Y0-BVP-945

WIND ENERGY

HANDBOOK

WIND ENERGY

HANDBOOK

Tony Burton
Wind Energy Consultant, Carno, UK

David Sharpe
CREST, Loughborough University, UK

Nick Jenkins
UMIST, Manchester, UK

Ervin Bossanyi
Garrad Hassan & Partners, Bristol, UK

JOHN WILEY & SONS, LTD
Chichester • New York • Weinheim • Brisbane • Singapore • Toronto

Telephone (+44) 1243 779777

Email (for orders and customer service enquiries): cs-books@wiley.co.uk
Visit our Home Page on www.wileyeurope.com or www.wiley.com

Reprinted December 2001, November 2002, March and September 2004, August 2005,
June and December 2006, April and September 2008
Reprinted with corrections January 2008

Other Wiley Editorial Offices

John Wiley & Sons Inc., 111 River Street, Hoboken, NJ 07030, USA

Jossey-Bass, 989 Market Street, San Francisco, CA 94103-1741, USA

Wiley-VCH Verlag GmbH, Boschstr. 12, D-69469 Weinheim, Germany

John Wiley & Sons Australia Ltd, 33 Park Road, Milton, Queensland 4064, Australia

John Wiley & Sons (Asia) Pte Ltd, 2 Clementi Loop #02-01, Jin Xing Distripark, Singapore 129809

John Wiley & Sons Canada Ltd, 22 Worcester Road, Etobicoke, Ontario, Canada M9W 1L1

Wiley also publishes its books in a variety of electronic formats. Some content that appears in print may not be available in electronic books.

Library of Congress Cataloging-in-Publication Data

Handbook of wind energy / Tony Burton, David Sharpe, Nick Jenkins and Ervin Bossanyi.
p. cm
Includes bibliographical references and index.
ISBN 0-471-48997-2
1. Wind power - Handbooks, manuals, etc. I. Burton, Tony, 1947.

TJ820.H35 2001
621.31'2136-dc21 2001024908

British Library Cataloguing in Publication Data

A catalogue record for this book is available from the British Library

ISBN 13: 978-0-471-48997-9 (H /B)

Typeset in 10/12 pt Palatino by Keytec Typesetting Ltd, Bridport, Dorset
Printed and bound in Great Britain by CPI Antony Rowe, Chippenham, Wiltshire

Contents

Acknowledgements

A large number of individuals have assisted the authors in a variety of ways in the preparation of this work. In particular, however, we would like to thank David Infield for providing some of the content of Chapter 4, David Quarton for scrutinising and commenting on Chapter 5, Mark Hancock, Martin Ansell and Colin Anderson for supplying information and guidance on blade material properties reported in Chapter 7, and Ray Hicks for insights into gear design. Thanks are also due to Roger Haines and Steve Gilkes for illuminating discussions on yaw drive design and braking philosophy, respectively, and to James Shawler for assistance and discussions about Chapter 3.

We have made extensive use of ETSU and Risø publications and record our thanks to these organisations for making documents available to us free of charge and sanctioning the reproduction of some of the material therein.

While acknowledging the help we have received from the organisations and individuals referred to above, the responsibility for the work is ours alone, so corrections and/or constructive criticisms would be welcome.

Extracts from British Standards reproduced with the permission of the British Standards Institution under licence number 2001/SK0281. Complete Standards are available from BSI Customer Services. (Tel +44 (0) 20 8996 9001).

List of Symbols

Note: This list is not exhaustive, and omits many symbols that are unique to particular chapters

a	axial flow induction factor
a'	tangential flow induction factor
a'_{t}	tangential flow induction factor at the blade tip
a_0	two-dimensional lift curve slope, $(\mathrm{d}C_1/\mathrm{d}\alpha)$
a_1	constant defining magnitude of structural damping
A, A_{D}	rotor swept area
A_∞, A_{W}	upstream and downstream stream-tube cross-sectional areas
b	face width of gear teeth
c	blade chord; Weibull scale parameter
$\hat{c}$	damping coefficient per unit length
c_i	generalized damping coefficient with respect to the ith mode
C	decay constant
$C(\nu)$	Theodorsen's function, where ν is the reduced frequency: $C(\nu) = F(\nu) + iG(\nu)$
C_{d}	sectional drag coefficient
C_{f}	sectional force coefficient (i.e., C_{d} or C_1 as appropriate)
C_1	sectional lift coefficient
C_n^m	coefficient of a Kinner pressure distribution
C_p	pressure coefficient
C_P	power coefficient
C_Q	torque coefficient
C_T	thrust coefficient; total cost of wind turbine
C_{TB}	total cost of baseline wind turbine
C_x	coefficient of sectional blade element force normal to the rotor plane
C_y	coefficient of sectional blade element force parallel to the rotor plane
$C(\Delta r, n)$	coherence—i.e., normalized cross spectrum – for wind speed fluctuations at points separated by distance s measured in the across wind direction
$C_{jk}(n)$	coherence—i.e., normalized cross spectrum – for longitudinal wind speed fluctuations at points j and k
d	streamwise distance between vortex sheets in a wake
d_1	pitch diameter of pinion gear

d_{PL}	pitch diameter of planet gear
D	drag force; tower diameter; rotor diameter; flexural rigidity of plate
E	energy capture, i.e., energy generated by turbine over defined time period; modulus of elasticity
$E\{\}$	time averaged value of expression within brackets
f	tip loss factor; Coriolis parameter
$f()$	probability density function
$f_j(t)$	blade tip displacement in jth mode
$f_{in}(t)$	blade tip displacement in ith mode at the end of the nth time step
$f_J(t)$	blade j first mode tip displacement
$f_T(t)$	hub displacement for tower first mode
F	force
F_X	load in x (downwind) direction
F_Y	load in y direction
F_t	force between gear teeth at right angles to the line joining the gear centres
$F(\mu)$	function determining the radial distribution of induced velocity normal to the plane of the rotor
$F()$	cumulative probability density function
g	vortex sheet strength; peak factor, defined as the number of standard deviations of a variable to be added to the mean to obtain the extreme value in a particular exposure period, for zero-up-crossing frequency, ν
g_0	peak factor as above, but for zero upcrossing frequency n_0
G	geostrophic wind speed; shear modulus; gearbox ratio
$G(t)$	t second gust factor
h	height of atmospheric boundary layer; duration of time step; thickness of thin-walled panel; maximum height of single gear tooth contact above critical root section
H	hub height
H_{jk}	elements of transformational matrix, $\mathbf{H}$, used in wind simulation
$H_i(n)$	complex frequency response function for the ith mode
I	turbulence intensity; second moment of area; moment of inertia; electrical current (shown in bold when complex)
I_B	blade inertia about root
I_0	ambient turbulence intensity
I_+	added turbulence intensity
I_{++}	added turbulence intensity above hub height
I_R	inertia of rotor about horizontal axis in its plane
I_u	longitudinal turbulence intensity
I_v	lateral turbulence intensity
I_w	vertical turbulence intensity
I_{wake}	total wake turbulence intensity
j	$\sqrt{-1}$
k	shape parameter for Weibull function; integer; reduced frequency, $(\omega c/2W)$
k_i	generalized stiffness with respect to the ith mode, defined as $m_i\omega_i^2$

K_P	power coefficient based on tip speed
K_{SMB}	size reduction factor accounting for the lack of correlation of wind fluctuations over structural element or elements
$K_{Sx}(n_1)$	size reduction factor accounting for the lack of correlation of wind fluctuations at resonant frequency over structural element or elements
$K_\nu()$	modified Bessel function of the second kind and order ν
$K(\chi)$	function determining the induced velocity normal to the plane of a yawed rotor
L	length scale for turbulence (subscripts and superscripts according to context); lift force
L_u^x	integral length scale for the along wind turbulence component, u, measured in the longitudinal direction, x
m	mass per unit length, integer
m_i	generalized mass with respect to the ith mode
m_{T1}	generalized mass of tower, nacelle and rotor with respect to tower first mode
M	moment; integer
$\overline{M}$	mean bending moment
M_T	teeter moment
M_X	blade in-plane moment (i.e., moment causing bending in plane of rotation); tower side-to-side moment
M_Y	blade out-of-plane moment (i.e., moment causing bending out of plane of rotation); tower fore-aft moment
M_Z	blade torsional moment; tower torsional moment
M_{YS}	low-speed shaft moment about rotating axis perpendicular to axis of blade 1
M_{ZS}	low-speed shaft moment about rotating axis parallel to axis of blade 1
M_{YN}	moment exerted by low-speed shaft on nacelle about (horizontal) y-axis
M_{ZN}	moment exerted by low-speed shaft on nacelle about (vertical) z-axis
n	frequency (Hz); number of fatigue loading cycles; integer
n_0	zero up-crossing frequency of quasistatic response
n_1	frequency (Hz) of 1st mode of vibration
N	number of blades; number of time steps per revolution; integer
$N(r)$	centrifugal force
$N(S)$	number of fatigue cycles to failure at stress level S
p	static pressure
P	aerodynamic power; electrical real (active) power
$P_n^m()$	associated Legrendre polynomial of the first kind
$q(r, t)$	fluctuating aerodynamic lift per unit length
Q	rotor torque; electrical reactive power
Q_a	aerodynamic torque
$\dot{Q}$	rate of heat flow
$\overline{Q}$	mean aerodynamic lift per unit length
Q_D	dynamic factor defined as ratio of extreme moment to gust quasistatic moment
Q_g	load torque at generator

Q_L	loss torque
$Q_n^m()$	associated Legrendre polynomial of the second kind
$Q_1(t)$	generalized load, defined in relation to a cantilever blade by Equation (A5.13)
r	radius of blade element or point on blade; correlation coefficient between power and wind speed; radius of tubular tower
r'	radius of point on blade
r_1, r_2	radii of points on blade or blades
R	blade tip radius; ratio of minimum to maximum stress in fatigue load cycle; electrical resistance
Re	Reynold's number
$R_u(n)$	normalized power spectral density, $n.S_u(n)/\sigma_u^2$, of longitudinal wind-speed fluctuations, u, at a fixed point
s	distance inboard from the blade tip; distance along the blade chord from the leading edge; separation between two points; Laplace operator; slip of induction machine
s_1	separation between two points measured in the along-wind direction
S	wing area; autogyro disc area; fatigue stress range
$\mathbf{S}$	electrical complex (apparent) power (bold indicates a complex quantity)
$S()$	uncertainty or error band
$S_{jk}(n)$	cross spectrum of longitudinal wind-speed fluctuations, u, at points j and k (single sided)
$S_M(n)$	single-sided power spectrum of bending moment
$S_{Q1}(n)$	single-sided power spectrum of generalized load
$S_u(n)$	single-sided power spectrum of longitudinal wind-speed fluctuations, u, at a fixed point
$S_u^0(n)$	single-sided power spectrum of longitudinal wind-speed fluctuations, u, as seen by a point on a rotating blade (also known as rotationally sampled spectrum)
$S_u^0(r_1, r_2, n)$	cross spectrum of longitudinal wind-speed fluctuations, u, as seen by points at radii r_1 and r_2 on a rotating blade or rotor (single sided)
$S_v(n)$	single-sided power spectrum of lateral wind speed fluctuations, v, at a fixed point
$S_w(n)$	single-sided power spectrum of vertical wind-speed fluctuations, w, at a fixed point
t	time; gear tooth thickness at critical root section; tower wall thickness
T	rotor thrust; duration of discrete gust; wind-speed averaging period
u	fluctuating component of wind speed in the x-direction; induced velocity in x-direction; in-plane plate deflection in x-direction; gear ratio
u^*	friction velocity in boundary layer
U_∞	free stream velocity
U, $U(t)$	instantaneous wind speed in the along-wind direction
$\overline{U}$	mean component of wind speed in the along-wind direction – typically taken over a period of 10 min or 1 h
U_{ave}	annual average wind speed at hub height

U_d	streamwise velocity at the rotor disc
U_w	streamwise velocity in the far wake
U_{e1}	extreme 3 s gust wind speed with 1 year return period
U_{e50}	extreme 3 s gust wind speed with 50 year return period
U_0	turbine upper cut-out speed
U_r	turbine rated wind speed, defined as the wind speed at which the turbine's rated power is reached
U_{ref}	reference wind speed defined as 10 min mean wind speed at hub height with 50 year return period
U_1	strain energy of plate flexure
U_2	in-plane strain energy
v	fluctuating component of wind speed in the y-direction; induced velocity in y-direction; in-plane plate deflection in y-direction
V	airspeed of an autogyro; longitudinal air velocity at rotor disc, $U_\infty(1 - a)$ (Section 7.1.9); voltage (shown in bold when complex)
$V(t)$	instantaneous lateral wind speed
VA	electrical volt-amperes
V_f	fibre volume fraction in composite material
V_t	blade tip speed
w	fluctuating component of wind speed in the z-direction; induced velocity in z-direction; out-of-plane plate deflection
W	wind velocity relative to a point on rotating blade; electrical power loss
x	downwind co-ordinate – fixed and rotating axis systems; downwind displacement
$x(t)$	stochastic component of a variable
x_n	length of near wake region
$\bar{x}_1$	first-mode component of steady tip displacement
X	electrical inductive reactance
y	lateral co-ordinate with respect to vertical axis (starboard positive) – fixed-axis system
y	lateral co-ordinate with respect to blade axis – rotating-axis system
y	lateral displacement
z	vertical co-ordinate (upwards positive) – fixed-axis system
z	radial co-ordinate along blade axis – rotating-axis system
Z	section modulus
$\mathbf{Z}$	electrical impedance (bold indicates a complex quantity)
z_0	ground roughness length
z_1	number of teeth on pinion gear
$z(t)$	periodic component of a variable

Greek

α	angle of attack – i.e., angle between air flow incident on the blade and the blade chord line; wind-shear power law exponent
β	inclination of local blade chord to rotor plane (i.e., blade twist plus pitch angle, if any)

χ	wake skew angle: angle between the axis of the wake of a yawed rotor and the axis of rotation of rotor
χ_{M1}	weighted mass ratio defined in Section 5.8.6
δ_a	logarithmic decrement of aerodynamic damping
δ_s	logarithmic decrement of structural damping
δ	logarithmic decrement of combined aerodynamic and structural damping; width of tower shadow deficit region
δ_3	angle between axis of teeter hinge and the line perpendicular to both the rotor axis and the low-speed shaft axis
ε_1, ε_2, ε_3	proportion of time in which a variable takes the maximum, mean or minimum values in a three-level square wave
ϕ	flow angle of resultant velocity W to rotor plane
γ	yaw angle; Euler's constant (= 0.5772)
γ_L	load factor
γ_{mf}	partial safety factor for material fatigue strength
γ_{mu}	partial safety factor for material ultimate strength
Γ	blade circulation; vortex strength
$\Gamma()$	gamma function
η	ellipsoidal parameter; shaft tilt; one eighth of Lock number (defined in Section 5.8.8)
φ	impact factor in DS 472, defined as ratio of extreme moment to the 10 min mean
κ	von Karman's constant
$\kappa_L(s)$	cross-correlation function between velocity components at points in space a distance s apart, in the direction parallel to the line joining them
$\kappa_T(s)$	cross-correlation function between velocity components at points in space a distance s apart, in the direction perpendicular to the line joining them
$\kappa_u(r, \tau)$	auto-correlation function for along-wind velocity component at radius r on stationary rotor
$\kappa_u^0(r, \tau)$	auto-correlation function for along-wind velocity component as seen by a point at radius r on a rotating rotor
$\kappa_u(r_1, r_2, \tau)$	cross-correlation function between along-wind velocity components at radii r_1 and r_2 (not necessarily on same blade), for stationary rotor
$\kappa_u^0(r_1, r_2, \tau)$	cross-correlation function between along-wind velocity components as seen by points (not necessarily on same blade) at radii r_1 and r_2 on a rotating rotor
λ	tip speed ratio; latitude; ratio of longitudinal to transverse buckle half wavelengths
λ_r	tangential speed of blade element at radius r divided by wind speed: local speed ratio
Λ	yaw rate
μ	non-dimensional radial position, r/R; viscosity; coefficient of friction
$\mu_i(r)$	mode shape of ith blade mode
$\mu_i(z)$	mode shape of ith tower mode
$\mu_T(z)$	tower first mode shape

$\mu_{\mathrm{TJ}}(r)$	normalized rigid body deflection of blade j resulting from excitation of tower first mode
μ_z	mean value of variable z
ν	ellipsoidal co-ordinate; mean zero up-crossing frequency
θ	wind-speed direction change; random phase angle; cylindrical panel co-ordinate; brake disc temperature
ρ	air density
$\rho_u^0(r_1, r_2, \tau)$	normalized cross-correlation function between along-wind velocity components as seen by points (not necessarily on same blade) at radii r_1 and r_2 on a rotating rotor (i.e., $\kappa_u^0(r_1, r_2, \tau)/\sigma_u^2$)
σ	blade solidity; standard deviation; stress
$\overline{\sigma}$	mean stress
σ_M	standard deviation of bending moment
σ_{M1}	standard deviation of first-mode resonant bending moment, at blade root for blade resonance, and at tower base for tower resonance
σ_{MB}	standard deviation of quasistatic bending moment (or bending moment background response)
σ_{Mh}	standard deviation of hub dishing moment
σ_{MT}	standard deviation of teeter moment for rigidly mounted, two-bladed rotor
$\sigma_{\overline{M}}$	standard deviation of mean of blade root bending moments for two-bladed rotor
σ_{Q1}	standard deviation of generalized load with respect to first mode
σ_{r}	rotor solidity
σ_{u}	standard deviation of fluctuating component of wind in along-wind direction
σ_v	standard deviation of wind speed in across-wind direction
σ_w	standard deviation of wind speed in vertical direction
σ_{x1}	standard deviation of first-mode resonant displacement, referred to blade tip for blade resonance, and to nacelle for tower resonance
τ	time interval; non-dimensional time; shear stress
υ	Poisson's ratio
ω	angular frequency (rad/s)
ω_{d}	demanded generator rotational speed
ω_i	natural frequency of ith mode (rad/s)
ω_{g}	generator rotational speed
ω_{r}	induction machine rotor rotational speed
ω_{s}	induction machine stator field rotational speed
Ω	rotational speed of rotor; earth's rotational speed
ξ	damping ratio
ψ	angle subtended by cylindrical plate panel
$\psi_{\mathrm{uu}}(r, r', n)$	real part of normalized cross spectrum (see Appendix 1, section A1.4)
Ψ	blade azimuth
ζ	teeter angle

Subscripts

a	aerodynamic

B	baseline
c	compressive
d	disc; drag; design
e1	extreme value with return period of 1 year
e50	extreme value with return period of 50 years
ext	extreme
f	fibre
i	mode i
j	mode j
J	blade J
k	characteristic
l	lift
m	matrix
M	moment
max	maximum value of variable
min	minimum value of variable
n	value at end of nth time step
Q	generalized load
R	value at tip radius, R
s	structural
t	tensile
T	thrust
u	downwind; ultimate
v	lateral
w	vertical
w	wake
x	deflection in along-wind direction

Superscripts

0	rotationally sampled (applied to wind-speed spectra)

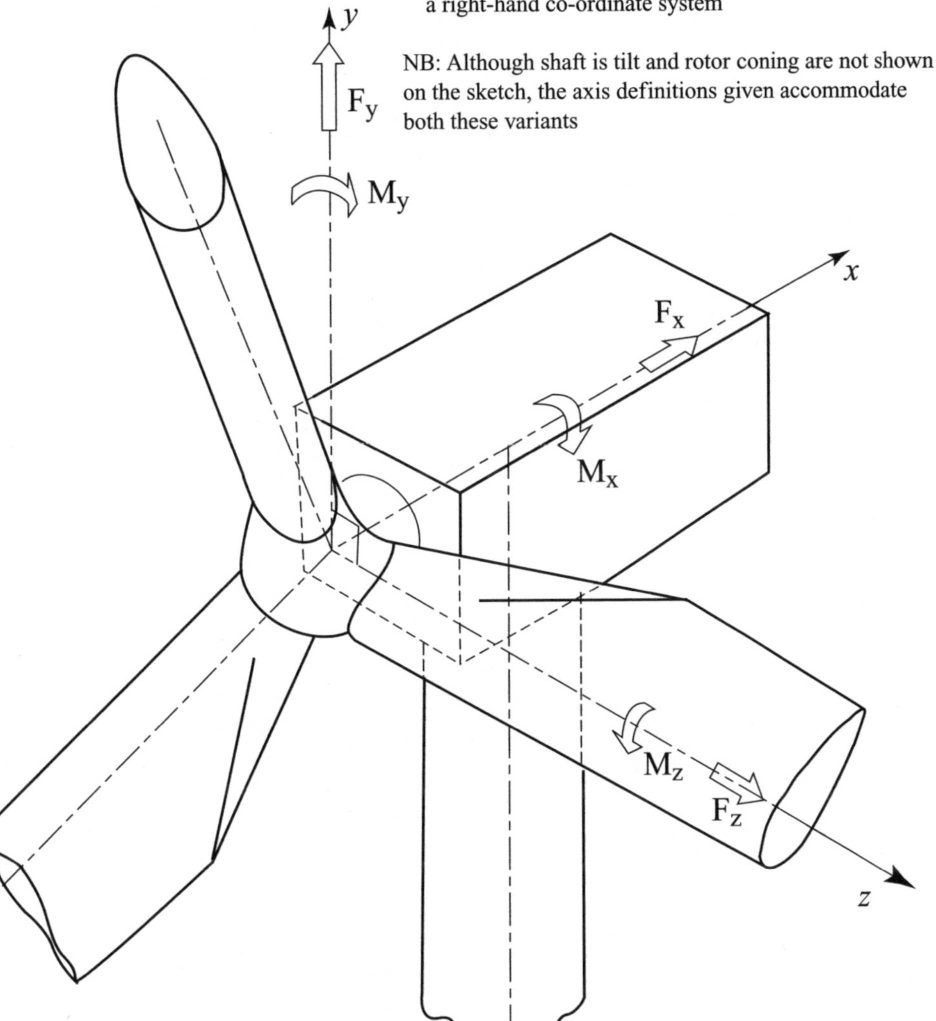

Figure C1 Co-ordinate System for Blade Loads, Positions and Deflections (rotates with blade) NB: Although shaft tilt and rotor coning are not shown on the sketch, the axis definitions given accommodate both these variants.

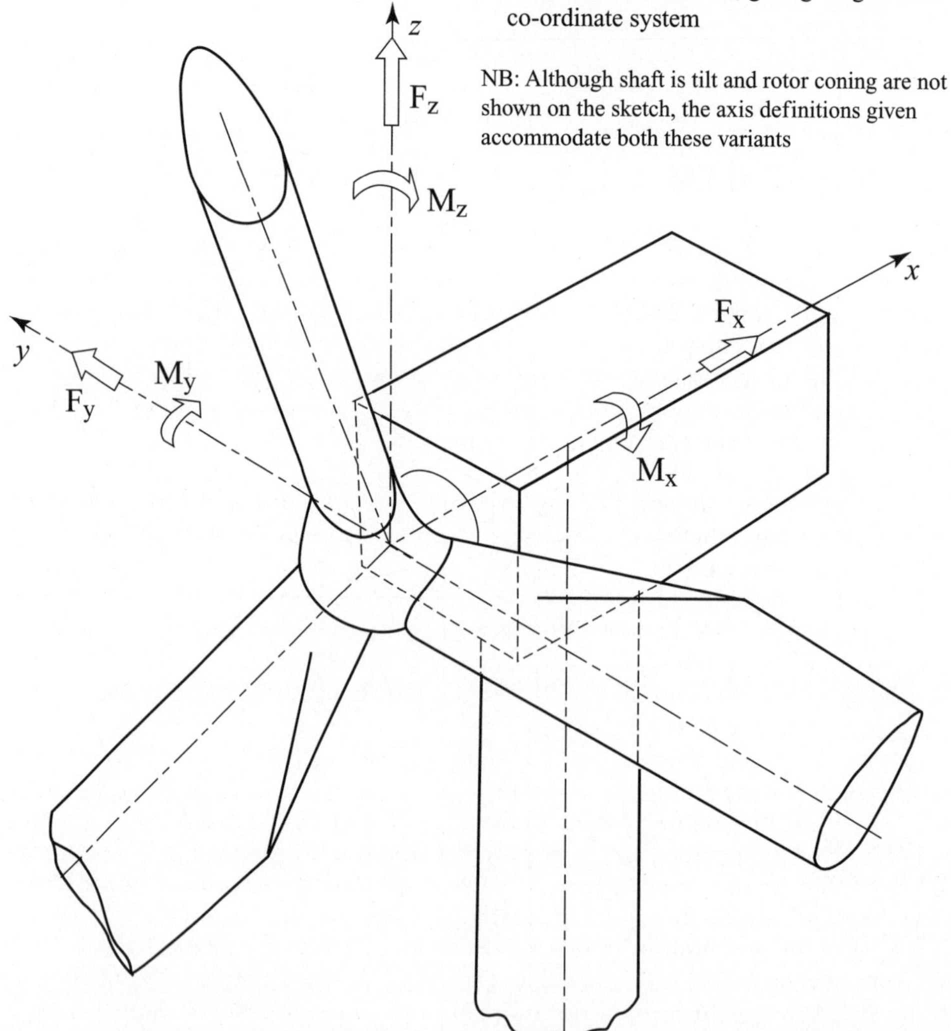

Figure C2 Fixed Co-ordinate System for Hub Loads and Deflections, and Positions with Respect to Hub NB: Although shaft tilt and rotor coning are not shown on the sketch, the axis definitions given accommodate both these variants.

1 Introduction

1.1 Historical Development

Windmills have been used for at least 3000 years, mainly for grinding grain or pumping water, while in sailing ships the wind has been an essential source of power for even longer. From as early as the thirteenth century, horizontal-axis windmills were an integral part of the rural economy and only fell into disuse with the advent of cheap fossil-fuelled engines and then the spread of rural electrification. The use of windmills (or wind turbines) to generate electricity can be traced back to the late nineteenth century with the 12 kW DC windmill generator constructed by Brush in the USA and the research undertaken by LaCour in Denmark. However, for much of the twentieth century there was little interest in using wind energy other than for battery charging for remote dwellings and these low-power systems were quickly replaced once access to the electricity grid became available. One notable exception was the 1250 kW Smith–Putnam wind turbine constructed in the USA in 1941. This remarkable machine had a steel rotor 53 m in diameter, full-span pitch control and flapping blades to reduce loads. Although a blade spar failed catastrophically in 1945, it remained the largest wind turbine constructed for some 40 years (Putnam, 1948).

Golding (1955) and Shepherd and Divone in Spera (1994) provide a fascinating history of early wind turbine development. They record the 100 kW 30 m diameter Balaclava wind turbine in the then USSR in 1931 and the Andrea Enfield 100 kW 24 m diameter pneumatic design constructed in the UK in the early 1950s. In this turbine hollow blades, open at the tip, were used to draw air up through the tower where another turbine drove the generator. In Denmark the 200 kW 24 m diameter Gedser machine was built in 1956 while Electricité de France tested a 1.1 MW 35 m diameter turbine in 1963. In Germany, Professor Hutter constructed a number of innovative, lightweight turbines in the 1950s and 1960s. In spite of these technical advances and the enthusiasm, among others, of Golding at the Electrical Research Association in the UK there was little sustained interest in wind generation until the price of oil rose dramatically in 1973.

The sudden increase in the price of oil stimulated a number of substantial Government-funded programmes of research, development and demonstration. In the USA this led to the construction of a series of prototype turbines starting with the 38 m diameter 100 kW Mod-0 in 1975 and culminating in the 97.5 m diameter 2.5 MW Mod-5B in 1987. Similar programmes were pursued in the UK, Germany and Sweden. There was considerable uncertainty as to which architecture might prove most cost-effective and several innovative concepts were investigated at full scale. In Canada, a 4 MW vertical-axis Darrieus wind turbine was constructed and this concept was also investigated in the 34 m diameter Sandia Vertical Axis Test Facility in the USA. In the UK, an alternative vertical-axis design using straight blades to give an 'H' type rotor was proposed by Dr Peter Musgrove and a 500 kW prototype constructed. In 1981 an innovative horizontal-axis 3 MW wind turbine was built and tested in the USA. This used hydraulic transmission and, as an alternative to a yaw drive, the entire structure was orientated into the wind. The best choice for the number of blades remained unclear for some while and large turbines were constructed with one, two or three blades.

Much important scientific and engineering information was gained from these Government-funded research programmes and the prototypes generally worked as designed. However, it has to be recognized that the problems of operating very large wind turbines, unmanned and in difficult wind climates were often under-

Figure 1.1 1.5 MW, 64 m diameter Wind Turbine (Reproduced by permission of NEG MICON, www.neg-micon.dk)

estimated and the reliability of the prototypes was not good. At the same time as the multi-megawatt prototypes were being constructed private companies, often with considerable state support, were constructing much smaller, often simpler, turbines for commercial sale. In particular the financial support mechanisms in California in the mid-1980s resulted in the installation of a very large number of quite small (< 100 kW) wind turbines. A number of these designs also suffered from various problems but, being smaller, they were in general easier to repair and modify. The so-called 'Danish' wind turbine concept emerged of a three-bladed, stall-regulated rotor and a fixed-speed, induction machine drive train. This deceptively simple architecture has proved to be remarkably successful and has now been implemented on turbines as large as 60 m in diameter and at ratings of 1.5 MW. The machines of Figures 1.1 and 1.2 are examples of this design. However, as the sizes of commercially available turbines now approach that of the large prototypes of the 1980s it is interesting to see that the concepts investigated then of variable-speed operation, full-span control of the blades, and advanced materials are being used increasingly by designers. Figure 1.3 shows a wind farm of direct-drive, variable-speed wind turbines. In this design, the synchronous generator is coupled directly to the aerodynamic rotor so eliminating the requirement for a gearbox. Figure 1.4 shows a more conventional, variable-speed wind turbine that uses a gearbox, while a small wind farm of pitch-regulated wind turbines, where full-span control of the blades is used to regulate power, is shown in Figure 1.5.

Figure 1.2 750 kW, 48 m diameter Wind Turbine, Denmark (Reproduced by permission of NEG MICON, www.neg-micon.dk)

Figure 1.3 Wind Farm of Variable-Speed Wind Turbines in Complex Terrain (Reproduced by permission of Wind Prospect Ltd., www.windprospect.com)

Figure 1.4 1 MW Wind Turbine in Northern Ireland (Reproduced by permission of Renewable Energy Systems Ltd., www.res-ltd.com)

The stimulus for the development of wind energy in 1973 was the price of oil and concern over limited fossil-fuel resources. Now, of course, the main driver for use of wind turbines to generate electrical power is the very low CO_2 emissions (over the entire life cycle of manufacture, installation, operation and de-commissioning)

Figure 1.5 Wind Farm of Six Pitch-regulated Wind Turbines in Flat Terrain (Reproduced by permission of Wind Prospect Ltd., www.windprospect.com)

and the potential of wind energy to help limit climate change. In 1997 the Commission of the European Union published its White Paper (CEU, 1997) calling for 12 percent of the gross energy demand of the European Union to be contributed from renewables by 2010. Wind energy was identified as having a key role to play in the supply of renewable energy with an increase in installed wind turbine capacity from 2.5 GW in 1995 to 40 GW by 2010. This target is likely to be achievable since at the time of writing, January 2001, there was some 12 GW of installed wind-turbine capacity in Europe, 2.5 GW of which was constructed in 2000 compared with only 300 MW in 1993. The average annual growth rate of the installation of wind turbines in Europe from 1993–9 was approximately 40 percent (Zervos, 2000). The distribution of wind-turbine capacity is interesting with, in 2000, Germany accounting for some 45 percent of the European total, and Denmark and Spain each having approximately 18 percent. There is some 2.5 GW of capacity installed in the USA of which 65 percent is in California although with increasing interest in Texas and some states of the midwest. Many of the California wind farms were originally

Table 1.1 Installed Wind Turbine Capacity Throughout the World, January 2001

Location	Installed capacity (MW)
Germany	5432
Denmark	2281
Spain	2099
Netherlands	444
UK	391
Total Europe	**11831**
California	1622
Total USA	**2568**
Total World	**16461**

Courtesy of Windpower Monthly News Magazine

constructed in the 1980s and are now being re-equipped with larger modern wind turbines.

Table 1.1 shows the installed wind-power capacity worldwide in January 2001 although it is obvious that with such a rapid growth in some countries data of this kind become out of date very quickly.

The reasons development of wind energy in some countries is flourishing while in others it is not fulfilling the potential that might be anticipated from a simple consideration of the wind resource, are complex. Important factors include the financial-support mechanisms for wind-generated electricity, the process by which the local planning authorities give permission for the construction of wind farms, and the perception of the general population particularly with respect to visual impact. In order to overcome the concerns of the rural population over the environmental impact of wind farms there is now increasing interest in the development of sites offshore.

1.2 Modern Wind Turbines

The power output, P, from a wind turbine is given by the well-known expression:

$$P = \frac{1}{2} C_P \rho A U^3$$

where ρ is the density of air (1.225 kg/m^3), C_P is the power coefficient, A is the rotor swept area, and U is the wind speed.

The density of air is rather low, 800 times less than that of water which powers hydro plant, and this leads directly to the large size of a wind turbine. Depending on the design wind speed chosen, a 1.5 MW wind turbine may have a rotor that is more than 60 m in diameter. The power coefficient describes that fraction of the power in the wind that may be converted by the turbine into mechanical work. It has a theoretical maximum value of 0.593 (the Betz limit) and rather lower peak values are achieved in practice (see Chapter 3). The power coefficient of a rotor varies with the tip speed ratio (the ratio of rotor tip speed to free wind speed) and is only a maximum for a unique tip speed ratio. Incremental improvements in the power coefficient are continually being sought by detailed design changes of the rotor and, by operating at variable speed, it is possible to maintain the maximum power coefficient over a range of wind speeds. However, these measures will give only a modest increase in the power output. Major increases in the output power can only be achieved by increasing the swept area of the rotor or by locating the wind turbines on sites with higher wind speeds.

Hence over the last 10 years there has been a continuous increase in the rotor diameter of commercially available wind turbines from around 30 m to more than 60 m. A doubling of the rotor diameter leads to a four-times increase in power output. The influence of the wind speed is, of course, more pronounced with a doubling of wind speed leading to an eight-fold increase in power. Thus there have been considerable efforts to ensure that wind farms are developed in areas of the highest wind speeds and the turbines optimally located within wind farms. In certain countries very high towers are being used (more than 60–80 m) to take advantage of the increase of wind speed with height.

In the past a number of studies were undertaken to determine the 'optimum' size of a wind turbine by balancing the complete costs of manufacture, installation and operation of various sizes of wind turbines against the revenue generated (Molly *et al.*, 1993). The results indicated a minimum cost of energy would be obtained with wind turbine diameters in the range of 35–60 m, depending on the assumptions made. However, these estimates would now appear to be rather low and there is no obvious point at which rotor diameters, and hence output power, will be limited particularly for offshore wind turbines.

All modern electricity-generating wind turbines use the lift force derived from the blades to drive the rotor. A high rotational speed of the rotor is desirable in order to reduce the gearbox ratio required and this leads to low solidity rotors (the ratio of blade area/rotor swept area). The low solidity rotor acts as an effective energy concentrator and as a result the energy recovery period of a wind turbine, on a good site, is less than 1 year, i.e., the energy used to manufacture and install the wind turbine is recovered within its first year of operation (Musgrove in Freris, 1990).

1.3 Scope of the Book

The use of wind energy to generate electricity is now well accepted with a large industry manufacturing and installing thousands of MWs of new capacity each year. Although there are exciting new developments, particularly in very large wind turbines, and many challenges remain, there is a considerable body of established knowledge concerning the science and technology of wind turbines. This book is intended to record some of this knowledge and to present it in a form suitable for use by students (at final year undergraduate or post-graduate level) and by those involved in the design, manufacture or operation of wind turbines. The overwhelming majority of wind turbines presently in use are horizontal-axis, land-based turbines connected to a large electricity network. These turbines are the subject of this book.

Chapter 2 discusses the wind resource. Particular reference is made to wind turbulence due to its importance in wind-turbine design. Chapter 3 sets out the basis of the aerodynamics of horizontal-axis wind turbines while Chapter 4 discusses their performance. Any wind-turbine design starts with establishing the design loads and these are discussed in Chapter 5. Chapter 6 sets out the various design options for horizontal-axis wind turbines with approaches to the design of some of the important components examined in Chapter 7. The functions of the wind-turbine controller are discussed in Chapter 8 and some of the possible analysis techniques described. In Chapter 9 wind farms and the development of wind-energy projects are reviewed with particular emphasis on environmental impact. Finally, Chapter 10 considers how wind turbines interact with the electrical power system.

The book attempts to record well-established knowledge that is relevant to wind turbines, which are currently commercially significant. Thus, it does not discuss a number of interesting research topics or where wind-turbine technology is still evolving rapidly. Although they were investigated in considerable detail in the 1980s, vertical-axis wind turbines have not proved to be commercially competitive and are not currently manufactured in significant numbers. Hence the particular issues of vertical-axis turbines are not dealt with in this text.

There are presently some two billion people in the world without access to mains electricity and wind turbines, in conjunction with other generators, e.g., diesel engines, may in the future be an effective means of providing some of these people with power. However, autonomous power systems are extremely difficult to design and operate reliably, particularly in remote areas of the world and with limited budgets. A small autonomous AC power system has all the technical challenges of a large national electricity system but, due to the low inertia of the plant, requires a very fast, sophisticated control system to maintain stable operation. Over the last 20 years there have been a number of attempts to operate autonomous wind–diesel systems on islands throughout the world but with only limited success. This class of installation has its own particular problems and again, given the very limited size of the market at present, this specialist area is not dealt with.

Installation of offshore wind turbines is now commencing. The few offshore wind farms already installed are in rather shallow waters and resemble land-based wind farms in many respects using medium sized wind turbines. Very large wind farms with multi-megawatt turbines located in deeper water, many kilometres offshore, are now being planned and these will be constructed over the coming years. However, the technology of offshore wind-energy projects is still evolving at too rapid a pace for inclusion in this text which attempts to present established engineering practice.

References

CEU, (1997). 'Energy for the future, renewable sources of energy – White Paper for a Community Strategy and Action Plan'. *COM (97) 559 final.*

Freris, L. L. (ed.), (1990). *Wind energy conversion systems*. Prentice Hall, New York, US.
Golding, E. W. (1955). *The generation of electricity from wind power*. E. & F. N. Spon (reprinted R. I. Harris, 1976).
Molly, J. P. Keuper, A. and Veltrup, M., (1993). 'Statistical WEC design and cost trends'. *Proceedings of the European Wind Energy Conference*, pp 57–59.
Putnam, G. C. (1948). *Power from the wind*. Van Nostrand Rheinhold, New York, USA.
Spera, D. A. (1994) *Wind-turbine technology, fundamental concepts of wind-turbine engineering*. ASME Press, New York, US.
Zervos, A. (2000) 'European targets, time to be more ambitious?'. *Windirections*, 18–19. European Wind Energy Association, www.ewea.org.

Bibliography

Eggleston, D. M. and Stoddard, F. S., (1987). *Wind turbine engineering design*. Van Nostrand Rheinhold, New York, USA.
Gipe, P., (1995). *Wind energy comes of age*. John Wiley and Sons, New York, USA.
Harrison, R., Hau, E. and Snel, H., (2000). *Large wind turbines, design and economics*. John Wiley and Sons.
Johnson, L., (1985). *Wind energy systems*. Prentice-Hall.
Le Gourieres, D., (1982). *Wind power plants theory and design*. Pergamon Press, Oxford, UK.
Twiddell, J. W. and Weir, A. D., (1986). *Renewable energy sources*. E. & F. N. Spon.

2
The Wind Resource

2.1 The Nature of the Wind

The energy available in the wind varies as the cube of the wind speed, so an understanding of the characteristics of the wind resource is critical to all aspects of wind energy exploitation, from the identification of suitable sites and predictions of the economic viability of wind farm projects through to the design of wind turbines themselves, and understanding their effect on electricity distribution networks and consumers.

From the point of view of wind energy, the most striking characteristic of the wind resource is its variability. The wind is highly variable, both geographically and temporally. Furthermore this variability persists over a very wide range of scales, both in space and time. The importance of this is amplified by the cubic relationship to available energy.

On a large scale, spatial variability describes the fact that there are many different climatic regions in the world, some much windier than others. These regions are largely dictated by the latitude, which affects the amount of insolation. Within any one climatic region, there is a great deal of variation on a smaller scale, largely dictated by physical geography – the proportion of land and sea, the size of land masses, and the presence of mountains or plains for example. The type of vegetation may also have a significant influence through its effects on the absorption or reflection of solar radiation, affecting surface temperatures, and on humidity.

More locally, the topography has a major effect on the wind climate. More wind is experienced on the tops of hills and mountains than in the lee of high ground or in sheltered valleys, for instance. More locally still, wind velocities are significantly reduced by obstacles such as trees or buildings.

At a given location, temporal variability on a large scale means that the amount of wind may vary from one year to the next, with even larger scale variations over periods of decades or more. These long-term variations are not well understood, and may make it difficult to make accurate predictions of the economic viability of particular wind-farm projects, for instance.

On time-scales shorter than a year, seasonal variations are much more predictable, although there are large variations on shorter time-scales still, which although reasonably well understood, are often not very predictable more than a few days ahead. These 'synoptic' variations are associated with the passage of weather systems. Depending on location, there may also be considerable variations with the

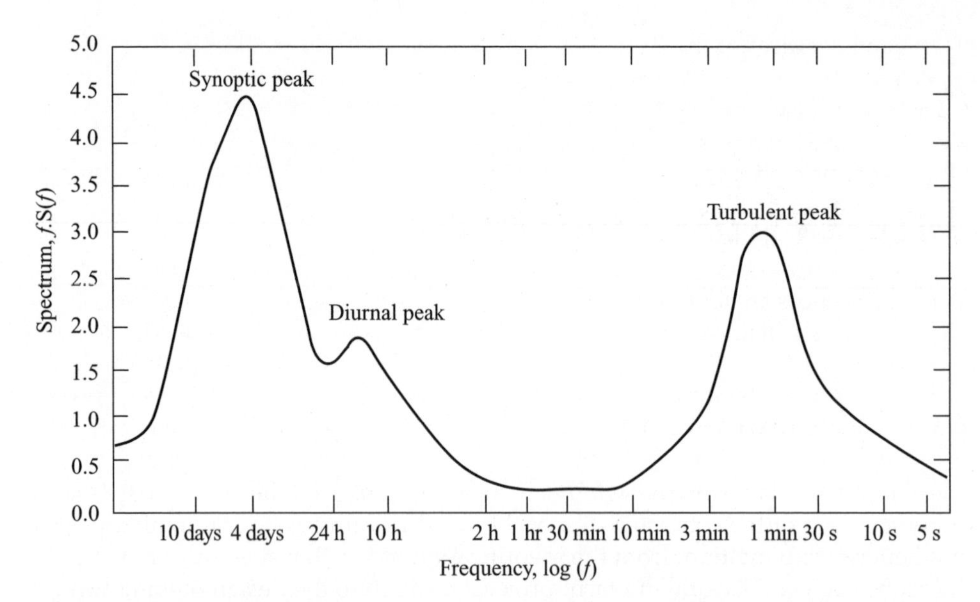

Figure 2.1 Wind Spectrum Farm Brookhaven Based on Work by van der Hoven (1957)

time of day (diurnal variations) which again are usually fairly predictable. On these time-scales, the predictability of the wind is important for integrating large amounts of wind power into the electricity network, to allow the other generating plant supplying the network to be organized appropriately.

On still shorter time-scales of minutes down to seconds or less, wind-speed variations known as turbulence can have a very significant effect on the design and performance of the individual wind turbines, as well as on the quality of power delivered to the network and its effect on consumers.

Van der Hoven (1957) constructed a wind-speed spectrum from long- and short-term records at Brookhaven, New York, showing clear peaks corresponding to the synoptic, diurnal and turbulent effects referred to above (Figure 2.1). Of particular interest is the so-called 'spectral gap' occurring between the diurnal and turbulent peaks, showing that the synoptic and diurnal variations can be treated as quite distinct from the higher-frequency fluctuations of turbulence. There is very little energy in the spectrum in the region between 2 h and 10 min.

2.2 Geographical Variation in the Wind Resource

Ultimately the winds are driven almost entirely by the sun's energy, causing differential surface heating. The heating is most intense on land masses closer to the equator, and obviously the greatest heating occurs in the daytime, which means that the region of greatest heating moves around the earth's surface as it spins on its axis. Warm air rises and circulates in the atmosphere to sink back to the surface in cooler areas. The resulting large-scale motion of the air is strongly influenced by coriolis forces due to the earth's rotation. The result is a large-scale global circulation pattern. Certain identifiable features of this such as the trade winds and the 'roaring forties' are well known.

The non-uniformity of the earth's surface, with its pattern of land masses and oceans, ensures that this global circulation pattern is disturbed by smaller-scale variations on continental scales. These variations interact in a highly complex and non-linear fashion to produce a somewhat chaotic result, which is at the root of the day-to-day unpredictability of the weather in particular locations. Clearly though, underlying tendencies remain which lead to clear climatic differences between regions. These differences are tempered by more local topographical and thermal effects.

Hills and mountains result in local regions of increased wind speed. This is partly a result of altitude – the earth's boundary layer means that wind speed generally increases with height above ground, and hill tops and mountain peaks may 'project' into the higher wind-speed layers. It is also partly a result of the acceleration of the wind flow over and around hills and mountains, and funnelling through passes or along valleys aligned with the flow. Equally, topography may produce areas of reduced wind speed, such as sheltered valleys, areas in the lee of a mountain ridge or where the flow patterns result in stagnation points.

Thermal effects may also result in considerable local variations. Coastal regions are often windy because of differential heating between land and sea. While the sea is warmer than the land, a local circulation develops in which surface air flows from the land to the sea, with warm air rising over the sea and cool air sinking over the land. When the land is warmer the pattern reverses. The land will heat up and cool down more rapidly than the sea surface, and so this pattern of land and sea breezes tends to reverse over a 24 h cycle. These effects were important in the early development of wind power in California, where an ocean current brings cold water to the coast, not far from desert areas which heat up strongly by day. An intervening mountain range funnels the resulting air flow through its passes, generating locally very strong and reliable winds (which are well correlated with peaks in the local electricity demand caused by air-conditioning loads).

Thermal effects may also be caused by differences in altitude. Thus cold air from high mountains can sink down to the plains below, causing quite strong and highly stratified 'downslope' winds.

The brief general descriptions of wind speed variations in Sections 2.1 to 2.5 are illustrative, and more detailed information can be found in standard meteorological texts. Section 9.1.3 describes how the wind regimes at candidate sites can be assessed, while wind forecasting is covered in Section 2.9.

Section 2.6 presents a more detailed description of the high-frequency wind fluctuations known as turbulence, which are crucial to the design and operation of wind turbines and have a major influence on wind turbine loads. Extreme winds are also important for the survival of wind turbines, and these are described in Section 2.8.

2.3 Long-term Wind speed Variations

There is evidence that the wind speed at any particular location may be subject to very slow long-term variations. Although the availability of accurate historical records is a limitation, careful analysis by, for example, Palutikoff, Guo and Halliday (1991) has demonstrated clear trends. Clearly these may be linked to long-term temperature variations for which there is ample historical evidence. There is also much debate at present about the likely effects of global warming, caused by human activity, on climate, and this will undoubtedly affect wind climates in the coming decades.

Apart from these long-term trends there may be considerable changes in windiness at a given location from one year to the next. These changes have many causes. They may be coupled to global climate phenomema such as *el niño*, changes in atmospheric particulates resulting from volcanic eruptions, and sunspot activity, to name a few. These changes add significantly to the uncertainty in predicting the energy output of a wind farm at a particular location during its projected lifetime.

2.4 Annual and Seasonal Variations

While year-to-year variation in annual mean wind speeds remains hard to predict, wind speed variations during the year can be well characterized in terms of a probability distribution. The Weibull distribution has been found to give a good representation of the variation in hourly mean wind speed over a year at many typical sites. This distribution takes the form

$$F(U) = \exp\left(-\left(\frac{U}{c}\right)^k\right) \tag{2.1}$$

where $F(U)$ is the fraction of time for which the hourly mean wind speed exceeds U. It is characterized by two parameters, a 'scale parameter' c and a 'shape parameter' k which describes the variability about the mean. c is related to the annual mean wind speed $\overline{U}$ by the relationship

$$\overline{U} = c\Gamma(1 + 1/k) \tag{2.2}$$

where Γ is the complete gamma function. This can be derived by consideration of the probability density function

$$f(U) = -\frac{\mathrm{d}F(U)}{\mathrm{d}U} = k\frac{U^{k-1}}{c^k}\exp\left(-\left(\frac{U}{c}\right)^k\right) \tag{2.3}$$

since the mean wind speed is given by

$$\overline{U} = \int_0^\infty Uf(U)\mathrm{d}U \tag{2.4}$$

A special case of the Weibull distribution is the Rayleigh distribution, with $k = 2$, which is actually a fairly typical value for many locations. In this case, the factor

$\Gamma(1 + 1/k)$ has the value $\sqrt{\pi}/2 = 0.8862$. A higher value of k, such as 2.5 or 3, indicates a site where the variation of hourly mean wind speed about the annual mean is small, as is sometimes the case in the trade wind belts for instance. A lower value of k, such as 1.5 or 1.2, indicates greater variability about the mean. A few examples are shown in Figure 2.2. The value of $\Gamma(1 + 1/k)$ varies little, between about 1.0 and 0.885 see (Figure 2.3).

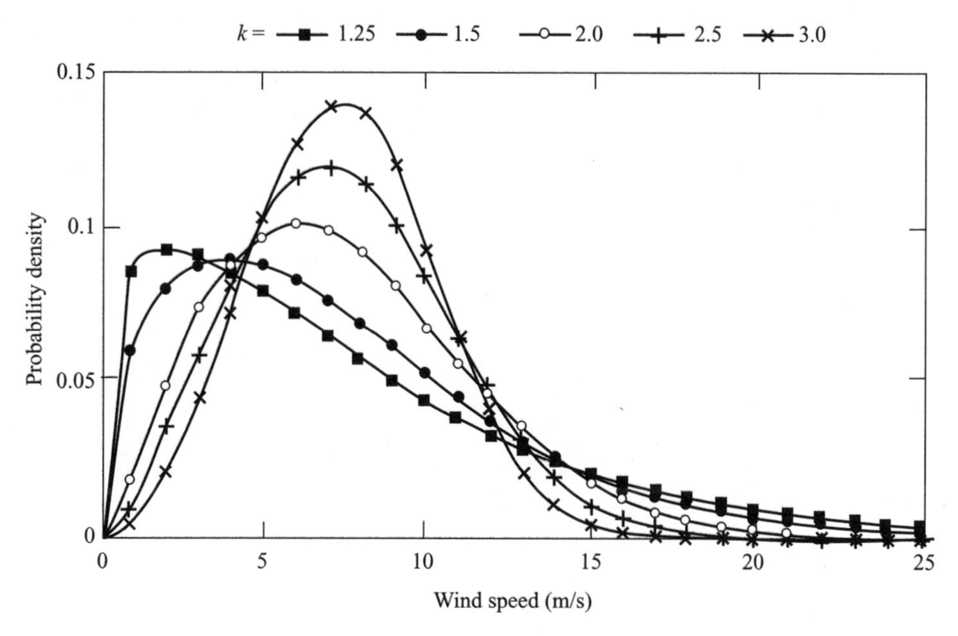

Figure 2.2 Example Weibull Distributions

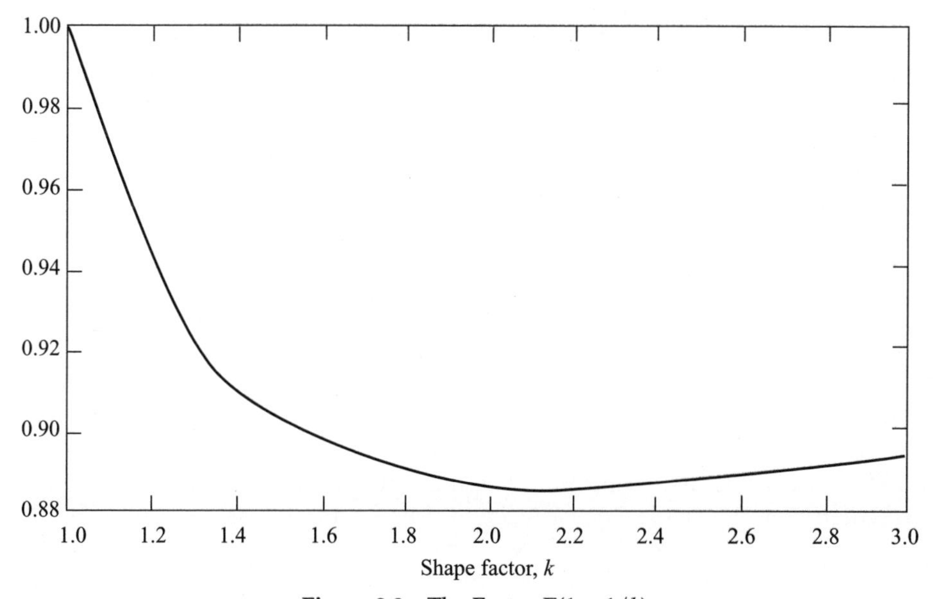

Figure 2.3 The Factor $\Gamma(1 + 1/k)$

The Weibull distribution of hourly mean wind speeds over the year is clearly the result of a considerable degree of random variation. However, there may also be a strong underlying seasonal component to these variations, driven by the changes in insolation during the year as a result of the tilt of the earth's axis of rotation. Thus in temperate latitudes the winter months tend to be significantly windier than the summer months. There may also be a tendency for strong winds or gales to develop around the time of the spring and autumn equinoxes. Tropical regions also experience seasonal phenomena such as monsoons and tropical storms which affect the wind climate. Indeed the extreme winds associated with tropical storms may significantly influence the design of wind turbines intended to survive in these locations.

Although a Weibull distribution gives a good representation of the wind regime at many sites, this is not always the case. For example, some sites showing distinctly different wind climates in summer and winter can be represented quite well by a double-peaked 'bi-Weibull' distribution, with different scale factors and shape factors in the two seasons, i.e.,

$$F(U) = F_1 \exp\left(-\left(\frac{U}{c_1}\right)^{k_1}\right) + (1 - F_1)\exp\left(-\left(\frac{U}{c_2}\right)^{k_2}\right) \tag{2.5}$$

Certain parts of California are good examples of this.

2.5 Synoptic and Diurnal Variations

On shorter time-scales than the seasonal changes described in Section 2.4, wind speed variations are somewhat more random, and less predictable. Nevertheless these variations contain definite patterns. The frequency content of these variations typically peaks at around 4 days or so. These are the 'synoptic' variations, which are associated with large-scale weather patterns such as areas of high and low pressure and associated weather fronts as they move across the earth's surface. Coriolis forces induce a circular motion of the air as it tries to move from high- to low-pressure regions. These coherent large-scale atmospheric circulation patterns may typically take a few days to pass over a given point, although they may occasionally 'stick' in one place for longer before finally moving on or dissipating.

Following the frequency spectrum to still higher frequencies, many locations will show a distinct diurnal peak, at a frequency of 24 h. This is usually driven by local thermal effects. Intense heating in the daytime may cause large convection cells in the atmosphere, which die down at night. This process is described in more detail in Section 2.6 as it also contributes significantly to turbulence, on time-scales representative of the size of the convection cells. Land and sea breezes, caused by differential heating and cooling between land and sea, also contribute significantly to the diurnal peak. The daily direction reversal of these winds would be seen as a 12 h peak in the spectrum of wind speed magnitude.

2.6 Turbulence

2.6.1 The nature of turbulence

Turbulence refers to fluctuations in wind speed on a relatively fast time-scale, typically less than about 10 min. In other words it corresponds to the highest frequency spectral peak in Figure 2.1. It is useful to think of the wind as consisting of a mean wind speed determined by the seasonal, synoptic and diurnal effects described above, which varies on a time-scale of one to several hours, with turbulent fluctuations superimposed. These turbulent fluctuations then have a zero mean when averaged over about 10 min. This description is a useful one as long as the 'spectral gap' in Figure 2.1 is reasonably distinct.

Turbulence is generated mainly from two causes: 'friction' with the earth's surface, which can be thought of as extending as far as flow disturbances caused by topographical features such as hills and mountains, and thermal effects which can cause air masses to move vertically as a result of variations of temperature, and hence of the density of the air. Often these two effects are interconnected, such as when a mass of air flows over a mountain range and is forced up into cooler regions where it is no longer in thermal equilibrium with its surroundings.

Turbulence is clearly a complex process, and one which cannot be represented simply in terms of deterministic equations. Obviously it does obey certain physical laws, such as those describing the conservation of mass, momentum and energy. However, in order to describe turbulence using these laws it is necessary to take account of temperature, pressure, density and humidity as well as the motion of the air itself in three dimensions. It is then possible to formulate a set of differential equations describing the process, and in principle the progress of the turbulence can be predicted by integrating these equations forward in time starting from certain initial conditions, and subject to certain boundary conditions. In practice, of course, the process can be described as 'chaotic' in that small differences in initial conditions or boundary conditions may result in large differences in the predictions after a relatively short time. For this reason it is generally more useful to develop descriptions of turbulence in terms of its statistical properties.

There are many statistical descriptors of turbulence which may be useful, depending on the application. These range from simple turbulence intensities and gust factors to detailed descriptions of the way in which the three components of turbulence vary in space and time as a function of frequency.

The turbulence intensity is a measure of the overall level of turbulence. It is defined as

$$I = \frac{\sigma}{\overline{U}} \tag{2.6}$$

where σ is the standard deviation of wind speed variations about the mean wind speed $\overline{U}$, usually defined over 10 min or 1 h. Turbulent wind speed variations can be considered to be roughly Gaussian, meaning that the speed variations are normally distributed, with standard deviation σ, about the mean wind speed $\overline{U}$. However, the tails of the distribution may be significantly non-Gaussian, so this approximation is not reliable for estimating, say, the probability of a large gust within a certain period.

The turbulence intensity clearly depends on the roughness of the ground surface and the height above the surface. However, it also depends on topographical features such as hills or mountains, especially when they lie upwind, as well as more local features such as trees or buildings. It also depends on the thermal behaviour of the atmosphere: for example, if the air near to the ground warms up on a sunny day, it may become buoyant enough to rise up through the atmosphere, causing a pattern of convection cells which are experienced as large-scale turbulent eddies.

Clearly as the height above ground increases, the effects of all these processes which are driven by interactions at the earth's surface become weaker. Above a certain height, the air flow can be considered largely free of surface influences. Here it can be considered to be driven by large-scale synoptic pressure differences and the rotation of the earth. This air flow is known as the geostrophic wind. At lower altitudes, the effect of the earth's surface can be felt. This part of the atmosphere is known as the boundary layer. The properties of the boundary layer are important in understanding the turbulence experienced by wind turbines.

2.6.2 The boundary layer

The principal effects governing the properties of the boundary layer are the strength of the geostrophic wind, the surface roughness, Coriolis effects due to the earth's rotation, and thermal effects.

The influence of thermal effects can be classified into three categories: stable, unstable and neutral stratification. Unstable stratification occurs when there is a lot of surface heating, causing warm air near the surface to rise. As it rises, it expands due to reduced pressure and therefore cools adiabatically. If the cooling is not sufficient to bring the air into thermal equilibrium with the surrounding air then it will continue to rise, giving rise to large convection cells. The result is a thick boundary layer with large-scale turbulent eddies. There is a lot of vertical mixing and transfer of momentum, resulting in a relatively small change of mean wind speed with height.

It the adiabatic cooling effect causes the rising air to become colder than its surroundings, its vertical motion will be suppressed. This is known as stable stratification. It often occurs on cold nights when the ground surface is cold. In this situation, turbulence is dominated by friction with the ground, and wind shear (the increase of mean wind speed with height) can be large.

In the neutral atmosphere, adiabatic cooling of the air as it rises is such that it remains in thermal equilibrium with its surroundings. This is often the case in strong winds, when turbulence caused by ground roughness causes sufficient mixing of the boundary layer. For wind energy applications, neutral stability is usually the most important situation to consider, particularly when considering the turbulent wind loads on a turbine, since these are largest in strong winds. Nevertheless, unstable conditions can be important as they can result in sudden gusts from a low level, and stable conditions can give rise to significant asymmetric loadings due to high wind shear. There can also be rapid changes in wind direction with height in this situation.

In the following sections, a series of relationships are presented which describe the properties of the atmospheric boundary layer, such as turbulence intensities, spectra, length scales and coherence functions. These relationships are partly based on theoretical considerations, and partly on empirical fits to a wide range of observations from many researchers taken in various conditions and in various locations.

In the neutral atmosphere, the boundary-layer properties depend mainly on the surface roughness and the Coriolis effect. The surface roughness is characterized by the roughness length z_o. Typical values of z_o are shown in Table 2.1.

The Coriolis parameter f is defined as

$$f = 2\Omega \sin(|\lambda|) \tag{2.7}$$

where Ω is the angular velocity of the earth's rotation, and λ is the latitude. This is zero at the equator, so the following description applies only to temperate latitudes. Here the height of the boundary layer is given by

$$h = u^*/(6f) \tag{2.8}$$

where u^* is known as the friction velocity, given by:

$$u^*/\overline{U}(z) = \kappa/[\ln(z/z_o) + \Psi] \tag{2.9}$$

where κ is the von Karman constant (approximately 0.4), z is the height above ground and z_o is the surface roughness length. Ψ is a function which depends on stability: it is negative for unstable conditions, giving rise to low wind shear, and positive for stable conditions, giving high wind shear. For neutral conditions, ESDU (1985) gives $\Psi = 34.5fz/u^*$, which is small compared to $\ln(z/z_o)$ for situations of interest here. If Ψ is ignored, the wind shear is then given by a logarithmic wind profile:

$$\overline{U}(z) \propto \ln(z/z_o) \tag{2.10}$$

Table 2.1 Typical Surface Roughness Lengths

Type of terrain	Roughness length z_o(m)
Cities, forests	0.7
Suburbs, wooded countryside	0.3
Villages, countryside with trees and hedges	0.1
Open farmland, few trees and buildings	0.03
Flat grassy plains	0.01
Flat desert, rough sea	0.001

A power law approximation,

$$\overline{U}(z) \propto z^{\alpha} \tag{2.11}$$

is often used, where the exponent α is typically about 0.14, but varies with the type of terrain. However, the value of α should also depend on the height interval over which the expression is applied, making this approximation less useful than the logarithmic profile.

If the surface roughness changes, the wind shear profile changes gradually downwind of the transition, from the original to the new profile. Essentially, a new boundary layer starts, and the boundary between the new and old boundary layers increases from zero at the transition point until the new boundary layer is fully established. The calculation of wind shear in the transition zone is covered by, for example, Cook (1985).

By combining Equations (2.8) and (2.9), we obtain the wind speed at the top of the boundary layer as

$$\overline{U}(h) = \frac{u^*}{\kappa}\left(\ln\left(\frac{u^*}{f z_0}\right) - \ln 6 + 5.75\right) \tag{2.12}$$

This is similar to the so-called geostrophic wind speed, G, which is the notional wind speed driving the boundary layer as calculated from the pressure field. The geostrophic wind speed is given by

$$G = \frac{u^*}{\kappa}\sqrt{\left[\ln\left(\frac{u^*}{f z_0}\right) - A\right]^2 + B^2} \tag{2.13}$$

where, for neutral conditions, $A = \ln 6$ and $B = 4.5$. This relationship is often referred to as the geostrophic drag law.

The effect of surface roughness is not only to cause the wind speed to decrease closer to the ground. There is also a change in direction between the 'free' pressure-driven geostrophic wind and the wind close to the ground. Although the geostrophic wind is driven by the pressure gradients in the atmosphere, coriolis forces act to force the wind to flow at right angles to the pressure gradient, causing a characteristic circulating pattern. Thus in the northern hemisphere, wind flowing from high pressure in the south to low pressure in the north will be forced eastwards by coriolis effects, in effect to conserve angular momentum on the rotating earth. The result is that the wind circulates anti-clockwise around low-pressure areas and clockwise around high-pressure areas, or the other way round in the southern hemisphere. Close to the ground, these flow directions are modified due to the effect of surface friction. The total direction change, α, from the geostrophic to the surface wind is given by

$$\sin\alpha = \frac{-B}{\sqrt{\left[\ln\left(\frac{u^*}{f z_0}\right)\right]^2 + B^2}} \tag{2.14}$$

2.6.3 Turbulence intensity

The turbulence intensity in the neutral atmosphere clearly depends on the surface roughness. For the longitudinal component, the standard deviation σ_u is approximately constant with height, so the turbulence intensity decreases with height. More precisely, the relationship $\sigma_u \approx 2.5u^*$ may be used to calculate the standard deviation, with the friction velocity u^* calculated as in the previous section. More recent work (ESDU, 1985) suggests a variation given by:

$$\sigma_u = \frac{7.5\eta(0.538 + 0.09\ln(z/z_0))^p u^*}{1 + 0.156\ln(u^*/f z_0)} \tag{2.15}$$

where

$$\eta = 1 - 6fz/u^* \tag{2.16}$$

$$p = \eta^{16} \tag{2.17}$$

This approximates to $\sigma_u = 2.5u^*$ close to the ground, but gives larger values at greater heights. The longitudinal turbulence intensity is then

$$I_u = \sigma_u/\overline{U} \tag{2.18}$$

The lateral (v) and vertical (w) turbulence intensities are given (ESDU, 1985) by

$$I_v = \frac{\sigma_v}{\overline{U}} = I_u\left(1 - 0.22\cos^4\left(\frac{\pi z}{2h}\right)\right) \tag{2.19}$$

$$I_w = \frac{\sigma_w}{\overline{U}} = I_u\left(1 - 0.45\cos^4\left(\frac{\pi z}{2h}\right)\right) \tag{2.20}$$

Note that specific values of turbulence intensity for use in design calculations are prescribed in some of the standards used for wind turbine design calculations, and these may not always correspond with the above expressions. For example, the Danish standard (DS472, 1992) specifies

$$I_u = 1.0/\ln(z/z_0) \tag{2.21}$$

with $I_v = 0.8I_u$ and $I_w = 0.5I_u$. The IEC standard (IEC, 1999) gives

$$I_u = I_{15}(a + 15/\overline{U})/(a + 1) \tag{2.22}$$

where $I_{15} = 0.18$ for 'higher turbulence sites' and 0.16 for 'lower turbulence sites', with corresponding values of a of 2 and 3 respectively. For the lateral and vertical components, a choice is allowed: either $I_v = 0.8I_u$ and $I_w = 0.5I_u$, or an isotropic model with $I_u = I_v = I_w$. The Germanischer Lloyd rules (GL, 1993) simply specify 20 percent turbulence intensity. Figure 2.4 shows example longitudinal turbulence intensities for the GL, IEC and Danish standards. The values for the Danish standard are given for 50 m height with roughness lengths of 0.3 and 0.03 m respectively.

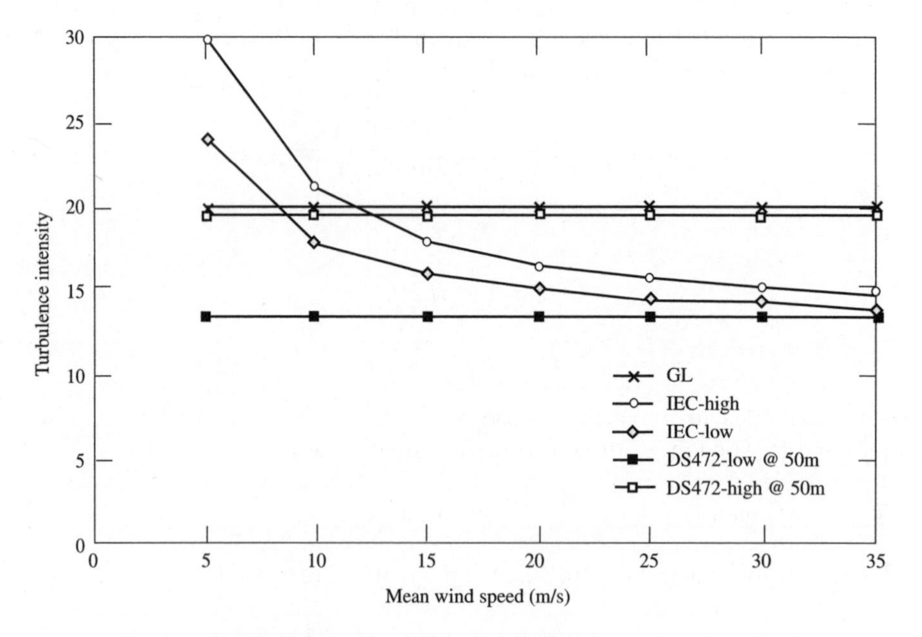

Figure 2.4 Turbulence Intensities According to Various Standards

2.6.4 Turbulence spectra

The spectrum of turbulence describes the frequency content of wind-speed variations. According to the Kolmogorov law, the spectrum must approach an asymptotic limit proportional to $n^{-5/3}$ at high frequency (here n denotes the frequency, in Hz). This relationship is based on the decay of turbulent eddies to higher and higher frequencies as turbulent energy is dissipated as heat.

Two alternative expressions for the spectrum of the longitudinal component of turbulence are commonly used, both tending to this asymptotic limit. These are the Kaimal and the von Karman spectra, which take the following forms:

$$\text{Kaimal:} \quad \frac{nS_u(n)}{\sigma_u^2} = \frac{4nL_{1u}/\overline{U}}{(1+6nL_{1u}/\overline{U})^{5/3}} \tag{2.23}$$

$$\text{von Karman:} \quad \frac{nS_u(n)}{\sigma_u^2} = \frac{4nL_{2u}/\overline{U}}{(1+70.8(nL_{2u}/\overline{U})^2)^{5/6}} \tag{2.24}$$

where $S_u(n)$ is the autospectral density function for the longitudinal component and L_{1u} and L_{2u} are length scales. In order for these two forms to have the same high-frequency asymptotic limit, these length scales must be related by the ratio $(36/70.8)^{-5/4}$, i.e., $L_{1u} = 2.329 L_{2u}$. The appropriate length scales to use are discussed in the next section.

According to Petersen *et al.* (1998), the von Karman spectrum gives a good description for turbulence in wind tunnels, although the Kaimal spectrum may give a better fit to empirical observations of atmospheric turbulence. Nevertheless the von Karman spectrum is often used for consistency with analytical expressions for the correlations. The length scale L_{2u} is identified as the integral length scale of the longitudinal component in the longitudinal direction, denoted ${}^xL_u$. The Kaimal spectrum has a lower, broader peak than the von Karman spectrum (see Figures 2.5 to 2.7).

Recent work suggests that the von Karman spectrum gives a good representation of atmospheric turbulence above about 150 m, but has some deficiencies at lower altitudes. Several modifications have been suggested (Harris, 1990) and a modified von Karman spectrum of the following form is recommended (ESDU, 1985):

$$\frac{nS_u(n)}{\sigma_u^2} = \beta_1 \frac{2.987nL_{3u}/\overline{U}}{(1+(2\pi nL_{3u}/\overline{U})^2)^{5/6}} + \beta_2 \frac{1.294nL_{3u}/\overline{U}}{(1+(\pi nL_{3u}/\overline{U})^2)^{5/6}} F_1 \tag{2.25}$$

All three of these spectra have corresponding expressions for the lateral and vertical components of turbulence. The Kaimal spectra have the same form as for the longitudinal component but with different length scales, L_{1v} and L_{1w} respectively. The von Karman spectrum for the i component ($i = v$ or w) is

$$\frac{nS_i(n)}{\sigma_i^2} = \frac{4(nL_{2i}/\overline{U})(1+755.2(nL_{2i}/\overline{U})^2)}{(1+283.2(nL_{2i}/\overline{U})^2)^{11/6}} \tag{2.26}$$

where $L_{2v} = {}^xL_v$ and $L_{2w} = {}^xL_w$. For the modified von Karman spectrum it is

$$\frac{nS_i(n)}{\sigma_i^2} = \beta_1 \frac{2.987(nL_{3i}/\overline{U})(1+\frac{8}{3}(4\pi nL_{3i}/\overline{U})^2)}{(1+(4\pi nL_{3i}/\overline{U})^2)^{11/6}} + \beta_2 \frac{1.294nL_{3i}/\overline{U}}{(1+(2\pi nL_{3i}/\overline{U})^2)^{5/6}} F_{2i} \tag{2.27}$$

2.6.5 Length scales and other parameters

To use the spectra defined above, it is necessary to define the appropriate length scales. Additional parameters β_1, β_2, F_1, and F_2 are also required for the modified von Karman model.

The length scales are dependent on the surface roughness z_0, as well as on the height above ground (z); proximity to the ground constrains the size of turbulent eddies and thus reduces the length scales. If there are many small obstacles on the ground of typical height z', the height above ground should be corrected for the effect of these by assuming that the effective ground surface is at a height $z' - 2.5z_0$ (ESDU, 1975). Far enough above the ground, i.e., for $z > z_i$, the turbulence is no longer constrained by the proximity of the surface and becomes isotropic. According to ESDU (1975), $z_i = 1000z_0^{0.18}$ and above this height ${}^xL_u = 280$ m, and ${}^yL_u = {}^zL_u = {}^xL_v = {}^zL_v = 140$ m. Even for very small roughness lengths z_0, the isotropic region is well above the height of a wind turbine and the following corrections for $z < z_i$ should be applied:

$$
\begin{aligned}
{}^xL_u &= 280(z/z_i)^{0.35} \\
{}^yL_u &= 140(z/z_i)^{0.38} \\
{}^zL_u &= 140(z/z_i)^{0.45} \\
{}^xL_v &= 140(z/z_i)^{0.48} \\
{}^zL_v &= 140(z/z_i)^{0.55}
\end{aligned}
\qquad (2.28)
$$

together with ${}^xL_w = {}^yL_w = 0.35z$ (for $z < 400$ m). Expressions for ${}^yL_v$ and ${}^zL_w$ are not given. The length scales ${}^xL_u$, ${}^xL_v$ and ${}^xL_w$ can be used directly in the von Karman spectra. For the Kaimal spectra we already have $L_{1u} = 2.329{}^xL_u$, and to achieve the same high frequency asymptotes for the other components we also have $L_{1v} = 3.2054{}^xL_v$, $L_{1w} = 3.2054{}^xL_w$.

Later work based on measurements for a greater range of heights (Harris, 1990; ESDU, 1985) takes into account an increase in length scales with the thickness of the boundary layer, h, which also implies a variation of length scales with mean wind speed. This yields more complicated expressions for the nine length scales in terms of z/h, σ_u/u^* and the Richardson number $u^*/(fz_0)$.

Note that some of the standards used for wind turbine loading calculations prescribe that certain turbulence spectra and/or length scales be used. These are often simplified compared to the expressions given above. Thus the Danish standard (DS 472, 1992) specifies a Kaimal spectrum with

$$
\begin{aligned}
L_{1u} &= 150 \text{ m}, \quad \text{or } 5z \text{ for } z < 30 \text{ m} \\
L_{1v} &= 0.3L_{1u} \\
L_{1w} &= 0.1L_{1u}
\end{aligned}
\qquad (2.29)
$$

while the IEC standard (IEC, 1999) recommends either a Kaimal model with

$$
\begin{aligned}
L_{1u} &= 170.1 \text{ m}, \quad \text{or } 5.67z \text{ for } z < 30 \text{ m} \\
L_{1v} &= 0.3333 L_{1u} \\
L_{1w} &= 0.08148 L_{1u}
\end{aligned} \tag{2.30}
$$

or an isotropic von Karman model with

$$
\begin{aligned}
{}^{x}L_u &= 73.5 \text{ m, or } 2.45z \text{ for } z < 30 \text{ m} \\
{}^{x}L_v &= {}^{x}L_w = 0.5^{x}L_u
\end{aligned} \tag{2.31}
$$

Eurocode 1 (1997) specifies a longitudinal spectrum of Kaimal form with $L_{1u} = 1.7L_i$, where

$$
L_i = 300(z/300)^{\varepsilon} \tag{2.32}
$$

for $z < 300$ m, with ε varying between 0.13 over open water to 0.46 in urban areas. This standard is used for buildings, but not usually for wind turbines.

Figure 2.5 compares these various longitudinal turbulence spectra at 30 m height, for a mean wind speed of 10 m/s. The surface roughness is 0.001 m, corresponding to very flat land or rough sea, and the latitude is 50°. There is reasonable agreement between the various spectra in this situation, apart from the Eurocode spectrum which is shifted to somewhat lower frequencies. Note the characteristic difference between the Kaimal and von Karman spectra, the latter being rather more sharply peaked. The improved von Karman spectrum (Equation 2.25) is intermediate in shape.

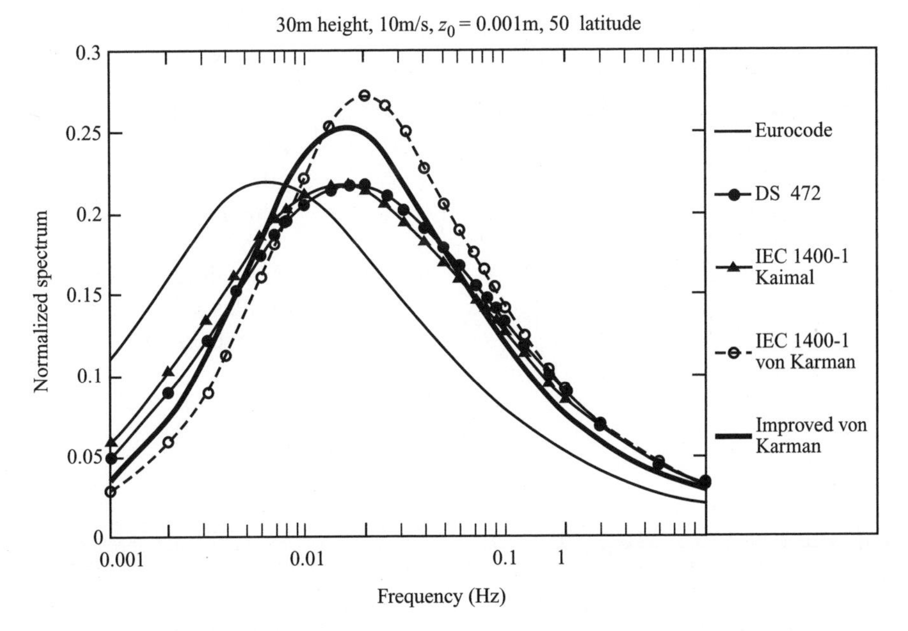

Figure 2.5 Comparison of Spectra over Smooth Terrain at 10 m/s

Figure 2.6 shows a similar figure for a roughness length of 0.1 m, typical of villages or farmland with hedges and trees. Only the Eurocode and improved von Karman spectra have moved, coming closer together.

A further comparison is shown in Figure 2.7. The intermediate roughness length of 0.01 m is used here, but the wind speed is increased to 25 m/s. Now the IEC and DS472 spectra have shifted further than the other two towards the higher frequencies. There is now a rather marked difference between the two groups of spectra.

2.6.6 Cross-spectra and coherence functions

The turbulence spectra presented in the preceding sections describe the temporal variation of each component of turbulence at any given point. However, as the wind-turbine blade sweeps out its trajectory, the wind-speed variations it experiences are not well represented by these single-point spectra. The spatial variation of turbulence in the lateral and vertical directions is clearly important, since this spatial variation is 'sampled' by the moving blade and thus contributes to the temporal variations experienced by it.

In order to model these effects, the spectral description of turbulence must be extended to include information about the cross-correlations between turbulent fluctuations at points separated laterally and vertically. Clearly these correlations decrease as the distance separating two points increases. The correlations are also smaller for high-frequency than for low-frequency variations. They can therefore be

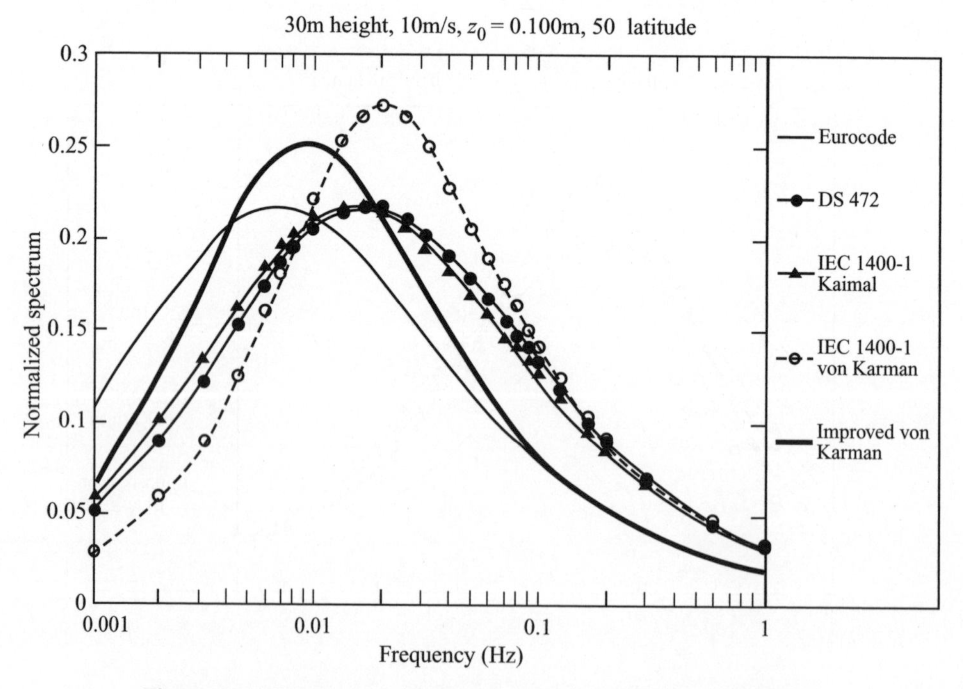

Figure 2.6 Comparison of spectra over Rough Terrain at 10 m/s

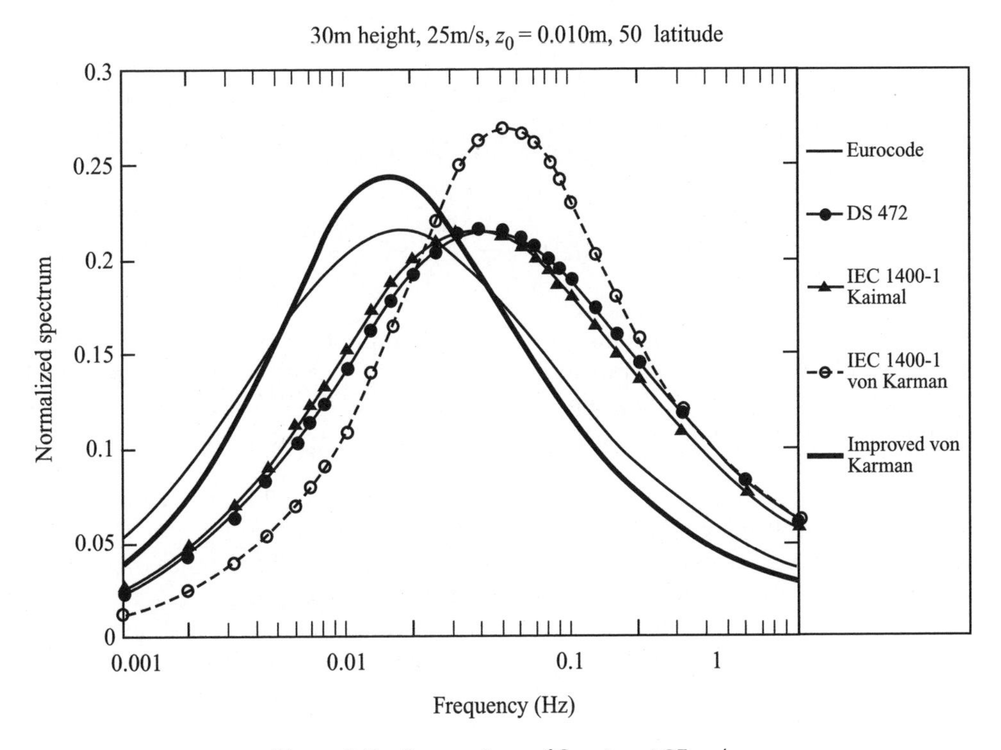

Figure 2.7 Comparison of Spectra at 25 m/s

described by 'coherence' functions, which describe the correlation as a function of frequency and separation. The coherence $C(\Delta r, n)$ is defined by

$$C(\Delta r, n) = \frac{|S_{12}(n)|}{\sqrt{S_{11}(n)S_{22}(n)}} \tag{2.33}$$

where n is frequency, $S_{12}(n)$ is the cross-spectrum of variations at the two points separated by Δr, and $S_{11}(n)$ and $S_{22}(n)$ are the spectra of variations at each of the points (usually these can be taken as equal).

Starting from von Karman spectral equations, and assuming Taylor's frozen turbulence hypothesis, an analytical expression for the coherence of wind-speed fluctuations can be derived. Accordingly for the longitudinal component at points separated by a distance Δr perpendicular to the wind direction, the coherence $C_u(\Delta r, n)$ is:

$$C_u(\Delta r, n) = 0.994(A_{5/6}(\eta_u) - \tfrac{1}{2}\eta_u^{5/3}A_{1/6}(\eta_u)) \tag{2.34}$$

Here $A_j(x) = x^j K_j(x)$ where K is a fractional order modified Bessel function, and

$$\eta_u = \Delta r \sqrt{\left(\frac{0.747}{L_u}\right)^2 + \left(c\frac{2\pi n}{\overline{U}}\right)^2} \tag{2.35}$$

with $c = 1$. L_u is a local length scale which can be defined as:

$$L_u(\Delta r, n) = 2f_u(n)\sqrt{\frac{({}^yL_u\Delta y)^2 + ({}^zL_u\Delta z)^2}{\Delta y^2 + \Delta z^2}} \tag{2.36}$$

where Δy and Δz are the lateral and vertical components of the separation Δr, and ${}^yL_u$ and ${}^zL_u$ are the lateral and vertical length scales for the longitudinal component of turbulence. Normally $f_u(n) = 1$, but ESDU (1975) suggests a modification at low frequencies where the wind becomes more anisotropic, with $f_u(n) = MIN(1.0, 0.04n^{-2/3})$.

The IEC (1999) standard gives an isotropic turbulence model for use with the von Karman spectrum, in which ${}^xL_u = 2{}^yL_u = 2{}^zL_u$, and then $L_u = {}^xL_u$, and $f_u(n) = 1$. The modified von Karman model described in Equation (2.25) also uses $f_u(n) = 1$, but the factor c in Equation (2.35) is modified instead (ESDU, 1985).

For the lateral and vertical components, the corresponding equations are as follows. The analytical derivation for the coherence, based as before on the von Karman spectrum and Taylor's hypothesis, is

$$C_i(\Delta r, n) = \frac{0.597}{2.869\gamma_i^2 - 1}[4.781\gamma_i^2 A_{5/6}(\eta_i) - A_{11/6}(\eta_i)] \tag{2.37}$$

for $i = u$ or v, where η_i is calculated as in Equation (2.35) but with L_u replaced by L_v or L_w respectively, and with $c = 1$. Also

$$\gamma_i = \frac{\eta_i L_i(\Delta r, n)}{\Delta r} \tag{2.38}$$

and L_v and L_w are given by expressions analogous to Equation (2.36).

The expressions for spatial coherence in Equations (2.34) and (2.37) above are derived theoretically from the von Karman spectrum, although there are empirical factors in some of the expressions for length scales for example. If a Kaimal rather than a von Karman spectrum is used as the starting point, there are no such relatively straightforward analytical expressions for the coherence functions. In this case a simpler and purely empirical exponential model of coherence is often used. The IEC (1999) standard, for example, gives the following expression for the coherence of the longitudinal component of turbulence:

$$C_u(\Delta r, n) = \exp\left(-8.8\Delta r\sqrt{\left(\frac{0.12}{L_u}\right)^2 + \left(\frac{n}{\overline{U}}\right)^2}\right)$$

$$\cong \exp(-1.4\eta_u) \tag{2.39}$$

with η_u as in Equation (2.35).

The standard also states that this may also be used with the von Karman model, as an approximation to Equation (2.34). However, the standard does not specify the coherence of the other two components to be used in conjunction with the Kaimal model.

The three turbulence components are usually assumed to be independent of one another. This is a reasonable assumption, although in practice Reynolds stresses may result in a small correlation between the longitudinal and vertical components near to the ground.

Clearly there are significant discrepancies between the various recommended spectra and coherence functions. Also these wind models are applicable to flat sites, and there is only limited understanding of the way in which turbulence characteristics change over hills and in complex terrain. Given the important effect of turbulence characteristics on wind turbine loading and performance, this is clearly an area in which there is scope for further research.

2.7 Gust Wind Speeds

It is often useful to know the maximum gust speed which can be expected to occur in any given time interval. This is usually represented by a gust factor G, which is the ratio of the gust wind speed to the hourly mean wind speed. G is obviously a function of the turbulence intensity, and it also clearly depends on the duration of the gust – thus the gust factor for a 1 s gust will be larger than for a 3 s gust, since every 3 s gust has within it a higher 1 s gust.

While it is possible to derive expressions for gust factors starting from the

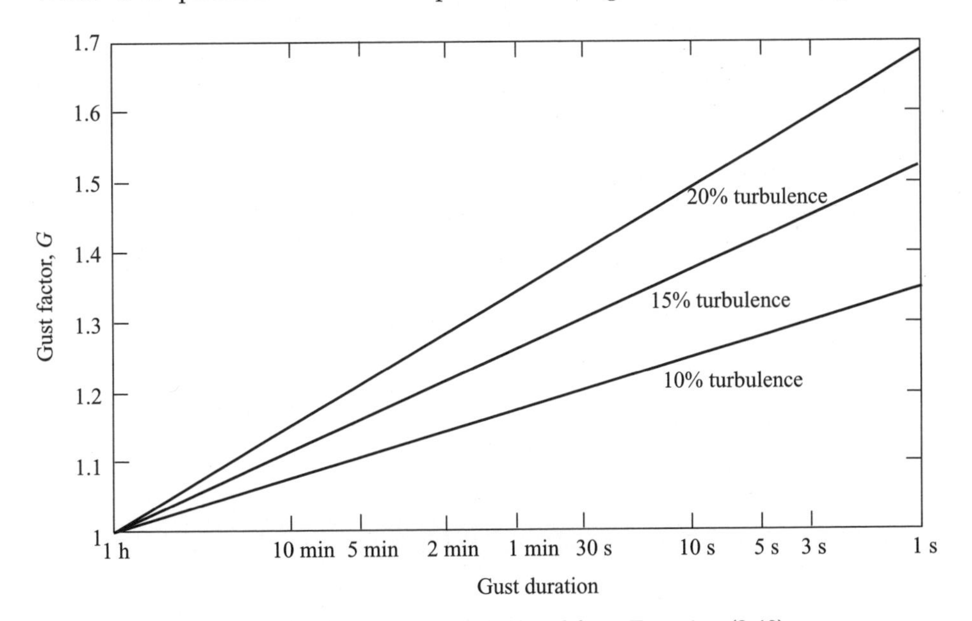

Figure 2.8 Gust Factors Calculated from Equation (2.40)

turbulence spectrum (Greenway, 1979; ESDU, 1983), an empirical expression due to Weiringa (1973) is often used as it is much simpler, and agrees well with theoretical results. Accordingly the t s gust factor is given by

$$G(t) = 1 + 0.42 I_u \ln \frac{3600}{t} \tag{2.40}$$

where I_u is the longitudinal turbulence intensity. Figure 2.8 shows the gust factors for several different turbulence intensities and gust durations calculated according to this expression.

2.8 Extreme Wind Speeds

In addition to the foregoing descriptions of the average statistical properties of the wind, it is clearly of interest to be able to estimate the long-term extreme wind speeds which might occur at a particular site.

A probability distribution of hourly mean wind speeds such as the Weibull distribution will yield estimates of the probability of exceeding any particular level of hourly mean wind speed. However, when used to estimate the probability of extreme winds, an accurate knowledge of the high wind speed tail of the distribution is required, and this will not be very reliable since almost all of the data which was used to fit the parameters of the distribution will have been recorded at lower wind speeds. Extrapolating the distribution to higher wind speeds cannot be relied upon to give an accurate result.

Fisher and Tippett (1928) and Gumbel (1958) have developed a theory of extreme values which is useful in this context. If a measured variable (such as hourly mean wind speed $\overline{U}$) conforms to a particular cumulative probability distribution $F(\overline{U})$, i.e., $F(\overline{U}) \to 1$ as $\overline{U}$ increases, then the peak values of hourly mean wind speed in a given period (a year, for example) will have a cumulative probability distribution of F^N, where N is the number of independent peaks in the period. In the UK, for example, Davenport (1964) has estimated that there are about 160 independent wind speed peaks per year, corresponding to the passage of individual weather systems. Thus if

$$F(\overline{U}) = 1 - \exp\left(-\left(\frac{\overline{U}}{c}\right)^k\right)$$

as for a Weibull distribution, the wind speed peaks in 1 year will have a cumulative probability distribution given approximately by

$$\left[1 - \exp\left(-\left(\frac{\overline{U}}{c}\right)^k\right)\right]^{160}$$

However, as indicated above, this is unlikely to give accurate estimates for extreme hourly means, since the high-wind tail of the distribution cannot be considered to be reliably known. However, Fisher and Tippett (1928) demonstrated that for any cumulative probability distribution function which converges towards unity at least exponentially (as is usually the case for wind speed distributions, including the Weibull distribution), the cumulative probability distribution function for extreme values $\hat{U}$ will always tend towards an asymptotic limit

$$F(\hat{U}) = \exp(-\exp(-a(\hat{U} - U')))$$

as the observation period increases. U' is the most likely extreme value, or the *mode* of the distribution, while $1/a$ represents the width or spread of the distribution and is termed the *dispersion*.

This makes it possible to estimate the distribution of extreme values based on a fairly limited set of measured peak values, for example a set of measurements of the highest hourly mean wind speeds $\hat{U}$ recorded during each of N storms. The N measured extremes are ranked in ascending order, and an estimate of the cumulative probability distribution function is obtained as

$$\tilde{F}(\hat{U}) \cong \frac{m(\hat{U})}{N+1}$$

where $m(\hat{U})$ is the rank, or position in the sequence, of the observation $\hat{U}$. Then a plot of $-\ln(-\ln(\tilde{F}(\hat{U})))$ against $\hat{U}$ is used to estimate the mode U' and dispersion $1/a$ by fitting a straight line to the datapoints. This is the method due to Gumbel. Lieblein (1974) has developed a numerical technique which gives a less biased estimate of U' and $1/a$ than a simple least squares fit to a Gumbel plot.

Having made an estimate of the cumulative probability distribution of extremes $F(\hat{U})$, the M year extreme hourly mean wind speed can be estimated as the value of $\hat{U}$ corresponding to the probability of exceedance $F = 1 - 1/M$.

According to Cook (1985), a better estimate of the probability of extreme winds is obtained by fitting a Gumbel distribution to extreme values of wind speed squared. This is because the cumulative probability distribution function of wind speed squared is closer to exponential than the distribution of wind speed itself, and it converges much more rapidly to the Gumbel distribution. Therefore by using this method to predict extreme values of wind speed squared, more reliable estimates can be obtained from a given number of observations.

2.8.1 Extreme winds in standards

The design of wind turbines must allow them to withstand extremes of wind speed, as well as responding well to the more 'typical' conditions described above. Therefore the various standards also specify the extremes of wind speed which must be designed for. This includes extreme mean wind speeds as well as various types of severe gust.

Extreme conditions may be experienced with the machine operating, parked or idling with or without various types of fault or grid loss, or during a particular operation such as a shut-down event. The extreme wind conditions may be characterized by a 'return time': for example a 50 year gust is one which is so severe that it can be expected to occur on average only once every 50 years. It would be reasonable to expect a turbine to survive such a gust, provided there was no fault on the turbine.

It is always possible that the turbine happens to be shut down on account of a fault when a gust occurs. If the fault impairs the turbine's ability to cope with a gust, for example if the yaw system has failed and the turbine is parked at the wrong angle to the wind, then the turbine may have to withstand even greater loads. However, the probability of the most extreme gusts occurring at the same time as a turbine fault is very small, and so it is usual to specify that a turbine with a fault need only be designed to withstand, for example, the annual extreme gust and not the 50 year extreme gust.

For this to be valid, it is important that the faults in question are not correlated with extreme wind conditions. Grid loss is not considered to be a fault with the turbine, and is actually quite likely to be correlated with extreme wind conditions.

Clearly the extreme wind speeds and gusts (both in terms of magnitude and shape) may be quite site-specific. They may differ considerably between flat coastal sites and rugged hill-tops for example.

The IEC (1999) standard, for example, specifies a 'reference wind speed' V_{ref} which is five times the annual mean wind speed. The 50 year extreme wind speed is then given by 1.4 times V_{ref} at hub height, and varying with height using a power law exponent of 0.11. The annual extreme wind speed is taken as 75 percent of the 50 year value.

The standard goes on to define a number of transient events which the turbine must be designed to withstand. These are described in more detail in Section 5.4.3 and include the following.

- *Extreme operating gust (EOG)*: a decrease in speed, followed by a steep rise, a steep drop, and a rise back to the original value. The gust amplitude and duration vary with the return period.

- *Extreme direction change (EDC)*: this is a sustained change in wind direction, following a cosine-shaped curve. The amplitude and duration of the change once again depend on the return period.

- *Extreme coherent gust (ECG)*: this is a sustained change in wind speed, again following a cosine-shaped curve with the amplitude and duration depending on the return period.

- *Extreme coherent gust with direction change (ECD)*: simultaneous speed and direction transients similar to EDC and ECG.

- *Extreme wind shear (EWS)*: a transient variation in the horizontal and vertical wind gradient across the rotor. The gradient first increases and then falls back to the initial level, following a cosine-shaped curve.

These transient events are deterministic gusts intended to represent the extreme turbulent variations which would be expected to occur at the specified return period. They are not intended to occur *in addition* to the normal turbulence described previously. Such deterministic coherent gusts, however, have little basis in terms of actual measured or theoretical wind characteristics, and are likely to be superseded in future standards by something more closely related to the actual characteristics of extreme turbulence.

2.9 Wind-speed Prediction and Forecasting

Because of the variable nature of the wind resource, the ability to forecast wind speed some time ahead is often valuable. Such forecasts fall broadly into two categories: predicting short-term turbulent variations over a time-scale of seconds to minutes ahead, which may be useful for assisting with the operational control of wind turbines or wind farms, and longer-term forecasts over periods of a few hours or days, which may be useful for planning the deployment of other power stations on the network.

Short-term forecasts necessarily rely on statistical techniques for extrapolating the recent past, whereas the longer-term forecasts can make use of meteorological methods. A combination of meteorological and statistical forecasts can give very useful predictions of wind farm power output.

2.9.1 Statistical methods

The simplest statistical prediction is known as a 'persistence' forecast: the prediction is set equal to the last available measurement. In other words the last measured value is assumed to persist into the future without any change:

$$\hat{y}_k = y_{k-1}$$

where y_{k-1} is the measured value at step $k-1$ and $\hat{y}_k$ is the prediction for the next step.

A more sophisticated prediction might be some linear combination of the last n measured values, i.e.,

$$\hat{y}_k = \sum_{i=1}^{n} a_i y_{k-i}$$

This is known as an nth order autoregressive model, or AR(n). We can now define the prediction error at step k by

$$e_k = \hat{y}_k - y_k$$

and then use the recent prediction errors to improve the prediction:

$$\hat{y}_k = \sum_{i=1}^{n} a_i y_{k-i} + \sum_{j=1}^{m} b_j e_{k-j}$$

This is known as an nth order autoregressive, mth order moving average model, or ARMA(n, m). This can be further extended to an ARMAX model, where the X stands for an 'exogenous' variable: another measured variable which is included in the prediction because it influences y.

The model parameters a_i, b_j can be estimated in various ways. A useful technique is the method of recursive least squares, or RLS (Ljung and Söderström, 1983). Estimates of the model parameters are updated on each timestep in such a way as to minimize the expected value of the sum of squares of the prediction errors. By including a so-called 'forgetting factor', the influence of older observations can be progressively reduced, leading to an adaptive estimation of the parameters, which will gradually change to accommodate variations in the statistical properties of the variable y.

Bossanyi (1985) investigated the use of ARMA models for wind-speed predictions from a few seconds to a few minutes ahead, obtaining reductions in rms prediction errors of up to 20 percent when compared to a persistence forecast. The best results were obtained when predicting 10 min ahead from 1 min data.

Kariniotakis, Nogaret and Stavrakakis (1997) compare ARMA methods against a selection of more recent techniques such as neural network, fuzzy logic and wavelet-based methods. The fuzzy logic method is tentatively selected as giving the best predictions over periods of 10 min to 2 h, with improvements of 10–18 percent compared to persistence.

Nielsen and Madsen (1999) use an ARX model with recursive least squares to predict wind-farm power output based on previous values of power output, and measured wind speed as an exogenous variable, supplemented by a function describing the diurnal variations of wind speed and by meteorological forecasts of wind speed and direction. Predictions up to 48 h ahead are considered, and the inclusion of meteorological forecasts is shown to improve the predictions significantly, especially for the longer period forecasts.

2.9.2 Meteorological methods

As indicated in the previous section, much better predictions can be made by using meteorological forecasts than by using purely statistical methods, when predictions over time-scales of a few hours or days are considered. Very sophisticated meteorological forecasts are available from highly detailed simulation models of the atmosphere, fed by many recorded observations of pressure, temperature, wind speed, etc. over wide areas of land and sea.

Landberg (1997, 1999) describes the use of such models to predict wind-farm output, by extrapolating the large-scale wind predictions produced by these models down the specific wind-farm site. The geostrophic drag law and the logarithmic wind shear profile (Section 2.6.2) are used to extrapolate the wind forecasts down to ground level. Modifications to the flow resulting from the topography, the physical geography and surface roughness conditions in the area surrounding the wind-farm are then modelled by the WA^{s}P program (Mortensen *et al.*, 1993). A turbine wake interaction model PARK (Sanderhoff, 1993) then takes account of wind direction in relation to the actual turbine positions to calculate wake losses, and finally a statistical model (as described in the previous section) combines the meteorological forecasts with recent measurements on the wind farm to give predictions of the energy output which are good enough to be useful in planning the deployment of other power stations on the network.

2.10 Turbulence in Wakes and Wind Farms

As it extracts energy from the wind, a turbine leaves behind it a wake characterized by reduced wind speeds and increased levels of turbulence. Another turbine operating in this wake, or deep inside a wind farm where the effects of a number of wakes may be felt simultaneously, will therefore produce less energy and suffer greater structural loading than a turbine operating in the free stream.

Immediately behind a turbine, its wake can crudely be considered as a region of reduced wind speed slightly larger in diameter than the turbine itself (Vermeulen, 1980). The reduction in velocity is directly related to the thrust coefficient of the turbine, since this determines the momentum extracted from the flow. As this reduced wind speed region convects downstream, the wind speed gradient between the wake and the free flow outside the wake results in additional shear-generated turbulence, which assists the transfer of momentum into the wake from the surrounding flow. Thus the wake and the surrounding flow start to mix, and the region of mixing spreads inwards to the centre of the wake, as well as outwards to make the width of the wake increase. In this way, the velocity deficit in the wake is eroded and the wake becomes broader but shallower until the flow has fully recovered far downstream. The rate at which this occurs is dependent on the ambient turbulence level.

In addition to this shear-generated turbulence, the turbine itself generates additional turbulence directly, as a result of the tip vortices shed by the blades and the general disturbance to the flow caused by the blades, nacelle and tower. This 'mechanical' component of turbulence is of relatively high frequency, and decays relatively quickly. Bossanyi (1983) develops a theoretical model which describes how this additional turbulence might decay: large eddies give rise to smaller ones, and so the turbulent energy moves to higher and higher frequencies until it is eventually dissipated as heat. The model predicts a faster rate of decay in low winds and in high ambient turbulence intensities.

Beyond the near wake region, once the shear-generated turbulence has reached the centre of the wake and started to erode the centreline velocity deficit, the mean velocity variation in the wake can be described well by an axisymmetric profile with a Gaussian cross-section. The development of the wake profile downstream of this point can be reasonably well predicted, for example using an eddy viscosity model (Ainslie, 1988). This is partly theoretical, derived from the Navier–Stokes equation of fluid flow, but with some empirical terms.

Theoretical models capable of predicting turbulence levels in the wake are less well developed. Quarton and Ainslie (1989) examined a number of different sets of wake turbulence measurements, both in wind tunnels using small wind turbine models or gauze simulators, and behind full-size turbines in the free stream. An empirical formula for added turbulence I_+ at a downstream distance x from the turbine was found to give a good fit to the various measurements:

$$I_+ = 4.8C_T^{0.7}I_0^{0.68}(x/x_n)^{-0.57}$$

where C_T is the turbine thrust coefficient, I_0 the ambient turbulence intensity, and x_n the length of the near wake region. Here I_0 and I_+ are expressed as percentages. On the basis of further work, an improved expression was subsequently proposed by Hassan (1992):

$$I_+ = 5.7C_T^{0.7}I_0^{0.68}(x/x_n)^{-0.96}$$

The added turbulence is defined as the square root of the additional wind speed variance normalized by the mean wind speed, i.e.,

$$I_+ = \sqrt{I_{\text{wake}}^2 - I_0^2}$$

where I_{wake} is the total wake turbulence intensity, at any given downstream distance.

The length of the near wake region, x_n, is calculated according to Vermeulen (1980) in terms of the rotor radius R and the thrust coefficient C_T as

$$x_n = \frac{nr_0}{\left(\frac{dr}{dx}\right)}$$

where

$$r_0 = R\sqrt{\frac{m+1}{2}}$$

$$m = \frac{1}{\sqrt{1-C_T}}$$

$$n = \frac{\sqrt{0.214+0.144m}(1-\sqrt{0.134+0.124m})}{(1-\sqrt{0.214+0.144m})\sqrt{0.134+0.124m}}$$

and $\mathrm{d}r/\mathrm{d}x$ is the wake growth rate:

$$\frac{\mathrm{d}r}{\mathrm{d}x} = \sqrt{\left(\frac{\mathrm{d}r}{\mathrm{d}x}\right)_\alpha^2 + \left(\frac{\mathrm{d}r}{\mathrm{d}x}\right)_m^2 + \left(\frac{\mathrm{d}r}{\mathrm{d}x}\right)_\lambda^2}$$

where

$$\left(\frac{\mathrm{d}r}{\mathrm{d}x}\right)_\alpha = 2.5I_0 + 0.005$$

is the growth rate contribution due to ambient turbulence,

$$\left(\frac{\mathrm{d}r}{\mathrm{d}x}\right)_m = \frac{(1-m)\sqrt{1.49+m}}{(1+m)9.76}$$

is the contribution due to shear-generated turbulence, and

$$\left(\frac{\mathrm{d}r}{\mathrm{d}x}\right)_\lambda = 0.012B\lambda$$

is the contribution due to mechanical turbulence, where B is the number of blades and λ is the tip speed ratio.

Deep inside a wind farm, the reduction in wind speed and the increase in turbulence intensity are the result of the superposition of wakes from many upwind turbines. Frandsen and Thøgersen (1999) propose a model based on the geostrophic drag law which takes into account the additional 'surface roughness' caused by the turbines themselves. This leads to a formula for added turbulence above hub height:

$$I_{++} = \frac{0.36}{1 + 0.2\sqrt{s_1 s / C_T}}$$

where s_1 and s are the inter-turbine spacings, normalized by rotor diameter, within a row and between rows. Since this does not apply below hub height, the average added turbulence intensity I_+ is then calculated as

$$I_+ = \tfrac{1}{2}\left(I_0 + \sqrt{I_0^2 + I_{++}^2}\right)$$

However, no consensus has yet emerged on a sufficiently well-validated formula for turbulence intensity within a wind farm for use in wind turbine design calculations.

2.11 Turbulence in Complex Terrain

Predicting the turbulence intensity and spectrum at a given point within an area of complex terrain is not straightforward. Hilly terrain upwind of the site in question will lead to generally higher turbulence levels, and some authors have suggested that this can be calculated from a 'regional roughness length' which takes the topography into account as well as the surface roughness (Tieleman, 1992). On the other hand, distortion of the flow by the local terrain may reduce the turbulence intensity. At heights above ground which are of importance for wind turbines, rapid distortion theory applies, which means that the variance of the turbulent fluctuations will not change much as the flow passes over terrain features such as hills. Therefore if there is acceleration of the flow as it passes over a hill, the turbulence intensity will decrease, and the length scale will increase, resulting in a shift of the turbulence spectrum towards lower frequencies, without any change of shape (Schlez, 2000). This effect is therefore easily estimated once a model such as WA^{S}P has been used to calculate the speed-up factor at a particular point. However, the effect is also accompanied by a shift of turbulent energy from the longitudinal to the lateral and vertical components of turbulence, causing the turbulence on hilltops to be more isotropic (Petersen *et al.*, 1998).

References

Ainslie, J. F., (1988). 'Calculating the flowfield in the wake of wind turbines.' *J. Wind Engng. Ind. Aerodyn.*, **27**, 213–224.

Bossanyi, E. A., (1983). 'Windmill wake turbulence decay – a preliminary theoretical model.' *SERI/TR-635-1280*. Solar Energy Research Institute, Colorado, USA.

Bossanyi, E. A., (1985). 'Short-term stochastic wind prediction and possible control applications.' *Proceedings of the Delphi Workshop on Wind Energy Applications*.

Cook, N. J., (1985). *The designer's guide to wind loading of building structures*, Part 1. Butterworths, UK.

Davenport, A. G., (1964). 'Note on the distribution of the largest value of a random function with application to gust loading.' *Proc. Inst. Civ. Eng.*, **28**, 187–196.

DS 472, (1992). 'Code of practice for loads and safety of wind turbine constructions.' *DS 472*, The Danish Society of Engineers and the Federation of Engineers.

ESDU, (1975). 'Characteristics of atmospheric turbulence near the ground. Part III: Variations in space and time for strong winds (neutral atmosphere).' *ESDU 75001*, Engineering Sciences Data Unit, UK.

ESDU, (1983). 'Strong winds in the atmospheric boundary layer. Part 2: Discrete gust speeds.' *ESDU 83045*. Engineering Sciences Data Unit, UK.

ESDU, (1985). 'Characteristics of atmospheric turbulence near the ground. Part II: Single point data for strong winds (neutral atmosphere).' *ESDU 85020* (amended 1993), Engineering Sciences Data Unit, UK.

Eurocode, (1997). 'Eurocode 1: Basis of design and actions on structures – part 2.4: Actions on structures – Wind actions'.

Fisher, R. A. and Tippett, H. C., (1928). 'Limiting forms of the frequency distribution of the largest or smallest member of a sample.' *Proc. Camb. Phil. Soc.*, **24**, 180–190.

Frandsen, S. and Thøgersen, M., (1999). 'Integrated fatigue loading for wind turbines in wind farms by combining ambient turbulence and wakes.' *Wind Engng.*, **23**, 6, 327–339.

Germanischer Lloyd, (1993). 'Rules and regulations IV – Non-marine technology, Part 1 – Wind Energy,' (supplemented 1994, 1998). *Germanischer Lloyd*.

Greenway, M. E., (1979). 'An analytical approach to wind velocity gust factors.' *J. Ind. Aerodyn.*, **5**, 61–91.

Gumbel, E. J., (1958). *Statistics of extremes*. Columbia University Press, New York.

Harris, R. I., (1990). 'Some further thoughts on the spectrum of gustiness in strong winds.' *J. Wind Eng. Ind. Aerodyn.*, **33**, 461–477.

Hassan, U., (1992). 'A wind tunnel investigation of the wake structure within small wind turbine farms.' *E/5A/CON/5113/1890*. UK Department of Energy, ETSU.

IEC, (1999). 'Wind turbine generator systems – Part 1: Safety requirements.' *International Standard 61400-1*, Second Edition. International Electrotechnical Commission.

Kariniotakis, G., Nogaret, E. and Stavrakakis, G., (1997). 'Advanced short-term forecasting of wind power production.' *Proceedings of the European Wind Energy Conference*, pp 751–754.

Landberg, L., (1997). 'Predicting the power output from wind farms.' *Proceedings of the European Wind Energy Conference*, pp 747–750.

Landberg, L., (1999). 'Operational results from a physical power prediction model.' *Proceedings of the European Wind Energy Conference*, pp 1086–1089.

Lieblein, J., (1974). 'Efficient methods of extreme-value methodology.' *NBSIR 74-602*. National Bureau of Standards, Washington.

Ljung, L. and Söderström, T., (1983). *Theory and practice of recursive identification*. MIT Press.

Mortensen, N. G. *et al.*, (1993). 'Wind atlas analysis and application program (WA^{S}P), User's guide.' *Risø-I-666-(EN)v.2)*. Risø National Laboratory, Roskilde, Denmark.

Nielsen, T. L. and Madsen, H., (1999). 'Experiences with statistical methods for wind power prediction.' *Proceedings of the European Wind Energy Conference*, pp 1066–1069.

Palutikoff, J. P., Guo, X. and Halliday, J. A., (1991). 'The reconstruction of long wind speed records in the UK.' *Proceedings of the Thirteenth British Wind Energy Association Conference*, Mechanical Engineering Publications, Bury St Edmunds, UK.

Petersen, E. L. *et al.*, (1998). 'Wind power meteorology. Part I: Climate and turbulence.' *Wind Energy*, **1**, 1, 2–22.

Quarton, D. C. and Ainslie, J. F., (1989). 'Turbulence in wind turbine wakes.' *Proceedings of the European Wind Energy Conference, BWEA/EWEA*. Peter Peregrinus, Bristol, UK.

Sanderhoff, P., (1993). 'PARK – User's guide.' *Risø-I-668(EN)*, Risø National Laboratory, Roskilde, Denmark.

Schlez, W., (2000). *Voltage fluctuations caused by groups of wind turbines*. PhD Thesis, Loughborough University.

Tieleman, H. W., (1992). 'Wind characteristics in the surface layer over heterogeneous terrain.' *J. Wind Eng. Ind. Aerodyn.*, **41–44**, 329–340.

Van der Hoven, I., (1957). 'Power spectrum of horizontal wind speed in the frequency range from 0.0007 to 900 cycles per hour.' *J. Met.*, **14**, 160–4.

Vermeulen, P. E. J., (1980). 'An experimental analysis of wind turbine wakes.' *Third International Symposium on Wind Energy Systems*. BHRA.

Weiringa, J., (1973). 'Gust factors over open water and built-up country.' *Boundary-layer Met.*, **3**, 424–441.

3

Aerodynamics of Horizontal-Axis Wind Turbines

To study the aerodynamics of wind turbines some knowledge of fluid dynamics in general is necessary and, in particular, aircraft aerodynamics. Excellent text books on aerodynamics are readily available, a bibliography is given at the end of this chapter, and any abbreviated account of the subject that could have been included in these pages would not have done it justice; recourse to text books would have been necessary anyway. Some direction on which aerodynamics topics are necessary for the study of wind turbines would, however, be useful to the reader.

For Sections 3.2 and 3.3 a knowledge of Bernoulli's theorem for steady, incompressible flow is required together with the concept of continuity. For Sections 3.4 and 3.10 an understanding of vortices is desirable and the flow field induced by vortices. The Biot–Savart law, which will be familiar to those with a knowledge of electric and magnetic fields, is used to determine velocities induced by vortices. The Kutta–Joukowski theorem for determining the force on a bound vortex should also be studied. For Sections 3.5, 3.6 and 3.7 to 3.10 a knowledge of the lift and drag of aerofoils is essential, including the stalled flow and so a brief introduction has been included in the Appendix at the end of this chapter.

3.1 Introduction

A wind turbine is a device for extracting kinetic energy from the wind. By removing some of its kinetic energy the wind must slow down but only that mass of air which passes through the rotor disc is affected. Assuming that the affected mass of air remains separate from the air which does not pass through the rotor disc and does not slow down a boundary surface can be drawn containing the affected air mass and this boundary can be extended upstream as well as downstream forming a long stream-tube of circular cross section. No air flows across the boundary and so the mass flow rate of the air flowing along the stream-tube will be the same for all stream-wise positions along the stream-tube. Because the air within the stream-tube slows down, but does not become compressed, the cross-sectional area of the stream-tube must expand to accommodate the slower moving air (Figure 3.1).

Although kinetic energy is extracted from the airflow, a sudden step change in velocity is neither possible nor desirable because of the enormous accelerations and

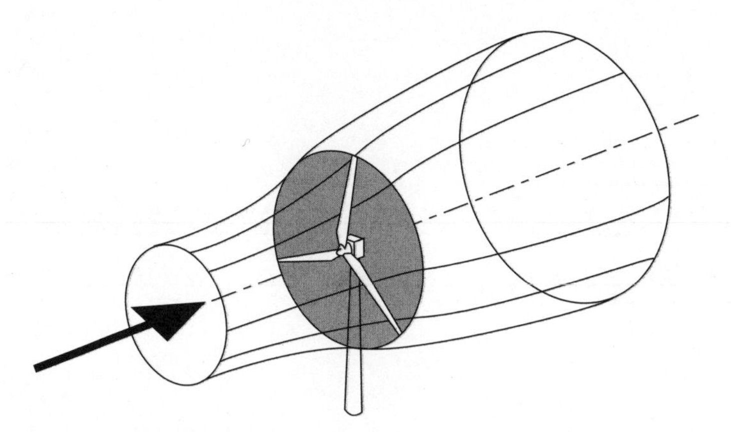

Figure 3.1 The Energy Extracting Stream-tube of a Wind Turbine

forces this would require. Pressure energy *can* be extracted in a step-like manner, however, and all wind turbines, whatever their design, operate in this way.

The presence of the turbine causes the approaching air, upstream, gradually to slow down such that when the air arrives at the rotor disc its velocity is already lower than the free-stream wind speed. The stream-tube expands as a result of the slowing down and, because no work has yet been done on, or by, the air its static pressure rises to absorb the decrease in kinetic energy.

As the air passes through the rotor disc, by design, there is a drop in static pressure such that, on leaving, the air is below the atmospheric pressure level. The air then proceeds downstream with reduced speed and static pressure – this region of the flow is called the *wake*. Eventually, far downstream, the static pressure in the wake must return to the atmospheric level for equilibrium to be achieved. The rise in static pressure is at the expense of the kinetic energy and so causes a further slowing down of the wind. Thus, between the far upstream and far wake conditions, no change in static pressure exists but there is a reduction in kinetic energy.

3.2 The Actuator Disc Concept

The mechanism described above accounts for the extraction of kinetic energy but in no way explains what happens to that energy; it may well be put to useful work but some may be spilled back into the wind as turbulence and eventually be dissipated as heat. Nevertheless, we can begin an analysis of the aerodynamic behaviour of wind turbines without any specific turbine design just by considering the energy extraction process. The general device that carries out this task is called an *actuator disc* (Figure 3.2).

Upstream of the disc the stream-tube has a cross-sectional area smaller than that of the disc and an area larger than the disc downstream. The expansion of the stream-tube is because the mass flow rate must be the same everywhere. The mass of air which passes through a given cross section of the stream-tube in a unit length of time is ρAU, where ρ is the air density, A is the cross-sectional area and U is the

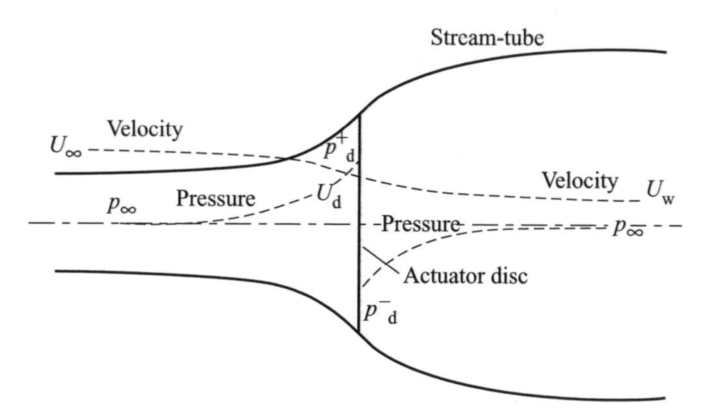

Figure 3.2 An Energy Extracting Actuator Disc and Stream-tube

flow velocity. The mass flow rate must be the same everywhere along the stream-tube and so

$$\rho A_\infty U_\infty = \rho A_d U_d = \rho A_w U_w \tag{3.1}$$

The symbol ∞ refers to conditions far upstream, d refers to conditions at the disc and w refers to conditions in the far wake.

It is usual to consider that the actuator disc induces a velocity variation which must be superimposed on the free-stream velocity. The stream-wise component of this induced flow at the disc is given by $-aU_\infty$, where a is called the axial flow induction factor, or the inflow factor. At the disc, therefore, the net stream-wise velocity is

$$U_d = U_\infty(1 - a) \tag{3.2}$$

3.2.1 Momentum theory

The air that passes through the disc undergoes an overall change in velocity, $U_\infty - U_w$ and a rate of change of momentum equal to the overall change of velocity times the mass flow rate:

$$\text{Rate of change of momentum} = (U_\infty - U_w)\rho A_d U_d \tag{3.3}$$

The force causing this change of momentum comes entirely from the pressure difference across the actuator disc because the stream-tube is otherwise completely surrounded by air at atmospheric pressure, which gives zero net force. Therefore,

$$(p_d^+ - p_d^-)A_d = (U_\infty - U_w)\rho A_d U_\infty(1 - a) \tag{3.4}$$

To obtain the pressure difference $(p_d^+ - p_d^-)$ Bernoulli's equation is applied separately to the upstream and downstream sections of the stream-tube; separate equations are necessary because the total energy is different upstream and downstream. Bernoulli's equation states that, under steady conditions, the total energy in the flow, comprising kinetic energy, static pressure energy and gravitational potential energy, remains constant provided no work is done on or by the fluid. Thus, for a unit volume of air,

$$\frac{1}{2}\rho U^2 + p + \rho g h = \text{constant}. \tag{3.5}$$

Upstream, therefore, we have

$$\frac{1}{2}\rho_\infty U_\infty^2 + \rho_\infty g h_\infty = \frac{1}{2}\rho_d U_d^2 + p_d^+ + \rho_d g h_d \tag{3.6}$$

Assuming the flow to be incompressible ($\rho_\infty = \rho_d$) and horizontal ($h_\infty = h_d$) then,

$$\frac{1}{2}\rho U_\infty^2 + p_\infty = \frac{1}{2}\rho U_d^2 + p_d^+ \tag{3.6a}$$

Similarly, downstream,

$$\frac{1}{2}\rho U_w^2 + p_\infty = \frac{1}{2}\rho U_d^2 + p_d^- \tag{3.6b}$$

Subtracting these equations we obtain

$$(p_d^+ - p_d^-) = \frac{1}{2}\rho(U_\infty^2 - U_w^2)$$

Equation (3.4) then gives

$$\frac{1}{2}\rho(U_\infty^2 - U_w^2)A_d = (U_\infty - U_w)\rho A_d U_\infty(1 - a) \tag{3.7}$$

and so

$$U_w = (1 - 2a)U_\infty \tag{3.8}$$

That is, half the axial speed loss in the stream-tube takes place upstream of the actuator disc and half downstream.

3.2.2 Power coefficient

The force on the air becomes, from Equation (3.4)

$$F = (p_d^+ - p_d^-)A_d = 2\rho A_d U_\infty^2 a(1 - a) \tag{3.9}$$

As this force is concentrated at the actuator disc the rate of work done by the force is FU_d and hence the power extraction from the air is given by

$$\text{Power} = FU_d = 2\rho A_d U_\infty^3 a(1-a)^2 \tag{3.10}$$

A *power coefficient* is then defined as

$$C_P = \frac{\text{Power}}{\frac{1}{2}\rho U_\infty^3 A_d} \tag{3.11}$$

where the denominator represents the power available in the air, in the absence of the actuator disc. Therefore,

$$C_P = 4a(1-a)^2 \tag{3.12}$$

3.2.3 The Betz limit

The maximum value of C_P occurs when

$$\frac{dC_P}{da} = 4(1-a)(1-3a) = 0$$

which gives a value of $a = \frac{1}{3}$.

Hence,

$$C_{P_{max}} = \frac{16}{27} = 0.593 \tag{3.13}$$

The maximum achievable value of the power coefficient is known as the Betz limit after Albert Betz the German aerodynamicist (119) and, to date, no wind turbine has been designed which is capable of exceeding this limit. The limit is caused not by any deficiency in design, for, as yet, we have no design, but because the stream-tube has to expand *upstream* of the actuator disc and so the cross section of the tube where the air is at the full, free-stream velocity is smaller than the area of the disc.

C_P could, perhaps, more fairly be defined as

$$C_P = \frac{\text{Power extracted}}{\text{Power available}} = \frac{\text{Power extracted}}{\frac{16}{27}\left(\frac{1}{2}\rho U_\infty^3 A_d\right)} \tag{3.14}$$

but this not the accepted definition of C_P.

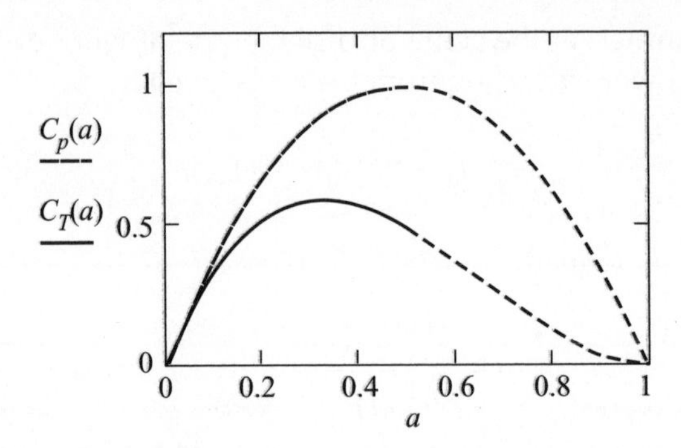

Figure 3.3 Variation of C_P and C_T with Axial Induction Factor a

3.2.4 The thrust coefficient

The force on the actuator disc caused by the pressure drop, given by Equation (3.9), can also be non- dimensionalized to give a *Coefficient of Thrust* C_T

$$C_T = \frac{\text{Power}}{\frac{1}{2}\rho U_\infty^2 A_d} \tag{3.15}$$

$$C_T = 4a(1-a) \tag{3.16}$$

A problem arises for values of $a \geqslant \frac{1}{2}$ because the wake velocity, given by $(1-2a)U_\infty$, becomes zero, or even negative; in these conditions the momentum theory, as described, no longer applies and an empirical modification has to be made (Section 3.5).

The variation of power coefficient and thrust coefficient with a is shown in Figure 3.3.

3.3 Rotor Disc Theory

The manner in which the extracted energy is converted into usable energy depends upon the particular turbine design. Most wind energy converters employ a rotor with a number of blades rotating with an angular velocity Ω about an axis normal to the rotor plane and parallel to the wind direction. The blades sweep out a disc and by virtue of their aerodynamic design develop a pressure difference across the disc, which, as discussed in the previous section, is responsible for the loss of axial momentum in the wake. Associated with the loss of axial momentum is a loss of energy which can be collected by, say, an electrical generator attached to the rotor shaft if, as well as a thrust, the rotor experiences a torque in the direction of rotation. The generator exerts a torque equal and opposite to that of the airflow which keeps the rotational speed constant. The work done by the aerodynamic torque on the generator is converted into electrical energy. The required aerodynamic design of the rotor blades to provide a torque as well as a thrust is discussed in Section 3.5.

3.3.1 Wake rotation

The exertion of a torque on the rotor disc by the air passing through it requires an equal and opposite torque to be imposed upon the air. The consequence of the reaction torque is to cause the air to rotate in a direction opposite to that of the rotor; the air gains angular momentum and so in the wake of the rotor disc the air particles have a velocity component in a direction which is tangential to the rotation as well as an axial component (Figure 3.4).

The acquisition of the tangential component of velocity by the air means an increase in its kinetic energy which is compensated for by a fall in the static pressure of the air in the wake in addition to that which is described in the previous section.

The flow entering the actuator disc has no rotational motion at all. The flow exiting the disc does have rotation and that rotation remains constant as the fluid progresses down the wake. The transfer of rotational motion to the air takes place entirely across the thickness of the disc (see Figure 3.5). The change in tangential velocity is expressed in terms of a tangential flow induction factor a'. Upstream of the disc the tangential velocity is zero. Immediately downstream of the disc the tangential velocity is $2\Omega ra'$. At the middle of the disc thickness, a radial distance r from the axis of rotation, the induced tangential velocity is $\Omega ra'$. Because it is produced in reaction to the torque the tangential velocity is opposed to the motion of the rotor.

An abrupt acquisition of tangential velocity cannot occur in practice. Figure 3.5 shows the flow accelerating in the tangential direction as it is 'squeezed' between the blades; the separation of the blades has been reduced for effect but it is the increasing solid blockage that the blades present to the flow as the root is approached that causes the high values of tangential velocity close to the root.

3.3.2 Angular momentum theory

The tangential velocity will not be the same for all radial positions and it may well also be that the axial induced velocity is not the same. To allow for variation of both induced velocity components consider only an annular ring of the rotor disc which is of radius r and of radial width δr.

The increment of rotor torque acting on the annular ring will be responsible for

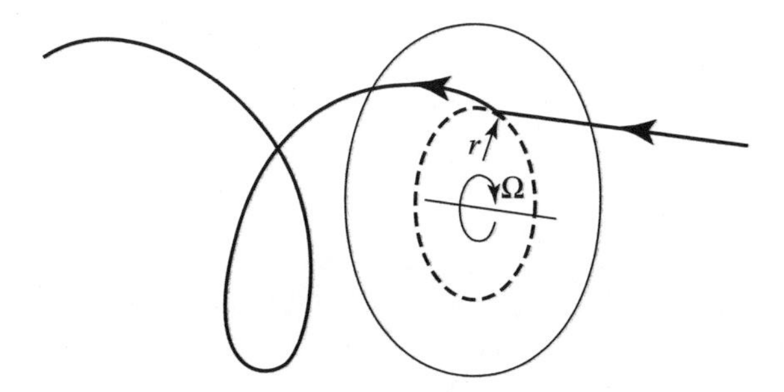

Figure 3.4 The Trajectory of an Air Particle Passing Through the Rotor Disc

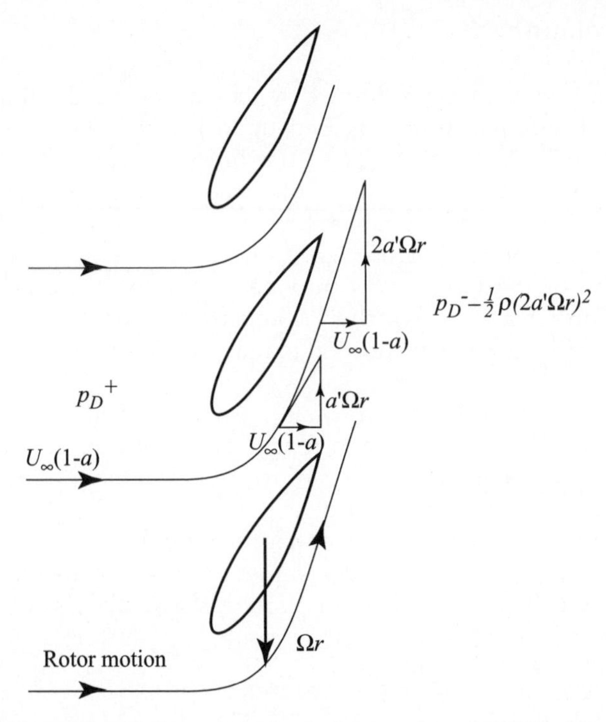

Figure 3.5 Tangential Velocity Grows Across the Disc Thickness

imparting the tangential velocity component to the air whereas the axial force acting on the ring will be responsible for the reduction in axial velocity. The whole disc comprises a multiplicity of annular rings and each ring is assumed to act independently in imparting momentum only to the air which actually passes through the ring.

The torque on the ring will be equal to the rate of change of angular momentum of the air passing through the ring. Thus,

torque = rate of change of angular momentum

= mass flow rate × change of tangential velocity × radius

$$\delta Q = \rho \delta A_{\rm d} U_\infty (1 - a) 2\Omega a' r^2 \tag{3.17}$$

where $\delta A_{\rm d}$ is taken as being the area of an annular ring.

The driving torque on the rotor shaft is also δQ and so the increment of rotor shaft power output is

$$\delta P = \delta Q \Omega$$

The total power extracted from the wind by slowing it down is therefore determined by the rate of change of axial momentum given by Equation (3.10) in Section 3.2.2

$$\delta P = 2\rho\delta A_{\mathrm{d}} U_\infty^3 a(1-a)^2$$

Hence

$$2\rho\delta A_{\mathrm{d}} U_\infty^3 a(1-a)^2 = \rho\delta A_{\mathrm{d}} U_\infty (1-a) 2\Omega^2 a' r^2$$

and

$$U_\infty^2 a(1-a) = \Omega^2 r^2 a'$$

Ωr is the tangential velocity of the spinning annular ring and so $\lambda_r = \Omega r/U_\infty$ is called the *local speed ratio*. At the edge of the disc $r = R$ and $\lambda = \Omega R/U_\infty$ is known at the *tip speed ratio*. Thus

$$a(1-a) = \lambda_r^2 a' \tag{3.18}$$

The area of the ring is $\delta A_{\mathrm{D}} = 2\pi r \delta r$ therefore the incremental shaft power is, from Equation (3.17),

$$\delta P = \mathrm{d}Q\Omega = \left(\frac{1}{2}\rho U_\infty^3 2\pi r \delta r\right) 4a'(1-a)\lambda_r^2$$

The term in brackets represents the power flux through the annulus, the term outside the brackets, therefore, is the efficiency of the blade element in capturing the power, or blade element efficiency:

$$\eta_r = 4a'(1-a)\lambda_r^2 \tag{3.19}$$

In terms of power coefficient

$$\frac{\mathrm{d}}{\mathrm{d}r} C_P = \frac{4\pi\rho U_\infty^3 (1-a) a' \lambda_r^2 r}{\frac{1}{2}\rho U_\infty^3 \pi R^2} = \frac{8(1-a)a'\lambda_r^2 r}{R^2}$$

$$\frac{\mathrm{d}}{\mathrm{d}\mu} C_P = 8(1-a)a'\lambda^2\mu^3 \tag{3.20}$$

where $\mu = r/R$.

Knowing how a and a' vary radially, Equation (3.20) can be integrated to determine the overall power coefficient for the disc for a given tip speed ratio, λ.

3.3.3 Maximum power

The values of a and a' which will provide the maximum possible efficiency can be determined by differentiating Equation (3.19) by either factor and putting the result equal to zero. Whence

$$\frac{\mathrm{d}}{\mathrm{d}a'} a = \frac{1-a}{a'} \tag{3.21}$$

From Equation (3.18)

$$\frac{\mathrm{d}}{\mathrm{d}a'}a = \frac{\lambda_r^2}{1-2a}$$

giving

$$a'\lambda_r^2 = (1-a)(1-2a) \tag{3.22}$$

The combination of Equations (3.18) and (3.21) gives the required values of a and a' which maximize the incremental power coefficient:

$$a = \frac{1}{3} \text{ and } a' = \frac{a(1-a)}{\lambda^2\mu^2} \tag{3.23}$$

The axial flow induction for maximum power extraction is the same as for the non-rotating wake case, that is, $a = \frac{1}{3}$ and is uniform over the entire disc. On the other hand a' varies with radial position.

From Equation (3.20) the maximum power is

$$C_P = \int_0^1 8(1-a)a'\lambda^2\mu^3\,\mathrm{d}\mu$$

Substituting for the expressions in Equations (3.23)

$$C_P = \int_0^1 8(1-a)\left[\frac{a(1-a)}{\lambda^2\mu^2}\right]\lambda^2\mu^3\,\mathrm{d}\mu = 4a(1-a)^2 = \frac{16}{27} \tag{3.24}$$

Which is precisely the same as for the non-rotating wake case.

3.3.4 Wake structure

The angular momentum imparted to the wake increases the kinetic energy in the wake but this energy is balanced by a loss of static pressure:

$$\Delta p_r = \frac{1}{2}\rho(2\Omega a' r)^2 \tag{3.25}$$

Substituting the expression for a' given by Equations (3.23)

$$\Delta p_r = \frac{1}{2}\rho U_\infty^2\left[2\frac{a(1-a)}{\lambda\mu}\right]^2 \tag{3.26}$$

The tangential velocity increases with decreasing radius (Equation (3.23)) and so the pressure decreases creating a radial pressure gradient. The radial pressure gradient balances the centrifugal force on the rotating fluid. The pressure drop across the disc caused by the rate of change of axial momentum as developed in Section 3.2.1 (Equation (3.9)) is additional to the pressure drop associated with the rotation of the wake and is uniform over the whole disc.

If the wake did not expand as it slows down the rotational wake structure together with the rotational pressure gradient would not change as the wake develops whereas the pressure loss caused by the change of axial momentum will gradually reduce to zero in the fully-developed wake, as shown in Figure 3.2. The pressure in the fully developed wake would therefore be atmospheric superimposed on which would be the pressure loss given by Equation (3.26). Consequently, the axial force on the fluid in the wake causing it to slow down would be only that caused by the uniform pressure drop across the disc given by Equation (3.9), as is assumed in the simple theory of Section 3.2.1. The rotational pressure drop does not contribute to the change of axial momentum.

In fact, the wake does expand and the full details of the analysis are given by Glauert (1935a). Glauert's analysis is applied to propellers where the flow is accelerated by the rotor but this is only a matter of reversing the signs of the flow induction factors. The inclusion of flow expansion and wake rotation in a fully integrated momentum theory shows that the axial induced velocity in the developed wake is greater than $2a$ but the effect is only significant at tip speed ratios less than about 1.5, which is probably outside of the operating range for most modern wind turbines. The analysis does, however, demonstrate that the kinetic energy of wake rotation is accounted for by reduced static pressure in the wake. Glauert's conclusion about wake expansion and its interaction with wake rotation is that its inclusion makes little difference to the results obtained from the simple axial momentum theory and so can be ignored. Where, in the same reference, Glauert deals with 'Windmills and Fans' (1935b) he adopts the simple momentum theory but then has to account for kinetic energy of wake rotation, which he does by assuming that it is drawn from the kinetic energy of the flow. The rotational kinetic energy of the wake is therefore regarded as a loss and reduces the level of the energy that can be extracted. Consequently, at low local speed ratios, the inboard sections of a rotor, the local aerodynamic efficiency falls below the Betz limit. Most authors since Glauert have assumed the same conclusion but, in fact, Glauert himself has demonstrated that the conclusion is wrong. The error makes very little difference to the final results for most modern wind turbines designed for the generation of electricity. For wind pumps, where a high starting torque and high solidity are required, the error would probably be very significant because they operate at very low tip speed ratios.

3.4 Vortex Cylinder Model of the Actuator Disc

3.4.1 Introduction

The momentum theory of Section 3.1 uses the concept of the actuator disc across which a pressure drop develops constituting the energy extracted by the rotor. In the rotor disc theory of Section 3.3 the actuator disc is depicted as being swept out by a multiplicity of aerofoil blades each with radially uniform bound circulation $\Delta\Gamma$. From the tip of each blade a helical vortex of strength $\Delta\Gamma$ convects downstream with the local flow velocity (Figure 3.6). If the number of blades is assumed to be very large but the solidity of the total is finite and small then the accumulation of helical tip vortices will form the surface of a tube. As the number of blades approaches infinity the tube surface will become a continuous tubular vortex sheet.

From the root of each blade, assuming it reaches to the axis of rotation, a line vortex of strength $\Delta\Gamma$ will extend downstream along the axis of rotation contributing to the total root vortex of strength Γ. The vortex tube will expand in radius as the flow of the wake inside the tube slows down. Vorticity is confined to the surface of the tube, the root vortex and to the bound vortex sheet swept by the multiplicity of blades to form the rotor disc; elsewhere in the wake and everywhere else in the entire flow field the flow is irrotational.

The nature of the tube's expansion cannot be determined by means of the momentum theory and so, as an approximation, the tube is allowed to remain cylindrical Figure 3.7. The Biot–Savart law is used to determine the induced velocity at any point in the vicinity of the actuator disc. The cylindrical vortex model allows the whole flow field to be determined and is accurate within the limitations of the non-expanding cylindrical wake.

3.4.2 Vortex cylinder theory

The vortex cylinder has surface vorticity which follows a helical path with a helix angle ϕ or, as it has been termed previously, the flow angle at the blade tip. The strength of the vorticity is $g = \mathrm{d}\Gamma/\mathrm{d}n$, where n is a direction in the tube surface normal to the direction of $\Delta\Gamma$, and has a component $g_\theta = g\cos\phi_t$ parallel to the rotor disc. Due to g_θ the axial (parallel to the axis of rotor rotation) induced velocity at the rotor plane is *uniform* over the rotor disc and can be determined by means of the Biot–Savart law as

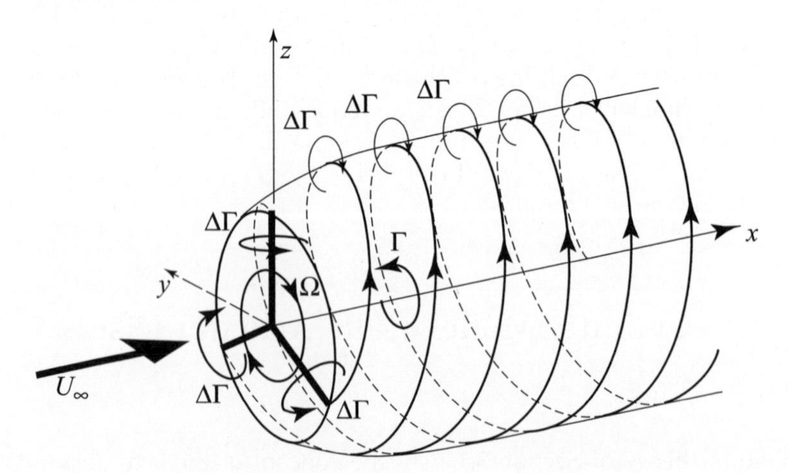

Figure 3.6 Helical Vortex Wake Shed by Rotor with Three Blades Each with Uniform Circulation $\Delta\Gamma$

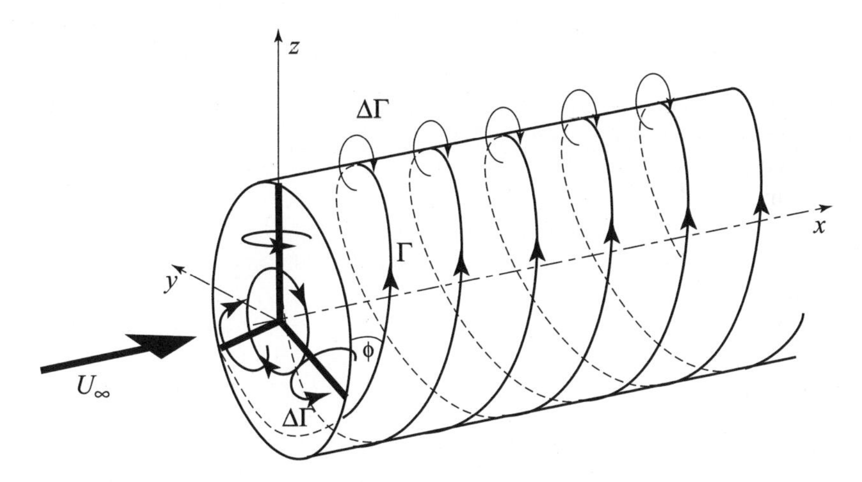

Figure 3.7 Simplified Helical Vortex Wake Ignoring Wake Expansion

$$u_{\rm d} = -\frac{g_\theta}{2} = -aU_\infty \tag{3.27}$$

In the far wake the axial induced velocity is also uniform within the cylindrical wake and is

$$u_{\rm w} = -g_\theta = -2aU_\infty \tag{3.28}$$

The ratio of the two induced velocities corresponds to that of the simple momentum theory and justifies the assumption of a cylindrical vortex sheet.

3.4.3 Relationship between bound circulation and the induced velocity

The total circulation on all of the multiplicity of blades is Γ which is shed at a uniform rate into the wake in one revolution. So, from Figure 3.8 in which the cylinder has been slit longitudinally and opened out flat,

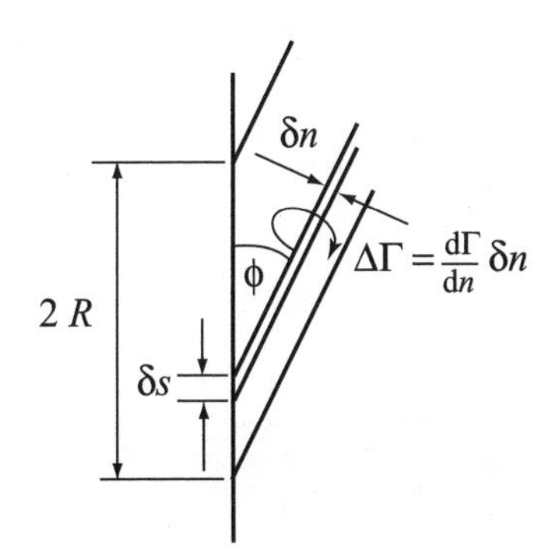

Figure 3.8 The Geometry of the Vorticity in the Cylinder Surface

$$g = \frac{\Gamma}{2\pi R \sin(\phi_t)} \tag{3.29}$$

hence

$$g_\theta = \frac{\Gamma}{2\pi R}\frac{\cos\phi_t}{\sin\phi_t} = \frac{\Gamma}{2\pi}\frac{\Omega R(1+a'_t)}{U_\infty(1-a)} \tag{3.30}$$

therefore

$$2aU_\infty = \frac{\Gamma}{2\pi R}\frac{\Omega R(1+a'_t)}{U_\infty(1-a)} \tag{3.31}$$

So, the total circulation is related to the induced velocity

$$\Gamma = \frac{4\pi U_\infty^2 a(1-a)}{\Omega(1+a'_t)} \tag{3.32}$$

3.4.4 Root vortex

Just as a vortex is shed from each blade tip a vortex is also shed from each blade root. If it is assumed that the blades extend to the axis of rotation, obviously not a practical option, then the root vortices will each be a line vortex running axially downstream from the centre of the disc. The direction of rotation of the all the root vortices will be the same forming a core, or root, vortex, of total strength Γ. The root vortex is primarily responsible for inducing the tangential velocity in the wake flow and in particular the tangential velocity on the rotor disc.

On the rotor disc surface the tangential velocity induced by the root vortex, given by the Biot–Savart law, is

$$\Omega r a' = \frac{\Gamma}{4\pi r}$$

$$a' = \frac{\Gamma}{4\pi r^2 \Omega} \tag{3.33}$$

This relationship can also be derived from the momentum theory: the rate of change of angular momentum of the air which passes through an annulus of the disc of radius r and radial width $\mathrm{d}r$ is equal to the torque increment imposed upon the annulus

$$\mathrm{d}Q = \rho U_\infty(1-a)2\pi r\,\mathrm{d}r 2a'\Omega r^2 \tag{3.34}$$

By the Kutta–Joukowski theorem the lift per unit radial width is

$$L = \rho(W \times \Gamma)$$

where $(W \times \Gamma)$ is a vector product.

$$\frac{\mathrm{d}}{\mathrm{d}r}Q = \rho W \times \Gamma r \sin\phi_t = \rho \Gamma r U_\infty (1 - a) \tag{3.35}$$

Equating the two expressions gives

$$a' = \frac{\Gamma}{4\pi r^2 \Omega}$$

Hence

$$a_t' = \frac{U_\infty^2 a(1-a)}{\Omega^2 R^2 (1 + a_t')} = \frac{a(1-a)}{\lambda^2 (1 + a_t')}$$

so

$$a_t'(1 + a_t') = \frac{a(1-a)}{\lambda^2} \tag{3.36}$$

Equation (3.36) is not quite the same as Equation (3.23) of Section 3.3.3 and it can be shown that this is a result of ignoring the wake expansion.

3.4.5 Torque and power

The torque on an annulus of radius r and radial width δr is

$$\frac{\mathrm{d}}{\mathrm{d}r}Q\delta r = \rho W \Gamma r \sin\phi_t \delta r = \frac{\rho 4\pi r U_\infty^3 a(1-a)^2}{\Omega(1 + a_t')}\delta r$$

$$\frac{\mathrm{d}}{\mathrm{d}r}Q = \frac{\frac{1}{2}\rho U_\infty^3 2\pi r 4a(1-a)^2}{\Omega(1 + a_t')} \tag{3.37}$$

Power

$$\frac{\mathrm{d}}{\mathrm{d}r}P = \Omega \frac{\mathrm{d}}{\mathrm{d}r}Q = \frac{\frac{1}{2}\rho U_\infty^3 2\pi r 4a(1-a)^2}{(1 + a_t')} \tag{3.38}$$

$$P = \frac{\frac{1}{2}\rho U_\infty^3 \pi R^2 4a(1-a)^2}{(1 + a_t')} \tag{3.39}$$

Power coefficient

$$C_P = \frac{4a(1-a)^2}{(1+a_t')} = 4a_t'(1-a)\lambda^2 \tag{3.40}$$

The reduced efficiency compared with the simple actuator disc result, $C_P = 4a(1-a)^2$, is caused by the energy required to spin the wake, as a rigid body, with an angular velocity $2a_t'\Omega$. It should be noted that any additional rotational energy is accounted for by the loss of static pressure caused by the pressure gradient which balances the centrifugal forces on the rotating fluid.

The general momentum theory of Glauert (1935a) includes the expansion of the wake and shows that no contribution at all is actually required from the kinetic energy of the free-stream flow to maintain wake rotation; all the kinetic energy of rotation is derived from static pressure energy.

3.4.6 Axial flow field

By means of the Biot-Savart law the induced velocity in the wind-wise (axial) direction at the actuator disc can be determined both upstream of the disc and downstream in the developing wake, as well as on the disc itself. The flow field (net velocity) is axisymmetric and a radial cross section is shown in Figure 3.9. Both radial and axial distances are divided by the disc radius with the axial distance being measured downstream from the disc and the radial distance being measured from the rotational axis. The velocity is divided by the wind speed.

The axial velocity within the wake is sharply lower than without and is radially uniform at the disc and in the far wake, just as the momentum theory predicts. There is a small acceleration of the flow immediately outside of the wake. The induced velocity on the wake cylinder itself is $\frac{1}{2}a$ at the disc and a in the far wake.

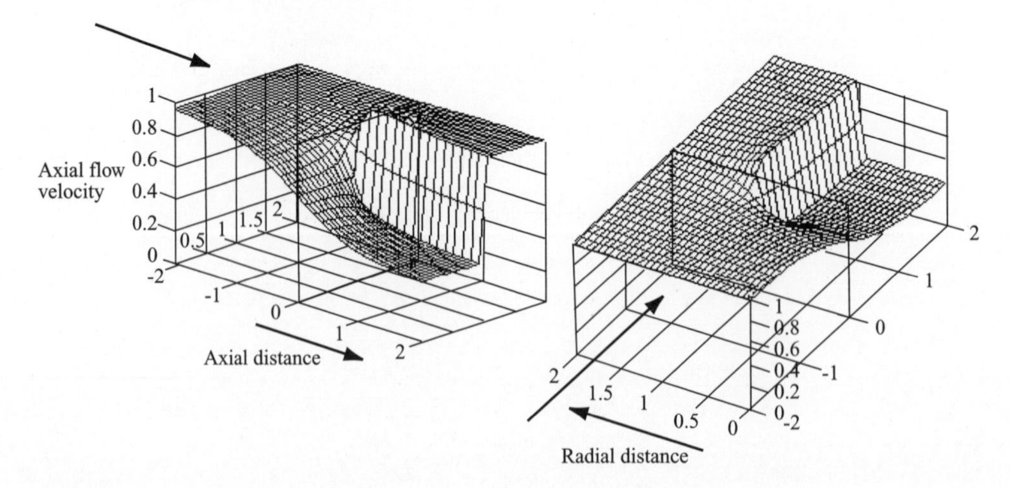

Figure 3.9 The Radial and Axial Variation of Axial Velocity in the Vicinity of an Actuator Disc, $a = 1/3$

3.4.7 Tangential flow field

The tangential induced velocity is determined not only by the root vortex but also by the component of vorticity $g \sin \phi$ on the wake cylinder and the bound vorticity on the rotor disc. At a radial distance equal to half the disc radius, as an example, the axial variation of the three contributions are shown in Figure 3.10.

The bound vorticity causes rotation in opposite senses upstream and downstream of the disc with a step change across the disc. The upstream rotation, which is in the same sense as the rotor rotation, is nullified by the root vortex, which induces rotation in the opposite sense to that of the rotor. The downstream rotation is in the same sense for both the root vortex and the bound vorticity the stream-wise variations of the two summing to give a uniform velocity in the stream-wise sense. The vorticity located on the surface of the wake cylinder makes a small contribution.

At the disc itself the bound vorticity induces no rotation, see Figure 3.10, the wake cylinder induces no rotation either and so it is only the root vortex which does induce rotation and that value is half the total induced generally in the wake. It is now clear why only half the rotational velocity is used to determine the flow angle at the disc.

The rotational flow is confined to the wake, that is, inside the cylinder. There is no rotational flow anywhere outside of the wake, neither upstream of the disc or outside the cylinder. The rotational velocity within the cylinder falls with increasing radius but is not zero at the outer edge of the wake, therefore there is an abrupt fall of rotational velocity across the cylindrical surface. The contributions of the three vorticity sources to the rotational flow at a radius of 101 percent of the disc radius are shown in Figure 3.11; the total rotational flow is zero at all axial positions but the individual components are not zero.

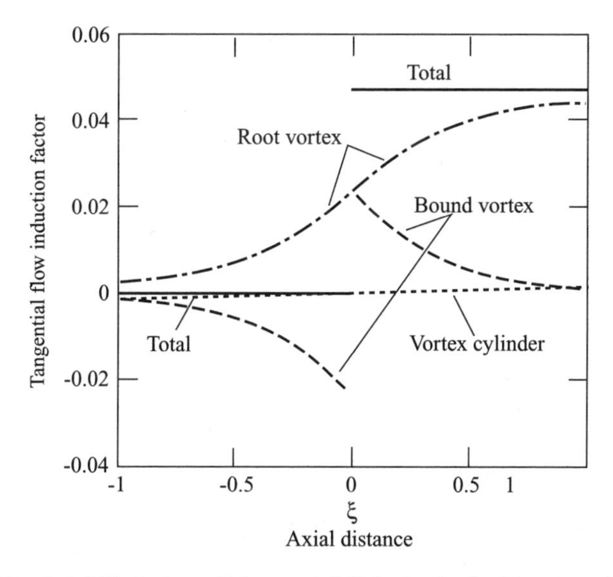

Figure 3.10 The Axial Variation of Tangential Velocity in the Vicinity of an Actuator Disc at 50% Radius, $a = 1/3, \lambda = 6$

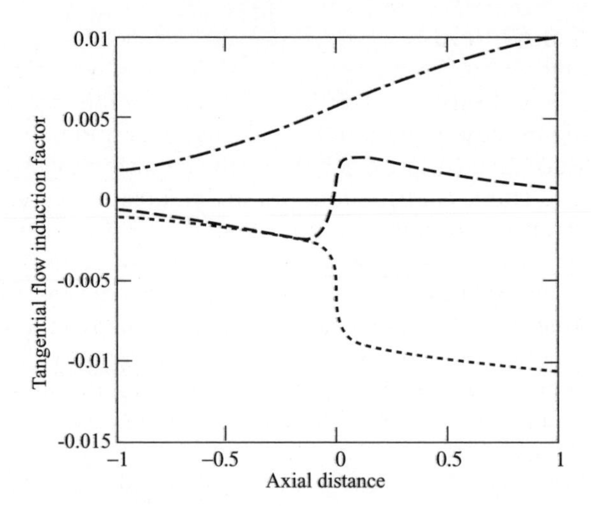

Figure 3.11 The Axial Variation of Tangential Velocity in the Vicinity of an Actuator Disc at 101 percent Radius, $a = 1/3, \lambda = 6$

3.4.8 Radial flow field

Although the vortex cylinder theory has been simplified by not allowing the cylinder to expand, nevertheless the theory predicts flow expansion. A radial velocity distribution is predicted by the theory as shown in Figure 3.12 which shows a longitudinal section of the flow field through the rotor disc.

The radial velocity, as calculated, is greatest as the flow passes through the rotor disc and, although not shown in Figure 3.12, it is infinite at the edge of the disc. The situation is very similar to the determination of the potential flow field around a flat, solid, circular disc which is normal to the oncoming flow; an infinite radial

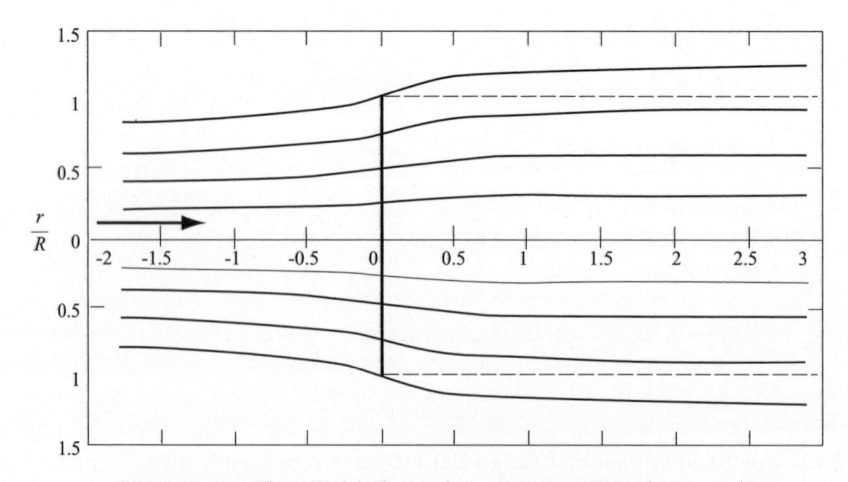

Figure 3.12 Flow Field Through an Actuator Disc for $a = 1/3$

velocity is predicted at the disc edge for the flow to pass around and continue radially inwards on the downstream side. In practice there would be insufficient static pressure available to fuel an infinite, or even a very high, velocity and so some discontinuity in the flow must occur. The presence of even the smallest amount of viscosity would produce a thick boundary layer towards the disc edge because of the high radial velocity. The viscosity in the boundary layer would absorb much of the available energy and dissipate it as heat so that as the flow accelerated around the disc edge the maximum velocity attainable would be limited by the static pressure approaching zero. Instead of the flow moving around the edge it would separate from the edge and continue downstream leaving a very low pressure region behind the disc with very low velocity—a stagnant region.

In the case of the permeable rotor disc there would be some flow through the disc, which would behave as described above, but the separation of the flow at the disc edge would produce an additional low pressure in the wake.

The problem of the infinite radial velocity at the rotor disc edge arises because of the assumption of an infinite number of rotor blades. If the theory is modified such that there are only a few blades the infinite radial velocity disappears. However, if, for a given rotor, the tip speed ratio is increased, with a consequent increase of the axial flow induction factor, the radial velocity at the tip rises sharply and the problem of edge separation returns, which is what actually occurs, see Section 3.6.

3.4.9 Conclusions

Despite the exclusion of wake expansion, the vortex theory produces results largely in agreement with the momentum theory and enlightens understanding of the flow through an energy extracting actuator disc.

3.5 Rotor Blade Theory

3.5.1 Introduction

The aerodynamic lift (and drag) forces on the span-wise elements of radius r and length δr of the several blades of a wind turbine rotor are responsible for the rate of change of axial and angular momentum of all of the air which passes through the annulus swept by the blade elements. In addition, the force on the blade elements caused by the drop in pressure associated with the rotational velocity in the wake must also be provided by the aerodynamic lift and drag. As there is no rotation of the flow approaching the rotor the reduced pressure on the downwind side of the rotor caused by wake rotation appears as a step pressure drop just like that which causes the change in axial momentum. Because the wake is still rotating in the far wake the pressure drop caused by the rotation is still present and so does not contribute to the axial momentum change.

3.5.2 Blade element theory

It is assumed that the forces on a blade element can be calculated by means of two-dimensional aerofoil characteristics using an angle of attack determined from the incident resultant velocity in the cross-sectional plane of the element; the velocity component in the span-wise direction is ignored. Three-dimensional effects are also ignored.

The velocity components at a radial position on the blade expressed in terms of the wind speed, the flow factors and the rotational speed of the rotor will determine the angle of attack. Having information about how the aerofoil characteristic coefficients C_d and C_d vary with the angle of attack the forces on the blades for given values of a and a' can be determined.

Consider a turbine with N blades of tip radius R each with chord c and set pitch angle β measured between the *aerofoil zero lift line* and the plane of the disc. Both the chord length and the pitch angle may vary along the blade span. Let the blades be rotating at angular velocity Ω and let the wind speed be U_∞. The tangential velocity Ωr of the blade element shown in Figure 3.13 combined with the tangential velocity of the wake $a'\Omega r$ means that the net tangential flow velocity experienced by the blade element is $(1 + a')\Omega r$. Figure 3.14 shows all the velocities and forces relative to the blade chord line at radius r.

From Figure 3.14 the resultant relative velocity at the blade is

$$W = \sqrt{U_\infty^2(1-a)^2 + \Omega^2 r^2(1+a')^2} \tag{3.41}$$

which acts at an angle ϕ to the plane of rotation, such that

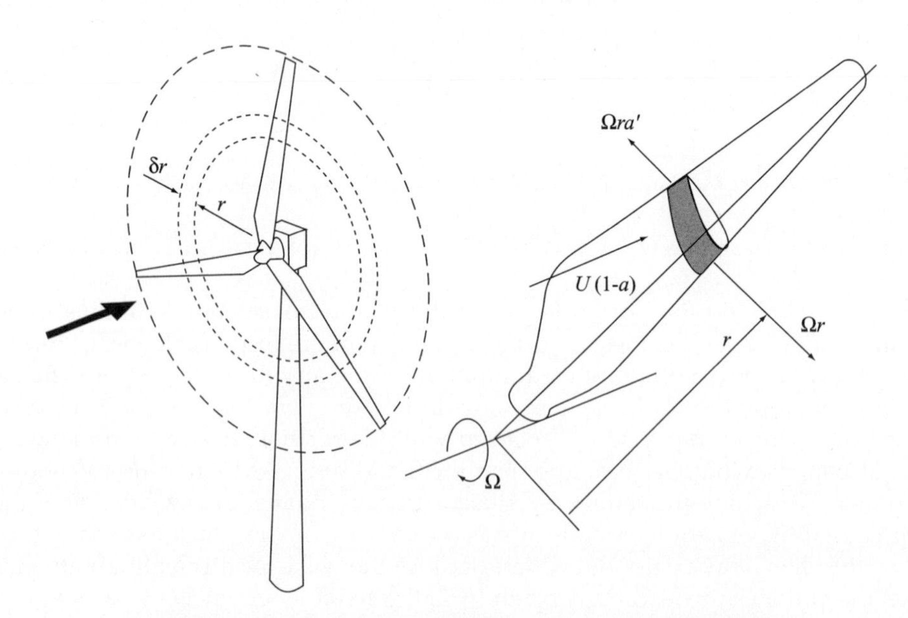

Figure 3.13 A Blade Element Sweeps Out an Annular Ring

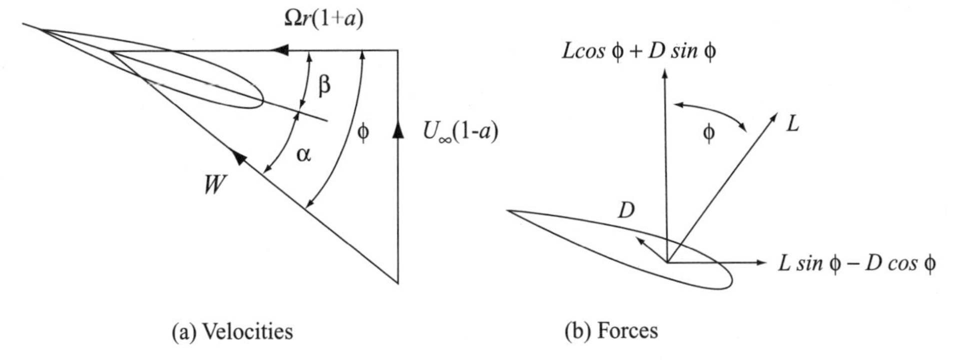

Figure 3.14 Blade Element Velocities and Forces

$$\sin\phi = \frac{U_\infty(1-a)}{W} \text{ and } \cos\phi = \frac{\Omega r(1+a')}{W} \tag{3.42}$$

The angle of attack α is then given by

$$\alpha = \phi - \beta \tag{3.43}$$

The lift force on a span-wise length δr of each blade, normal to the direction of W, is therefore

$$\delta L = \frac{1}{2}\rho W^2 c C \delta r \tag{3.44}$$

and the drag force parallel to W is

$$\delta D = \frac{1}{2}\rho W^2 c C_d r \delta r \tag{3.45}$$

3.5.3 The blade element – momentum (BEM) theory

The basic assumption of the BEM theory is that the force of a blade element is solely responsible for the change of momentum of the air which passes through the annulus swept by the element. It is therefore to be assumed that there is no radial interaction between the flows through contiguous annuli—a condition that is, strictly, only true if the axial flow induction factor does not vary radially. In practice, the axial flow induction factor is seldom uniform but experimental examination of flow through propeller discs by Lock (1924) shows that the assumption of radial independence is acceptable.

The component of aerodynamic force on N blade elements resolved in the axial direction is

$$\delta L\cos\phi + \delta D\sin\phi = \frac{1}{2}\rho W^2 N c(C_L\cos\phi + C_d\sin\phi)\delta r \tag{3.46}$$

The rate of change of axial momentum of the air passing through the swept annulus is

$$\rho U_\infty(1-a)2\pi r\delta r 2aU_\infty = 4\pi\rho U_\infty^2 a(1-a)r\delta r \tag{3.47}$$

The drop in wake pressure caused by wake rotation is equal to the increase in dynamic head, which is

$$\frac{1}{2}\rho(2a'\Omega r)^2$$

Therefore the additional axial force on the annulus is

$$\frac{1}{2}\rho(2a'\Omega r)^2 2\pi r\delta r$$

Thus

$$\frac{1}{2}\rho W^2 Nc(C_l\cos\phi + C_d\sin\phi)\delta r = 4\pi\rho[U_\infty^2 a(1-a) + (a'\Omega r)^2]r\delta r$$

Simplifying,

$$\frac{W^2}{U_\infty^2}N\frac{c}{R}(C_l\cos\phi + C_d\sin\phi) = 8\pi(a(1-a) + (a'\lambda\mu)^2)\mu \tag{3.48}$$

The element of axial rotor torque caused by aerodynamic forces on the blade elements is

$$(\delta L\sin\phi - \delta D\cos\phi)r = \frac{1}{2}\rho W^2 Nc(C_l\sin\phi - C_d\cos\phi)r\delta r \tag{3.49}$$

The rate of change of angular momentum of the air passing through the annulus is

$$\rho U_\infty(1-a)\Omega r 2a' r 2\pi r\delta r = 4\pi\rho U_\infty(\Omega r)a'(1-a)r^2\delta r$$

Equating the two moments

$$\frac{1}{2}\rho W^2 Nc(C_l\sin\phi - C_d\cos\phi)r\delta r = 4\pi\rho U_\infty(\Omega r)a'(1-a)r^2\delta r \tag{3.50}$$

Simplifying,

$$\frac{W^2}{U_\infty^2}N\frac{c}{R}(C_l\sin\phi - C_d\cos\phi) = 8\pi\lambda\mu^2 a'(1-a) \tag{3.50a}$$

where the parameter $\mu = r/R$.

It is convenient to put

$$C_l \cos\phi + C_d \sin\phi = C_x$$

and

$$C_l \sin\phi - C_d \cos\phi = C_y$$

Solving Equations 3.48 and (3.50a) to obtain values for the flow induction factors a and a' using two-dimensional aerofoil characteristics requires an iterative process. The following equations, derived from (3.48) and (3.50a), are convenient in which the right-hand sides are evaluated using existing values of the flow induction factors yielding simple equations for the next iteration of the flow induction factors.

$$\frac{a}{1-a} = \frac{\sigma_r}{4\sin\phi^2}\left[(C_x) - \frac{\sigma_r}{4\sin^2\phi}C_y^2\right] \tag{3.51}$$

$$\frac{a}{1+a'} = \frac{\sigma_r C_y}{4\sin\phi\cos\phi} \tag{3.52}$$

Blade solidity σ is defined as total blade area divided by the rotor disc area and is a primary parameter in determining rotor performance. Chord solidity σ_r is defined as the total blade chord length at a given radius divided by the circumferential length at that radius.

$$\sigma_r = \frac{N}{2\pi}\frac{c}{r} = \frac{N}{2\pi\mu}\frac{c}{R} \tag{3.53}$$

It is argued by Wilson and Lissaman (1974) that the drag coefficient should not be included in Equations (3.51) and (3.52) because the velocity deficit caused by drag is confined to the narrow wake which flows from the trailing edge of the aerofoil. Furthermore, Wilson and Lissaman reason, the drag-based velocity deficit is only a feature of the wake and does not contribute to the velocity deficit upstream of the rotor disc. The basis of the argument for excluding drag in the determination of the flow induction factors is that, for attached flow, drag is caused only by skin friction and does not affect the pressure drop across the rotor. Clearly, in stalled flow the drag is overwhelmingly caused by pressure. In attached flow it has been shown by Young and Squire (1938) that the modification to the inviscid pressure distribution around an aerofoil caused by the boundary layer has an affect both on lift and drag. The ratio of pressure drag to total drag at zero angle of attack is approximately the same as the thickness to chord ratio of the aerofoil and increases as the angle of attack increases.

One last point about the BEM theory: the theory is strictly only applicable if the blades have uniform circulation, i.e., if a is uniform. For non-uniform circulation there is a radial interaction and exchange of momentum between flows through adjacent elemental annular rings. It cannot be stated that the only axial force acting on the flow through a given annular ring is that due to the pressure drop across the disc. However, in practice, it appears that the error involved in relaxing the above constraint is small for tip speed ratios greater than 3.

3.5.4 Determination of rotor torque and power

The calculation of torque and power developed by a rotor requires a knowledge of the flow induction factors, which are obtained by solving Equations (3.51) and (3.52). The solution is usually carried out iteratively because the two-dimensional aerofoil characteristics are non-linear functions of the angle of attack.

To determine the complete performance characteristic of a rotor, i.e., the manner in which the power coefficient varies over a wide range of tip speed ratio, requires the iterative solution. The iterative procedure is to assume a and a' to be zero initially, determining ϕ, C_p and C_d on that basis, and then to calculate new values of the flow factors using Equations (3.51) and (3.52). The iteration is repeated until convergence is achieved.

From Equation (3.50) the torque developed by the blade elements of span-wise length δr is

$$\delta Q = 4\pi\rho U_\infty(\Omega r)a'(1-a)r^2\delta r$$

If drag, or part of the drag, has been excluded from the determination of the flow induction factors then its effect must be introduced when the torque caused by drag is calculated from blade element forces, see Equation (3.49),

$$\delta Q = 4\pi\rho U_\infty(\Omega r)a'(1-a)r^2\delta r - \frac{1}{2}\rho W^2 N c C_d \cos(\phi) r\delta r$$

The complete rotor, therefore, develops a total torque Q:

$$Q = \frac{1}{2}\rho U_\infty^2 \pi R^3 \lambda \left[\int_0^R \mu^2 \left[8a'(1-a)\mu - \frac{W}{U_\infty}\frac{N\frac{c}{R}}{\pi}C_d(1+a')\right]d\mu\right] \tag{3.54}$$

The power developed by the rotor is $P = Q\Omega$. The power coefficient is

$$C_P = \frac{P}{\frac{1}{2}\rho U_\infty^3 \pi R^2}$$

Solving the BEM Equations (3.51) and (3.52) for a given, suitable blade geometrical and aerodynamic design yields a series of values for the power and torque coefficients which are functions of the tip speed ratio. A typical performance curve for a modern, high-speed wind turbine is shown in Figure 3.15.

The maximum power coefficient occurs at a tip speed ratio for which the axial flow induction factor a, which in general varies with radius, approximates most

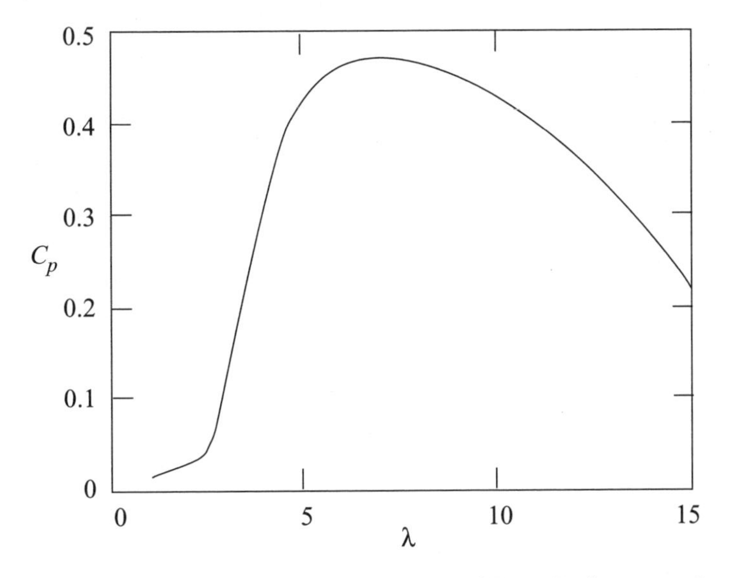

Figure 3.15 Power Coefficient – Tip Speed Ratio Performance Curve

closely to the Betz limit value of $\frac{1}{3}$. At lower tip speed ratios the axial flow induction factor can be much less than $\frac{1}{3}$ and aerofoil angles of attack are high leading to stalled conditions. For most wind turbines stalling is much more likely to occur at the blade root because, from practical constraints, the built-in pitch angle β of a blade is not large enough in that region. At low tip speed ratios blade stalling is the cause of a significant loss of power, as demonstrated in Figure 3.15. At high tip speed ratios a is high, angles of attack are low and drag begins to predominate. At both high and low tip speed ratios, therefore, drag is high and the general level of a is non-optimum so the power coefficient is low. Clearly, it would be best if a turbine can be operated at all wind speeds at a tip speed ratio close to that which gives the maximum power coefficient.

3.6 Breakdown of the Momentum Theory

3.6.1 Free-stream/wake mixing

For heavily loaded turbines, when a is high, the momentum theory predicts a reversal of the flow in the wake. Such a situation cannot actually occur so what happens is that the wake becomes turbulent and, in doing so, entrains air from outside the wake by a mixing process which re-energizes the slow moving air which has passed through the rotor.

A rotor operating at increasingly high tip speed ratios presents a decreasingly permeable disc to the flow. Eventually, when λ is high enough for the axial flow factor to be equal to one, the disc effectively becomes a solid plate.

The flow past a solid disc, because of viscosity, separates at the disc's edge. A boundary layer develops as the flow over the front of the disc spreads out radially and by the time it reaches the edge viscosity has sapped much of the kinetic energy. As the boundary layer flows around the disc edge it accelerates causing a large drop in static pressure (Bernoulli). To flow around the disc edge would require very high velocity and there is insufficient static pressure to provide the necessary kinetic energy. The flow, therefore, separates from the disc and continues in the general stream-wise direction. In the region directly behind the disc there is slow moving, almost stagnant, air at the low static pressure of the flow separating at the disc edge. At the front of the disc, at the very centre, the flow is brought to rest and so there is a large increase in static pressure as the kinetic energy is converted to pressure energy. Elsewhere on the front surface the flow moves radially with a velocity, outside the boundary layer, which increases towards the disc edge. The static pressure is generally higher on the front of the disc than on the rear and so the disc experiences a pressure drag force.

A similar process happens with a spinning rotor at high tip speed ratios. The air which does not pass through the rotor disc moves radially outwards and separates at the disc edge causing a low static pressure to develop behind the disc; the drop in static pressure caused by the separation increases as the tip speed ratio rises and the axial flow factor increases. The air which does pass through the rotor emerges into a low pressure region and is moving slowly. There is insufficient kinetic energy to provide the rise in static pressure necessary to achieve the ambient atmospheric pressure that must exist in the far wake. The air can only achieve atmospheric pressure by gaining energy from the mixing process in the turbulent wake. The shear layer in the flow between the free-stream air and the wake air is what becomes of the boundary layer that develops on the front of the disc. The shear layer is unstable and breaks up into the turbulence that causes the mixing and re-energization of the wake air.

3.6.2 Modification of rotor thrust caused by flow separation

The low static pressure downstream of the rotor disc caused by the separation of the free-stream flow at the edge of the disc and the high static pressure at the stagnation point on the upstream side causes a large thrust on the disc, much larger than that predicted by the momentum theory. Some experimental results reported by Glauert (1926) for a whole rotor can be seen in Figure 3.16 where the simple expression for the thrust force coefficient, as derived from the momentum theory ($C_T = 4a(1 - a)$), is given for comparison.

The thrust (or drag) coefficient for a simple, flat circular plate is given by Hoerner (1965) as 1.17 but, as demonstrated in Figure 3.16, the thrust on the rotating disc is higher. It might have been expected that when $a = 1$ the rotor would have the same thrust coefficient as the circular plate. The principal difference between the circular plate and the rotor is that the latter is rotating and, as Hoerner also describes, this causes energy to be dissipated in a thicker, rotating boundary layer on the upstream surface of the rotor disc giving rise to an even lower pressure on the downstream side.

It would follow from the above arguments that for high values of the axial induction factor most of the pressure drop across the disc is not associated with

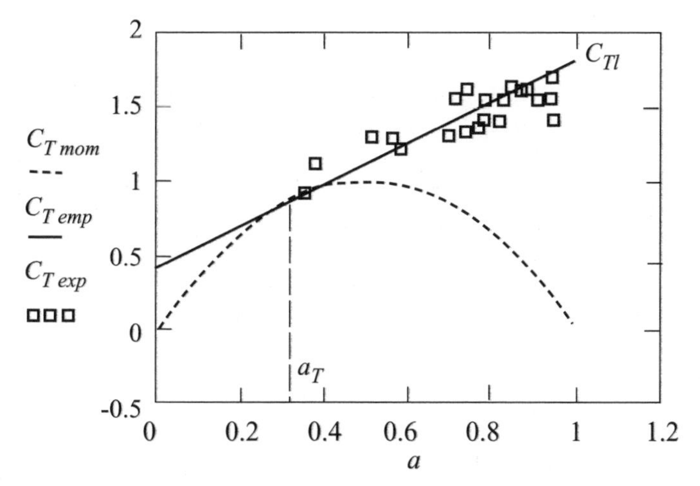

Figure 3.16 Comparison of Theoretical and Measured Values of C_T

blade circulation, there is no circulation present in the case of the circular plate. Circulation would cause a small pressure drop similar to that given by the momentum theory because it would be determined by the very low axial velocity of the flow which actually permeates the disc.

3.6.3 Empirical determination of thrust coefficient

A suitable straight line through the experimental points would appear to be possible, although Glauert proposed a parabolic curve, and provides an empirical solution to the problem of the thrust on a heavily loaded turbine (a rotor operating at a high value of the axial flow induction factor).

Most authors assume that the entire thrust on the rotor disc causes axial momentum change. Therefore, for the empirical line to be useful it must be assumed that it applies not only to the whole rotor but also to each separate stream-tube. Let C_1 be the empirical value of C_t when $a = 1$. Then, as the straight line must be a tangent to the momentum theory parabola at the transition point, the equation for the line is

$$C_T = C_{T1} - 4(\sqrt{C_{T1}} - 1)(1 - a) \tag{3.55}$$

and the value of a at the transition point is

$$a_T = 1 - \frac{1}{2}\sqrt{C_{T1}} \tag{3.56}$$

By inspection of Figure 3.16, C_{T1} must lie between 1.6 and 2; $C_{T1} = 1.816$ would appear to be the best fit to the experimental data of Figure 3.16, whereas Wilson and Lissaman (1974) favour the lower value of $C_{T1} = 1.6$. Glauert fits a parabolic curve to the data giving much higher values of C_{T1} at high values of a but he was considering the case of an *airscrew* in the windmill brake state where the angles of attack are negative.

The flow field through the turbine under heavily loaded conditions cannot be modelled easily and the results of this empirical analysis must be regarded as being only approximate at best. They are, nevertheless, better than those predicted by the momentum theory. For most practical designs the value of axial flow induction factor rarely exceeds 0.6 and for a well-designed blade will be in the vicinity of 0.33 for much of its operational range.

For values of a greater than a_T it is common to replace the momentum theory thrust in Equation (3.16) with Equation (3.55), in which case Equation (3.51) is replaced by

$$(1-a)^2 \frac{\sigma_r}{\sin\phi^2} C_x + 4(\sqrt{C_{T1}} - 1)(1-a) - C_{T1} = 0 \tag{3.51a}$$

in which the pressure drop caused by wake rotation in ignored as it is very small. However, as the additional pressure drop is caused by edge flow separation then this course of action is questionable and it may be more appropriate to retain Equation (3.51).

3.7 Blade Geometry

3.7.1 Introduction

The purpose of most wind turbines is to extract as much energy from the wind as possible and each component of the turbine has to be optimized for that goal. Optimal blade design is influenced by the mode of operation of the turbine, that is, fixed rotational speed or variable rotational speed and, ideally, the wind distribution at the intended site. In practice engineering compromises are made but it is still necessary to know what would be the best design.

Optimizing a blade design means maximizing the power output and so a suitable solution to blade element – momentum Equations (3.51) and (3.52) is necessary.

3.7.2 Optimal design for variable-speed operation

A turbine operating at variable speed can maintain the constant tip speed ratio required for the maximum power coefficient to be developed regardless of wind speed. To develop the maximum possible power coefficient requires a suitable blade geometry the conditions for which will now be derived.

For a chosen tip speed ratio λ the torque developed at each blade station is maximized if

$$\frac{d}{da'} 8\pi\lambda\mu^2 a'(1-a) = 0$$

(see Equation 3.50a), giving

$$\frac{\mathrm{d}}{\mathrm{d}a'}a = \frac{1-a}{a'} \tag{3.57}$$

From Equations (3.48) and (3.50a) a relationship between the flow induction factors can be obtained. Dividing equations 3.48 and 3.50a

$$\frac{\dfrac{C_l}{C_d}\tan\phi - 1}{\dfrac{C_l}{C_d} + \tan\phi} = \frac{\lambda\mu a'(1-a)}{a(1-a) + (a'\lambda\mu)^2} \tag{3.58}$$

The flow angle ϕ is given by

$$\tan\phi = \frac{1-a}{\lambda\mu(1+a')} \tag{3.59}$$

Substituting Equation (3.59) into Equation (3.58) gives

$$\frac{\dfrac{C_l}{C_d}\dfrac{1-a}{\lambda\mu(1+a')} - 1}{\dfrac{C_l}{C_d} + \dfrac{1-a}{\lambda\mu(1+a')}} = \frac{\lambda\mu a'(1-a)}{a(1-a) + (a'\lambda\mu)^2}$$

Simplifying,

$$\frac{\dfrac{C_l}{C_d}(1-a) - \lambda\mu(1+a')}{\lambda\mu(1+a')\dfrac{C_l}{C_d} + (1-a)} = \frac{\lambda\mu a'(1-a)}{a(1-a) + (a'\lambda\mu)^2}$$

$$\left[\frac{C_l}{C_d}(1-a) - \lambda\mu(1+a')\right][a(1-a) + (a'\lambda\mu)^2] = \left[\lambda\mu(1+a')\frac{C_l}{C_d} + (1-a)\right]\lambda\mu a'(1-a) \tag{3.60}$$

At this stage the process is made easier to follow if drag is ignored, Equation (3.60) then reduces to

$$a(1-a) - \lambda^2\mu^2 a' = 0 \tag{3.60a}$$

Differentiating Equation (3.60a) with respect to a' gives

$$(1-2a)\frac{\mathrm{d}}{\mathrm{d}a'}a - \lambda^2\mu^2 = 0 \tag{3.61}$$

and substituting Equation (3.57) into (3.61)

$$(1-2a)(1-a) - \lambda^2\mu^2 a' = 0 \tag{3.62}$$

Equations (3.60a) and (3.62), together, give the flow induction factors for optimized operation

$$a := \frac{1}{3} \text{ and } a' = \frac{a(1-a)}{\lambda^2\mu^2} \tag{3.63}$$

which agree exactly with the momentum theory prediction (Equation 3.23) because no losses, such as aerodynamic drag, have been included and the number of blades is assumed to be large; every fluid particle which passes through the rotor disc interacts with a blade resulting in a uniform axial velocity over the area of the disc.

To achieve the optimum conditions the blade design has to be specific and can be determined from either of the fundamental Equations (3.48) and (3.50). Choosing Equation (3.50), because it is the simpler, and ignoring the drag, the torque developed in optimized operation is

$$\delta Q = 4\pi\rho U_\infty(\Omega r)a'(1-a)r^2\delta r = 4\pi\rho\frac{U_\infty^3}{\Omega}a(1-a)^2 r\delta r$$

The component of the lift per unit span in the tangential direction is therefore

$$L\sin\phi = 4\pi\rho\frac{U_\infty^3}{\Omega}a(1-a)^2$$

By the Kutta–Joukowski theorem the lift per unit span is

$$L = \rho W\Gamma$$

where Γ is the sum of the individual blade circulations.

Consequently

$$\rho W\Gamma\sin\phi = \rho\Gamma U_\infty(1-a) = 4\pi\rho\frac{U_\infty^3}{\Omega}a(1-a)^2 \tag{3.64}$$

so

$$\Gamma = 4\pi\frac{U_\infty^2}{\Omega}a(1-a) \tag{3.65}$$

The circulation is therefore uniform along the blade span and this is a condition for optimized operation.

To determine the blade geometry, that is, how should the chord size vary along the blade and what pitch angle β distribution is necessary, we must return to Equation (3.50a):

$$\frac{W^2}{U_\infty^2}N\frac{c}{R}C_l\sin\phi = 8\pi\lambda\mu^2 a'(1-a)$$

substituting for $\sin\phi$ gives

$$\frac{W}{U_\infty} N \frac{c}{R} C_l(1-a) = 8\pi\lambda\mu^2 a'(1-a) \tag{3.66}$$

From which is derived

$$\frac{N}{2\pi}\frac{c}{R} = \frac{4\lambda\mu^2 a'}{\frac{W}{U_\infty} C_l}$$

The only unknown on the right-hand side of the above equation is the value of the lift coefficient C_l and so it is common to include it on the left-side of the equation with the chord solidity as a blade geometry parameter. The lift coefficient can be chosen as that value which corresponds to the maximum lift/drag ratio C_l/C_d as this will minimize drag losses; even though drag has been ignored in the determination of the optimum flow induction factors and blade geometry it cannot be ignored in the calculation of torque and power. Blade geometry also depends upon the tip speed ratio λ so it is also included in the blade geometry parameter. Hence

$$\sigma_r \lambda C_l = \frac{N}{2\pi}\frac{c}{R}\lambda C_l = \frac{4\lambda^2\mu^2 a'}{\sqrt{(1-a)^2 + (\lambda\mu(1+a'))^2}} \tag{3.67}$$

Introducing the optimum conditions of Equations (3.63)

$$\sigma_r \lambda C_l = \frac{\frac{8}{9}}{\sqrt{\left(1-\frac{1}{3}\right)^2 + \lambda^2\mu^2\left[1+\frac{2}{9(\lambda^2\mu^2)}\right]^2}} \tag{3.67a}$$

The parameter $\lambda\mu$ is called the local speed ratio and is equal to the tip speed ratio where $\mu = 1$.

If, for a given design, C_l is held constant then Figure 3.17 shows the blade planform for increasing tip speed ratio. A high design tip speed ratio would require a long, slender blade (high aspect ratio) whilst a low design tip speed ratio would need a short, fat blade. The design tip speed ratio is that at which optimum performance is achieved. Operating a rotor at other than the design tip speed ratio gives a less than optimum performance even in ideal drag free conditions.

In off-optimum operation the axial inflow factor is not uniformly equal to $\frac{1}{3}$, in fact it is not uniform at all.

The local inflow angle ϕ at each blade station also varies along the blade span as shown in Equation (3.68) and Figure 3.18

$$\tan\phi = \frac{1-a}{\lambda\mu(1+a')} \tag{3.68}$$

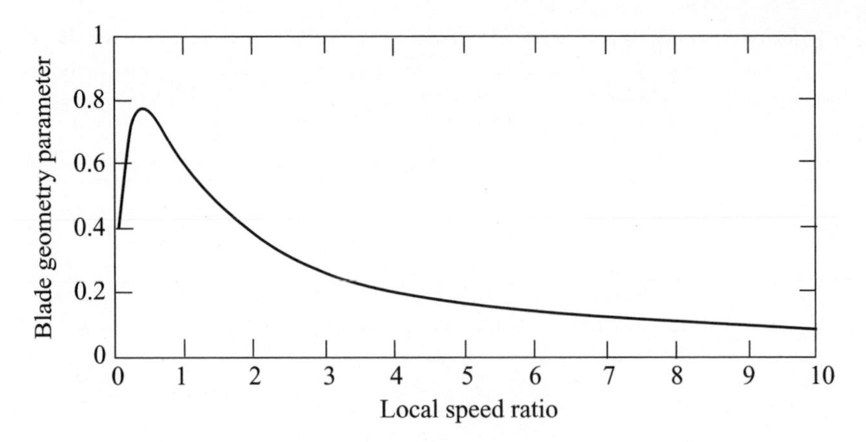

Figure 3.17 Variation of Blade Geometry Parameter with Local Speed Ratio

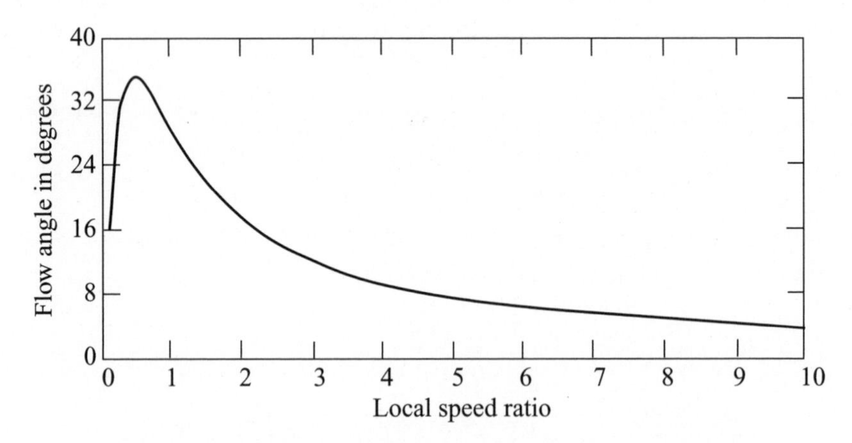

Figure 3.18 Variation of Inflow Angle with Local Speed Ratio

which, for optimum operation, is

$$\tan\phi = \frac{1 - \dfrac{1}{3}}{\lambda\mu\left(1 + \dfrac{2}{3\lambda^2\mu^2}\right)} \tag{3.68a}$$

Close to the blade root the inflow angle is large which could cause the blade to stall in that region. If the lift coefficient is to be held constant such that drag is minimized everywhere then the angle of attack α also needs to be uniform at the appropriate value. For a prescribed angle of attack variation the design pitch angle $\beta = \phi - \alpha$ of the blade must vary accordingly.

As an example, suppose that the blade aerofoil is NACA 4412, popular for hand-built wind turbines because the bottom (high pressure) side of the profile is almost flat which facilitates manufacture. At a Reynolds number of about 5×10^5 the maximum lift/drag ratio occurs at a lift coefficient of about 0.7 and an angle of attack of about 3°. Assuming that both C_l and α are to be held constant along each blade and there are to be three blades operating at a tip speed ratio of 6 then the blade design in plan-form and pitch (twist) variation are shown in Figures 3.19 (a) and (b), respectively.

3.7.3 A practical blade design

The blade design of Figure 3.19 is efficient but complex to build, and therefore costly. Suppose the plan-form was prescribed to have a uniform taper such that the outer part of the blade corresponds closely to Figure 3.19 (a). A straight line drawn through the 70 percent and 90 percent span points as shown Figure 3.20 not only simplifies the plan-form but removes a lot of material close to the root.

The expression for the new plan-form is

$$\frac{c_u}{R} = \frac{8}{9\lambda 0.8}\left(2 - \frac{\lambda\mu}{\lambda 0.8}\right)\frac{2\pi}{C_l\lambda N} \tag{3.69}$$

The 0.8 in Equation 3.69 refers to the 80 percent point, midway between the target points.

Equations (3.67a) and (3.69) can be combined to give the required span-wise variation of C_l for optimal operation.

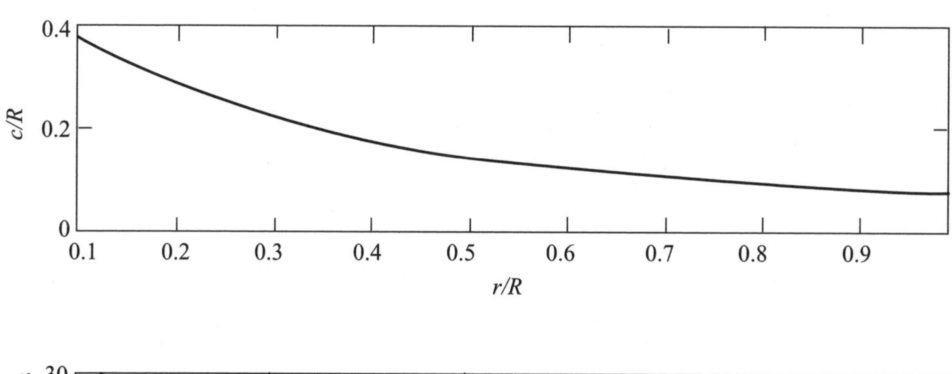

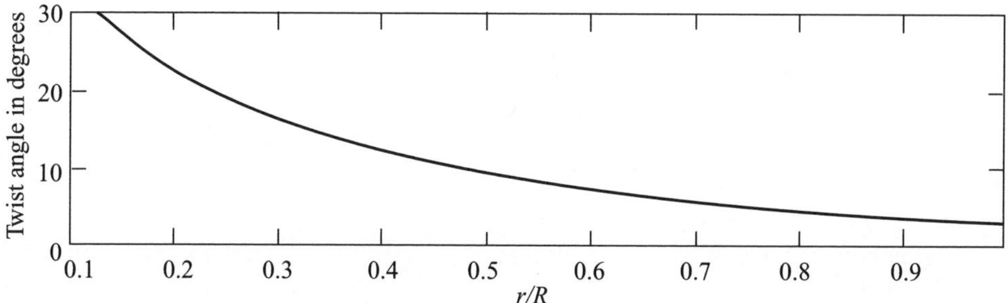

Figure 3.19 Optimum Blade Design for Three Blades and $\lambda = 6$

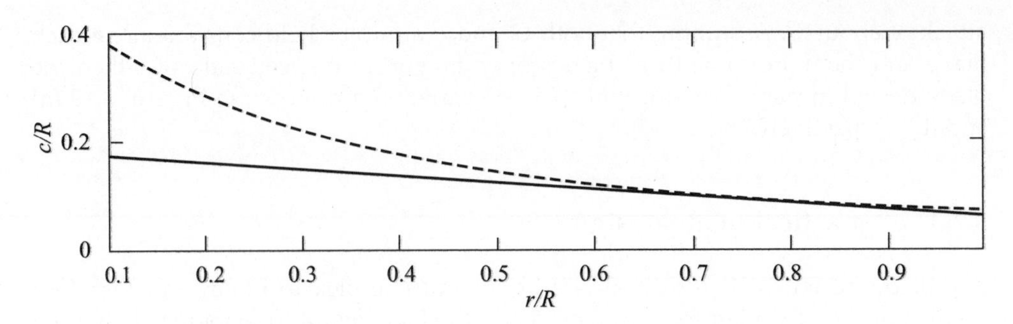

Figure 3.20 Uniform Taper Blade Design for Optimal Operation

$$C_l = \frac{8}{9} \frac{1}{\frac{Nc_u\lambda}{2\pi}\sqrt{\left(1-\frac{1}{3}\right)^2 + \lambda^2\mu^2\left[1+\frac{2}{9(\lambda^2\mu^2)}\right]^2}}$$

Close to the blade root the lift coefficient approaches the stalled condition and drag is high but the penalty is small because the adverse torque is small in that region.

Assuming that stall does not occur and that for the aerofoil in question, which has a 4 percent camber, the lift coefficient is given approximately by

$$C_l = 0.1(\alpha + 4\text{deg})$$

where α is in degrees, so

$$\alpha = \frac{C_l}{0.1} - 4\text{deg}$$

The blade twist distribution can now be determined from Equations (3.68a) and (3.43).

The twist angle close to the root is still high but lower than for the constant C_l blade.

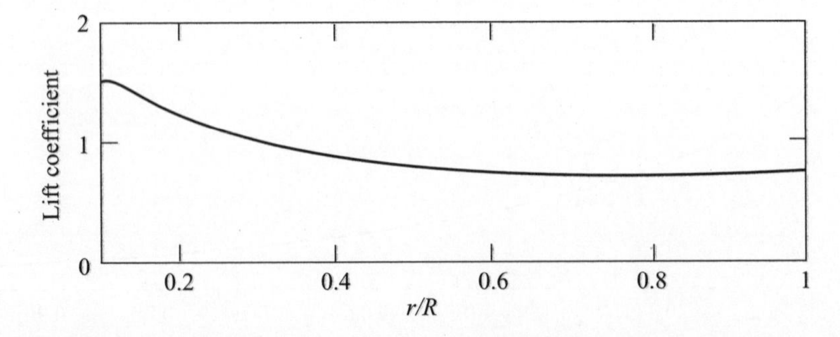

Figure 3.21 Span-wise Distribution of the Lift Coefficient Required for the Linear Taper Blade

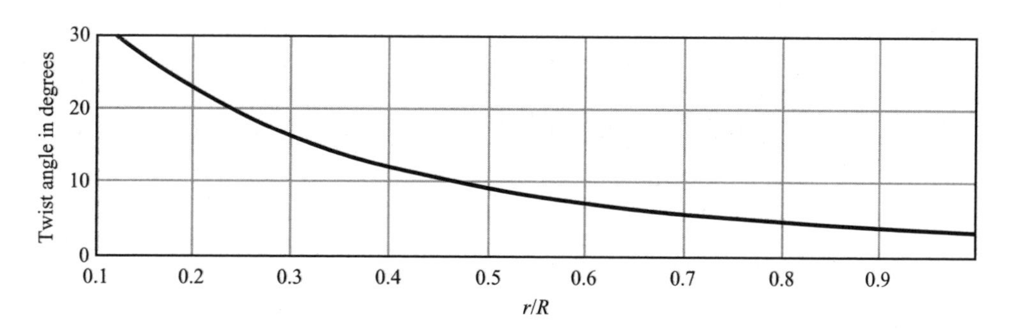

Figure 3.22 Span-wise Distribution of the Twist Required for the Linear Taper Blade

3.7.4 Effects of drag on optimal blade design

If, despite the views of Wilson and Lissaman (1974), see Section 3.5.3, the effects of drag are included in the determination of the flow induction factors we must return to Equation (3.46) and follow the same procedure as described for the drag free case.

In the current context the effects of drag are dependent upon the magnitude of the lift/drag ratio which, in turn, depends on the aerofoil profile but largely on Reynolds number and on the surface roughness of the blade. A high value of lift/drag ratio would be about 150, whereas a low value would be about 40.

Unfortunately, with the inclusion of drag, the algebra of the analysis is complex. Polynomial equations have to be solved for both a and a'. The details of the analysis are left for the reader to discover.

In the presence of drag, the axial flow induction factor for optimal operation is not uniform over the disc as it is in the hypothetical drag free situation. However, the departure of the axial flow distribution from uniformity is not great, even when the lift/drag ratio is low provided the flow around a blade remains attached.

The radial variation of the axial and tangential flow induction factors is shown in Figure 3.23 for zero drag and for a lift/drag ratio of 40. The tangential flow

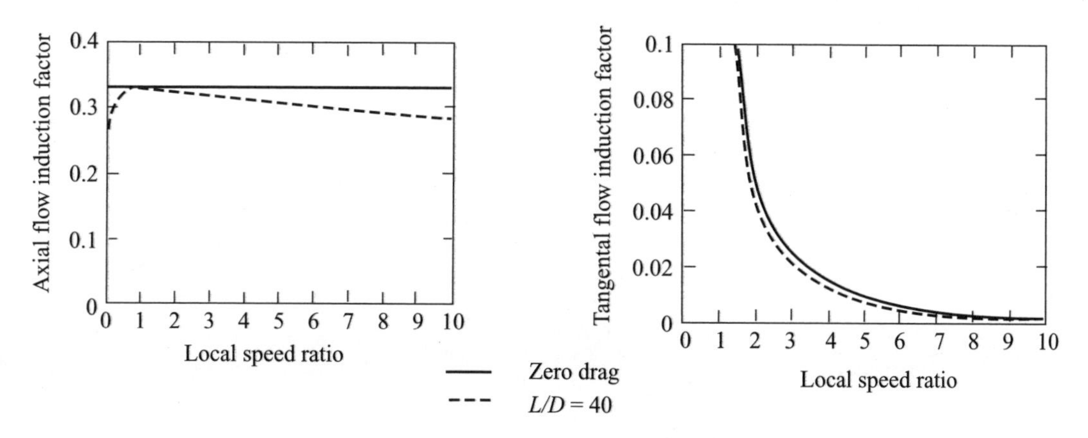

Figure 3.23 Radial Variation of the Flow Induction Factors with and without Drag

induction factor is lower in the presence of drag than without because the blade drags the fluid around in the direction of rotation, opposing the general rotational reaction to the shaft torque.

From the torque/angular momentum Equation (3.50) the blade geometry parameter becomes

$$\frac{N}{2\pi}\frac{c}{R}\lambda C_l = \frac{4\lambda^2\mu^2 a'(1-a)}{\dfrac{W}{U_\infty}\left[(1-a) - \dfrac{C_d}{C_l}\lambda\mu(1+a')\right]} \tag{3.70}$$

Figure 3.24 compares the blade geometry parameter distributions for zero drag and a lift/drag ratio of 40 and, as is evident, drag has very little effect on blade

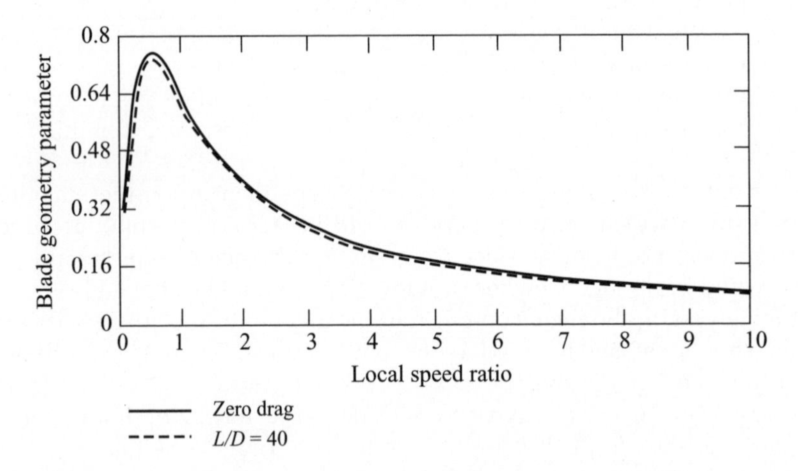

Figure 3.24 Span-wise Variation of the Blade Geometry Parameter with and without Drag

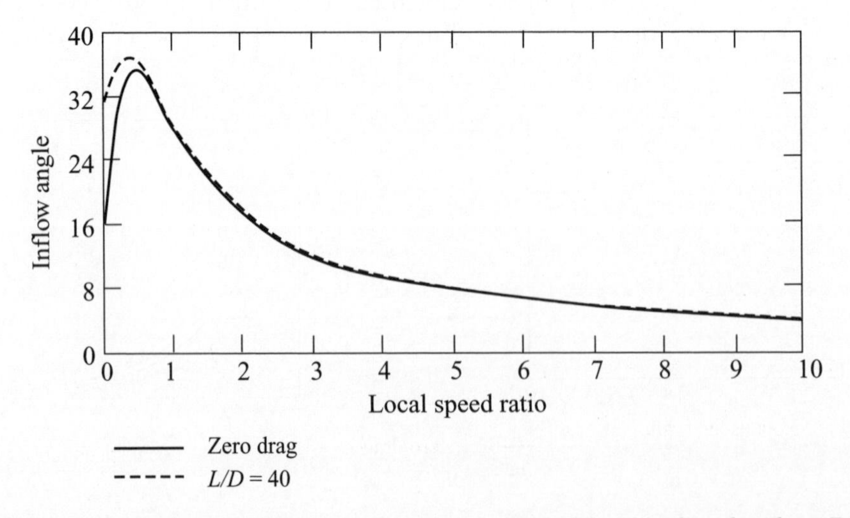

Figure 3.25 Variation of Inflow Angle with Local Speed Ratio with and without Drag

optimal design. A similar result is apparent for the inflow angle distribution where drag also has little influence, see Figure 3.25. As far as blade design for optimal operation is concerned drag can be ignored, greatly simplifying the process.

The results of Equation (3.54) showing the maximum power coefficients for a range of design tip speed ratios and several lift/drag ratios is shown in Figure 3.26. The flow induction factors have been determined without drag (Equations (3.63)), as have the blade designs (Equations 3.67a and 3.68a), but the torque has been calculated using Equation (3.54), which includes drag. The losses caused by drag are significant and increase with increasing design tip speed ratio. As will be shown later, when tip losses are also taken into account, the losses at low tip speed ratios are even greater.

3.7.5 Optimal blade design for constant-speed operation

If the rotational speed of a turbine is maintained at a constant level as is common for grid connected turbines then the tip speed ratio is continuously changing and a blade optimized for a fixed tip speed ratio would not necessarily be appropriate.

No simple technique is available for the optimal design of a blade operating at constant rotational speed. A non-linear programming method could be applied by maximizing energy capture at a site with a specified wind-speed distribution. Alternatively, a design tip speed ratio could be chosen which corresponds to the wind speed at the specified site which contains the most energy or, more practically, the pitch angle for the whole blade can be adjusted to maximize energy capture.

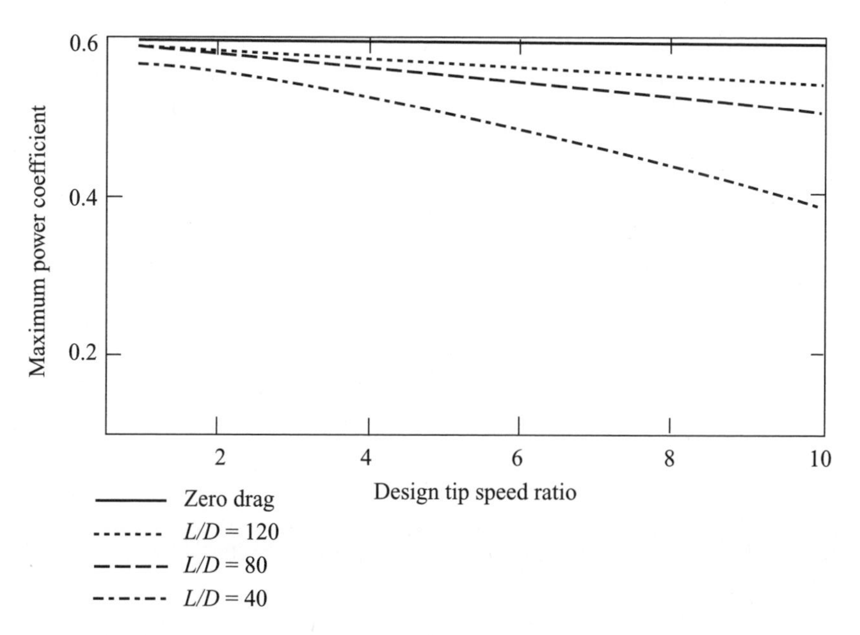

Figure 3.26 The Variation of Maximum C_P with Design λ for Various Lift/Drag Ratios

3.8 The Effects of a Discrete Number of Blades

3.8.1 Introduction

The analysis described in the previous Sections of this chapter assumes that there is a sufficient number of blades on the rotor for every fluid particle passing through the rotor disc to interact with a blade, i.e., that all fluid particles undergo the same loss of momentum. With a small number of blades some fluid particles will interact with them but most will pass between the blades and, clearly, the loss of momentum by a particle will depend on its proximity to a blade as the particle passes through the rotor disc. The axial induced velocity will, therefore, at any instant, vary around the disc, the average value determining the overall axial momentum of the flow and the larger value local to a blade determining the forces on the blade.

3.8.2 Tip-losses

If the axial flow induction factor a is large at the blade position then, by Equation (3.42), the inflow angle ϕ will be small and the lift force will be almost normal to the rotor plane. The component of the lift force in the tangential direction will be small and so will be its contribution to the torque. A reduced torque means reduced power and this reduction is known as tip loss because the effect occurs only at the outermost parts of the blades.

In order to account for tip losses, the manner in which the axial flow induction factor varies azimuthally needs to be known but, unfortunately, this requirement is beyond the abilities of the blade element–momentum theory.

Just as a vortex trails from the tip of an aircraft wing so does a vortex trail from the tip of a wind turbine blade. Because the blade tip follows a circular path it leaves a trailing vortex of a helical structure, as is shown in Figure 3.27, which convects downstream with the wake velocity. For a two-blade rotor, unlike an aircrafts wings, the bound circulations on the two blades are opposite in sign and so combine to shed a straight line vortex along the rotational axis with strength equal to the blade circulation times the number of blades.

For a single vortex to be shed from the blade tip the circulation strength along the blade span must be uniform and, as has been shown, uniform circulation is a requirement for optimized operation. However, the uniform circulation requirement assumes that the axial flow induction factor is uniform across the disc and, as has been argued above, with discrete blades rather than a uniform disc the flow factor is not uniform.

In the case of Figure 3.27, very close to the blade tips the tip vortex causes very high values of the flow factor a such that, locally, the net flow past the blade is in the upstream direction. The average value of a, azimuthally, is radially uniform which means that if high values occur in the vicinity of the blades then low values occur elsewhere. The azimuthal variation of a at various radial positions is shown in Figure 3.28 for a three blade rotor operating at a top speed ratio of 6. The

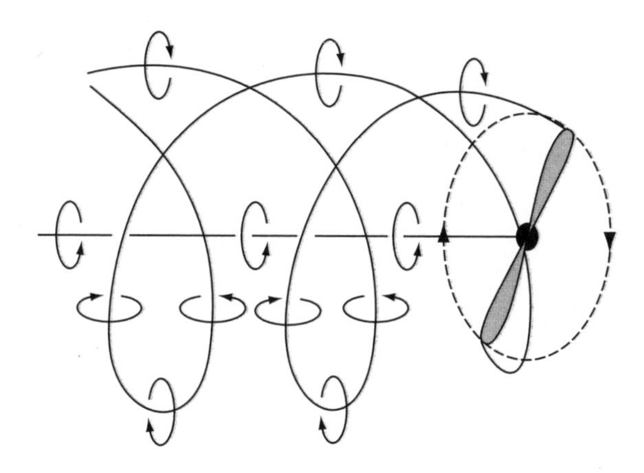

Figure 3.27 Helical Trailing Tip Vortices of a Horizontal Axis Turbine Wake

calculation for Figure 3.28 assumes a discrete vortex for each blade with a constant pitch and constant radius helical wake.

The ratio of the average value of a to that at a blade position is shown in Figure 3.29 and only near the tip does the ratio begin to fall to zero. The ratio is called 'the tip-loss factor'.

From Equation (3.20) and in the absence of tip-loss and drag the contribution of each blade element to the overall power coefficient is

$$\delta C_P = 8\lambda^2\mu^3 a'(1-a)\delta\mu \tag{3.71}$$

Substituting for a' from Equation (3.23) gives

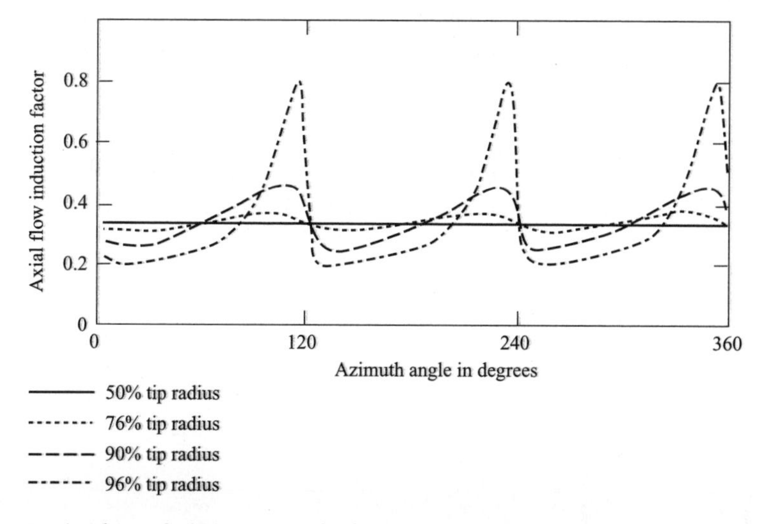

Figure 3.28 Azithimuthal Variation of a for Various Radial Positions for a Three-blade Rotor with Uniform Blade Circulation Operating at a Tip Speed Ratio of 6. The blades are at 120°, 240° and 360°.

$$\delta C_P = 8\mu a(1-a)^2\delta\mu \tag{3.72}$$

Whereas, from the Kutta–Joukowski theorem, the circulation Γ on the blade, which is uniform, provides a torque per unit span of

$$\frac{\delta Q}{\delta r} = \rho(W \times \Gamma)\sin\phi_r\, r$$

where the angle ϕ_r is determined by the flow velocity local to the blade.

The strength of the total circulation for all three blades is given by Equation (3.65) and so, in the presence of tip-loss the increment of power coefficient from a blade element is

$$\delta C_P = 8\mu a(1-a)(1-a_r)\delta\mu \tag{3.73}$$

where $a = 1/3$ is the average axial flow induction factor and a_r is the value local to the blade.

The results from Equations (3.72) and (3.73) are plotted in Figure 3.30 and clearly show the effect of tip-loss. Equation (3.72) assumes that $a = 1/3$ uniformly over the whole disc, Equation (3.73) recognizes that a is not uniform. The azimuthally averaged value of a is equal to 1/3 at every radial position but the azimuth variation gives rise to the tip-loss. The blade does not extract energy from the flow efficiently because a varies. Imagine the disc comprising a myriad of elemental discs, each with its own independent stream-tube, and not all of them operating at the Betz limit. Note that the power loss to the wind is exactly the same as that extracted by the blades, there is no effective drag associated with tip-loss.

With uniform circulation the azimuthal average value of a is also radially uniform but that implies a discontinuity of axial velocity at the wake boundary with a corresponding discontinuity in pressure. Whereas such discontinuities are acceptable in the idealized actuator disc situation they will not occur in practice with a finite number of blades. If it is assumed that a is zero outside of the wake then a must fall to zero in a regular fashion towards the blade tips and, consequently, the bound circulation must also fall to zero. The manner in which the circulation varies at the tip will be governed by the blade tip design, that is, the chord and pitch variation, and there will be a certain design which will minimize the tip-loss.

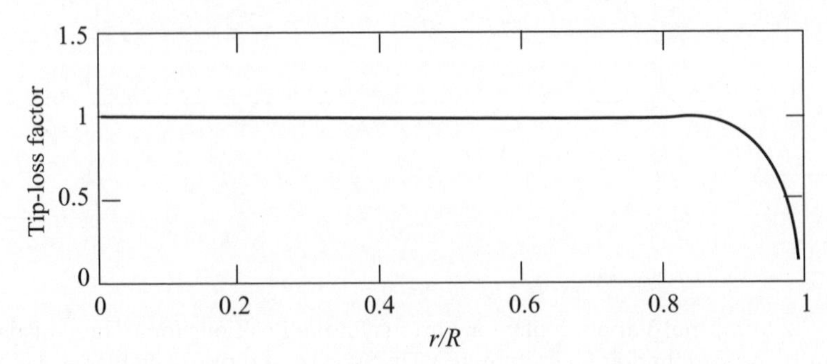

Figure 3.29 Span-wise Variation of the Tip-loss Factor for a Blade with Uniform Circulation

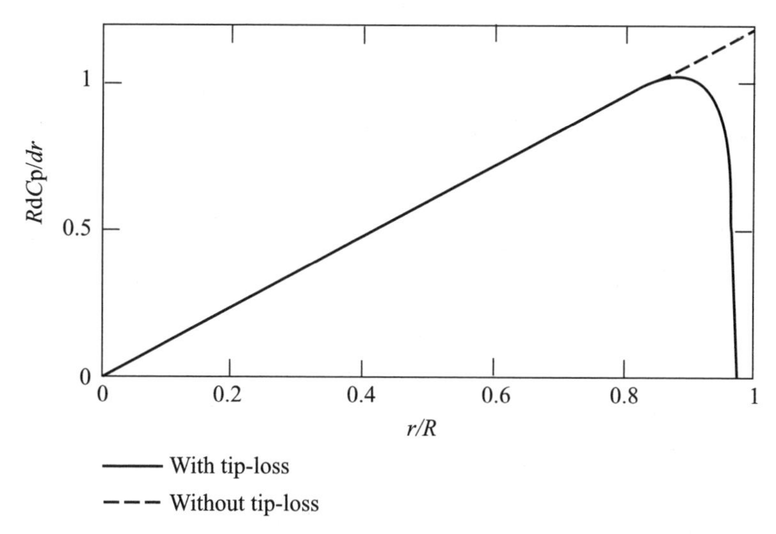

Figure 3.30 Span-wise Variation of Power Extraction in the Presence of Tip-loss for a Blade with Uniform Circulation on a Three-blade Turbine Operating at a Tip Speed Ratio of 6

If the circulation varies along the blade span vorticity is shed into the wake in a continuous fashion from the trailing edge. Therefore, each blade sheds a helicoidal sheet of vorticity, as shown in Figure 3.31, rather than a single helical vortex as shown in Figure 3.27. The helicoidal sheets convect with the wake velocity and so there can be no flow across the sheets which can therefore be regarded as impermeable. The intensity of the vortex sheets is equal to the rate of change of bound circulation along the blade span and so increases rapidly towards the blade tips. There is flow around the blade tips because of the pressure difference between the blade surfaces which means that on the upwind surface of the blades the flow moves towards the tips and on the downwind surface the flow moves towards the root. The flows from either surface leaving the trailing edge of a blade will not be

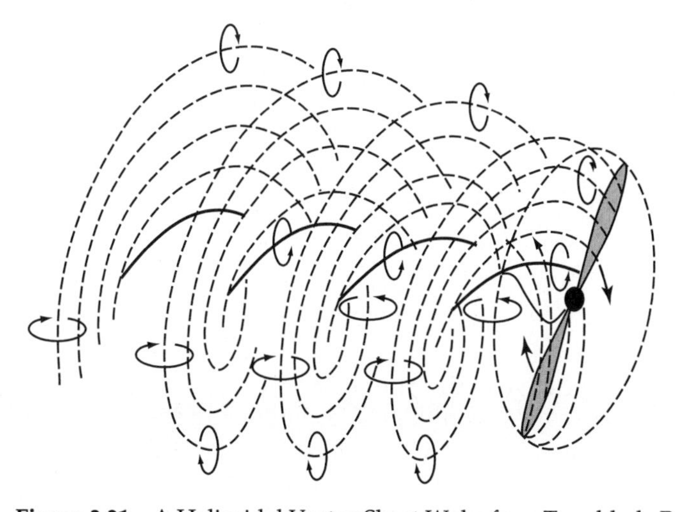

Figure 3.31 A Helicoidal Vortex Sheet Wake for a Two-blade Rotor

parallel to one another and will form a surface of discontinuity of velocity in a radial sense within the wake; the axial velocity components will be equal. The surface of discontinuity is called a vortex sheet. A similar phenomenon occurs with aircraft wings and a textbook of aircraft aerodynamics will explain it in greater detail.

A deeper understanding of the mechanism of tip-loss can be obtained by following the path of air particles. An air particle approaches the spinning rotor, 'senses' high pressure ahead and slows down accordingly. The high pressure on the upwind side of the rotor blades is effectively smeared around the whole disc. Slowing down also causes the particle to move outwards to maintain the mass flow rate. When the particle reaches the rotor plane it will either be close to a blade or not and its axial velocity will be affected accordingly, as shown in Figure 3.28. If the particle passes through the rotor plane close to blade then it will also be strongly affected by the blade's pressure field.

A particle which passes close to and in front of a blade will leave the trailing edge having accelerated in the tangential direction; it will then pass downstream, on the upwind side of the vortex sheet being shed from the trailing edge and so will also be moving radially outwards. The particle, therefore, migrates outward to the edge of the vortex sheet around which it is swept on to the downwind side and migrates inward with a radial velocity which reduces to zero at a radial point on the sheet where the shed vorticity is zero. The particle then continues downstream with the velocity of the axial and tangential velocities of the vortex sheet.

A second particle which passes a blade close to the downwind, low pressure, surface is accelerated tangentially in the opposite direction to the blade motion and then slows down, leaving the trailing edge with the same axial and tangential velocity components as the first particle but on the downwind side of the vortex sheet so it will have, in addition, a radially inwards velocity. The second particle will, depending on its radial position, migrate inwards until the radial velocity becomes zero.

A third particle which passes between two blades will be moving axially at a greater velocity than the first two particles, will not be strongly affected by the pressure fields of the blades but, because of the solid blockage presented by the blades (see Figure 3.5), will be directed into a helical path. Being faster, axially, than the vortex sheet ahead the particle will begin to catch up, as it does so the influence of the vortex sheet will move it outwards, around the edge of the sheet and then inwards, just like the first particle. Unlike the first particle, however, the third particle will still be moving faster than the vortex sheet and so will move axially away from the sheet, approaching the next sheet downstream and repeating the motion around the edge of that sheet. The particle will proceed downstream overtaking and hopping around each vortex sheet in turn.

The third particle does not lose as much axial momentum as particles one and two and is therefore affected by the so-called tip-loss. The affect is greater the closer the third particle is to the edge of the rotor disc as it passes through the disc.

A fourth particle passes between the blades but at a radial position, closer to the axis of rotation, where its axial velocity is equal to that of the vortex sheets. If the particle passes midway, say, between two blades then it remains midway between the two corresponding vortex sheets as it moves downstream and does not undergo any radial motion other than the general expansion caused by the slowing down of the flow. The fourth particle is totally unaffected by the fact that there is a finite number of blades and follows the same progress as if it were passing through a uniform actuator disc.

The axial flow induction factor varies, therefore, not only azimuthally but also radially, is a function of both r and θ. The azimuthally averaged value of $a(r) = a_b(r)f(r)$, where $f(r)$ is known as the tip-loss factor, has a value of unity inboard (particle 4) and falls to zero at the edge of the rotor disc. The value $a_b(r)$ is the level of axial flow induction factor that occurs locally at a blade element and is the velocity with which the vortex sheet convects downstream. If $a_b(r)$ can be held radially uniform then the vortex sheets will be radially flat, as shown in Figure 3.31, but if $a_b(r)$ is not uniform the vortex sheets will warp.

In the application of the blade element–momentum theory it is argued that the rate of change of axial momentum is determined by the azimuthally averaged value of axial flow induction factor, whereas the blade forces are determined by the value of the flow factor which occurs locally at the blade element, that experienced by the first and second particles.

$$\text{The mass flow rate through an annulus} = \rho U_\infty(1 - a_b(r)f(r))2\pi r\delta r$$

$$\text{The azimuthally averaged overall change of axial velocity} = 2a_b(r)f(r)U_\infty$$

$$\text{The rate of change of axial momentum} = 4\pi r\rho U_\infty^2(1 - a_b(r)f(r))a_b(r)f(r)\delta r$$

$$\text{The blade element forces} = \frac{1}{2}\rho W^2 N c C_l \text{ and } \frac{1}{2}e W^2 N_c C_d$$

where W, C_l and C_d are determined using $a_b(r)$.

The pressure force caused by the rotation of the wake is also calculated using an azimuthally averaged value of the tangential flow induction factor $2a'_b(r)f(r)$.

3.8.3 Prandtl's approximation for the tip-loss factor

The function for the tip-loss factor $f(r)$ is shown in Figure 3.29 for a blade with uniform circulation operating at a tip speed ratio of 6 and is not readily obtained by analytical means for any desired tip speed ratio. Sidney Goldstein did analyse the tip-loss problem for application to propellers in 1929 and achieved a solution in terms of Bessel functions but neither that nor the Biot–Savart solution used above is suitable for inclusion in the blade element–momentum theory. Fortunately, in 1919, Ludwig Prandtl (reported by Betz, 1919) had already developed an ingenious approximate solution which does yield a relatively simple analytical formula for the tip-loss function.

Prandtl's approximation was inspired by the fact that, being impermeable (particles one and three in the above description pass around the outer edge of a sheet but not through it), the vortex sheets could be replaced by material sheets which, provided they move with the velocity dictated by the wake, would have no effect upon the wake flow. The theory applies only to the developed wake. In order to simplify his analysis Prandtl replaced the helicoidal sheets with a succession of discs, moving with the uniform, central wake velocity $U_\infty(1-a)$ and separated by the same distance as the normal distance between the vortex sheets. Conceptually, the discs, travelling axially with velocity $U_\infty(1-a)$ would encounter the unattenuated free-stream velocity U_∞ at their outer edges. The fast flowing free-stream air would tend to weave in and out between successive discs. The wider apart successive discs the deeper, radially, the free-stream air would penetrate. Taking any line parallel to the rotor axis at a radius r, somewhat smaller than the wake radius (rotor radius), the average axial velocity along that line would be greater than $U_\infty(1-a)$ and less than U_∞. Let the average velocity be $U_\infty(1-af(r))$, where $f(r)$ is the tip-loss function, has a value less than unity and falls to zero at the wake boundary. At a distance from the wake edge the free-stream fails to penetrate and there is little or no difference between the induced velocity at the blade and that in the wake, i.e., $f(r)=1$.

A particle path, as shown in Figure 3.32, is very similar to that described for particle three, above, and may be interpreted as that of the *average* particle passing through the rotor disc at a given radius in the actual situation: the azimuthal variations of particle velocities at various radii are shown in Figure 3.28 and a 'Prandtl particle' would have a velocity equal to the average of the variation. Figure 3.32 depicts the developed wake. Prandtl's approximation defines quite well the downstream behaviour of particle three above, which passes the rotor plane between two blades.

The mathematical detail of Prandtl's analysis (see Glauert (1935a)) is beyond the scope of this text but, unlike Goldstein's theory, the result can be expressed in closed solution form; the tip-loss factor is given by

$$f(r)=\frac{2}{\pi}\cos^{-1}[\mathrm{e}^{-\pi(R_W/d-r/d)}] \tag{3.74}$$

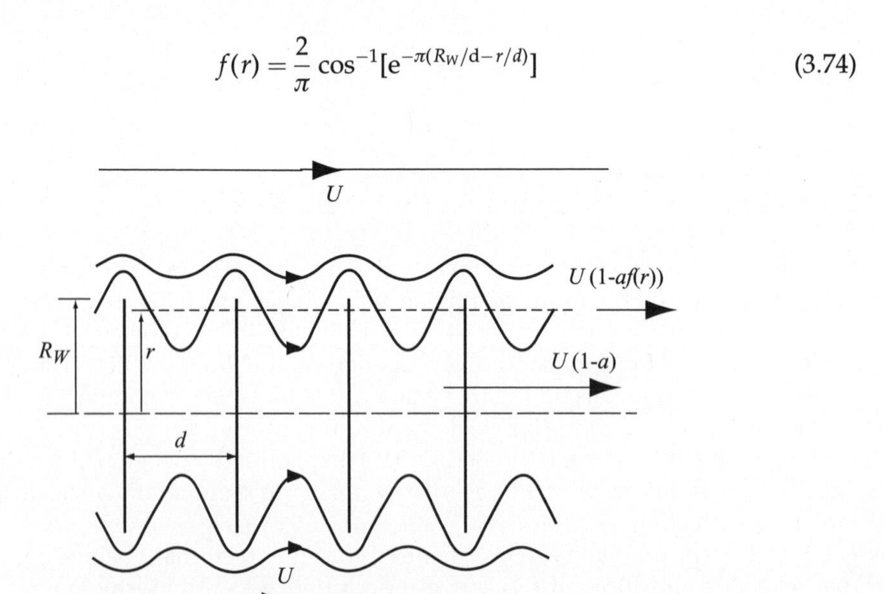

Figure 3.32 Prandtl's Wake-disc Model to Account for Tip-losses

$R_W - r$ is a distance measured from the wake edge. Distance d between the discs should be that of the distance travelled by particle three between successive vortex sheets. Glauert (1935) takes d as being the normal distance between successive helicoidcal vortex sheets.

The helix angle of the vortex sheets is the flow angle ϕ_S and so with N sheets intertwining from N blades

$$d = \frac{2\pi R_W}{N}\sin\phi_S = \frac{2\pi R_W}{N}\frac{U_\infty(1-a)}{W_S} \tag{3.75}$$

Prandtl's model has no wake rotation but the discs may spin at the rotor speed without affecting the flow at all, as it is inviscid, thus a' is zero and W_S is the resultant velocity (not including the radial velocity) at the edge of a disc. Glauert (1935a) argues that $R_W/W_S \approx r/W$ which is much more convenient to use.

$$W = \sqrt{[U_\infty(1-a)]^2 + (\Omega r)^2}$$

so

$$\pi\left(\frac{R_W - r}{r}\right) = \frac{N}{2}\left(\frac{R-r}{d}\right)\sqrt{1 + \frac{(\Omega r)^2}{[U_\infty(1-a)]^2}}$$

and

$$f(\mu) = \frac{2}{\pi}\cos[e^{((N/2)(1-\mu)/\mu)\sqrt{1+(\lambda\mu)^2/(1-a)^2}}] \tag{3.76}$$

The Prandtl tip-loss factor for a three-blade rotor operating at a tip speed ratio of 6 is compared with the tip-loss factor of the helical vortex wake in the Figure 3.33.

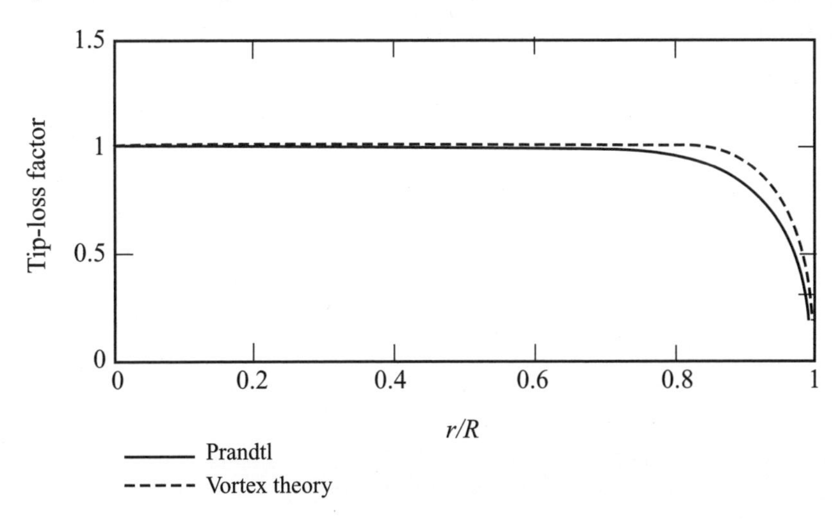

Figure 3.33 Comparison of Prandtl Tip-loss Factor with that Predicted by a Vortex Theory for a Three-blade Turbine Optimized for a Tip Speed Ratio of 6

It should also be pointed out that the vortex theory of Figure 3.28 also predicts that the tip-loss factor should be applied to the tangential flow induction factor.

It is now useful to know what is the variation of circulation along the blade. For the previous analysis, which disregarded tip-losses, the blade circulation was uniform (Equation (3.65)). Following the same procedure from which Equation (3.64) was developed

$$\rho(W \times \Gamma)\sin\phi = \rho\Gamma U_{\infty}(1 - a_b(r)) = 4\pi\rho\frac{U_{\infty}^3}{\Omega}a(r)(1 - a(r))^2$$

Recall that $a_b(r)$ is the flow factor local to the blade at radius r, which is equal to a and $a(r)$ is the average value of the flow factor at radius r. Therefore

$$\Gamma(r) = \frac{4\pi}{1 - a}\frac{U_{\infty}^2}{\Omega}af(r)(1 - af(r))^2 \qquad (3.77)$$

$\Gamma(r)$ is the total circulation for all blades and is shown in Figure 3.34 and, as can be seen, it is almost uniform except near to the tip. The dashed vertical line shows the effective blade length (radius) $R_e = 0.975$ if the circulation is assumed to be uniform at the level that pertains at the inboard section of the blade.

The Prandtl tip-loss factor appears to offer an acceptable, simple solution to a complex problem; not only does it account for the effects of discrete blades but it also allows the induction factors to fall to zero at the edge of the rotor disc.

3.8.4 Blade root losses

At the root of a blade the circulation must fall to zero as it does at the blade tip and so it can be presumed that a similar process occurs. The blade root will be at some distance from the rotor axis and the airflow through the disc inside the blade root

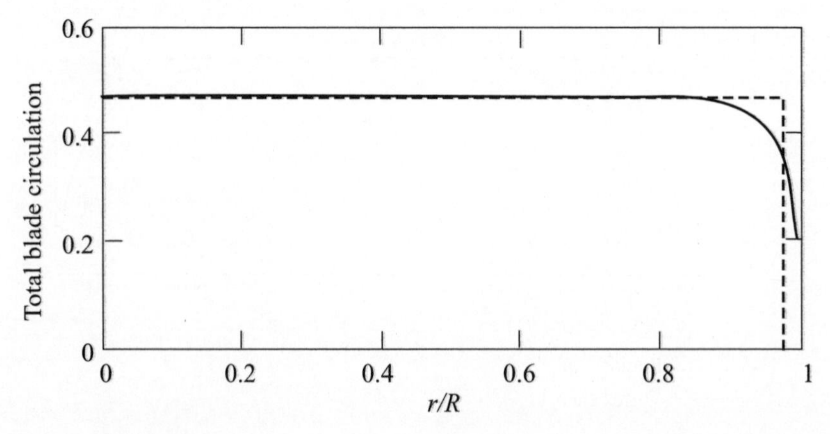

Figure 3.34 Span-wise Variation of Blade Circulation for a Three-blade Turbine Optimized for a Tip Speed Ratio of 6

radius will be at the free-stream velocity. Actually, the vortex theory of Section 3.4 can be extended to show that the flow through the root disc is somewhat higher than the free-stream velocity. It is usual, therefore, to apply the Prandtl tip-loss function at the blade root as well as at the tip.

If μ_R is the normalized root radius then the root loss factor can be determined by modifying the tip-loss factor of Equation (3.76).

$$f_R(\mu) = \frac{2}{\pi}\cos[\mathrm{e}^{-N/2(\mu-\mu_R/\mu)\sqrt{1+(\lambda\mu)^2/(1-a)^2}}] \tag{3.78}$$

If Equation (3.76) is now termed $f_T(r)$ the complete tip/root loss factor is

$$f(\mu) = f_T(\mu)f_R(\mu) \tag{3.79}$$

3.8.5 Effect of tip-loss on optimum blade design and power

With no tip-loss the optimum axial flow induction factor is uniformly 1/3 over the whole swept rotor. The presence of tip-loss changes the optimum value of the average value of a which reduces to zero at the edge of the wake but, local to the blade tends to increase in the tip region. If $a(r)$ is taken as the azimuthal average at radius r then locally, at the blade at that radius, the flow factor will be $a(r)/f(r)$. The inflow angle ϕ at the blade is then, from Equation (3.57),

$$\tan\phi = \frac{1}{\lambda\mu}\left(\frac{1-\dfrac{a}{f}}{1+\dfrac{a'}{f}}\right) \tag{3.80}$$

but Equation (3.56), which is the ratio of the non-dimensional rate of change of

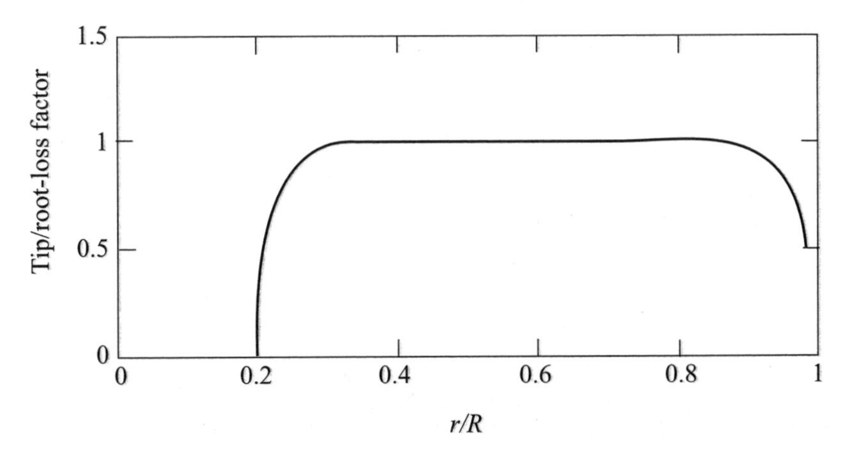

Figure 3.35 Span-wise Variation of Combined Tip/root-loss Factor for a Three-blade Turbine Optimized for a Tip Speed Ratio of 6 and with a Blade Root at 20% Span

angular momentum to the non-dimensional rate of change of axial momentum, is not changed because it deals with the whole flow through the disc and so uses average values. If drag is ignored for the present, Equation (3.59) becomes

$$\tan\phi = \frac{\lambda\mu a'(1-a)}{a(1-a)+(a'\lambda\mu)^2} \tag{3.81}$$

Hence

$$\frac{(1-a)\lambda\mu a'}{a(1-a)+(\lambda\mu a')^2} = \frac{\left(1-\dfrac{a}{f}\right)}{\left[\lambda\mu\left(1+\dfrac{a'}{f}\right)\right]}$$

which becomes

$$\lambda^2\mu^2\frac{(f-1)}{f}a'^2 - \lambda^2\mu^2(1-a)a' + a(1-a)\left(1-\frac{a'}{f}\right) = 0 \tag{3.82}$$

A great simplification can be made to Equation (3.82) by ignoring the first term because, clearly, it disappears for much of the blade, where $f = 1$, and for the tip region the value of a'^2 is very small. For tip speed ratios greater than 3 neglecting the first term makes negligible difference to the result.

$$\lambda^2\mu^2 a' = a\left(1-\frac{a}{f}\right) \tag{3.83}$$

As before, Equation (3.57) still applies

$$\frac{\mathrm{d}}{\mathrm{d}a'}a = \frac{1-a}{a'}$$

From Equation (3.83)

$$\frac{\mathrm{d}}{\mathrm{d}a}a' = \frac{1}{\lambda^2\mu^2}\left(1-2\frac{a}{f}\right)$$

Consequently

$$(1-a)\left(1-2\frac{a}{f}\right) = \lambda^2\mu^2 a'$$

which combined with Equation (3.83) gives

$$a^2 - \frac{2}{3}(f+1)a + \frac{1}{3}f = 0$$

so

$$a = \frac{1}{3} + \frac{1}{3}f - \frac{1}{3}\sqrt{1 - f + f^2} \tag{3.84}$$

The radial variation of the average value of a, as given by Equation (3.84), and the value local to the blade a/f is shown in Figure 3.36. An exact solution would also have the local induced velocity falling to zero at the blade tip.

Clearly, the required blade design for optimal operation would be a little different to that which corresponds to the Prandtl tip-loss factor because a/f, the local flow factor, does not fall to zero at the blade tip. The use of the Prandtl tip-loss factor leads to an approximation, but that was recognized from the outset.

The blade design, which gives optimum power output, can now be determined by adapting Equations (3.66) and (3.67) accordingly

$$\sigma_r \lambda C_l = \frac{4\lambda^2\mu^2 a'}{\sqrt{\left(1 - \frac{a}{f}\right)^2 + \left[\lambda\mu\left(1 + \frac{a'}{f}\right)\right]^2}} \left(\frac{1 - a}{1 - \frac{a}{f}}\right)$$

Introducing Equation (3.83) gives

$$\sigma_r \lambda C_l = \frac{4a(1 - a)}{\sqrt{\left(1 - \frac{a}{f}\right)^2 + \left[\lambda\mu\left[1 + \frac{a\left(1 - \frac{a}{f}\right)}{\lambda^2\mu^2 f}\right]\right]^2}} \tag{3.85}$$

The blade geometry parameter given by Equation (3.85) is shown in Figure 3.37 compared with the design which excludes tip-loss. As can be seen, only in the tip region is there any difference between the two designs.

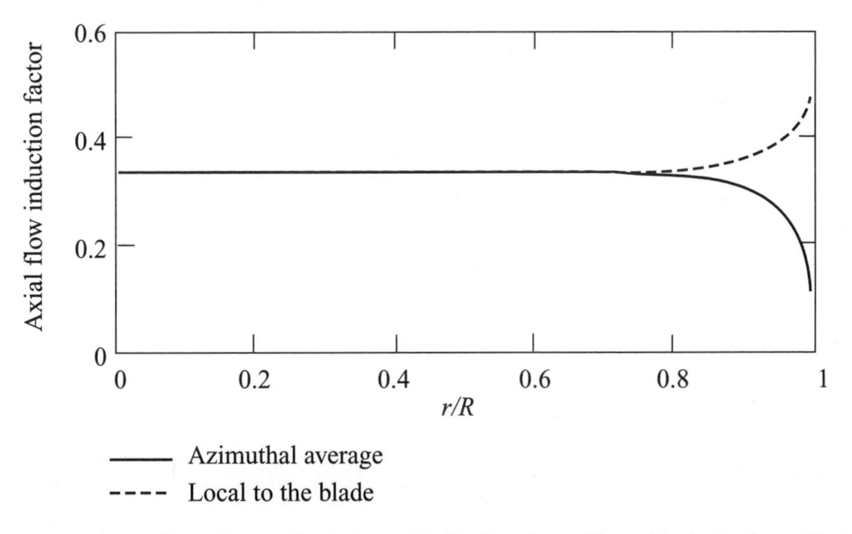

Figure 3.36 Axial Flow Factor Variation with Radius for a Three-blade Turbine Optimized for a Tip Speed Ratio of 6

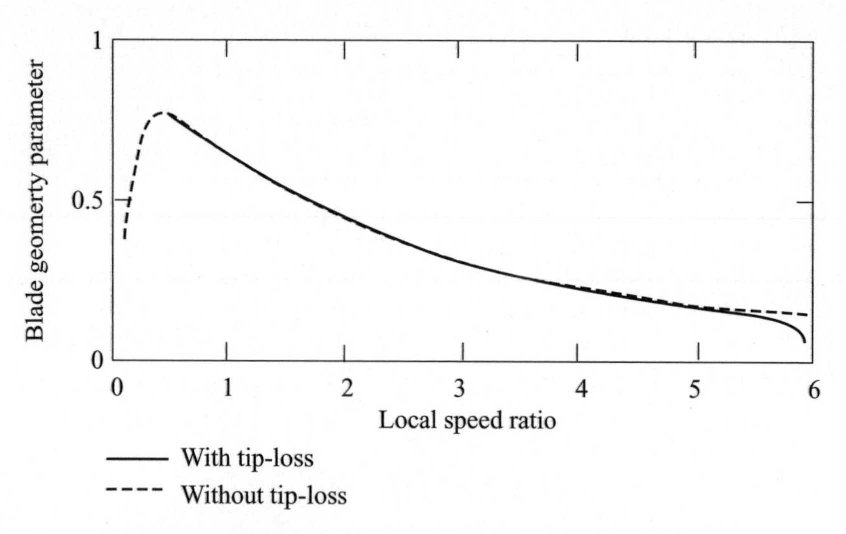

Figure 3.37 Variation of Blade Geometry Parameter with Local Speed Ratio, with and without Tip-loss for a Three-blade Rotor with a Design Tip Speed Ratio of 6

Similarly, the inflow angle distribution, shown in Figure 3.38, can be determined by suitably modifying Equation (3.66)

$$\tan\phi = \frac{1 - \dfrac{a}{f}}{\lambda\mu\left[1 + \dfrac{a\left(1 - \dfrac{a}{f}\right)}{\lambda^2\mu^2 f}\right]} \tag{3.86}$$

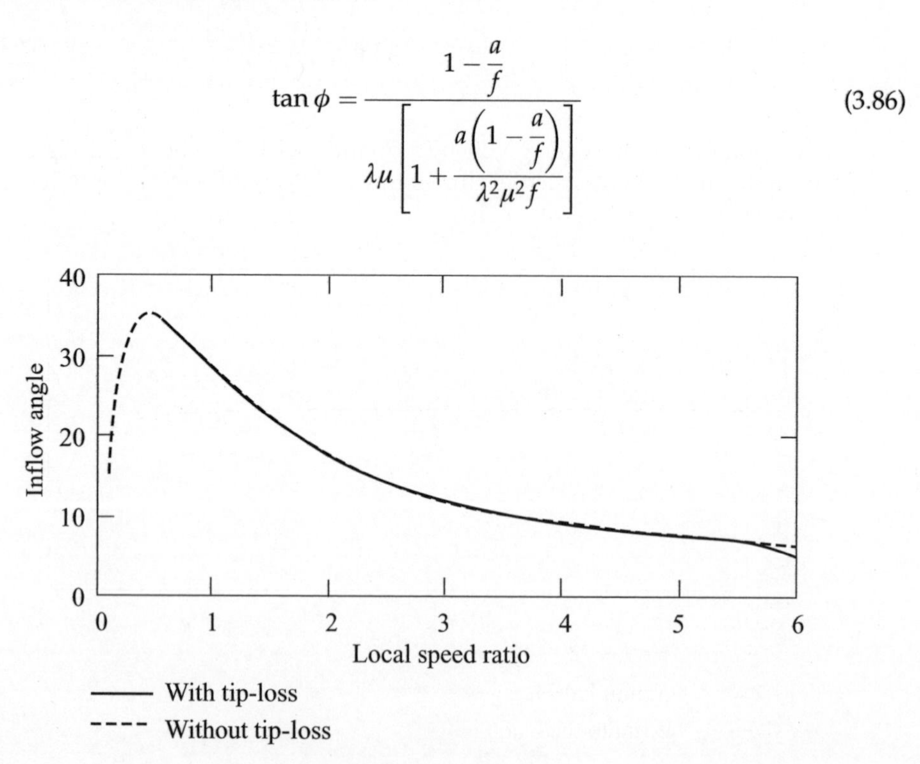

Figure 3.38 Variation of Inflow Angle with Local-Speed Ratio, with and without Tip-loss for a Three-blade Rotor with a Design Tip Speed Ratio of 6

Again, the effects of tip-loss are confined to the blade tip.

The power coefficient for an optimized rotor, operating at the design tip speed ratio, without drag and tip-losses is equal to the Betz limit 0.593 but with tip-loss there is obviously a reduced optimum power coefficient. Equation (3.20) determines the power coefficient distribution along the blade, see Figure 3.39.

The power coefficient

$$C_P = \frac{P}{\frac{1}{2}\rho U_\infty^3 \pi R^2} = 8\lambda^2 \int_0^1 a'(1-a)\mu^3 \, d\mu \tag{3.87}$$

for which a' and a are obtained from Equations (3.83) and (3.84).

The maximum power coefficient that can be achieved in the presence of both drag and tip-loss is significantly less than the Betz limit at all tip speed ratios. As is shown in Figure 3.42 drag reduces the power coefficient at high tip speed ratios but the effect of tip-loss is most significant at low tip speed ratios because the separation of the helicoidal vortex sheets is large.

3.8.6 Incorporation of tip-loss for non-optimal operation

The blade element–momentum Equations (3.51), (3.51a) and (3.52) are used to determine the flow induction factors for non-optimal operation. With tip-loss included the BEM equations have to be modified. The necessary modification

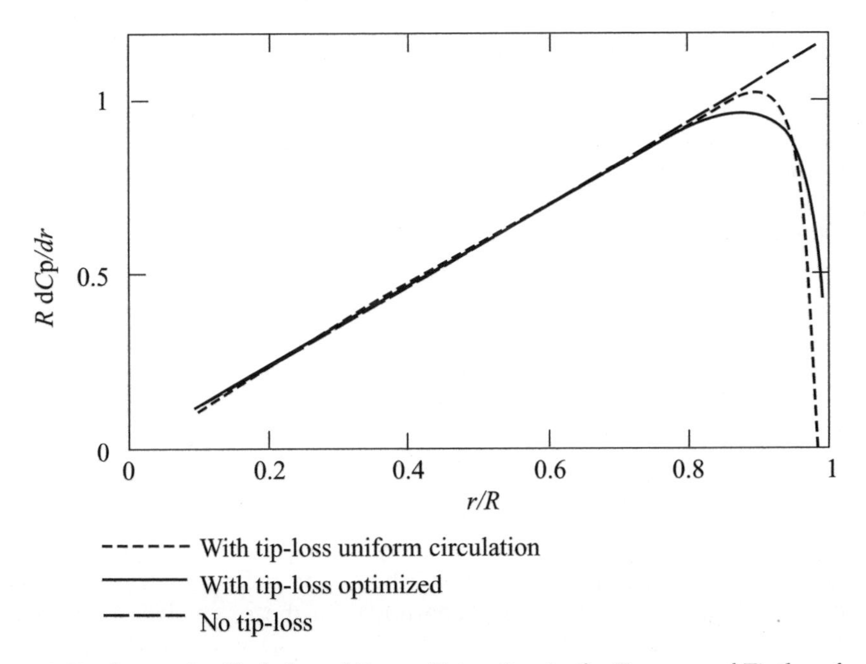

Figure 3.39 Span-wise Variation of Power Extraction in the Presence of Tip-loss for Three Blades with Uniform Circulation and of Optimized Design for a Tip Speed Ratio of 6

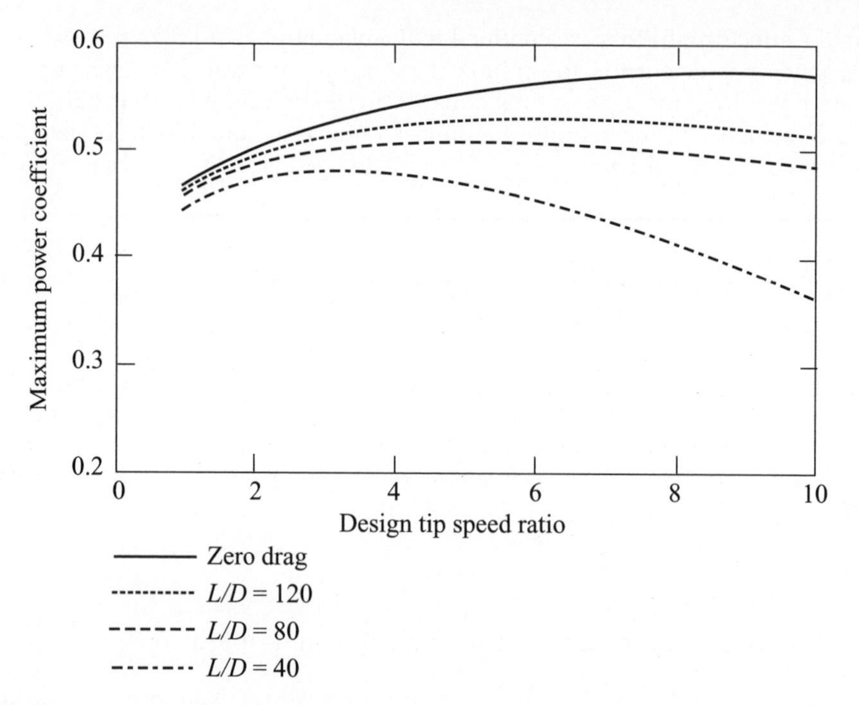

Figure 3.40 The Variation of Maximum C_P with Design λ for Various Lift/Drag Ratios and Including Tip-losses for a Three-bladed Rotor

depends upon whether the azimuthally averaged values of the flow factors are to be determined or the maximum (local to a blade element) values. If the former alternative is chosen then, in the momentum terms the flow factors remain unmodified but in the blade element terms the flow factors must appear as the average values divided by the tip-loss factor. Choosing to determine the maximum values of the flow factors means that they are not modified in the blade element terms but are multiplied by the tip-loss factor in the momentum terms. The latter choice allows the simplest modification of Equations (3.51) and (3.52).

$$\frac{af}{1-a} = \frac{\sigma_r}{4\sin^2\phi}\left(C_x - \frac{\sigma_r}{4\sin^2\phi}C_y^2\right)\frac{1-a}{1-af} \tag{3.51b}$$

$$\frac{a'f}{1+a'} = \frac{\sigma_r}{4\sin\phi\cos\phi}\frac{1-a}{1-af} \tag{3.52a}$$

There remains the problem of the breakdown of the momentum theory when wake mixing occurs. The helicoidal vortex sheets may not exist and so Prandtl's approximation is not physically appropriate. Nevertheless particles which pass between blades will no doubt still lose less momentum than those which interact with a blade and so the application of a tip-loss factor is necessary. Prandtl's approximation is the only practical method available and so is commonly used. In view of the manner in which the experimental results of Figure 3.16 were gathered it is the average value of a which should determine at which stage the momentum theory breaks down.

3.9 Calculated Results for an Actual Turbine

The blade design of a turbine operating at constant uniform rotational speed and fixed pitch is given in Table 3.1 and the aerofoil characteristics are shown in Figure 3.41.

The complete $C_P - \lambda$ curve for the design is given in Figure 3.15.

Using the above data the following results were obtained.

The blade is designed for optimum performance at a tip speed ratio of about 6 and, ideally, the angle of attacks uniform along the span at the level for which the lift/drag ratio is a maximum, about 7° for the aerofoil concerned. At the lowest tip speed ratio shown in Figure 3.42 the entire blade is stalled and for a rotational speed of 60 r.p.m. the corresponding wind speed will be 26 m/s which is about the cut-out speed. For the highest tip speed ratio shown the corresponding wind speed will be 4.5 m/s, the cut-in speed. Maximum power is developed at a tip speed ratio of 4 at a wind speed of 13 m/s and, clearly, much of the blade is stalled.

The axial flow induction factor is not uniform along the span at any tip speed ratio, indicating that the blade design is an engineering compromise, but at the tip speed ratio of 6 there is a value a little higher than 1/3. The flow factors shown in Figure 3.43 are those local to the blade and so the average value of axial flow factor will be close to 1/3 at a tip speed ratio of 6.

Generally, the axial flow factor increases with tip speed ratio while the tangential

Table 3.1 Blade Design of a 17m-diameter Rotor

Radius r (mm)	$\mu = r/R$	Chord c (mm)	Pitch β (degree)	Thickness/chord ratio of blade (%)
1700	0.20	1085	15.0	24.6
2125	0.25	1045	12.1	22.5
2150	0.30	1005	9.5	20.7
2975	0.35	965	7.6	19.5
3400	0.40	925	6.1	18.7
3825	0.45	885	4.9	18.1
4250	0.50	845	3.9	17.6
4675	0.55	805	3.1	17.1
5100	0.60	765	2.4	16.6
5525	0.65	725	1.9	16.1
5950	0.70	685	1.5	15.6
6375	0.75	645	1.2	15.1
6800	0.80	605	0.9	14.6
6375	0.85	565	0.6	14.1
7225	0.90	525	0.4	13.6
8075	0.95	485	0.2	13.1
8500	1.00	445	0.0	12.6

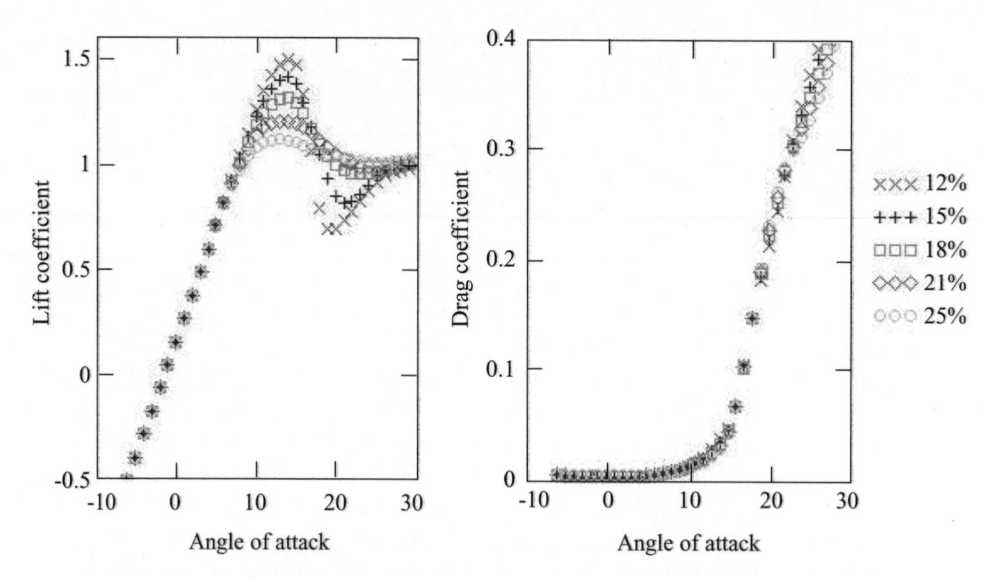

Figure 3.41 The Aerodynamic Characteristics of the NACA632XX Aerofoil Series

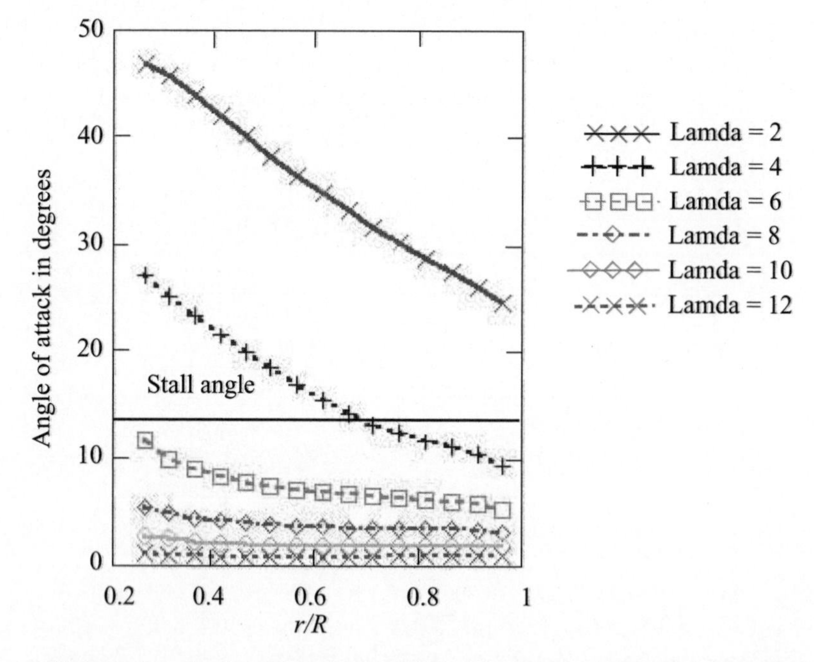

Figure 3.42 Angle of Attack Distribution for a Range of Tip Speed Ratios

flow factor decreases with tip speed ratio. The angular velocity of the wake increases sharply with decreasing radius because it is determined by the root vortex.

The importance of the outboard section of the blade is clearly demonstrated in Figure 3.44. The dramatic effect of stall is shown in the difference in torque distribution between the tip speed ratio of 4 and the tip speed ratio of 2. Note, also,

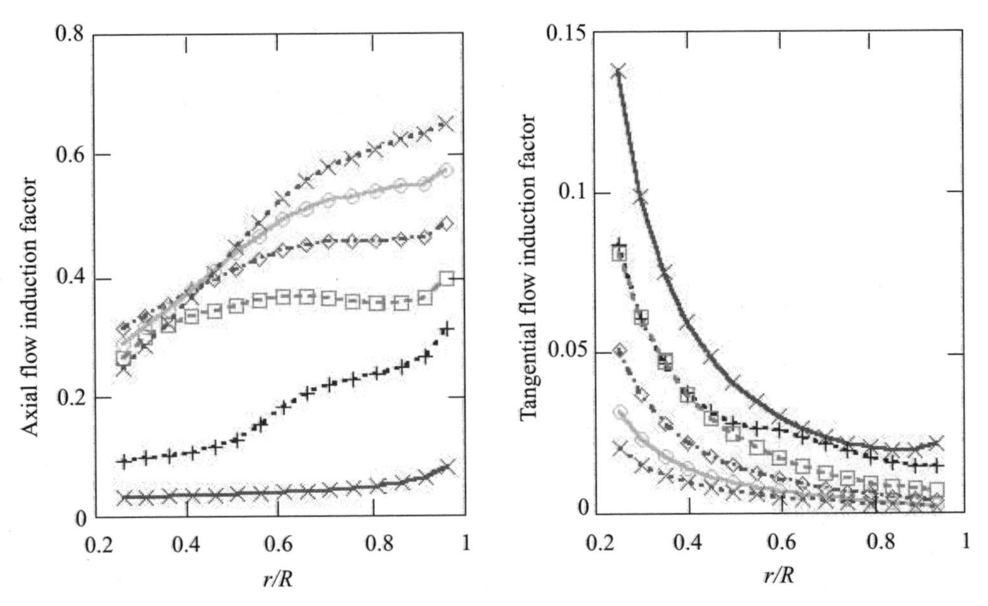

Figure 3.43 Distribution of the Flow Induction Factors for a Range of Tip Speed Ratios

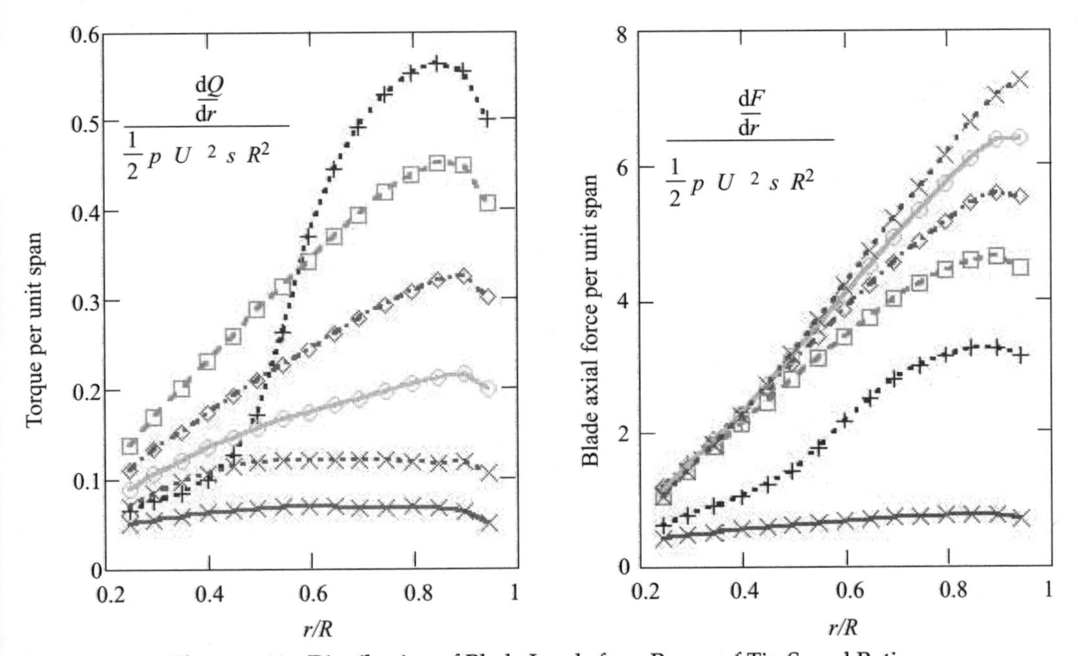

Figure 3.44 Distribution of Blade Loads for a Range of Tip Speed Ratios

the flat distribution of torque at the high tip speed ratio of 12; this is caused by the effect of drag which reduces torque as the square of the local speed ratio and with the low angle of attack at $\lambda = 12$ drag causes a significant loss of power.

Although the blade axial force *coefficient* increases with tip speed ratio it must be

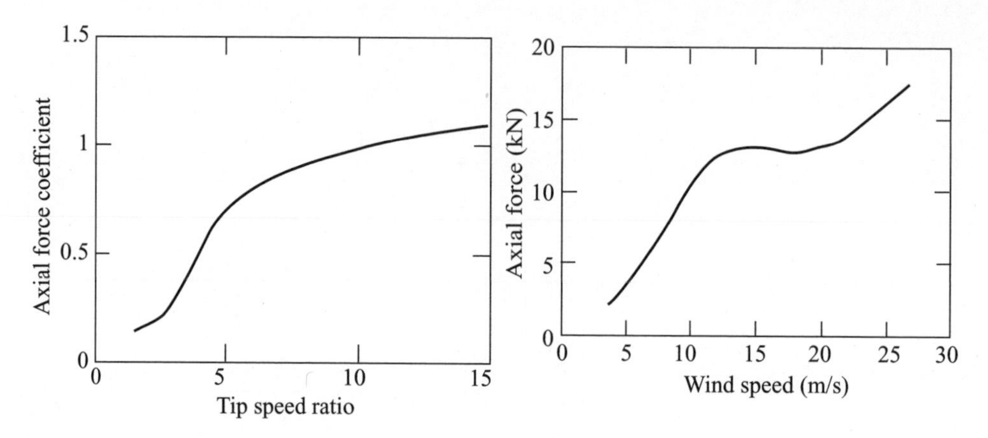

Figure 3.45 Axial Force Coefficient and the Variation of the Actual Force with Wind Speed

remembered that the actual thrust force increases with wind speed as is demonstrated in Figure 3.45.

3.10 The Aerodynamics of a Wind Turbine in Steady Yaw

The rotor axis of a wind turbine rotor is usually not aligned with the wind because the wind is continuously changing direction; the rotor is not capable of following this variability and so spends most of its time in a yawed condition. The yawed rotor is less efficient than the non-yawed rotor and so it is vital to assess the efficiency for purposes of energy production estimation.

In the yawed condition, even in a steady wind, the angle of attack on each blade is continuously changing as it rotates and so the loads on the rotor blades are fluctuating, causing fatigue damage. The changes in angle of attack mean that the blade forces cause not only a thrust in the axial direction but also moments about the yaw (z) axis and the tilt axis.

Even if the rotor is operating with a uniform induced velocity over the rotor disc when aligned with a steady wind, once the rotor is misaligned the induced velocity varies both azimuthally and radially which makes its determination much more difficult (Figure 3.46).

3.10.1 Momentum theory for a turbine rotor in steady yaw

The application of the momentum theory to an actuator disc representing a yawed rotor is somewhat problematical. The momentum theory is only capable of determining an average induced velocity for the whole rotor disc but, although in the non-yawed case the restriction was relaxed to allow some radial variation, it would not be appropriate to do this in the yawed case because the blade circulation is also changing with azimuth position. If it is assumed that the force on the rotor disc,

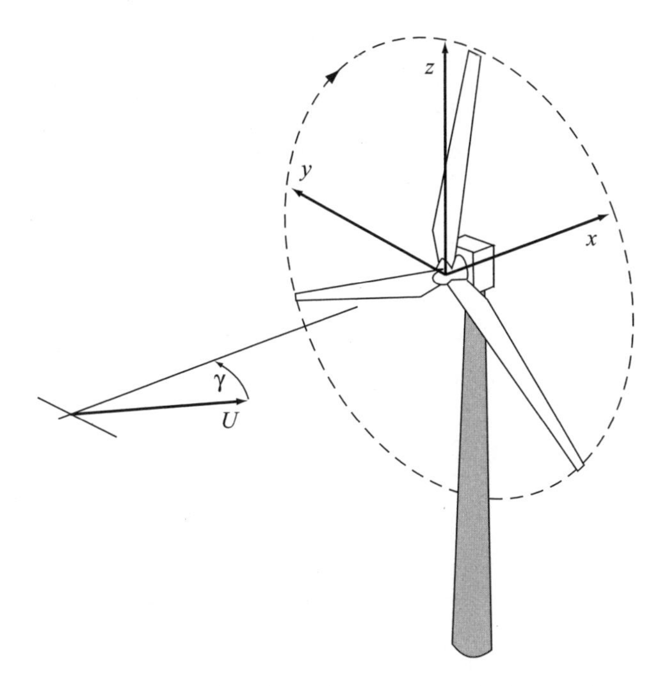

Figure 3.46 A Wind Turbine Yawed to the Wind Direction

which is a pressure force and so normal to the disc, is responsible for the rate of change of momentum of the flow then the average induced velocity must also be in a direction at right angles to the disc plane, i.e., in the *axial* direction. The wake is therefore deflected to one side because a component of the induced velocity is at right angles to the wind direction. As in the non-yawed case the average induced velocity at the disc is half that in the wake.

Let the rotor axis be held at an angle of yaw γ to the steady wind direction (Figure 3.47) then, assuming that the rate of change of momentum in the axial

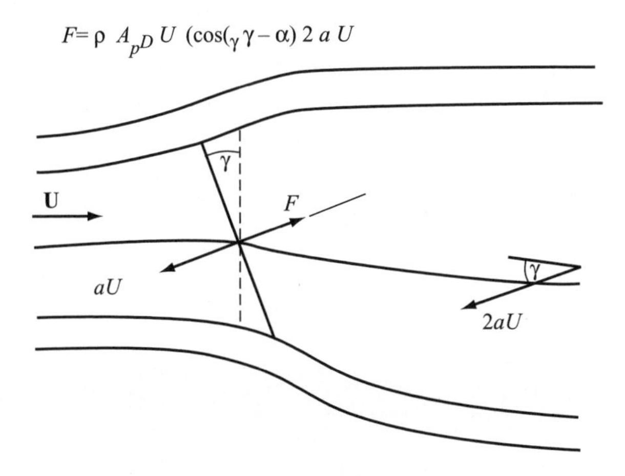

Figure 3.47 Deflected Wake of a Yawed Turbine and Induced Velocities

direction is equal to the mass flow rate through the rotor disc times the change in velocity normal to the plane of the rotor

$$F = \rho A_d U_\infty \quad (\cos\gamma - a)2aU_\infty \tag{3.88}$$

Therefore the thrust coefficient is

$$C_T = 4a(\cos\gamma - a) \tag{3.89}$$

and the power developed is

$$FU_\infty(\cos\gamma - a)$$

$$C_P = 4a(\cos\gamma - a)^2 \tag{3.90}$$

Figure 3.48 shows the decrease in power as the yaw angle increases.

To find the maximum value of C_P differentiate Equation (3.90) with respect to a and set equal to zero, whence

$$a = \frac{\cos\gamma}{3} \quad \text{and} \quad C_{P_{\max}} = \frac{16}{27}\cos^3\gamma \tag{3.91}$$

This $\cos^3\gamma$ rule is commonly adopted for power assessment in yawed flow.

A question remains: is it legitimate to apply the momentum theory in the above manner to the yawed rotor? Transverse pressure gradients which cause the wake to skew sideways may well also contribute to the net force on the flow in the axial direction, influencing the axial induced velocity. The above analysis might be satisfactory for determining the average axial induced velocity but there is even less justification to apply the momentum theory to each blade element position than there is in the non-yawed case. If a theory is going to be of any use in design it must

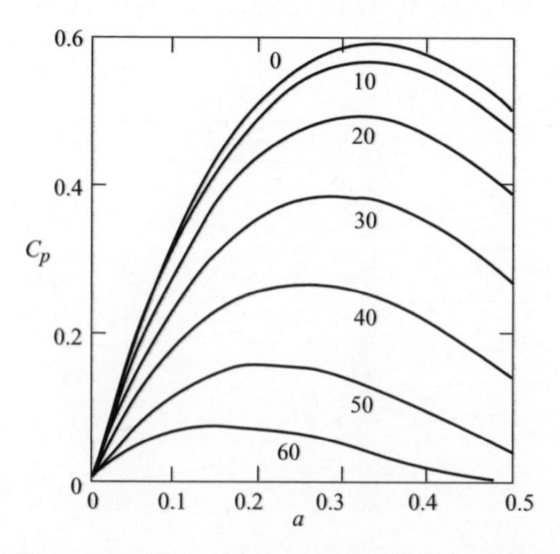

Figure 3.48 Power Coefficient Variation with Yaw Angle and Axial Flow Factor

be capable of determining the induced velocity at each blade element position to a satisfactory accuracy. The satisfactory calculation of blade forces is as important as the estimation of power.

3.10.2 Glauert's momentum theory for the yawed rotor

Glauert (1926) was primarily interested in the autogyro which is an aircraft with a rotor to provide lift and a conventional propeller to provide forward thrust. The lifting rotor has a rotational axis which inclines backwards from the vertical and by virtue of the forward speed of the aircraft air flows through the rotor disc causing it to rotate and to provide an upward thrust. Thus, the autogyro rotor is just like a wind turbine rotor in yaw, when in forward flight. At high forward speeds the yaw angle is large but in a power off vertical descent the yaw angle is zero.

Glauert maintained that at high forward speed the rotor disc, which is operating at a high tip speed ratio, is like a wing of circular plan-form at a small angle of attack (large yaw angle) and so the thrust on the disc is the lift on the circular wing. Simple lifting line wing theory (see Prandtl and Tietjens, 1957) states that the down-wash (induced velocity) at the wing, caused by the trailing vortex system, is uniform over the wing span (transverse diameter of the disc) for a wing with an elliptical plan-form and this would include the circular plan-form of the autogyro rotor.

The theory gives the uniform (average) induced velocity as

$$u = \frac{2L}{\pi(2R)^2\rho V} \tag{3.92}$$

where L is the lift and V is the forward speed of the aircraft.

The lift is in a direction normal to the effective incident velocity W (see Figure 3.49) and so is not vertical but leans backwards. The vertical component of the lift

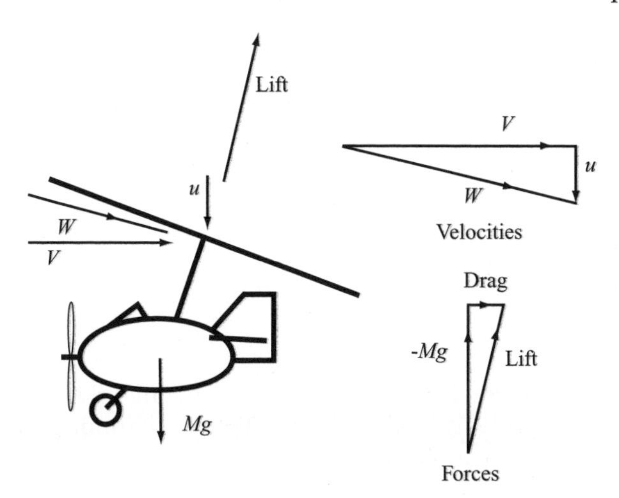

Figure 3.49 Velocities and Forces on an Autogyro in Fast Forward Flight

supports the weight of the aircraft and the horizontal component constitutes drag. In horizontal flight the vertical component of the lift does no work but the drag does do work.

The vector triangles of Figure 3.49 show that

$$\frac{D}{L} = \frac{u}{V} \tag{3.93}$$

In the wake of the aircraft the induced velocity u_{w} caused by the trailing vortices is greater than that at the rotor. A certain mass flow rate of air, ρVS, where S is an area, normal to velocity V, yet to be determined, undergoes a downward change in velocity u_{w} in the far wake. By the momentum theory the rate of change of downward momentum is equal to the lift, therefore

$$L = \rho VSu_{\mathrm{w}} \tag{3.94}$$

The rate of work done by the drag DV must be equal to the rate at which kinetic energy is created in the wake $\frac{1}{2}\rho u_{\mathrm{w}}^2 VS$, because the ambient static pressure in the wake of the aircraft is the same as the pressure ahead of the aircraft.

$$DV = \frac{1}{2}\rho u_{\mathrm{w}}^2 VS \tag{3.95}$$

Combining Equations (3.93), (3.94) and (3.95) gives

$$D = \frac{L^2}{2\rho V^2 S} \tag{3.96}$$

and

$$u_{\mathrm{w}} = 2u \tag{3.97}$$

Equation (3.97) should look familiar. Combining Equation (3.92), the lifting line theory's assessment of the induced velocity at the rotor, with Equation (3.93) gives

$$D = \frac{2L^2}{\rho V^2 \pi (2R)^2} \tag{3.98}$$

Comparing Equations (3.96) and (3.98) leads to an estimate of the required area S

$$S = \pi R^2 \tag{3.99}$$

S has the same area as the rotor disc but is normal to the flight direction.

Note that the above analysis has been simplified by assuming that the angle of attack is small. Actually, the trailing vortices from the rotor are influenced by their own induced velocity and so trail downwards behind the rotor. The induced velocity must therefore have a forward component, which means that the air undergoes a rate of change of momentum in the forward direction as well, thus balancing the drag. The drag is termed induced drag as it comes about by the backward tilting of the lift force caused by the induced velocity and has nothing to do with viscosity, it is entirely a pressure drag. Equation (3.92) should also be modified to replace V by W, the resultant velocity at the disc, and the area S will be in a plane normal to W. Also W has a direction which lies close to the plane of the rotor and so the lift force L will be almost the same as the thrust force F, which is normal to the plane of the rotor, and by the same argument the induced velocity is almost normal to the plane of the rotor

$$u = \frac{2F}{\pi(2R)^2 \rho W} \tag{3.100}$$

It can be assumed that a wind turbine rotor at high angles of yaw behaves just like the autogyro rotor.

At zero yaw the thrust force on the wind turbine rotor disc, given by the momentum theory, is

$$F = \pi R^2 \frac{1}{2} \rho 4u(U_\infty - u) \tag{3.101}$$

where U_∞ now replaces V, so the induced velocity is

$$u = \frac{2F}{\pi(2R)^2 \rho (U_\infty - u)} \tag{3.102}$$

The area S now coincides in position with the rotor disc.

Putting $W = U_\infty - u$ to represent the resultant velocity of the flow at the disc in Equation (3.102) then gives exactly the same Equation as (3.100) which is for a large angle of yaw. On the basis of this argument Glauert assumed that Equation (3.100), which is the simple momentum theory, could be applied at all angles of yaw, area $S = \pi R^2$, through which the mass flow rate is determined, always lying in a plane normal to the resultant velocity. The rotation of the area S is a crucially different assumption to that of the theory of Section 3.10.1 (which will now be referred as the axial momentum theory) and allows for part of the thrust force to be attributable to an overall lift on the rotor disc.

Thus

$$F = \rho \pi R^2 W 2u \tag{3.103}$$

where

$$W = \sqrt{U_\infty^2 \sin^2 \gamma + (U_\infty \cos \gamma - u)^2} \tag{3.104}$$

Thrust is equal to the mass flow rate times the change in velocity in the direction of the thrust. Both F and u are assumed to be normal to the plane of the disc. The thrust coefficient is then

$$C_T = 4a\sqrt{1 - a(2\cos\gamma - a)} \tag{3.105}$$

The power developed is a scalar quantity and so is the scalar product of the thrust force and the resultant velocity at the disc W. Hence, the power coefficient is

$$C_P = 4a\sqrt{1 - a(2\cos\gamma - a)}(\cos\gamma - a) \tag{3.106}$$

However, as some of the thrust is attributable to lift on the rotor disc acting as a circular wing that lift will not extract power from the wind because the net velocity field associated with the lift does not give rise to a flow through the rotor disc. Only that proportion of the thrust which arises from net flow through the disc will extract energy from the flow. Consequently, the axial momentum theory is more likely to estimate the power extraction correctly, whereas the Glauert theory is more likely to estimate the thrust correctly.

One very useful concept that emerges from Glauert's autogyro theory is that he predicted that the induced velocity through the rotor would not be uniform. The flow through the yawed rotor is depicted in Figure 3.50 and a simplification of the contributions to the velocity normal to the plane of the rotor along the rotor diameter parallel to the flight direction are shown. The mean induced velocity through the rotor, as determined by Equation (3.100), is shown as u_0, the normal component of the forward velocity of the aircraft is $U_\infty \cos\gamma$, also uniform over the disc, but, to account for the flow pattern shown, there needs to be a non-uniform component which decreases the normal induced velocity at the leading edge of the rotor disc and increases it at the rear. From symmetry, the induced velocity along the disc diameter normal to the flight direction (normal to the plane of the diagram) is uniform. The simplest form of the non-uniform component of induced velocity would be

$$u_1(r, \psi) = u_1 \frac{r}{R} \sin\psi \tag{3.107}$$

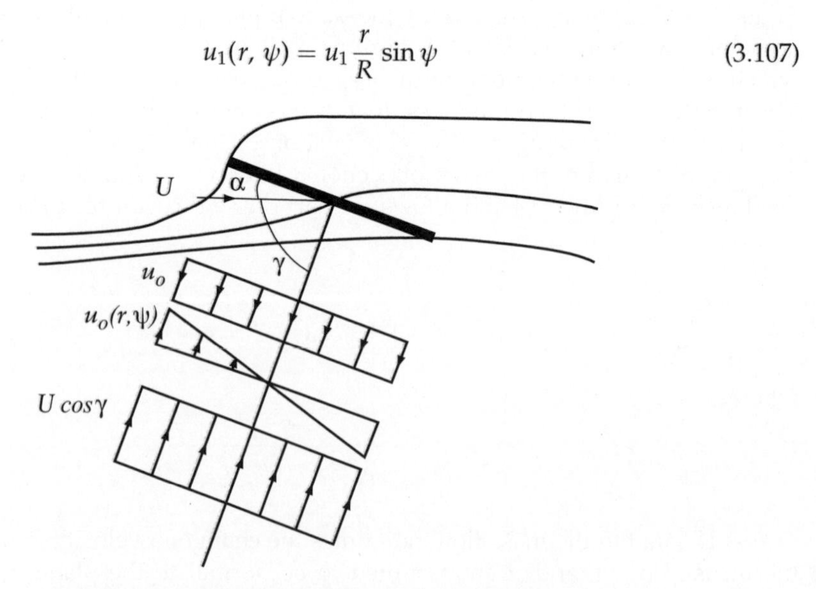

Figure 3.50 Velocities Normal to the Yawed Rotor

where ψ is the blade azimuth angle measured in the direction of rotation, $0°$ being when the blade is normal to the flight direction (or when the wind turbine blade is vertically upwards), and u_1 is the amplitude of the non-uniform component which is dependent on the yaw angle. There would, of course, need to be induced velocities parallel to the plane of the rotor disc but these are of secondary importance; the normal induced velocity has a much greater influence on the blade angle of attack than the in-plane component and therefore a much greater influence on blade element forces.

The value of u_1 in Equation (3.107) cannot be determined from momentum theory but Glauert suggested that it would be of the same order of magnitude as u_0. The total induced velocity, normal to the rotor plane, may then be written as

$$u = u_0\left(1 + K\frac{r}{R}\sin\psi\right) \tag{3.108}$$

The value of K must depend upon the yaw angle.

3.10.3 Vortex cylinder model of the yawed actuator disc

The vortex theory for the non-yawed rotor given in Section 3.4 was demonstrated to be equivalent to the momentum theory in its main results but, in addition, was shown to give much more detail about the flow-field. As the momentum theories of Sections 3.10.1 and 3.10.2 yield very limited results using the vortex approach for the yawed rotor may also prove to be useful, giving more flow structure detail than the momentum theory and, perhaps, a means of allying it with the blade element theory.

The wake of a yawed rotor is skewed to one side because the thrust F on the disc is normal to the disc plane and so has a component normal to the flow direction. The force on the flow therefore is in the opposite sense to F causing the flow to accelerate both upwind and sideways. The centre line of the wake will be at an angle χ to the axis of rotation (axis normal to the disc plane) known as the wake skew angle. The skew angle will be greater than the yaw angle (Figure 3.51). The

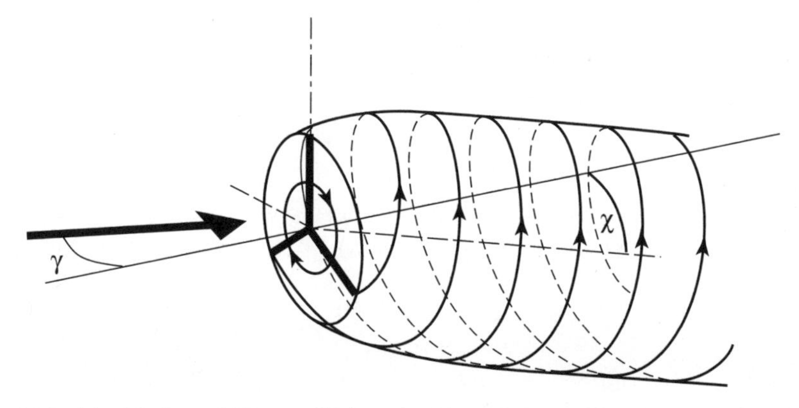

Figure 3.51 The Deflected Vortex Wake of a Yawed Rotor Showing the Shed Vortices of Three Blades

same basic theory as in Section 3.4 can be carried out for an actuator disc with a wake skewed to the rotor axis by an angle χ. There is an important proviso however: it must be assumed that the bound circulation on the rotor disc is radially and azimuthally uniform. As will be demonstrated the angle of attack of the blades is changing cyclically and so it would be impossible for the uniform circulation condition ever to be valid. What must be assumed is that the variation of circulation around a mean value has but a small effect on the induced velocity and the wake is therefore dominated by the vorticity shed from the blade tips by the mean value of circulation.

The expansion of the wake again imposes a difficulty for analysis and so, as before, it will be ignored (Figure 3.52).

The analysis of the yawed rotor was first carried out for purposes of understanding a helicopter rotor in forward flight by Coleman *et al.* (1945) but it can readily be applied to a wind turbine rotor by reversing the signs of the circulation and the induced velocities. An infinite number of blades is assumed as in the

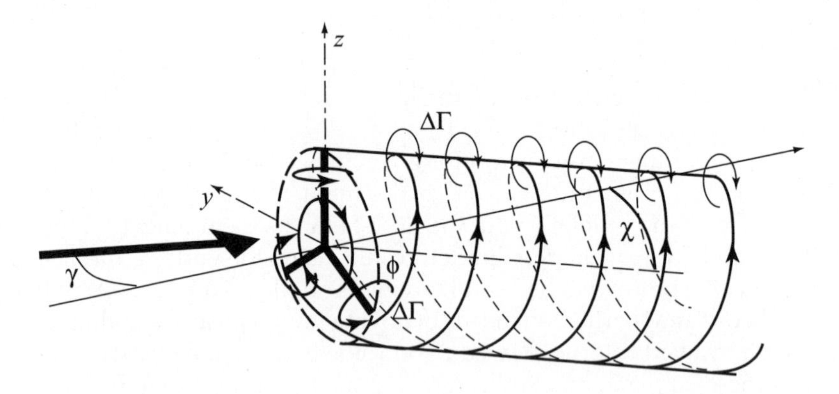

Figure 3.52 A Yawed Rotor Wake without Wake Expansion

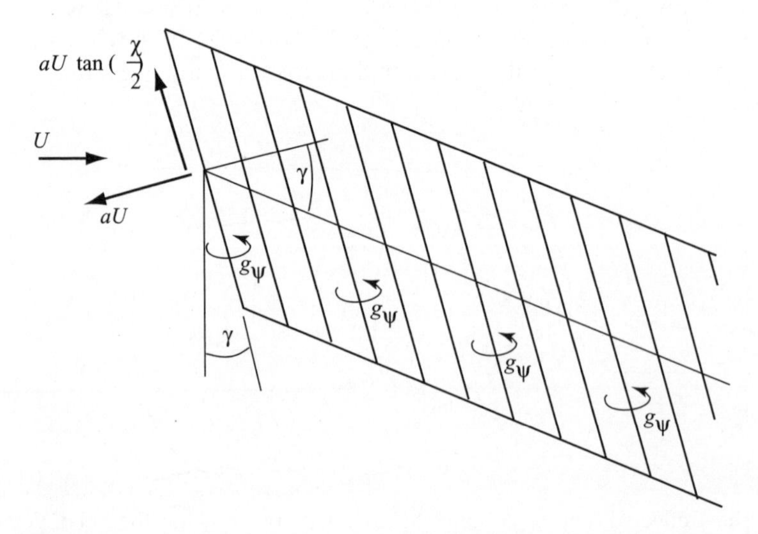

Figure 3.53 A Yawed Actuator Disc and the Skewed Vortex Cylinder Wake

analysis of Section 3.4. The vorticity g_ψ has a direction which remains parallel to the yawed disc and assuming it to be uniform (not varying with the azimuth angle), using the Biot–Savart law, induces an *average* velocity at the disc of $aU_\infty \sec\chi/2$ in a direction which bisects the skew angle between the central axis of the skew angle, as shown in Figure 3.54. The average axial induced velocity, normal to the rotor plane, is aU_∞, as in the unyawed case. In the fully developed wake the induced velocity is twice that at the rotor disc.

Because the average induced velocity at the disc is not in the rotor's axial direction, as is assumed for the momentum theory of Sections 3.10.1 and 3.10.2, the force F on the disc, which must be in the axial direction, cannot be solely responsible for the overall rate of change of momentum of the flow; there is a change of momentum in a direction normal to the rotor axis.

The velocity components at the rotor disc define the skew angle:

$$\tan\chi = \frac{U_\infty\left(\sin\gamma - a\tan\dfrac{\chi}{2}\right)}{U_\infty(\cos\gamma - a)} = \frac{2\tan\dfrac{\chi}{2}}{1 - \tan\dfrac{\chi^2}{2}} \tag{3.109}$$

From which it can be shown that a close, approximate relationship between χ, γ and a is

$$\chi = (0.6a + 1)\gamma \tag{3.110}$$

Using the velocities shown in Figure 3.54 a fresh analysis can be made of the flow. The average force on the disc can be determined by applying Bernoulli's equation to both the upwind and downwind regions of the flow.

Upwind

$$p_\infty + \frac{1}{2}\rho U_\infty^2 = p_d^+ + \frac{1}{2}\rho U_d^2$$

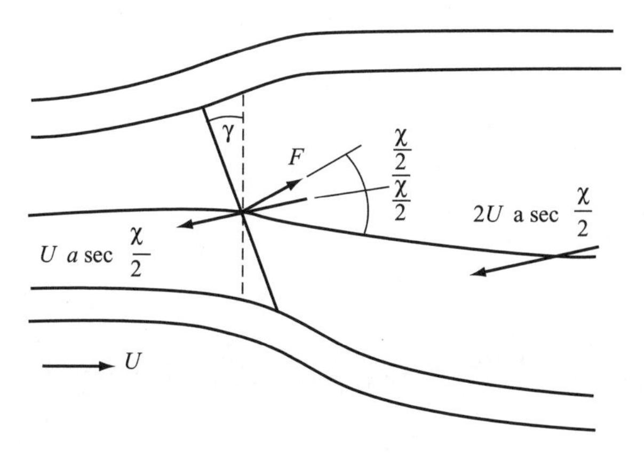

Figure 3.54 Average Induced Velocities Caused by a Yawed Actuator Disc

Downwind

$$p_d^- + \frac{1}{2}\rho U_d^2 = p_\infty + \frac{1}{2}\rho U_\infty^2 \left[(\cos\gamma - 2a)^2 + \left(\sin\gamma - 2a\tan\frac{\chi}{2} \right)^2 \right]$$

where U_d is the resultant velocity at the disc.

Subtracting the two equations to obtain the pressure drop across the disc

$$p_d^+ - p_d^- = \frac{1}{2}\rho U_\infty^2 \left[4a \left(\cos\gamma + \tan\frac{\chi}{2}\sin\gamma - a\sec^2\frac{\chi}{2} \right) \right]$$

The coefficient of thrust on the disc is therefore,

$$C_T = 4a \left(\cos\gamma + \tan\frac{\chi}{2}\sin\gamma - a\sec^2\frac{\chi}{2} \right) \tag{3.111}$$

and the power coefficient is

$$C_P = 4a \left(\cos\gamma + \tan\frac{\chi}{2}\sin\gamma - a\sec^2\frac{\chi}{2} \right)(\cos\gamma - a) \tag{3.112}$$

In a similar manner to the Glauert theory, it is not clear how much of the thrust in Equation (3.111) is capable of extracting energy from the flow and so the expression for power in Equation (3.112) will probably be an over estimate. A comparison of the maximum C_P values derived from the three theories, as a function of the yaw angle, is shown in Figure 3.55.

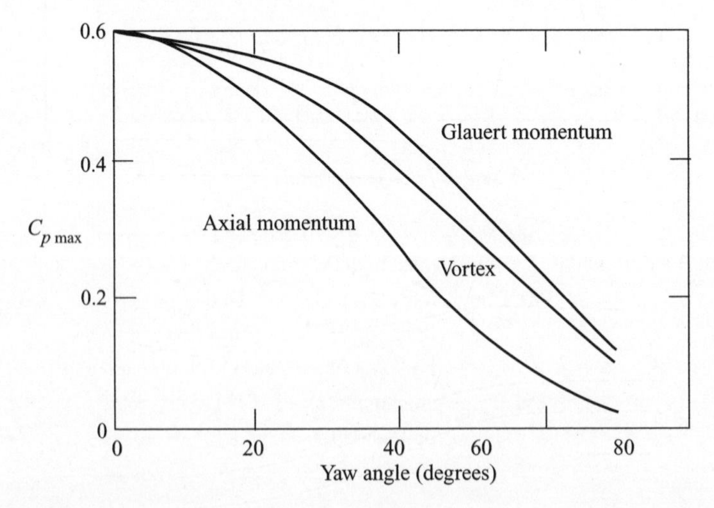

Figure 3.55 Maximum Power Coefficient Variation with Yaw Angle, Comparison of Momentum and Vortex Theories

3.10.4 Flow expansion

The induced velocity component parallel to the skewed axis of the wake is uniform over the disc with a value aU_∞, as can be deduced from Figure 3.54. The induced horizontal velocity, normal to the skewed axis of the wake is also uniform over the area of the disc with a value $a\tan(\chi/2)U_\infty$.

In addition to the uniform induced velocities of Figure 3.54 the expansion of the flow gives rise to velocities in the y and z directions, respectively (i.e., directions in a vertical plane at the skew angle χ to the rotor plane, see Figure 3.56). When resolved into the rotor plane the flow expansion velocities will give rise to a normal induced velocity of the type predicted by Glauert in Equation (3.108).

At a point on the disc at radius r and azimuth angle ψ, defined in Figure 3.56, the induced flow expansion velocities are non-simple functions of r and ψ. Across the horizontal diameter, where $\psi = \pm 90°$, Coleman *et al.* (1945) obtained an analytical solution for the flow expansion velocity in the y direction that involves complete elliptic integrals; the solution is not very practicable because numerical evaluation requires calculating the difference between two large numbers. Simplification of the analytical solution leads to the following expression for the horizontal flow expansion velocity which removes that difficulty but it is not in closed form.

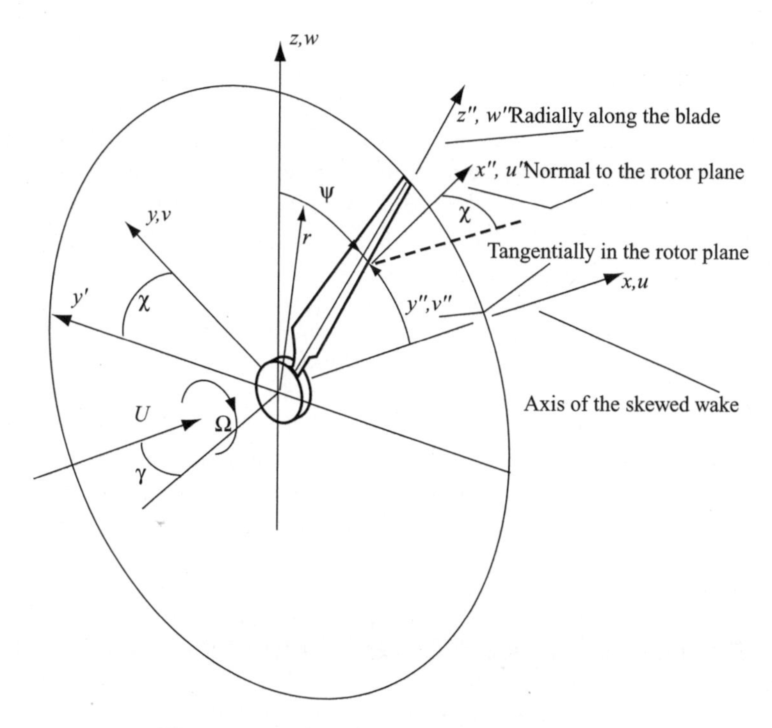

Figure 3.56 Axis System for a Yawed Rotor

$$\nu(\chi, \mu) =$$

$$\frac{-2\mu \sin\theta a U_\infty}{\pi}\int_0^{\frac{\pi}{2}} \frac{\sin^2 2\varepsilon}{\sqrt{(1+\mu)^2 - 4\mu\sin^2\varepsilon}}\frac{1}{[(\mu+\cos 2\varepsilon)^2]\cos^2\chi + \sin^2 2\varepsilon}\mathrm{d}\varepsilon \quad (3.113)$$

where $\mu = r/R$, ε is a parameter arising from the elliptic integrals, which is eliminated from the function by the definite integral, and aU_∞ is the average induced velocity as previously defined. An important feature of Equation (3.113) is that the flow expansion velocity is proportional to the average axial flow induction factor. Furthermore, if Equation (3.113) is divided by $\sec(\chi/2)^2$ the result is almost independent of the skew angle χ. Let

$$\frac{\nu(\chi, \mu)}{\sec\dfrac{\chi}{2}\sin\psi\, a\, U_\infty}$$

be defined as the flow expansion function $F(\mu)$ which is shown in Figure 3.57, clearly demonstrating how little $F(\mu)$ changes over a range of skew angle of 0° to 60°.

At all skew angles the value of the flow expansion function is infinite at the edge of the rotor disc, indicating a singularity in the flow which, of course, does not occur in practice but is a result of assuming uniform blade circulation. Circulation must fall to zero at the disc edge in a smooth fashion.

No analytical expressions for the flow expansion velocity components for values of ψ other than $\pm 90°$ were developed by Coleman *et al.* (1945) but numerical evaluations of the flow expansion velocities can be made using the Biot–Savart law.

The radial variation of the vertical flow expansion velocity across the vertical diameter of the rotor disc is much the same as $F(\mu)$ for skew angles between $\pm 45°$ but outside of this range the vertical velocity increases more sharply than the horizontal velocity at the disc edge. As will be shown, the vertical expansion

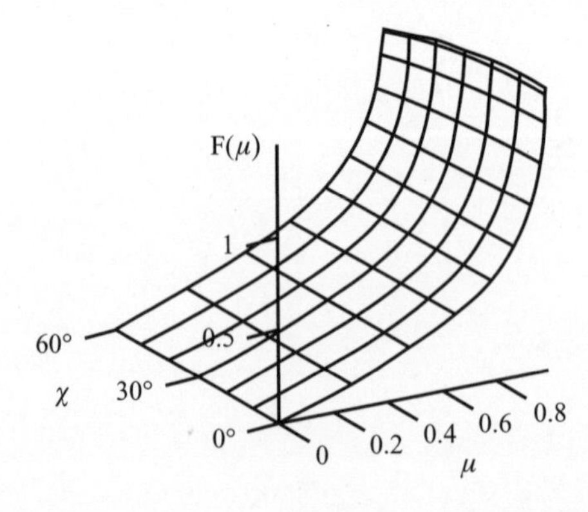

Figure 3.57 Flow Expansion Function Variation with Radial Position and Skew Angle

velocity is of less importance than the horizontal velocity in determining the aerodynamic behaviour of the yawed rotor.

The variation of the horizontal and vertical flow expansion velocities along radial lines on the rotor disc surface at varying azimuth angles (a radius sweeping out the disc surface as it rotates about the yawed rotor axis) shows that some further simplifications can be made for small skew angles.

Figure 3.58 shows the variation of the flow expansion velocities across the rotor disc for a skew angle of 30°. It should be emphasized that the velocity components lie in planes that are normal to the skewed axis of the wake. Inspection of the variations leads to simple approximations for the two velocity components.

$$v(\chi, \mu, \psi) = aU_\infty F(\mu)\sec^2\frac{\chi}{2}\sin\psi \tag{3.114}$$

$$w(\chi, \mu, \psi) = aU_\infty F(\mu)\sec^2\frac{\chi}{2}\sin\psi \tag{3.115}$$

where

$$F(\mu) = \frac{2\mu}{\pi}\int_0^{\frac{\pi}{2}} \frac{\sin^2 2\varepsilon}{\sqrt{(1+\mu)^2 - 4\mu\sin^2\varepsilon}}\,\frac{1}{(\mu^2 + 2\mu\cos(2\varepsilon+1)}\,\mathrm{d}\varepsilon \tag{3.116}$$

The drawback of Equations (3.114) and (3.115) is the singularity in the flow expansion function (3.116) at the outer edge of the disc. If the actuator disc is replaced with a rotor which has a small number of blades then the flow expansion function changes very significantly. Conducting a calculation using the Biot–Savart law for a non-yawed, single-bladed rotor represented by a lifting line vortex of radially uniform strength the flow expansion function can be determined numerically. It is found that the flow expansion velocity along the radial lifting line is a function of the helix (flow) angle of the discrete line vortex shed from the tip of the lifting line (blade). The vortex wake is assumed to be rigid in that the helix angle and the wake diameter are fixed everywhere at the values which pertain at the rotor. The solutions for a single-blade rotor can be used to determine the flow fields

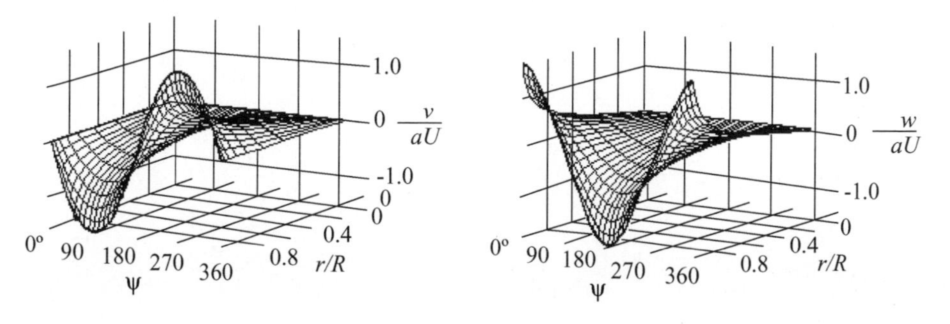

Figure 3.58 Azimuthal and Radial Variation of Horizontal (v) and Vertical (w) Velocities on the Rotor Plane for a Skew Angle of 30°

for multi-blade rotors by a simple process of superposition. The resulting flow expansion functions $F(\mu)_N$ are depicted in Figure 3.59 for one-, two- and three-blade rotors.

The radial variations in Figure 3.59 have been extended beyond the rotor radius to show the continuity which exists for the discrete blade situation as compared with the singularity that occurs for the actuator disc. There are two striking features of the flow expansion functions of Figure 3.59: the function is heavily modified by the value of the helix angle ϕ and the negative values (flow contraction) that can occur for the single-blade rotor.

An analytical expression which approximates the form of the diagrams shown in Figure 3.59 for two- and three-bladed rotors is

$$F_a(\mu, \phi_t, N) = \frac{F(\mu)}{\sqrt{1 + 50\dfrac{\tan^2\phi_t}{N^2} F(\mu)^2 (F(\mu)\mu(2-\mu))^{0.05/\tan\phi_t}\left(\dfrac{1}{\tan\phi_t} + 8\right)}} \tag{3.117}$$

where

$$\tan\phi_t = \frac{1-a}{\lambda(1+a')}$$

is the tangent of the flow angle at which the tip vortex is shed from the blade tips. The approximation of Equation (3.117) is shown in Figure 3.60 for two- and three-bladed rotors.

When transformed as components of velocity with respect to axes rotating about the rotor axis (x'', y'' and z'' axes as shown in Figure 3.56) the flow expansion velocities of Equations (3.114) and (3.116) are resolved into the components that are normal and tangential to the blade element (see Figure 3.63).

The normal component is

$$u'' = aU_\infty\left(1 + F(\mu) 2\tan\frac{\chi}{2}\sin\psi\right) \tag{3.118}$$

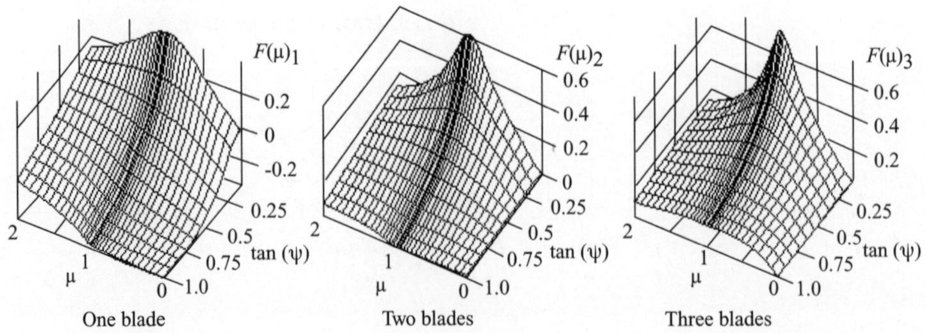

Figure 3.59 Flow Expansion Functions for One-, Two- and Three-blade Rotors

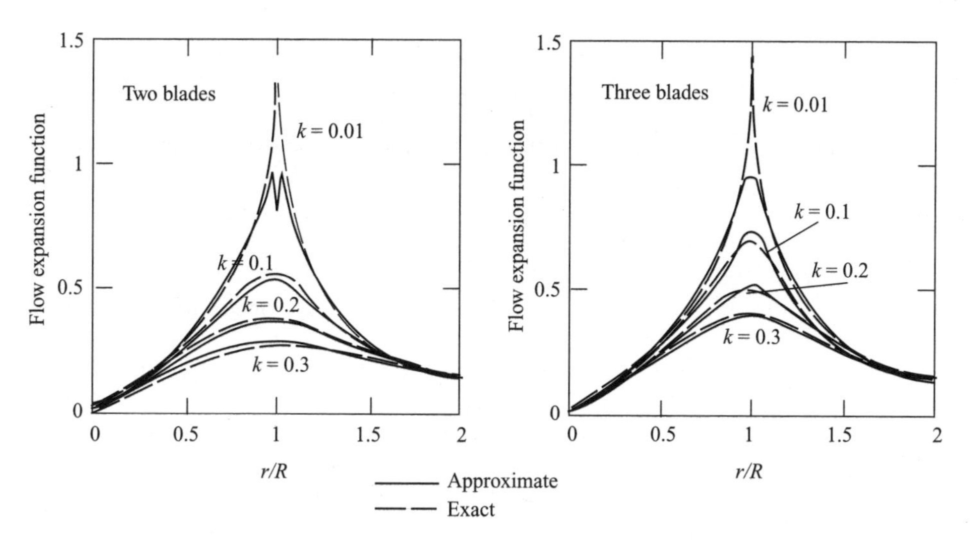

Figure 3.60 Approximate Flow Expansion Functions for Two- and Three-blade Rotors

And the tangential component is

$$v'' = aU_\infty \cos\psi \tan\frac{\chi}{2}\left(1 + F(\mu)2\tan\frac{\chi}{2}\sin\psi\right) \tag{3.119}$$

to which must be added the components of the wind velocity U_∞

the normal component

$$U'' = \cos\gamma\, U_\infty \tag{3.120}$$

and the tangential component

$$V = \cos\psi\sin\gamma\, U_\infty \tag{3.121}$$

There is a radial (span-wise) velocity component but this will not influence the angle of attack so can be ignored.

Clearly, from Equation (3.118), the Coleman theory determines the function $K(\chi)$, see equation 3.108, as being

$$K_C(\chi) = 2\tan\frac{\chi}{2} \tag{3.122}$$

In addition there is the tangential velocity Ωr caused by blade rotation and also the induced wake rotation but the latter will be ignored initially.

The velocities of Equations (3.118) to (3.121) will produce a lower angle of attack when the azimuth angle ψ is positive, see Figure 3.57, than when it is negative and so the angle of attack will vary cyclically. When ψ is positive the incident normal velocity u'' lies closer to the radial axis of the blade than when ψ is negative. The difference in angle of attack can be attributed to flow expansion as depicted in Figure 3.61.

The variation of the angle of attack makes the flow about a blade aerofoil unsteady and so the lift will have a response of the kind discussed in Section 3.1. The blade circulation will therefore vary during the course of a revolution, which means that the vortex model is incomplete because it is derived from the assumption that the circulation is constant.

There is clearly additional vorticity in the wake being shed from the blades' trailing edges influencing the induced velocity, which is not accounted for in the theory. The additional induced velocity would be cyclic so would probably not affect the average induced velocity normal to the rotor disc but would affect the amplitude and phasing of the angle of attack.

Further numerical analysis of the Coleman vortex theory reveals that at skew angles greater than ±45° higher harmonics than just the one per revolution term in Equation (3.108) become significant in the flow expansion induced velocities. Only odd harmonics are present, reflecting the anti-symmetry about the yaw axis.

3.10.5 Related theories

A number of refinements to the Glauert and Coleman theories have been proposed by other researchers, mostly addressing helicopter aerodynamics but some have been directed specifically at wind turbines. In particular, Øye (1992) undertook the same analysis as Coleman and proposed a simple curve-fit to Equation (3.116)

$$F_{\phi}(\mu) = \frac{1}{2}(\mu + 0.4\mu^3 + 0.4\mu^5) \tag{3.123}$$

Øye has clearly avoided the very large values that Equation (3.116) produces close to the outer edge of the disc and Equation (3.122) is in general accordance with the flow expansion functions shown in Figure 3.59.

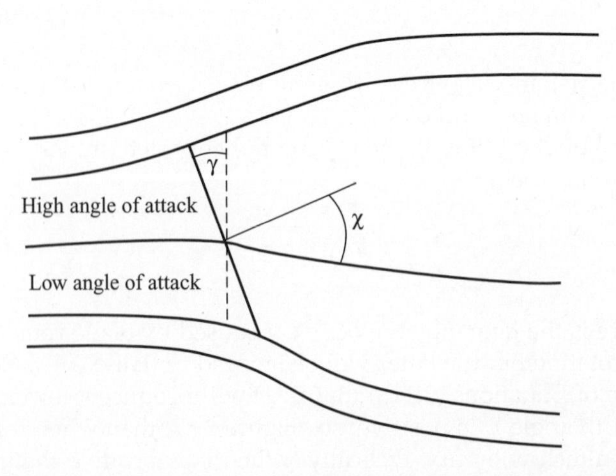

Figure 3.61 Flow Expansion Causes a Differential Angle of Attack

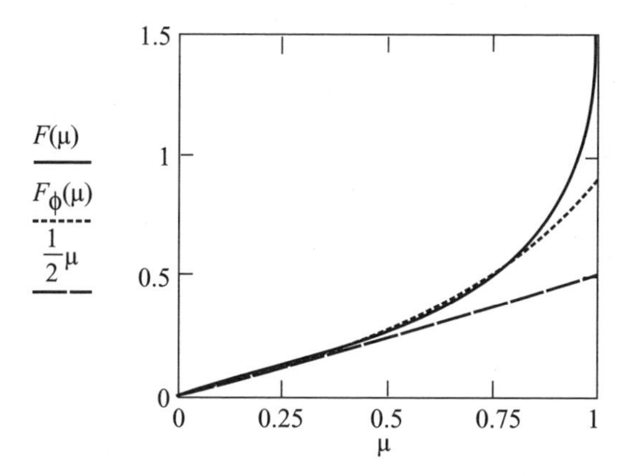

Figure 3.62 Øye's Curve-fit to Coleman's Flow Expansion Function

Meijer Drees (1949) has extended the Coleman *et al.* vortex model to include a cosinusoidal variation of blade circulation. The main result is a modification to the function $K(\chi)$ but Meijer-Drees retained Glauert's assumption of linear variation of normal induced velocity with radius

$$u'' = -aU_\infty \left\{ 1 + \frac{4}{3}\mu \left[1 - 1.8 \left(\frac{\sin\gamma}{\lambda} \right)^2 \right] \tan\frac{\chi}{2} \sin\psi \right\} \tag{3.124}$$

(see Schepers and Snel, 1995).

3.10.6 Wake rotation for a turbine rotor in steady yaw

Wake rotation is, of course, present in the wake flow but cannot be related only to the torque. The vortex theory needs also to include a root vortex, which will lie in the wake along the skewed wake axis. The rotation in the wake will therefore be about the skewed wake axis and not about the axis of rotation and the wake rotation velocity will lie in a plane normal to the skewed wake axis.

To determine the wake rotation velocity the rate of change of angular momentum about the skewed wake axis will be equated to the moment about the axis produced by blade forces.

If the wake rotation velocity is described, as before, in terms of the angular velocity of the rotor then

$$v''' = \Omega r''' a' h(\psi) \tag{3.125}$$

where the triple prime denotes an axis system rotating about the wake axis and $h(\psi)$ is a function which determines the intensity of the root vortex's influence. In the non-yawed case the root vortex induces a velocity at the rotor which is half of that it induces in the far wake at the same radial distance and the same would apply to a disc normal to the skewed axis with a centre located at the same position as the actual rotor disc. The distance upstream or downstream of a point on the actual rotor disc from the plane of the disc normal to the wake axis determines the value of the root vortex influence function $h(\psi)$. The value of $h(\psi)$ will lie between 0.0 and 2.0, being equal to 1.0 at points on the vertical diameter.

The velocity induced by a semi-infinite line vortex of strength Γ lying along the x-axis from zero to infinity at a point with cylindrical co-ordinates (x''', ψ''', r''') is, using the Biot–Savart law,

$$\overrightarrow{V'''} = \frac{\Gamma}{4\pi r'''}\begin{bmatrix} 0 \\ \left(1 + \dfrac{x'''}{\sqrt{x'''^2 + r'''^2}}\right) \\ 0 \end{bmatrix} = \begin{bmatrix} 0 \\ v''' \\ 0 \end{bmatrix} \tag{3.126}$$

The induced velocity when $x''' = \infty$ is twice that when $x''' = 0$ and is zero when $x''' = -\infty$.

For a point on the rotor disc $(0, \psi, r)$ the corresponding co-ordinates (x''', ψ''', r''') in normal disc axes are

$$x''' = -y''' \sin\chi = r \sin\psi \sin\chi,\ r''' = r\sqrt{\cos\psi^2 + \cos\chi^2 \sin\psi^2}$$

and

$$\cos\psi''' = \frac{r}{r'''}\cos\psi, \qquad \sin\psi''' = \frac{r}{r'''}\sin\psi\cos\chi \tag{3.127}$$

the induced velocity at the same point is

$$v''' = \Omega r''' a'\left(1 + \frac{x'''}{\sqrt{x'''^2 + r'''^2}}\right) \tag{3.128}$$

So, transforming the velocity of (3.126) to the rotating axes in the plane of the rotor disc

$$\overrightarrow{V''} = \begin{bmatrix} 1 & 0 & 0 \\ 0 & \cos\psi & \sin\psi \\ 0 & -\sin\psi & \cos\psi \end{bmatrix}\begin{bmatrix} \cos\chi & \sin\chi & 0 \\ -\sin\chi & \cos\chi & 0 \\ 0 & 0 & 1 \end{bmatrix}\begin{bmatrix} 1 & 0 & 0 \\ 0 & \cos\psi''' & -\sin\psi''' \\ 0 & \sin\psi''' & \cos\psi''' \end{bmatrix}\begin{bmatrix} 0 \\ v''' \\ 0 \end{bmatrix} \tag{3.129}$$

Substituting (3.127) and (3.128) into (3.129) gives

$$\overrightarrow{V''} = \begin{bmatrix} \cos\psi \sin\chi \\ \cos\chi \\ 0 \end{bmatrix}\Omega r a'(1 + \sin\psi\sin\chi) \tag{3.130}$$

Thus the wake rotation produces two velocity components, one in the rotor plane and one normal to the rotor plane; there is no radial component.

3.10.7 The blade element theory for a turbine rotor in steady yaw

There is doubt about the applicability of the blade element theory in the case of a yawed turbine because the flow, local to a blade element, is unsteady and because the theory representing the vortex half of the equation, which replaces the momentum theory, is incomplete in this respect. However, it is not clear how large or significant are the unsteady forces. If the unsteady forces are large then the blade element theory is inapplicable and the results of applying the theory will bear no relation to measured results. If the unsteady forces are small then there should be some correspondence with actual values. In a steady yawed condition the flow velocities at a point on the rotor disc do not change with time, if an infinity of blades is assumed, and so there is no added mass term to consider. However, the change of angle of attack with time at a point on the blade does mean that the two dimensional lift force should really be modified by a lift deficiency function similar to that determined by Theodorsen (1935) for the rectilinear wake of sinusoidally pitching aerofoil.

Neglecting the effects of shed vorticity the net velocities in the plane of a local blade element are shown in Figure 3.63. The radial (span-wise) velocity component is not shown in Figure 3.63 but it is neglected as it is not considered to have any influence on the angle of attack and therefore on the lift force.

The flow angle ϕ is then determined by the components of velocity shown in Figure 3.63.

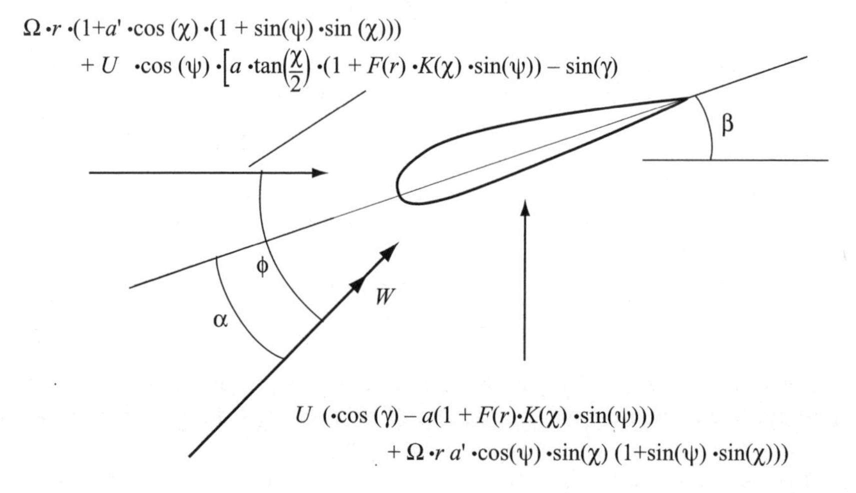

Figure 3.63 The Velocity Components in the Plane of a Blade Cross-Section

$$\tan\phi = \frac{\begin{bmatrix} U_\infty(\cos\gamma - a(1 + F(\mu)K(\chi)\sin\psi)) \cdots \\ + \Omega r a' \cos\psi \sin\chi(1 + \sin\psi\sin\chi) \end{bmatrix}}{\begin{array}{l}(\Omega r(1 + a'\cos\chi(1 + \sin\psi\sin\chi))) \cdots \\ + U_\infty\cos(\psi)\left[a\tan\dfrac{\chi}{2}(1 + F(\mu)K(\chi)\sin\psi) - \sin\gamma\right]\end{array}} \tag{3.131}$$

where r is measured radially from the axis of rotor rotation.

The angle of attack α is found from

$$\alpha = \phi - \beta \tag{3.132}$$

Lift and drag coefficients taken from two-dimensional experimental data, just as for the non-yawed case, are determined from the angle of attack calculation for each blade element (each combination of μ and ψ).

3.10.8 The blade element–momentum theory for a rotor in steady yaw

The forces on a blade element can be determined via Equations (3.131) and (3.132) for given values of the flow induction factors.

The thrust force will be calculated using Equation (3.46) in Section 3.5.3, which is for a complete annular ring of radius r and radial thickness δr.

$$\delta L\cos\phi + \delta D\sin\phi = \frac{1}{2}\rho W^2 Nc(C_l\cos\phi + C_d\sin\phi)\delta r \tag{3.46}$$

For an elemental area of the annular ring swept out as the rotor turns through an angle $\delta\psi$ the proportion of the force is

$$\delta F_b = \frac{1}{2}\rho W^2 Nc(C_l\cos\phi + C_d\sin\phi)\delta r\frac{\delta\psi}{2\pi}$$

putting $C_x = C_l\cos\phi + C_d\sin\phi$ and

$$\sigma_r = \frac{Nc}{2\pi r}$$

then

$$\delta F_b = \frac{1}{2}\rho W^2\sigma_r C_x r\delta r\delta\psi \tag{3.133}$$

The values of C_l and C_d should really include unsteady effects because of the ever changing blade circulation with azimuth angle which will depend upon the level of the reduced frequency of the circulation fluctuation.

If it is chosen to ignore drag, or use only that part of the drag attributable to pressure, then Equation (3.133) should be modified accordingly.

The rate of change of momentum will use either Equation (3.105), Glauert's theory, or Equation (3.111), the vortex cylinder theory; in both equations the flow induction factor a should be replaced by af to account for Prandtl tip loss.

For Glauert's theory

$$\delta F_m = \frac{1}{2}\rho U_\infty^2 4af\sqrt{1 - af(2\cos\gamma - af)}r\delta\psi\delta r \tag{3.134a}$$

Or, for the vortex theory,

$$\delta F_m = \frac{1}{2}\rho U_\infty^2 4af\left(\cos\gamma + \tan\frac{\chi}{2}\sin\gamma - af\sec^2\frac{\chi}{2}\right)r\delta\psi\delta r \tag{3.134b}$$

The Equations (3.134) do not include the drop in pressure caused by wake rotation but to do this would require greater detail about wake rotation velocities from the vortex theory. The algebraic complexity of estimating the wake rotation velocities is great and even then fluctuation of bound circulation is ignored. The drop in pressure caused by wake rotation, however, is shown to be small in the non-yawed case and so it is assumed that it can safely be ignored in the yawed case.

The moment of the blade element force about the wake axis is

$$\delta M_b = \frac{1}{2}\rho W^2 Nc(\cos\psi\sin\chi C_x + \cos\chi C_y r\delta r\frac{\delta\psi}{2\pi}$$

where

$$C_y = C_l\sin\phi - C_d\cos\phi$$

therefore

$$\delta M_b = \frac{1}{2}\rho W^2\sigma_r(\cos\psi\sin\chi C_x + \cos\chi C_y)r^2\delta r\delta\psi \tag{3.135}$$

The rate of change of angular momentum is the mass flow rate through an elemental area of the disc times the tangential velocity times radius.

$$\delta M_m = \rho U_\infty(\cos\gamma - af)r\delta\psi\delta r 2a'f\Omega r'''^2$$

where

$$r'''^2 = r^2(\cos^2\psi + \cos^2\chi\sin^2\psi)$$

therefore

$$\delta M_m = \frac{1}{2}\rho U_\infty^2\lambda\mu 4a'f(\cos\gamma - af)(\cos^2\psi + \cos^2\chi\sin^2\psi)r^2\delta r\delta\psi \tag{3.136}$$

The momentum theory, as developed, applies only to the whole rotor disc where the flow induction factor a is the average value for the disc. However, it may be argued that it is better to apply the momentum equations to an annular ring, as in the non-yawed case, to determine a distribution of the flow induction factors varying with radius, reflecting radial variation of circulation. Certainly, for the angular momentum case the tangential flow factor a' will not vary with azimuth position because it is generated by the root vortex and although, in fact, the axial flow factor a does vary with azimuth angle it is consistent to use an annular average for this factor as well.

To find an average for an annular ring the elemental values of force and moment must be integrated around the ring.

For the axial momentum case, taking the vortex method as an example,

$$\int_0^{2\pi} \frac{1}{2}\rho U_\infty^2 4af\left(\cos\gamma + \tan\frac{\chi}{2}\sin\gamma - af\sec^2\frac{\chi}{2}\right) r\delta r\,\mathrm{d}\psi$$

$$= \int_0^{2\pi} \frac{1}{2}\rho W^2 \sigma_r C_x r\delta r\,\mathrm{d}\psi$$

Therefore

$$8\pi af\left(\cos\gamma + \tan\frac{\chi}{2}\sin\gamma - af\sec^2\frac{\chi}{2}\right) = \sigma_r \int_0^{2\pi} \frac{W^2}{U_\infty^2} C_x\,\mathrm{d}\psi \tag{3.137}$$

The resultant velocity W and the normal force coefficient C_x are functions of ψ. And for the angular momentum case

$$\int_0^{2\pi} \frac{1}{2}\rho U_\infty^2 \lambda\mu 4a' f(\cos\gamma - af)(\cos^2\psi + \cos^2\chi\sin^2\psi) r^2\delta r\,\mathrm{d}\psi$$

$$= \int_0^{2\pi} \frac{1}{2}\rho W^2 \sigma_r(\cos\psi\sin\chi C_x + \cos\chi C_y) r^2\delta r\,\mathrm{d}\psi$$

which reduces to

$$4a' f(\cos\gamma - af)\lambda\mu\pi(1 + \cos^2\chi) = \sigma_r \int_0^{2\pi} \frac{W^2}{U_\infty^2}(\cos\psi\sin\chi C_x + \cos\chi C_y)\mathrm{d}\psi \tag{3.138}$$

The non-dimensionalized resultant velocity relative to a blade element is given by

$$\frac{W^2}{U_\infty^2} = \begin{bmatrix} (\cos\gamma - a)\cdots \\ +\lambda\mu a'\cos\psi\sin\chi(1+\sin\psi\sin\chi) \end{bmatrix}^2 \cdots \\ + \begin{bmatrix} (\lambda\mu(1+a'\cos\chi(1+\sin\psi\sin\chi)))\cdots \\ +\cos\psi\left(a\tan\frac{\chi}{2} - \sin\gamma\right) \end{bmatrix}^2 \tag{3.139}$$

Note that the flow expansion terms, those terms that involve $F(\mu)K(\chi)$, have been excluded from the velocity components in Equation (3.139) because flow expansion is not present in the wake and so there is no associated momentum change. The blade force, which arises from the flow expansion velocity, is balanced in the wake by pressure forces acting on the sides of the stream-tubes, which have a stream-wise component because the stream-tubes are expanding.

Equations (3.137) and (3.138) can be solved by iteration, the integrals being determined numerically. Initial values are chosen for a and a', usually zero. For a given blade geometry, at each blade element position μ and at each blade azimuth position ψ, the flow angle ϕ is calculated from Equation (3.131), which have been suitably modified to remove the flow expansion velocity, in accordance with Equation (3.139). Then, knowing the blade pitch angle β at the blade element, the local angle of attack can be found. Lift and drag coefficients are obtained from tabulated aerofoil data. Once an annular ring (constant μ) has been completed the integrals are calculated. The new value of axial flow factor a is determined from Equation (3.137) and then the tangential flow factor a' is found from Equation (3.138). Iteration proceeds for the same annular ring until a satisfactory convergence is achieved before moving to the next annular ring (value of μ).

Although the theory supports only the determination of azimuthally averaged values of the axial flow induced velocity, once the averaged tangential flow induction factors have been calculated the elemental form of the momentum equation (3.134) and the blade element force (Equation (3.133)) can be employed to yield values of a which vary with azimuth.

For the determination of blade forces the flow expansion velocities must be included. The total velocity components, normal and tangential to a blade element, are then as shown in Figure 3.63 and the resultant velocity is

$$\frac{W^2}{U_\infty^2} = \begin{bmatrix} (\cos\gamma - a(1+F(\mu)K(\chi)\sin\psi))\cdots \\ +\lambda\mu a'\cos\psi\sin\chi(1+\sin\psi\sin\chi) \end{bmatrix}^2 \cdots \\ + \begin{bmatrix} (\lambda\mu(1+a'\cos\chi(1+\sin\psi\sin\chi)))\cdots \\ +\cos\psi\left[a\tan\frac{\chi}{2}(1+F(\mu)K(\chi)\sin\psi) - \sin\gamma\right] \end{bmatrix}^2 \tag{3.139a}$$

3.10.9 Calculated values of induced velocity

The measurement of the induced velocities of a wind turbine rotor in yaw has been undertaken at Delft University of Technology, (Snel and Schepers, 1995). The tests were carried out using a small wind tunnel model so that a steady yaw could be maintained in a steady wind with no tower shadow and no wind shear. The rotor had two blades of 1.2 m diameter which were twisted but had a uniform chord length of 80 mm. The blade root was at a radius of 180 mm and the blade twist was 9° at the root varying linearly with radius to 4° at 540 mm radius and remaining at 4° from there to the tip. The blade aerofoil profile was NACA 0012. The rotor speed was kept constant at 720 rev/min and the wind speed was held constant at 6.0 m/s. Tests were carried out at 10°, 20° and 30° of yaw angle.

Calculated induced velocities using the vortex momentum equation for the Delft turbine are shown in Figure 3.64; these are the average values for each annulus obtained using Equation (3.137) and (3.138).

The component velocities at each blade element, as defined in Figure 3.63, are shown in Figure 3.65. Because of the rotational speed of the blades the tangential

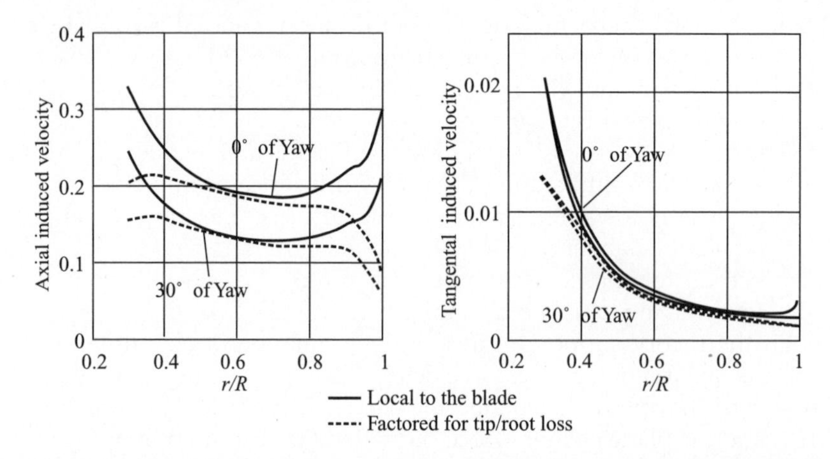

Figure 3.64 Azimuthally Averaged Induced Velocity Factors for the Delft Turbine

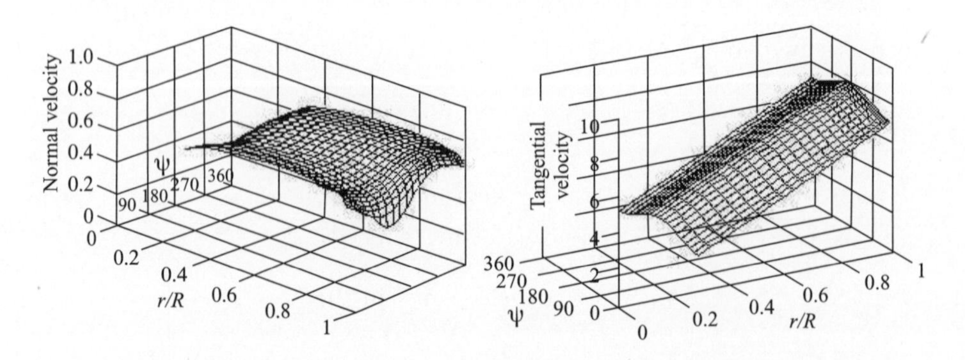

Figure 3.65 Component Velocities, Normalized with Wind Speed, at 30° of Yaw

velocity is much greater than the normal velocity but it is the latter which most influences the variation in angle of attack at the important, outboard sections of the blades, shown in Figure 3.66.

At the inboard sections of the blades it is the variation in tangential velocity which mostly influences the angle of attack variation and this is largely as a result of the changing geometry with azimuth angle rather than the effect of induced velocity.

3.10.10 Blade forces for a rotor in steady yaw

Once the flow induction factors have been determined blade forces can then be calculated. Although the flow expansion velocity is excluded from the determination of the flow induction factors, on the grounds that the consequent blade forces do not cause any change in the momentum of the flow because flow expansion is not present in the developed wake, it must be included when the blade forces are calculated. The flow expansion velocity should be dependent on an overall average value of the axial flow induction factor but it is more convenient to use the annular average value as determined by Equations (3.137) and (3.138).

The flow angle and the angle of attack need to be determined anew at each blade element position μ and at each blade azimuth position θ because the flow expansion velocity must now be included, so Equation (3.131) is used in its unmodified form. Drag must also be included in the determination of forces even if it was not in the calculation of the induced velocities.

The blade force per unit span normal to the plane of rotation is

$$\frac{\mathrm{d}}{\mathrm{d}r}F_x = \frac{1}{2}\rho W^2 c C_x \tag{3.140}$$

which will vary with the azimuth position of the blade. The total force normal to the rotor plane can be obtained by integrating Equation (3.138) along the blade length for each of the blades, taking account of their azimuthal separation, and summing the results. The total normal force will also vary with rotor azimuth.

Similarly, the tangential blade force per unit span is

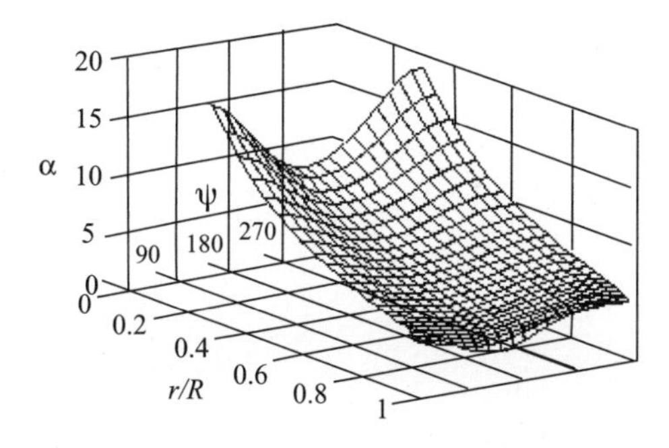

Figure 3.66 Angle of Attack Variation at 30° of Yaw

$$\frac{\mathrm{d}}{\mathrm{d}r}F_x = \frac{1}{2}\rho W^2 c C_y$$

and the blade torque contribution about the axis of rotation is

$$\frac{\mathrm{d}}{\mathrm{d}r}Q = \frac{1}{2}\rho W^2 c r C_y \tag{3.141}$$

The total torque is found by integrating along each blade and summing over all the blades, just as for the normal force. Again, the torque on the rotor will vary with azimuth position so to find the average torque will require a further integration with respect to azimuth.

3.10.11 Yawing and tilting moments in steady yaw

The asymmetry of the flow through a yawed rotor, caused by the flow expansion, means that a blade sweeping upwind has a higher angle of attack than when it is sweeping downwind, as shown in Figure 3.61. The blade lift upwind will therefore be greater than the lift downwind and a similar differential applies to the forces normal to the rotor plane. It can be seen, therefore, that there is a net moment about the yaw (vertical axis) in a direction which will tend to restore the rotor axis to a position aligned with the wind direction. The yawing moment is obtained from the normal force of Equation (3.140)

$$\frac{\mathrm{d}}{\mathrm{d}r}M_z = \frac{1}{2}\rho W^2 c r \sin\psi C_x \tag{3.142}$$

which will also vary with the azimuth position of the blade. The total single-blade yawing moment at each azimuth position is obtained by integrating Equation (3.142) along the length of the blade. Summing the moments for all blades, suitably separated in phase, will result in the yawing moment on the rotor.

A similar calculation can be made for the tilting moment, the moment about the horizontal diametral axis (y-axis) of the rotor:

$$\frac{\mathrm{d}}{\mathrm{d}r}M_z = \frac{1}{2}\rho W^2 c r \cos\psi C_x \tag{3.143}$$

Measured results of rotor yaw moment for the Delft turbine are shown in Figure 3.67 and the corresponding calculated yawing moments are shown in Figure 3.68.

The measured yawing moments were derived from strain gauge readings of the flat-wise bending strain close to the root of the blade at 129 mm radius. Flat-wise, or flap-wise bending causes only displacements normal to the rotor plane. The calculated yawing moments are determined at the same radial position on the blade and are, therefore, not the true yawing moments about the actual yaw axis.

The comparison between the measured and calculated yaw moments is quite good taking into account the limitations of the theory. At 30° of yaw the calculated values underestimate the measurements significantly whereas at the two lower

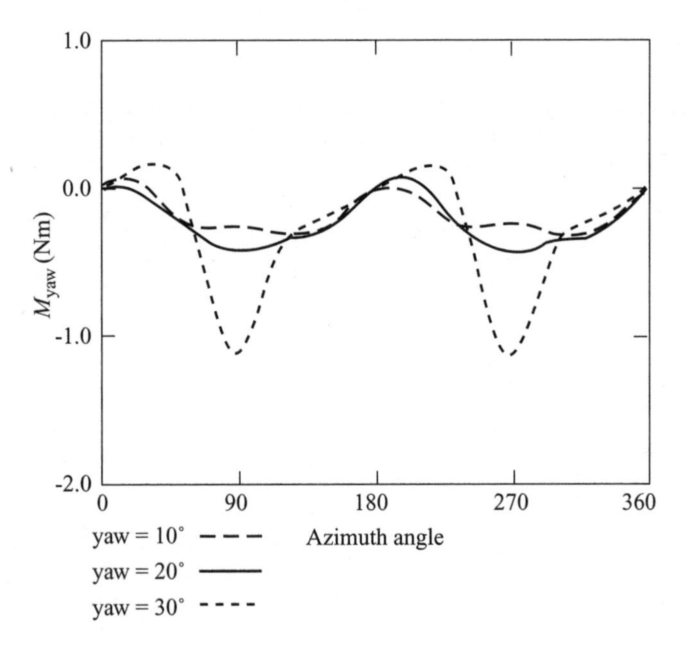

Figure 3.67 Measured Yaw Moments on the Delft Turbine, (Snel and Schepers, 1995)

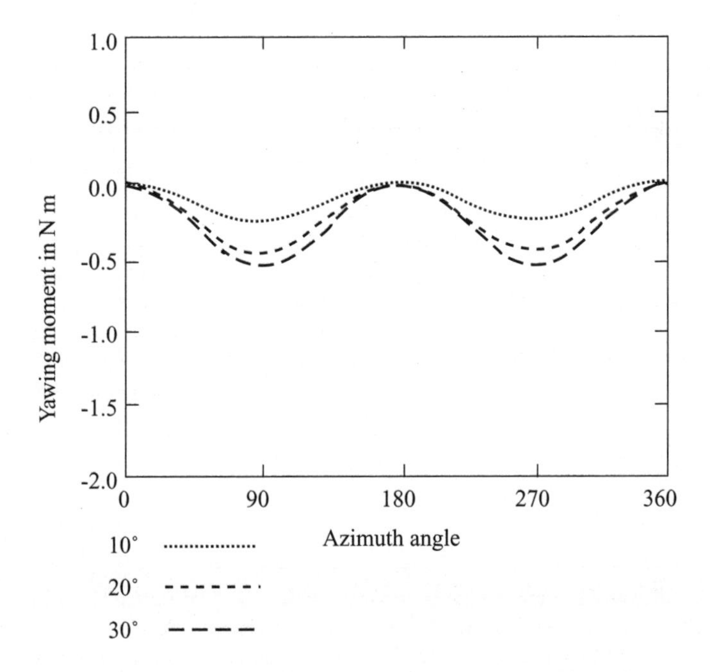

Figure 3.68 Calculated Yaw Moments on the Delft Turbine

angles the correspondence is much closer. It should be noted that the mean yawing moment is not zero and that the sign of the moment, being negative, means that it endeavours to restore the rotor axis to alignment with the wind direction.

The yawing moment comparison is a test of the usefulness of the theory developed in this section and it would seem that for general engineering purposes it passes the test.

The measured tilting moments (Figure 3.69) appear to be of about the amplitude for all three yaw angles whereas the calculated moments (Figure 3.70) increase with

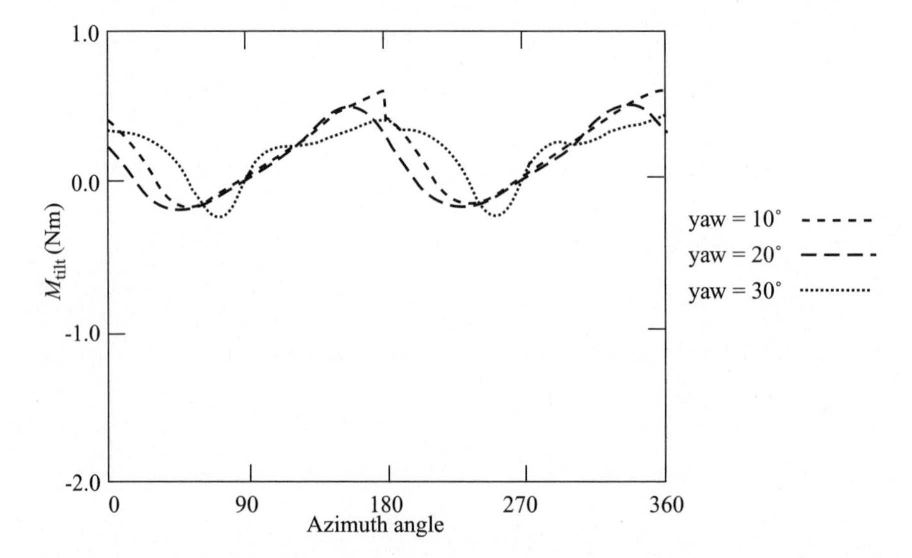

Figure 3.69 Measured Tilt Moments on the Delft Turbine, (Snel and Schepers, 1995)

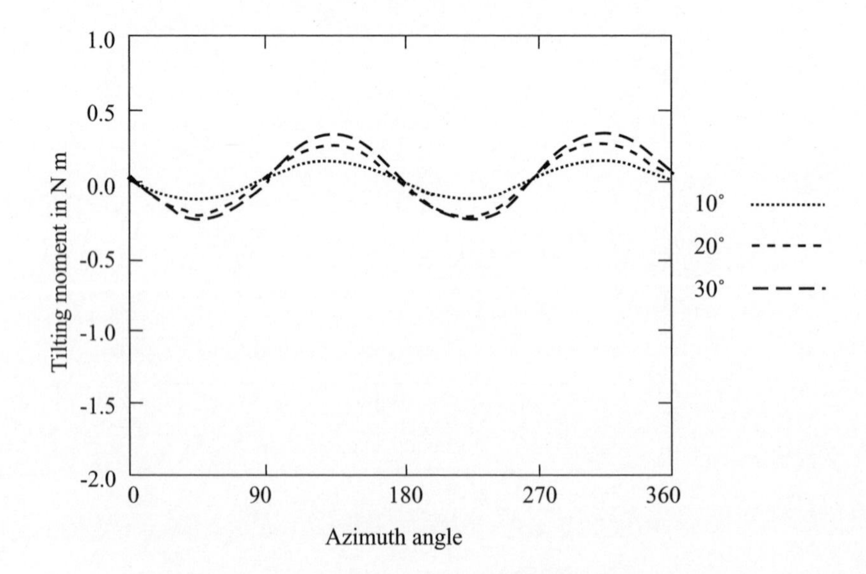

Figure 3.70 Calculated Tilt Moments on the Delft Turbine

yaw angle. For 30° of yaw the magnitudes of the measured and calculated tilting moments are comparable. The measured mean tilting moment is quite definitely non-zero and positive but the calculated mean moment is much smaller although still positive. A positive tilt rotation would displace the upper part of the rotor disc in the downwind direction. In the theory the small mean tilting moment is caused by the wake rotation velocities.

A theory based upon computational fluid mechanics should provide a much more accurate prediction of the aerodynamics of a wind turbine in yaw. However, the severe computational time limitations associated with CFD solutions precludes their use in favour of the simple theory outlined in these pages.

3.11 The Method of Acceleration Potential

3.11.1 Introduction

An aerodynamic model that is applied to the flight performance of helicopter rotors, and which can also be applied to wind turbine rotors that are lightly loaded, is that based upon the idea of acceleration potential. The method allows distributions of the pressure drop across an actuator disc that are more general than the, strictly, uniform pressure distribution of the momentum theory. The model has been expounded by Kinner (1937), inspired by Prandtl, who has developed expressions for the pressure field in the vicinity of an actuator disc, treating it as a circular wing. To regard a rotor as a circular wing requires an infinity of very slender blades so that the solidity remains small.

Kinner's theory, which is derived from the Euler equations, assumes that the induced velocities are small compared with the general flow velocity. If u, v and w are the velocities induced by the actuator disc in the x-, y- and z-directions, respectively, and which are very much smaller than the free-stream velocity in the x-direction U_∞, then the rate of change of momentum in the x-direction of a unit volume of air will be in response to the pressure gradient in that direction

$$\rho\left[(U_\infty + u)\frac{\partial(U_\infty + u)}{\partial x} + v\frac{\partial(U_\infty + u)}{\partial y} + w\frac{\partial(U_\infty + u)}{\partial z}\right] = -\frac{\partial p}{\partial x} \tag{3.144}$$

The free-stream velocity U_∞ does not change with position therefore, for example, $\partial(U_\infty)/\partial x = 0$. Also, $U_\infty \gg (u, v, w)$ and so, for example, $v(\partial u/\partial y)$ can be ignored in comparison with $U_\infty(\partial u/\partial x)$. The momentum equation in the x-direction then simplifies to

$$\rho U_\infty \frac{\partial u}{\partial x} = -\frac{\partial p}{\partial x} \tag{3.145a}$$

Similarly, in the y- and z-directions, the momentum equations are also simplified

$$\rho U_\infty \frac{\partial v}{\partial w} = -\frac{\partial p}{\partial y} \tag{3.145b}$$

and

$$\rho U_\infty \frac{\partial w}{\partial x} = -\frac{\partial p}{\partial z} \tag{3.145c}$$

Differentiating each momentum equation with respect to its particular direction and adding together the results gives

$$\rho U_\infty \frac{\partial}{\partial x}\left(\frac{\partial u}{\partial x} + \frac{\partial v}{\partial y} + \frac{\partial w}{\partial z}\right) = -\left(\frac{\partial^2 p}{\partial x^2} + \frac{\partial^2 p}{\partial y^2} + \frac{\partial^2 p}{\partial z^2}\right)$$

but, for continuity of the flow,

$$\frac{\partial u}{\partial x} + \frac{\partial v}{\partial y} + \frac{\partial w}{\partial z} = 0,$$

therefore

$$\frac{\partial^2 p}{\partial x^2} + \frac{\partial^2 p}{\partial y^2} + \frac{\partial^2 p}{\partial z^2} = 0 \tag{3.146}$$

which is the Laplace equation governing the pressure field on and surrounding the actuator disc. Given the boundary conditions at the actuator disc Equation (3.146) can be solved for the pressure field and, in particular, the pressure distribution at the disc. The pressure is continuous everywhere except across the disc surfaces where there is the usual pressure discontinuity, or pressure drop, in the wind turbine case.

In Coleman's analysis (1945) the pressure drop distribution across the disc is uniform (it is only as a result of combining the theory with the blade element theory that a non-uniform pressure distribution can be achieved) but falls to zero, abruptly, at the disc edge. Kinner assumes that the pressure drop is zero at the disc edge and changes in a continuous manner as radius decreases.

The simplified Euler Equations (3.145) allow pressure to be regarded as the potential field from which the acceleration field can be obtained, by differentiation, and thence the velocity field, by integration. Commencing upstream where the known free-stream conditions apply the velocity components can be determined by progressive integration towards the disc.

3.11.2 The general pressure distribution theory of Kinner

Kinner's solution (1937) is mathematically complex and is achieved by means of a co-ordinate transformation. The Cartesian co-ordinates centred in the rotor plane

(x'', y', z), as defined in Figure 3.56, are transformed to, what is termed, an ellipsoidal co-ordinate system (ν, η, ψ), ψ is the azimuth angle.

$$\frac{x''}{R} = \upsilon\eta, \ \frac{y'}{R} = \sqrt{1-\upsilon^2}\sqrt{1+\eta^2}\sin\psi \ \text{and} \ \frac{z}{R} = \sqrt{1-\upsilon^2}\sqrt{1+\eta^2}\cos\psi \tag{3.147}$$

On the surface of the rotor disc $\eta = 0$ and $r/R = \mu = \sqrt{1-\upsilon^2}$ or, conversely, $\upsilon = \sqrt{1-\mu^2}$

The transformation separates the variables and allows the pressure field to be expressed as the product of three functions

$$p(\upsilon, \eta, \psi) = \Phi_1(\upsilon)\Phi_2(\eta)\Phi_3(\psi) \tag{3.148}$$

each separate function being the solution of a separate, ordinary differential equation,

$$\frac{d}{d\upsilon}\left[(1-\upsilon^2)\frac{d}{d\upsilon}\Phi_1(\upsilon)\right] + \left[n(n+1) - \frac{m^2}{1-\upsilon^2}\right]\Phi_1(\upsilon) = 0 \tag{3.149a}$$

$$\frac{d}{d\eta}\left[(1-\eta^2)\frac{d}{d\eta}\Phi_2(\eta)\right] + \left[\frac{m^2}{1-\eta^2} - n(n+1)\right]\Phi_2(\eta) = 0 \tag{3.149b}$$

$$\frac{d^2}{d\psi^2}\Phi_3(\psi) + m^2\Phi_3(\psi) = 0 \tag{3.149c}$$

where m and n are positive integers.

Equations (3.149a) and (3.149b) have the form of Legendre's associated differential equations which has solutions which are called associated Legendre polynomials of the first and second kinds, respectively (see van Bussel, 1995).

If $m = 0$ then Equations (3.149a) and (3.149b) are reduced to Legendre's differential equations the solutions for which are

$$\Phi_1(\upsilon) = P_n(\upsilon) \tag{3.150a}$$

and

$$\Phi_1(\upsilon) = Q_n(\upsilon) \tag{3.150b}$$

where $P_n(\upsilon)$ is a Legendre polynomial of the first kind and $Q_n(\eta)$ is a Legendre polynomial of the second kind.

$$P_n(\upsilon) = \frac{1}{2^n n!}\frac{d^n}{d\upsilon^n}(\upsilon^2 - 1)^n \tag{3.151}$$

Although the polynomials extend beyond the range $\upsilon^2 \leqslant 1$, over that interval the polynomials are mutually orthogonal

$$\int_{-1}^{1} P_n(v)P_k \mathrm{d}v = 0 \quad n \neq k \tag{3.152}$$

For $n = 0$ the Legendre polynomial of the second kind is

$$Q_0(v) = \frac{1}{2}\ln\left(\frac{1+v}{1-v}\right) \tag{3.153}$$

For $n > 0$ the Legendre polynomials of the second kind $Q_n(v)$ can be obtained from the polynomials of the first kind. For $n = 1$ to 4

$$\begin{aligned}
Q_1(v) &= (P_1(v)Q_0(v)) - 1 \\
Q_2(v) &= (P_2(v)Q_0(v)) - \frac{3}{2}v \\
Q_3(v) &= (P_3(v)Q_0(v)) - \frac{5}{2}v^2 + \frac{2}{3} \\
Q_4(v) &= (P_4(v)Q_0(v)) - \frac{35}{8}v^3 + \frac{55}{24}v
\end{aligned} \tag{3.154}$$

The solutions for Equation (3.149b), with $m = 0$, are the same as for (3.149a) but with imaginary arguments, i.e.,

$$\Phi_2(\eta) = P_n(i\eta) \tag{3.155a}$$

and

$$\Phi_2(\eta) = Q_n(i\eta) \tag{3.155b}$$

where

$$Q_0(i\eta) = i\tan^{-1}(\eta) \quad \text{for } \eta < 1$$

and

$$Q_0(i\eta) = i\left(\frac{\pi}{2} - \tan^{-1}\eta\right) \quad \text{for } \eta > 1$$

For non-zero values of the integer m the solutions of Equation (3.149a) become

$$P_n^m(v) = (1 - v^2)^{m/2}\frac{\mathrm{d}^m}{\mathrm{d}v^m}P_n(v) \tag{3.156}$$

If $m > n$ then $P_n^m(v) = 0$.

$$Q_n^m(\upsilon) = (1 - \upsilon^2)^{m/2} \frac{\mathrm{d}^m}{\mathrm{d}\upsilon^m} Q_n(\upsilon)$$

But, from Equation (3.153), $Q_n^m(\upsilon) = \infty$ at $\upsilon^2 = 1$ which is physically inapplicable and so these functions are excluded from the solution.

For solutions of Equation (3.149b) for non-zero values of m

$$P_n^m(i\eta) = (1 + \eta^2)^{m/2} \frac{\mathrm{d}^m}{\mathrm{d}i\eta^m} P_n(i\eta)$$

Inspection reveals that $P_n^m(i\eta) \to \infty$ as $\eta \to \infty$ which means that pressure would be infinite in the far field which is not physically acceptable therefore these terms will also not be included.

$$Q_n^m(i\eta) = (1 + \eta^2)^{m/2} \frac{\mathrm{d}^m}{\mathrm{d}i\eta^m} Q_n(i\eta) \tag{3.157}$$

Equations (3.156) and (3.157) are known as associated Legendre polynomials.

The solution to differential Equation (3.149c) is more straight forward than for the other two governing equations

$$\Phi_3(\psi) = \cos m\psi, \sin m\psi \tag{3.158}$$

The complete solution, therefore, for the pressure field surrounding a rotor disc is

$$p(\upsilon, \eta, \psi) = \sum_{m=0}^{M} \sum_{n=m}^{N} P_n^m(\upsilon) Q_n^m(i\eta)(C_n^m \cos m\psi + D_n^m \sin m\psi) \tag{3.159}$$

The upper limits M and N can have any positive integer value.

The polynomial $Q_n^m(i\eta)$ is imaginary for odd values of m and real for even values therefore the arbitrary constants C_n^m and D_n^m must be real or imaginary, accordingly, in order that the pressure field be real.

Any combination of terms in Equation (3.159) can be used, whatever suits the conditions. For there to be a pressure discontinuity across the disc, but continuously varying pressure elsewhere the solutions must be restricted to those for which $n + m$ is odd. Of course, limiting the number of terms, other than has been described, may result in an approximate solution.

The pressure discontinuity across the rotor disc will be as shown in Figure 3.2. The magnitude of the step in pressure will be twice the pressure level (above the far field level) that occurs just upstream of the disc. The pressure gradient, however, normal to the rotor disc, will be continuous.

3.11.3 The axi-symmetric pressure distributions

For the wind turbine rotor disc the simplest situation is for $m = 0$ which means that the pressure distribution is axisymmetric. The permitted values of n must be odd.

For $n = 1$ the polynomials are

$$P_1^0(v) = v = \sqrt{1 - \mu^2} \tag{3.160a}$$

and

$$Q_1^0(i\eta) = \eta \tan^{-1}\frac{1}{\eta} - 1 \tag{3.160b}$$

So, on the disc, where $\eta = 0$, $Q_1^0(i0) = -1$.

Therefore, the pressure distribution is

$$p(\mu) = -C_1^0\sqrt{1 - \mu^2} \tag{3.161}$$

If the pressure in Equation (3.161) is non-dimensionalized using the free-stream dynamic pressure $(1/2)\rho U_\infty^2$ the value of C_1^0 can be related to the thrust coefficient by integrating the pressure distribution of Equation (3.161) over the disc area

$$\pi R^2 C_T = -R^2 C_1^0 \int_0^{2\pi}\int_0^1 \sqrt{1 - \mu^2}\mu \, d\mu \, d\psi = -\frac{2}{3}\pi R^2 C_1^0$$

Therefore,

$$C_1^0 = -\frac{3}{2}C_T \tag{3.162}$$

and so the pressure step across the disc is

$$p_1(\mu) = C_T\frac{3}{2}\sqrt{1 - \mu^2} \tag{3.163}$$

All the remaining polynomials (for $m = 0$ and odd values of $n > 1$) produce zero thrust. To modify the pressure distribution to suit the boundary conditions an appropriate linear combination of solutions can be added to that of Equation (3.163).

The application to helicopter rotors leads to a requirement for the pressure and the radial pressure gradient to be zero at the rotor axis as these conditions correspond to the pressure on actual rotors. The above pressure distribution does not have zero pressure at the rotor axis and so needs to be combined with at least one other solution. The second axisymmetric solution, $n = 3$, is

$$P_3^0(v) = \frac{1}{2}v(5v^2 - 3) = \frac{1}{2}\sqrt{1 - \mu^2}(2 - 5\mu^2) \tag{3.164a}$$

and

$$Q_3^0(i\eta) = -\frac{\eta}{2}(5\eta^2 + 3)\tan^{-1}\frac{1}{\eta} + \frac{5}{2}\eta^2 + \frac{2}{3}, \text{ so } Q_3^0(i0) = \frac{2}{3} \tag{3.164b}$$

The second pressure distribution is, therefore,

$$p_2(\mu) = \frac{1}{3} C_3^0 \sqrt{1 - \mu^2}(2 - 5\mu^2) \tag{3.165}$$

The sum of the two pressure distributions must be zero where $\mu = 0$, so

$$C_3^0 = -\frac{9}{4} C_T$$

and the combination of the two distributions is

$$p_{1-2}(\mu) = \frac{15}{4} C_T \mu^2 \sqrt{1 - \mu^2} \tag{3.166}$$

The three distributions are shown in Figure 3.71.

As most modern wind turbines are designed to achieve as uniform a pressure distribution as practicable, to maximize efficiency, the form chosen for the helicopter rotor might need to be modified. A uniform pressure distribution can be formed by combining solutions but, because the pressure discontinuity must itself be discontinuous at the disc edge, it would mean that a great many solutions would be required. Tip-loss effects would require zero pressure at both the blade tips and at the hub but for most of the blade span the pressure should be uniform. It should be pointed out that the *blade loading* caused by uniform pressure does increase linearly with radius.

The induced velocity field caused by the axisymmetric pressure distribution has to be obtained from the pressure field by integrating Equations (3.145) commencing far upstream where free-stream conditions are assumed to apply. The upstream

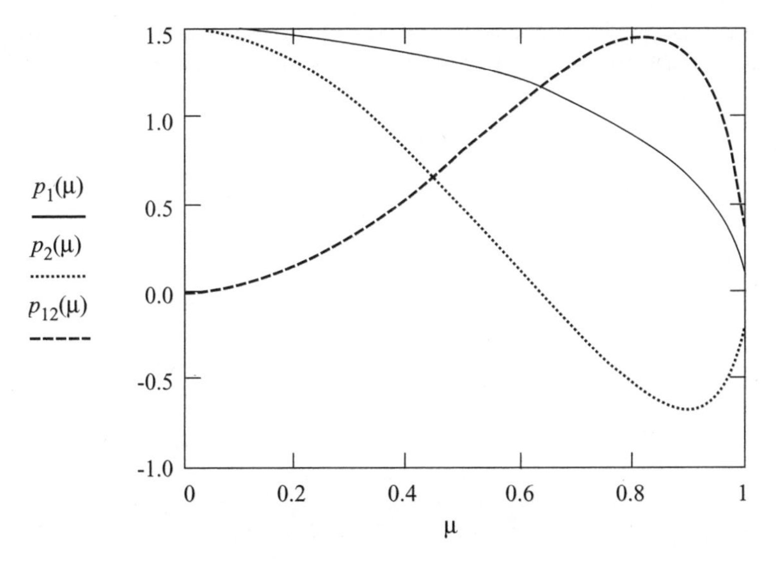

Figure 3.71 Radial Pressure Distributions of the First Two Solutions and their Combination to Satisfy the Requirements at the Rotor Axis

conditions also depend upon the angle of yaw of the disc. The integration continues until a point on the disc is reached where the induced velocity is to be determined.

The particular induced velocity component that is most important for determining the angle of attack on a blade element is normal to the rotor disc, i.e., the axial induced velocity. Mangler and Squire (1950) calculated the axial induced velocity distribution as a function of yaw angle by expressing the velocity as a Fourier series of the azimuth angle ψ.

$$\frac{u}{U_\infty} = C_T\left(\frac{A_0(\mu,\gamma)}{2} + \sum_{k=1}^{\infty} A_k(\mu,\chi)\sin k\psi\right) \tag{3.167}$$

For the pressure distribution of Equation (3.166) the Fourier coefficients are

$$A_0(\mu,\gamma) = -\frac{15}{8}\mu^2(1-\mu^2)^{1/2} \tag{3.168a}$$

$$A_1(\mu,\gamma) = -\frac{15\pi}{256}\mu(9\mu^2-4)\tan\frac{\gamma}{2} \tag{3.168b}$$

$$A_3(\mu,\gamma) = \frac{45\pi}{256}\mu^3\tan\left(\frac{\gamma}{2}\right)^3 \tag{3.168c}$$

Higher-order odd terms are zero. There are also even terms which have the general form

$$A_k = (-1)^{(k-2)/2}\frac{3}{4}\left[\frac{k+v}{k^2-1}\left(\frac{9v^2+k^2-6}{k^2-9}\right)+\frac{3v}{k^2-9}\right]\left(\frac{1-v}{1+v}\right)^{k/2}\tan\frac{\gamma}{2}^{k/2}$$

where $v^2 = 1-\mu^2$ and k is an even integer greater than zero.

The average value of the axial induced flow factor is independent of yaw angle and is given by

$$a_0 = -\frac{u_0}{U_\infty} = \frac{1}{4}C_T$$

where u_0 is the average axial induced velocity.

Thus, the *average* value of the axial flow induced flow factor is related to the thrust coefficient by

$$C_T = 4a_0 \tag{3.169}$$

compared with the momentum theory $C_T = 4a_0(1-a_0)$ or compared with any of the expressions developed for yawed conditions (Equations (3.91), (3.106) and (3.112)).

Because of the assumption that the induced velocity is small compared with the flow velocity, a_0 is small compared with 1. Clearly, the acceleration potential method only applies if the value of C_T is much less than 1.

The once per revolution term in Equation (3.167) will cause an angle of attack variation and, hence, a lift variation that will cause a yawing moment on the disc. However, the pressure distribution, being axi-symmetric, cannot cause a yawing moment. The situation is much the same as for the vortex theory of Coleman, Feingold and Stempin (1945).

Pitt and Peters (1981) use, or rather, impose Glauert's assumption (Equation (3.108)) for the variation of the axial induced flow factor:

$$a = a_0 + a_S \mu \sin \psi \tag{3.170}$$

The value of a_S is obtained by equating the first moment about the yaw axis of Equation (3.170) with the first moment of Equation (3.167) using the Mangler and Squire velocity distributions of Equations (3.168).

$$\int_0^{2\pi}\int_0^1 \mu \sin \psi (a_0 + a_S \mu \sin \psi) 2\pi\mu \, d\mu \, d\psi$$

$$= \int_0^{2\pi}\int_0^1 \mu \sin \psi C_T \left(\frac{A_0(\mu, \gamma)}{2} + \sum_{k=1}^{\infty} A_k(\mu, \gamma) \sin k\psi \right) \mu \, d\mu \, d\psi \tag{3.171}$$

All terms, apart from that containing A_1, vanish on integration, giving

$$a_S = \frac{15\pi}{128} \tan \frac{\gamma}{2} C_T \tag{3.172}$$

Hence, using Equation (3.169), the axial induced velocity becomes

$$a = a_0 \left(1 + \frac{15\pi}{32} \mu \tan \frac{\gamma}{2} \sin \psi \right) \tag{3.173}$$

Which, apart from the use of the yaw angle instead of the wake skew angle, has the same form as Equations (3.108) and (3.118) and so there is some consistency in the various methods for dealing with yawed flow.

3.11.4 The anti-symmetric pressure distributions

As has been determined in Section 3.10.11, there is a moment about the vertical diameter of a yawed wind turbine rotor disc, the restoring yaw moment. An axi-symmetric pressure distribution, however, is not capable of producing a yaw moment and so more terms from the series solution of Equation (3.159) need to be included.

The only terms in Equation (3.159) which will yield a yawing moment are those for which $m = 1$ and for which $D_n^1 \neq 0$. Terms for which $m = 1$ and $C_n^1 \neq 0$ will cause a tilting moment. Recalling that $m + n$ must be odd to achieve a pressure discontinuity across the disc the values of n that may be combined with $m = 1$ must be even.

Because of the nature of the Legendre polynomials only one term in the series of Equation (3.159) will produce a net thrust and only one term will produce a yawing moment, which is a first moment. Similarly only one term will produce a second moment, and so on.

The unique term in Equation (3.159) which yields a yawing moment is that for which $m = 1$, $n = 2$ and $C_n^1 \neq 0$, therefore

$$P_2^1(v) = 3v\sqrt{1 - v^2} = 3\mu\sqrt{1 - \mu^2} \tag{3.174}$$

and

$$Q_2^1(i\eta) = 3i\eta\sqrt{1 + \eta^2}\tan^{-1}\frac{1}{\eta} - 3i\sqrt{1 + \eta^2} + \frac{i}{\sqrt{1 + \eta^2}}, \tag{3.175}$$

so

$$Q_2^1(i0) = -2i \tag{3.175a}$$

A zero pressure gradient at the rotor axis is not appropriate in this case because the pressure distribution is anti-symmetric about the yaw axis, therefore,

$$p(\mu, \psi) = P_2^1(\mu)Q_2^1(i0)D_2^1 \sin\psi = -6iD_2^1\mu\sqrt{1 - \mu^2}\sin\psi \tag{3.176}$$

The pressure distribution is shown in Figure 3.72.

The yawing moment coefficient is defined by

$$C_{mz} = \frac{M_z}{\frac{1}{2}\rho U_\infty^2 \pi R^3} \tag{3.177}$$

As before, if the pressure in Equation (3.176) is non-dimensionalized by the free-stream dynamic pressure $(1/2)\rho U_\infty^2$, then

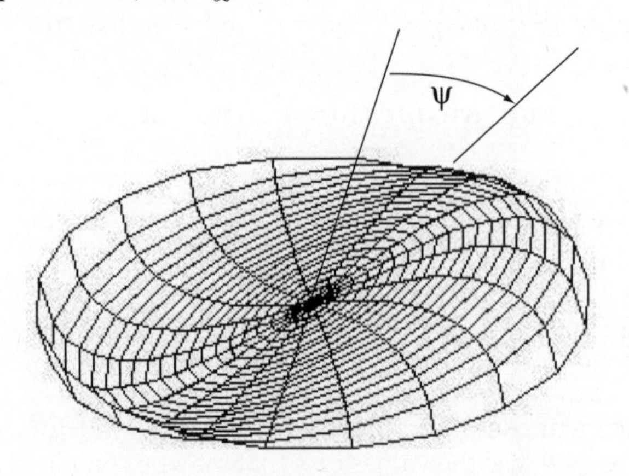

Figure 3.72 The Form of the Pressure Distribution which Yields a Yawing Moment

$$C_{mz} = \frac{1}{\pi}\int_0^{2\pi}\int_0^1 \mu \sin\psi p(\mu, \psi)\mu \, d\mu \, d\psi = \frac{-1}{\pi} 6iD_2^1 \int_0^1 \mu^3\sqrt{1-\mu^2}\, d\mu \int_0^{2\pi} \sin^2\psi \, d\psi \quad (3.178)$$

which gives

$$iD_2^1 = -\frac{5}{4}C_{mz} \quad (3.179)$$

To establish a relationship between the yawing moment coefficient and the axial velocity induced by the pressure distribution of Equation (3.176) the velocity distribution has to be obtained by integrating Equations (3.145). Unfortunately, no analytical solution has been determined for the anti-symmetric case, as Mangler and Squire have done for the symmetric case. Numerical values of induced velocities need to be calculated from Equations (3.145) using the pressure distribution defined by Equations (3.174) and (3.175).

Pitt and Peters (1981) have determined the axial velocity distribution for values of the yaw angle from 0° to 90°: the yaw angle fixes the far upstream conditions where the integration commences. The velocity distribution found corresponds to that of Equation (3.167) for the axi-symmetric case. Pitt and Peters again impose the form of Equation (3.170) and determine the average value of the axial induced velocity a_0 and the value of a_s, using the same method of Equation (3.171); in both cases, of course, numerical integration is necessary.

The values of a_0, are not zero, as might have been expected from the anti-symmetric pressure distribution, but are equal and opposite to the values of a_s found for the corrected axi-symmetric pressure distribution of Equation (3.166). The variation of the two coefficients a_0 and a_s with yaw angle γ is determined numerically but, using the Mangler and Squire analytical forms for guidance, analytical variations can be inferred. Pitt and Peters found that the linearized axial induced velocity distribution is

$$a_0 = -\frac{15}{128}\pi \tan\frac{\gamma}{2}C_{mz} \quad (3.180)$$

and

$$a_S = -\left(1 - \tan^2\frac{\gamma}{2}\right)C_{mz} \quad (3.181)$$

Pitt and Peters also include a cosine term in the linearized axial induced flow factor representation of Equation (3.170) which will only arise if $C_2^1 \neq 0$

$$a = a_0 + a_S\mu \sin\psi + a_c\mu \cos\psi \quad (3.182)$$

In which case there will be an additional pressure distribution given by

$$p(\mu, \psi) = P_2^1(\mu)Q_2^1(i0)C_2^1 \cos\psi = 6iC_2^1\mu\sqrt{1-\mu^2}\cos\psi \quad (3.183)$$

The tilting moment coefficient is given by

$$C_{my} = \frac{1}{\pi}\int_0^{2\pi}\int_0^1 \mu\cos\psi p(\mu,\psi)\mu\,d\mu\,d\psi = \frac{-1}{\pi}6iC_2^1\mu^3\sqrt{1-\mu^2}\,d\mu\int_0^{2\pi}\cos^2\psi\,d\psi \quad (3.184)$$

Therefore

$$iC_2^1 = -\frac{5}{4}C_{my} \quad (3.185)$$

The axial induced velocity distribution resulting from the pressure field of Equation (3.183) is calculated by integration of Equations (3.145) and is then matched with the linear velocity distribution of Equation (3.182) using the same method as for Equation (3.171).

$$\int_0^{2\pi}\int_0^1 \mu\cos\psi(a_0 + a_c\mu\cos\psi)2\pi\mu\,d\mu\,d\psi$$

$$= \int_0^{2\pi}\int_0^1 \mu\cos\psi C_T\left(A_0\frac{(\mu,\gamma)}{2} + \sum_{k=1}^{\infty}A_k(\mu,\gamma)\cos k\psi\right)\mu\,d\mu\,d\psi \quad (3.186)$$

The functions $A_k(\mu,\gamma)$ being determined numerically. Again, using the Mangler and Squire results as guidance, an expression for a_c is found.

$$a_c = -\sec^2\frac{\gamma}{2}C_{my} \quad (3.187)$$

3.11.5 The Pitt and Peters model

Pitt and Peters (1981) have developed the linear theory that relates the axial induced flow factors to the thrust and moment coefficients given in Equations (3.169), (3.172), (3.180), (3.181) and (3.187) which collect together in matrix form

$$\begin{bmatrix} a_0 \\ a_c \\ a_S \end{bmatrix} \begin{bmatrix} \frac{1}{4} & 0 & -\frac{15}{128}\pi\tan\frac{\gamma}{2} \\ 0 & -\sec^2\frac{\gamma}{2} & 0 \\ \frac{15}{128}\pi\tan\frac{\gamma}{2} & 0 & -\left(1-\tan^2\frac{\gamma}{2}\right) \end{bmatrix} \begin{bmatrix} C_T \\ C_{my} \\ C_{mz} \end{bmatrix} \quad (3.188a)$$

$$(a) = [L](C) \quad (3.188b)$$

The procedure for using Equation (3.188) is to assume initial values for (a) from which the values of (C) can be calculated from the blade element theory. New values of (a) are then found from Equation (3.188) and an iteration proceeds.

For the wind turbine the value of a_0 may not be small compared with 1 and so the above procedure will converge on values of a_0 which are too small compared with what the momentum theory would deliver.

To produce more realistic results that is, results in line with Glauert's momentum theory where

$$C_T = 4a\sqrt{1 - a(2\cos\gamma - a)} = 4aA_G(a) \tag{3.105}$$

or the Coleman theory, where

$$C_T = 4a\left(\cos\gamma + \tan\frac{\chi}{2}\sin\gamma - a\sec^2\frac{\chi}{2}\right) = 4aA_C(a) \tag{3.111}$$

Also, the wake skew angle should be used in matrix $[L]$ instead of the yaw angle.

The matrix $[L]$ should then be modified to become

$$[L] = \begin{bmatrix} \dfrac{1}{4A(a_0)} & 0 & -\dfrac{15}{128}\pi\tan\dfrac{\chi}{2} \\ 0 & -\sec^2\dfrac{\chi}{2} & 0 \\ \dfrac{15}{128A(a_0)}\pi\tan\dfrac{\chi}{2} & 0 & -\left(1 - \tan^2\dfrac{\chi}{2}\right) \end{bmatrix} \tag{3.189}$$

where $A(a_0)$ is chosen according to which momentum theory is to be used.

The Pitt and Peters method does not include any determination of induced velocities in the plane of the rotor disc and as a consequence it is not possible to account for wake rotation. However, it is possible that the Kinner solutions $Q_n^m(\nu)$ that were excluded from the analysis because they give infinite pressure at $\nu^2 = 1$, which lies along the axis of rotation, may give velocity distributions which provide for wake rotation; the momentum theory of Section 3.3 also predicts an infinite pressure at the axis of rotation because of wake rotation. In practice, of course, the rotor disc would not extend to the axis of rotation and the singularity would not occur.

With or without wake rotation a flow angle ϕ can be determined from which a torque can be found. If the normal force on an element of the rotor disc is equal to $\delta L\cos\phi$ then the tangential force will be $\delta L\sin\phi$.

3.11.6 The general acceleration potential method

Peters with a number of associates has developed the theory further and a reading of these references (Pitt and Peters, 1981, Goanker and Peters, 1988, HaQuang and Peters, 1988) is recommended. The acceleration potential method has been developed specifically for wind turbines by van Bussell (1995) where a much more comprehensive account of the theory is given.

3.11.7 Comparison of methods

A project to compare existing methods of predicting yaw behaviour, among other aspects of the aerodynamic behaviour of wind turbines, was reported upon (Snel and Schepers, 1995). Figure 3.73 shows results obtained by various methods for predicting the yawing moment of the 2 MW, three-blade turbine at Tjæreborg in Denmark at a yaw angle of 32° and a wind speed of 8.5 m/s.

Most of the theoretical predictions in Figure 3.73 have the correct phasing and about the correct mean yawing moment but the amplitude of the yawing moment varies. In this example the second method bears the closest comparison with the measured data. Generally, the amplitude of the yawing moment variation is underestimated by the theoretical predictions, whereas the mean yawing moment is quite well predicted.

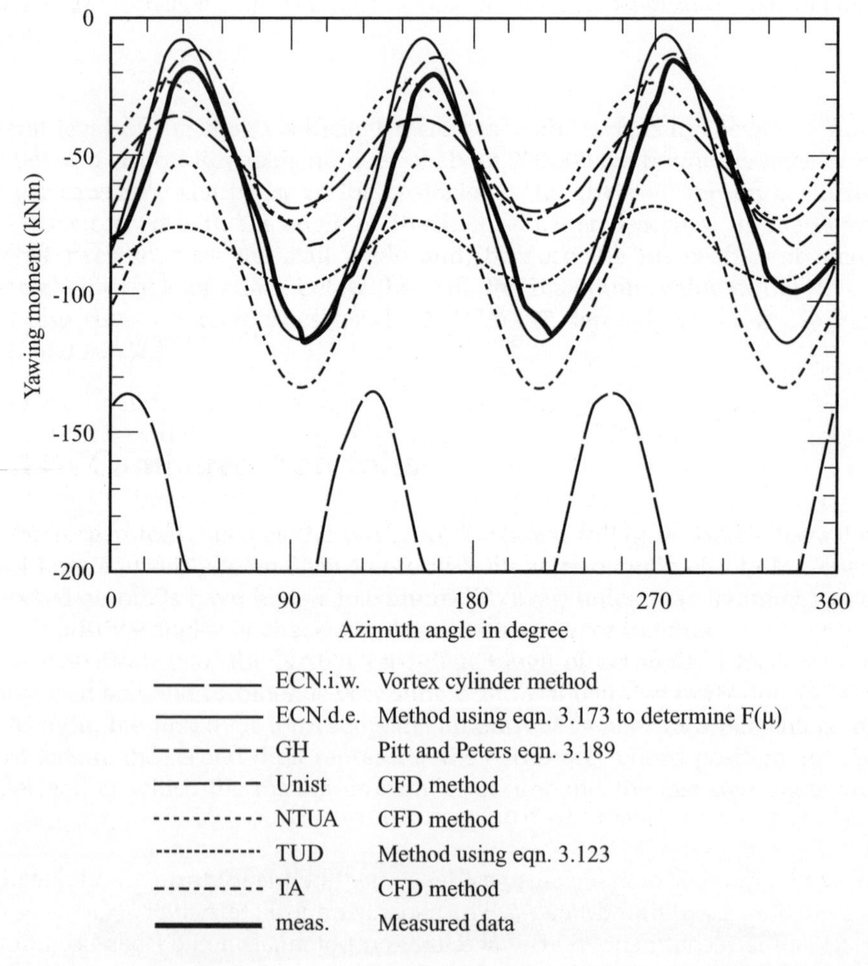

Figure 3.73 Yawing Moment on the Tjæreborg Turbine at 32° Yaw and 8.5 m/s, (Srel and Schepers 1995)

3.12 Stall Delay

A phenomenon first noticed on propellers by Himmelskamp (1945) is that of lift coefficients being attained at the inboard section of a rotating blade which are significantly in excess of the maximum value possible in two-dimensional static tests. In other words the angle of attack at which stall occurs is greater for a rotating blade than for the same blade tested statically. The power output of a rotor is measurably increased by the stall-delay phenomenon and, if included, improves the comparison of theoretical prediction with measured output. It is noticed that the effect is greater near the blade root and decreases with radius.

The reason for the stall delay has been the cause of much discussion but a convincing physical process has not yet been established. What is agreed is that, for whatever reason, the adverse pressure gradient experienced by the flow passing over the downwind surface of the blade is reduced by the blade's rotation. The adverse pressure gradient slows down the flow as it approaches the trailing edge of the blade after the velocity peak reached close the the leading edge. In the boundary layer viscosity also slows down the flow and the combination of the two effects, if sufficiently large, can bring the boundary layer flow to a standstill (relative to the blade surface) or even cause a reversal of flow direction. When flow reversal takes place the flow separates from the blade surface and stall occurs giving rise to loss of lift and a dramatic increase in pressure drag.

Aerodynamic analyses (Wood (1991) and Snel *et al.* (1993)) of rotating blades using computational fluid dynamic techniques, which include the effects of viscosity, do show a decrease in the adverse pressure gradient but it is not obvious from these numerical calculations as to what exactly is occurring physically.

It is also agreed that the parameter that influences stall delay predominantly is the local blade solidity $c(r)/r$. The evidence which does exist shows that for attached flow conditions, below what would otherwise be the static (non-rotating) stall angle of attack, there is little difference between two-dimensional flow conditions and rotating conditions. When stall does occur, however, the air in the separated region, which is moving very slowly with respect to the blade surface, is rotating with the blade and so is subject to centrifugal force causing it to flow radially outwards. Prior to stalling taking place, centrifugal forces on the fluid in the boundary layer, again causing radial flow, may reduce the displacement thickness and so increase the resistance to separation.

Blade surface pressures have been measured by Ronsten (1991) on a blade while static and while rotating. Figure 3.74 shows the comparison of surface pressure coefficients for similar angles of attack in the static and rotating conditions (tip speed ratio of 4.32) for three span-wise locations. At the 30% span location the estimated angle of attack at 30.41° is well above the static stall level which is demonstrated by the static pressure coefficient distribution. The rotating pressure coefficient distribution at 30% span shows a high leading edge suction pressure peak with a uniform pressure recovery slope over the rear section of the upper surface of the chord. The gradual slope of the pressure recovery indicates a reduced adverse pressure gradient with the effect on the boundary layer that it is less likely to separate. The level of the leading edge suction peak, however, is very much less

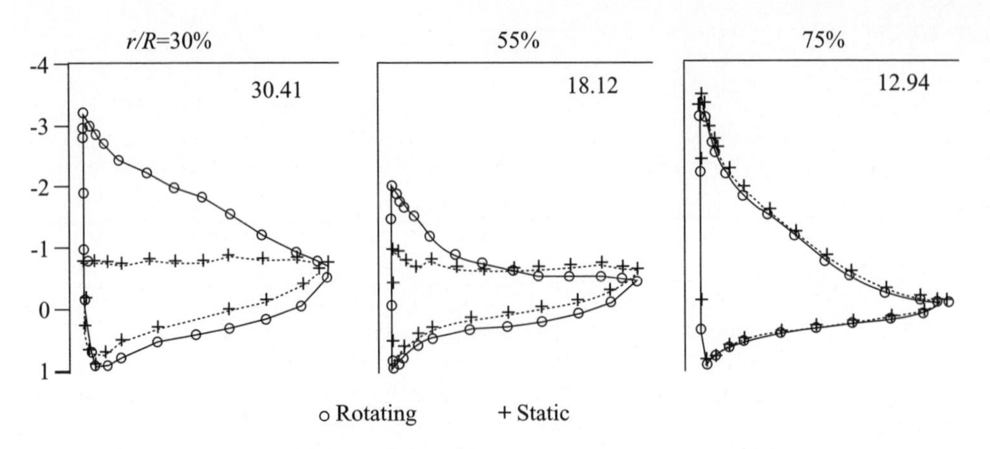

Figure 3.74 Pressure Measurements on the Surface of a Wind Turbine Blade while Rotating and while Static by Ronsten (1991)

than it would be if, in the non-rotating situation, it were possible for flow still to be attached at 30.41°.

The situation at the 55% span-wise location is similar to that at 30%; the static pressures indicate that the section has stalled but the rotating pressures show leading edge suction peak which is small but significant. At the 75% span location there is almost no difference between static and rotating blade pressure coefficient distributions at an angle of attack of 12.94°, which is below the static stall level; the leading edge suction pressure peak is a little higher than that at 30% span, much higher than that at 55% but the pressure recovery slope is much steeper. The flow appears to be attached at the 30% and 55% span locations on the rotating blade, but the suction pressures are too low for that actually to be the case, so stall appears to be greatly delayed and the low adverse pressure gradient shown by the gentle slope of the pressure recovery appears to indicate the reason for the delay. At 30% span the ratio $c/r = 0.374$, $c/r = 0.161$ at 55% span and at the 75% location $c/r = 0.093$. The increased lift also occurs in the post stall region and is attributed to the radial flow in the separated flow regions.

Snel, *et al.* 1993, have proposed a simple, empirical modification to the usually available two dimensional, static aerofoil lift coefficient data which fits the measured lift coefficients by Ronsten (1991) and computed results using a three-dimensional computational fluid dynamics code.

Table 3.2 Summary of Ronsten's measurements of lift coefficient and lift coefficients corrected to rotating conditions using Equation (3.190)

$r/R^{*}100$	30%	55%	75%
c/R	0.374	0.161	0.093
Angle of attack α	30.41°	18.12°	12.94°
C_l static (measured)	0.8	0.74	1.3
C_l rotating (measured)	1.83	0.93	1.3
C_l rotating (Snel, 1995)	1.87	0.84	1.3

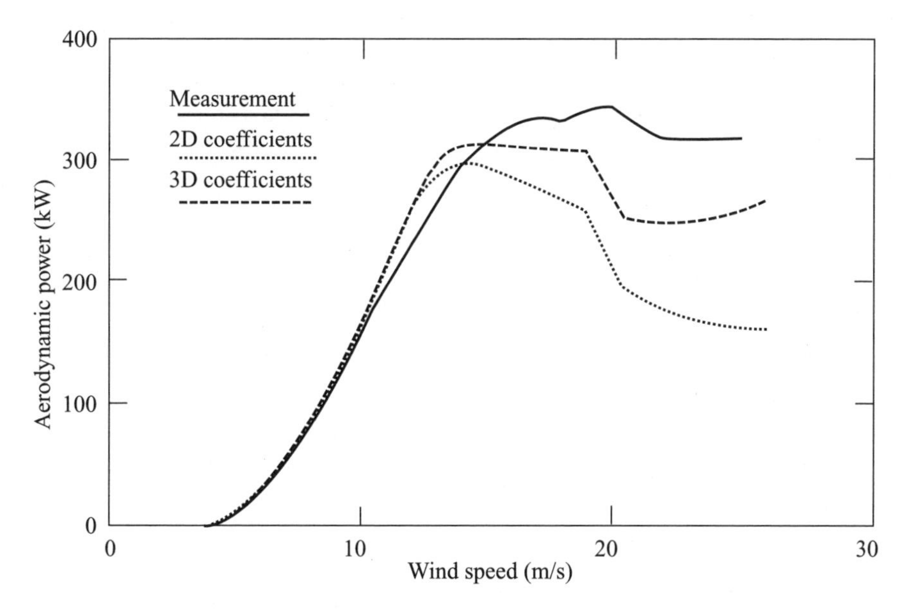

Figure 3.75 A Comparison of Measured and Snel's Predicted Power Curves for a NORTANK 300 kW Turbine

If the linear part of the static, two-dimensional, $C_l - \alpha$ curve is extended beyond the stall then let ΔC_l be the difference between the two curves. Then the correction to the two-dimensional curve to account for the rotational, three-dimensional, effects is $3(c/r)^2 \Delta C_l$.

$$C_{l_{3\text{-D}}} = C_{l_{2\text{-D}}} + 3\left(\frac{c}{r}\right)^2 \Delta C_l \qquad (3.190)$$

Table 3.2 compares the measured static ($C_{l_{2\text{-D}}}$) and rotating ($C_{l_{3\text{-D}}}$) lift coefficients with the calculated values for the rotating values using Snel's correction of Equation (3.190). The correction is quite good and is very simple to apply. An example of the correction is given by Snel *et al.* (1993) and is shown in Figure 3.75.

3.13 Unsteady flow – Dynamic inflow

3.13.1 Introduction

Natural winds are almost never steady in either strength or direction and so it is seldom that the conditions for the momentum theory apply. It takes a finite time for the wind to travel from far upwind of a rotor to far downwind and in that time wind conditions will change so an equilibrium state is never achieved. Even if the 'average' wind speed changes only slowly small-scale turbulence will cause a continuous unsteadiness in the velocities impinging on a rotor blade.

Several approximate solutions offer themselves for the determination of the dynamic flow conditions at the rotor disc. It could be assumed that the induced velocity remains fixed at the level determined by the average wind speed blowing over a set period of time, that may be quite short. The wake remains frozen while the unsteady component of the wind passes through the rotor disc unattenuated by the presence of the rotor. The unsteady forces, that would impose zero net force on the rotor, would be determined by the blade element theory. Alternatively, the induced velocity through the rotor disc could be determined from the instantaneous wind velocity as if that velocity was steady. The induced velocity will change as the wind speed changes but it must be assumed that the entire wake changes instantaneously to remain in step. Equilibrium in the wake is maintained at all times. The truth lies somewhere between the two scenarios given above, both of which rely on simple assumptions about the state of the wake.

The acceleration potential method avoids reference to the wake and allows the flow conditions at the rotor disc to be determined by the upwind flow field, which is much simpler to determine than that of the wake.

In steady-flow conditions the velocity at a fixed point in the upwind flow field is constant; acceleration of the flow from point to point takes place (e.g., $U_\infty(\partial u/\partial x)$ in the x-direction, assuming u, the induced velocity, is much smaller than U_∞) but no rate of change of velocity with time (e.g., $\partial u/\partial t$) occurs at a single point. In unsteady flow, conditions at a fixed point do change with time and the total acceleration in the x-direction is then $(\partial u/\partial t) + U_\infty(\partial u/\partial x)$. The additional acceleration requires an additional inertia force the reaction to which will change the force on the rotor disc. The additional force is often termed the added mass force because if the unsteadiness in the relative flow past a blade can be attributed not to flow turbulence but to an unsteady motion of the blade itself some of the air will be forced to move (accelerate) with the blade, effectively adding to the mass of the blade.

3.13.2 Adaptation of the acceleration potential method to unsteady flow

If the unsteady acceleration terms are added to Equations (3.145), which are simplified to account for the induced velocities being very much smaller than the wind velocity, then those equations become

$$\begin{aligned}
\rho\left(\frac{\partial u}{\partial t} + U_\infty\frac{\partial u}{\partial x}\right) &= -\frac{\partial p}{\partial x}\\
\rho\left(\frac{\partial v}{\partial t} + U_\infty\frac{\partial v}{\partial x}\right) &= -\frac{\partial p}{\partial y}\\
\rho\left(\frac{\partial w}{\partial t} + U_\infty\frac{\partial w}{\partial x}\right) &= -\frac{\partial p}{\partial z}
\end{aligned} \tag{3.191}$$

As before, differentiating each equation with respect to its particular direction and adding together the results gives

$$\rho\left[\frac{\partial}{\partial t}\left(\frac{\partial u}{\partial x}+\frac{\partial v}{\partial y}+\frac{\partial w}{\partial z}\right)+U_\infty\frac{\partial}{\partial x}\left(\frac{\partial u}{\partial x}+\frac{\partial v}{\partial y}+\frac{\partial w}{\partial z}\right)\right]=-\left(\frac{\partial^2 p}{\partial x^2}+\frac{\partial^2 p}{\partial y^2}+\frac{\partial^2 p}{\partial z^2}\right)$$

but, for continuity of the flow,

$$\frac{\partial u}{\partial x}+\frac{\partial v}{\partial y}+\frac{\partial w}{\partial z}=0$$

Therefore the condition

$$\frac{\partial^2 p}{\partial x^2}+\frac{\partial^2 p}{\partial y^2}+\frac{\partial^2 p}{\partial z^2}=0$$

applies together with the Kinner pressure distributions of Section 3.11.2.

The accelerations

$$\left(\frac{\partial u}{\partial t},\frac{\partial v}{\partial t},\frac{\partial w}{\partial t}\right)$$

can be determined directly from Equations (3.191) without integration being necessary but it is only the component $\partial u/\partial t$ that is required because it is normal to the rotor disc and so will give rise to a normal force.

The solutions for

$$\rho U_\infty\frac{\partial u}{\partial x}=-\frac{\partial p}{\partial x}$$

have already been obtained in Sections 3.11.2–3.11.4 and so it remains to determine the solutions for

$$\rho\frac{\partial u}{\partial t}=-\frac{\partial p}{\partial x} \tag{3.192}$$

Equation (3.192) cannot be solved for the velocity u because that is the solution of the complete equation that is the first of Equations (3.191). What can be determined from Equation (3.192) is the acceleration $\partial u/\partial t$ for which it is necessary to differentiate the chosen pressure distribution. The Kinner pressure distributions, that are solutions of Equations (3.149), are given as functions of the ellipsoidal co-ordinates (ν, η, ψ) so to obtain the derivative with respect to x a co-ordinate transformation is required. The relationships between the ellipsoidal co-ordinates and the Cartesian co-ordinates (x, y, z) are given in Equations (3.157) from which can be obtained the derivatives $\partial x/\partial v$, $\partial x/\partial \eta$, $\partial x/\partial \psi$, etc., but what are really needed are the inverses of these derivatives.

We can find by appropriate differentiations of Equations (3.147)

$$\frac{\partial}{\partial v}=\frac{\partial x''}{\partial v}\frac{\partial}{\partial x''}+\frac{\partial y'}{\partial v}\frac{\partial}{\partial y'}+\frac{\partial z}{\partial v}\frac{\partial}{\partial z}$$

for example, and so the complete Jacobian can be determined:

$$\begin{bmatrix} \dfrac{\partial}{\partial \upsilon} \\ \dfrac{\partial}{\partial \eta} \\ \dfrac{\partial}{\partial \psi} \end{bmatrix} = \begin{bmatrix} \dfrac{\partial x''}{\partial \upsilon} & \dfrac{\partial y'}{\partial \upsilon} & \dfrac{\partial z}{\partial \upsilon} \\ \dfrac{\partial x''}{\partial \eta} & \dfrac{\partial y'}{\partial \eta} & \dfrac{\partial z}{\partial \eta} \\ \dfrac{\partial x''}{\partial \psi} & \dfrac{\partial y'}{\partial \psi} & \dfrac{\partial z}{\partial \psi} \end{bmatrix} \begin{bmatrix} \dfrac{\partial}{\partial x''} \\ \dfrac{\partial}{\partial y'} \\ \dfrac{\partial}{\partial z} \end{bmatrix} \tag{3.193}$$

the inverse of which is what is required:

$$\begin{bmatrix} \dfrac{\partial}{\partial x''} \\ \dfrac{\partial}{\partial y'} \\ \dfrac{\partial}{\partial z} \end{bmatrix} = \begin{bmatrix} \dfrac{\partial \upsilon}{\partial x''} & \dfrac{\partial \eta}{\partial x''} & \dfrac{\partial \psi}{\partial x''} \\ \dfrac{\partial \upsilon}{\partial y'} & \dfrac{\partial \eta}{\partial y'} & \dfrac{\partial \psi}{\partial y'} \\ \dfrac{\partial \upsilon}{\partial z} & \dfrac{\partial \eta}{\partial z} & \dfrac{\partial \psi}{\partial z} \end{bmatrix} \begin{bmatrix} \dfrac{\partial}{\partial \upsilon} \\ \dfrac{\partial}{\partial \eta} \\ \dfrac{\partial}{\partial \psi} \end{bmatrix} \tag{3.194}$$

The Jacobian matrix of Equation (3.193) can be determined algebraically from Equations (3.147) and this can then be inverted algebraically to give the inverse Jacobian of Equation (3.194). From the inverse Jacobian it is found that

$$\frac{\partial}{\partial x''} = \frac{\eta(1-\upsilon^2)}{R(\eta^2+\upsilon^2)}\frac{\partial}{\partial \upsilon} + \frac{\upsilon(1+\eta^2)}{R(\eta^2+\upsilon^2)}\frac{\partial}{\partial \eta} \tag{3.195}$$

However, only the acceleration at the rotor disc itself is required and there the value of η is zero so

$$\frac{\partial}{\partial x''} = \frac{1}{R\upsilon}\frac{\partial}{\partial \eta} \tag{3.196}$$

If the corrected axi-symmetric pressure drop distribution of Equation (3.166) is chosen, to conform with the steady flow case then, for the whole flow field,

$$p(\upsilon, \eta) = \frac{15}{32}\upsilon C_{\mathrm{TD}} \begin{bmatrix} \left[-7\eta \tan^{-1}\dfrac{1}{\eta} + 4(1-\upsilon^2)\right] \cdots \\ + 15\upsilon^2\eta^2\left(\eta \tan^{-1}\dfrac{1}{\eta} - 1\right) \cdots \\ + 9\eta\left[\eta + (\upsilon^2-\eta^2)\tan^{-1}\dfrac{1}{\eta}\right] \end{bmatrix} \tag{3.197}$$

in which the pressure is normalized by $(1/2)\rho U_\infty^2$. The term C_{TD} is the contribution to the total thrust coefficient of the dynamic acceleration $\partial u/\partial t$. Note that, as explained at the end of Section 3.11.1, the pressure level just upwind of the rotor disc, as given by Equation (3.197), is half the magnitude of the pressure drop across the disc given by Equation (3.166).

By means of Equation (3.195), at the rotor plane, where $\eta = 0$, the pressure gradient is found to be

$$\frac{\partial p}{\partial x''} = \frac{1}{R}\frac{15\pi}{64} C_{TD}(9v^2 - 7)$$

Therefore, in terms of parameter μ, from Equation (3.192)

$$\rho\frac{\partial u}{\partial t} = -\frac{\partial p}{\partial x''}\frac{1}{2}\rho U_\infty^2 = \frac{1}{R}\frac{15\pi}{64} C_{TD}(9\mu^2 - 2)\frac{1}{2}\rho U_\infty^2 \tag{3.198}$$

It should be noted that the axial acceleration distribution is axi-symmetric and independent of the yaw angle.

The mean value of axial acceleration over the area of the disc is

$$\frac{\partial u_o}{\partial t} = \frac{75\pi}{256}\frac{U_\infty^2}{R} C_{TD} \tag{3.199}$$

The non-dimensional form of the acceleration can be expressed as

$$\frac{\partial a_o}{\partial \tau} = \frac{R}{U_\infty^2}\frac{\partial u_o}{\partial t} = \frac{75\pi}{256} C_{TD} \tag{3.200}$$

where $a_o = u_o/U_\infty$, axial flow factor and $\tau = tR/U_\infty$ which is called non-dimensional time.

The axial force on the disc is

$$F_x = \frac{1}{2}\rho U_\infty^2 \pi R^2 C_{TD}$$

Substituting for C_{TD} from Equation (3.199) gives

$$F_x = \frac{128}{75}\rho R^3 \frac{\partial u_o}{\partial t} \tag{3.201}$$

The added mass is, therefore, $(128/75)\rho R^3$.

The added mass term associated with a solid disc is 8/3 (Tuckerman, 1925), compared with 128/75 given in Equation (3.201), is in agreement with the value that is given by the uncorrected axi-symmetric Kinner pressure distribution of Equations (3.160). Although Pitt and Peters (1981) determine the value 128/75 in subsequent papers by Peters and other workers the value 8/3 is recommended and has come to be generally accepted (see Schepers and Snel, 1995). The use of the so-called 'corrected' pressure distribution for wind turbines has already been questioned in Section 3.11.3; there is no need to impose a zero pressure difference on the rotor disc at the rotation axis.

3.13.3 Unsteady yawing and tilting moments

For unsteady flow in yaw the normal unsteady acceleration distribution on the disc is required to have the same form of linear variation as the velocity, given in Equation (3.182). In terms of flow factors

$$\frac{\partial a}{\partial \tau} = \frac{\partial a_0}{\partial \tau} + \frac{\partial a_{\mathrm{s}}}{\partial \tau}\mu \sin\psi + \frac{\partial a_{\mathrm{c}}}{\partial \tau}\mu \cos\psi \tag{3.202}$$

The condition that causes a yawing moment arises from the anti-symmetric pressure distribution of Section 3.11.4 and can be obtained from Equations (3.175). For the whole field surrounding the rotor disc the pressure distribution is

$$p(v, \eta, \psi) = -\frac{3}{2}D_2^1 v\sqrt{1-v^2}\left(3i\eta\sqrt{1+\eta^2}\tan^{-1}\frac{1}{\eta} - 3i\sqrt{1+\eta^2} + \frac{i}{\sqrt{1+\eta^2}}\right)\sin\psi$$

which, on the disc, produces the pressure shown in Figure 3.72. The coefficient D_2^1 is related to the yawing moment coefficient in Equation (3.179)

$$iD_2^1 = -\frac{5}{4}C_{mz} \tag{3.179}$$

Therefore

$$p(v, \eta, \psi) = \frac{15}{8}C_{mz} v\sqrt{1-v^2}\left(3\eta\sqrt{1+\eta^2}\tan^{-1}\frac{1}{\eta} - 3\sqrt{1+\eta^2} + \frac{1}{\sqrt{1+\eta^2}}\right)\sin\psi \tag{3.203}$$

As before, the pressure in Equation (3.203) is non-dimensionalized by the free-stream dynamic pressure $(1/2)\rho U_\infty^2$.

Applying the differential operator given in Equation (3.196) to Equation (3.203), from Equation (3.192) we get at the rotor disc, where $\eta = 0$,

$$\frac{\partial u_{\mathrm{s}}}{\partial t} = \frac{45}{32}\pi\frac{U_\infty^2}{R}C_{mz\mathrm{D}}\mu \sin\psi \tag{3.204}$$

In terms of non-dimensional time and velocity

$$\frac{\partial a_{\mathrm{s}}}{\partial \tau} = \frac{45}{32}\pi C_{mz\mathrm{D}}\mu \sin\psi \tag{3.205}$$

Similarly, if there is a tilting moment then the corresponding acceleration is

$$\frac{\partial a_{\mathrm{c}}}{\partial \tau} = \frac{45}{32}\pi C_{my\mathrm{D}}\mu \cos\psi \tag{3.206}$$

The radial variation is linear and so no linearization adjustment is necessary. Again, the acceleration is independent of yaw angle. The mean acceleration is zero and so there is no coupling between the cases.

The relationship between accelerations and force coefficients is therefore

$$\begin{bmatrix} \frac{16}{3\pi} & 0 & 0 \\ 0 & \frac{32}{45\pi} & 0 \\ 0 & 0 & \frac{32}{45\pi} \end{bmatrix} \begin{bmatrix} \frac{\partial a_0}{\partial \tau} \\ \frac{\partial a_c}{\partial \tau} \\ \frac{\partial a_s}{\partial \tau} \end{bmatrix} = \begin{bmatrix} C_T \\ C_{my} \\ C_{mz} \end{bmatrix}_D \tag{3.207a}$$

$$[M]\left\{\frac{\partial a}{\partial \tau}\right\} = \{C\}_D \tag{3.207b}$$

The complete equation of motion combines Equation (3.207) and the steady yaw Equation (3.188). The combination is achieved by adding the corresponding force coefficients, which means that both equations must be inverted.

$$[M]\left\{\frac{\partial a}{\partial \tau}\right\} + [L]^{-1}\{a\} = \{C\}_D + \{C\}_S \tag{3.208}$$

The right-hand side of Equation (3.208) can also be determined from blade element theory and will be a time-dependent function of the inflow factor. The blade forces will vary in a manner determined by the time-varying velocity of the oncoming wind and consequent dynamic structural deflections of the necessarily elastic rotor. Equation (3.208) applies to the whole rotor disc and the blade element forces need to be integrated along the blade lengths.

Numerical solutions to Equation (3.208) require a procedure for dealing with first-order differential equations and the tried and tested fourth-order Runge-Kutta method is recommended. Starting with a steady-state solution the progress in time of the induced velocity as an unsteady flow passes through the rotor can be tracked. However, non-dimensionalizing with respect to wind speed is not very useful if wind speed is changing dynamically and it is common to work directly in terms of induced velocity rather than flow factors.

Equation (3.208) really applies to the whole rotor and the only spatial variation of the induced velocity and acceleration that is permitted is as defined in Equations (3.182) and (3.202). However, a relaxation of the strict approach has been adopted by several workers (see, for example, Schepers and Snel, 1995) where the induced velocities are determined for separate annular rings, as described in Section 3.10.8. The added mass term for an annular ring can be taken as a proportion of the whole added mass according to the appropriate acceleration distribution, Equations (3.198), (3.204) and (3.206).

Figure 3.76 shows measured and calculated flap-wise (out of the rotor plane) blade root bending moments for the Tjæreborg turbine caused by a pitch change from 0.070° to 3.716° with the reversed change 30 seconds later. The turbine was not in yaw and the wind speed was 8.7 m/s. The calculated results were made according to the equilibrium wake method and with a differential equation method

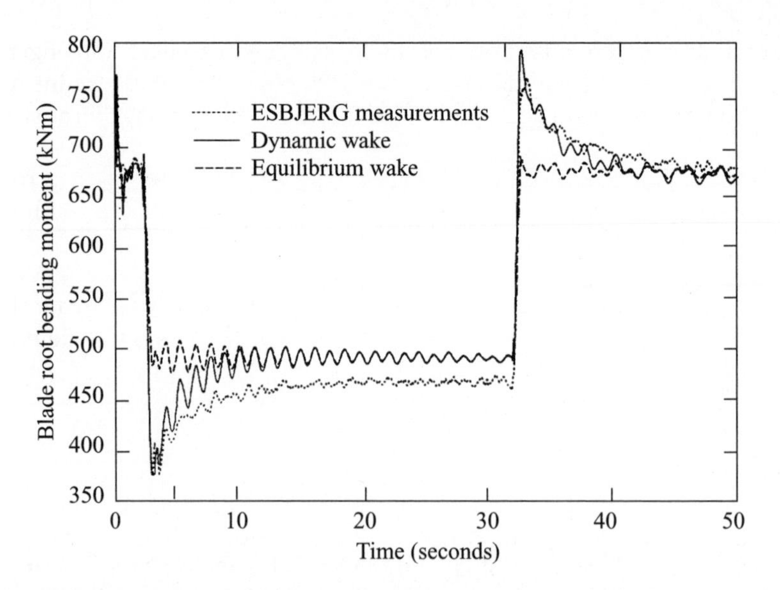

Figure 3.76 Measured and Calculated Blade Root Bending Moment Responses to Blade Pitch Angle Changes on the Tjæreborg Turbine (from Lindenburg, 1996).

similar to that of Equation 3.206. The comparison with the measured results clearly shows that the dynamic analysis predicts the initial overshoot in bending moment whereas the equilibrium wake method does not. Neither theory predicts the steady state bending moment achieved betweeen the pitch changes. Figure 3.76 is taken from Reference (27) which describes the PHATAS III aero-elastic code developed at ECN in the Netherlands. The Tjæreborg turbine is sited near Esbjerg in Denmark, details of which can be obtained from Snel and Schepers, 1995.

The solution procedure requires the time varying blade element force to determine the right-hand side of Equation (3.208), but calculating the lift and drag forces on a blade element in unsteady flow conditions is not a straight forward process. The lift force on a blade element is dependent upon the circulation around the element but after a change in conditions the circulation takes time to settle at a new level and in the interim the instantaneous lift cannot be determined *via* the instantaneous angle of attack. In a continuously changing situation the lift is not in phase with the angle of attack and does not have a magnitude that can be determined using static, two-dimensional aerofoil lift *versus* angle of attack data.

3.13.5 Quasi-steady aerofoil aerodynamics

When the oncoming flow relative to an aerofoil is unsteady the angle of attack is continuously changing and so the lift also is changing with time. The simple, but incorrect, way of dealing with this problem is to assume that the instantaneous angle of attack corresponds to the same lift coefficient as if that angle of attack were to be constantly applied. The angle of attack is determined by the oncoming flow velocity and the velocity of the blade's motion. If the blade motion includes a torsional (pitching) component then the angle of attack will vary along the chord length: thin aerofoil theory (see Anderson, 1991) shows that the point 3/4 of the chord length from the leading edge is where the angle of attack must be determined.

The velocities that determine the quasi-steady angle of attack for a rotor blade element are shown in Figure 3.77, the dot represents differentiation with respect to time t.

The flow velocity $W(t)$, which includes the rotational speed of the blade element, varies in magnitude and direction ($\alpha_W(t)$) with the unsteady wind. $W(t)$ also includes the induced velocities caused by the rotor disc as might be determined by Equation (3.208). The elastic deflection velocities (subscript e) caused by blade vibration also influence the quasi-steady angle of attack, which is

$$\alpha(t) = \alpha_W(t) - \left[v_e(t) - \frac{\partial \beta_e}{\partial t}\left(\frac{3}{4} - h\right)c \right] \frac{1}{W(t)} \tag{3.209}$$

The structural velocity caused by chord-wise (edge-wise) deflections of the blade will also influence the angle of attack but by a very small amount. The non-dimensional parameter h defines the position of the pitching axis (flexural axis, shear centre position) of the blade element.

Assuming the structural deflection velocities to be small the lift force per unit span is then

$$L_c(t) = \frac{1}{2}\rho W(t)^2 c \frac{dC_l}{d\alpha} \sin \alpha(t) \tag{3.210}$$

The lift-curve slope $dC_l/d\alpha$ is assumed to be the same as for the static case.

3.14.6 Aerodynamic forces caused by aerofoil acceleration

In addition to the circulatory forces there are forces on the aerofoil caused by the inertia of the surrounding air that is accelerated as the aerofoil accelerates in its motion. The additional terms are added mass forces. The added mass per unit span of blade can be shown to be that of a circular cylinder of air of diameter equal to the aerofoil chord $(\pi c^2/4)\rho$. There are two components to the added mass force (Fung, 1969).

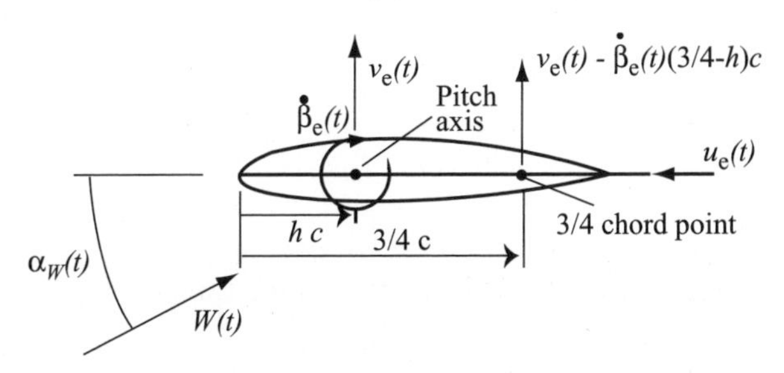

Figure 3.77 Unsteady Flow and Structural Velocities Adjacent to a Rotor Blade

(1) A lift force with the centre of pressure at the mid-chord point of an amount equal to the apparent mass times the acceleration normal to the chord line of the mid-chord point:

$$L_{m1}(\tau) = -\frac{1}{4}\pi c^2 \rho \left[\frac{\partial v_e}{\partial t} - c\left(\frac{1}{2} - h\right)\frac{\partial^2 \beta_e}{\partial t^2}\right] \tag{3.211}$$

(2) A lift force with the centre of pressure at the 3/4 chord point, of the nature of a centrifugal force, of an amount equal to the apparent mass times $W(t)(\partial\beta_e/\partial t)$

$$L_{m2}(\tau) = -\frac{1}{4}\pi c^2 \rho W(t)\frac{\partial \beta_e}{\partial t} \tag{3.212}$$

There is also a nose-down pitching moment equal to the apparent moment of inertia $(\pi/128)c^4\rho$ (which, actually, is only a quarter of the moment of inertia per unit length of the cylinder of air of diameter c) times the pitching acceleration $\partial^2\beta_e/\partial t^2$

$$M_m = -\frac{1}{128}\pi \rho c^4 \frac{\partial^2 \beta_e}{\partial t^2} \tag{3.213}$$

The added masses are determined by a process similar to that of Section 3.13.2.

3.13.7 The effect of the wake on aerofoil aerodynamics in unsteady flow

If the angle of attack of the flow relative to an aerofoil changes, the strength of the circulation also changes, but the process is not instantaneous because the circulation can only develop gradually. To determine how the lift on an aerofoil actually develops with time after an impulsive change of angle of attack occurs it is necessary to include the wake in the analysis. The sudden change of a causes a build up of circulation around the aerofoil that is matched by an equal and opposite vorticity being shed into the wake.

The bound circulation on an aerofoil is actually distributed along the chord but, for simplicity, can be assumed to be a concentrated vortex Γ at the aerodynamic centre 1/4 chord point). In steady flow conditions, the velocity induced by the vortex, normal to the chord-line, at the 3/4 chord point is exactly equal and opposite to the component of the flow velocity normal to the chord-line. The two opposed velocities ensure that no flow passes through the aerofoil at the 3/4 chord point, a condition which, of course, must be true everywhere along the chord-line but the 3/4 chord point is used as a control point. The simplified situation assumes that the aerofoil can be represented geometrically by its chord-line, this known as thin aerofoil representation.

In unsteady flow conditions the velocity induced at the 3/4 chord point (often referred to as downwash) is caused jointly by the bound vortex and the wake vorticity, see Figure 3.78, but must still be equal and opposite to the component of the flow velocity normal to the chord-line.

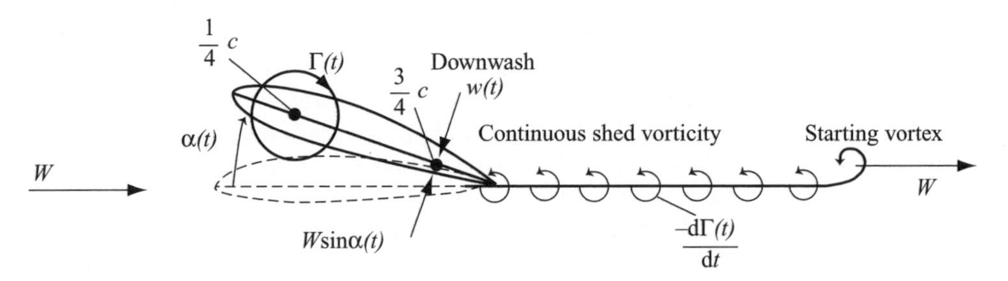

Figure 3.78 Wake Development after an Impulsive Change of Angle of Attack

After the impulsive change of angle of attack there is a sudden change in upwash ($W \sin \alpha$) which must be matched by a sudden change of downwash. The change of upwash implies an impulsive acceleration of the mass of the air that causes an added mass force on the aerofoil. The sudden change of downwash must come from a sudden increase in circulation that must be matched by an equal and opposite starting vortex being shed into the wake and then convected downstream. The influence of the starting vortex on the downwash gets gradually weaker as the vortex moves away so the bound vortex must increase in strength to maintain that the downwash matches the upwash. The increasing strength of the bound vortex means that, to keep the overall angular momentum contained in the vorticity zero (there was none before the impulse), continuous vorticity of the opposite sense must be shed into the wake and this also contributes to the downwash.

The rate of increase of the bound vortex strength gradually reduces with a corresponding reduction of the strength of the shed vorticity and eventually, asymptotically, the steady-state bound circulation strength is developed.

The analytical solution to the problem was developed by Wagner (1925); it is complex and expressed in terms of Bessel functions but several approximations to the Wagner function exist, the most accurate of which is given by Jones (1945).

$$\frac{L_c(\tau)}{\frac{1}{2}\rho W^2 c \frac{\mathrm{d}C_l}{\mathrm{d}\alpha}\sin(\alpha)} = \Phi(\tau) = 1 - 0.165\,\mathrm{e}^{-0.0455\tau} - 0.335\,\mathrm{e}^{-0.30\tau} \tag{3.214}$$

where $\tau = tc/2W$ is the non-dimensional time based upon the half chord length $c/2$ of the aerofoil and $\mathrm{d}C_l/\mathrm{d}\alpha$ is the slope of the static lift *versus* angle of attack characteristic of the aerofoil. τ can also be regarded as the number of half-chord lengths travelled downstream by the starting vortex after a time t has elapsed since the impulsive change of angle of attack. Equation (3.214) is an example of an indicial equation.

Figure 3.79 shows the progression of the growth of the lift as time proceeds from the original impulsive change of angle of attack. The added mass lift gradually dies away as the circulatory lift develops. Eventually, the steady state, full circulatory lift is achieved. In the situation where the angle of attack is continuously changing, which is the case for the wind turbine blade, the circulation never reaches an equilibrium state and the added mass lift never dies away.

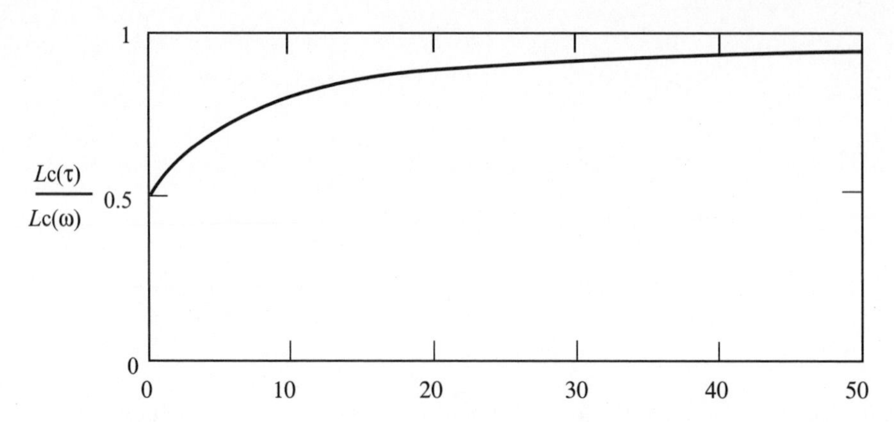

Figure 3.79 Lift Development after an Impulsive Change of Angle of Attack

In an unsteady wind, if for each change of wind velocity over an increment of time δt there is a corresponding impulsive angle attack change then the lift on the aerofoil will subsequently be influenced by that change in the manner of Equation (3.214) and Figure 3.79. The accumulation of all such changes in a continuous manner will determine the unsteady lift force on the aerofoil.

Assume the flow has been in progress for a long time t_0 and let t be any time prior to t_0, the lift at time t is then given by, see Fung (1969),

$$L_c(t) = \frac{1}{2}\rho\frac{dC_l}{d\alpha}c\int_0^{\tau} W(\tau_0)\Phi(\tau-\tau_0)\frac{d}{d\tau_0}w(\tau_0)d\tau_0 \tag{3.215}$$

where $\delta w = (d/d\tau_0)w(\tau_0)\delta\tau_0$ is the change in upwash (downwash) determined by the change in $W(\tau)$ and the changes in blade motion during the time interval. The use of non-dimensional time in the above equation poses a problem and it is more convenient to use actual time in a numerical integration.

Theodorsen (1935) solved Equation (3.215) for the case of an aerofoil oscillating sinusoidally in pitch and heave (flapping motion) at fixed frequency and immersed in a steady oncoming wind U. The unsteady lift on the aerofoil is also sinusoidal but not in phase with the angle of attack variation nor is the amplitude of the lift variation related to the amplitude of the angle of attack by static aerofoil characteristics.

Theodorsen's solution shows that the circulatory lift on the aerofoil equals the quasi-steady lift of Equation (3.209) multiplied by Theodorsen's function $C(k)$ that has both real and imaginary parts. $k = \omega c/2U$ is called the *reduced frequency* and $\omega t = k\tau$. In addition there is the added mass lift given by Equations (3.211) and (3.212).

$$C(k) = \frac{1}{1+A(k)} = \frac{1}{1+\left(\dfrac{Y_0(k)+iJ_0(k)}{J_1(k)-iY_1(k)}\right)} \tag{3.216}$$

where $J_n(k)$ and $Y_n(k)$ are Bessel functions of the first and second kind, respectively and n is an integer. Like the Legendre polynomials the Bessel functions are the solutions to a second-order ordinary differential equation called Bessel's equation

$$k^2 \frac{\mathrm{d}^2}{\mathrm{d}k^2} y + k \frac{\mathrm{d}}{\mathrm{d}k} y + (k^2 - n^2) y = 0$$

Unlike the Legendre polynomials the Bessel functions cannot be expressed in closed form but only as an infinite series.

Theodorsen's function is often divided into two functions, one describing the real part and the other the imaginary part:

$$C(k) = F(k) + iG(k) \tag{3.217}$$

From Jones' approximation to the Wagner function, Equation (3.214), an approximation to Theodorsen's function is obtained

$$C(k) = 1 - \frac{0.165}{1 - \dfrac{0.0455}{k} i} - \frac{0.335}{1 - \dfrac{0.30}{k} i} = F(k) = iG(k) \tag{3.218}$$

The exact and approximate parts of $C(k)$ are shown in Figures 3.80(a) and (b).

The real part of $C(k)$ gives the lift that is in phase with the angle of attack defined in Equation (3.209) and the imaginary part gives the lift that is 90° out of phase with the angle of attack.

The drawback of the Theodorsen function for rotor blade application is that the wake streams away from the blade in a straight line whereas the rotor blade wake is

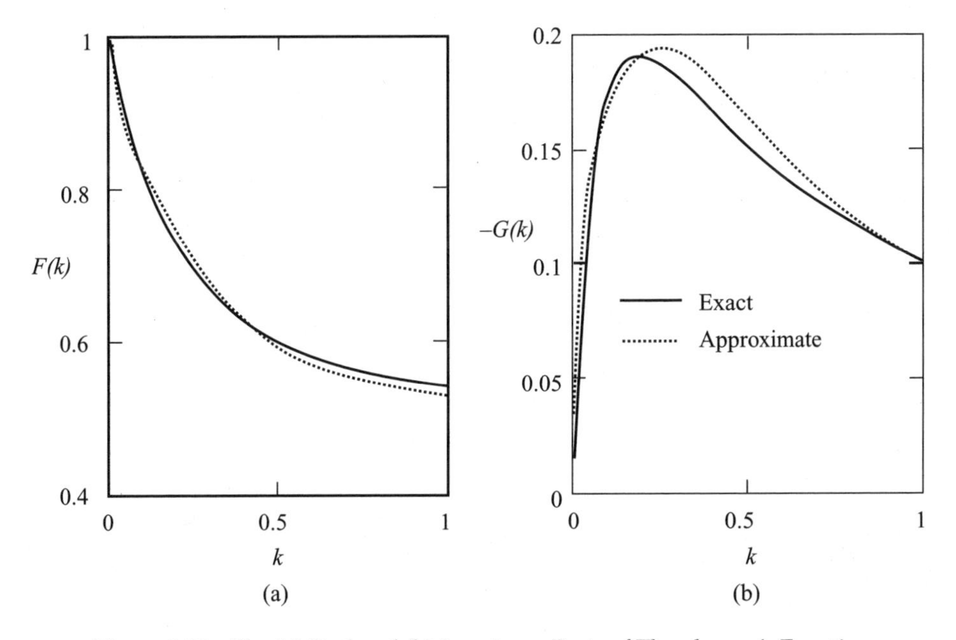

Figure 3.80 The (a) Real and (b) Imaginary Parts of Theodorsen's Function

helical and the wakes of other blades will also be present. Loewy (1957) developed a theory for a rotor blade that accounts for the repeated wake in a similar manner to Prandtl (see Section 3.8.3). As did Theodorsen, Loewy used two-dimensional, thin aerofoil theory and produced a modification to Theodorsen's function. In Equation (3.216) the Bessel function of the first kind, $J_n(k)$, is multiplied by $(1 + W(k))$ where $W(k)$ is called the Loewy wake-spacing function.

$$W(k) = \frac{1}{e^{(2[d/c]k+i2\pi)} - 1} \tag{3.219}$$

d is the wake spacing defined in Equation (3.75) and c is the chord of the aerofoil. Miller (1964) arrived at a very similar result to Loewy by using a discrete vortex wake model.

Loewy's and Miller's theories apply only to the non-yawed rotor but Peters, Boyd and He (1989) have developed a much more extensive theory based upon the method of acceleration potential. A sufficient number of Kinner pressure distributions are required to model both the radial and azimuthal pressure distribution on a helicopter rotor such that the pressure spikes of individual blades are present. The theory obviates the use of blade element theory and includes automatically unsteady effects and tip losses; modelling of the blade geometry by this method does present some problems, however. Suzuki and Hansen (1999) have applied the theory of Peters, Boyd and He to wind turbine rotors and make comparisons with the blade-element/momentum theory. Van Bussel's theory (1995) is very similar to that of Peters, Boyd and He but is intended for application to wind turbines.

References

Betz, A., (1919). *Schraubenpropeller mit geringstem energieverlust*. Gottinger Nachr., Germany.

Coleman, R. P., Feingold, A. M. and Stempin, C. W., (1945). 'Evaluation of the induced velocity field of an idealised helicopter rotor'. *N.A.C.A. A.R.R,. No. L5E10*.

Glauert, H., (1926a). 'The analysis of experimental results in the windmill brake and vortex ring states of an airscrew'. *ARCR R&M No. 1026*.

Glauert, H., (1926b). 'A general theory of the autogyro'. *ARCR R&M No. 1111*.

Glauert, H., (1935a). 'Airplane propellers'. *Aerodynamic theory* (ed. Durand, W. F.). Julius Springer, Berlin, Germany.

Glauert, H., (1935b). 'Windmills and fans'. *Aerodynamic theory* (ed. Durand, W. F.). Julius Springer, Berlin, Germany.

Goankar, G. H. and Peters, D. A., (1988). 'Review of dynamic inflow modelling for rotorcraft flight dynamics'. *Vertica*, **2**, 3, 213–242.

Goldstein, S., (1929). 'On the vortex theory of screw propeller'. *Proc. Roy. Soc. (A)*, **123**, 440.

HaQuang, N. and Peters, D. A., (1988). 'Dynamic inflow for practical applications'. *J. Am. Heli. Soc.*, Technical Note.

Himmelskamp, H., (1945). *Profile investigations on a rotating airscrew*. Ph.D. Thesis, Gottingen University, Germany.

Hoerner, S. F., (1965). 'Pressure drag on rotating bodies'. *Fluid dynamic drag* (ed. Hoerner, S. F.), pp. 3–14.

Jones, W. P., (1945). 'Aerodynamic forces on wings in non-uniform motion', *ARCR R&M 2117.*

Kinner, W., (1937). 'The principle of the potential theory applied to the circular wing'. *Ing. Arch.*, **VIII**, 47–80. (Translated by Flint, M., R.T.P. Translation No. 2345.)

Lindenburg, C., (1996). 'Results of the PHATAS-III development', International Energy Agency, 28th Meeting of Experts, Lyngby, Denmark.

Lock, C. N. H., (1924). 'Experiments to verify the independence of the elements of an airscrew blade'. *ARCR R&M No. 953.*

Loewy, R. G., (1957). 'A two-dimensional approach to the unsteady aerodynamics of rotary wings'. *J. Aerospace Sci.*, **24**, 2.

Mangler, K. W. and Squire, H. B., (1950). 'The induced velocity field of a rotor'. *ARCR R&M No. 2642.*

Meijer Drees, J., (1949). A theory of airflow through rotors and its application to some helicopter problems'. *J. Heli. Ass. G.B.*, **3**, 2, 79–104.

Miller, R. H., (1964). 'Rotor blade harmonic air loading'. *AIAA Journal*, **2**, 7.

Øye, S. (1992). 'Induced velocities for rotors in yaw'. *Proceedings of the Sixth IEA Symposium.* ECN, Petten, Holland.

Peters, D. A., Boyd, D. D. and He, C. J., (1989). 'Finite state induced flow model for rotors in hover and forward flight'. *J. Amer. Heli. Soc.*, **34**, 4, 5–17.

Pitt, D. M. and Peters, D. A., (1981). 'Theoretical prediction of dynamic inflow derivatives'. *Vertica*, **5**, 21–34.

Ronsten, G., (1991). 'Static pressure measurements in a rotating and a non-rotating 2.35 m wind turbine blade'. *Proceedings of the EWEC Conference.*

Schepers, J. G. and Snel, H., (1995). 'Joint investigation of dynamic inflow effects and implementation of an engineering method'. *Report: ECN-C-94-107.* ECN, Petten, Holland.

Snel, H. *et al.*, (1993). 'Sectional prediction of three-dimensional effects for stalled flow on rotating blades and comparison with measurements'. *Proceedings of the EWEC Conference.*

Suzuki, A. and Hansen, A. C., (1999). 'Generalized dynamic wake model for Yawdyn'. *AIAA Wind Symposium, AIAA-99-0041.*

Theodorsen, T., (1935). 'General theory of aerodynamic instability and the mechanism of flutter'. *NACA Report 496.*

Tuckerman, L. B., (1925). 'Inertia factors of ellipsoids for use in airship design', *NACA Report Number 210.*

Van Bussel, G. J. W., (1995). *The aerodynamics of horizontal-axis wind-turbine rotors explored with asymptotic expansion methods*, Ph.D. Thesis, Delft University of Technology, Holland.

Wagner, H., (1925). 'Über die Entstahung des dynamischen Auftriebes von Tragflügel', *Zeischrift für angewandte Mathematik und Mechanik*, **5**, 1.

Wilson, R. E. and Lissaman, P. B. S., (1974). 'Applied aerodynamics of wind-power machines'. *NTIS: PB-238-595*, Oregon State University, USA.

Wood, D. H., (1991). 'A three-dimensional analysis of stall-delay on a horizontal-axis wind turbine'. *J. Wind Eng. Indl. Aerodyn.*, **37**, 1–14.

Young, A. D. and Squire, H. B., (1938). *R&M No. 1838.*

Bibliography

Abbott, I. H. and von Doenhoff, A. E., (1959). *Theory of wing sections.* Dover, New York, USA.

Anderson, J. D., (1991). *Fundamentals of aerodynamics, Second edition.* McGraw-Hill, Singapore.

Barnard, R. H. and Philpot, D. R., (1989). *Aircraft flight - a description of the physical principles of aircraft flight.* Longmans, Singapore.

Duncan, W. J., Thom, A. S. and Young, A. D., (1970). *Mechanics of fluids, Second edition.* Edward Arnold, London, UK.
Eggleston, D. M. and Stoddard, F. S., *Wind Turbine Engineering Design*, Van Nostrand Reinhold Co., New York, USA.
Fung, Y. C., (1969). *An introduction to the theory of aeroelasticity*. Dover, New York, USA.
Katz, J. and Plotkin, A., (1991). *Low-speed aerodynamics - from wing theory to panel methods.* McGraw-Hill, Singapore.
Prandtl, L. and Tietjens, O. G., (1957). *Applied hydro- and aeromechanics*. Dover, New York, USA.
Stepniewski, W. Z. and Keys, C. N., (1984). *Rotary-wing aerodynamics*. Dover, New York, USA.

Appendix: Lift and Drag of Aerofoils

The forces acting on a body immersed in a fluid moving fluid can be resolved into stream-wise (drag) and normal (lift) components. Neglecting buoyancy, the fluid mechanics which give rise to lift and drag are associated with the boundary layer of slow-moving fluid close to the body's surface. The fluid force on the surface of the body can be either parallel to the surface (viscous or skin friction force) or normal to the surface (pressure force).

For a thorough understanding of the phenomena of lift and drag an aerodynamics text should be consulted but for the purposes of wind-turbine aerodynamics the basic results are given below.

A3.1 Definition of Drag

The drag on a body immersed in an oncoming flow is defined as the force on the body in a direction parallel to the flow direction. In a very slow-moving fluid the drag on a body may be directly attributable to the viscous, frictional shear stresses set up in the fluid due to the fact that, at the body wall, there is no relative motion. This type of flow is known as Stokes' flow after Sir George Stokes.

Two centuries before Stokes, Isaac Newton showed that that the shear stress t at a boundary wall, or between two layers of fluid moving relative to one another, is proportional to the transverse velocity gradient at the boundary, or between the two layers:

$$\tau = \mu \frac{\mathrm{d}u}{\mathrm{d}y} \tag{A3.1}$$

where the constant of proportionality is μ the fluid viscosity.

Using Newton's theory, Stokes determined the drag force on a sphere in creeping flow (Figure A3.1).

$$\text{Drag} = 3\pi\mu U d \tag{A3.2}$$

where d is the sphere diameter and U is the general flow velocity.

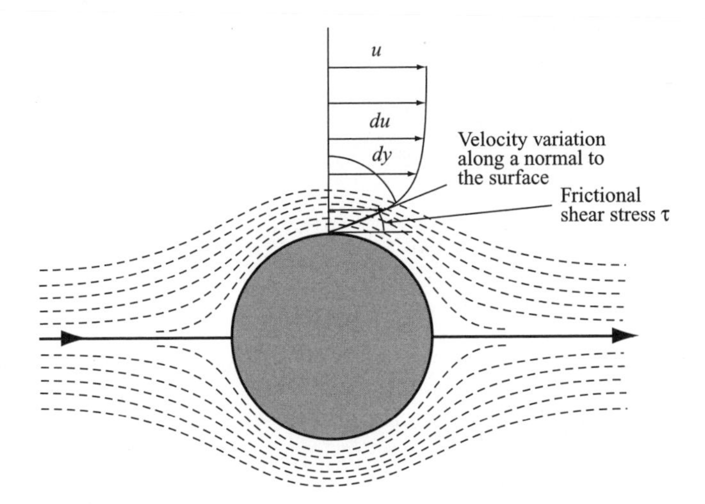

Figure A3.1 Creeping Flow Past a Circular Cylinder

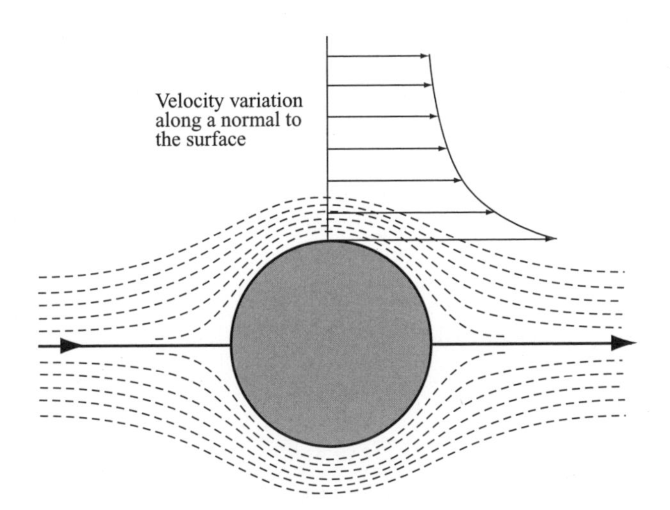

Figure A3.2 Inviscid Flow Pattern Around a Cylinder

The inviscid flow pattern around a cylinder (Figure A3.2) appears very similar to that of creeping flow but the nature of the flow is very different indeed. By definition inviscid flow causes no viscous drag but it also causes no pressure drag, that is, drag caused by pressure forces aggregated over the whole surface area. The pressure distribution for the inviscid flow past a cylinder is shown in Figure A3.3, where the atmospheric pressure p_∞ has been subtracted from the pressure around the surface. The symmetry of the pressure distribution fore and aft shows clearly that no pressure drag arises. At the nose of the body the flow is brought exactly to rest and this is called the stagnation point. Another stagnation point occurs at the rear of the body.

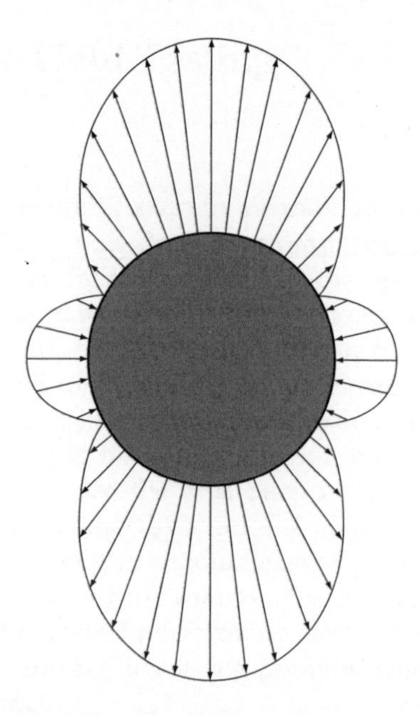

Figure A3.3 Inviscid Flow Pressure Distribution Around a Cylinder

In a real fluid, when the viscosity is low and the velocity is relatively high, the drag force that exists is due primarily to an asymmetric pressure distribution, fore and aft (Figure A3.5). This is caused by the fact that the fluid does not follow the boundary of the body but separates from it leaving low pressure, stagnant fluid in the wake (Figure A3.4). On the upstream side the flow remains attached and the pressure is high.

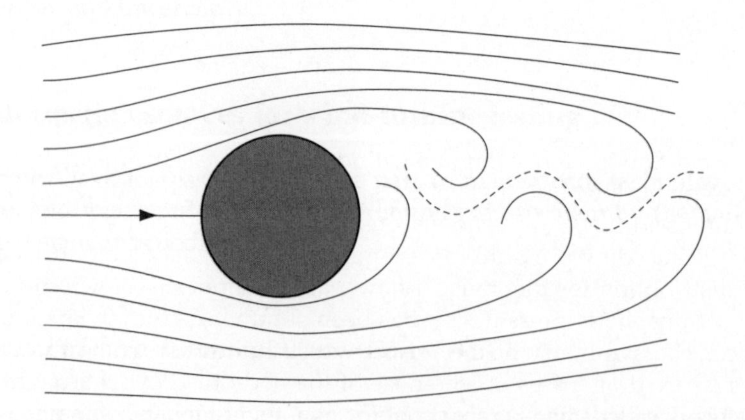

Figure A3.4 Separated Flow Pattern Around a Cylinder

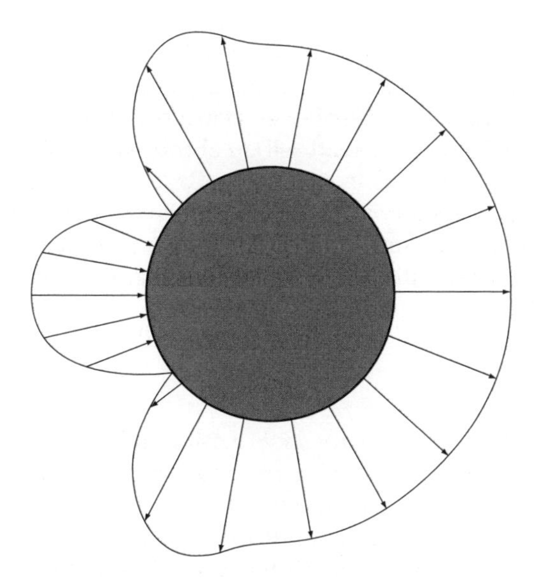

Figure A3.5 Separated Flow Pressure Distribution Around a Cylinder

A3.2 Drag coefficient

If Stokes' drag Equation (A3.2) for the sphere is re-arranged, giving

$$\text{Drag} = 3\pi\mu Ud = \left[24\left(\frac{\mu}{\rho Ud}\right)\right]\left(\frac{1}{2}\rho U^2\right)\left(\frac{\pi d^2}{4}\right) \tag{A3.3}$$

it is then in the standard form of drag coefficient (C_d)× dynamic pressure ($\frac{1}{2}\rho U^2$) × frontal area (A). The drag coefficient is then defined as

$$C_d = \frac{\text{Drag}}{\frac{1}{2}\rho U^2 A} \tag{A3.4}$$

Note that, $\rho Ud/\mu$ is known as the Reynolds number (Re) and represents the ratio of the inertia force acting on a unit volume of fluid, as it is accelerated by a pressure gradient, and the viscous force on the same volume of fluid which is resisting the motion of the fluid. For high Reynolds numbers viscous forces are low and *vice versa*. The drag coefficient term in Equation (A3.3) is $C_d = 24/Re$ and is clearly a function only of the Reynolds number; this turns out to be valid for all bodies in incompressible flow but the functional relationship is not usually as simple as in the above case. However, it can be stated, generally, that C_d falls with increasing Reynolds number.

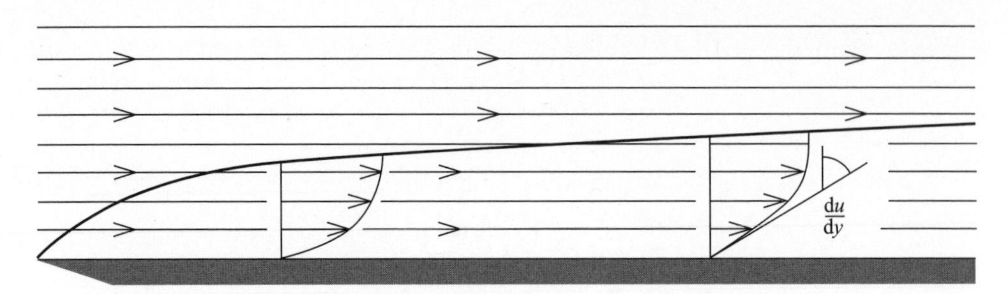

Figure A3.6 Boundary Layer Showing the Velocity Profile

A3.3 The Boundary Layer

The reason for the separated flow at the higher Reynolds numbers is the existence of a thin boundary layer of slow moving fluid, close to the body surface, within which viscous forces predominates. Outside this layer the flow behaves almost inviscidly. The drag on the body caused directly by viscosity is quite small but the effect on the flow pattern is profound.

The drag on an aerofoil can be attributed both to pressure and viscous sources and the drag coefficient varies significantly with both angle of attack and Reynolds number.

A3.4 Boundary-layer Separation

Referring to Figure A3.3, the inviscid flow pressure distribution around a cylinder, fore and aft the pressure is high above and below the pressure is low. The fluid on the downstream side is slowing down against an adverse pressure gradient and, at the wall boundary, it slows down exactly to a standstill at the rear stagnation point.

In the real flow the boundary layer, which has already been slowed down by viscosity, comes to a halt well before the stagnation point is reached and the flow

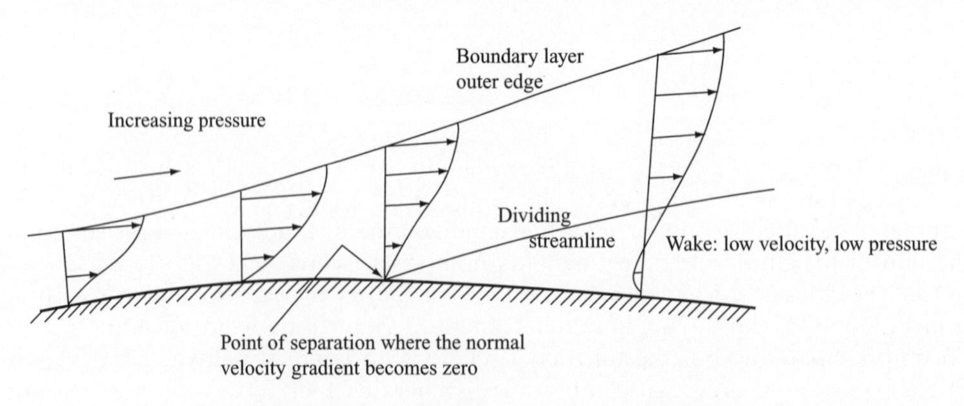

Figure A3.7 Separation of a Boundary Layer

begins to reverse under the action of the adverse pressure. At this point, where the pressure is still low, the boundary layer separates from the body surface forming a wake of stagnant, low-pressure fluid (Figure A3.7), the resulting pressure distribution is thereby dramatically altered as shown in Figure A3.5. The high pressure acting at and around the forward stagnation point is no longer balanced by the high pressure at the rear and so a drag-wise pressure force is exerted.

A3.5 Laminar and Turbulent Boundary Layers

A boundary layer grows in thickness from the forward stagnation point, or leading edge. Initially, the flow in the layer is ordered and smooth (laminar) but, at a critical distance l from the stagnation point, characterized by $Re_{\text{crit}} = \rho Ul/\mu$, the flow begins to become turbulent (Figure A3.8). This turbulence causes mixing of the boundary layer with the faster moving fluid outside resulting in re-energization and delaying of the point of separation. The result is to reduce the pressure drag, because the low-pressure stagnant rear area is reduced, to increase the viscous (frictional) drag, because the velocity gradient at the surface is increased, and increase the boundary-layer thickness.

The coefficient of drag, therefore, varies with *Re* in a complex fashion (Figure A3.9). For small bodies at low speeds the critical *Re* is never reached and so separation takes place early. For large bodies, or high speeds, turbulence develops quickly and separation is delayed.

Turbulence can be artificially triggered by roughening the body surface or simply by using a 'trip wire'. General flow turbulence tends to produce turbulent boundary layers at Reynolds numbers ostensibly below the critical value and this certainly seems to happen in the case of wind-turbine blades. A sharp edge on a body will *always* cause separation. For a flat plate broad side on to the flow (Figure A3.10) the boundary layer separates at the sharp edges and C_d is almost independent of *Re*, but *is* dependent upon the plate's aspect ratio.

So-called streamlined bodies such as an aerfoil taper gently in the aft region so that the adverse pressure gradient is small and separation is delayed until very close to the trailing edge. This produces a very much narrower wake and a very low drag because it is largely caused by skin friction rather than pressure (Figure A3.11).

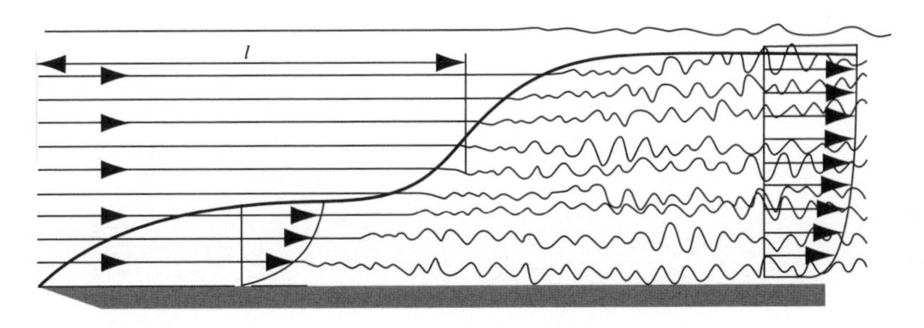

Figure A3.8 Laminar and Turbulent Boundary Layers

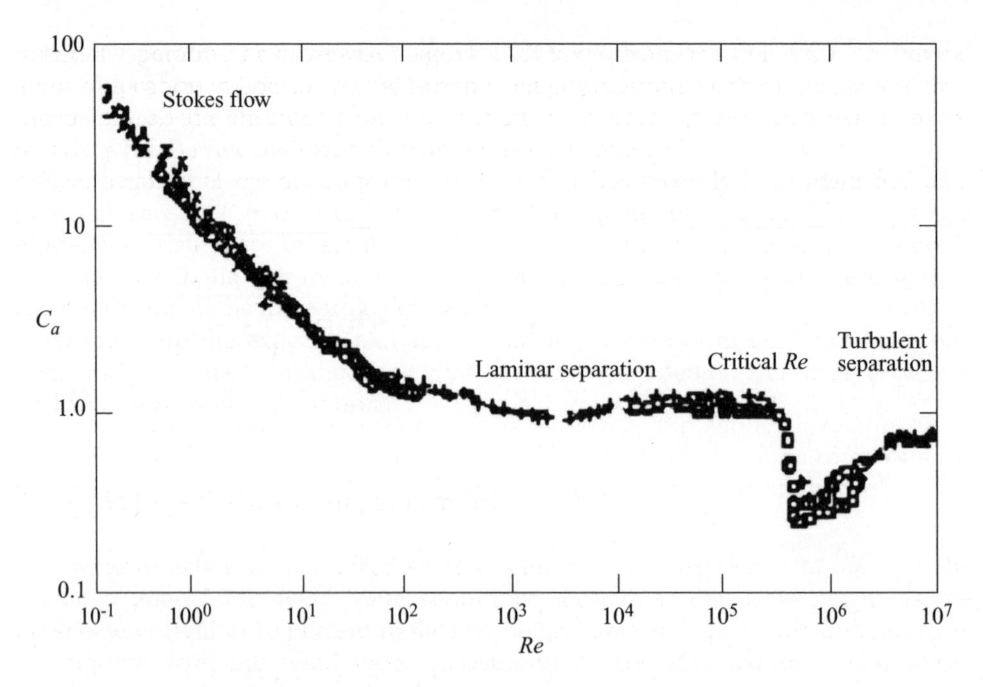

Figure A3.9 Variation of C_d with *Re* for a Long Cylinder

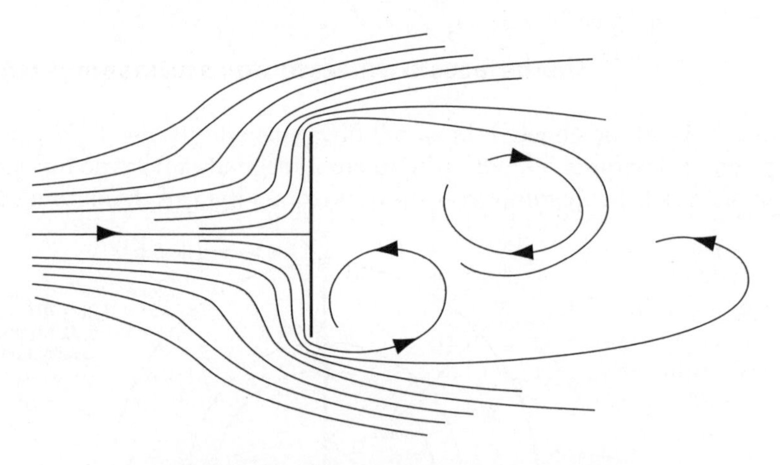

Figure A3.10 Separated Flow Past a Flat Plate

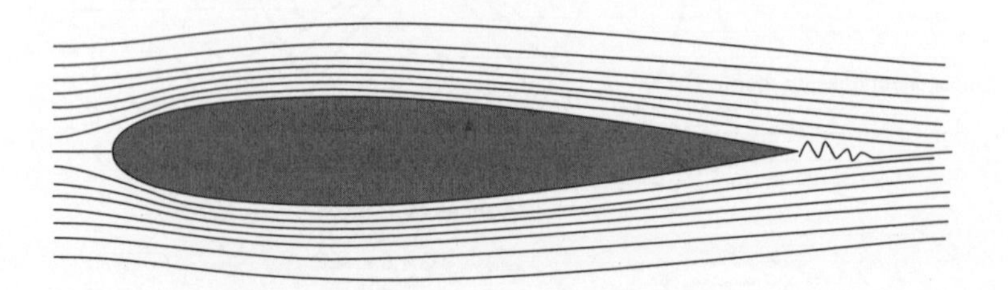

Figure A3.11 Flow Past a Streamlined Body

A3.6 Definition of Lift and its Relationship to Circulation

The lift on a body immersed is defined as the force on the body in a direction normal to the flow direction. Lift will only be present if the flow incorporates a circulatory flow about the body such as that which exists about a spinning circular cylinder. If the fluid also has a uniform velocity U past the cylinder, the resulting flow field is as shown in Figure A3.12. The velocity above the cylinder is increased, and so the static pressure there is reduced. Conversely, the velocity beneath is slowed down, giving an increase in static pressure. There is clearly a normal force upwards on the cylinder, a lift force.

The phenomenon is known as the Magnus effect after its original discoverer and explains, for example, why spinning tennis balls veer in flight. The circulatory flow, shown in Figure A3.13, is generated by skin friction and has the same structure as that of a vortex.

The lift force is given by the Kutta-Joukowski theorem called after the two pioneering aerodynamicists who, independently, realized that this was the key to the understanding of the phenomenon of lift:

$$L = \rho(\Gamma \times U) \tag{A3.5}$$

where Γ is the circulation, or vortex strength, around the cylinder, defined as the integral

$$\Gamma = \int v \, \mathrm{d}s \tag{A3.6}$$

around any path enclosing the cylinder and v is the velocity tangential to the path s. For convenience choosing a circular path of radius r around the cylinder, and

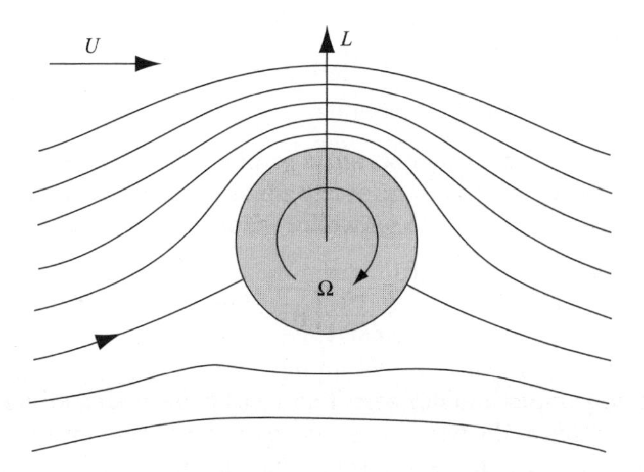

Figure A3.12 Flow Past a Rotating Cylinder

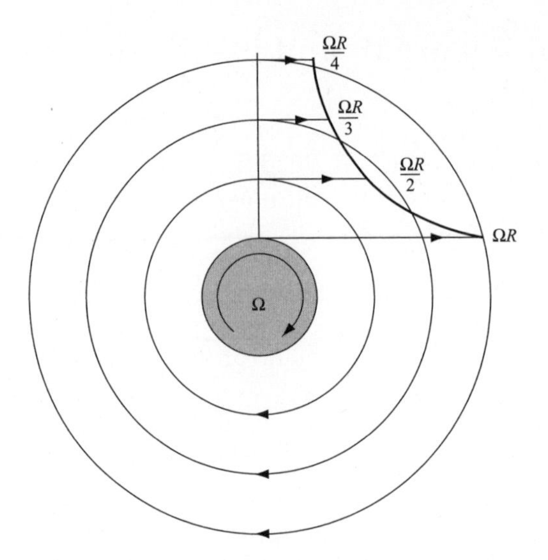

Figure A3.13 Circulatory Flow Round a Rotating Cylinder

ignoring the general flow velocity U, then it can be shown that $v = k/r$, where k is a constant. At the cylinder wall $r = R$, so $v_R = \Omega R = k/R$. Therefore, $k = \Omega R^2$.

The circulation Γ, which is the same for every path enclosing the cylinder, is given by

$$\Gamma = \int \frac{\Omega R^2}{r} \, \mathrm{d}s = \int_0^{2\pi} \Omega R^2 \, \mathrm{d}\psi = 2\pi\Omega R^2 \tag{A3.7}$$

To achieve a circulatory flow about a non-rotating body it must have a sharp trailing edge like an aerofoil cross-sectional shape or a thin plate. An aerofoil works in a similar manner to the spinning cylinder and does so because of its sharp trailing edge. Consider an aerofoil at a small angle of attack α to the oncoming flow. The inviscid flow pattern around the aerofoil, in which no boundary layer forms, is as shown in Figure A3.14(a). The theoretical inviscid flow condition is such that no force on the aerofoil exists at all.

In real flow (Figure A3.14(c)) boundary-layer separation occurs at the sharp trailing edge, causing the flow to leave the edge smoothly. The separation leaves no low-pressure wake so the flow remains attached everywhere else and the flow pattern is now altered such that there is a net circulation around the aerofoil (Figure A3.14(b)) increasing the velocity over the top and reducing it below, resulting in a lift force. There can be no flow around the sharp trailing edge because this would require very high local velocities that are precluded by the boundary layer. The drag is very low because, in the absence of a wake, it is attributable largely to skin friction caused by the shearing stresses in the boundary layer. The situation, which is imposed by the sharp trailing edge, is known as the Kutta condition.

In a manner similar to the Magnus effect, a pressure difference occurs across the aerofoil and the overall circulation Γ can be shown to be $\pi U c \sin\alpha$, where c is the chord length of the aerofoil and α is termed the *angle of attack*. Although the

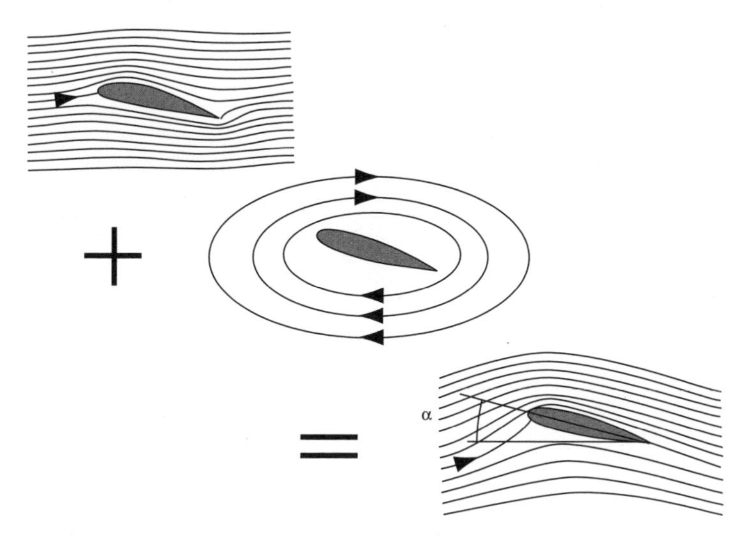

Figure A3.14 Flow Past an Aerofoil at a Small Angle of Attack

velocities and pressures above and below the aerofoil at the trailing edge must be the same, the particles which meet there are not the same ones that parted company at the leading edge; the particle which travelled above the aerofoil reaches the trailing edge first because it is speeded up by the circulatory flow.

The pressure variation (minus the ambient static pressure of the undisturbed flow) around an aerofoil is shown in Figure A3.15. The upper surface is subject to suction (with the ambient pressure subtracted) and is responsible for most of the lift force. The pressure distribution is calculated without the presence of the boundary layer.

Figure A3.16 shows the same distribution with the pressure coefficient ($C_p = p - p_\infty/[1/2]\rho U^2$) plotted against the chord-wise co-ordinate of the aerofoil profile: the full line shows the pressure distribution if the effects of the boundary layer are ignored, and the dashed line shows the actual distribution.

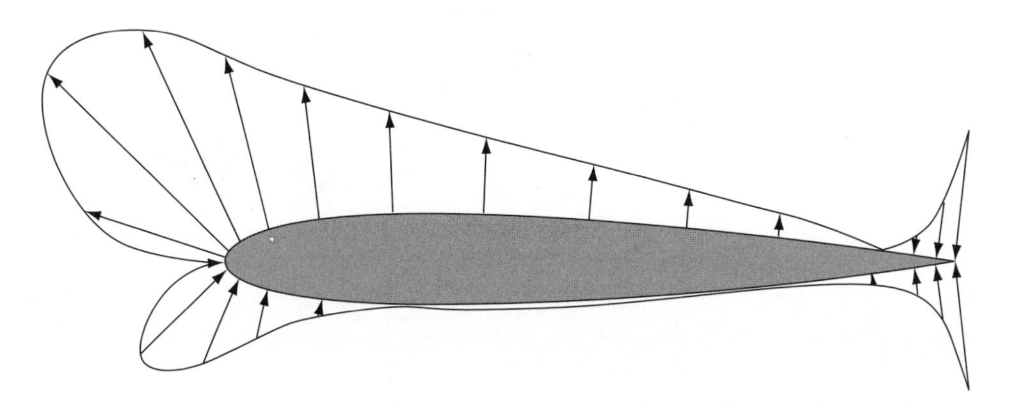

Figure A3.15 The Pressure Distribution Around the NACA0012 Aerofoil at $\alpha = 5°$

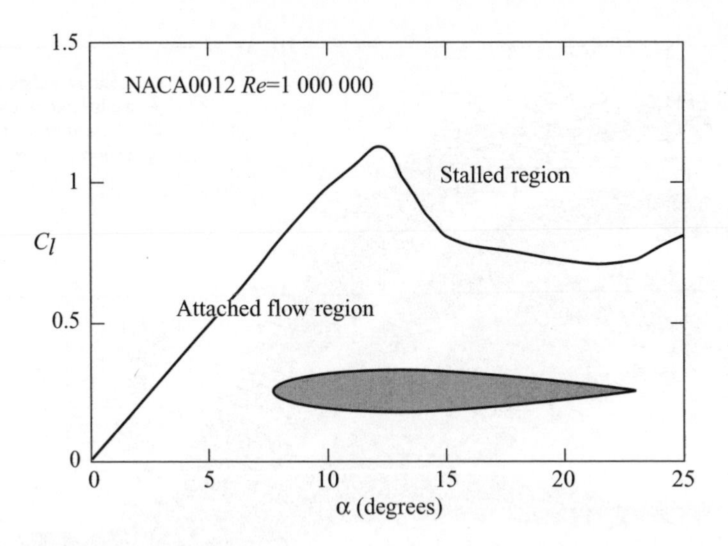

Figure A3.16 The Pressure Distribution Around the NACA0012 Aerofoil at $\alpha = 5°$

The effect of the boundary layer is to modify the pressure distribution at the rear of the aerofoil such that lower pressure occurs there than if the boundary is ignored. The modified pressure distribution gives rise to pressure drag which is added to the skin friction drag, also caused by the boundary layer.

A3.7 The Stalled Aerofoil

If the angle of attack exceeds a certain critical value (10° to 16°, depending on the *Re*), separation of the boundary layer on the upper surface takes place. This causes a wake to form from above the aerofoil, reduces the circulation, reduces the lift and increases the drag. The flow past the aerofoil has then stalled (Figure A3.17). A flat plate will also develop circulation and lift but will stall at a very low angle of attack because of the sharp leading edge. Arching the plate will improve the stalling

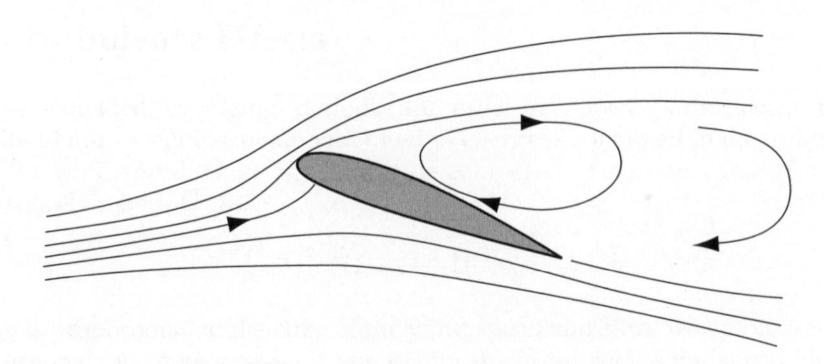

Figure A3.17 Stalled Flow Around an Aerofoil

behaviour but a much greater improvement can be obtained by giving thickness to the aerofoil together with a well-rounded leading edge.

A3.8 The Lift Coefficient

The lift coefficient is defined as

$$C_l = \frac{\text{Lift}}{\frac{1}{2}\rho U^2 A} \tag{A3.8}$$

U is the flow speed and A is the plan area of the body. For a long body, such as an aircraft wing or a wind turbine blade, the lift per unit span is used in the definition and the plan area is replaced by the chord length.

$$C_l = \frac{\text{Lift/unit span}}{\frac{1}{2}\rho U^2 c} = \frac{\rho(\Gamma \times U)}{\frac{1}{2}\rho U^2 c} = \frac{\rho\pi U c \sin\alpha U}{\frac{1}{2}\rho U^2 c} = 2\pi \sin\alpha \tag{A3.9}$$

In practice,

$$C_l = a_0 \sin\alpha \tag{A3.10}$$

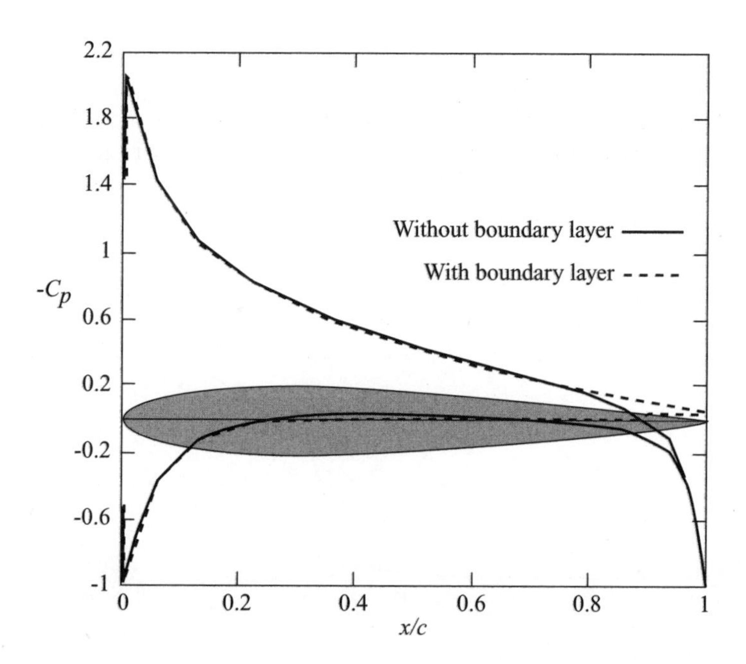

Figure A3.18 $C_l - \alpha$ Curves for a Symmetrical Aerofoil

where a_0, called the lift-curve slope $dC_l/d\alpha$, is about 5.73 (0.1/degree), rather than 2π. Note that a_0 should not be confused with the flow induction factor.

Lift, therefore, depends on two parameters, the angle of attack α and the flow speed U. The same lift force can be generated by different combinations of α and U. The variation of C_l with the angle of attack α is shown in Figure A3.15 for a typical *symmetrical* aerofoil (NACA0012). Notice that the simple relationship of Equation (A3.6) is only valid for the pre-stall region, where the flow is attached, and because the angle of attack is small ($< 16°$) the equation is often written as

$$C_l = a_0\alpha \tag{A3.10a}$$

A3.9 Aerofoil Drag Characteristics

The definition of the drag coefficient for an aircraft wing or a wind-turbine blade is based not on the frontal area but on the plan area, for reasons that will become clear later. The flow past a body which has a large span normal to the flow direction is basically two-dimensional and in such cases the drag coefficient can be based upon the drag force per unit span using the stream-wise chord length for the definition:

$$C_d = \frac{\text{Drag/unit span}}{\frac{1}{2}\rho U^2 c} \tag{3.11}$$

For a wing of large span the value of C_d is roughly 0.01, at moderate Reynolds numbers.

The drag coefficient of an aerofoil also varies with angle of attack. Figure A3.16 shows that on the upper surface pressure is rising as the flow moves towards the trailing edge, this is called an adverse pressure gradient and seeks to slow the air down. If the air is slowed to a standstill stall will occur and the pressure drag will rise

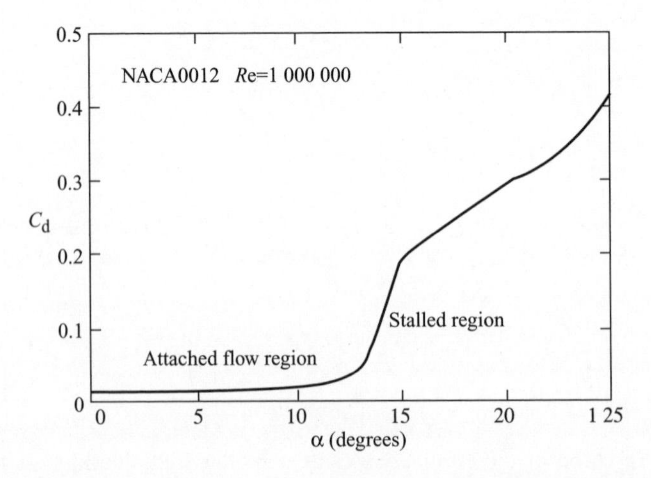

Figure A3.19 Variation of C_d with α for the NACA0012 Aerofoil

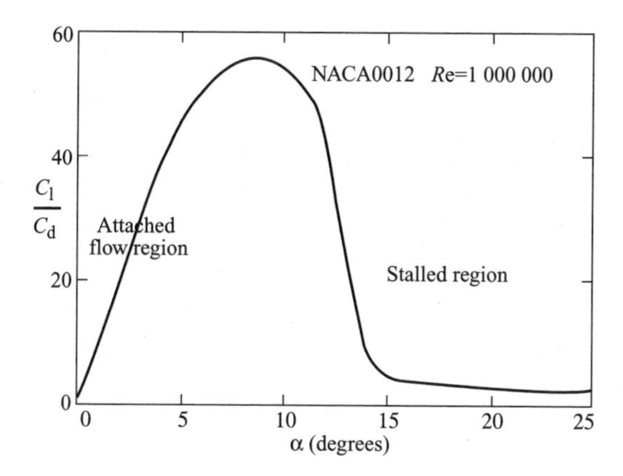

Figure A3.20 Lift/Drag Ratio Variation for the NACA0012 Aerofoil

sharply. The strength of the adverse pressure gradient increases with angle of attack and so it can be expected that the drag will rise with angle of attack. Figure A3.19 shows the variation of C_d with α also for the symmetrical NACA0012 aerofoil.

The lift/drag ratio (shown in Figure A3.20) has a significant affect upon the efficiency of a wind turbine and it is desirable that a turbine blade operates at the maximum ratio.

A3.10 Variation of Aerofoil Characteristics with Reynolds Number

The nature of the flow pattern around an aerofoil is determined by the Reynolds number and this significantly affects the values of the lift and drag coefficients. The

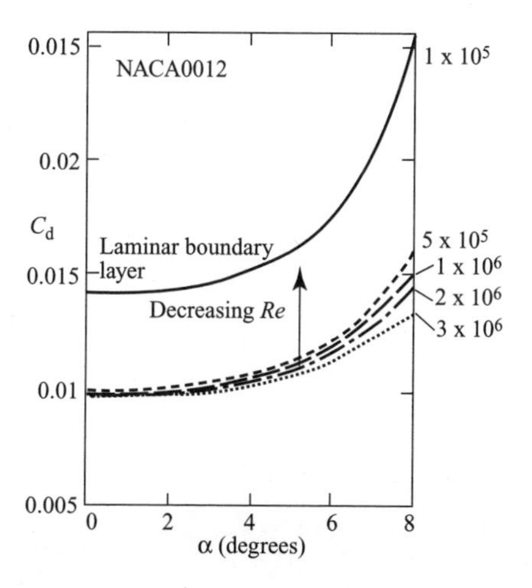

Figure A3.21 Variation of the Drag Coefficient with Reynolds Number at Low Angles of Attack

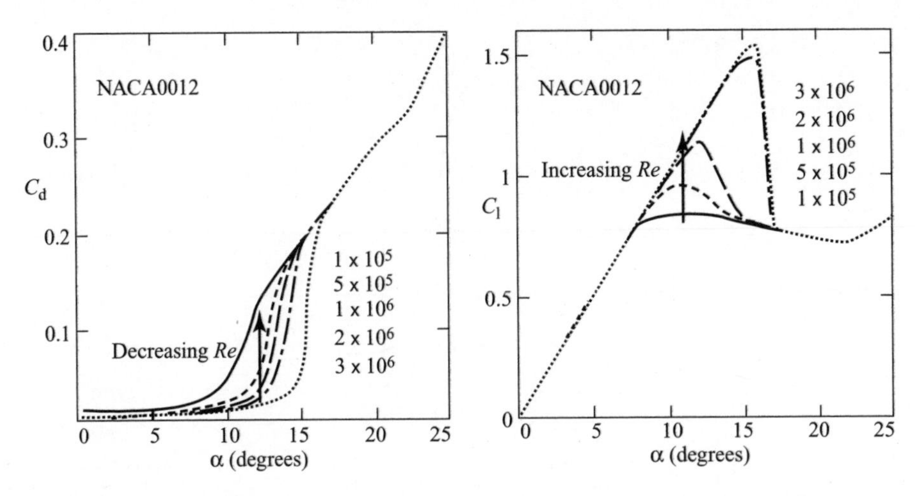

Figure A3.22 Variation of the Drag and Lift Coefficients with Reynolds Number in the Stall Region

general level of the drag coefficient increases with decreasing Reynolds number and below a critical Reynolds number of about 200 000 the boundary layer remains laminar causing a sharp rise in the coefficient. The affect on the lift coefficient is largely concerned with the angle of attack at which stall occurs. As the Reynolds number rises so does the stall angle and, because the lift coefficient increases linearly with angle of attack below the stall, the maximum value of the lift coefficient also rises. Characteristics for the NACA0012 aerofoil are shown in Figures A3.21 and A3.22.

A3.11 Cambered Aerofoils

Cambered aerofoils, such as the NACA4412, shown in Figure A3.23, have curved chord lines and this allows them to produce lift at zero angle of attack. Generally, cambered aerofoils have higher maximum lift/drag ratios than symmetrical aerofoils for positive angles of attack and this is the reason for their use.

The classification of the NACA four-digit range of aerofoils, which were commonly used on wind turbines, is very simple and is illustrated in Figure A3.24: from left to right, the first digit represents the amount of camber as a percentage of the chord length, the second digit represents the percentage chord position, in units of 10 percent, at which the maximum camber occurs and the last two digits are the

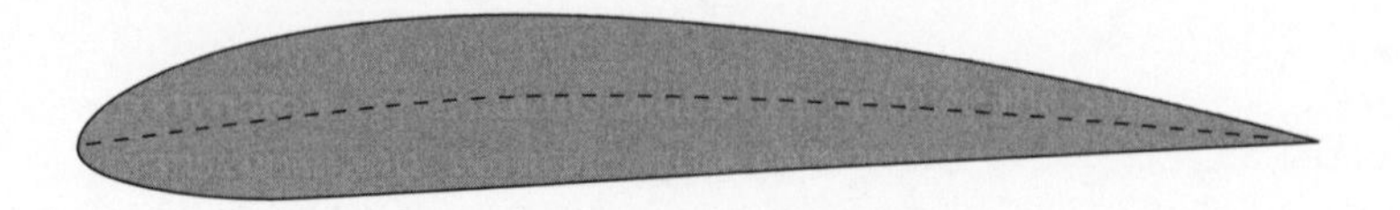

Figure A3.23 The Profile of the NACA4412 Aerofoil

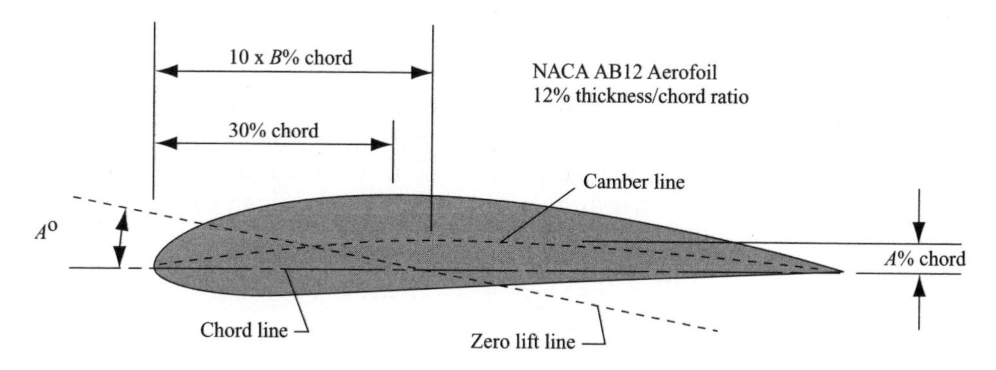

Figure A3.24 Classification of the NACAXXXX Aerofoil Range

maximum thickness to chord ratio, as a percentage of the chord length, which, in this family of aerofoils, is at the 30 percent chord position. The cambered chordline, now called the camber line, comprises two parabolic arcs that join smoothly at the point of maximum camber. For other ranges of aerofoils the reader should refer to Abbott and von Doenhoff (1959).

The angle of attack α is measured form the chord line which is now defined as the straight line joining the ends of the camber line. Note that the lift at zero angle of attack is no longer zero; zero lift occurs at a small negative angle of attack. With most cambered aerofoils the zero lift line is approximately at $-A^{\circ}$ where A is the percentage camber. The behaviour of the NACA4412 aerofoil is shown in Figure A3.25 for angles of attack below and just above the stall. Note that the lift at zero

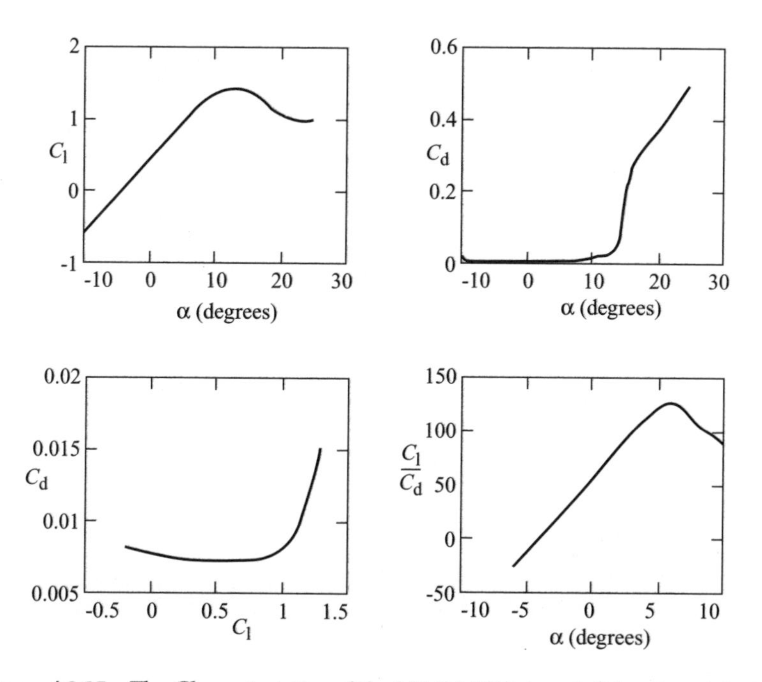

Figure A3.25 The Characteristics of the NACA4412 Aerofoil for $Re = 1.5 \cdot 10^6$

angle of attack is no longer zero; zero lift occurs at a small negative angle of attack of approximately 4°.

The centre of pressure, which is at the $\frac{1}{4}$ chord position on symmetrical aerofoils lies aft of the $\frac{1}{4}$ chord position on cambered aerofoils and moves towards the trailing edge with increasing angle of attack. However, if a fixed chordwise position is chosen then the resultant force through that point is accompanied by a pitching moment (nose-up positive, by convention). If a pitching moment coefficient is defined as

$$C_m = \frac{\text{Pitching/unit span}}{\frac{1}{2}\rho U^2 c} \tag{3.12}$$

then there will be a position, called the *aerodynamic centre*, for which $dC_m/dC_l = 0$. Theoretically, the aerodynamic centre lies at the $\frac{1}{4}$ chord position and is close to this point for most practical aerofoils. The value of C_m depends upon the degree of camber but for the NACA4412 the value is -0.1, note that pitching moments are always negative in practice (nose down) despite the sign convention. Above the stall there is no aerodynamic centre, as defined, and so the pre-stall position continues to be used to determine the pitching moment coefficient, which then becomes dependent upon α.

4
Wind-turbine Performance

4.1 The Performance Curves

The performance of a wind turbine can be characterized by the manner in which the three main indicators—power, torque and thrust—vary with wind speed. The power determines the amount of energy captured by the rotor, the torque developed determines the size of the gear box and must be matched by whatever generator is being driven by the rotor. The rotor thrust has great influence on the structural design of the tower. It is usually convenient to express the performance by means of non-dimensional, characteristic performance curves from which the actual performance can be determined regardless of how the turbine is operated, e.g., at constant rotational speed or some regime of variable rotor speed. Assuming that the aerodynamic performance of the rotor blades does not deteriorate the non-dimensional aerodynamic performance of the rotor will depend upon the tip speed ratio and, if appropriate, the pitch setting of the blades. It is usual, therefore, to display the power, torque and thrust coefficients as functions of tip speed ratio.

4.1.1 The $C_P - \lambda$ performance curve

The theory described in Chapter 3 gives the wind turbine designer a means of examining how the power developed by a turbine is governed by the various design parameters. The usual method of presenting power performance is the non-dimensional $C_P - \lambda$ curve and the curve for a typical, modern, three-blade turbine is shown in Figure 4.1.

The first point to notice is that the maximum value of C_P is only 0.47, achieved at a tip speed ratio of 7, which is much less than the Betz limit. The discrepancy is caused, in this case, by drag and tip losses but the stall also reduces the C_P at low values of the tip speed ratio (Figure 4.2).

Even with no losses included in the analysis the Betz limit is not reached because the blade design is not perfect.

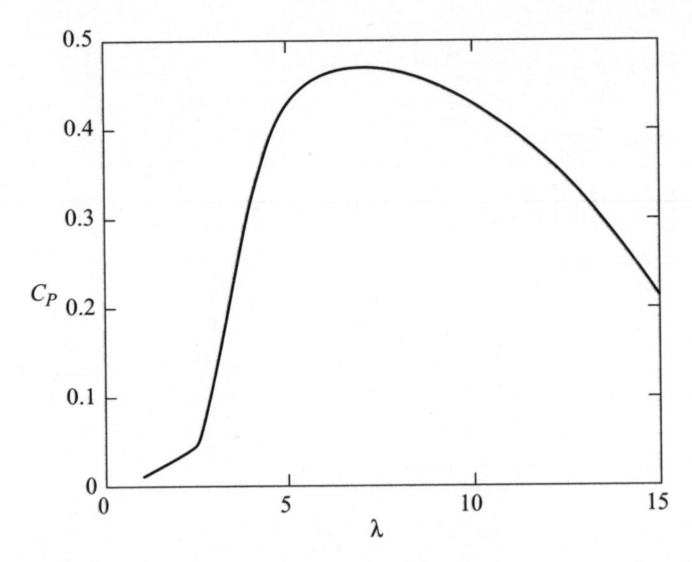

Figure 4.1 $C_P - \lambda$ Performance Curve for a Modern Three-blade Turbine

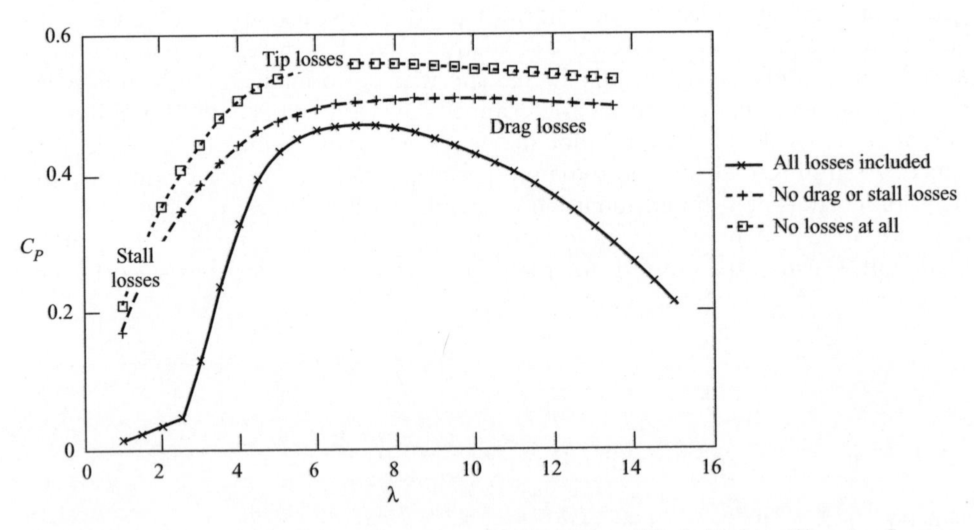

Figure 4.2 $C_P - \lambda$ Performance Curve for a Modern Three-blade Turbine Showing Losses

4.1.2 The effect of solidity on performance

At this stage, the other principal parameter to consider is the solidity, defined as total blade area divided by the swept area. For the three-blade machine above the solidity is 0.0345 but this can be altered readily by changing its number of blades. The solidity could also have been changed by changing the blade chord.

The main effects to observe of changing solidity are as follows, see Figure 4.3.

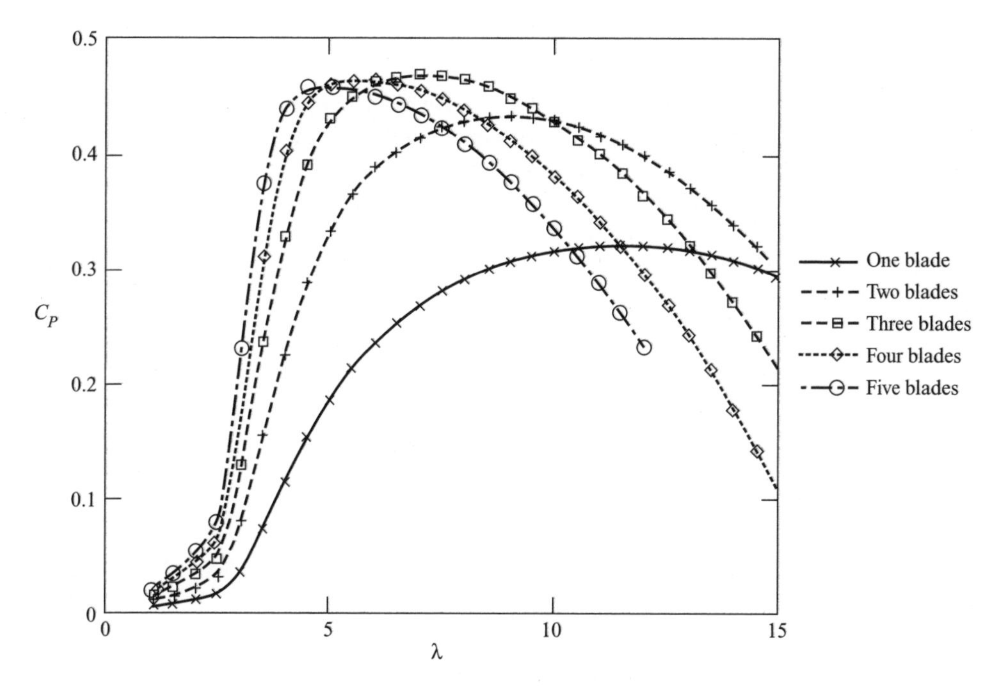

Figure 4.3 Effect of Changing Solidity

(1) Low solidity produces a broad, flat curve which means that the C_P will change very little over a wide tip speed ratio range but the maximum C_P is low because the drag losses are high (drag losses are roughly proportional to the cube of the tip speed ratio).

(2) High solidity produces a narrow performance curve with a sharp peak making the turbine very sensitive to tip speed ratio changes and, if the solidity is too high, has a relatively low maximum C_P. The reduction in $C_{P_{\max}}$ is caused by stall losses.

(3) An optimum solidity appears to be achieved with three blades, but two blades might be an acceptable alternative because although the maximum C_P is a little lower the spread of the peak is wider and that might result in a larger energy capture.

It might be argued that a good solution would be to have a large number of blades of small individual solidity but this greatly increases production costs and results in blades which are structurally weak and very flexible.

There are applications which require turbines of relatively high solidity, one is the directly driven water pump and the other is the very small turbine used for battery charging. In both cases it is the high starting torque (high torque at very low tip speed ratios) which is of importance and this also allows small amounts of power to be developed at very low wind speeds, ideal for trickle charging batteries.

4.1.3 The $C_Q - \lambda$ curve

The torque coefficient is derived from the power coefficient simply by dividing by the tip speed ratio and so it does not give any additional information about the turbine's performance. The principal use of the C_Q–λ curve is for torque assessment purposes when the rotor is connected to a gear box and generator.

Figure 4.4 shows how the torque developed by a turbine rises with increasing solidity. For modern high-speed turbines designed for electricity generation as low a torque as possible is desirable in order to reduce gearbox costs. On the other hand the multi-bladed, high-solidity turbine, developed in the nineteenth century for water pumping, rotates slowly and has a very high starting torque coefficient necessary for overcoming the torque required to start a positive displacement pump.

The peak of the torque curve occurs at a lower tip speed ratio than the peak of the power curve. For the highest solidity shown in Figure 4.4 the peak of the curve occurs while the blade is stalled.

4.1.4 The $C_T - \lambda$ curve

The thrust force on the rotor is directly applied to the tower on which the rotor is supported and so considerably influences the structural design of the tower.

Generally, the thrust on the rotor increases with increasing solidity (Figure 4.5).

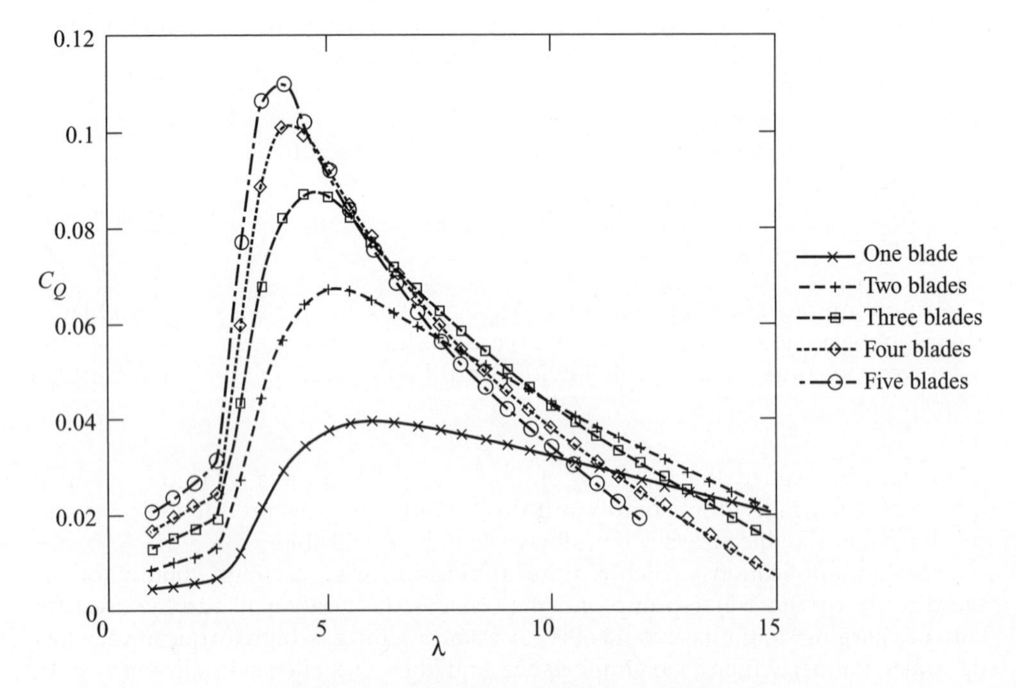

Figure 4.4 The Effect of Solidity on Torque

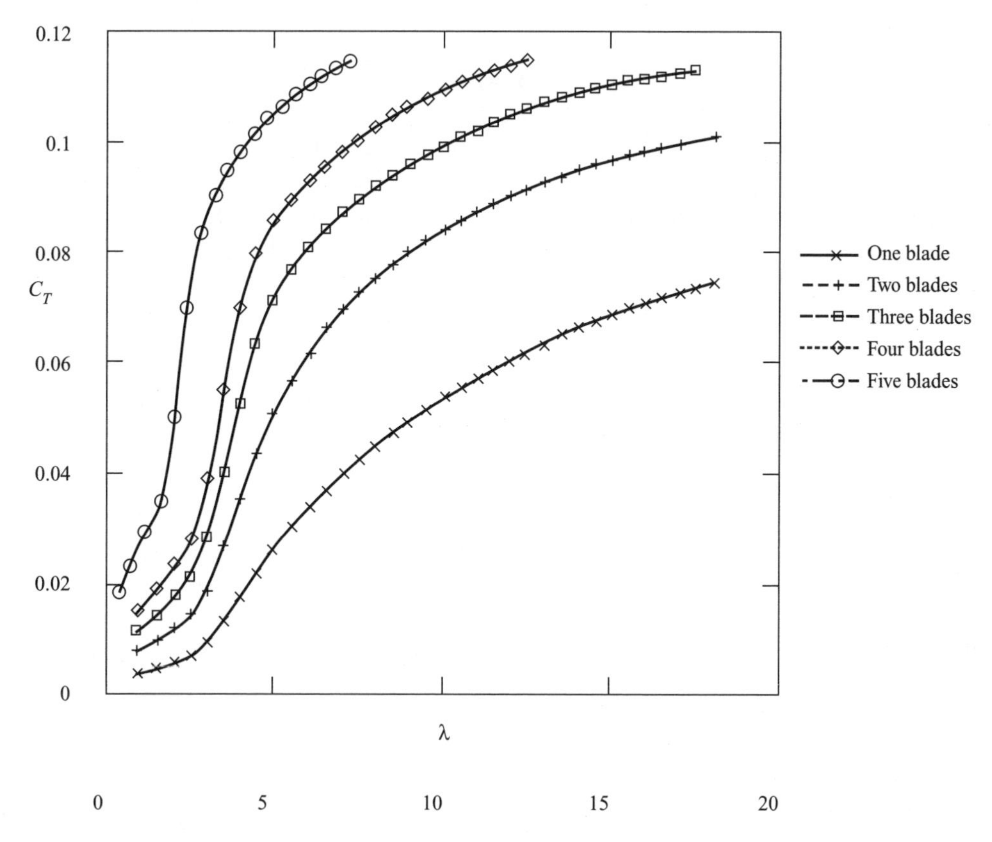

Figure 4.5 The Effect of Solidity on Thrust

4.2 Constant Rotational Speed Operation

The majority of wind turbines currently installed generate electricity. Whether or not these turbines are grid connected they need to produce an electricity supply which is of constant frequency or else many common appliances will not function properly. Consequently, the most common mode of operation for a wind turbine is constant rotational speed. Connected to the grid a constant speed turbine is automatically controlled whereas a stand-alone machine needs to have speed control and a means of dumping excess power.

4.2.1 The $K_P - 1/\lambda$ curve

An alternative performance curve can be produced for a turbine controlled at constant speed. The $C_P - \lambda$ curve shows, non-dimensionally, how the power would vary with rotational speed if the wind speed was held constant. The $K_P - 1/\lambda$ curve describes, again non-dimensionally, how the power would change with wind speed when constant rotational speed is enforced. K_P is defined as

$$K_P = \frac{\text{Power}}{\frac{1}{2}\rho(\Omega R)^3 A_d} = \frac{C_P}{\lambda^3} \tag{4.1}$$

The $C_P - \lambda$ and $K_P - 1/\lambda$ curves for a typical fixed-pitch wind turbine are shown in Figure 4.6. The $K_P - 1/\lambda$ curve, as stated above, has the same form as the power – wind speed characteristic of the turbine. The efficiency of the turbine (given by the $C_P - \lambda$ curve) varies greatly with wind speed, a disadvantage of constant speed operation, but it should be designed such that the maximum efficiencies are achieved at wind speeds where there is the most energy available.

4.2.2 Stall regulation

An important feature of this $K_P - 1/\lambda$ curve is that the power, initially, falls off once stall has occurred and then gradually increases with wind speed. This feature provides an element of passive power output regulation, ensuring that the generator is not overloaded as the wind speed increases. Ideally, the power should rise with wind speed to the maximum value and then remain constant regardless of the increase in wind speed; this is called perfect stall regulation. However, stall regulated turbines do not exhibit the ideal, passive stall behaviour.

Stall regulation provides the simplest means of controlling the maximum power generated by a turbine to suit the sizes of the installed generator and gearbox and until recently, at the time of writing, is the most commonly adopted control method. The principal advantage of stall control is simplicity but there are significant disadvantages. The power *versus* wind speed curve is fixed by the aerodynamic characteristics of the blades, in particular the stalling behaviour. The post stall power output of a turbine varies very unsteadily and in a manner which, so far, defies prediction, see Figure 4.13, for example. The stalled blade also exhibits low vibration damping because the flow about the blade is unattached to the low pressure surface and blade vibration velocity has little effect on the aerodynamic forces. The low damping can give rise to large vibration displacement amplitudes

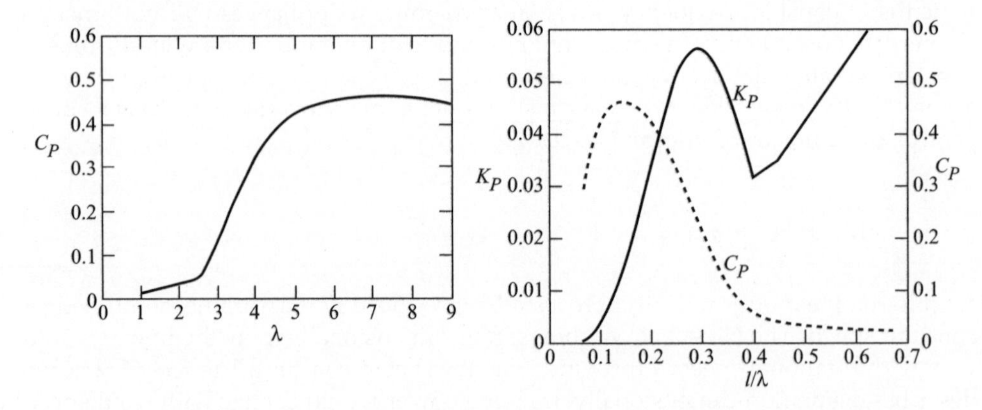

Figure 4.6 Non-dimensional Performance Curves for Constant Speed Operation

which will inevitably be accompanied by large bending moments and stresses, causing fatigue damage. When parked in high, turbulent winds a rotor with fixed pitch blades may well be subject to large aerodynamic loads which cannot be alleviated by adjusting (feathering) the blade pitch angle. Consequently, a fixed-pitch, stall-regulated turbine experiences more severe blade and tower loads than a pitch regulated turbine.

4.2.3 Effect of rotational speed change

The power output of a turbine running at constant speed is strongly governed by the chosen, operational rotational speed. If a low rotation speed is used the power reaches a maximum at a low wind speed and consequently it is very low. To extract energy at wind speeds higher than the stall peak the turbine must operate in a stalled condition and so is very inefficient. Conversely, a turbine operating at a high speed will extract a great deal of power at high wind speeds but at moderate wind speeds it will be operating inefficiently because of the high drag losses. Figure 4.7 demonstrates the sensitivity to rotation speed of the power output – a 33 percent increase in r.p.m. from 45 to 60 results in a 150 percent increase in peak power, reflecting the increased wind speed at which peak power occurs at 60 r.p.m.

At low wind speeds, on the other hand there is a marked fall in power with increasing rotational speed as shown in Figure 4.8. In fact, the higher power available at low wind speeds if a lower rotational speed is adopted has led to two-speed turbines being built. Operating at one fixed speed which maximizes energy capture at wind speeds at, or above, the average level will result in a rather high cut-in wind speed, the lowest wind speed at which generation is possible. Employ-

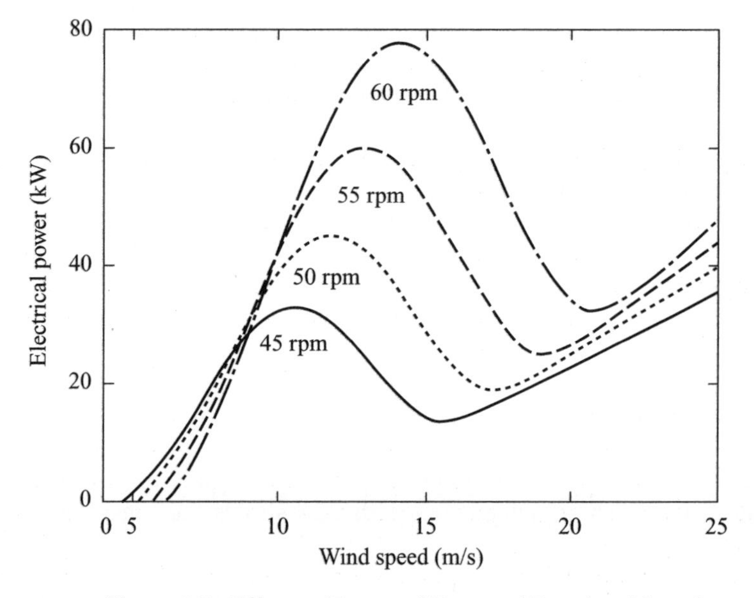

Figure 4.7 Effect on Extracted Power of Rotational Speed

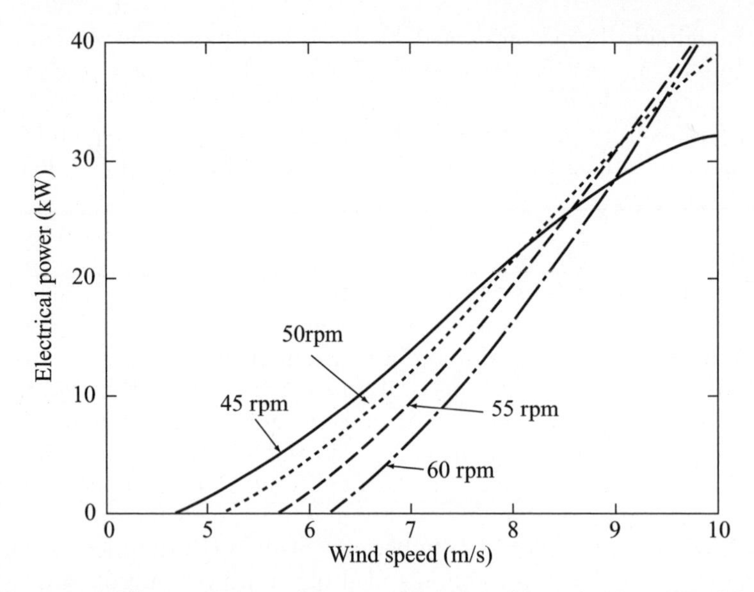

Figure 4.8 Effect on Extracted Power of Rotational Speed at Low Wind Speeds

ing a lower rotational speed at low wind speeds reduces the cut-in wind speed and increases energy capture. The increased energy capture is, of course, offset by the cost of the extra machinery.

4.2.4 Effect of blade pitch angle change

Another parameter which affects the power output is the pitch setting angle of the blades β_s. Blade designs almost always involve twist but the blade can be set at the root with an overall pitch angle. The effects of a few degrees of pitch are shown in Figure 4.9.

Small changes in pitch setting angle can have a dramatic effect on the power output. Positive pitch angle settings increase the design pitch angle and so decrease the angle of incidence. Conversely, negative pitch angle settings increase the angle of incidence and may cause stalling to occur as shown in Figure 4.9. A turbine rotor designed to operate optimally at a given set of wind conditions can be suited to other conditions by appropriate adjustments of blade pitch angle and rotational speed.

4.2.5 Pitch regulation

Many of the shortcomings of fixed pitch/passive stall regulation can be overcome by providing active pitch angle control. Figure 4.9 shows the sensitivity of power output to pitch angle changes.

The most important application of pitch control is for power regulation but pitch control has other advantages. By adopting a large positive pitch angle a large

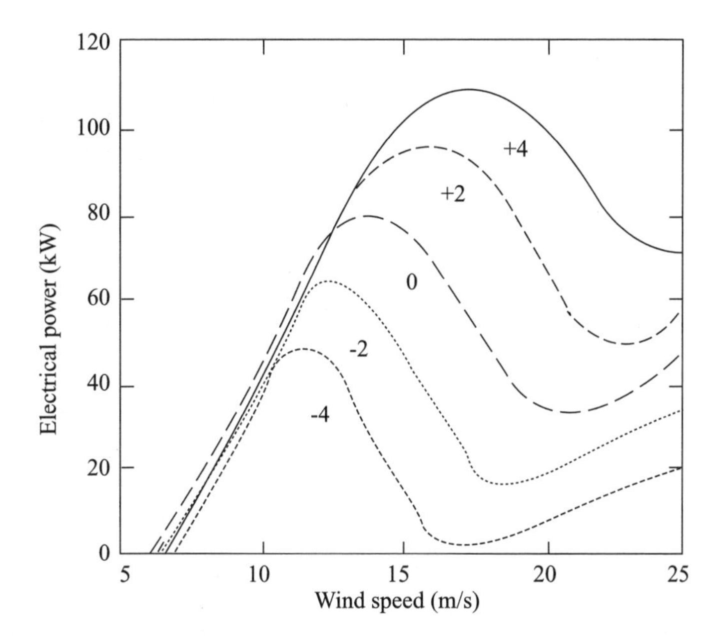

Figure 4.9 Effect on Extracted Power of Blade Pitch Set Angle

starting torque can be generated as a rotor begins to turn. A 90° pitch angle is usually used when shutting down because this minimises the rotor idling speed at which the parking brake is applied. At 90° of positive pitch the blade is said to be 'feathered'. The principal disadvantages of pitch control are reliability and cost. Power regulation can be achieved either by pitching to promote stalling or pitching to feather which reduces the lift force on the blades by reducing the angle of attack.

4.2.6 Pitching to stall

Figure 4.9 shows the power curves for a turbine rated at 60 kW, which is achieved at 12 m/s. At wind speeds below the rated level the blade pitch angle is kept at 0°. As rated power is reached only a small negative pitch angle, initially of about 2°, is necessary to promote stalling and so to limit the power to the rated level. As the wind speed increases small adjustments in both the positive and negative directions are all that are needed to maintain constant power. The small sizes of the pitch angle adjustments make pitching to stall very attractive to designers but the blades have the same damping and fatigue problems as fixed pitch turbines.

4.2.7 Pitching to feather

By increasing the pitch angle as rated power is reached the angle of attack can be reduced. A reduced angle of attack will reduce the lift force and the torque. The

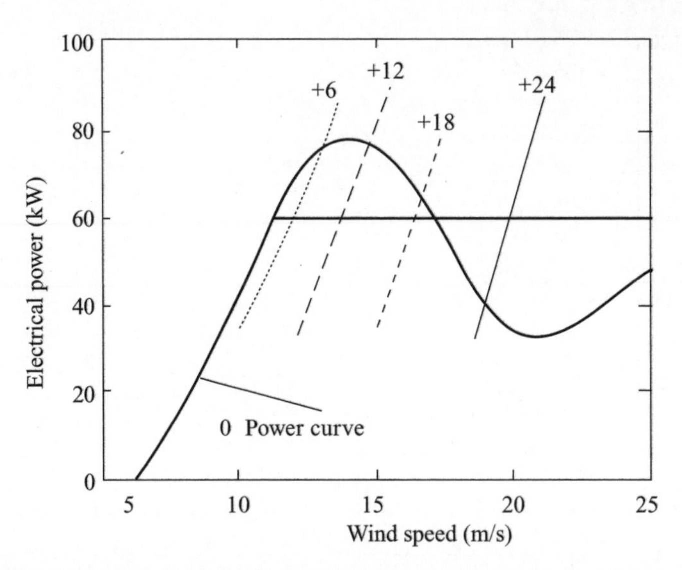

Figure 4.10 Pitching to Feather Power Regulation Requires Large Changes of Pitch Angle

flow around the blade remains attached. Figure 4.10 is for the same turbine as Figure 4.9 but only the zero degree power curve is shown below the rated level. Above the rated level fragments of power curves for higher pitch angles are shown as they cross the rated power line; the crossing points give the necessary pitch angles to maintain rated power at the corresponding wind speeds. As can be seen in Figure 4.10, the required pitch angles increase progressively with wind speed and are generally much larger than is needed for the pitching to stall method. In gusty conditions large pitch excursions are needed to maintain constant power and the inertia of the blades will limit the speed of the control system's response.

Because the blades remain unstalled if large gusts occur at wind speeds above the rated level large changes of angle of attack will take place with associated large changes in lift. Gust loads on the blades can therefore be more severe than for stalled blades. The advantages of the pitching to feather method are that the flow around the blade remains attached, and so well-understood, and provides good, positive damping. Feathered blade parking and assisted starting are also available. Pitching to feather has been the preferred pitch control option mainly because the blade loads can be predicted with more confidence than for stalled blades.

4.3 Comparison of Measured with Theoretical Performance

The turbine considered in this section is run at constant rotational speed, the most common mode of operation, because this allows electricity to be generated at constant frequency. More detail about this method of operation will be discussed in the next section but the main feature is that there is, theoretically, a unique power output for a given wind speed.

When the turbine was under test the chosen rotational speed was 44 r.p.m. Energy output and wind speed were measured over 1 min time intervals and the average power and wind speed determined. The test was continued until a sufficient range of wind speeds had been covered. The results were then sorted in 'bins' 0.5 m/s of wind speed wide and a fairly smooth power *versus* wind-speed curve was obtained as shown in Figure 4.11. The turbine has a diameter of 17 m and would be expected to produce rather more power than shown above if operated at a higher rotational speed.

From the data in Figure 4.11 the $C_P - \lambda$ curve can be derived. The tip speed of the blades is $44 \times \pi/30$ rad/s $\times$ 8.5m = 39.2 m/s, the swept area is $\pi \times 8.5^2 = 227$ m^2 and the air density was measured (from air pressure and temperature readings) at 1.19 kg/m^3. Therefore

$$\lambda = \frac{39.2}{\text{Windspeed}} \quad \text{and} \quad C_P = \frac{\text{Power}\,\lambda^3}{\frac{1}{2}1.19\;39.2^3\;227}$$

The mechanical and electrical losses were estimated at 5.62 kW and this value was used to adjust the theoretical values of C_P. The resulting comparison of measured and theoretical results are shown in Figure 4.12.

This comparison looks reasonable and shows that the theory is reliable but the quality of the theoretical predictions really relies upon the quality of the aerofoil data. The blade and aerofoil design are the same as given in Section 3.9.

One last point should be made before classifying the theory as complete: it would be as well to look at the raw, 1 min averages data, which was reduced down by the binning process and is shown in Figure 4.12. In the post stall region there seems to be a much more complex process taking place than the simple theory predicts and this could be caused by unsteady aerodynamic effects.

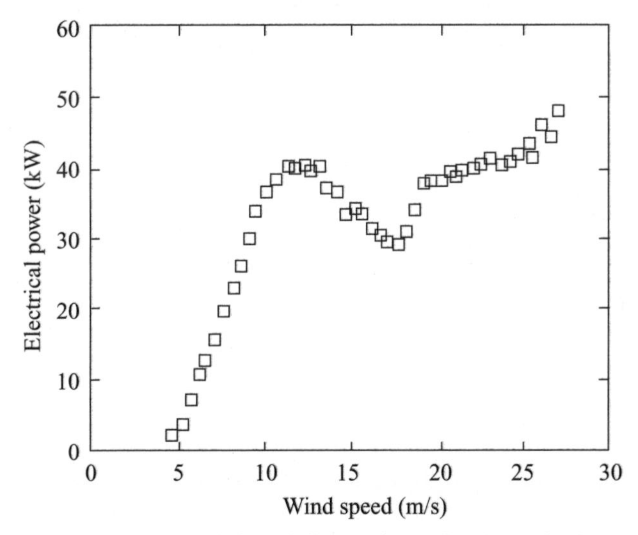

Figure 4.11 Power *versus* Wind Speed Curve from the Binned Measurements of a Three-blade Stall Regulated Turbine

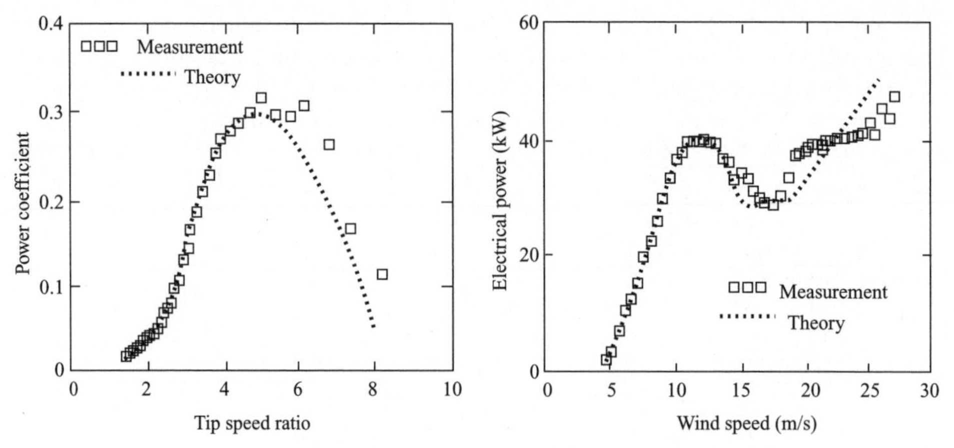

Figure 4.12 Comparison of Measured and Theoretical Performance Curves

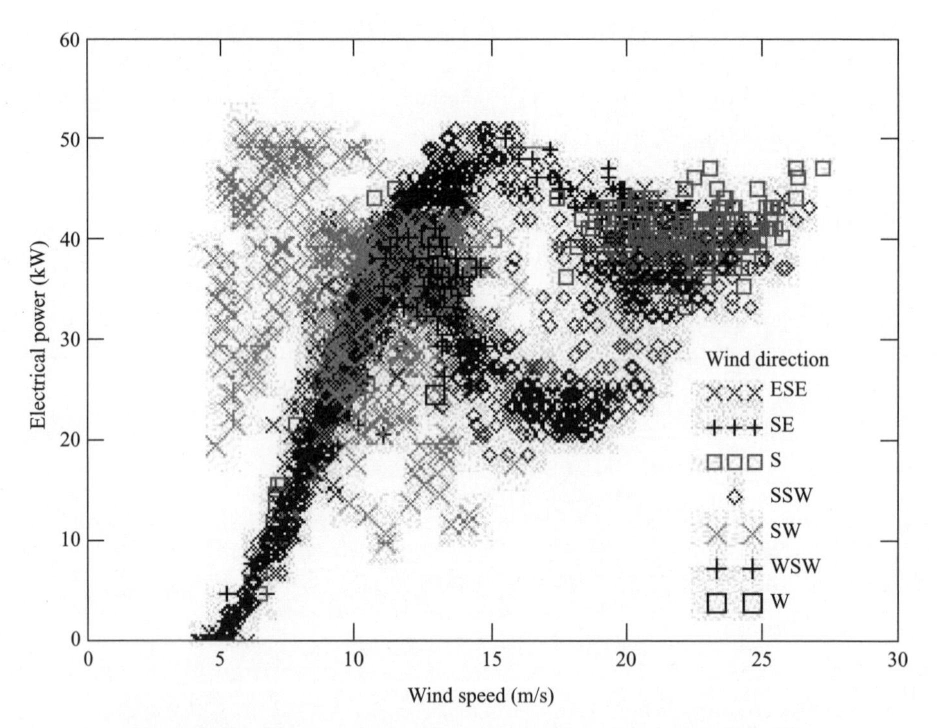

Figure 4.13 Measured Raw Results of a Three-blade Wind Turbine

4.4 Variable-speed Operation

If the speed of the rotor can be continuously adjusted such that the tip speed ratio remains constant at the level which gives the maximum C_P then the efficiency of the turbine will be significantly increased. Active pitch control is necessary to maintain a constant tip speed ratio but only in the process of adjustment of rotational speed; the pitch angle should always return to the optimum setting for highest efficiency. Pitch control regulation is also required in conditions above the rated wind speed when the rotational speed is kept constant.

4.5 Estimation of Energy Capture

The quantity of energy that can be captured by a wind turbine depends upon the power *versus* wind speed characteristic of the turbine and the wind-speed distribution at the turbine site. The wind-speed distribution at a site can be represented by the Weibull function: the probability that the wind speed will exceed a value U is

$$F(U) = e^{-(U/c)^k} \tag{4.2}$$

where c, called the scale factor, is a characteristic speed related to the average wind speed at the site by

$$c = \frac{\overline{U}}{\Gamma\left(1 + \dfrac{1}{k}\right)} \tag{4.3}$$

(Γ being the gamma function and k is a shape parameter, see also Section 2.4). Let $U/\overline{U} = u$ a normalized wind speed.

The wind speed distribution density is then the modulus of the derivative of Equation (4.2) with respect to u:

$$f(u) = k\left(\frac{\overline{U}}{c}\right)^k u^{k-1} e^{-\left(\frac{\overline{U}}{c}\right) u^k} \tag{4.4}$$

i.e., the probability that the wind speed lies between u and $u + \delta u$ is $f(u)\delta u$. Alternatively, Equation (4.4) gives the proportion of time for which the wind speed u will occur.

The performance curve shown in Figure 4.14 is for a turbine designed with an optimum tip speed ratio of 7. As an example, assume that the turbine is stall-regulated and operates at a fixed rotational speed at a site where the average wind speed is 6 m/s and the Weibell shape factor $k = 1.8$ then, from Equation (4.3), the scale factor $c = 6.75$ m/s.

Figure 4.15 shows the $K_P - 1/\lambda$ curve for the turbine; from inspection of that curve the tip speed ratio at which stall (maximum power) occurs is 3.7 and the corresponding C_P is 0.22.

The required maximum electrical power of the machine is 500 kW, the transmission loss is 10 kW, the mean generator efficiency is 90 percent and the availability of the turbine (amount of time for which it is available to operate when maintenance and repair time is taken into account) is 98 percent.

The maximum rotor shaft power (aerodynamic power) is then

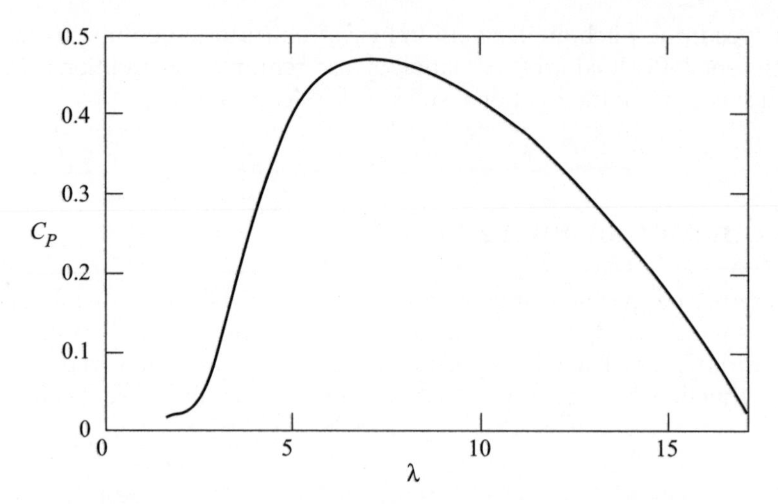

Figure 4.14 $C_P - \lambda$ Curve for a Design Tip Speed Ratio of 7

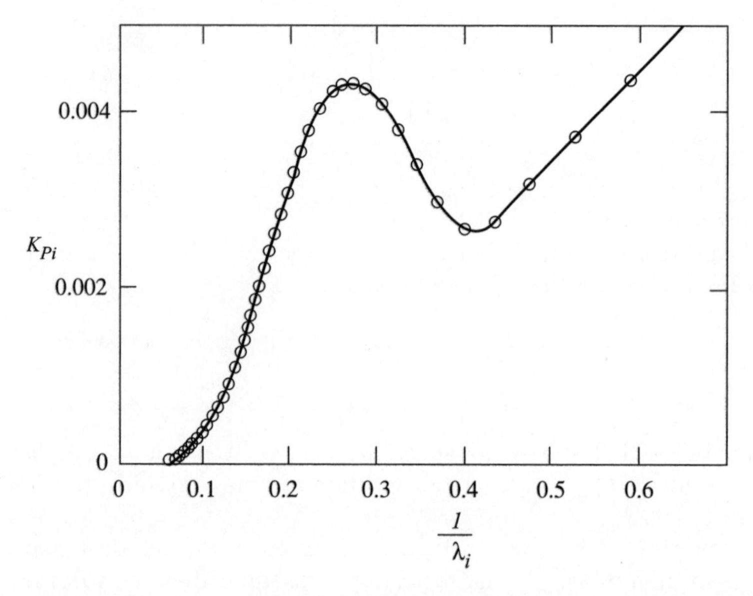

Figure 4.15 $K_P - 1/\lambda$ Curve for a Fixed Speed, Stall-regulated Turbine

$$P_s = 500/0.9 + 10 = 567 \text{ kW}$$

The wind speed at which maximum power is developed is 13 m/s, therefore the rotor swept area must be, assuming an air density of 1.225 kg/m^3,

$$567\,000/(1/2 \times 1.225 \times 13^3 \times 0.22) = 1.92 \times 10^3 \text{ m}^2$$

The rotor radius is therefore 24.6 m.

The tip speed of the rotor will be 3.7×13 m/s $= 48.1$ m/s and so the rotational speed will be $48.1/24.6$ rad/s $= 1.96$ rad/s, which is $1.96 \times 60/2\pi$ rev/min $=$ 18.7 rev/min.

The power *versus* wind-speed curve for the turbine can then be obtained from Figure 4.15.

$$\text{Power (electrical)} = (K_P \times \tfrac{1}{2} \times 1.225\ \text{kg/m}^3 \times (48.1\ \text{m/s})^3 \times 1.92 \times 10^3\ \text{m}^2 - 10 \times 1000\ \text{W}) \times 0.9$$

$$\text{Wind speed} = 48.1\ \text{m/s}/\lambda$$

To determine the energy capture of the turbine over a time period T multiply the power by $f(u) \times T$ (Equation 4.4); because $f(u)$ is the proportion of time T spent at a normalized wind speed u.

$$f(u)\delta u = \frac{\delta T}{T} \tag{4.5}$$

and

$$\int_0^\infty f(u)\mathrm{d}u = 1 \tag{4.6}$$

Plot power against u and integrate over the operational wind-speed range of the turbine to give the energy capture.

The operational speed range will be between the cut-in speed and the cut-out speed. The cut-in speed is determined by the transmission losses and is the wind speed at which the turbine begins to generate power. The cut-in speed is usually

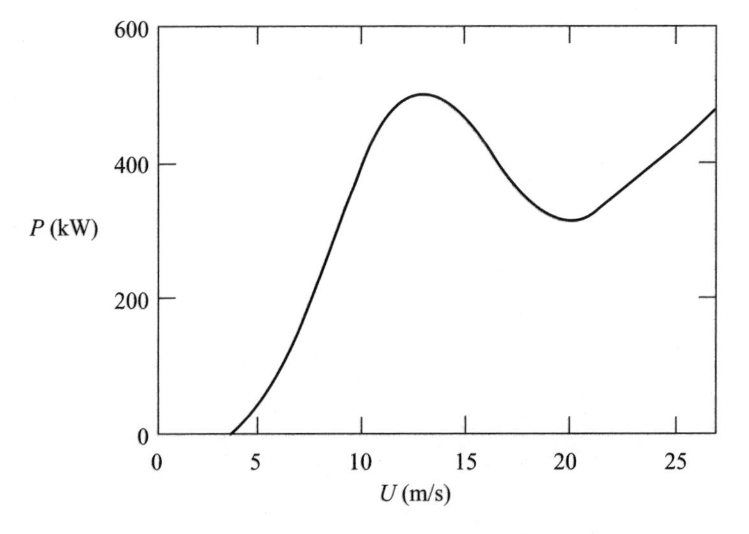

Figure 4.16 Power *versus* Wind Speed

chosen to be somewhat higher than the zero power speed, in the present case, say 4 m/s.

The cut-out speed is chosen to protect the turbine from high loads, usually about 25 m/s.

The energy captured over a time period T (ignoring down time) will be

$$T\int_{U_{ci}/\overline{U}}^{U_{co}/\overline{U}} P(u)f(u)\mathrm{d}u = E \tag{4.7}$$

which is the area under the curve of Figure 4.17 times the time period T. Unfortunately, the integral does not have a closed mathematical form in general and so a numerical integration is required, such as the trapezoidal rule or, for better accuracy, Simpson's rule.

For a time period of 1 year $T = 365 \times 24$ h. Therefore, for the 10 data points shown in Figure 4.18, the energy capture will be, using the trapezoidal rule,

$$E = 0.98T\sum_{i=1}^{9}(P_{i+1}f(u_{i+1}) + P_i f(u_i))\frac{(u_{i+1} - u_i)}{2} \tag{4.8}$$

where

$$E = 4.5413 \times 10^5 \text{ kWh}$$

Even though the upper limit of integration $u_{co} = 4.17$ is greater than highest value of u shown in Figure 4.18 it is clear that almost no energy is captured between those speeds.

A turbine which has pitch control would be able to capture more energy but at the expense of providing the control system and the concomitant reduction in

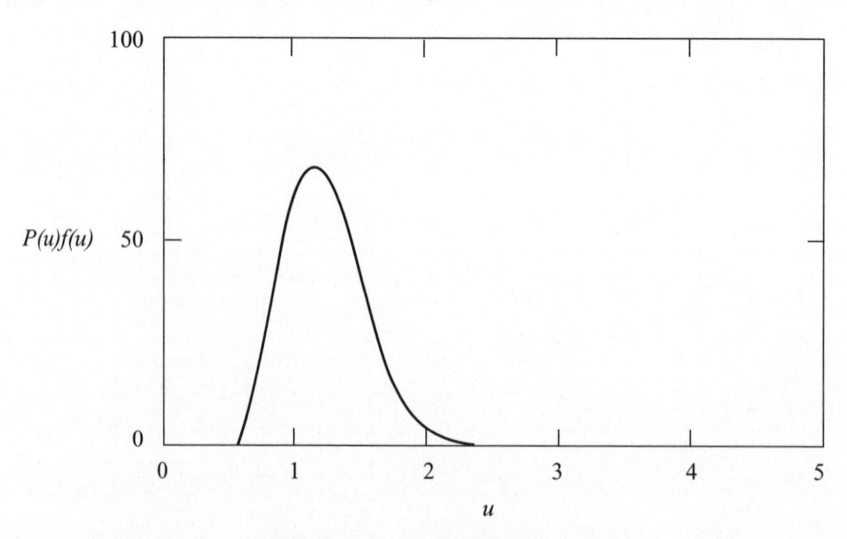

Figure 4.17 Energy Capture Curve

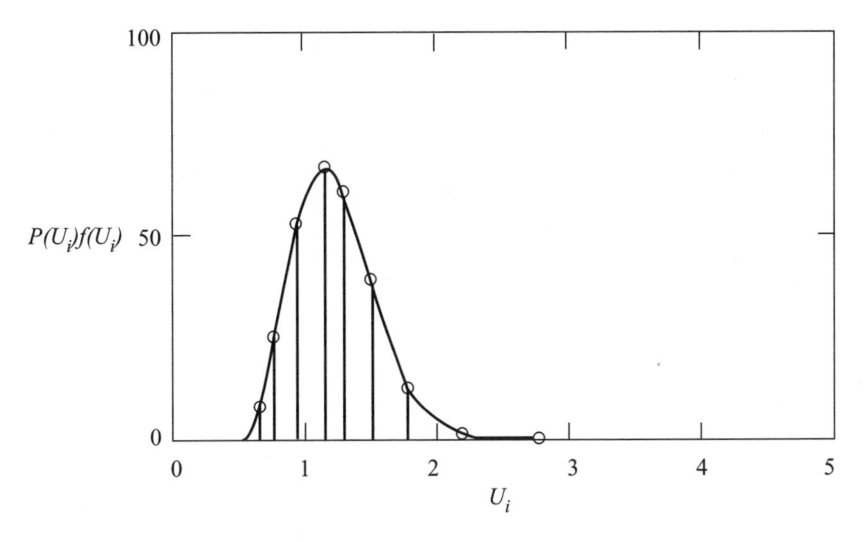

Figure 4.18 Energy Capture Curve for Numerical Integration

reliability. A turbine operating at variable speed (constant tip speed ratio) until maximum power is reached and thence at constant speed, constant power and pitch control, see Figure 4.19, would capture the maximum possible amount of energy in a given time.

The annual energy capture would be

$$E = 4.8138 \times 10^5 \text{ kWh}$$

which is a 6 percent increase in energy capture compared with the fixed-speed, stall-regulated machine. Whether or not the increase in captured energy is economically worthwhile is a matter for debate.

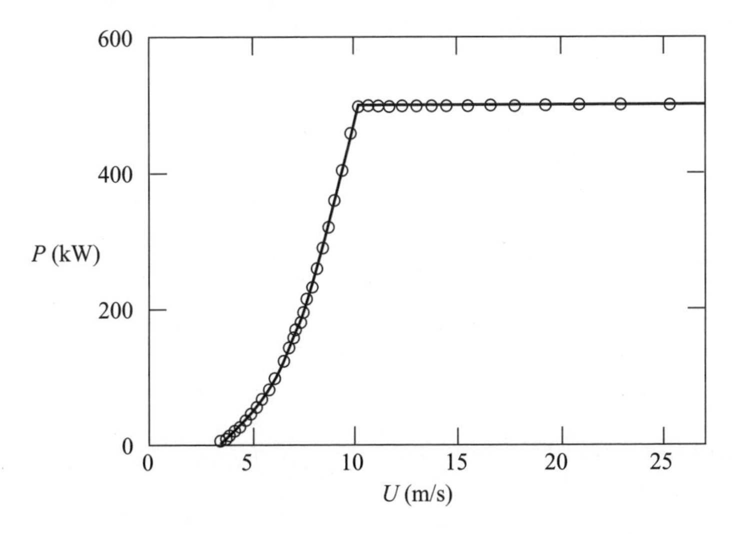

Figure 4.19 Power Curve for Variable Speed Turbine

4.6 Wind-turbine Field Testing

4.6.1 Introduction

Wind-turbine field testing is undertaken in the main for two different reasons. First, as part of the development of new designs, manufacturers and researchers undertake a wide range of measurements to check on the operation of a given machine and in some instances to validate wind turbine models used in the design process. The second, and perhaps the most common, reason for testing is to establish the performance of a given wind turbine for commercial reasons. Because in the second case the objectives are better defined and since many of the problems are common to both sorts of tests, there will be a concentration on performance measurement. Performance measurement is also the area best covered by agreed standards and recommendations, although some difficulties and inconsistencies still remain. It should be mentioned that testing is also increasingly undertaken, in a commercial context by the operators, to verify the manufacturer's performance warranty. Such tests will tend to follow the agreed performance testing methodology although the context is often more demanding, involving possibly complex terrain and also the influence of adjacent turbines.

There are other specialized tests that are of importance, such as for determining the acoustic emission characteristics, and the assessment of noise at a given site. Other areas such as the evaluation of fatigue, including the associated mechanical loads, and power quality can involve direct measurement.

This section will provide an overview of field testing, and in particular will:

- describe the reasons for undertaking field testing,
- identify recognized testing procedures,
- examine practical aspects of transducers and data loggers,
- discuss the difficulties associated with aerodynamic performance assessment, and
- look at errors and uncertainty.

4.6.2 Information sources for wind-turbine testing

Wind turbines have increasingly become part of mainstream technology over the last decade, and this is reflected by the attention paid to them by the national and international standards bodies.

International Electro-technical Commission (IEC)

The IEC is currently developing a range of standards specifically applicable to wind turbines. The work is being undertaken in the main by its TC88 Technical Committee, and covers power performance, acoustics, blade testing, mechanical loads and power quality. Power performance testing is covered by the published IEC standard 61400-12 (1998). The IEC standard also exists as a British and European Standard BS EN 61400-12:1998. All national standards bodies have a responsibility which extends to any significant technology. Accordingly, national standards committees have been formed in most countries to cover wind energy technology. In many instances these merely formalize the national participation in international standards formation, and tend to adopt acceptable international standards as national standards.

CENELEC

CENELEC is a European committee, concerned with electro-technical matters, which has established a task force to manage the harmonization of safety and testing standards in support of various European Union Directives. The task force is presently assessing the adequacy of evolving IEC standards and may, or may not, issue its own standards relevant to wind turbines in the future.

International Energy Agency (IEA)

In the early days of wind energy, prior to the involvement of the standards organizations, the IEA took the lead in drawing up recommendations, or informal codes of practice. The IEA codes of practice have been highly influential, and covered power performance, acoustics, power quality and fatigue.

MEASNET

Most western European countries have established national wind test sites. There has been a tradition of close co-operation between these sites and, supported by the European Commission, many useful documents have been developed by ECWETS (the European Community Wind Energy Test Station). More recently, the MEASNET network for the promotion of quality, common interpretation and mutual recognition in testing standards, was created.

MEASNET is a European collaboration of experts concerned with establishing rigorous standards for all measurement procedures related to wind energy. In particular, they have drawn up draft guidelines for 'power performance measurement procedure'. Although heavily reliant on the IEC Standard, they are in some areas more precise.

4.7 Wind-turbine Performance Measurement

In the final analysis, wind-turbine performance is concerned with the estimation of long-term energy production expected on a given site. The wind resource is described by the probability distribution (usually annual) of 1 h (sometimes, 10 min) mean wind speeds. To calculate the average energy production for a given probability distribution of wind speeds a relationship between wind speed and wind power is needed—this is the power curve of the wind turbine. As dynamic effects are not of interest for long-term performance, averaging of the measured wind speed and wind turbine power is carried out which improves the correlation between them and attenuates the effects of wind turbulence. This is not to say that site turbulence is irrelevant—a point which will be dealt with later.

It is also important to note the power from the wind turbine which is of relevance here is the net power, defined as the power available from the wind turbine less power needed for control, monitoring, display or maintaining operation. In other words it is the power available to the user and is measured at the point of connection to the network.

4.7.1 Field testing methodology

Although a few years ago, the International Energy Agency (IEA) recommended practices for wind turbine testing which were the nearest thing to an agreed procedure for wind turbine evaluation. Now, as mentioned, an IEC standard is available: IEC 61400-12: Wind-turbine Generator Systems, Part 12:- Wind-turbine Power Performance Testing (1998). It is interesting to contrast it with Volume 1 of the IEA recommendations (1982) and subsequent editions (1990), which deal with power performance testing. These notes will draw on both the IEC standard, which forms the basis of European and national standards, and its precursor IEA documents. It is essential that the IEC standard, and any amendments which may have been agreed, be obtained before contemplating power performance measurements.

The wind turbine should ideally be located on a site free from obstructions and local topographic features which could affect the measurements; the IEC standard specifies relevant criteria. Site calibration is needed for non-ideal sites, and the IEC document also includes an approach, included as Annex B (designated 'informative' in contrast to the compulsory, 'normative' content of the Standard). This is to facilitate the evaluation of machine performance *in situ*, within a commercial wind farm, which is unlikely to conform to the test site requirements. However, MEASNET find that this calibration procedure is not rigorous enough, and have recommended an alternative method. Testing should be conducted under 'natural' conditions which excludes towing tests, testing while it is raining (or other precipitation) and when ice accumulates on the turbine, all of which affect wind turbine performance and should be avoided. If conditions like rain do occur during the tests the data affected should be appropriately demarcated. Clearly, modifications or adjustments should not be made to the wind turbine during the period of data collection.

The test procedure consists of taking a series of measurements of wind speed, wind direction, atmospheric pressure and temperature, net power and in some circumstances rotor speed. These measurements should be taken over as wide a range of wind speeds as possible. All data should be checked for accuracy and consistency and if any of the variables have been found to be misread the sample should be discarded. Data collected while the anemometer is in the wake of the wind turbine or in the wake of the anemometer tower, or other obstacle, should also be discarded.

4.7.2 Wind-speed measurement

Wind speed is the most critical parameter to be measured so considerable emphasis should be placed on its accuracy. According to the IEA (1982) the anemometer should have an accuracy of 5 percent or better over the range of relevant wind speeds and according to the revised IEA recommendation (1990) it should be accurate to ± 0.1 m/s or less for wind speeds between 4 and 25 m/s. Finally the IEC have opted to eschew a stated precision, and require instead calibration against a traceable instrument. The instrument should be calibrated, before and after the test, so as to establish that its accuracy has been maintained throughout the test (MEASNET have documented a specified calibration procedure). To avoid problems, it is advisable to run in a new anemometer for a period of about 2 months before use, to allow the bearings to ease. Another characteristic of an anemometer is its distance constant, which the IEC states, should be 5 m or less. The distance constant is an indication of the response of the anemometer and is defined as the length of wind run which must pass the anemometer for its output to reach $(1 - 1/e) = 0.63$ of its final value. Large distance constants can give rise to a significant over speeding effect because the cup anemometer responds more quickly to increases in wind speed than decreases, and this is the reason for the 5 m cut off in the standard. Some believe that, even with a distance constant of 5 m, the instrument should be assessed to evaluate the likely over-speeding error, and a correction applied if necessary. The IEC allow this as the accuracy can be shown to be improved.

The wind speed that is measured should be as representative as possible of wind which would have been present in the plane of the rotor in the absence of the wind turbine. The desired velocity never exists and so a suitable upstream velocity is selected instead. The anemometer location is generally chosen so as to minimize any interference from the wind turbine rotor itself whilst maintaining a reasonable correlation between the measured wind speed and the output from the wind turbine. Between 2 and 4 rotor diameters from the wind turbine is stated in the IEC standard, which recommends 2.5 diameters as the optimum. This compares with the 2 and 6 diameters recommended by the IEA. If the anemometer were placed significantly nearer than 2 diameters, correction for the velocity deficit caused by the rotor would have to be made.

Although corrections can in theory be made to take account of wind shear for anemometers not at hub height, this is discouraged and whenever possible the wind speed should be measured at the same height (relative to ground level) as the hub of the wind turbine rotor. The IEC specify a height within 2.5 percent of the turbine hub height.

As already mentioned, the anemometer must not be located in the wake of the turbine, or other significant obstacles on the site, including of course other wind turbines, operating or otherwise. Figure 4.20, taken from the Standard, shows the allowed location in relation to the turbine being measured, with the exclusion zone dependent on the distance from the turbine. A precise specification exists in the Standard for the calculation of all other exclusion zones.

Poor location of the anemometer on the tower has recently been identified as a potential cause of error and for the first time with the IEC standard, precise guidelines exist. These reflect the need to avoid mast wake effects and any significant blockage in the vicinity of the instrument. An ideal location is on a vertical tube clear of the top of the meteorological mast.

To speed up the experimental assessment it is common to use more than one meteorological mast, arranged so that at least one anemometer is free of any exclusion zone at any given time.

4.7.3 Wind-direction measurement

The wind direction is monitored so as to eliminate wind-speed data taken in the excluded zones. The wind vane should be located at the same height as the anemometer (within 10 percent of the hub height) and in its proximity but not so as to interfere with the wind-speed measurement. The IEC requires an absolute accuracy better than 5 degrees for the direction measurement.

4.7.4 Air temperature and pressure measurement

For a given wind velocity the energy in the wind depends on the air density. So as to be able to correct for changes in air density, the air temperature and pressure should be measured. At high temperatures it is recommended that relative humid-

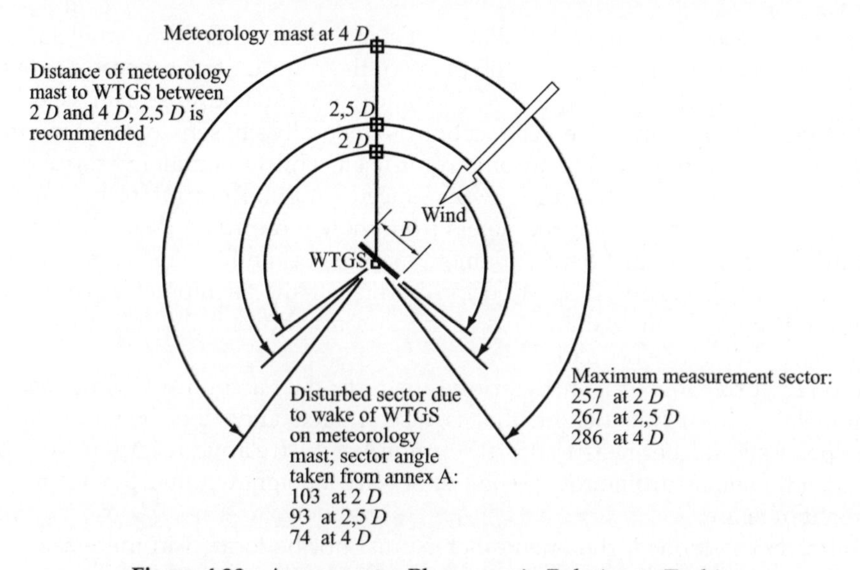

Figure 4.20 Anemometer Placement in Relation to Turbine

ity also be measured, and corrected for. Accuracy should be such as to give the air density to a precision of ±1 percent. The IEC Draft states that pressure should be measured at hub height or corrected to that height using ISO 2533. The IEC require that the data be normalized to two different reference air densities, the average measured air density at the test site (rounded to the nearest 0.05 kg/m^3), and standard conditions at sea level defined as 15°C and 1013.3 mbar which, for dry air, corresponds to a density of 1.225 kg/m^3.

The standard requires that the measured power is corrected by multiplying it by the ratio of the standard air density and the test air density, calculated from the temperature and pressure, but only when the average air density lies outside the standard value, plus or minus 0.05 kg/m^3.

The above procedure is appropriate for stall-regulated machines, but where pitch regulation is active the results of the correction can be misleading. In these circumstances it is more appropriate to correct the wind speed using the one third power of the ratio of the test air density to the standard density. This approach was adopted by the IEA (1990) and is specified for power levels above 70 percent of rated. The IEC standard follows this line, and the formulae to be used for the correction according to the standard are given below.

Density corrections

For stall-regulated wind turbines each 10 min averaged power value should be corrected as follows. The test air density ρ_T, is given by:

$$\rho_T = 1.225\left[\frac{288.15}{T}\right]\left[\frac{B}{1013.3}\right]$$

where T is the test air temperature in degrees absolute and B is the barometric pressure in mbar. The power corrected to standard condition, ρ_S is given by

$$P_S = P_T\left[\frac{\rho_S}{\rho_T}\right]$$

where P_T is the measured power and ρ_S is the standard air density of 1.225 kg/m^3.

For pitch-regulated wind turbines the above approach is applied below 70 percent of rated power but above that value the following correction should be used:

$$U_S = U_T\left[\frac{\rho_T}{1.225}\right]^{1/3}$$

Here the averaged wind-speed value is corrected rather than the power.

Where U_T is the measured wind speed in m/s and U_S is the value corrected to standard conditions. After considerable debate, the exponent of 1/3 appearing in Equation (4.13) has been adopted in the IEC standard, following the IEA precedent.

4.7.5 Power measurement

As it is the net power which is of interest the power transducer should be located downstream of any auxiliary loads. It is generally assumed that the wind turbine will be operating at a nominally fixed speed. For variable speed operation, the IEA indicated that the rotor speed must also be measured to enable changes in kinetic energy to be calculated and compensated for, and this should be done to within 1 percent of the nominal rotor speed. No such prescription is included in the international standard.

Generally the electrical output will be three-phase, 50/60 Hz with voltage in the range 380–415 V. The recommended approaches are the '3 watt meter' method and the '2 watt meter' method where no neutral connection exists. Both of these take account of load imbalance between the phases. The IEC standard refers to IEC 60688 and recommends a transducer of class 0.5 or better (which means a maximum error of 0.5 percent at rated power); the current transformers, and voltage transformers if used, should reach the equivalent standard (IEC 60044-1 and 60186 respectively). Usually these transducers will have an analogue output.

Alternative power measurement equipment, such as kWh meters equipped to produce pulse outputs, can be used provided an equivalent accuracy can be established. Whatever transducer is chosen a calibration should be obtained. The transducers must be able to cope with the power range −50 percent to 200 percent of the turbine rated capacity.

4.7.6 Wind-turbine status

At least one output should be measured which indicates the operational status of the wind turbine system. MEASNET make clear that this should not be a sensor showing whether the turbine is connected to the grid, but rather showing that the turbine is available. This should be used to determine the time periods for which the measured power data should be selected for performance analysis.

4.7.7 Data acquisition system

An automatic digital data acquisition system capable of taking analogue signals (and pulse train inputs where appropriate) should be used. Raw data from all channels should be stored and preferably the system should be able to collect data continuously over the measurement period. Commercially available equipment now enables quite sophisticated data logging systems to be built around a standard micro-computer. A typical arrangement is shown in Figure 4.21. Some data processing such as the application of calibrations and averaging can be done on-line with more complex analysis left to be done later. It is required that the resolution of the data acquisition system does not reduce the accuracy of the data collected, indeed measurement uncertainty should be minimal compared to the sensors used. Care should also be taken to ensure that the signals are free from spurious noise.

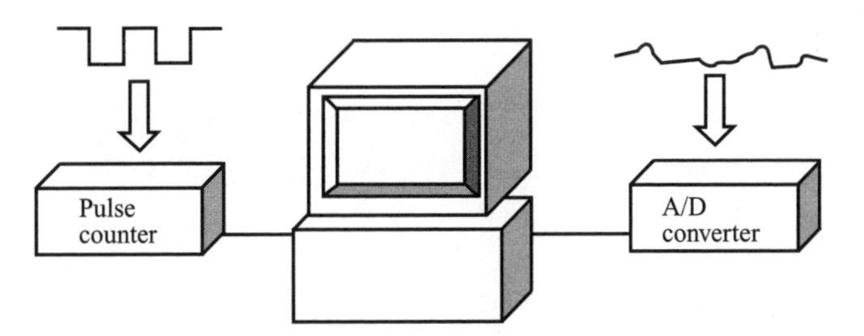

Figure 4.21 Data Logging Arrangement

4.7.8 Data acquisition rate

For the purpose of power performance estimation the collected data are averaged to increase the correlation between wind speed and power. Consequently high rates of data sampling are not required. Where pulse generating instruments are used the logging interval should be chosen long enough to provide an acceptable resolution. For example, an anemometer might give 20 pulses/m of wind run. If this is sampled at 0.5 Hz at a wind speed of 5 m/s the resolution error will be 1 in 200 or 0.5 percent which is adequate. Analogue measurements are more likely and the international standard specifies a minimum sampling rate of 0.5 Hz.

4.8 Analysis of Test Data

Both the IEA and the IEC standard use a 10 min averaging time. This corresponds approximately to the 'spectral gap' (Section 2.1) and means that wind distributions of either 10 min or 1 h means can be used with reasonable confidence to estimate annual energy production. Once erroneous data have been eliminated and any corrections applied, 10 min averages of wind speed and wind power should be calculated. Scatter plots should be presented as shown in Figure 4.22. The data are then analysed using the 'method of bins' (Akins, 1978). According to this procedure the wind speed range is divided into a series of intervals (known as bins). The IEC standard requires 0.5 m/s bins throughout the range. Data sets are distributed into the bins according to wind speed and the ensemble average of the data sets in each bin calculated as follows:

$$
\begin{aligned}
U_i &= \frac{1}{N_i}\sum_{j=1}^{N_j} U_{ij} \\
P_i &= \frac{1}{N_i}\sum_{j=1}^{N_i} P_{ij}
\end{aligned}
\tag{4.9}
$$

where U_{ij} is the jth 10 min average of wind speed in the ith bin; P_{ij} is the jth 10 min average of power in the ith bin; and N_i is the number of data sets in the ith bin. The ensemble averages (U_i, P_i) are then plotted and a curve drawn through the plotted

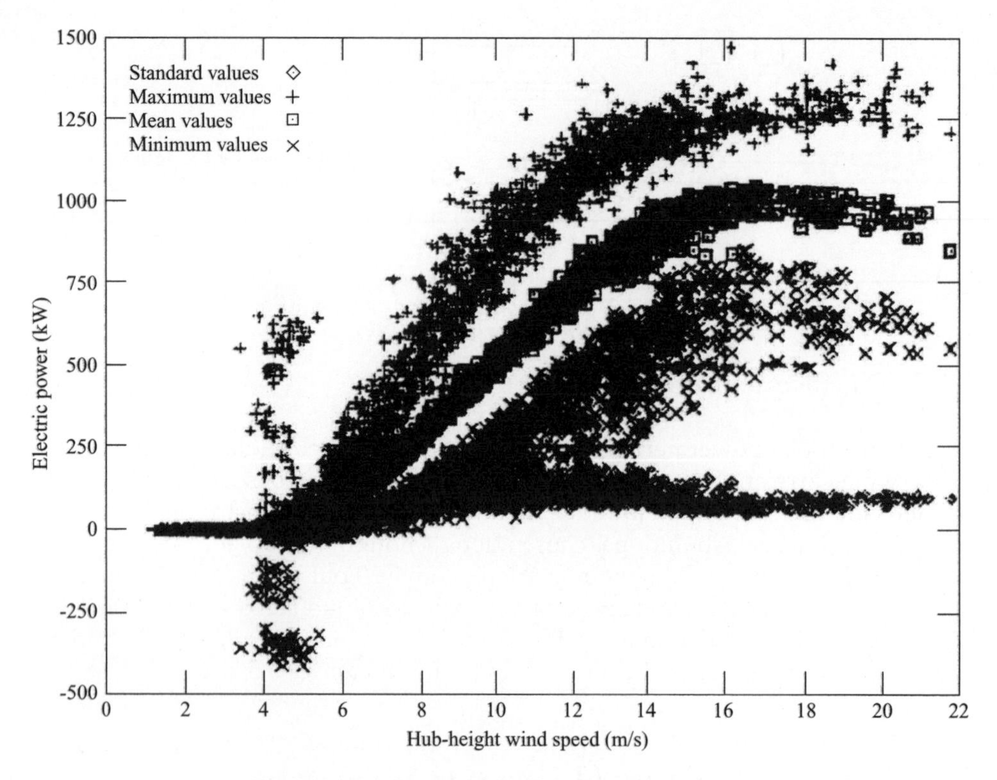

Figure 4.22 Scatter Plots Showing 10 min Data

points. This curve is defined to be the *power curve* for the wind turbine. Figure 4.23 gives an example power curve, also showing errors bars representing the total uncertainty. How these are calculated is dealt with in Section 4.11.

The IEC standard also provides a detailed prescription as to how annual energy yield should be calculated from the power curve and site wind statistics which are assumed to follow the Rayleigh probability distribution.

4.9 Turbulence Effects

Because extended averaging periods are used in power performance testing (usually 10 min) a significant amount of wind energy is contained in the turbulence.

Let the wind speed, U, at any instant be composed of a mean value $\overline{U}$ and the fluctuation U^* about $\overline{U}$. Then

$$U = \overline{U} + U^* \tag{4.10}$$

Energy is proportional to the cube of the wind speed and since we are interested in the mean energy, an expression is wanted for the mean cube wind speed, denoted $\overline{U^3}$. From Equation (4.10)

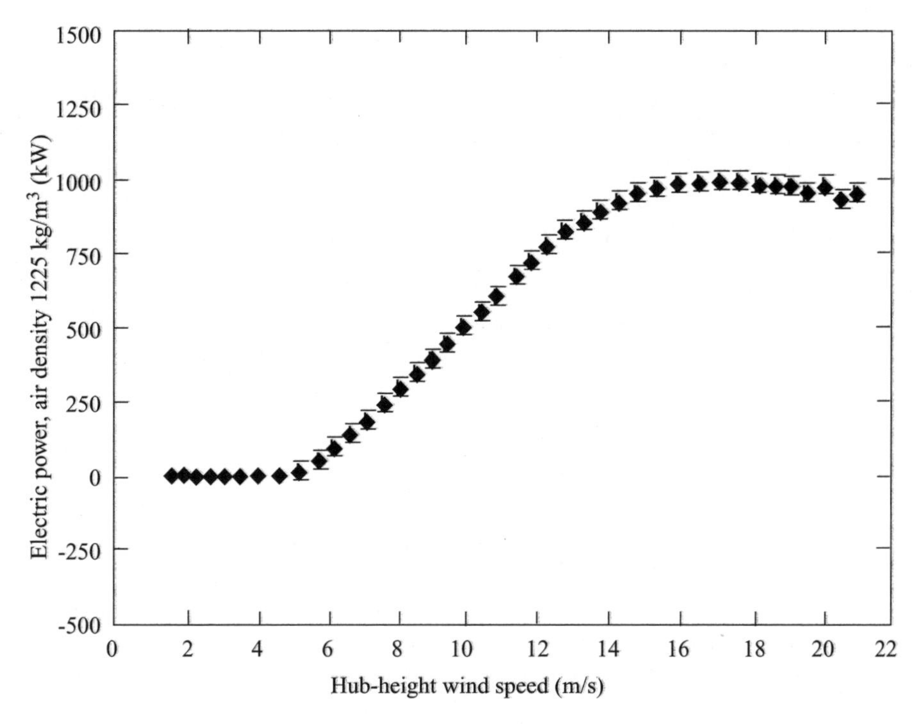

Figure 4.23 Turbine Power Curve with Error Bars

$$\overline{U^3} = \overline{(\overline{U} + U^*)^3}$$

$$\overline{\overline{U}^3} + \overline{3\overline{U}^2 U^*} + \overline{3U^{*2}\overline{U}} + \overline{U}^{*3}$$

The fluctuations U^* are assumed symmetrical about $\overline{U}$ and so the mean values of U^* and U^{*3} are zero and hence this simplifies to:

$$\overline{U^3} = \overline{U}^3 + 3\overline{U^{*2}}\overline{U} \tag{4.11}$$

By definition the mean square fluctuation about the mean is just the variance, σ^2, and so Equation (4.11) can be written as

$$\overline{U^3} = \overline{U}^3 + 3\sigma^2\overline{U}$$

Also by definition the turbulence intensity, I is simply $\sigma/\overline{U}$, and hence

$$\overline{U^3} = \overline{U}^3(1 + 3I^2) \tag{4.12}$$

Consequently there is more energy in the wind than indicated by the average value of the wind speed. Corrections to the power curve have been suggested on the basis of Equation (4.12). This can, however, be misleading as it assumes that the wind turbine maintains a constant C_P across its output range and this is far from true. Christensen and Dragt (1983) discuss this issue in detail and conclude that such simple corrections should not be made. An example calculated in the same report does show that a turbulence intensity of 13 percent could give rise to an error in the predicted energy yield of a particular wind turbine of 1.6 percent. Higher turbulence levels can be encountered and could give rise to errors of 5 to 15 percent. If the proposed site has a turbulence intensity similar to that on which the wind turbine was tested then no need for correction will arise. What is needed then is a record of the turbulence present on the site during the testing, and this is indeed specified in the international standard. Specifically, plots of mean wind speed and turbulence intensity as a function of wind direction should be presented for each data set selected.

4.10 Aerodynamic Performance Assessment

There is often a need for assessing the aerodynamic or instantaneous performance of a wind turbine. Detailed features, such as the stall characteristic, will tend to be smoothed out by the 10 min averaging employed in power performance assessment. Consequently only short averaging periods can be used, but this introduces a further limitation of the method of bins. It has been shown that poor correlation between power and wind speed results in a systematic distortion of the binned relationship (Christensen and Dragt, 1986; Dragt, 1983) and shorter averaging times result in poorer correlation. The effect is to rotate the power curve about the point where the wind speed probability is highest as shown in Figure 4.24. This can be understood as follows. Consider a short gust of high wind at the anemometer. The likelihood is that at this instant, the wind speed at the turbine will be lower (i.e. nearer the mean), and consequently the power output measured will be less than would have been expected from the power curve with the wind speed as measured at the anemometer. This will bias the measured power curve down at higher than average wind speeds. A similar argument shows that the curve will be biased upwards at lower than average wind speeds, as illustrated in Figure 4.24.

Dragt (1983) has developed a formula for correcting the measured wind speed based on the statistics of the sample data set. The corrected wind speed, U^*, is given by:

$$U^* = U - (1 - r)(U - \overline{U}) \tag{4.13}$$

Here U is the measured wind speed, $\overline{U}$ is the sample mean, and r is the correlation coefficient between power and wind speed. This correction was derived on the basis of a normal distribution for the wind speed variations, and care should be taken to check that this is applicable to the sample data set before carrying out the correction. It could be that data are collected so as to cover the wind-speed range evenly. If this has been done effectively then no systematic distortion should occur.

Another approach to reducing the averaging time whilst maintaining a high

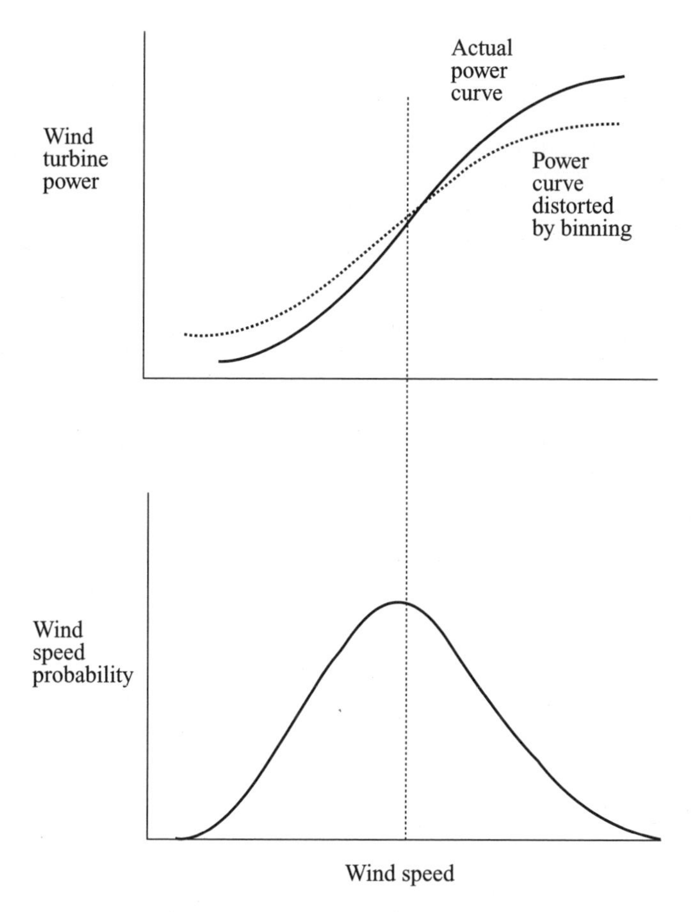

Figure 4.24 Biasing Effect with Binning

correlation is to measure the wind much closer to the wind turbine. Some years ago, at Rutherford Appleton Laboratory (RAL) in the UK, for a downwind machine, the wind speed was measured by using a boom mounted anemometer located only one radius upwind of the rotor. As close as one radius distance from the wind turbine some retarding (velocity deficit) will be apparent. The easiest way to take account of this is to determine experimentally the relationship between the boom anemometer reading and the measured free wind speed. Figure 4.25 shows the uncorrected power curve based on the boom anemometer readings and shows that the results for the different averaging times are in close agreement, as expected, since high correlations were achieved.

Figures 4.26 to 4.29 show the effect of applying Dragt's correction directly to measurements made on a 17 m, stall-regulated wind turbine at RAL. It is evident that 1 min averaged data, when corrected, is preferable to 10 min averaging due to the extended parameter ranges achieved. The similarity of the corrected results derived from the differently averaged data lends confidence to the technique.

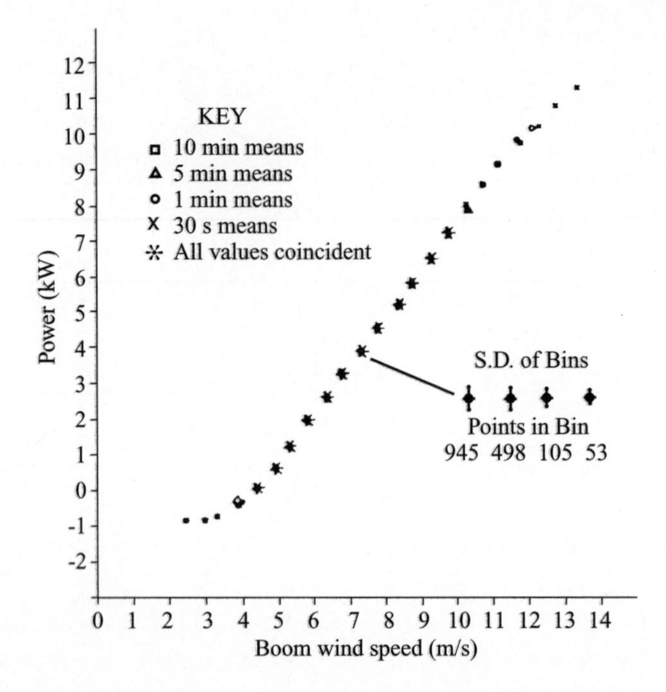

Figure 4.25 Power Curve Based on Boom Wind Speed Measurement

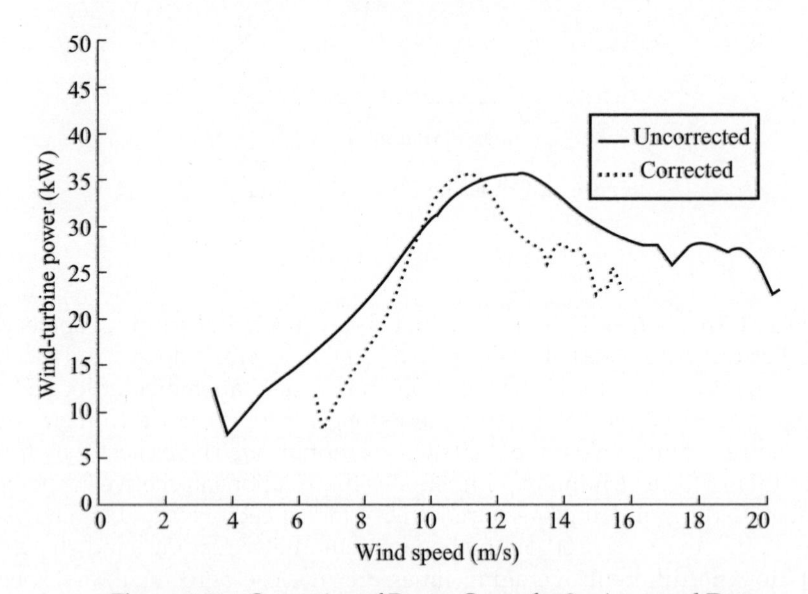

Figure 4.26 Correction of Power Curve for 2 s Averaged Data

Detailed aerodynamic experiments will often require the simultaneous measurement of the transverse and vertical wind components, in addition to the longitudinal component measure by the cup anemometer. Very fast response sonic anemometers are well suited to this purpose.

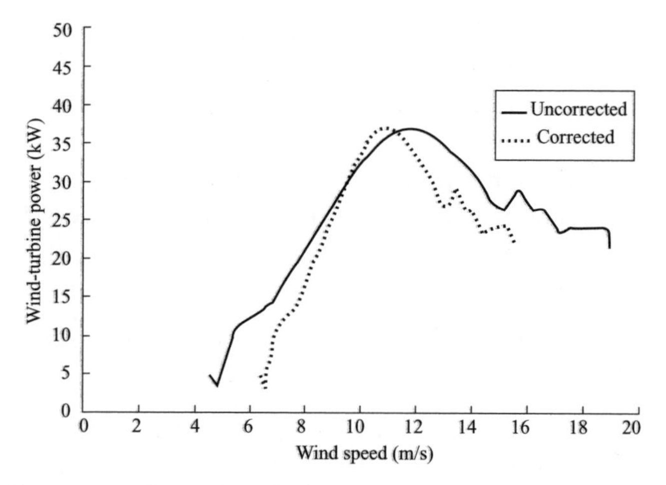

Figure 4.27 Correction of Power Curve for 30 s Averaged Data

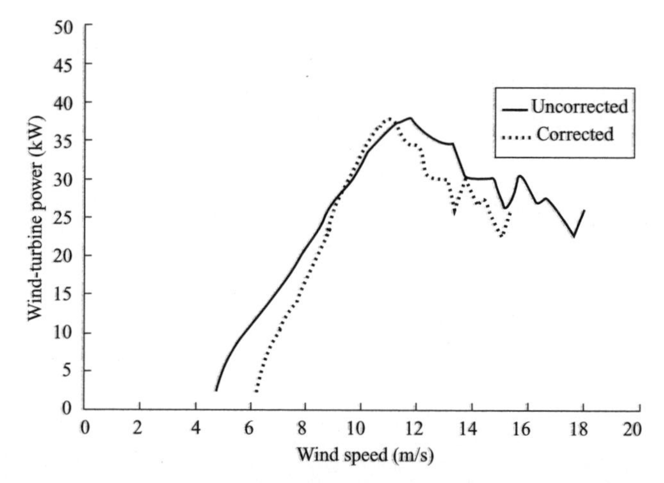

Figure 4.28 Correction of Power Curve for 1 min Averaged Data

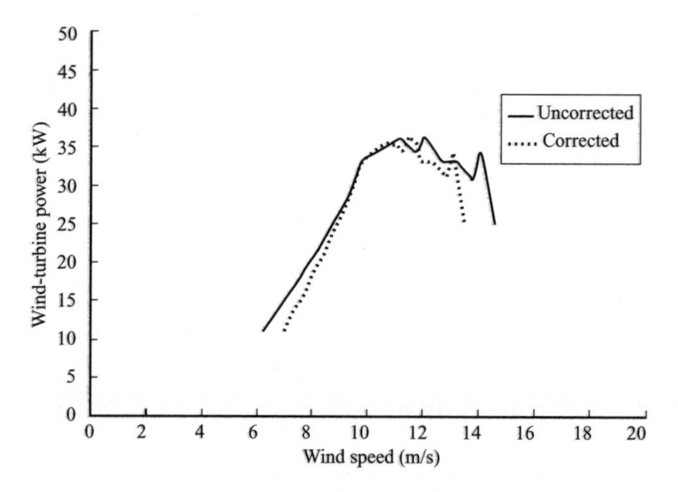

Figure 4.29 Correction of Power Curve for 10 min Averaged Data

4.11 Errors and Uncertainty

The testing guidelines and subsequent Standards were drawn up with a key objective of reducing errors and uncertainty. Despite this, inaccuracies will remain. For some years this has been an area of concern, and the recent IEC standard now makes clear how these should be assessed.

A comprehensive study funded by the CEC (Christensen and Dragt, 1986) identified the major sources of inaccuracy. As well as measurement and data analysis error, uncertainties are also introduced by other factors such as operating conditions (rain, ice, turbulence etc), blade condition/roughness, generator and gearbox temperature, yaw error, unmeasured wind velocity components and wind shear. These cannot be directly assessed, although some will be reflected in an increase of the scatter in the results which is taken into account in the proposed IEC analysis.

Work at Riso National Laboratory (Pedersen and Peterson, 1988) used a frequency domain model to quantify the uncertainty associated with factors such as averaging time, anemometer position and wind turbulence. Accuracy was expressed in terms of the precision index which is defined by:

$$S_i = \frac{\sigma_i}{\sqrt{N_i}} \tag{4.14}$$

where N_i is the number of samples and σ_i is the standard deviation of power measurements for the ith bin. The uncertainty in annual energy production $S(E)$ resulting from a measured power curve was calculated by using the precision index for each bin:

$$S(E) = 8760 \sum_i S(P)_i f(U_i) \tag{4.15}$$

Where $S(P)_i$ is the power error band (computed from S_i) for a given confidence level and $f(U_i)$ is the annual probability of the wind being in the ith bin. It should be noted that this treatment only deals with random errors (referred to as category A uncertainties by the IEC) and that the effect of bias errors associated with the instruments and other factors (category B uncertainties) should be included separately. The RISO work concluded that the testing procedures were not accurate enough to identify with confidence differences in performance of less than 5 percent.

The current IEC standard presents a detailed methodology for analysing how errors propagate through the power curve to the energy yield calculations. Because of its importance, the approach is outlined here.

4.11.1 Evaluation of uncertainty

The measurands are the power curve, or more precisely the individual bin averages which constitute the curve, and the estimated annual energy production. Uncertainties in the measurements are converted into uncertainty in the measurands by means of sensitivity factors following the ISO methodology (1995).

Table 4.1 lists the minimum selection of parameters which must be included in the analysis; all are category B except the variability of electrical power which is determined from the scatter in the bins.

4.11.2 Sensitivity factors

The sensitivity factors indicate how changes in a particular measured parameter affect the relevant measurand. For example, temperature measurements are used to calculate the air density used in the power curve calculation through correction of the wind speed or power. We are interested in the rate of change of power (the measurand) with temperature, i.e. the gradient $\partial P/\partial K$. From the correction formula, this factor is $P_i/288.15$ (kW/K). Similarly the sensitivity factor for air pressure measurement is $P_i/1013$ (kW/hPa).

The sensitivity factor for the impact of wind-speed error on the power curve is given directly by the local gradient of the power curve, calculated from:

$$\frac{P_i - P_{i-1}}{U_j - U_{i-1}} \tag{4.16}$$

Table 4.1 Uncertainty Components

Measured parameter	Uncertainty component
Electric power	Current transformers
	Voltage transformers
	Power transducer
	Data acquisition system
	Variability of electric power
Wind speed	Anemometer calibration
	Operational characteristics
	Mounting effects
	Data acquisition system
	Flow distortion due to terrain
Air temperature	Temperature sensor
	Radiation shielding
	Mounting effects
	Data acquisition system
Air pressure	Pressure sensor
	Mounting effects
	Data acquisition system
Data acquisition system	Signal transmission
	System accuracy
	Signal conditioning

The same sensitivity factor applies to all other influences on wind-speed measurement such as flow distortion and anemometer mounting effects.

Measurement uncertainties directly affecting power such as those associated with current transformers, power traducers etc, all take a sensitivity factor of unity. Annex D (ISO, 1995) gives a comprehensive specification of all the sensitivity factors to be used.

4.11.3 Estimating uncertainties

Category A uncertainties are based on the standard deviation of the scatter in each bin, calculated in the conventional way. The main uncertainties in this category are for power variation and the relevant standard deviation is $\sigma_{P,i}$. Statistical theory (the central limit theorem) requires that the standard uncertainty also reflects the number of points in the bin. The appropriate expression is:

$$S_i = S_{P,i} = \frac{\sigma_{P,i}}{\sqrt{N_i}} \tag{4.17}$$

This is the same as the sensitivity factor defined in Equation (4.14). Other statistical uncertainties such as climatic variation could in theory be calculated by experiments designed to isolate these effects. In practice such an approach is unlikely.

Category B uncertainties must be estimated from knowledge of the instrument. If for example a sensor has an accuracy of $\pm U$, it is reasonable to assume that the real value is equally likely to take any value within this interval. Such a rectangular probability distribution implies that the standard uncertainty $\sigma = U/\sqrt{3}$. If the probability distribution is better represented by a triangular distribution, then $\sigma = U/\sqrt{6}$.

4.11.4 Combining uncertainties

Uncertainty components are the individual contributions to the overall uncertainty associated with particular measurements. The general expression for the combined uncertainty in the ith bin, $u_{c,i}$, is given by:

$$u_{c,i}^2 = \sum_{k=1}^{M} \sum_{l=1}^{M} c_{k,i} u_{k,i} c_{i,j} u_{l,j} \rho_{k,l,i,j} \tag{4.18}$$

where: $c_{k,i}$ is the sensitivity factor of component k in bin i, $u_{k,i}$ is the standard uncertainty of component k in bin i, M is the number of uncertainty components in each bin, and $\rho_{k,l,i,j}$ is the correlation between uncertainty component k in bin i and uncertainty component l in bin j.

Note that in Equation (4.17), components k and l both are both in bin i. The correlation coefficients are in practice almost impossible to estimate and assumptions are usually made. For example, that different components are independent

($\rho = 0$). This will pick out only the terms with $k = 1$, for which $\rho = 1$. If in addition we divide the M uncertainty components into subsets of M_A category A components and M_B category B components, Equation (4.17) simplifies to:

$$u_{c,i}^2 = \sum_{k=1}^{M_A} c_{k,i}^2 s_{k,i}^2 + \sum_{l=1}^{M_B} c_{k,i}^2 u_{k,i}^2 = s_i^2 + u_i^2 \tag{4.19}$$

where s_i and u_i are the combined uncertainties for categories A and B respectively.

Varying the bin size from the standard value will affect the overall uncertainty calculated. This is because $s_{P,i}$ depends on the number of data points in each bin, according to Equation (4.17).

The combined uncertainty for annual energy production, u_{AEP}, is given by:

$$u_{AEP}^2 = N_h^2 \sum_{i=1}^{N} f_i^2 \sum_{k=1}^{M_A} c_{k,i}^2 s_{k,i}^2 + N_h^2 \sum_{k=1}^{M_B} \left(\sum_{i=1}^{N} f_i c_{k,i} u_{k,i} \right)^2 \tag{4.20}$$

where f_i is the annual wind-speed probability associated with bin i and N_h is the number of hours in the year. This expression differs from Equation (4.15), in that the second term is an algebraic sum rather than a sum of squares which reflects the fact that the category B uncertainty components are fully correlated across the different bins.

References

Akins, R. E., (1978). 'Performance evaluation of wind-energy conversion systems using the method of bins - current status'. *Internal Report SAND-77-1375*. Sandia Laboratory, Alberquerque, USA.

Christensen, C. J. and Dragt, J. B. (eds), (1986). 'Accuracy of power-curve measurements'. *Riso-M-2632*. Riso National Laboratory, Roskilde, Denmark.

Dragt, J. B., (1983). 'On the systematic errors in the measurement of the aerodynamic performance of a WECS by the method of bins'. ECN.

International Energy Agency, (1982). *Recommended practices for wind-turbine testing and evaluation: 1 Power performance testing*. DTI, London, UK.

International Energy Agency, (1990). *Recommended practices for wind-turbine testing and evaluation: 1 Power performance testing, Second edition.* (Currently available in third edition). DTI, London, UK.

IEC, (1998). 'Wind-turbine generator systems. Part 12: Wind-turbine power performance testing'. *IEC 61400-12*.

ISO, (1995). 'Guide to the expression of uncertainty in measurement'. ISO Information Publication.

Pedersen, T. F. and Peterson, S. M., (1988). 'Evaluation of power performance testing procedures by means of measurements on a 90 kW wind turbine'. *Proceedings of the EWEC*. H. S. Stephens, Bedford, UK.

5

Design Loads for Horizontal-Axis Wind Turbines

5.1 National and International Standards

5.1.1 Historical development

The preparation of national and international standards containing rules for the design of wind turbines began in the 1980s. The first publication was a set of regulations for certification drawn up by Germanischer Lloyd in 1986. These initial rules were subsequently considerably refined as the state of knowledge grew, leading to the publication by Germanischer Lloyd of the *Regulation for the Certification of Wind Energy Conversion Systems* in 1993. This was further amended by supplements issued in 1994 and 1998. Meanwhile national standards were published in The Netherlands (NEN 6096, Dutch Standard, 1988) and Denmark (DS 472, Danish Standard, 1992).

The International Electrotechnical Commission (IEC) began work on the first international standard in 1988, leading to the publication of IEC 1400-1 *Wind turbine generator systems – Part 1 Safety Requirements* in 1994 (Second Edition IEC, 1997). A revised edition containing some significant changes appeared in 1999, bearing the new number IEC 61400-1. The following sections describe the scope of the IEC 61400-1, Germanischer Lloyd and Danish requirements in outline.

5.1.2 IEC 61400-1

IEC 61400-1 *Wind turbine generator systems – Part 1 Safety Requirements* identifies four different classes of wind turbines to suit differing site wind conditions, with increasing class designation number corresponding to reducing wind speed. The wind speed parameters for each class are given in Table 5.1.

The reference wind is defined as the 10 min mean wind speed at hub-height with a 50 year return period. To allow for sites where conditions do not conform to any of these classes, a fifth class is provided for in which the basic wind parameters are to be specified by the manufacturer. The normal value of air density is specified as 1.225 kg/m^3.

A crucial parameter for wind turbine design is the turbulence intensity, which is

Table 5.1 Wind Speed Parameters for Wind Turbine Classes

Parameters	Class I	Class II	Class III	Class IV
Reference wind speed, U_{ref} (m/s)	50	42.5	37.5	30
Annual average wind speed, U_{ave} (m/s)	10	8.5	7.5	6
50 year return gust speed, 1.4 U_{ref} (m/s)	70	59.5	52.5	42
1 year return gust speed, 1.05 U_{ref} (m/s)	52.5	44.6	39.4	31.5

defined as the ratio of the standard deviation of wind speed fluctuations to the mean. The standard specifies two levels of turbulence intensity, designated category A (higher) and category B (lower), which are independent of the wind speed classes above. In each case the turbulence varies with hub height mean wind speed, $\overline{U}$, according to the formula

$$I_u = I_{15}(a + 15/\overline{U})/(a + 1)$$

(Section 2.6.3) where I_{15} is the turbulence intensity at a mean wind speed of 15 m/s, defined as 18 percent for category A and 16 percent for category B. The constant a takes the values 2 and 3 for categories A and B respectively.

The standard then proceeds to the definition of external wind and other environmental conditions on the one hand, and turbine normal operational states and fault situations on the other. The selection of certain combinations of these results in the specification of some 17 different ultimate load cases and five fatigue load cases which require consideration in the design of the turbine. The standard does not extend to the prescription of particular methods of loading analysis. Subsequent sections cover the control and protection systems, the electrical system, installation, commissioning, operation and maintenance.

5.1.3 Germanischer Lloyd rules for certification

Germanischer Lloyd's *Regulation for the Certification of Wind Energy Conversion Systems*, commonly referred to as the GL rules, adopts the same classification of wind turbines as IEC 61400-1, but specifies a single value of hub-height turbulence intensity of 20 percent. A larger number of load cases are specified, but many of them parallel cases in IEC 61400-1. However, the GL rules also provide a simplified fatigue spectrum for aerodynamic loading and simplified design loads for turbines with three non-pitching blades.

The GL rules then go on to describe the design processes required for each component of the turbine in turn – beginning with the blades and ending with the foundation. This includes design load definition, analysis methods, material strengths and fatigue properties. The level of detail provided here sets the GL rules apart from the IEC and Danish standards, and is a consequence of their role in defining the design documentation required for certification.

There is a rigorous treatment of the requirements for the control and safety systems, and for the associated protection and monitoring devices. The centrality of these systems to the overall design process is emphasized by placing them at the start of the document. Final sections deal with operation and maintenance, noise and lightning protection.

5.1.4 Danish Standard DS 472

DS 472 bases the derivation of design-extreme wind speeds on four terrain classes, ranging from the very smooth (expanses of water) to the very rough (e.g., built-up areas). The base wind velocity is taken to be the same all over Denmark, so the result is four alternative profiles of wind speed variation with height. The philosophy behind the selection of design load cases in the Danish standard is similar to that in IEC-1400 and the GL rules, although the number of load cases is fewer. Similarly, the requirements for the control and safety systems are again clearly set out. DS 472 is distinctive in that it includes detailed treatments of the derivation of simplified fatigue load spectra for a three-bladed, stall-regulated machine of up to 25 m diameter and a method of calculating gust response factors for the blades and the tower.

5.2 Basis for Design Loads

5.2.1 Sources of loading

The sources of loading to be taken into account may be catagorized as follows:

- aerodynamic loads,
- gravitational loads,
- inertia loads (including centrifugal and gyroscopic effects), and
- operational loads arising from actions of the control system (e.g., braking, yawing, blade-pitch control, generator disconnection).

5.2.2 Ultimate loads

The load cases selected for ultimate load design must cover realistic combinations of a wide range of external wind conditions and machine states. It is common practice to distinguish between normal and extreme wind conditions on the one hand, and between normal machine states and fault states on the other. The load cases for design are then chosen from:

- normal wind conditions in combination with normal machine states,
- normal wind conditions in combination with machine fault states, or
- extreme wind conditions in combination with normal machine states.

Extreme and normal wind conditions are generally defined in terms of the worst condition occurring with a 50 year and 1 year return period respectively. It is assumed that machine fault states arise only rarely and are uncorrelated with extreme wind conditions, so that the occurrence of a machine fault in combination with an extreme wind condition is an event with such a high return period that it need not be considered as a load case. However, IEC 61400-1 wisely stipulates that if there *is* some correlation between an extreme external condition and a fault state, then the combination should be considered as a design case.

5.2.3 Fatigue loads

A typical wind turbine is subjected to a severe fatigue loading regime. The rotor of a 600 kW machine will rotate some 2×10^8 times during a 20 year life, with each revolution causing a complete gravity stress reversal in the low speed shaft and in each blade, together with a cycle of blade out-of-plane loading due to the combined effects of wind shear, yaw error, shaft tilt, tower shadow and turbulence. It is therefore scarcely surprising that the design of many wind turbine components is often governed by fatigue rather than by ultimate load.

The design fatigue load spectrum should be representative of the loading cycles experienced during power production over the full operational wind speed range, with the numbers of cycles weighted in accordance with the proportion of time spent generating at each wind speed. For completeness, load cycles occurring at start-up and shut-down and, if necessary, during shut-down, should also be included. It is generally assumed that the extreme load cases occur so rarely that they will not have a significant effect on fatigue life.

5.2.4 Partial safety factors for loads

Limit-state design requires the design load for a component to be calculated as the sum of the products of each characteristic load and the appropriate partial load factor. The partial safety factors for ultimate loads stipulated by IEC 61400-1, the GL rules and DS 472 are given in Table 5.2:

Note that the GL rules treat machine fault conditions as abnormal cases with a partial safety factor of unity. This implies that the fault conditions are considered to be very rare events, which may be questionable in practice. On the other hand, IEC 61400-1 classifies load cases involving a machine fault as normal. IEC 61400-1 observes that in many cases, especially where varying loads result in dynamic load effects, the loads from various sources cannot be evaluated separately. In these situations, the standard requires the use of a single partial load factor equal to the highest of the factors in the Table for the relevant design situation. The partial safety factor for fatigue loads is taken as unity.

Table 5.2 Partial Safety Factors for Loads

Source of loading	Unfavourable loads						Favourable loads		
	Types of loading								
	Normal and extreme				Abnormal				
	IEC	GL		DS	IEC	GL and DS	IEC	GL	DS
		Normal	Extreme						
Aerodynamic	1.35	1.2	1.5	1.3	1.1	1.0	0.9	–	–
Operational	1.35	1.35	1.2	1.3	1.1	1.0	0.9	–	–
Gravity	1.1*	1.1*	1.1*	1.0	1.1	1.0	0.9	1.0	–
Inertia	1.25	1.1*	1.1*	1.0	1.1	1.0	0.9	1.0	–

*Factor increased to 1.35 if masses are not determined by weighing.

5.2.5 Functions of the control and safety systems

A primary function of the control system is to maintain the machine operating parameters within their normal limits. The purpose of the safety system (referred to as 'protection system' in IEC 61400-1) is to ensure that, should a critical operating parameter exceed its normal limit as a result of a fault or failure in the wind turbine or the control system, the machine is maintained in a safe condition. Normally the critical operating parameters are:

- turbine rotational speed,
- power output,
- vibration level,
- twist of pendant cables running up into nacelle.

For each parameter it is necessary to set an activation level at which the safety system is triggered. This has to be set at a suitable margin above the normal operating limit to allow for overshooting by the control system, but sufficiently far below the maximum safe value of the parameter to allow scope for the safety system to rein it in. The rotor speed at which the safety system is activated is a key input to the design-load case involving rotor overspeed.

5.3 Turbulence and Wakes

Fluctuation of the wind speed about the short-term mean, or turbulence, naturally has a major impact on the design loadings, as it is the source of both the extreme gust loading and a large part of the blade fatigue loading. The latter is exacerbated by the gust slicing effect in which a blade will slice through a localized gust repeatedly in the course of several revolutions.

The nature of free stream turbulence, and its mathematical descriptions in statistical terms, form the subject of Section 2.6. Within a wind farm, turbines operating in the wakes of other turbines experience increased turbulence and reduced mean velocities. In general, a downwind turbine will lie off-centre with respect to the wake of the turbine immediately upwind, leading to horizontal wind shear. Models describing velocity deficits and the increase in turbulence intensity due to turbine wakes are reported in Section 2.10 but, as is pointed out, no consensus has yet emerged as regards their use for wind turbine design calculations.

5.4 Extreme Loads

5.4.1 Non-operational load cases—normal machine state

A non-operational machine state is defined as one in which the machine is neither generating power, nor starting up, nor shutting down. It may be stationary, i.e., 'parked', or idling.

The design wind speed for this load case is commonly taken as the gust speed with a return period of 50 years. The magnitude of the 50 year return gust depends on the gust duration chosen, which in turn should be based on the size of the loaded area. For example, British Standard CP3 Chapter V, Part 2 (1972) *Code of basic data for the design of buildings: Wind Loading*, states that a 3 s gust can envelope areas up to 20 m across, but advises that for larger areas up to 50 m across, a 5 s gust is appropriate. Nevertheless, IEC 61400-1 and the GL rules specify the use of gust durations of 3 s and 5 s respectively, regardless of the turbine size. In each case, the 50 year return gust value is defined as 1.4 times the 50 year return 10 min mean (the 'reference wind speed'), despite the fact that other authorities estimate the 5 s gust to be between 5 percent (CP3) and 2 percent (ASCE, 1993) smaller than the 3 s gust.

By contrast, DS 472 bases extreme loads on the dynamic pressure resulting from the extreme 10 min mean wind speed rather than a 3 or 5 s gust. These loads are augmented by an 'impact factor' (see Equation (5.15)), which takes into account both wind gusting and the excitation of resonant oscillations thereby. This approach is also followed in Eurocode 1 (1997).

Care is required in selecting the turbine configurations to be considered in the investigation of this case. IEC 61400-1 indicates that the possibility of grid failure is to be considered as part of this load case, which would prevent the yaw system tracking any subsequent changes in wind direction. This case is considered in more detail in Section 5.12.1. Where slippage of the yaw brake is a possibility, it will be necessary to consider a load case in which the turbine becomes backwinded, even though winds are unlikely to change direction by more than 90° and still remain at storm force.

5.4.2 Non-operational load cases—machine fault state

Examples of load cases in this category are ones involving the failure of the yaw or pitch mechanisms. On the assumption that there is no correlation between such a failure and extreme winds, the design wind for this load case is normally taken as the gust speed with a return period of 1 year. GL rules lay down that this is to be taken as 80 percent of the 50 year return gust speed, whereas IEC 61400-1 specifies the lower ratio of 75 percent.

5.4.3 Operational load cases—normal machine state

Several load cases have to be investigated in this category, so that the effects of extremes of gust loading, wind direction change and wind shear can be evaluated in turn. Two types of deterministic discrete gust models are used for the gust loading: the 'rising gust' and the 'rising and falling gust'. In the former case, the wind speed rises sinusoidally over the time interval $t = 0$ to $t = T/2$ to a new value, according to the formula $U(t) = \overline{U} + \Delta U(1 - \cos 2\pi t/T)$, and remains there. In the simple version of the 'rising and falling gust', the sinusoidal wind speed variation continues over the full cycle until the wind speed has returned to its original value. However, IEC 61400-1 defines a more sophisticated version, incorporating a brief dip in the wind speed before and after the rising and falling gust, the complete cycle being defined by

$$U(z, t) = \overline{U}(z) - 0.37\beta\left(\frac{\sigma_u}{1 + 0.1(D/\Lambda_1)}\right)\sin 3\pi t/T[1 - \cos 2\pi t/T] \tag{5.1}$$

where σ_u is the standard deviation of the turbulent wind speed fluctuations, and the factor β takes the values 4.8 and 6.4 for gusts with recurrence periods of 1 and 50 years respectively. D/Λ_1 is the ratio of the rotor diameter to the turbulence scale parameter (see load case 1.7, below). The duration of the gust, T, is specified as 10.5 s for the 1 year and 14 s for the 50 year return gust. The gust profile is illustrated in Figure 5.1.

The ultimate-load cases during normal running defined in IEC 61400-1 are described below. The standard points out that the cases listed are the minimum that should be investigated, and that other cases may need to be considered in relation to specific turbine designs. Note that an inclination of the mean air flow of up to 8° to the horizontal is to be considered in each case. The acronyms in capitals are those used by the code to identify the types of gust and/or direction change.

Load case 1.1: Hub-height wind speed equal to U_r or U_o, with turbulence (Section 5.1.2), where U_r is the rated wind speed, defined as the wind speed at which the turbine's rated power is reached, and U_o is the upper cut-out speed.

(*Load case 1.2* is a Fatigue Case).

Load case 1.3: Gust and direction change (ECD). Hub-height wind speed equal to U_r

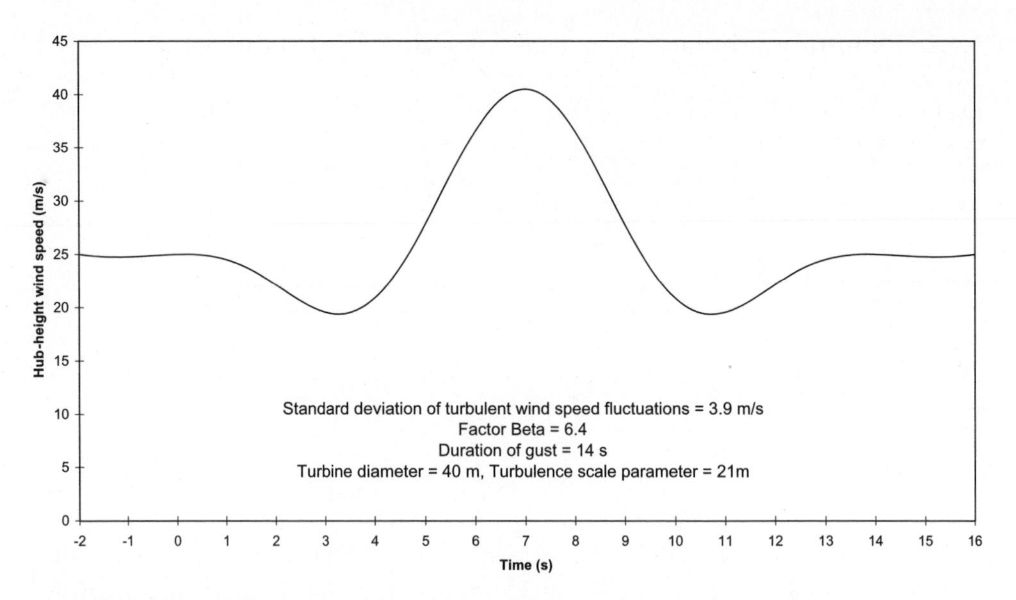

Figure 5.1 IEC 61400-1 Extreme Rising and Falling Gust with 50 year Return Period for Steady Wind of 25 m/s and Category A Turbulence

plus a 15 m/s rising gust, in conjunction with a simultaneous direction change of $720/U_r$ degrees—i.e., 60° for a rated wind speed of 12 m/s, for example. The gust rise time and the period over which the direction change takes place are both specified as 10 s. Wind shear is to be included according to the 'Normal wind profile model', with $U(z) \propto z^{0.2}$. This is referred to as normal wind shear in the load cases below.

Load case 1.4: External electrical fault, with hub height wind speed equal to U_r or U_o. Normal wind shear is included, but turbulence is not.

Load case 1.5: Loss of load; this case is considered separately in Section 5.4.4.

Load case 1.6: 50 year return rising and falling gust, as defined by Equation (5.1), superimposed on hub-height wind speed of U_r or U_o with normal wind shear (EOG_{50}). The duration of the gust, T, is specified as 14 s. See illustration in Figure 5.1 for the case of $U_o = 25$ m/s with category A turbulence for a 40 m diameter rotor.

Load case 1.7: Extreme wind shear (EWS). Additional vertical or horizontal transient wind shear superimposed on the 'Normal wind profile model', for hub-height wind speeds of U_r or U_o. The additional wind shears are specified as:

$$\left(\frac{z - z_{\text{hub}}}{D}\right)\left(2.5 + 0.2\beta\sigma_{\text{u}}\left(\frac{D}{\Lambda_1}\right)^{0.25}\right)\left(1 - \cos\left(\frac{2\pi t}{T}\right)\right)$$

$$\text{m/s for } 0 < t < T, \text{ for vertical shear} \quad (5.2)$$

$$\left(\frac{y}{D}\right)\left(2.5 + 0.2\beta\sigma_{\mathrm{u}}\left(\frac{D}{\Lambda_1}\right)^{0.25}\right)\left(1 - \cos\left(\frac{2\pi t}{T}\right)\right)$$

$$\text{m/s for } 0 < t < T, \text{ for horizontal shear} \quad (5.3)$$

where z is the height above ground, y is the lateral co-ordinate with respect to the hub, D is the rotor diameter, β and σ_u are as defined above, Λ_1 is the turbulence scale parameter of $0.7z_{\mathrm{hub}}$, or 21 m, whichever is the lesser, and T is the duration of the transient wind shear, set at 12 s. The two shears are to be applied independently as separate cases, not simultaneously. In the case of a 40 m diameter machine operating at a 25 m/s cut-out speed, the resulting maximum additional wind speed at the tip of a blade is 8.37 m/s, assuming category A turbulence.

Load case 1.8: 50 year return direction change for steady hub-height wind speed, U_{hub}, of U_r or U_o, with normal wind shear (EDC_{50}). The direction change, θ_{e50}, is defined as:

$$\theta_{e50} = \pm\beta \arctan\left(\frac{\sigma_{\mathrm{u}}}{U_{\mathrm{hub}}[1 + 0.1(D/\Lambda_1)]}\right) \quad (5.4)$$

with the direction varying over time according to the relation:

$$\theta(t) = 0.5\theta_{e50}\{1 - \cos(2\pi t/T)\} \text{ for } 0 < t < T/2 \quad (5.5)$$

The direction change takes place over a period $T/2$ of 6 s. For a 40 m diameter machine operating at a 25 m/s cut-out speed, the direction change is 48°, assuming category A turbulence.

Load case 1.9: 15 m/s rising gust superimposed on hub-height wind speed of U_r with normal wind shear (ECG). The gust rise time is specified as 10 s.

Blade loadings arising from the above load cases are compared in Section 7.1.8.

5.4.4 Operational load cases—loss of load

If the connection to the grid is lost, then the aerodynamic torque will no longer meet with any resistance from the generator—which therefore experiences 'loss of load'—and so the rotor will begin to accelerate until the braking systems are brought into action. Depending on the speed of braking response, this load case may well result in critical rotor loadings.

Grid loss is likely to be caused by a fault on the utility network and subsequent circuit breaker operation, and thus may happen at any time. Accordingly, the concurrent machine state is taken to be normal and so the grid loss case has to be considered in combination with extreme wind conditions. The IEC 61400-1 grid loss case is as described below.

Load case 1.5: Grid loss, with a rising and falling 1 year return gust, as defined by

Equation (5.1), superimposed on hub-height wind speed of U_r or U_o (EOG_1). The duration of the gust, T, is specified as 10.5 s.

5.4.5 Operational load cases—machine fault states

As noted in Section 5.2.2, it is commonly assumed that the incidence of machine faults is uncorrelated with extreme winds, so only normal wind conditions need to be considered in the relevant load cases. IEC 61400-1 specifies two load cases as follows.

Load case 2.1: Control system fault, with steady hub-height wind speed equal to U_r or U_o and normal wind shear. Partial safety factor: normal.

Load case 2.2: Protection system fault or preceding internal electrical fault, with steady hub-height wind speed equal to U_r or U_o and normal wind shear. Partial safety factor: abnormal.

5.4.6 Start-up and shut-down cases

IEC 61400-1 also specifies start-up and shut-down cases with a 1 year return rising and falling gust, a start up case with a 1 year return direction change and an emergency shut-down case. In each, hub-height wind speeds of U_r and U_o are to be considered.

5.4.7 Blade/tower clearance

In addition to checking the acceptability of the stresses arising from the above load cases, the designer should also check that none can result in a collision between the blade and the tower.

5.5 Fatigue Loading

5.5.1 Synthesis of fatigue load spectrum

The complete fatigue load spectrum for a particular wind turbine component has to be built up from separate load spectra derived for turbine operation at different wind speeds and from the load cycles experienced at start-up, normal and emergency shut-down and while the machine is parked or idling. First the cycle counts for each stress range for 1 h operation in a particular wind speed band are calculated and scaled-up by the predicted number of hours of operation in that band over the machine lifetime. IEC 61400-1 specifies that this prediction is to be based on the Rayleigh distribution (Section 2.4), with the annual mean wind speed set according to the turbine class (see Section 5.1.2). Finally, the lifetime cycle counts obtained for operation in the different wind speed bands are combined and added to those calculated for start-ups, shut-downs and periods of non-operation.

5.6 Stationary Blade Loading

5.6.1 Lift and drag coefficients

Maximum blade loadings are in the out-of-plane direction and occur when the wind direction is either approximately normal to the blade, giving maximum drag, or at an angle of between 12° and 16° to the plane of the blade when the angle of attack is such as to give maximum lift.

In the absence of data on drag coefficients for airflow normal to the blade, designers formerly utilized the drag coefficient for an infinitely long flat plate of 2.0, with an adjustment downwards based on the aspect ratio. Thus, on a typical blade with a mean chord equal to one fifteenth of the radius, the length to width ratio would be taken as 30, because free flow cannot take place around the inboard end of the blade. Following CP3 (British Standard, 1972), the British code for wind loading on buildings, this would give a drag coefficient of 1.68. More recently, field measurements have shown that such an approach is unduly conservative, with drag coefficients of 1.24 being reported for the LM 17.2 m blade (Rasmussen, 1984) and 1.25 for the Howden HWP-300 blade (Jamieson and Hunter, 1985). The Danish Standard DS 472 (1992) stipulates a minimum value of 1.3 for the drag coefficient.

The choice of lift coefficient value is more straightforward, because aerofoil data for low-angle attacks is more generally available and is, in any case, required for assessing rotor performance. The maximum lift coefficient rarely exceeds 1.6, but values down to as low as 1.1 will obtain on the thicker, inboard portion of the blade. The minimum value of lift coefficient of 1.5 specified in DS 472 for the calculation of blade out-of-plane loads is therefore probably conservative.

5.6.2 Critical configuration for different machine types

It was shown in the preceding section that the maximum lift coefficient is likely to exceed the maximum drag coefficient for a wind turbine blade, so consequently the maximum loading on a stationary blade will occur when the air flow is in a plane perpendicular to the blade axis and the angle of attack is such as to produce maximum lift. For a stall-regulated machine, this will be the case when the blade is vertical and the wind direction is 70°–80° to the nacelle axis, whereas for a pitch-regulated machine the blade only needs to be approximately vertical with a wind direction at 10°–20° to the nacelle axis.

5.6.3 Dynamic response

Tip displacement

Wind fluctuations at frequencies close to the first flapwise mode blade natural frequency excite resonant blade oscillations and result in additional, inertial loadings over and above the quasistatic loads that would be experienced by a completely rigid blade. As the oscillations result from fluctuations of the wind speed about the mean value, the standard deviation of resonant tip displacement can be expressed in terms of the wind turbulence intensity and the normalized power spectral density at the resonant frequency, $R_u(n_1) = n.S_u(n_1)/\sigma_u^2$, as follows:

$$\frac{\sigma_{x1}}{\bar{x}_1} = 2\frac{\sigma_u}{\overline{U}}\frac{\pi}{\sqrt{2\delta}}\sqrt{R_u(n_1)}\sqrt{K_{Sx}(n_1)} \tag{5.6}$$

Here $\bar{x}_1$ is the first mode component of the steady tip displacement, $\overline{U}$ is the mean wind speed (usually averaged over 10 min), δ is the logarithmic decrement of damping and $K_{Sx}(n_1)$ is a size reduction factor, which results from the lack of correlation of the wind along the blade at the relevant frequency. Note that the dynamic pressure, $\frac{1}{2}\rho U^2 = \frac{1}{2}\rho(\overline{U}+u)^2 = \frac{1}{2}\rho(\overline{U}^2 + 2\overline{U}u + u^2)$, is linearized to $\frac{1}{2}\rho\overline{U}(\overline{U}+2u)$ to simplify the result. See Sections A5.2–4 in the Appendix for the derivations of Equation (5.6) and the expression for $K_{Sx}(n_1)$.

Damping

It is evident from Equation (5.6) that a key determinant of resonant tip response is the value of damping present. Generally the damping consists of two components, aerodynamic and structural. In the case of a vibrating blade flat on to the wind, the fluctuating aerodynamic force per unit length is given by $\frac{1}{2}\rho(\overline{U}-\dot{x})^2 C_d.c(r) - \frac{1}{2}\rho\overline{U}^2 C_d.c(r) \cong \rho\overline{U}\dot{x}C_d.c(r)$, where $\dot{x}$ is the blade flatwise velocity, C_d the drag coefficient and $c(r)$ the local blade chord. Hence the aerodynamic damping per unit length, $\hat{c}_a(r)$, is $\rho\overline{U}C_d c(r)$, and the first mode aerodynamic damping ratio,

$$\xi_{a1} = c_{a1}/2m_1\omega_1 = \int_0^R \hat{c}_a(r)\mu_1^2(r)\,dr/2m_1\omega_1$$

is given by

$$\xi_{a1} = \rho\overline{U}C_d\int_0^R \mu_1^2(r)c(r)\,dr/2m_1\omega_1$$

Here $\mu_1(r)$ is the first mode shape,

$$m_1 = \int_0^R m(r)\mu_1^2(r)\,dr$$

is the generalized mass, and ω_1 is the first mode natural frequency in radians/s. The logarithmic decrement is obtained by multiplying the damping ratio by 2π.

When the wind direction is angled to the blade so as to generate maximum lift, the blade will be approaching stall, with the result that the aerodynamic damping is effectively zero. In this situation tip deflections are limited only by the blade structural damping. Structural damping is discussed in Section 5.8.4, and values for typical blade materials given.

Root bending moment

The standard deviation of tip displacement in combination with the blade mode shape yields an inertial loading distribution from which the standard deviation of the resulting bending moment at any position along the blade may be calculated. In particular, the standard deviation of the root bending moment may be expressed in terms of the mean root bending moment as follows:

$$\frac{\sigma_{M1}}{\overline{M}} = 2\frac{\sigma_u}{\overline{U}}\frac{\pi}{\sqrt{2\delta}}\sqrt{R_u(n_1)}\sqrt{K_{Sx}(n_1)}.\lambda_{M1} = \frac{\sigma_{x1}}{\bar{x}_1}\lambda_{M1} \tag{5.7}$$

where

$$\lambda_{M1} = \frac{\int_0^R m(r)\mu_1(r)r\,\mathrm{d}r}{m_1\int_0^R c(r)r\,\mathrm{d}r}\int_0^R c(r)\mu_1(r)\,\mathrm{d}r \tag{5.8}$$

(See Section A5.5 in the Appendix for the derivation of the expression for λ_{M1}.)

The standard deviation of the quasistatic root bending moment fluctuation, or root bending moment *background* response is expressed in terms of the mean root bending moment by

$$\frac{\sigma_{MB}}{\overline{M}} = 2\frac{\sigma_u}{\overline{U}}\sqrt{K_{SMB}} \tag{5.9}$$

where K_{SMB} is a size reduction factor to take account of lack of correlation of wind fluctuations along the blade. As shown in Section A5.6 of the Appendix, K_{SMB} is usually only slightly less than unity because the blade length is small compared with the integral length scale of longitudinal turbulence measured in the across wind direction.

The variance of the total root bending moment fluctuations is equal to the sum of the resonant and background response variances, i.e.,

$$\sigma_M^2 = \sigma_{M1}^2 + \sigma_{MB}^2$$

Hence

$$\frac{\sigma_M}{\overline{M}} = 2\frac{\sigma_u}{\overline{U}}\sqrt{K_{SMB} + \frac{\pi^2}{2\delta}R_u(n_1)K_{Sx}(n_1)\lambda_{M1}^2} \tag{5.10}$$

The design extreme root bending moment is typically calculated as that due to the 50 year return, 10 min mean wind speed plus the number of standard deviations of the root bending moment fluctuations corresponding to the likely peak excursion in a 10 min period. Thus

$$M_{\max} = \overline{M} + g.\sigma_M \tag{5.11}$$

where g is known as the peak factor, and depends on the number of cycles of root bending moment fluctuations in 10 min, according to the formula

$$g = \sqrt{2\ln(600\nu)} + \frac{0.577}{\sqrt{2\ln(600\upsilon)}} \tag{5.12}$$

Here, ν, is the mean zero-upcrossing frequency (i.e., the number of times per second the moment fluctuation changes from negative to positive) of the root bending moment fluctuations, which will be intermediate between that of the quasistatic wind loading and the blade natural frequency, n_1 (see Section A5.7 of the Appendix). (Note that, as g varies relatively slowly with frequency, it is a reasonable approximation to set g at an upper limit of 3.9, which corresponds to a frequency of about 1.9 Hz, as is suggested in DS 472.)

Substituting Equation (5.10) into Equation (5.11) yields

$$M_{\max} = \overline{M}\left[1 + g\frac{\sigma_M}{\overline{M}}\right] = \overline{M}\left[1 + g\left(2\frac{\sigma_u}{\overline{U}}\right)\sqrt{K_{SMB} + \frac{\pi^2}{2\delta}R_u(n_1)K_{Sx}(n_1)\lambda_{M1}^2}\right] \tag{5.13}$$

The expression in square brackets corresponds to the formula for the impact factor φ in Annex B of DS 472:

$$\varphi = 1 + 2g\frac{\sigma_u}{\overline{U}}\sqrt{k_b + k_r} \tag{5.14}$$

It is often necessary to express the maximum moment in terms of the quasistatic moment due to the 50 year return gust speed, U_{e50}. In order to do this, we equate the latter quantity to the quasistatic component of Equation (5.13), obtaining

$$C_f.\tfrac{1}{2}\rho U_{e50}^2\int_0^R c(r)r\,\mathrm{d}r = \overline{M}\left(1 + g_0.2\frac{\sigma_u}{\overline{U}}\sqrt{K_{SMB}}\right) \tag{5.15}$$

Here the peak factor, g takes a lower value, g_0, corresponding to the lower frequency of the quasistatic root bending moment fluctuations. Equation (5.15) can then be combined with Equation (5.13) to yield

$$M_{\max} = C_f.\tfrac{1}{2}\rho U_{e50}^2 \int_0^R c(r) r\,dr.Q_D \tag{5.16}$$

where Q_D is a dynamic factor given by

$$Q_D = \frac{1 + g\left(2\dfrac{\sigma_u}{\overline{U}}\right)\sqrt{K_{SMB} + \dfrac{\pi^2}{2\delta} R_u(n_1) K_{Sx}(n_1) \lambda_{M1}^2}}{1 + g_0\left(2\dfrac{\sigma_u}{\overline{U}}\right)\sqrt{K_{SMB}}} \tag{5.17}$$

There is considerable advantage in starting with the extreme gust speed and calculating the extreme root moment as the product of $C_f \frac{1}{2}\rho U_{e50}^2 \int_0^R c(r) r\,dr$ and Q_D, because it eliminates most of the error associated with linearizing the formula for dynamic pressure. For example, if the extreme gust is 1.4 times the extreme 10 min mean wind speed, as postulated in IEC 61400-1 (which implies that the product $g_0(\sigma_u/\overline{U})$ is 0.4), then the dynamic pressure due to the gust will be $1.4^2 = 1.96$ times that due to the 10 min mean, rather than 1.8 times as given by the formula $1 + g_0(2\{\sigma_u/\overline{U}\})$.

Example 5.1
Evaluate the dynamic factor, Q_D, for the blade root bending moment for a 20 m long stationary blade under extreme loading.

Consider a trial 20 m long blade design (designated Blade TR) utilizing NACA 632XX aerofoil sections with the chord and thickness distributions shown in Figure 5.2(a). The thickness distribution has a pronounced knee near mid span to minimize the thickness to chord ratio in the outer half of the span. Assuming a uniform skin thickness along the blade, apart from local thickening at the root, the resulting mass and stiffness distributions are as shown in Figure 5.2(b). Modal analysis as described in Section 5.8.2 yields the first and second mode shapes shown in

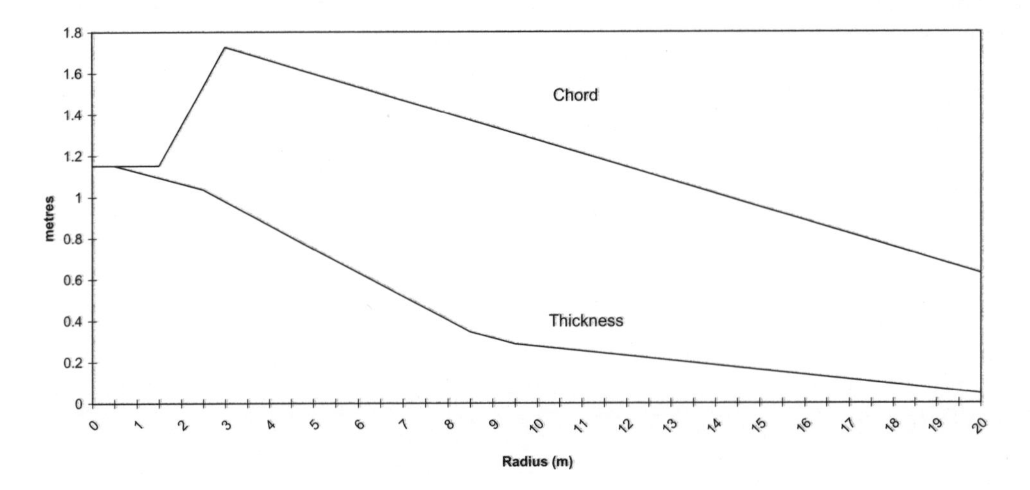

Figure 5.2(a) Blade 'TR' Chord and Thickness Distributions

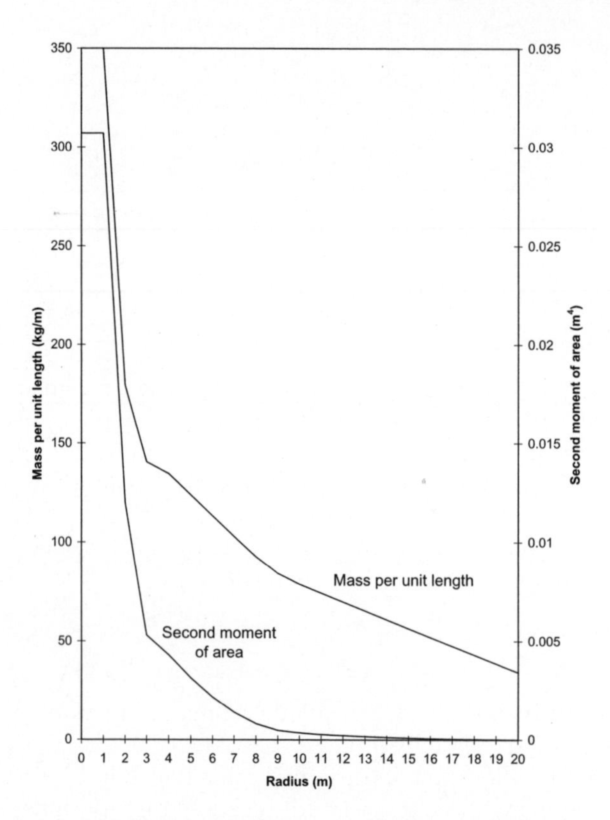

Figure 5.2(b) Blade 'TR' Mass and Stiffness Distributions

Figure 5.3. The first mode shape for a blade of constant cross section (i.e., a uniform cantilever) is also shown for comparison purposes, and it is evident that the high stiffness of the inboard portion of the tapered blade results in dramatically reduced deflections there as a proportion of tip deflection. For a fibreglass blade, typical values of the Young's modulus and material density would be 40 000 N/mm^2 and 1.7 Tonnes/m^3 respectively, resulting in a first mode natural frequency of 1.65 Hz, and a second mode natural frequency of 5.72 Hz.

Values of the other parameters assumed are:

Blade height, z	35 m	
50 year return 10 min mean wind speed at blade height, $\overline{U}$	45 m/s	
Eurocode 1 Terrain Category	I	(Roughness length, $z_o = 0.01$ m)
Turbulence intensity, $I(z) = 1/\ln(z/z_0)$	0.1225	

The corresponding integral length scale for longitudinal turbulence is 227 m according to Eurocode 1 (1997).

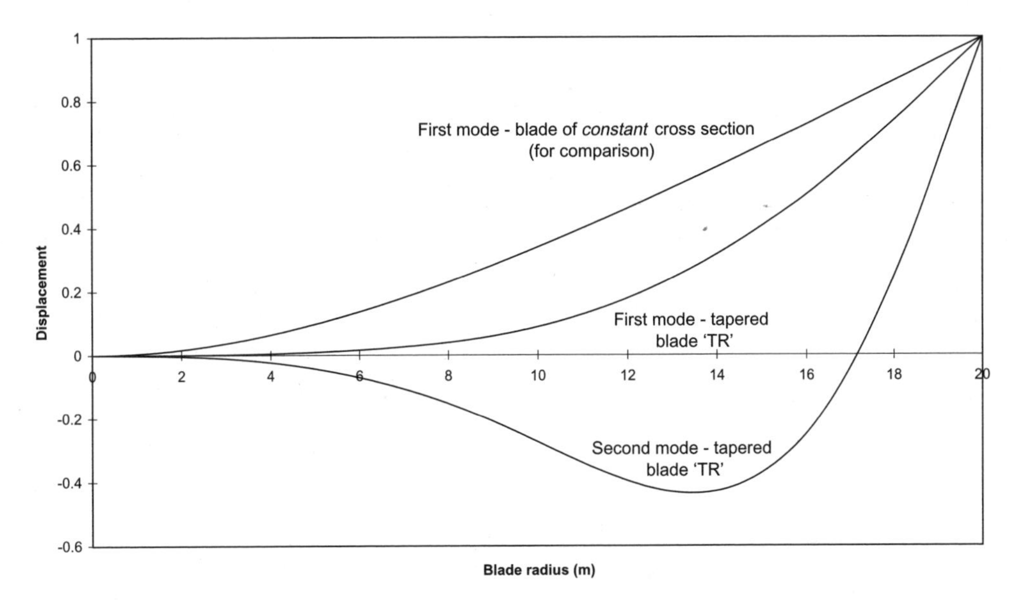

Figure 5.3 Blade 'TR' 1st and 2nd Mode Shapes

The values of the parameters in Equation (5.6) governing the resonant tip response are determined as follows.

(a) The aerodynamic damping is assumed to be zero, so the damping logarithmic decrement is taken as 0.05, corresponding to the structural damping value for fibreglass.

(b) The non-dimensional power spectral density of longitudinal wind turbulence, $R_u(n) = nS_u(n_1)/\sigma_u^2$, is calculated at the blade first mode natural frequency according the Kaimal power spectrum defined in Eurocode 1 (Appendix, Equation A5.8) as 0.0339.

(c) A value of 9.2 is taken for the non-dimensional decay constant in the exponential expression for the normalized co-spectrum used in the derivation of the size reduction factor, $K_{Sx}(n_1)$, in Equation (A5.25).

The various stages in the derivation of the extreme root bending moment and the dynamic factor, Q_D, are set out below. The figures in square brackets are the corresponding values obtained using the method of Annex B of DS 472, which are included for comparison.

Size reduction factor for resonant response, $K_{Sx}(n_1)$	0.426	(Equation (A5.25))	[0.312]

Ratio of standard deviation of resonant tip displacement to the first mode component of steady tip displacement,

$\frac{\sigma_{y1}}{\bar{y}_1} = 2\frac{\sigma_u}{\bar{U}}\frac{\pi}{\sqrt{2\delta}}\sqrt{R_u(n_1)}\sqrt{K_{Sx}(n_1)}$ $= 2 \times 0.1225 \times 9.935 \times \sqrt{0.0339} \times \sqrt{0.426}$ $= 2 \times 0.1225 \times 9.935 \times 0.184 \times 0.652$ $= 2 \times 0.1225 \times 1.192$	0.292	(Equation (5.6))	[N/A]
Root moment factor, λ_{M1}	0.579	(Equation (5.8))	[N/A]
Ratio of standard deviation of resonant root moment to mean value, $\frac{\sigma_{M1}}{\bar{M}} = \frac{\sigma_{x1}}{\bar{x}_1}\lambda_{M1}$	0.169	(Equation (5.7))	[0.338]
Size reduction factor for quasistatic or background response, K_{SMB}	0.926	(Equation (A5.40))	[0.78]
Ratio of standard deviation of quasistatic root moment response to mean value, $\frac{\sigma_{MB}}{\bar{M}} = 2\frac{\sigma_u}{\bar{U}}\sqrt{K_{SMB}} = 2 \times 0.1225 \times 0.962$	0.236	(Equation (5.9))	[0.216]
Ratio of standard deviation of total root moment response to mean value, $\frac{\sigma_M}{\bar{M}} = \sqrt{\left(\frac{\sigma_{MB}}{\bar{M}}\right)^2 + \left(\frac{\sigma_{M1}}{\bar{M}}\right)^2} = \sqrt{0.236^2 + 0.169^2}$	0.290		[0.402]
Zero up-crossing frequency of quasistatic response, n_0	0.41 Hz	(Equation (A5.57))	[N/A]
Zero up-crossing frequency of total root moment response, ν	1.02 Hz	(Equation (A5.54))	[N/A]
Peak factor, g, based on ν	3.74	(Equation (5.12))	[3.9]
Ratio of extreme moment to mean value, $\frac{M_{max}}{\bar{M}} = 1 + g\left(\frac{\sigma_M}{\bar{M}}\right) = 1 + 3.74(0.290)$	2.087	(Equation (5.13))	[2.57]
Peak factor, g_0, based on n_0	3.49		[3.5]
Ratio of quasistatic component of extreme moment to mean value $= 1 + g_0\frac{\sigma_{MB}}{\bar{M}} = 1 + 3.49(0.236)$	1.823	(Equation (5.15))	[1.76]
Dynamic factor, $Q_D = 2.087/1.823$	1.145	(Equation (5.17))	[1.46]

It is apparent that the DS 472 method yields a significantly larger value of the extreme root bending moment. However, the DS 472 ratio of extreme to mean bending moment is intended to apply at all points along the blade, so a conservative value at the root is inescapable, as is shown in the next section which examines the variation of bending moment along the blade.

Spanwise variation of bending moment

The resonant and quasistatic components of bending moments at intermediate positions along the blade can be related to those at the root in a straightforward way.

As far as the quasistatic bending moment fluctuations are concerned, the variation along the blade follows closely the bending moment variation due to the steady loading, although slight changes in the size reduction factor have a small effect. The bending moment diagram for the resonant oscillations is, however, of a very different shape, because of the dominance of the inertia loading on the tip. An

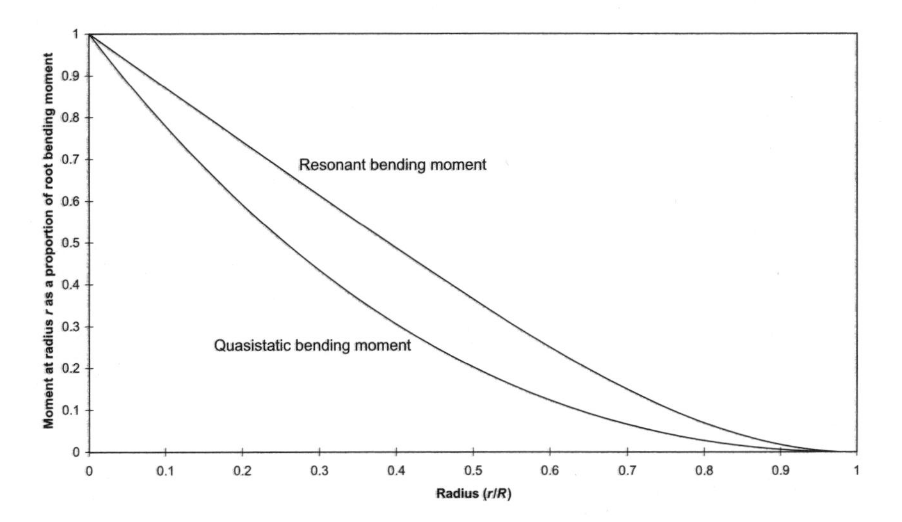

Figure 5.4 Spanwise Variation of Resonant and Quasistatic Bending Moments – Blade 'TR'

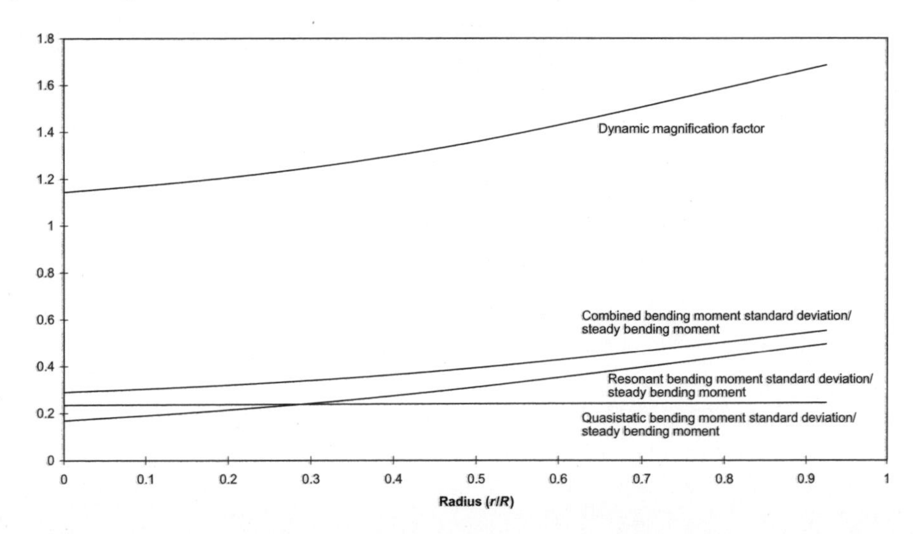

Figure 5.5 Spanwise Variation of (a) Fluctuating Bending Moment Standard Deviations in Terms of Local Steady Bending Moment, and (b) Dynamic Magnification Factor – for Blade 'TR'

expression for the resonant bending moment variation along the blade is given in Section A5.8 of the Appendix, and it is plotted out for the example above in Figure 5.4, with the quasistatic bending moment variation alongside for comparison. It is seen that the resonant bending moment diagram is closer to linear than the quasistatic one, which approximates to a parabola.

A consequence of the much slower decay of the resonant bending moment out towards the tip is an increase in the ratio of the resonant bending moment standard deviation to the local steady moment with radius. This results in an increase in the dynamic magnification factor, Q_D, from 1.145 at the root to 1.69 at the tip for the example above (see Figure 5.5).

5.7 Blade Loads During Operation

5.7.1 Deterministic and stochastic load components

It is normal to separate out the loads due to the steady wind on the rotating blade from those due to wind speed fluctuations and analyse them in different ways. The periodic loading on the blade due to the steady spatial variation of wind speed over the rotor swept area is termed the deterministic load component, because it is uniquely determined by a limited number of parameters – i.e., the hub-height wind speed, the rotational speed, the wind shear, etc. On the other hand, the random loading on the blade due to wind speed fluctuations (i.e., turbulence) has to be described probabilistically, and is therefore termed the stochastic load component.

In addition to wind loading, the rotating blade is also acted on by gravity and inertial loadings. The gravity loading depends simply on blade azimuth and mass distribution, and is thus deterministic, but the inertial loadings may be affected by turbulence – as, for example, in the case of a teetering rotor – and so will sometimes contain stochastic as well as deterministic components.

5.7.2 Deterministic Aerodynamic Loads

Steady, uniform flow perpendicular to plane of rotor

The application of momentum theory to a blade element, which is described in Section 3.5.3, enables the aerodynamic forces on the blade to be calculated at different radii. Equations (3.51) or (3.51a) and (3.52) are solved iteratively for the flow induction factors, a and a', at each radius, enabling the flow angle, ϕ, the angle of attack, α, and hence the lift and drag coefficients to be determined. The solution of the equations is normally simplified by omitting the C_y^2 term in Equation (3.51) – an approximation which is justifed, because $\sigma_r C_y^2/C_x$ is negligibly small away from the root area.

For loadings on the outboard portion of the blade, allowance for tip loss must be made, so Equations (3.51) and (3.52) are replaced by Equations (3.51b) and (3.52a) in Section 3.8.5, (with the omission of the C_y^2 term in Equation (3.51b) again being permissible). These equations can be arranged to give the following expressions for the forces per unit length on an element perpendicular to the plane of rotation and in the direction of blade motion, known as the out-of-plane and in-plane forces respectively:

$$\text{Out-of-plane force per unit length: } F_X = C_x \tfrac{1}{2}\rho W^2 c = 4\pi\rho U_\infty^2 (1 - af) a \frac{f}{N} r \quad (5.18)$$

$$\text{In-plane force per unit length: } -F_Y = C_y \tfrac{1}{2}\rho W^2 c = 4\pi\rho\Omega U_\infty (1 - af) a' \frac{f}{N} r^2 \quad (5.19)$$

The parameters in the expressions are as defined in Chapter 3 (f is the tip loss factor, and N is the number of blades), while the x and y directions are as defined in Figure C1.

The variation of the in-plane and out-of-plane forces with radius is shown in Figure 5.6 for a typical machine operating in a steady 10 m/s wind speed. The 40 m stall-regulated turbine considered in this example is fitted with three 'TR' blades as described in Example 5.1 and rotates at 30 r.p.m. The blade twist distribution is linear, and selected to produce the maximum energy yield for an annual mean wind speed of 7 m/s. It is evident that the out-of-plane load per unit length increases approximately linearly with radius, in spite of the reducing blade chord until the effects of tip loss are felt beyond about 75 percent of tip radius. Note that the form of the variation would be the same for any combination of rotational speed, wind speed and tip radius yielding the same tip speed ratio, because it is the tip speed ratio that determines the radial distribution of flow angle ϕ, and of the induction factors a and a'.

Integration of these forces along the blade then yields in-plane and out-of-plane

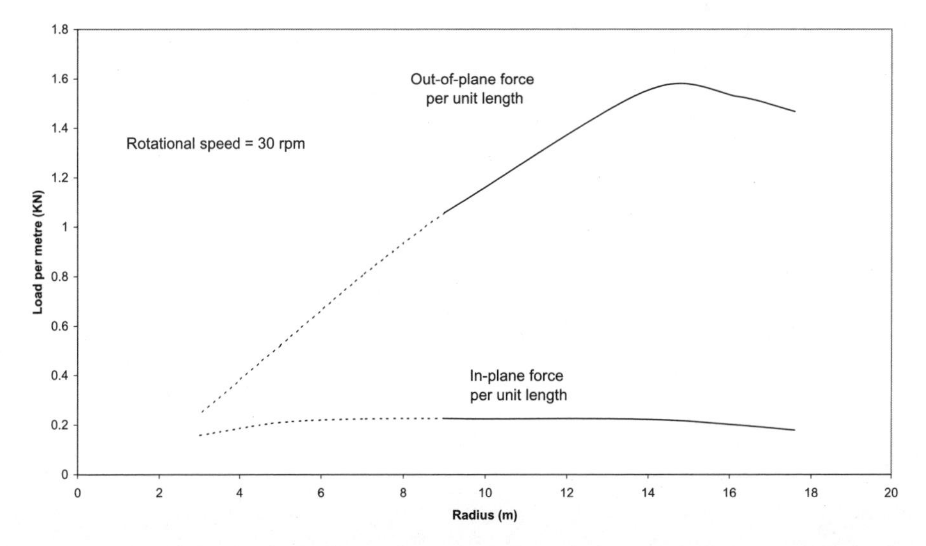

Figure 5.6 Distribution of Blade In-plane and Out-of-plane Aerodynamic Loads during Operation of Typical 40 m Diameter Stall-regulated Machine in a Steady, Uniform 10 m/s Wind

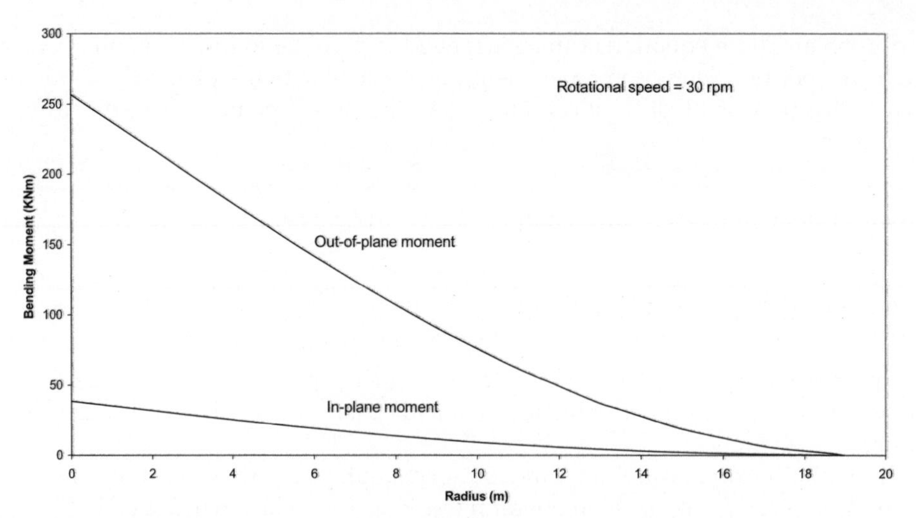

Figure 5.7 Blade In-plane and Out-of-plane Aerodynamic Bending Moment Distributions for Typical 40 m Diameter Stall-regulated Machine Operating in a Steady, Uniform 10 m/s Wind

aerodynamic blade bending moments. The variation of these moments with radius is shown in Figure 5.7 for the example above. The blade bending moments effectively decrease linearly with increasing radius over the inboard third of the blade because of the concentration of loading outboard.

The variation of the blade root out-of-plane bending moment with wind speed is illustrated in Figure 5.8 for the 40 m diameter example machine described above.

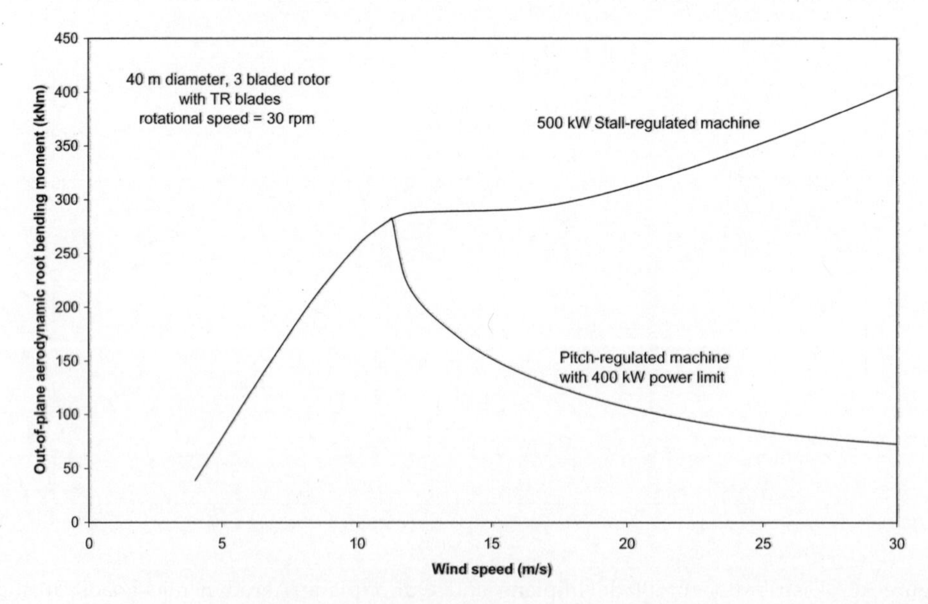

Figure 5.8 Blade Out-of-plane Root Bending Moment During Operation in Steady, Uniform Wind – Variation with Wind Speed for Similar Stall-regulated and Pitch-regulated Machines

As explained in Section 3.12, the phenomenon of stall delay results in significantly increased values of the lift coefficient at higher wind speeds on the inboard section of the rotating blade than predicted by static aerofoil data, such as that reproduced in Figure 3.41. Accordingly, Figure 5.8 and the other figures referred to in this section have been derived using realistic aerofoil data for a rotating LM-19.0 blade reported in Petersen *et al.* (1998), which is based on an empirical modification of static or two-dimensional aerofoil data. The modified data are reproduced in Figure 5.9, and display much higher lift coefficients for the thicker, inboard blade sections at high angles of attack than for the thinner, outboard blade sections because of stall delay at the inboard sections.

Figure 5.8 shows the blade root out-of-plane bending moment increasing nearly linearly with wind speed at first and then levelling off, becoming almost constant for winds between 12 m/s and 16 m/s, as the blade goes into stall. Thereafter the root moment increases again, but much more gently than before.

Also shown on Figure 5.8 is the variation of blade root out-of-plane bending moment with wind speed for the same machine with pitch regulation to limit the power output to 400 kW. It is evident that the bending moment drops away rapidly at wind speeds above rated.

Yawed flow

The application of blade element–momentum theory to steady yawed flow is decribed in Section 3.10.8. This methodology has been used to derive Figure 5.10,

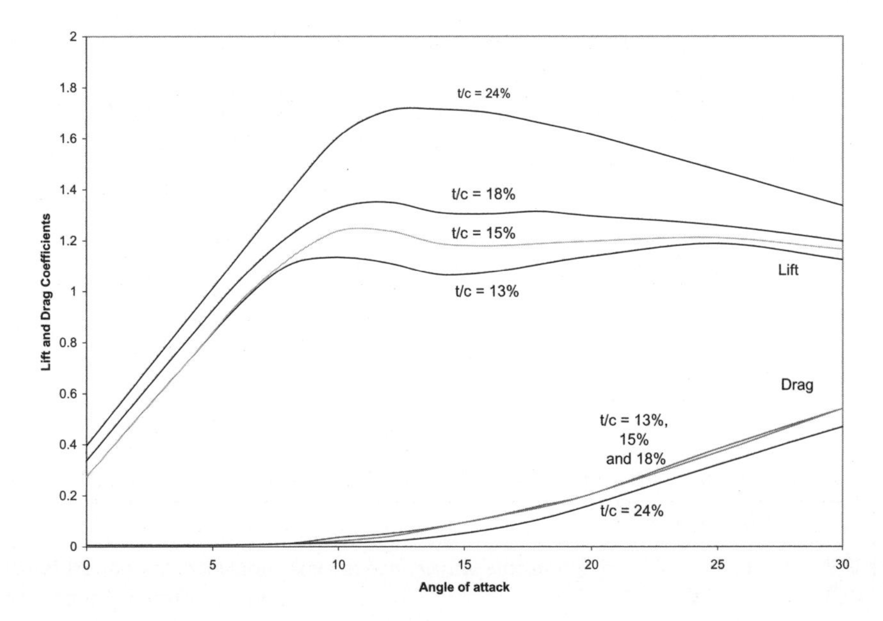

Figure 5.9 Aerofoil Data for LM-19.0 Blade for Various Thickness/Chord Ratios (from Petersen *et al.* (1998))

which shows the variation of the blade root out-of-plane and in-plane moments with azimuth for the 40 m diameter stall-regulated machine described above, operating at a steady yaw angle of +30°. Note that the blade azimuth is measured in the direction of blade rotation, from a zero value at top dead centre, and the yaw angle is defined as positive when the lateral component of air flow with respect to the rotor disc is in the same direction as the blade movement at zero azimuth.

Figure 5.10 reveals a distinct difference between the behaviour at 10 m/s and 20 m/s. In the latter case, the bending moment variation is sinusoidal with a maximum value at 180° azimuth, indicating that the variation is dominated by the effect of the fluctuation of the air velocity relative the blade, W. At 10 m/s, however, the maximum out-of-plane bending moment occurs at about 240° azimuth, suggesting that the non-uniform component of induced velocity, u_1 (Equation (3.107)) is also significant. As wind speed increases, of course, the induction factor, a, becomes small, reducing the impact of u_1.

Shaft tilt

Upwind machines, i.e., wind turbines with the rotor positioned between the tower and the oncoming wind, normally have the rotor shaft tilted upwards by several degrees in order to increase the clearance between the rotor and the tower. Thus, as for the case of yaw misalignment, the flow is inclined to the rotor shaft axis, but tilted upwards rather than sideways, so the treatment of shaft tilt mirrors that of yawed flow.

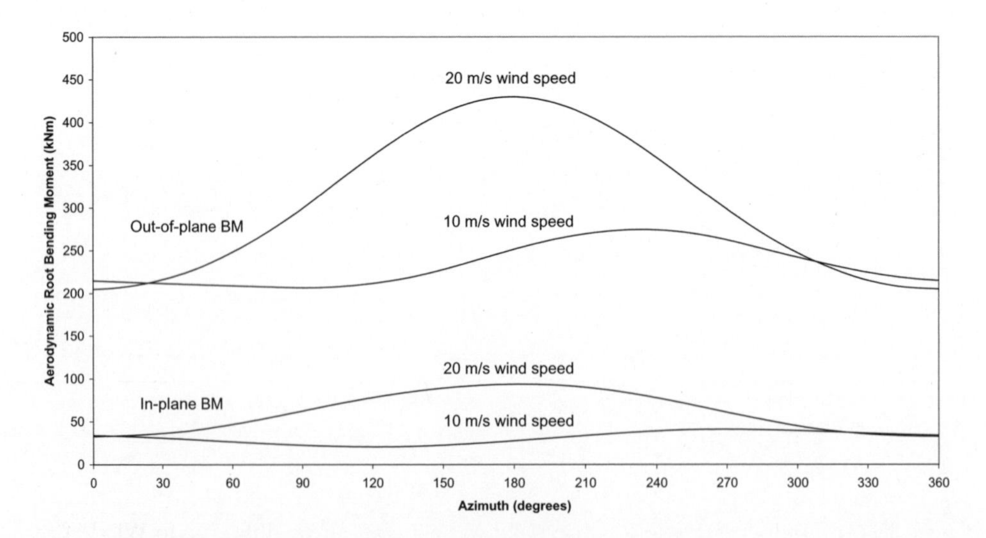

Figure 5.10 Variation of Blade Root Bending Moments with Azimuth, for Typical 40 m Diameter Stall-regulated Machine Operating at a Steady 30° Yaw

Wind shear

The increase of wind speed with height is known as wind shear. The theoretical logarithmic profile, $U(z) \propto \ln(z/z_0)$, is usually approximated by the power law, $U(z) \propto (z/z_{ref})^{\alpha}$ for wind turbine design purposes. The appropriate value of the exponent α increases with the surface roughness, z_0, with a figure of 0.14 typically quoted for level countryside, although the speed up of airflow close to the ground over rounded hills usually results in a lower value at hill tops. As already noted, IEC 61400-1 specifies a conservative value of 0.20.

In applying momentum theory to this case, the velocity component at right angles to the plane of rotation is expressed as $U_\infty(1 + r\cos\psi/z_{hub})^{\alpha}(1 - a)$. The variation of blade root bending moments with azimuth due to wind shear is illustrated in Figure 5.11 for the example 40 m diameter stall-regulated machine, taking the exponent as 0.20, the hub height as 40 m and considering hub-height wind speeds of 10 m/s and 15 m/s. In the former case, the variation is nearly sinusoidal, but in the latter case the root bending moments are effectively constant, as the blade is in stall.

Tower shadow

Blocking of the air flow by the tower results in regions of reduced wind speed both upwind and downwind of the tower. This reduction is more severe for tubular

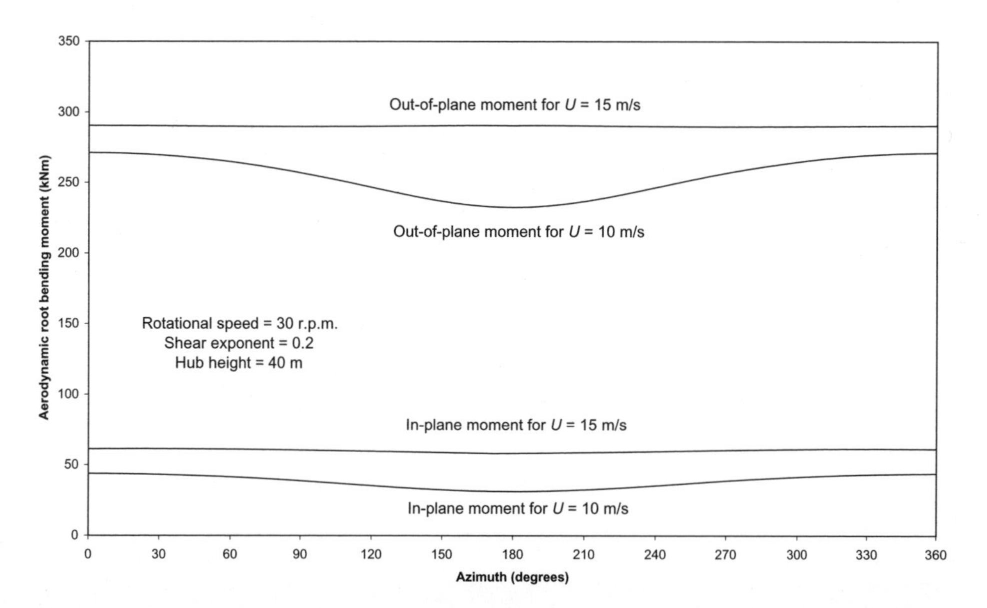

Figure 5.11 Variation of Blade Root Bending Moments with Azimuth due to Wind Shear, for Typical 40 m Diameter Stall-regulated Machine Operating in Steady Hub-height Winds of 10 m/s and 15 m/s (shear exponent = 0.2)

towers than for lattice towers and, in the case of tubular towers, is larger on the downwind side because of flow separation. As a consequence, designers of downwind machines usually position the rotor plane well clear of the tower to minimize the interference effect.

The velocity deficits upwind of a tubular tower can be modelled using potential flow theory. The flow around a cylindrical tower is derived by superposing a doublet, i.e., a source and sink at very close spacing, on a uniform flow, U_∞, giving the stream function:

$$\psi = U_\infty y\left(1 - \frac{(D/2)^2}{x^2 + y^2}\right) \tag{5.20}$$

where D is the tower diameter, and x and y are the longitudinal and lateral coordinates with respect to the tower centre (see Figure 5.12). Differentiation of ψ with respect to y yields the following expression for the flow velocity in the x direction:

$$U = U_\infty\left(1 - \frac{(D/2)^2(x^2 - y^2)}{(x^2 + y^2)^2}\right) \tag{5.21}$$

The second term within the brackets, which is the velocity deficit as a proportion of the undisturbed wind speed, is plotted out against the lateral co-ordinate, y, divided by tower diameter, for a range of upwind distances, x, in Figure 5.13. The velocity deficit on the flow axis of symmetry is equal to $U_\infty(D/2x)^2$ and the total width of the deficit region is twice the upwind distance. Consequently the velocity gradient encountered by a rotating blade decreases rapidly as the upwind distance, x, increases.

The effect of tower shadow on blade loading can be estimated by setting the local velocity component at right angles to the plane of rotation equal to $U(1-a)$ in place of $U_\infty(1-a)$, and applying blade element theory as usual. Results for blade root bending moments for the example 40 m diameter stall-regulated machine are given in Figure 5.14, assuming a tower diameter of 2 m and ignoring dynamic effects. The plots show the variation of in-plane and out-of-plane root moments with azimuth during operation in wind speeds of 10 m/s and 15 m/s, for a blade-tower clearance equal to the tower radius i.e., for $x/D = 1$. Note that the dip in out-of-plane bending moment is more severe at the lower wind speed. Also shown are 10 m/s plots for $x/D = 1.5$, which exhibit a much less severe disturbance.

In the case of downwind turbines, the flow separation and generation of eddies which take place are less amenable to analysis, so empirical methods are used to estimate the mean velocity deficit. Commonly the profile of the velocity deficit is assumed to be of cosine form, so that

$$U = U_\infty\left(1 - k\cos^2\left(\frac{\pi y}{\delta}\right)\right) \tag{5.22}$$

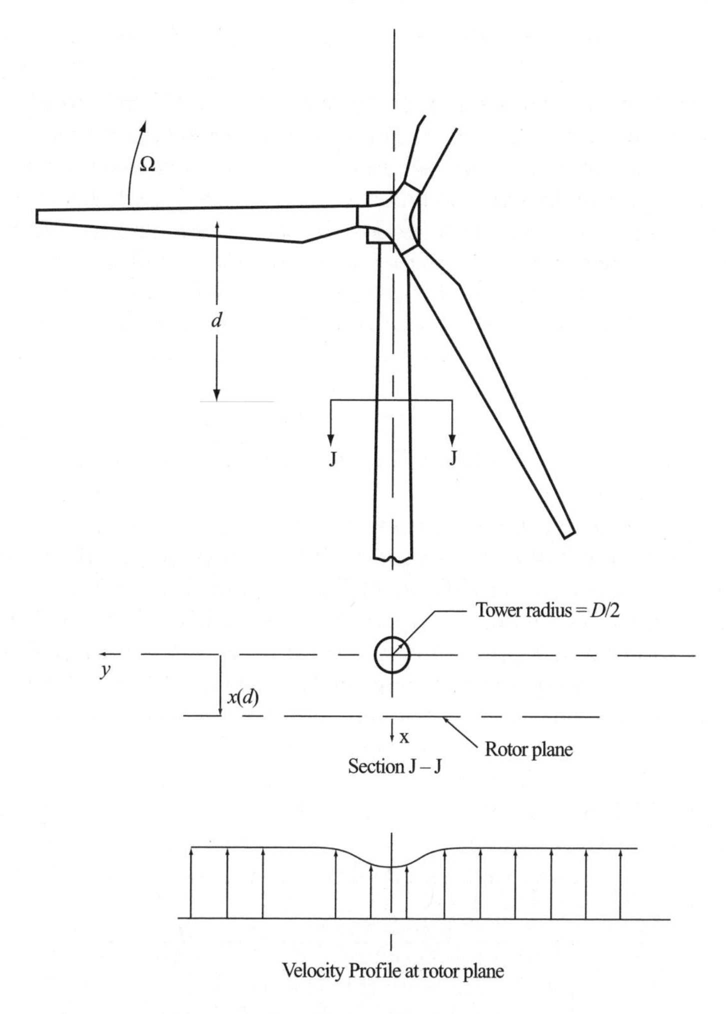

Figure 5.12 Tower Shadow Parameters

where δ is the total width of the deficit region. The slight enhancement of velocities beyond the deficit region is usually ignored (see also Section 6.13.2).

The sharp dip in blade loading caused by tower shadow is more prone to excite blade oscillations than the smooth variations in load due to wind shear, shaft tilt and yaw, and this aspect is considered in the section on blade dynamic response.

Wake effects

Within a wind-farm it is common for one turbine to be operating wholly or partly in the wake of another. In the latter case, which is more severe, the downwind turbine is effectively subjected to horizontal wind shear, and the blade load fluctuations can be analysed accordingly.

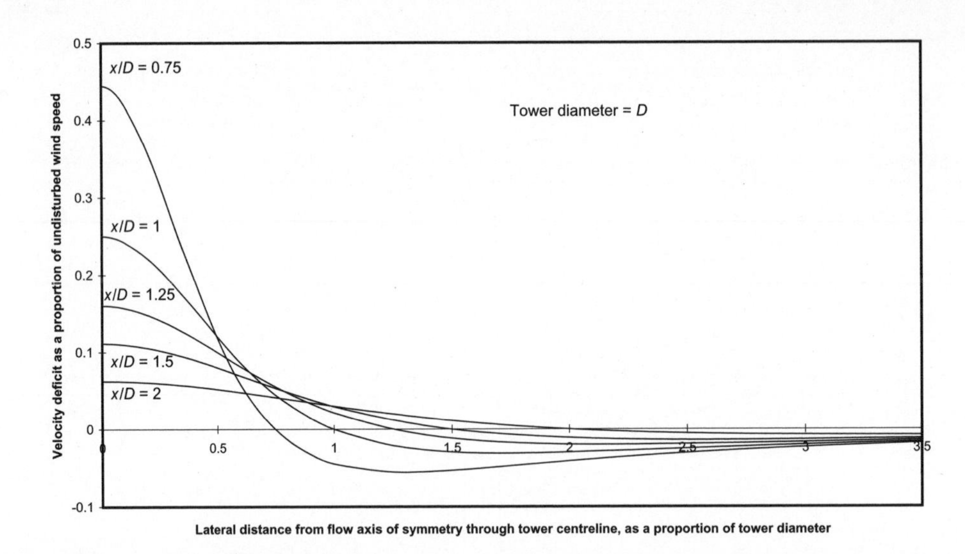

Figure 5.13 Profile of Velocity Deficit due to Tower Shadow at Different Distances x/D Upwind of Tower Centreline

5.7.3 Gravity loads

Gravity loading on the blade results in a sinusoidally varying edgewise bending moment which reaches a maximum when the blade is horizontal, and which changes sign from one horizontal position to the other. It is thus a major source of fatigue loading. For the blade 'TR' (see Example 5.1), the maximum gravity moment, $\int_0^R m(r)r\,dr$ is 134 kNm, so the edgewise bending moment range due to gravity is 268 kNm. This dwarfs the variations in edgewise moment due to yaw or wind shear, which are typically one tenth this value or less. The spanwise distribution of gravity bending moment is shown in Figure 5.15 for blade 'TR'.

5.7.4 Deterministic inertia loads

Centrifugal loads

For a rigid blade rotating with its axis perpendicular to the axis of rotation, the centrifugal forces generate a simple tensile load in the blade which at radius r^* is given by the expression $\Omega^2 \int_{r^*}^R m(r)r\,dr$. As a result, the fluctuating stresses in the blade arising from all loading sources always have a tensile bias during operation. For blade 'TR' rotating at 30 r.p.m., the centrifugal force at the root amounts to 134 kN – approximately seven times its weight.

Thrust loading causes flexible blades to deflect downwind, with the result that the centrifugal forces generate blade out-of-plane moments in opposition to those due to the thrust. This reduction of the moment due to thrust loading is known as

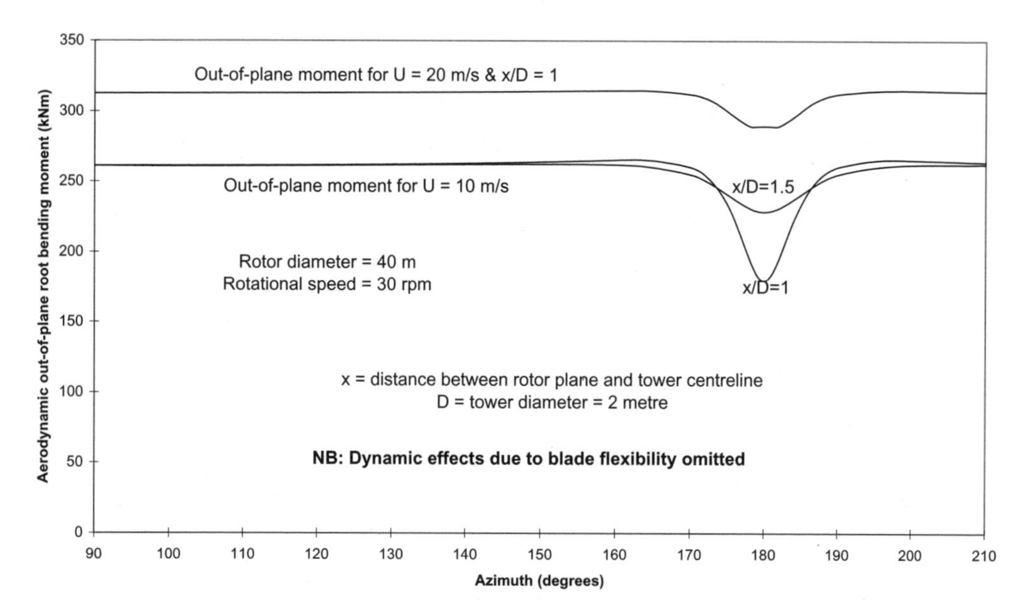

Figure 5.14 Variation of Blade Root Bending Moments with Azimuth due to Tower Shadow, for Typical 40 m Diameter Stall-regulated Upwind Machine Operating in Steady, Uniform Winds of 10 m/s and 20 m/s

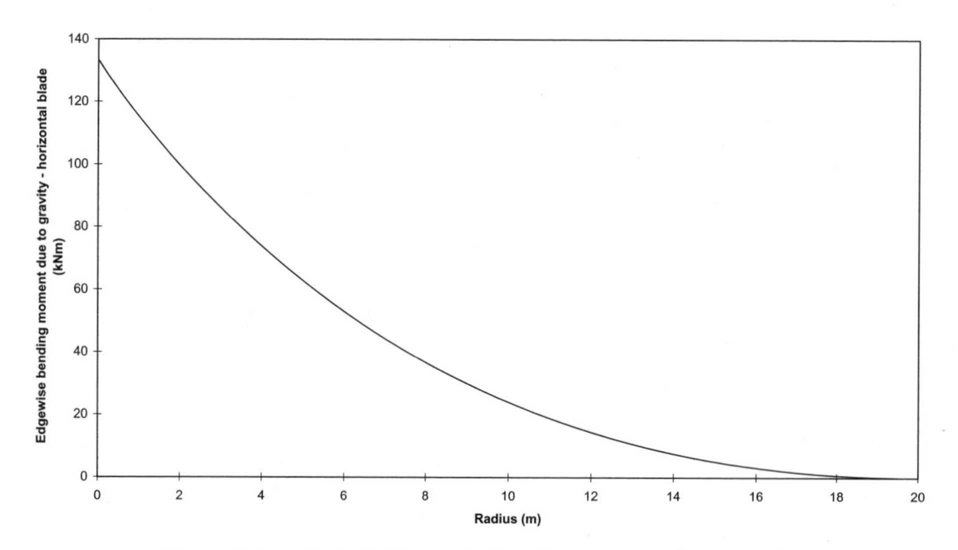

Figure 5.15 Blade 'TR' Gravity Bending Moment Distribution

centrifugal relief. The phenomenon is non-linear, so iterative techniques are required to arrive at a solution. Greater centrifugal relief can be obtained by coning the rotor so that the blades are inclined downwind in the first place. A balance can be struck so that the maximum forward out-of-plane moment due to centrifugal loads in very low wind is approximately equal to the maximum rearward out-of-plane moment due to the thrust loading in combination with centrifugal loads during operation in rated wind.

Gyroscopic loads

When an operating machine yaws, the blades experience gyroscopic loads perpendicular to the plane of rotation. Consider the point A on a rotor rotating clockwise at a speed of Ω rad/s, as illustrated in Figure 5.16. The instantaneous horizontal velocity component of point A due to rotor rotation is Ωz, where z is the height of the point above the hub. If the machine is yawing clockwise in plan at a speed of Λ rad/s, then it can be shown that point A accelerates at $2\Omega\Lambda z$ towards the wind, assuming the rotor is rigid. Integrating the resulting inertial force over the blade length gives the following expression for blade root out-of-plane bending moment

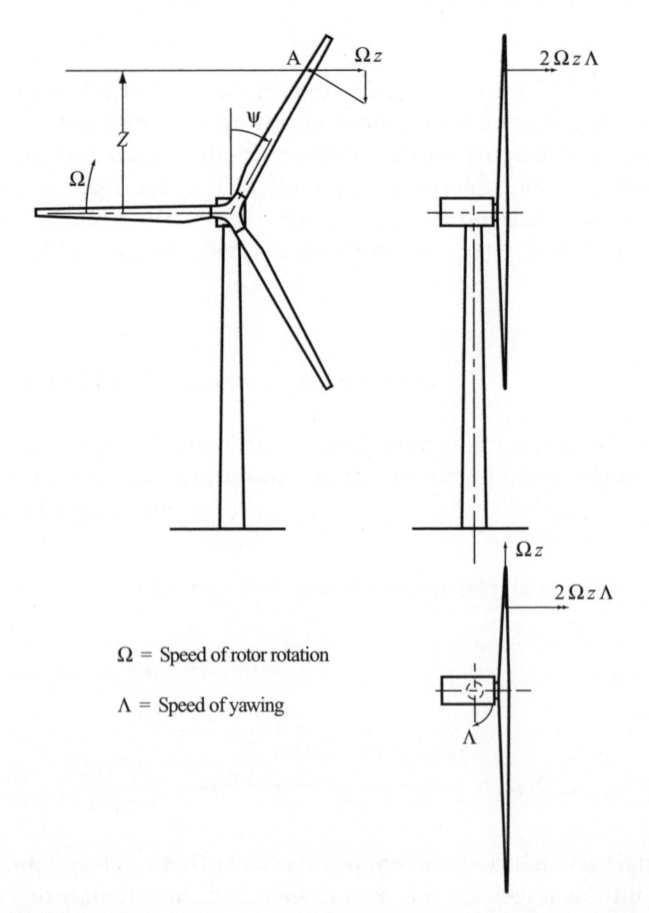

Figure 5.16 Gyroscopic Acceleration of a Point on a Yawing Rotor

$$M_Y = \int_0^R 2\Omega\Lambda zrm(r)\,\mathrm{d}r = 2\Omega\Lambda\cos\psi\int_0^R r^2 m(r)\,\mathrm{d}r = 2\Omega\Lambda\cos\psi I_B \tag{5.23}$$

where I_B is the blade inertia about the root.

As an example, consider a 40 m machine with 'TR' blades yawing at one degree per second during operation at 30 r.p.m. The blade root inertia is 153 Tm^2, so the maximum value of M_Y is $2\pi(0.0175)153 = 17$ kNm. This is only about 5 percent of the maximum out-of-plane moment due to aerodynamic loads.

Braking loads

Rotor deceleration due to mechanical braking introduces edgewise blade bending moments which are additive to the gravity moments on a descending blade.

Teeter loads

Blade out-of-plane root bending moments can be eliminated entirely by mounting each blade on a hinge so that it is free to rotate in the fore–aft direction. Although centrifugal forces are effective in controlling the cone angle of each blade at normal operating speeds, the need for alternative restraints during start-up and shut-down means that such hinges are rarely used. However, in the case of two bladed machines, it is convenient to mount the whole rotor on a single shaft hinge allowing fore–aft rotation or 'teetering', and this arrangement is frequently adopted in order to reduce out-of-plane bending moment fluctuations at the blade root, and to prevent the transmission of blade out-of-plane moments to the low speed shaft. As teetering is essentially a dynamic phenomenon, consideration of teeter behaviour is deferred to Section 5.8.

5.7.5 Stochastic aerodynamic loads – analysis in the frequency domain

As noted in Section 5.7.1, the random loadings on the blade due to short-term wind speed fluctuations are known as stochastic aerodynamic loads. The wind speed fluctuations about the mean at a *fixed* point in space are characterized by a probability distribution – which, for most purposes, can be assumed to be normal – and by a power spectrum which describes how the energy of the fluctuations is distributed between different frequencies (see Sections 2.6.3 and 2.6.4).

The stochastic loads are most conveniently analysed in the frequency domain but, in order to facilitate this, it is usual to assume a linear relation between the fluctuation, u, of the wind speed incident on the aerofoil and the resultant loadings. This is a reasonable assumption at high tip speed ratio, as will be shown. The fluctuating quasisteady aerodynamic lift per unit length, L, is $\frac{1}{2}\rho W^2 C_l c$, where W is the air velocity relative to the blade, C_l is the lift coefficient and the drag term is ignored. Because the flow angle, ϕ, is small at high tip speed ratio, λ, the relative air velocity, W, can be assumed to be changing much more slowly with the wind speed than C_l, so that $\mathrm{d}W/\mathrm{d}u$ can be ignored. As a result,

$$\frac{\mathrm{d}L}{\mathrm{d}u} = \tfrac{1}{2}\rho W^2 c \frac{\mathrm{d}C_l}{\mathrm{d}\alpha}\frac{\mathrm{d}\alpha}{\mathrm{d}u} \tag{5.24}$$

where α, the angle of attack, is equal to $(\phi - \beta)$.

If the blades are not pitching, then the local blade twist, β, is constant, so that $\mathrm{d}\alpha/\mathrm{d}u = \mathrm{d}\phi/\mathrm{d}u$. To preserve linearity, it is necessary to assume that the rate of change of lift coefficient with angle of attack, $\mathrm{d}C_l/\mathrm{d}\alpha$ is constant, which is tenable only if the blade remains unstalled. Assuming, for simplicity, that the wake is frozen, i.e., that the induced velocity, $\overline{U}a$, remains constant, despite the wind speed fluctuations, u, we obtain

$$\tan\phi \cong (\overline{U}(1-a) + u)/\Omega r,$$

so that, for ϕ small, $\mathrm{d}\varphi/\mathrm{d}u \cong 1/\Omega r$ and $W \cong \Omega r$, leading to

$$\Delta L = L - \overline{L} = u\frac{\mathrm{d}L}{\mathrm{d}u} = \tfrac{1}{2}\rho(\Omega r)^2 c\frac{\mathrm{d}C_l}{\mathrm{d}\alpha}\frac{u}{\Omega r} = \tfrac{1}{2}\rho\Omega r c\frac{\mathrm{d}C_l}{\mathrm{d}\alpha}u \tag{5.25}$$

Hence

$$\sigma_L = \left(\tfrac{1}{2}\rho\Omega\frac{\mathrm{d}C_l}{\mathrm{d}\alpha}\right) r c \sigma_u$$

Normally $\mathrm{d}C_l/\mathrm{d}\alpha$ is equal to 2π.

If the turbulence integral length scale is large compared to the blade radius, then the expression for the standard deviation of the blade root bending moment (assuming a completely rigid blade) approximates to

$$\sigma_M = \int_0^R \sigma_L r\,\mathrm{d}r = \tfrac{1}{2}\rho\Omega\frac{\mathrm{d}C_l}{\mathrm{d}\alpha}\sigma_u\int_0^R c(r)r^2\,\mathrm{d}r \tag{5.26}$$

where σ_u is the standard deviation of the wind speed incident on the rotor disc which, by virtue of the 'frozen wake' assumption, equates to the standard deviation of the wind speed in the undisturbed flow. If, as will be the case in practice, the longitudinal wind fluctuations are not perfectly correlated along the length of the blade, then

$$\sigma_M^2 = \left(\tfrac{1}{2}\rho\Omega\frac{\mathrm{d}C_l}{\mathrm{d}\alpha}\right)^2\int_0^R\int_0^R \kappa_u(r_1, r_2, 0)c(r_1)c(r_2)r_1^2 r_2^2\,\mathrm{d}r_1\,\mathrm{d}r_2 \tag{5.27}$$

where $\kappa_u(r_1, r_2, 0)$ is the cross correlation function $\kappa_u(r_1, r_2, \tau)$ between the wind fluctuations at radii r_1 and r_2 with the time lag τ set equal to zero, i.e.,

$$\kappa_u(r_1, r_2, 0) = \left[\frac{1}{T}\int_0^T u(r_1, t)u(r_2, t)\,\mathrm{d}t\right] \tag{5.28}$$

In reality, of course, the blade will not be completely rigid, so the random wind loading will excite the natural modes of blade vibration. In order to quantify these excitations, it is first necessary to know the energy content of the incident wind fluctuations *as seen by each point on the rotating blade* at the blade natural frequencies – information which is provided by the 'rotationally sampled spectrum'. This spectrum is significantly different from the fixed point spectrum, because a rotating blade will often slice through an individual gust (defined as a volume of air travelling at above average speed) several times, as the gust dimensions are frequently large compared with the distance travelled by the air in one turbine revolution. This phenomenon, known as 'gust slicing', considerably enhances the frequency content at the rotational frequency, and to a lesser extent, at its harmonics also.

The method for deriving the rotational spectrum is described below. The dynamic response of a flexible blade to random wind loading is explored in Section 5.8.

Rotationally sampled spectrum

The derivation of the power spectrum of the wind seen by a point on a rotating blade is based on the Fourier transform pairs:

$$S_u(n) = 4\int_0^\infty \kappa_u(\tau)\cos 2\pi n\tau\,\mathrm{d}\tau \tag{5.29}$$

$$\kappa_u(\tau) = \int_0^\infty S_u(n)\cos 2\pi n\tau\,\mathrm{d}n \tag{5.30}$$

where $S_u(n)$ is the single sided spectrum of wind speed fluctuations in terms of frequency in Hz. First, the latter equation is used to obtain the autocorrelation function, $\kappa_u(\tau)$, for the along wind turbulent fluctuations at a fixed point in space from the corresponding power spectrum. Second, $\kappa_u(\tau)$ is used to derive the related autocorrelation function, $\kappa_u^o(r, \tau)$, for a point on the rotating blade at radius r. Finally this function is transformed using Equation (5.29) to yield the rotationally sampled spectrum. The three steps are set out in more detail below. Note that three key simplifying assumptions are made: that the turbulence is homogeneous and isotropic, and that the flow is incompressible.

Step 1 – Derivation of the autocorrelation function at a fixed point: The von Karman spectrum

$$\frac{S_u(n)}{\sigma_u^2} = \frac{4L_u^x}{\overline{U}(1 + 70.8(nL_u^x/\overline{U})^2)^{\frac{5}{6}}} \tag{5.31}$$

is chosen as the input power spectrum of the along-wind wind speed fluctuations at a fixed point in space. It can be shown that Equation (5.30) yields the following expression for the corresponding auto correlation function:

$$\kappa_u(\tau) = \frac{2\sigma_u^2}{\Gamma\left(\frac{1}{3}\right)}\left(\frac{\tau/2}{T'}\right)^{\frac{1}{3}} K_{\frac{1}{3}}\left(\frac{\tau}{T'}\right) \tag{5.32}$$

where T' is related to the integral length scale, L_u^x, by the formula

$$T' = \frac{\Gamma\left(\frac{1}{3}\right)}{\Gamma\left(\frac{5}{6}\right)\sqrt{\pi}}\frac{L_u^x}{\overline{U}} \cong 1.34\frac{L_u^x}{\overline{U}} \tag{5.33}$$

$\Gamma()$ is the Gamma function and $K_{\frac{1}{3}}(x)$ is a modified Bessel function of the second kind and order $v = \frac{1}{3}$. The general definition of $K_v(x)$ is:

$$K_v(x) = \frac{\pi}{2\sin\pi v}\sum_{m=0}^{\infty}\frac{(x/2)^{2m}}{m!}\left[\frac{(x/2)^{-v}}{\Gamma(m-v+1)} - \frac{(x/2)^{v}}{\Gamma(m+v+1)}\right] \tag{5.34}$$

Step 2 – Derivation of the autocorrelation function at a point on the rotating blade: This derivation makes use of Taylor's 'frozen turbulence' hypothesis, by which the instantaneous wind speed at point C at time $t = \tau$ is assumed to be equal to that at a point B a distance $\overline{U}\tau$ upwind of C at time $t = 0$, $\overline{U}$ being the mean wind speed. Thus, referring to Figure 5.17, the autocorrelation function $\kappa_u^o(r, \tau)$ for the along-wind wind fluctuations seen by a point Q at radius r on the rotating blade is equal to the cross correlation function $\kappa_u(\vec{s}, 0)$ between the simultaneous along-wind wind fluctuations at points A and B. Here A and C are the positions of point Q at the beginning and end of time interval τ respectively, B is $\overline{U}\tau$ upwind of C and $\vec{s}$ is the vector BA. (Note that the superscript $^\circ$ denotes that the autocorrelation function relates to a point on a rotating blade rather than a fixed point. The same convention will be adopted in relation to power spectra.)

Batchelor (1953) has shown that, if the turbulence is assumed to be homogeneous and isotropic, the cross correlation function, $\kappa_u(\vec{s}, 0)$ is given by:

$$\kappa_u(\vec{s}, 0) = (\kappa_L(s) - \kappa_T(s))\left(\frac{s_1}{s}\right)^2 + \kappa_T(s) \tag{5.35}$$

where $\kappa_L(s)$ is the cross correlation function between velocity components at points A and B, s apart, in a direction *parallel* to AB (v_L^A and v_L^B in Figure 5.17), and $\kappa_T(s)$ is

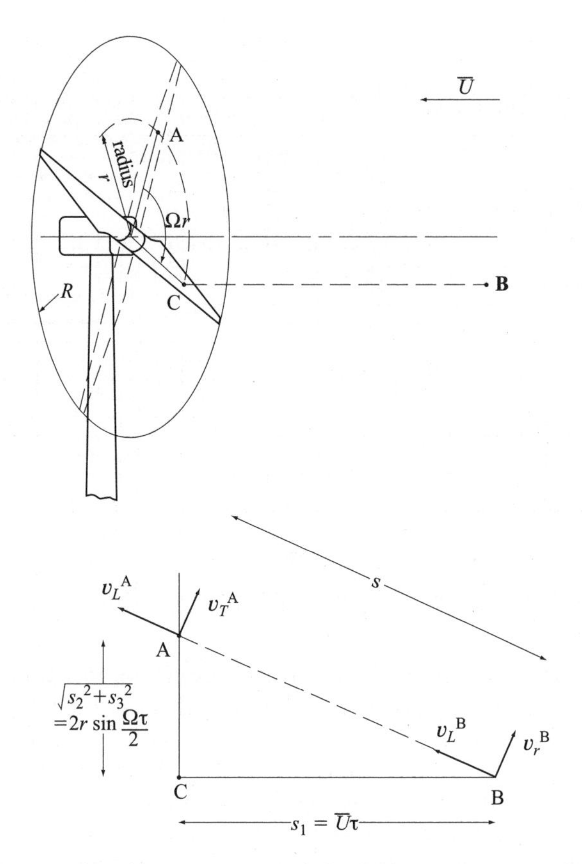

Figure 5.17 Geometry for the Derivation of the Velocity Autocorrelation Function for a Point on a Rotating Blade

the corresponding function for velocity components (v_T^A and v_T^B) in a direction *perpendicular* to AB. s_1 is the separation of points A and B measured in the along-wind direction, $\overline{U}\tau$. Noting that the distance between points A and C on the rotor disc is $2r\sin(\Omega\tau/2)$, we have

$$s^2 = \overline{U}^2\tau^2 + 4r^2\sin^2(\Omega\tau/2) \tag{5.36}$$

Hence

$$\kappa_u(\vec{s}, 0) = \kappa_L(s)\left(\frac{\overline{U}\tau}{s}\right)^2 + \kappa_T(s)\left[1 - \left(\frac{\overline{U}\tau}{s}\right)^2\right] = \kappa_L(s)\left(\frac{\overline{U}\tau}{s}\right)^2 + \kappa_T(s)\left(\frac{2r\sin(\Omega\tau/2)}{s}\right)^2 \tag{5.37}$$

For incompressible flow, it can also be shown (Batchelor, 1953) that

$$\kappa_T(s) = \kappa_L(s) + \frac{s}{2}\frac{d\kappa_L(s)}{ds} \tag{5.38}$$

Substitution of Equation (5.38) in Equation (5.37) gives

$$\kappa_u(\vec{s}, 0) = \kappa_L(s) + \frac{s}{2}\frac{d\kappa_L(s)}{ds}\left(\frac{2r\sin(\Omega\tau/2)}{s}\right)^2 \tag{5.39}$$

When the vector $\vec{s}$ is in the along-wind direction, $\kappa_L(s)$ translates to $\kappa_u(s_1)$, which, by Taylor's 'frozen turbulence' hypothesis, equates to the autocorrelation function at a fixed point, $\kappa_u(\tau)$ (Equation (5.32)), with $\tau = s_1/\overline{U}$. Thus

$$\kappa_L(s_1) = \frac{2\sigma_u^2}{\Gamma\left(\frac{1}{3}\right)}\left(\frac{s_1/2}{T'\overline{U}}\right)^{\frac{1}{3}} K_{\frac{1}{3}}\left(\frac{s_1}{T'\overline{U}}\right) \tag{5.40}$$

Because the turbulence is assumed to be isotropic, $\kappa_L(s)$ is independent of the direction of the vector $\vec{s}$, so we can write, with the aid of Equation (5.33),

$$\kappa_L(s) = \frac{2\sigma_u^2}{\Gamma\left(\frac{1}{3}\right)}\left(\frac{s/2}{T'\overline{U}}\right)^{\frac{1}{3}} K_{\frac{1}{3}}\left(\frac{s}{T'\overline{U}}\right) = \frac{2\sigma_u^2}{\Gamma\left(\frac{1}{3}\right)}\left(\frac{s/2}{1.34L_u^x}\right)^{\frac{1}{3}} K_{\frac{1}{3}}\left(\frac{s}{1.34L_u^x}\right) \tag{5.41}$$

Noting that

$$\frac{d}{dx}[x^v K_v(x)] = x^v K_{(1-v)}(x)$$

the following expression for the autocorrelation function for the along-wind fluctuations at a point at radius r on the rotating blade is obtained by substituting Equation (5.41) in Equation (5.39):

$$\kappa_u^o(r, \tau) = \kappa_u(\vec{s}, 0)$$

$$= \frac{2\sigma_u^2}{\Gamma\left(\frac{1}{3}\right)}\left(\frac{s/2}{1.34L_u^x}\right)^{\frac{1}{3}}\left[K_{1/3}\left(\frac{s}{1.34L_u^x}\right) + \frac{s}{2(1.34L_u^x)}K_{2/3}\left(\frac{s}{1.34L_u^x}\right)\left(\frac{2r\sin(\Omega\tau/2)}{s}\right)^2\right] \tag{5.42}$$

where s is defined in terms of τ by Equation (5.36) above.

Step 3 – Derivation of the power spectrum seen by a point on the rotating blade: The rotationally sampled spectrum is obtained by taking the Fourier transform of $\kappa_u^o(r, \tau)$ from Equation (5.42):

$$S_u^o(n) = 4\int_0^{\infty} \kappa_u^o(r, \tau)\cos 2\pi n\tau \, d\tau$$

$$= 2\int_{-\infty}^{\infty} \kappa_u^o(r, \tau)\cos 2\pi n\tau \, d\tau \qquad \text{as } \kappa_u^o(r, \tau) = \kappa_u^o(r, -\tau) \tag{5.43}$$

As no analytical solution has been found for the integral, a solution has to be obtained numerically using a discrete Fourier transform (DFT). First the limits of integration are reduced to $-T/2$, $+T/2$, as $\kappa_u^o(r, \tau)$ tends to zero for large τ. Then the limits of integration are altered to 0, T with $\kappa_u^o(r, \tau)$ set equal to $\kappa_u^o(r, T-\tau)$ for $\tau > T/2$, as $\kappa_u^o(r, \tau)$ is now assumed to be periodic with period T. Thus

$$S_u^o(n) = 2\int_0^T \kappa_u^{*o}(r, \tau)\cos 2\pi n\tau \, d\tau \tag{5.44}$$

where the asterisk denotes that $\kappa_u^o(r, \tau)$ is 'reflected' for $T > T/2$. The discrete Fourier transform then becomes

$$S_u^o(n_k) = 2T\left[\frac{1}{N}\sum_{p=0}^{N-1} \kappa_u^{*o}(r, pT/N)\cos 2\pi kp/N\right] \tag{5.45}$$

Here, N is the number of points taken in the time series of $\kappa_u^{*o}(r, pT/N)$, and the power spectral density is calculated at the frequencies $n_k = k/T$ for $k = 0$, $1, 2 \ldots N-1$. The expression in square brackets can be evaluated using a standard fast Fourier transform (FFT), provided N is chosen equal to a power of 2. Clearly N should be as large as possible if a wide range of frequencies is to be covered at high resolution. Just as $\kappa_u^{*o}(r, \tau)$ is symmetrical about $T/2$, the values of $S_u^o(n_k)$ obtained from the FFT are symmetrical about the mid-range frequency of $N/(2T)$, and the values above this frequency have no real meaning. Moreover, the values of power spectral density calculated by the DFT at frequencies approaching $N/(2T)$ will be in error as a result of aliasing, because these are falsely distorted by frequency components above $N/(2T)$ which contribute to the $\kappa_u^{*o}(r, pT/N)$ series. Assuming that the calculated spectral densities are valid up to a frequency of $N/(4T)$, then the selection of $T = 100$ s and $N = 1024$ would enable the FFT to give useful results up to a frequency of about 2.5 Hz at a frequency interval of 0.01 Hz.

Example 5.2

As an illustration, results have been derived for points on a 20 m radius blade rotating at 30 r.p.m. in a mean wind speed of 8 m/s. Following IEC 61400-1, the integral length scale L_u^x is taken as 73.5 m. Figure 5.18 shows how the normalized autocorrelation function, $\rho_u^o(r, \tau)\,(= \kappa_u^o(r, \tau)/\sigma_u^2)$, for the longitudinal wind fluctuations varies with the number of rotor revolutions at 20 m, 10 m and 0 m radii. For $r = 10$ m, and even more so for $r = 20$ m, these curves display pronounced peaks after each full revolution, when the blade may be thought of as encountering the initial gust or lull once more.

Figure 5.19 shows the corresponding rotationally sampled power spectral density function, $R_u^o(r, n)(= nS_u^o(r, n)/\sigma_u^2)$ plotted out against frequency, n, using a logarithmic scale for the latter. It is clear that there is a substantial shift of the frequency content of the spectrum to the frequency of rotation and, to a lesser degree to its harmonics, with the extent of the shift increasing with radius. Note that the spectral density is shown as increasing above a frequency of about 3 Hz. This is an error

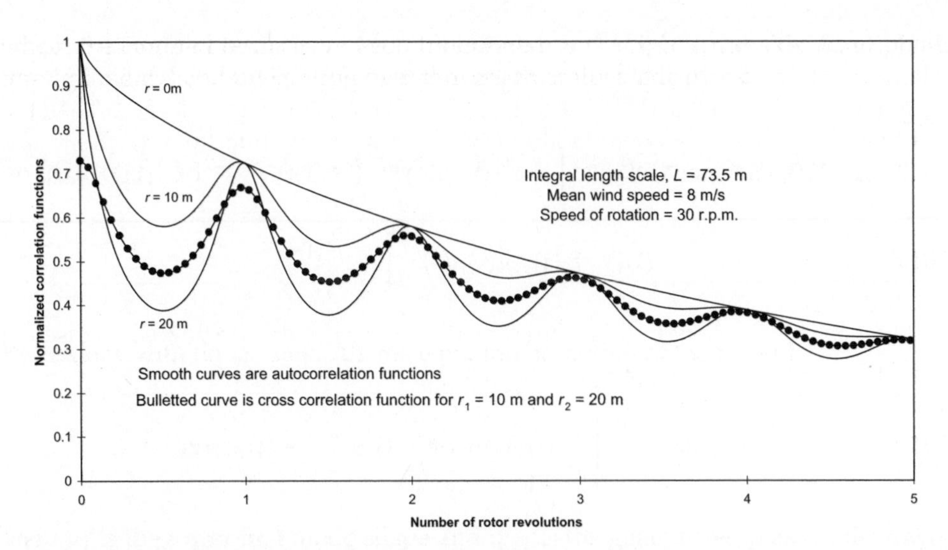

Figure 5.18 Normalized Autocorrelation and Cross Correlation Functions for Along-wind Wind Fluctuations for Points on a Rotating Blade at Different Radii

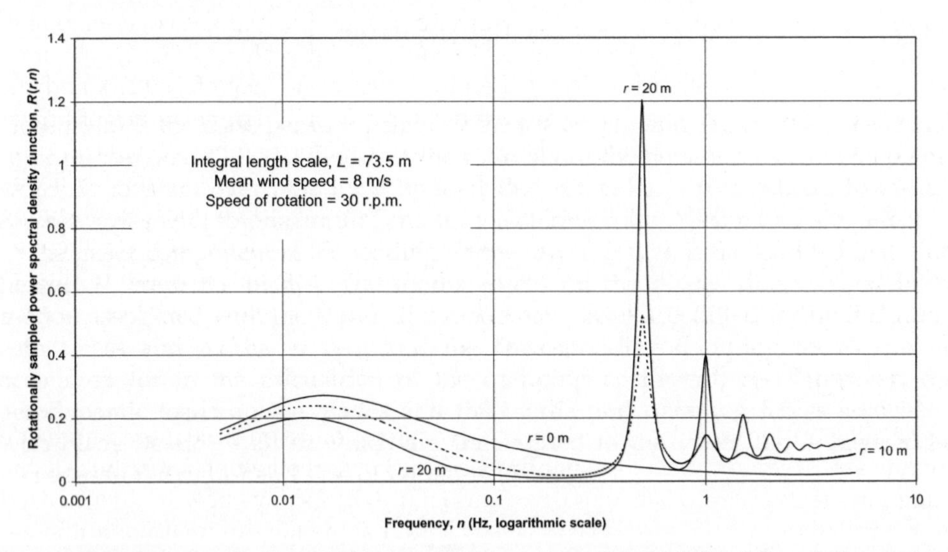

Figure 5.19 Rotationally Sampled Power Spectra of Longitudinal Wind Speed Fluctuations at Different Radii

which arises from the aliasing effect described above. Figure 5.20 is a repeat of Figure 5.19, but with a logarithmic scale used on both axes.

It is instructive to consider how the various input parameters affect the shift of energy to the rotational frequency. As $\kappa_u^o(r, \tau) = \kappa_u^o(\vec{s}, 0)$ decreases monotonically with increasing s, Equation (5.36) indicates that the depths of the troughs in this function – and hence the transfer of energy to the rotational frequency – increases

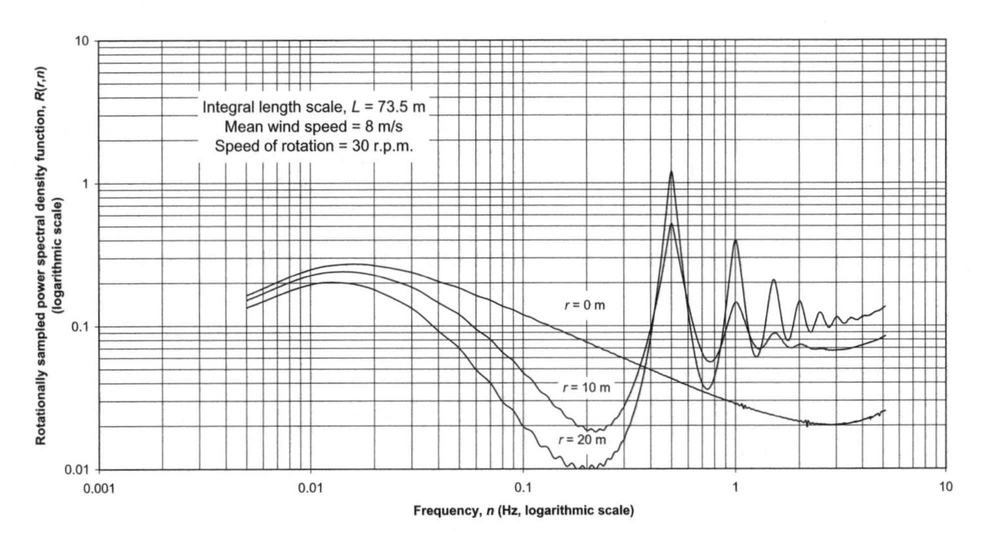

Figure 5.20 Rotationally Sampled Power Spectra of Longitudinal Wind Speed Fluctuations at Different Radii: log–log Plot

roughly in proportion to the tip speed ratio, $\Omega r/\overline{U}$, and will thus be most significant for fixed-speed two-bladed machines (which generally rotate faster than three-bladed ones) in low wind speeds.

Rotationally sampled cross spectra

The expressions for the spectra of blade bending moments and shears are normally functions of entities known as rotationally sampled cross spectra for pairs of points along the blade, which are analogous to the rotationally sampled ordinary spectra for single points described above. The cross spectrum for a pair of points at radii r_1 and r_2 on a rotating blade is thus related to the corresponding cross correlation function by the Fourier transform pair

$$S_u^o(r_1, r_2, n) = 4\int_0^\infty \kappa_u^o(r_1, r_2, \tau)\cos 2\pi n\tau \,\mathrm{d}\tau \tag{5.46a}$$

$$\kappa_u^o(r_1, r_2, \tau) = \int_0^\infty S_u^o(r_1, r_2, n)\cos 2\pi n\tau \,\mathrm{d}n \tag{5.46b}$$

Setting $\tau = 0$ in Equation (5.46b) gives

$$\kappa_u^o(r_1, r_2, 0) = \int_0^\infty S_u^o(r_1, r_2, n)\,\mathrm{d}n \tag{5.47}$$

which, when substituted into the expression for the standard deviation of the blade root bending moment in Equation (5.27) gives

$$\sigma_M^2 = \left(\tfrac{1}{2}\rho\Omega\frac{dC_l}{d\alpha}\right)^2 \int_0^R \int_0^R \left[\int_0^\infty S_u^o(r_1, r_2, n)\,dn\right] c(r_1)c(r_2)r_1^2 r_2^2\,dr_1\,dr_2 \tag{5.48}$$

From this, it can be deduced that the power spectrum of the blade root bending moment is

$$S_M(n) = \left(\tfrac{1}{2}\rho\Omega\frac{dC_l}{d\alpha}\right)^2 \int_0^R \int_0^R S_u^o(r_1, r_2, n)c(r_1)c(r_2)r_1^2 r_2^2\,dr_1\,dr_2 \tag{5.49}$$

The derivation of the rotationally sampled cross spectrum, $S_u^o(r_1, r_2, n)$, exactly parallels the derivation of the rotationally sampled single point spectrum given above, with the cross correlation function $\kappa_u^o(r_1, r_2, \tau)$ between the longitudinal wind fluctuations at points at radii r_1 and r_2 on the rotating blade replacing the autocorrelation function in step 2. Here the expression for the separation distance, s, given in Equation (5.36), is replaced by

$$s^2 = \overline{U}^2\tau^2 + r_1^2 + r_2^2 - 2r_1 r_2 \cos\Omega\tau \tag{5.50}$$

The expression for the cross-correlation function thus becomes:

$$\kappa_u^o(r_1, r_2, \tau) = \frac{2\sigma_u^2}{\Gamma(\frac{1}{3})}\left(\frac{s/2}{1.34L_u^x}\right)^{\frac{1}{3}} \left[K_{1/3}\left(\frac{s}{1.34L_u^x}\right) + \frac{s}{2(1.34L_u^x)} K_{2/3}\left(\frac{s}{1.34L_u^x}\right)\left(\frac{r_1^2 + r_2^2 - 2r_1 r_2\cos\Omega\tau}{s^2}\right)\right] \tag{5.51}$$

with s defined by Equation (5.50).

The form of the resulting normalized cross-correlation function, $\rho_u^o(r_1, r_2, \tau) = \kappa_u^o(r_1, r_2, \tau)/\sigma_u^2$, is illustrated in Figure 5.18 for the case considered in Example 5.2, taking $r_1 = 10$ m and $r_2 = 20$ m. In Figure 5.21, the rotationally sampled cross spectrum for this case is compared with the rotationally sampled single point spectra or 'autospectra' at these radii. It can be seen that the form of the cross spectrum curve is similar to that of the autospectra with a pronounced peak at the rotational frequency roughly midway between the peaks of the two autospectra. At higher frequencies, however, the cross spectrum falls away much more rapidly.

The evaluation of the the power spectrum of the blade root bending moment is, in practice, carried out using summations to approximate to the integrals in Equation (5.49), as follows:

$$S_M(n) = \left(\tfrac{1}{2}\rho\Omega\frac{dC_l}{d\alpha}\right)^2 \sum_j \sum_k S_u^o(r_j, r_k, n)c(r_j)c(r_k)r_j^2 r_k^2(\Delta r)^2 \tag{5.52}$$

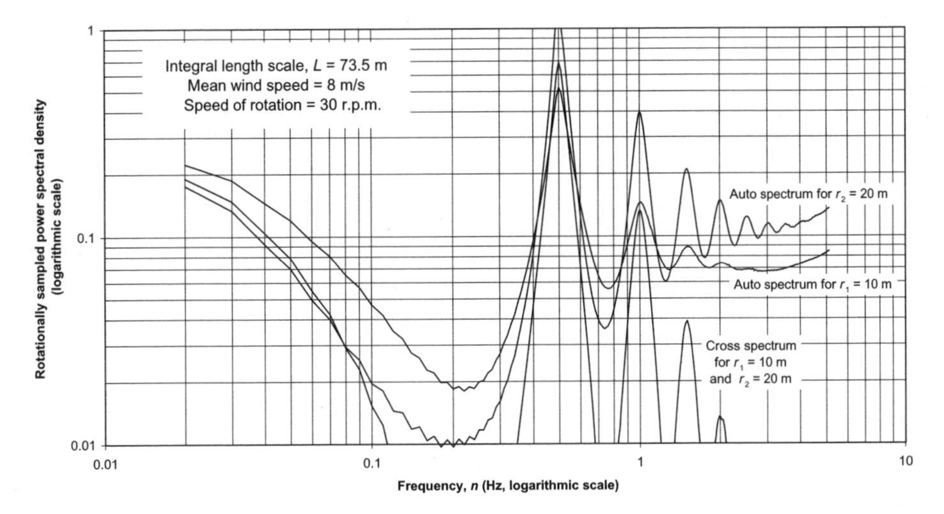

Figure 5.21 Rotationally Sampled Cross Spectrum of Longitudinal Wind Speed Fluctuations at 10 m and 20 m Radii Compared with Auto Spectra: log–log Plot

Limitations of analysis in the frequency domain

As noted at the beginning of this section, analysis of stochastic aerodynamic loads in the frequency domain depends for its validity on a *linear* relationship between the incident wind speed and the blade loading. Thus the method becomes increasingly inaccurate for pitch regulated machines as winds approach the cut-out value and breaks down completely for stall-regulated machines once the wind speed is high enough to cause stall. In order to avoid these limitations, it is necessary to carry out the analysis in the time domain.

5.7.6 Stochastic aerodynamic loads – analysis in the time domain

Wind simulation

The analysis of stochastic aerodynamic loads in the time domain requires, as input, a simulated wind field extending over the area of the rotor disc and over time. Typically, this is obtained by generating simultaneous time histories at points over the rotor disc, which have appropriate statistical properties both individually, and in relation to each other. Thus, the power spectrum of each time history should conform to one of the standard power spectra (e.g., Von Karman or Kaimal), and the normalized cross spectrum (otherwise known as the coherence function) of the time histories at two different points should equate to the coherence function corresponding to the chosen power spectrum and the distance separating the points. For example, the coherence of the longitudinal component of turbulence corresponding to the Kaimal power spectrum for points j and k separated by a distance Δs_{jk} perpendicular to the wind direction is:

$$C_{jk}(n) = \frac{S_{jk}(n)}{S_u(n)} = \exp\left(-8.8\Delta s_{jk}\sqrt{\left(\frac{n}{\overline{U}}\right)^2 + \left(\frac{0.12}{L}\right)^2}\right) \tag{5.53}$$

(Note that coherence is sometimes termed coherency, and that some authors define coherence as the *square* of the normalized cross spectrum.) See Section 2.6.6 for details of the coherence corresponding to the Von Karman spectrum.

Three distinct approaches have been developed for generating simulation time histories:

(1) the transformational method, based on filtering Gaussian white noise signals;

(2) the correlation method, in which the velocity of a small body of air at the end of a time step is calculated as the sum of a velocity correlated with the velocity at the start of the time step and a random, uncorrelated increment;

(3) the harmonic series method, involving the summation of a series of cosine waves at different frequencies with amplitudes weighted in accordance with the power spectrum.

This last method is probably now the one in widest use, and is described in more detail below. The description is based on that given in Veers (1988).

Wind simulation by the harmonic series method

The spectral properties of the wind-speed fluctuations at N points can be described by a spectral matrix, $\mathbf{S}$, in which the diagonal terms are the *double-sided* single point power spectral densities at each point, $S_{kk}(n)$, and the off-diagonal terms are the cross-spectral densities, $S_{jk}(n)$, also double sided. This matrix is equated to the product of a triangular transformation matrix, $\mathbf{H}$, and its transpose, $\mathbf{H}^T$:

$$\begin{bmatrix} S_{11} & S_{21} & S_{31} & \ldots \\ S_{21} & S_{22} & S_{32} & \ldots \\ S_{31} & S_{32} & S_{33} & \ldots \\ \ldots & \ldots & \ldots & S_{NN} \end{bmatrix} = \begin{bmatrix} H_{11} & & & \\ H_{21} & H_{22} & & \\ H_{31} & H_{32} & H_{33} & \\ \ldots & \ldots & \ldots & H_{NN} \end{bmatrix} \begin{bmatrix} H_{11} & H_{21} & H_{31} & \ldots \\ & H_{22} & H_{32} & \ldots \\ & & H_{33} & \ldots \\ & & & H_{NN} \end{bmatrix}$$

resulting in a set of $N(N+1)/2$ equations linking the elements of the $\mathbf{S}$ matrix to the elements of the $\mathbf{H}$ matrix:

$$\begin{aligned} &S_{11} = H_{11}^2 \quad S_{21} = H_{21}.H_{11} \quad S_{22} = H_{21}^2 + H_{22}^2 \quad S_{31} = H_{31}.H_{11} \\ &S_{32} = H_{31}.H_{21} + H_{32}.H_{22} \quad S_{33} = H_{31}^2 + H_{32}^2 + H_{33}^2 \\ &S_{jk} = \sum_{l=1}^{k} H_{jl}.H_{kl} \quad S_{kk} = \sum_{l=1}^{k} H_{kl} \end{aligned} \tag{5.54}$$

As with the elements of the **S** matrix, the elements of the **H** matrix are all double-sided functions of frequency, n.

Noting that the expression for the power spectral density S_{kk} resembles that for the variance of the sum of group of k independent variables, it is apparent that the elements of the **H** matrix can be considered as the weighting factors for the linear combination of N independent, unit magnitude, white noise inputs to yield N correlated outputs with the correct spectral matrix. Thus the elements in the jth row of **H** are the weighting factors for the inputs contributing to the output at point j. The formula for the linear combination is

$$u_j(n) = \sum_{k=1}^{j} H_{jk}(n)\Delta n \exp(-i\theta_k(n)) \tag{5.55}$$

where $u_j(n)$ is the complex coefficient of the discretized frequency component at n Hz of the simulated wind speed at point j. The frequency bandwidth is Δn. $\theta_k(n)$ is the phase angle associated with the n Hz frequency component at point k, and is a random variable uniformly distributed over the interval $0 - 2\pi$.

The values of the weighting factors, H_{jk}, which are $N(N+1)/2$ in number are derived from the Equations (5.54), giving:

$$\mathrm{H}_{11} = \sqrt{S_{11}} \qquad H_{21} = S_{21}/H_{11} \qquad H_{22} = \sqrt{S_{22} - H_{21}^2} \qquad H_{31} = S_{31}/H_{11}$$

$$\text{etc.} \tag{5.56}$$

Hence

$$u_1(n) = \sqrt{S_{11}(n)}\Delta n \exp(-i\theta_1(n))$$

$$u_2(n) = \sqrt{S_{22}(n)}\Delta n[\mathrm{C}_{21}(n)\exp(-i\theta_1(n)) + \sqrt{1 - \mathrm{C}_{21}^2(n)}\exp(-i\theta_2(n))]$$

$$\text{etc.} \tag{5.57}$$

Time series for the wind-speed fluctuations are obtained by taking the inverse discrete Fourier transform of the coefficients $u_j(n)$ at each point j. Lateral and vertical wind-speed fluctuations can also be simulated, if desired, using the same method. As an illustration, examples of time series derived by this method for two points 10 m apart are shown in Figure 5.22, based on the Von Karman spectrum.

In his 1988 paper, Veers pointed out that computation time required can be reduced by arranging for the simulated wind speed to be calculated at each point only at those times when a blade is passing, i.e., at a frequency of $\Omega B/2\pi$, where B is the number of blades. This is achieved by applying a phase shift to each frequency component at each point of $\psi_j n 2\pi/\Omega$, where ψ_j is the azimuth angle of point j.

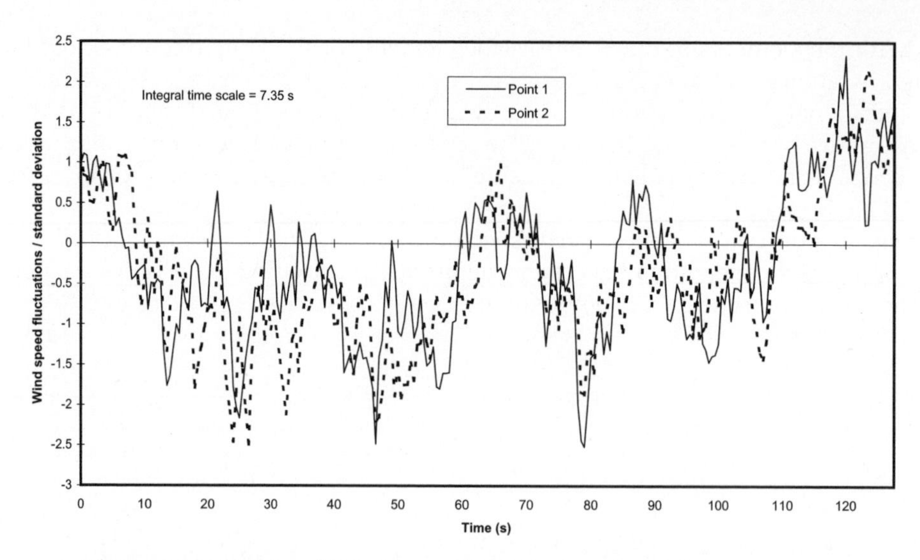

Figure 5.22 Simulated Wind Speed Time Series at Two Points 10 m Apart

Blade load time histories

Once the simulated wind speed time histories have been generated across the grid, the calculation of blade load histories at different radii can begin. If the wake is assumed to be 'frozen', then the axial induced velocity, $a\overline{U}$, and the tangential induced velocity, $a'\Omega r$, are taken as remaining constant over time, at each radius, at the values calculated for a steady wind speed of $\overline{U}$. The instantaneous value of the flow angle, ϕ, and hence the values of the lift and drag coefficients, may then be calculated directly from the instantaneous value of the wind-speed fluctuation (including lateral and vertical components) by means of the velocity diagram.

Alternatively, an equilibrium wake may be assumed. In this case, the induced velocities are taken to vary continuously so that the momentum equations are satisfied at each blade element at all times. Obviously, this requires that these equations are solved afresh at each time step, which is computationally much more demanding.

Neither the equilibrium wake model nor the frozen wake model provide an accurate description of wake behaviour. A better model is provided by dynamic inflow theory, which assumes that there is some delay before induced velocities react to changes in the incident wind field (see Section 3.13.1).

Note that, if desired, the spatial wind variations causing deterministic loading can be included in the simulated wind field, enabling the combined deterministic and stochastic loading on the blade to be calculated in a single operation.

5.7.7 Extreme loads

The derivation of extreme loads should properly take into account dynamic effects, which form the subject of the next section. However, in the interests of clarity, this section will be restricted to the consideration of extreme loads in the absence of dynamic effects.

As described in Section 5.4, it is customary for wind turbine design codes to specify extreme operating load cases in terms of deterministic gusts. The extreme blade loadings are then evaluated at intervals over the duration of the gust, using blade element and momentum theory as described in Section 5.7.2.

Although the discrete gust models prescribed by the codes have the advantage of clarity of definition, they are essentially arbitrary in nature. The alternative approach of adopting a stochastic representation of the wind provides a much more realistic description of the wind itself, but is dependant on assumptions of linearity as far as the calculation of loads is concerned.

Normally the loading under investigation, for example the blade root bending moment, will contain both periodic and random components. Although it is straightforward to predict the extreme values of each component independently, the prediction of the extreme value of the combined signal is quite involved. Madsen *et al.* (1984) have proposed the following simple, approximate approach, and have demonstrated that it is reasonably accurate.

The periodic component, $z(t)$, is considered as an equivalent 3 level square wave, in which the variable takes the maximum, mean (μ_z), and minimum values of the original waveform, for proportions ε_1, ε_2 and ε_3 of the wave period respectively. It is easy to show that:

$$\varepsilon_1 = \frac{\sigma_z^2}{(z_{max} - \mu_z)(z_{max} - z_{min})} \qquad \varepsilon_3 = \frac{\sigma_z^2}{(\mu_z - z_{min})(z_{max} - z_{min})} \tag{5.58}$$

Extreme values of the combined signal are only assumed to occur during the proportion of the time, ε_1, for which the square wave representation of the periodic component is at the maximum value, z_{max}.

Davenport (1964) gives the following formula for the extreme value of a random variable over a time interval T:

$$\frac{x_{max}}{\sigma_x} = \sqrt{2\ln(\nu T)} + \frac{\gamma}{\sqrt{2\ln(\nu T)}} \tag{5.59}$$

where ν is the zero up-crossing frequency (i.e., the number of times per second the variable changes from negative to positive) given by Equation (A5.46) and $\gamma = 0.5772$ (Euler's constant). Thus the extreme value of the combined periodic and random components is taken to be

$$z_{max} + x_{max} = z_{max} + \sigma_x\left(\sqrt{2\ln(\nu\varepsilon_1 T)} + \frac{\gamma}{\sqrt{2\ln(\nu\varepsilon_1 T)}}\right) = z_{max} + g_1.\sigma_x \tag{5.60}$$

where g_1 is termed the peak factor.

The variation of x_{max}/σ_x with exposure time, T, is shown Table 5.3 for a zero up-crossing frequency of 1 Hz. The periodic component is assumed to be a simple sinusoid, giving $\varepsilon_1 = 0.25$.

The method for determining the extreme load described above has to be applied with caution when the wind fluctuations exceed the rated wind speed. In the case

Table 5.3 Extreme Values of Random Component for Different Exposure Times

T	1 min	10 min	1 h	10 h	100 h	1000 h	1 year
T (s)	60	600	3600	36 000	360 000	3 600 000	3 153 600
$\varepsilon_1 T$ (s)	15	150	900	9000	90 000	900 000	7 884 000
x_{max}/σ_x	2.57	3.35	3.84	4.40	4.90	5.35	5.74

of a stall-regulated machine, the linearity assumption breaks down completely, invalidating the method. With pitch-regulated machines, however, the blade pitch will respond to wind fluctuations at frequencies below the rotor rotational frequency in order to limit power, causing a parallel reduction in blade loading. This will modify the spectrum of blade loading dramatically, effectively removing the frequency components below the pitch system cut-off frequency, and consequently reducing the magnitude of σ_x to be substituted in Equation (5.60).

To illustrate the method, the procedure for calculating the extreme flapwise blade root bending moment of a pitch-regulated machine operating at rated wind speed is described below.

(1) Equation (5.48) for the standard deviation of the random component of blade root bending moment is first modified to eliminate the contribution of frequencies below half the rotational speed to account for the blade pitching response, and then discretized to give:

$$\sigma_M^2 = \left(\tfrac{1}{2}\rho\Omega\frac{dC_l}{d\alpha}\right)^2 \sum_{j=1}^{m}\sum_{k=1}^{m}\left[\int_{\Omega}^{\infty} S_u^o(r_j, r_k, n)\,dn\right] c(r_j)c(r_k)r_j^2 r_k^2.(\Delta r)^2 \tag{5.61}$$

Here the blade is assumed to be divided up into m sections of equal length $\Delta r = R/m$.

(2) After evaluation of the integrals of the $m(m+1)/2$ different curtailed rotational spectra, the standard deviation of the blade root bending moment is obtained from Equation (5.61).

(3) The time, T, that the machine spends in a wind speed band centred on the rated wind speed is estimated using the Weibull curve, and multiplied in turn by the factor ε_1 appropriate to the waveform of the periodic component of blade root bending moment and by the zero up-crossing frequency of the random root bending moment fluctuations, to give the effective number of peaks, $\nu\varepsilon_1 T$.

(4) The predicted extreme value of the total moment is calculated by substituting the standard deviation of the blade root bending moment, $\sigma_M (= \sigma_x)$, the effective number of peaks, $\nu\varepsilon_1 T$, and the extreme value of the periodic moment into Equation (5.60).

In the case of a machine with a rated wind speed of 13 m/s operating at 30 r.p.m. at a site with an annual mean of 7 m/s, the expected proportion of the time spent operating within a 2 m/s wide band centred on the rated wind speed is 5.6 percent. Taking the machine lifetime as 20 years and a zero up-crossing frequency of 1.2 Hz, this results in a peak factor, g_1, of 5.8. For the 40 m diameter machine considered in Section 5.7.2 and a turbulence intensity of 20 percent, this translates to a peak value of the random component of blade root bending moment of about 230 kNm, compared to the extreme value of the periodic component, including wind shear, of about 300 kNm. It should be emphasized that the peak value of the random component quoted is a theoretical one, i.e., it assumes the linearity assumptions are maintained even for the large wind speed fluctuation needed to generate this moment. In practice, a machine operating in a steady wind speed equal to rated is usually not all that far from stall, so the larger fluctuations may induce stall. In this example, the square root of the weighted mean of the integrals of all the curtailed rotational spectra is about $0.5\sigma_u$, so the idealized uniform wind speed fluctuation equivalent to the extreme root moment is about $0.20 \times 13 \times 0.5 \times 5.8 = 7.5$ m/s.

The method outlined above has more validity at higher wind speeds, when the blades are pitched back, and are operating further away from stall. However, it is important to note that the other linearity assumption used in deriving Equation (5.25), namely that ϕ is small, becomes increasingly in error.

It will be now be evident that the calculation of stochastic extreme loads is fraught with difficulties because non-linearities are likely to arise as the extremes are approached. In so far as lift forces 'saturate' due to stall, or even drop back, as wind speed increases, a crude and simple approach to extreme out-of-plane operational loads is to calculate an upper-bound based on the maximum lift coefficient for the local aerofoil section and the relative air velocity, W. The induction factors will be small, and can be ignored.

The most sophisticated approach, however, is to analyse the loads generated by a simulated wind field. As computing costs normally restrict the length of simulated 'campaigns' to a few hundred seconds or less, statistical methods have to be used to extrapolate from the extreme values of loadings calculated during the campaign to the extreme values to be expected over the machine design life.

One method, which is discussed by Thomsen and Madsen (1997), is to use Equation (5.60) with T set equal to the appropriate exposure period over the machine design life, and values of z_{max} and σ_x abstracted from the simulation time history with the aid of azimuthal binning to separate the periodic and stochastic components. The danger of this approach with simulations of short duration is that the azimuthal binning process treats some load fluctuations due to the slicing of low frequency gusts as periodic rather than stochastic, so that the standard deviation of the stochastic component, σ_x, is underestimated.

5.8 Blade Dynamic Response

5.8.1 Modal analysis

Although dynamic loads on the blades will, in general, also excite the tower dynamics, tower head motion will initially be excluded from consideration in order to focus on the blade dynamic behaviour itself. The treatment is further limited to the response of blades in unstalled flow because of the inherent difficulty in predicting stalled behaviour.

The equation of motion for a blade element at radius r subject to a time varying load $q(r, t)$ per unit length in the out-of-plane direction is

$$m(r)\ddot{x} + \hat{c}(r)\dot{x} + \frac{\partial^2}{\partial r^2}\left[EI(r)\frac{\partial^2 x}{\partial r^2}\right] = q(r, t) \tag{5.62}$$

where the terms on the left-hand side are the loads on the element due to inertia, damping and flexural stiffness respectively. $I(r)$ is the second moment of area of the blade cross section about the weak principal axis (which for this purpose is assumed to lie in the plane of rotation) and x is the out-of-plane displacement. The expressions $m(r)$ and $\hat{c}(r)$ denote mass per unit length and damping per unit length respectively.

The dynamic response of a cantilever blade to the fluctuating aerodynamic loads upon it is most conveniently investigated by means of modal analysis, in which the the excitations of the various different natural modes of vibration are computed separately and the results superposed:

$$x(t, r) = \sum_{j=1}^{\infty} f_j(t)\mu_j(r) \tag{5.63}$$

where $\mu_j(r)$ is the jth mode shape, arbitrarily assumed to have a value of unity at the tip, and $f_j(t)$ is the variation of tip displacement with time. Equation (5.62) then becomes

$$\sum_{j=1}^{\infty}\left\{m(r)\mu_j(r)\ddot{f}_j(t) + \hat{c}(r)\mu_j(r)\dot{f}_j(t) + \frac{\mathrm{d}^2}{\mathrm{d}r^2}\left[EI(r)\frac{\mathrm{d}^2\mu_j(r)}{\mathrm{d}r^2}\right]f_j(t)\right\} = q(r, t) \tag{5.64}$$

For low levels of damping the beam natural frequencies are given by

$$m(r)\omega_j^2\mu_j(r) = \frac{\mathrm{d}^2}{\mathrm{d}r^2}\left[EI(r)\frac{\mathrm{d}^2\mu_j(r)}{\mathrm{d}r^2}\right] \tag{5.65}$$

so Equation (5.64) becomes

$$\sum_{j=1}^{\infty}\{m(r)\mu_j(r)\ddot{f}_j(t) + \hat{c}(r)\mu_j(r)\dot{f}_j(t) + m(r)\omega_j^2\mu_j(r)f_j(t)\} = q(r, t) \tag{5.66}$$

Multiplying both sides by $\mu_i(r)$, and integrating over the length of the blade, R, gives:

$$\sum_{j=1}^{\infty}\left\{\int_0^R m(r)\mu_i(r)\mu_j(r)\ddot{f}_j(t)\,\mathrm{d}r+\int_0^R \hat{c}(r)\mu_i(r)\mu_j(r)\dot{f}_j(t)\,\mathrm{d}r\right.$$

$$\left.+\int_0^R m(r)\omega_j^2\mu_i(r)\mu_j(r)f_j(t)\,\mathrm{d}r\right\}=\int_0^R \mu_i(r)q(r,\,t)\,\mathrm{d}r \tag{5.67}$$

The undamped mode shapes are orthogonal as a result of Betti's law, so they satisfy the orthogonality condition:

$$\int_0^R m(r)\mu_i(r)\mu_j(r)\,\mathrm{d}r=0 \quad \text{for } i\neq j \tag{5.68}$$

If we assume that the variation of the damping per unit length along the blade, $\hat{c}(r)$, is proportional to the variation in mass per unit length, $m(r)$, i.e., $\hat{c}(r)=a.m(r)$, then

$$\int_0^R \hat{c}(r)\mu_i(r)\mu_j(r)\,\mathrm{d}r=0 \quad \text{for } i\neq j \tag{5.69}$$

As a result, all the cross terms on the left-hand side of Equation (5.67) drop out, and it reduces to

$$m_i\ddot{f}_i(t)+c_i\dot{f}_i(t)+m_i\omega_i^2 f_i(t)=\int_0^R \mu_i(r)q(r,\,t)\,\mathrm{d}r \tag{5.70}$$

where $m_i=\int_0^R m(r)\mu_i^2(r)\,\mathrm{d}r$ and is known as the generalized mass, $c_i=\int_0^R \hat{c}(r)\mu_i^2(r)\,\mathrm{d}r$ and $\int_0^R \mu_i(r)q(r,\,t)\,\mathrm{d}r=Q_i(t)$ is termed the generalized fluctuating load with respect to the ith mode. Equation (5.70) is the fundamental equation governing modal response to time varying loading.

Blade flexural vibrations occur in both the flapwise and edgewise directions (i.e., about the weak and strong principal axes respectively). Blades are typically twisted some 15°, so the weak principal axis does not, in general, lie in the plane of rotation as assumed above. Consequently blade flexure about one principal axis inevitably results in some blade movement perpendicular to the other. This is illustrated in Figure 5.23, in which the maximum blade twist near the root has been exaggerated for clarity. Point P represents the undeflected position of the blade tip, point Q represents the deflected position as a result of flexure about the weak principal axis, and the line between them is built up of the contributions to the tip deflection made by flexure of each element along the blade, $M(R-r)\Delta r/EI$.

The interaction between flexure about the two principal axes can be explored with the help of some simplifying assumptions. If the moment M varies as $(R-r)$ for the first mode and I varies as $(\mathrm{R}-r)^2$, then each of the tip deflection contributions referred to above are equal, so that, for a linear twist distribution, the line PQ is the arc of a circle. If the twist varies between zero at the tip and a maximum value of β towards the root, then the tip deflection, δ_{12}, in the direction of the weak principal axis, at the blade section with maximum twist, is $\beta/2$ times the tip deflection, δ_{11}, perpendicular to this axis. Hence in the case of $\beta=15°$, the ratio

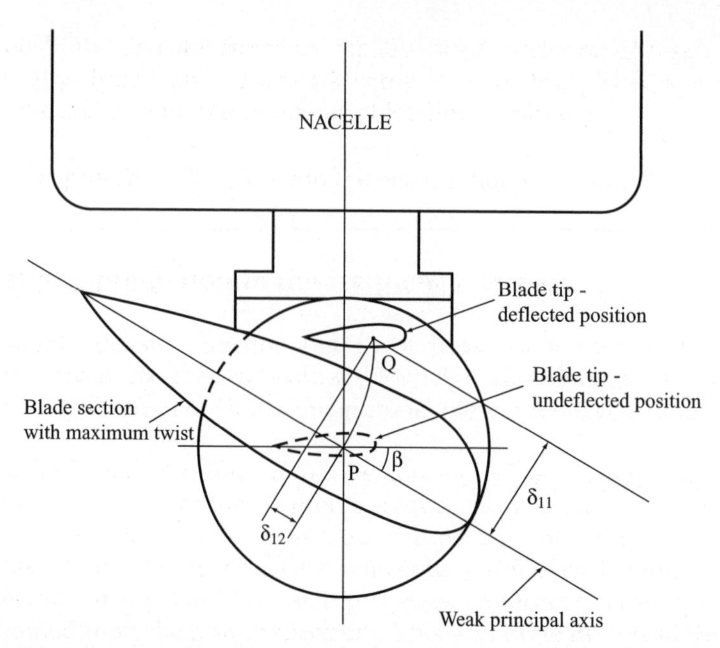

Figure 5.23 Deflection of Tip Due to Flapwise Bending of Twisted Blade (Viewed Along Blade Axis)

δ_{12}/δ_{11} approximates to 0.13, with the result that blade first mode flapwise oscillations will result in some relatively small simultaneous edgewise inertia loadings. These will not excite significant edgewise oscillations, because the edgewise first mode natural frequency is typically about double the flapwise one.

It can be seen from the above that the effects of interaction between flapwise and edgewise oscillations are generally minor, so they will not be considered further.

Blades will also be subject to torsional vibrations. However, these can generally be ignored, because both the exciting loads are small, and the high torsional stiffness of a typical hollow blade places the torsional natural frequencies well above the exciting frequencies.

Finally, in the case of a blade hinged at the root, the whole blade will experience oscillations involving rigid body rotation about the hinge. This phenomenon is considered in Section 5.8.8.

5.8.2 Mode shapes and frequencies

The mode shape and frequency of the first mode can be derived by an iterative technique called the Stodola method after its originator. Briefly, this consists of assuming a plausible mode shape, calculating the inertia loads associated with it for an arbitrary frequency of 1 rad/s, and then computing the beam deflected profile resulting from these inertia loads. This profile is then normalized, typically by dividing the deflections by the tip deflection, to obtain the input mode shape for the second iteration. The process is repeated until the mode shape converges, and the first mode natural frequency is calculated from the formula:

$$\omega_1 = \sqrt{\frac{\text{Tip deflection input to last iteration}}{\text{Tip deflection output from last iteration}}}$$

If the mode shapes are orthogonal, advantage can be taken of this property to simplify the derivation of the mode shapes and frequencies of the higher modes, provided this is carried out in ascending order. A trial mode shape is assumed as before, but before using it to calculate the inertia loadings, it is 'purified' so that it does not contain any lower mode content. For example, 'purification' of a second mode trial mode shape, $\mu_{2T}(r)$, of first mode content is achieved by subtracting

$$\mu_{2C}(r) = \mu_1(r)\frac{\int_0^R \mu_1(r)\mu_{2T}(r)m(r)\,\mathrm{d}r}{\int_0^R \mu_1^2(r)m(r)\,\mathrm{d}r} = \mu_1(r)\frac{\int_0^R \mu_1(r)\mu_{2T}(r)m(r)\,\mathrm{d}r}{m_1} \tag{5.71}$$

from it. The modified second mode trial mode shape, $\mu_{2M}(r) = \mu_{2T}(r) - \mu_{2C}(r)$, then satifies the orthogonality condition

$$\int_0^R m(r)\mu_1(r)\mu_{2M}(r)\,\mathrm{d}r = 0$$

After 'purification' of the trial mode shape, the Stodola method can be applied exactly as before. Further 'purification' before succeeding iterations should not be necessary if the lower mode shapes used for the initial 'purification' are accurate enough. See Clough and Penzien (1993) for a rigorous treatment of the method.

5.8.3 Centrifugal stiffening

When a rotating blade deflects either in its plane of rotation or perpendicular to it, the centrifugal force on each blade element exerts a restoring force which has the effect of stiffening the blade and thereby increasing the natural frequency compared with the stationary value. The centrifugal forces act radially outwards perpendicular to the axis of rotation, so in the case of an out-of-plane blade deflection, they are parallel to the undeflected blade axis and act at greater lever arms to the inboard part of the blade than they do in the case of in-plane blade deflection. This is illustrated in Figure 5.24.

In order to take account of the effects of centrifugal loads, the equation of motion for a blade element loaded in the out-of-plane direction is modified by the addition of an additional term to become

$$m(r)\ddot{x} + \hat{c}(r)\dot{x} - \frac{\partial}{\partial r}\left[N(r)\frac{\partial x}{\partial r}\right] + \frac{\partial^2}{\partial r^2}\left[EI\frac{\partial^2 x}{\partial r^2}\right] = q(r,\,t) \tag{5.72}$$

where the centrifugal force at radius r, $N(r)$, is the summation of the forces acting on each blade element outboard of radius r, that is $N(r) = \sum_{r=r}^{r=R} m(r)\Omega^2 r\Delta r$.

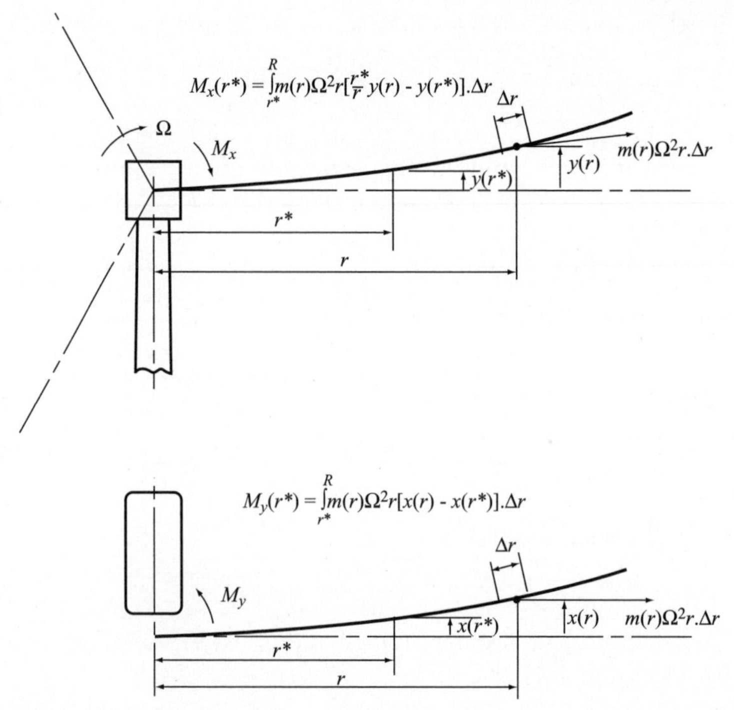

Figure 5.24 Restoring Moments due to Centrifugal Force for In-plane and Out-of-plane Blade Deflections

The Stodola method for deriving blade mode shapes and frequencies described in the preceding section can be modified to take account of centrifugal effects. In the case of out-of-plane modes, the procedure is:

(1) assume plausible trial mode shape;

(2) 'Purify' trial mode shape of any lower mode content;

(3) assume trial value for frequency, ω_j^2;

(4) calculate bending moment distribution due to lateral inertia forces according to:

$$M_{\text{Y.Lat}}(r^*) = \int_{r^*}^{R} m(r)\omega_j^2 \mu(r)[r - r^*]\,\mathrm{d}r \tag{5.73}$$

(5) calculate bending moment distribution due to centrifugal forces according to:

$$M_{\text{Y.CF}}(r^*) = -\int_{r^*}^{R} m(r)\Omega^2 r[\mu(r) - \mu(r^*)]\,\mathrm{d}r \tag{5.74}$$

(6) calculate combined bending moment distribution;

(7) calculate new deflected profile resulting from this bending moment distribution;

(8) calculate revised estimate of natural frequency from:

$$\omega_j' = \omega_j \sqrt{\frac{\text{Trial tip deflection}}{\text{Tip deflection calculated for new deflected profile}}}$$

(9) repeat steps (2)–(8) with revised mode shape and frequency until calculated mode shape converges.

It is important to note that the lateral loads and deflections of a centrifugally loaded beam do not conform to Betti's Law, so, as a consequence, the mode shapes are *not* orthogonal. It is for this reason that the 'purification' stage has been included in each cycle of iteration. When convergence of the calculated mode shape has occurred, it will be found that it differs significantly from the 'purified' mode shape input into each iteration, indicating that a true solution has not been obtained. It is then necessary to use a trial and error approach to modify the magnitudes of the 'purifying' corrections applied until the output mode shape and input 'purified' mode shape match. A few further iterations will be required until the natural frequency settles down.

A quick estimate of the first mode frequency of a rotating blade can be derived using the Southwell formula reported by Putter and Manor (1978) as follows

$$\omega_1 = \sqrt{\omega_{1,0}^2 + \phi_1 \Omega^2} \tag{5.75}$$

in which $\omega_{1,0}$ is the corresponding frequency for the non-rotating blade. The value of ϕ_1 depends on the blade mass and stiffness distribution, and Madsen *et al.* (1984) suggest the value 1.73 for wind-turbine blade out-of-plane oscillations. In the case of Blade TR rotating at 30 r.p.m., this yields a percentage increase in first mode frequency due to centrifugal stiffening of 7.7 percent compared to the correct value of 8.1 percent. Typically, centrifugal stiffening results in an increase of the first mode frequency for out-of-plane oscillations of between 5 percent and 10 percent. For higher modes, the magnitude of the centrifugal forces is less in proportion to the lateral inertia forces, so the percentage increase in frequency due to centrifugal stiffening becomes progressively less.

The procedure for deriving the blade first mode shape and frequency in the case of in-plane oscillations is the same as that described above for out-of-plane vibrations, except that the formula for the bending moment distribution due to the centrifugal forces has to be modified to:

$$M_{X.CF}(r^*) = \int_{r^*}^{R} m(r)\Omega^2 r \left[\frac{r^*}{r}\mu(r) - \mu(r^*)\right] \mathrm{d}r \tag{5.76}$$

The first mode frequency for in-plane oscillations of Blade TR in the absence of centrifugal force is 3.13 Hz. This is approximately double the corresponding frequency for out-of-plane oscillations of 1.65 Hz, and so the relative effect of the centrifugal loads is much reduced, even before allowance is made for the smaller lever arms at which they act (Figure 5.24). In fact the increase in the first mode frequency for in-plane oscillations due to centrifugal force is only 0.5 percent – probably small enough to be ignored.

5.8.4 Aerodynamic and structural damping

Blade motion is generally resisted by two forms of viscous damping, aerodynamic and structural, which are considered in turn.

An approximate expression for the aerodynamic damping per unit length in the flapwise direction can be derived by a method analagous to that used in Section 5.7.5 to derive the linear relation

$$q = \tfrac{1}{2}\rho\Omega r c(r)\frac{\mathrm{d}C_l}{\mathrm{d}\alpha}u \tag{5.25}$$

between blade load fluctuations per unit length, q, and fluctuations in the incident wind, u. The wind-speed fluctuation, u, is simply replaced by the blade flapwise velocity, $-\dot{x}$, giving

$$\hat{c}_a(r) = \frac{q}{-\dot{x}} = \tfrac{1}{2}\rho\Omega r\, c(r)\frac{\mathrm{d}C_l}{\mathrm{d}\alpha} \tag{5.77}$$

The rate of change of lift coefficient with angle of attack, $\mathrm{d}C_l/\mathrm{d}\alpha$, is constant and equal to 2π before the blade goes into stall, but can become negative post-stall, leading to the risk of instability (see Section 7.1.9).

It can be seen that the aerodynamic damping per unit length, $\hat{c}_a(r)$, varies spanwise as the product of radius and blade chord, and is therefore not very close to being proportional to the mass per unit length, as is required to satisfy the orthogonality condition. This will result in some aerodynamic coupling of modes, which is not accounted for in normal modal analysis.

The aerodynamic damping ratio for the ith mode, defined as $\xi_{ai} = c_{ai}/2m_i\omega_i = \int_0^R \hat{c}_a(r)\mu_i^2(r)\mathrm{d}r/2m_i\omega_i$, can be calculated using Equation (5.77) as follows:

$$\xi_{ai} = \frac{\tfrac{1}{2}\rho\Omega\dfrac{\mathrm{d}C_l}{\mathrm{d}\alpha}\displaystyle\int_0^R r\, c(r)\mu_i^2(r)\,\mathrm{d}r}{2\omega_i\displaystyle\int_0^R m(r)\mu_i^2(r)\,\mathrm{d}r} \tag{5.78}$$

In the case of fibreglass Blade TR described in Example 5.1, this yields values of 0.16 and 0.04 for the first and second modes respectively. These high values are a consequence of the lightness of the blade in relation to its width in the vicinity of the tip, an area which dominates the integrals thanks to the mode shape weighting. The corresponding first mode logarithmic decrement ($= 2\pi\xi_a$) is thus 1.0.

Structural damping can be considered as an internal resistance opposing the rate of strain, i.e., as an additional moment which results in an additional lateral load. Equation (5.62) can thus be written

$$m(r)\ddot{x} + \hat{c}_a(r)\dot{x} + \frac{\partial^2}{\partial r^2}\left[a_1 EI(r)\frac{\partial}{\partial t}\left\{\frac{\partial^2 x}{\partial r^2}\right\}\right] + \frac{\partial^2}{\partial r^2}\left[EI(r)\frac{\partial^2 x}{\partial r^2}\right] = q(r,\,t) \tag{5.79}$$

where $\hat{c}_a(r)$ is the aerodynamic damping per unit length, and a_1 is a constant defining the magnitude of the structural damping. Inserting $x(t,\,r) = \sum_{j=1}^{\infty} f_j(t)\mu_j(r)$ as before, and using Equation (5.65), the structural damping term becomes $\sum_{J=1}^{\infty} a_1 m(r)\omega_j^2\mu_j(r)\dot{f}_j(t)$. Thus the structural damping per unit length for the jth mode is $a_1 m(r)\omega_j^2\mu_j(r)$, and therefore varies as the mass per unit length as assumed in the Section 5.8.1. Continuing with the same procedure as in the Section 5.8.1, a modified modal response equation is obtained as follows

$$m_i\ddot{f}_i(t) + \{c_{ai} + a_1 m_i\omega_i^2\}\dot{f}_i(t) + m_i\omega_i^2 f_i(t) = \int_0^R \mu_i(r)p(r,\,t)\,dr \tag{5.80}$$

Thus the structural damping ratio for the ith mode, defined as $\xi_{si} = c_{si}/2m_i\omega_i = a_1 m_i\omega_i^2/2m_i\omega_i$, becomes $\xi_{si} = a_1\omega_i/2$, i.e., it increases in proportion to the modal frequency. Values for the structural damping logarithmic decrement, $\delta_s = 2\pi\xi_s$ at the fundamental natural (i.e., first mode) frequency are given in DS 472 for several different materials, and these are reproduced in Table 5.4 below. Similar values are given in Eurocode 1 for welded steel and concrete. Note that the first mode structural damping ratio for a fibreglass blade is much smaller than the aerodynamic damping ratio for Blade TR derived above.

It is instructive to evaluate the combined damping ratio for the first and second flapwise modes of Blade TR. These are presented in Table 5.5.

Table 5.4 Values of First Mode Structural Damping Logarithmic Decrements for Different Materials

Material	Logarithmic decrement, δ_s	Structural damping ratio, ξ_s
Concrete	0.05	0.008
Steel – welded	0.02	0.003
Steel – bolted	0.05	0.008
GRP	0.05	0.008
Timber	0.05	0.008

Table 5.5 Comparison of Blade TR Damping Ratios for First Two Modes

	First mode	Second mode
Natural frequency including effect of centrifugal stiffening	1.78 Hz	5.88 Hz
Aerodynamic damping ratio	0.16	0.04
Structural damping ratio (proportional to frequency)	0.008	0.03
Combined damping ratio	0.17	0.07

It is seen that the damping ratio for the second mode is under half that for the first.

5.8.5 Response to deterministic loads—step-by-step dynamic analysis

As set out in Section 5.8.1, blade dynamic response to time varying loading is best analysed in terms of the separate excitation of each blade mode of vibration, for which, under the assumptions of unstalled flow and mass-proportional aerodynamic damping , the governing equation is

$$m_i\ddot{f}_i(t) + c_i\dot{f}_i(t) + m_i\omega_i^2 f_i(t) = \int_0^R \mu_i(r)q(r, t)\,dr = Q_i(t) \tag{5.70}$$

where $f_i(t)$ and $\mu_i(r)$ are the tip displacement and mode shape for the ith mode respectively. Starting with the initial tip displacement, velocity and acceleration arbitrarily set at zero, this equation can be used to derive values for these quantities at successive time steps over a complete blade revolution by numerical integration. The procedure is then repeated for several more revolutions until the cyclic blade response to the periodic loading becomes sensibly invariant from one revolution to the next.

Linear acceleration method

The precise form of the equations linking the tip displacement, velocity and acceleration at the end of a time step to those at the beginning depends on how the acceleration is assumed to vary over the time step. Newmark has classified alternative assumptions in terms of a parameter β which measures the relative weightings placed on the initial and final accelerations in deriving the final displacement The simplest assumption is that the acceleration takes a constant value equal to the average of the initial and final values ($\beta = 1/4$). Clough and Penzien (1993), however, recommend that the acceleration is assumed to vary linearly between the initial and final values, as this will be a closer approximation to the actual variation. Step-by-step integration with this assumption is known as either the linear acceleration method or the Newmark $\beta = 1/6$ method.

Expressions for the tip displacement, velocity and acceleration at the end of the first time step – f_{i1}, $\dot{f}_{i1}$ and $\ddot{f}_{i1}$ respectively – are derived in terms of the initial values – f_{i0}, $\dot{f}_{i0}$ and $\ddot{f}_{i0}$ – as follows. The acceleration at time t during the time step of total duration h is

$$\ddot{f}_i(t) = \ddot{f}_{i0} + \left(\frac{\ddot{f}_{i1} - \ddot{f}_{i0}}{h}\right)t \tag{5.81}$$

This can be integrated to give the velocity at the end of the time step as

$$\dot{f}_{i1} = \dot{f}_{i0} + \ddot{f}_{i0}h + (\ddot{f}_{i1} - \ddot{f}_{i0})h/2 \tag{5.82}$$

Equation (5.81) can be integrated twice to give an expression for the displacement at the end of the time step, which, after rearrangement yields the following expression for the corresponding acceleration:

$$\ddot{f}_{i1} = \frac{6}{h^2}(f_{i1} - f_{i0}) - \frac{6}{h}\dot{f}_{i0} - 2\ddot{f}_{i0} \tag{5.83}$$

Substituting Equation (5.83) into Equation (5.82) yields

$$\dot{f}_{i1} = \frac{3}{h}(f_{i1} - f_{i0}) - 2\dot{f}_{i0} - \ddot{f}_{i0}h/2 \tag{5.84}$$

Equation (5.70) can be written as

$$m_i\ddot{f}_{in} + c_i\dot{f}_{in} + m_i\omega_i^2 f_{in} = \int_0^R \mu_i(r)q_n(r)\,\mathrm{d}r = Q_{in} \tag{5.85}$$

where the suffix n refers to the state at the end of the nth time step. Substituting Equations (5.83) and (5.84) into Equation (5.85) with $n = 1$ and collecting terms yields the displacement at the end of the first time step as:

$$f_{i1} = \frac{Q_{i1} + m_i\left(\frac{6}{h^2}f_{i0} + \frac{6}{h}\dot{f}_{i0} + 2\ddot{f}_{i0}\right) + c_i\left(\frac{3}{h}f_{i0} + 2\dot{f}_{i0} + \frac{h}{2}\ddot{f}_{i0}\right)}{m_i\omega_i^2 + \frac{3c_i}{h} + \frac{6m_i}{h^2}} \tag{5.86}$$

The velocity and acceleration at the end of the first time step are then obtained by substituting f_{i1} in Equations (5.84) and (5.83) respectively.

The full procedure for obtaining the blade dynamic response to a periodic loading using the Newmark $\beta = 1/6$ method (which is just one of many available) may be summarized as follows:

(1) calculate the blade mode shapes, $\mu_i(r)$;

(2) select the number of time steps, N, per complete revolution, then the time step, $h = 2\pi/N\Omega$;

(3) calculate the blade element loads, $q(r, \psi_n) = q_n(r)$, at blade azimuth positions corresponding to each time step (i.e., at $2\pi/N$ intervals) using momentum theory (here, the suffix n denotes the number of the time step);

(4) calculate the generalized load with respect to each mode, $Q_{in} = \int_0^R \mu_i(r)q_n(r)\,\mathrm{d}r$, for each time step;

(5) assume initial values of blade tip displacement, velocity and acceleration;

(6) calculate first mode blade tip displacement, velocity and acceleration at end of first time step, using Equations (5.86), (5.84) and (5.83) respectively (with $i = 1$);

(7) repeat Stage 6 for each successive time step over several revolutions until convergence achieved;

(8) calculate cyclic blade moment variation at radii of interest by multiplying the cyclic tip displacement variation by appropriate factors derived from the modal analysis;

(9) repeat Stages 6–8 for higher modes;

(10) combine the responses from different modes to obtain the total response.

Figure 5.25 shows some results of the application of the above procedure to the derivation of the out-of-plane root bending moment response of Blade TR to tower shadow loading. The case chosen is for a mean wind speed of 12 m/s, uniform across the rotor disc, and an x/D ratio of 1 (where x is the distance between the blade and the tower centreline, and D is the tower diameter), giving a maximum reduction in the blade root bending moment for a rigid blade of 70 kNm. Centrifugal stiffening is included in the derivation of the mode shapes and frequencies, and the damping ratios for the first and second modes are taken as 0.17 and 0.07 respectively. It is evident from Figure 5.25 that the tower shadow gives the blade a sharp 'kick' away from the tower, but the duration is too short in relation to the duration of the first mode half cycle for Blade TR to 'feel' the root bending moment reduction that would be experienced by a completely rigid blade. The blade

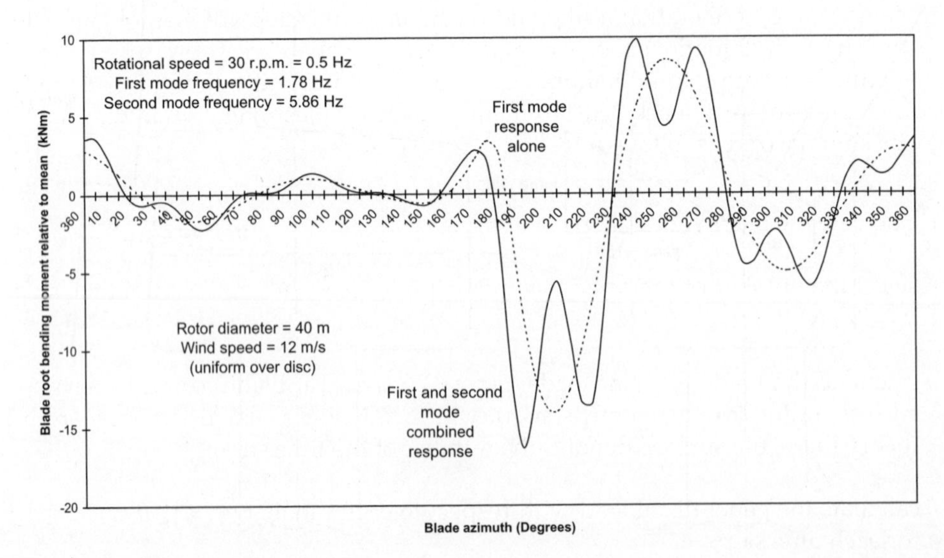

Figure 5.25 Blade TR Out-of-plane Root Bending Moment Dynamic Response to Tower Shadow

oscillations have largely died away after a complete revolution because of the relatively high levels of damping.

The response of the Blade TR out-of-plane root bending moment to tower shadow combined with wind shear is shown in Figure 5.26 for a hub-height wind speed of 12 m/s. Also plotted is the corresponding bending moment for a completely rigid blade. The wind shear loading is approximately sinusoidal (see Figure 5.11), and, consequently, the response is also. However, it is worth noting that the amplitude of the dominating first mode response to wind shear is the result of two effects working against each other – in other words the increase due to the dynamic magnification factor of about 9 percent is largely cancelled out by the reduction due to centrifugal stiffening.

Avoidance of resonance: the Campbell diagram

In the course of blade design, it is important to avoid the occurrence of a resonant condition, in which a blade natural frequency equates to the rotational frequency or a harmonic with a significant forcing load. This is often done with the aid of a Campbell diagram, in which the blade natural frequencies are plotted out against rotational frequency together with rays from the origin representing integer multiples of the rotational frequency. Then any intersections of the rays with a blade natural frequency over the turbine rotational speed operating range represent possible resonances. An example of a Campbell diagram is shown in Figure 5.27.

Clearly blade periodic loading is dominated by the loading at rotational frequency from wind shear, yawed flow and shaft tilt (Section 5.7.2), gravity (Section 5.7.3) and gust slicing (Section 5.7.5). However, the short-lived load relief resulting from tower shadow will be dominated by higher harmonics.

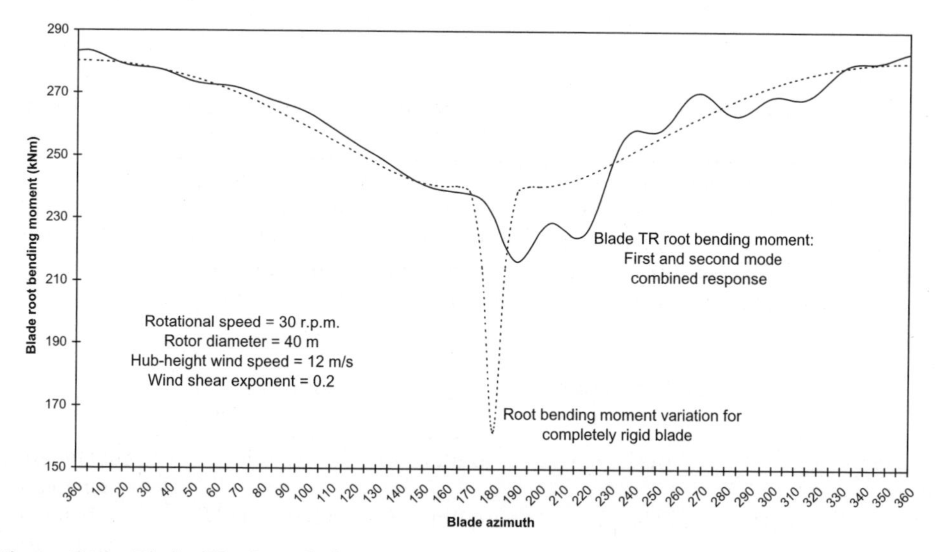

Figure 5.26 Blade TR Out-of-plane Root Bending Moment Dynamic Response to Tower Shadow and Wind Shear

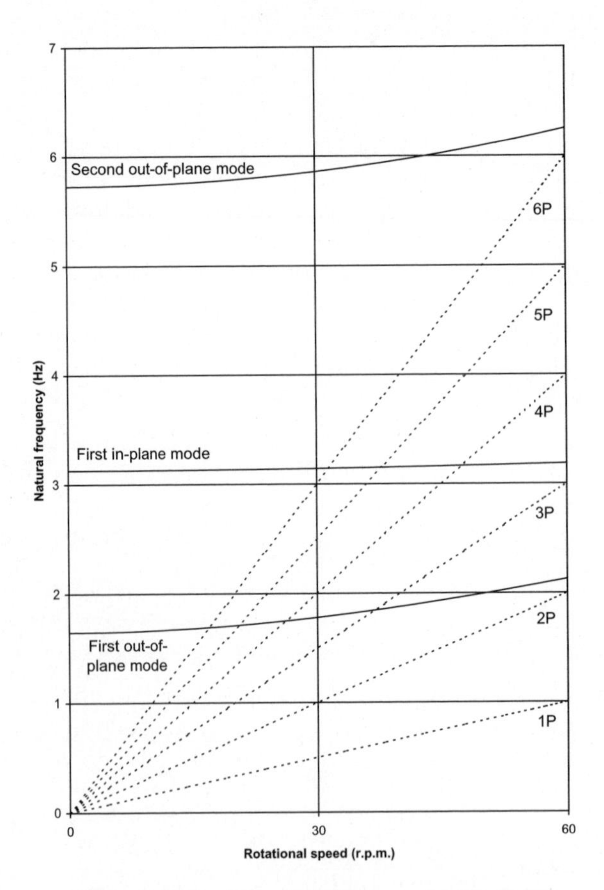

Figure 5.27 Campbell Diagram for Blade TR

5.8.6 Response to stochastic loads

The analysis of stochastic loads in the frequency domain has already been described in Section 5.7.5 for a rigid blade, and this will now be extended to cover the dynamic response of the different vibration modes of a flexible blade using the governing equation, Equation (5.70). Note that the restriction to an unstalled blade operating at a relatively high tip speed ratio still applies.

Power spectrum of generalized blade loading

The generalized fluctuating load with respect to the ith mode is $Q_i = \int_0^R \mu_i(r)q(r)\,dr$, where

$$q(r) = \tfrac{1}{2}\rho\Omega\frac{dC_l}{d\alpha}u(r,\,t)c(r)r \tag{5.25}$$

Hence

$$Q_i = \tfrac{1}{2}\rho\Omega \frac{dC_l}{d\alpha} \int_0^R \mu_i(r)u(r, t)c(r)\, dr \tag{5.87}$$

An expression for the standard deviation of Q_i, σ_{Q_i}, can be derived by a method analagous to that given in Section A5.4 of the Appendix for a non-rotating blade, yielding

$$\sigma_{Q_i}^2 = \left(\tfrac{1}{2}\rho\Omega \frac{dC_l}{d\alpha}\right)^2 \int_0^R \int_0^R \left[\int_0^\infty S_u^o(r_1, r_2, n)\, dn\right] \mu_i(r_1)\mu_i(r_2)c(r_1)c(r_2)r_1 r_2\, dr_1\, dr_2 \tag{5.88}$$

Here $S_u^0(r_1, r_2, n)$ is the rotationally sampled cross spectrum for a pair of points on the rotating blade at radii r_1 and r_2. Equation (5.88) is parallel to Equation (A5.16) in the Appendix with r and r' replaced by r_1 and r_2 and $(\rho\overline{U}C_F)^2$ replaced by $(\frac{1}{2}\rho\Omega(dC_l)/(d\alpha))^2$. From this it can be deduced that the power spectrum of the generalized load with respect to the ith mode is

$$S_{Q_i}(n) = \left(\tfrac{1}{2}\rho\Omega \frac{dC_l}{d\alpha}\right)^2 \int_0^R \int_0^R S_u^o(r_1, r_2, n)\mu_i(r_1)\mu_i(r_2)c(r_1)c(r_2)r_1 r_2\, dr_1\, dr_2 \tag{5.89}$$

In practice, this expression is evaluated using summations to approximate to the integrals.

Power spectrum of tip deflection

The expression for the amplitude of the ith mode blade tip response in response to excitation by a harmonically varying generalized load is given by Equation (A5.4) in the Appendix. Hence the power spectrum of the tip displacement is related to the power spectrum of the generalized load by

$$S_{xi}(n) = \frac{S_{Q_i}(n)}{k_i^2} \frac{1}{[(1 - n^2/n_i^2)^2 + 4\xi_i^2 n^2/n_i^2]} \tag{5.90}$$

This can be written $S_{xi}(n) = (S_{Q_i}(n)/k_i^2)[DMR]^2$ where DMR stands for the dynamic magnification ratio. n_i is the ith mode natural frequency in Hz.

Figure 5.28 shows the power spectrum of first mode tip deflection, $S_{x1}(n)$, for Blade TR operating at 30 r.p.m. in a mean wind of 8 m/s. A lower damping ratio of 0.1 has been selected so that the effect of dynamic magnification is emphasized and the turbulence intensity has been arbitrarily set at 12.5% so that $\sigma_u = 1$ m/s. Also shown is the first mode tip deflection spectrum ignoring dynamic magnification, $S_{Q1}(n)/k_i^2$, which, when multiplied by the square of the dynamic magnification ratio (also plotted), yields the $S_{x1}(n)$ curve. The standard deviation of first mode tip deflection, $\sigma_{x1} = \int_0^\infty S_{x1}(n)\, dn$, comes to 54 mm, a 24 percent increase compared with the value without dynamic magnification. The former is at a minimum,

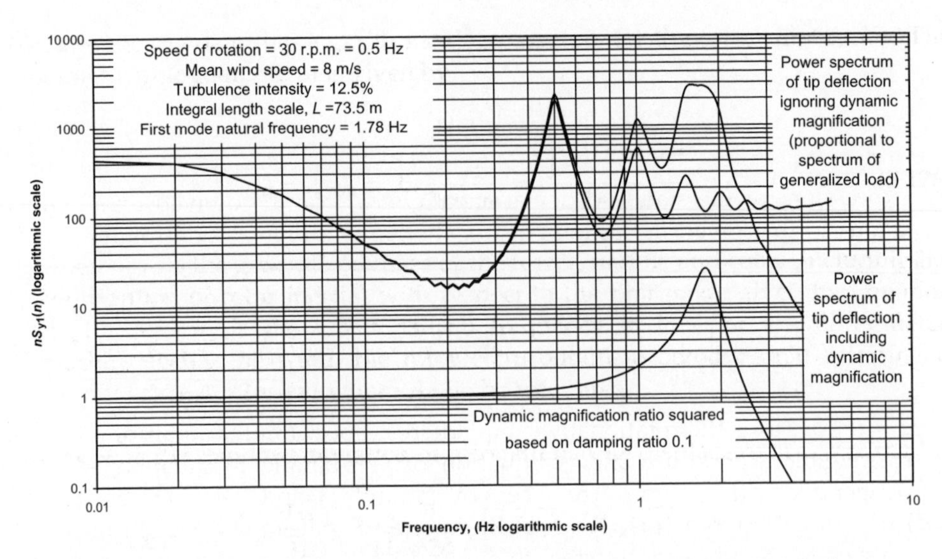

Figure 5.28 Power Spectrum of Blade TR First Out-of-plane Mode Tip Deflection

because the blade TR first mode natural frequency of 1.78 Hz is approximately midway between the third and fourth harmonics of the rotational frequency. However, it is found that, even if the first mode natural frequency coincided with the third harmonic of the rotational frequency, σ_{x1} would only increase by 4 percent. This increase is small, because the peak of $S_{x1}(n)$ at the third harmonic is not very pronounced, and because the peak in the dynamic magnification ratio is relatively broad.

Power spectrum of blade root bending moment

If the amplitude of tip deflection due to excitation of the blade resonant frequency is defined as $x_R(n_1)$, the amplitude of the corresponding blade root bending moment, $M_Y(n_1)$ is given by

$$M_Y(n_1) = \omega_1^2 x_R(n_1) \int_0^R m(r)\mu_1(r) r \, \mathrm{d}r \tag{5.91a}$$

Noting that $\omega_1^2 = k_1/m_1$, this becomes

$$\frac{M_Y(n_1)}{x_R(n_1)} = k_1 R \frac{\displaystyle\int_0^R m(r)\mu_1(r)(r/R)\mathrm{d}r}{m_1} = k_1 R \chi_{M1} \tag{5.91b}$$

This relationship applies at all exciting frequencies, because the right-hand side is essentially a function of mode shape. Hence the power spectrum of blade root bending moment due to excitation of the first mode is given by

$$S_{My1}(n) = (k_1 R\chi_{M1})^2 S_{x1}(n) = (R\chi_{M1})^2 S_{Q1}(n) \frac{1}{[(1 - n^2/n_1^2)^2 + 4\xi_1^2 n^2/n_1^2]} \tag{5.91c}$$

For Blade TR, the ratio χ_{M1} takes a value of 1.4.

5.8.7 Response to simulated loads

The blade dynamic response to time varying loading derived from wind simulation (Section 5.7.6) can be obtained by a step-by-step dynamic analysis such as that described for use with deterministic loads in Section 5.8.5. The procedure is essentially the same, except that it is more important to select realistic values for the initial blade tip displacement, velocity and acceleration, unless the results from the first few rotation cycles are to be discarded.

5.8.8 Teeter motion

When the rotor is rigidly mounted on the shaft, out-of-plane aerodynamic loads on the blades result in fluctuating bending moments in the low speed shaft additional to those due to gravity. In the case of two bladed machines, the transfer of blade out-of-plane aerodynamic moments to the shaft can be eliminated and blade root bending moments reduced by mounting the rotor on a hinge with its axis perpendicular to both the low speed shaft and the axis of the rotor. This allows the rotor to teeter to and fro in response to differential aerodynamic loads on each blade.

The restoring moment is generated by the lateral components of the centrifugal force acting on each blade element (see Figure 5.29). It is given by

$$M_R = \int_0^R r.m(r)\Omega^2 r.\zeta \, dr = I\Omega^2\zeta \tag{5.92}$$

where ζ is the teeter angle and I is the rotor moment of inertia about its centre. The equation of motion for free teeter oscillations is thus $I\ddot{\zeta} + I\Omega^2\zeta = 0$ (omitting the aerodynamic damping term for the moment), indicating that the natural frequency of the teeter motion with the teeter hinge perpendicular to the rotor axis is equal to the rotational frequency. Since both the deterministic and stochastic components of the exciting moment are dominated by this frequency, it is clear that the system operates at resonance, with aerodynamic damping alone controlling the magnitude of the teeter excursion.

The magnitude of teeter excursions would clearly be reduced if the teeter natural frequency were moved away from the rotational frequency. This can be done by rotating the teeter hinge axis relative to the rotor in the plane of rotation, as illustrated in Figure 5.29, so that teeter motion results in a change of blade pitch – positive in one blade and negative in the other – known as Delta 3 coupling. Consider the case of blade A slicing through a gust. The increased thrust on the blade will cause it to move in the downwind direction, by rotating about the teeter

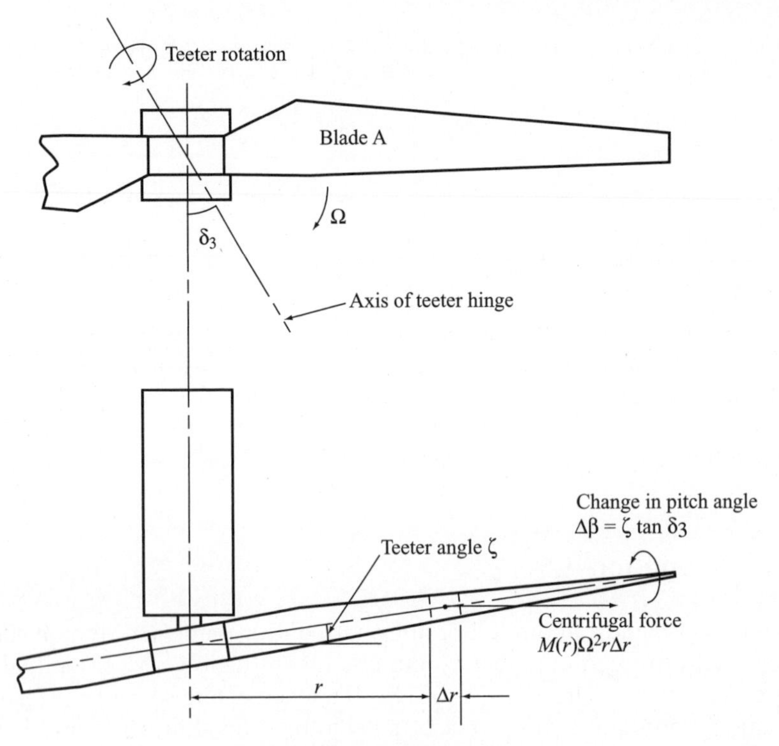

Figure 5.29 Teeter Geometry

hinge. If the teeter angle, defined as the rotation of the blade in its own radial plane, is ζ, then the increase in blade A's pitch angle will be $\zeta \tan \delta_3$, where δ_3 is as defined in Figure 5.29. The increase in the pitch angle of blade A will reduce the angle of attack, α, and thereby reduce the thrust loading on it. The net result of this and a simultaneous increase in the thrust loading on blade B is to introduce a restoring moment augmenting that provided by centrifugal force.

The first stage for the exploration of teeter response to different loadings is the derivation of the complete equation of motion. It is assumed that the blades are unstalled and are operating at a relatively high tip speed, so that the linear relations adopted in the derivation of Equation (5.25) in Section 5.7.5 can be retained. The various contributions to the change in the aerodynamic force on a blade element relative to the steady-state situation are therefore:

$$\tfrac{1}{2}\rho\Omega rc\frac{\mathrm{d}C_l}{\mathrm{d}\alpha}(u - \dot{\zeta}r) - \tfrac{1}{2}\rho(\Omega r)^2 c\frac{\mathrm{d}C_l}{\mathrm{d}\alpha}\Delta\theta \tag{5.93}$$

where the three terms result from the fluctuation of the incident wind, teeter motion and Delta 3 coupling respectively. Multiplication of these terms by radius, integration over the length of the blade and addition of the centrifugal and inertia hub moment terms yields the following equation of motion for the teeter response:

$$I\ddot{\zeta} + \tfrac{1}{2}\rho\Omega\frac{dC_l}{d\alpha}\left[\int_{-R}^{R} r^3 c(r)\,dr\right]\dot{\zeta} + \tfrac{1}{2}\rho\Omega\frac{dC_l}{d\alpha}\left[\int_{-R}^{R} r^3 c(r)\,dr\right]\Omega(\tan\delta_3)\zeta + I\Omega^2\zeta$$

$$= \tfrac{1}{2}\rho\Omega\frac{dC_l}{d\alpha}\int_{-R}^{R} u(r,\, t)c(r)r|r|dr \tag{5.94}$$

assuming a frozen wake. By dividing through by the moment of inertia and writing

$$\eta = \tfrac{1}{2}\frac{\rho}{I}\frac{dC_l}{d\alpha}\left[\int_{-R}^{R} r^3 c(r)\,dr\right] \tag{5.95}$$

this can be simplified to

$$\ddot{\zeta} + \eta\Omega\dot{\zeta} + (1 + \eta\tan\delta_3)\Omega^2\zeta = \tfrac{1}{2}\rho\frac{\Omega}{I}\frac{dC_l}{d\alpha}\int_{-R}^{R} u(r,\, t)c(r)r|r|\,dr \tag{5.96}$$

η is a measure of the ratio of aerodynamic to inertial forces acting on the blade, and is one eighth of the Lock number.

Delta 3 coupling thus raises the natural frequency, ω_n, of the teeter motion from Ω to $\Omega\sqrt{1+\eta\tan\delta_3}$. For a 40 m diameter rotor consisting of two TR blades mounted on a teeter hinge set at a δ_3 angle of 30°, $\eta = 0.888$ and $\tan\delta_3 = 0.577$, so the increase in natural frequency due to the δ_3 angle is 23 percent. The corresponding damping ratio, given by $\xi = (\eta/2)\sqrt{1+\eta\tan\delta_3}$ is quite high at 0.36.

Teeter response to deterministic loads

The teeter response to deterministic loads can be found using the same step-by-step integration procedure set out in Section 5.8.5. However, as the loadings due to wind shear and yaw are both approximately sinusoidal, an estimate of the maximum teeter angle for these cases may be obtained by using the standard solution for forced oscillations. For a harmonically varying teeter moment, $M_T = M_{T0}\cos\Omega t$ due to wind shear, the teeter angle is given by

$$\zeta = \frac{M_{TO}}{I\omega_n}\frac{\cos(\Omega t - \vartheta)}{\sqrt{(1-(\Omega/\omega_n)^2)^2 + (2\xi\Omega/\omega_n)^2}} \tag{5.97}$$

where $\vartheta = \tan^{-1}((2\xi\Omega/\omega_n)/(1-(\Omega/\omega_n)^2)) = 90° - \delta_3$ is the phase lag with respect to the excitation.

For the two-bladed turbine described above, rotating at 30 r.p.m. in a wind with a hub-height mean of 12 m/s and a shear exponent of 0.2, the teeter moment amplitude, M_o, is approximately 50 kNm (see Figure 5.11, which gives the blade root bending moment variation with azimuth for a fixed hub machine, based on momentum theory). Taking the rotor moment of inertia as 307 000 kg m^2 for TR blades, and $\omega_n = 1.23\pi$ rad/s, the maximum teeter angle comes to 0.9° for $\delta_3 = 30°$. This increases by about 16 percent to 1.05° if the δ_3 angle is reduced to zero.

If the wind speed variation due to wind shear is assumed to be linear with height, i.e., $u = \overline{U}(kr/R)$, and the teeter moment is calculated from the expression on the right-hand side of Equation (5.94), which assumes a frozen wake instead of the equilibrium wake resulting from momentum theory, a very simple expression for the teeter angle results in the case of zero δ_3 angle. The teeter moment becomes

$$M_T = M_{TO} \cos \Omega t = \tfrac{1}{2}\rho\Omega \frac{\mathrm{d}C_l}{\mathrm{d}\alpha} \frac{\overline{U}k}{R} \int_{-R}^{R} c(r) r^3 \,\mathrm{d}r \cos \Omega t \tag{5.98}$$

Substitution of Equation (5.98) in Equation (5.97), with ω_n set equal to Ω for the case of a zero δ_3 angle, results in the following expression for the teeter angle

$$\zeta = \frac{\overline{U}k}{\Omega R} \cos(\Omega t - \pi/2)) \tag{5.99}$$

Thus the teeter response lags the excitation by 90° and the magnitude of the teeter excursion is simply equal to the velocity gradient divided by the rotational speed. For a hub height of 35 m, the equivalent uniform velocity gradient over the rotor disc for the case above is $0.125\overline{U}/R = 0.075$ m/s per m, giving a teeter excursion of 0.024 radians or 1.4°. This differs from the earlier value of 1.05° because of the frozen wake assumption.

Teeter response to stochastic loads

As usual, it is convenient to analyse the response to the stochastic loads in the frequency domain. The teeter moment providing excitation is given by the right-hand side of Equation (5.94). By following a similar method to that used for the generalized load in Section 5.8.6, the following expression for the power spectrum of the teeter moment can be derived:

$$S_{MT}(n) = \left(\tfrac{1}{2}\rho\Omega \frac{\mathrm{d}C_l}{\mathrm{d}\alpha}\right)^2 \int_{-R}^{R} \int_{-R}^{R} S_u^o(r_1, r_2, n) c(r_1) c(r_2) r_1 r_2 |r_1||r_2| \,\mathrm{d}r_1 \,\mathrm{d}r_2 \tag{5.100}$$

where $S_u^o(r_1, r_2, n)$ is the rotationally sampled cross spectrum. In practice, $S_u^o(r_1, r_2, n)$ is evaluated for a few discrete radius values, and the integrals replaced by summations.

The power spectrum of the teeter angle response is related to the teeter moment power spectrum by a formula analagous to Equation (5.90), as follows:

$$S_\zeta(n) = \frac{S_{MT}(n)}{(I\omega_n^2)^2} \frac{1}{[(1 - (2\pi n/\omega_n)^2)^2 + (2\xi.2\pi n/\omega_n)^2]} \tag{5.101}$$

This can be written $S_\zeta(n) = (S_{MT}(n)/(I\omega_n^2)^2)[DMR]^2$ where DMR stands for the dynamic magnification ratio.

Figure 5.30 shows the teeter angle power spectrum, $S_\zeta(n)$, for a two-bladed rotor

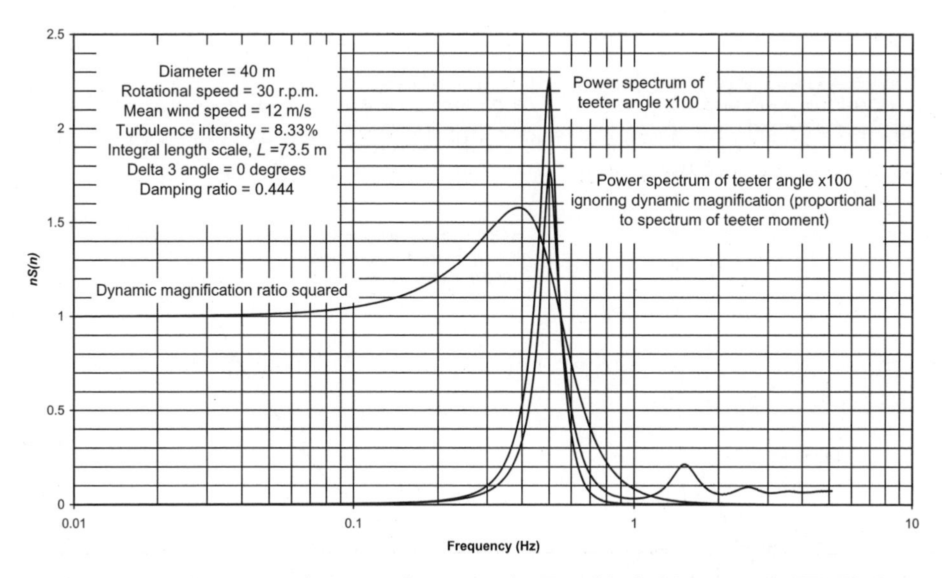

Figure 5.30 Teeter Angle Power Spectrum for Two-bladed Rotor with 'TR' Blades

with TR blades and zero δ_3 angle operating at 30 r.p.m. in a mean wind of 12 m/s. The turbulence intensity is arbitrarily taken as 8.33 percent, to give $\sigma_u = 1$ m/s, and the damping ratio, $\xi = \eta/2$, is 0.444, calculated from Equation (5.95). Also shown in the figure is the teeter angle power spectrum ignoring dynamic magnification, $S_{MT}(n)/(I\omega_n^2)^2$, which, when multiplied by the square of the dynamic magnification ratio (also plotted), yields the $S_\zeta(n)$ curve. The resulting teeter angle standard deviation, obtained by taking the square root of the area under the power spectrum, is 0.46°.

Having calculated the teeter angle standard deviation, the extreme value over any desired exposure period can be predicted from Equation (5.59). As is evident from Figure 5.30, the teeter angle power spectrum is all concentrated about the rotational frequency, Ω, so the zero upcrossing frequency, ν, can be set equal to it. Thus, for a machine operating at 30 r.p.m., a 1 h exposure period gives, $\nu T = 1800$ and $\zeta_{max}/\sigma_\zeta = 4.02$. Taking a turbulence intensity of 17 percent, the predicted maximum teeter angle due to stochastic loading over a 1 h period for the case above is $4.02 \times (12 \times 0.17) \times 0.46° = 3.8°$, which reduces to 3.2° if a δ_3 angle of 30° is introduced.

As already mentioned, teetering relieves blade root bending moments as well as those in the low speed shaft. The reduction of the stochastic component of root bending moment can be derived in terms of the standard deviations of blade root bending moment and hub teeter moment for a rigid hub two-blade machine. Integration of Equation (5.100) yields the following expression for the latter:

$$\sigma_{MT}^2 = \left(\frac{1}{2}\rho\Omega\frac{\mathrm{d}C_l}{\mathrm{d}\alpha}\right)^2 \int_{-R}^{R}\int_{-R}^{R} \kappa_u^o(r_1, r_2, 0)c(r_1)c(r_2).r_1 r_2 |r_1||r_2| \,\mathrm{d}r_1\,\mathrm{d}r_2 \tag{5.102}$$

where, as before, r_1 and r_2 take *negative* values on the second blade. $\kappa_u^o(r_1, r_2, 0)$ is the cross correlation function between the longitudinal wind fluctuations between points at radii r_1 and r_2 on the rotating rotor and is given by the right-hand side of Equation (5.51), with $\Omega\tau$ set equal to zero when r_1 and r_2 define points on the same blade, and replaced by π when r_1 and r_2 define points on different blades. Defining $\rho_u^o(r_1, r_2, 0)$ as the normalized cross correlation function, $\kappa_u^o(r_1, r_2, 0)/\sigma_u^2$, Equation (5.102) can be rewritten as:

$$\sigma_{MT}^2 = \sigma_u^2\left(\tfrac{1}{2}\rho\Omega\frac{dC_l}{d\alpha}\right)^2\int_{-R}^{R}\int_{-R}^{R}\rho_u^o(r_1, r_2, 0)c(r_1)c(r_2).r_1 r_2|r_1||r_2|\,dr_1\,dr_2 \tag{5.102a}$$

The corresponding expression for the standard deviation of the mean of the two-blade root bending moments is:

$$\sigma_{\overline{M}}^2 = \tfrac{1}{4}\sigma_u^2\left(\tfrac{1}{2}\rho\Omega\frac{dC_l}{d\alpha}\right)^2\int_{-R}^{R}\int_{-R}^{R}\rho_u^o(r_1, r_2, 0)c(r_1)c(r_2).r_1^2 r_2^2\,dr_1\,dr_2 \tag{5.103}$$

By inspection of the integrals, it is easily shown that:

$$\tfrac{1}{4}\sigma_{MT}^2 + \sigma_{\overline{M}}^2 = \sigma_u^2\left(\tfrac{1}{2}\rho\Omega\frac{dC_l}{d\alpha}\right)^2\int_{0}^{R}\int_{0}^{R}\rho_u^o(r_1, r_2, 0)c(r_1)c(r_2).r_1^2 r_2^2\,dr_1\,dr_2 = \sigma_M^2 \tag{5.104}$$

where σ_M is the standard deviation of root bending moment for a rigidly mounted blade. Thus, if the rotor is allowed to teeter, the standard deviation of the blade root bending moment will drop from σ_M to $\sigma_{\overline{M}}$ where σ_M is given by the equation above. The extent of the reduction is driven primarily by the ratio of rotor diameter to the integral length scale of the wind turbulence. For a two-bladed rotor with TR blades and an integral length scale of 73.5 m, the reduction is 11 percent.

5.8.9 Tower coupling

In the preceding sections, consideration of the dynamic behaviour of the blade has been based on the assumption that the nacelle is fixed in space, i.e., that the tower is rigid. In practice, of course, no tower is completely rigid, so fluctuating loads on the rotor will result in fore–aft flexure of the tower, which, in turn, will affect the blade dynamics. This section explores the effect the coupling of the blade and tower motions has on blade response.

The application of standard modal analysis techniques to the dynamic behaviour of the system comprising the tower and rotating rotor treated as a single entity is complicated by the system's continually changing geometry, which means that the mode shapes and frequencies of the structure taken as a whole would have to be re-evaluated at each succeeding rotor azimuth position.

An alternative approach is to base the analysis on the mode shapes and frequencies of the different elements of the structure considered separately, with the displacements arising from each set of modes superposed. Thus the tower modes are calculated on the basis of a completely rigid rotor, and the blade modes are calculated as if the blades were cantilevered from a rigidly mounted shaft, i.e., in the same way as before. The blade modes are not orthogonal to the tower modes, so the equations of motion for the different modes are no longer independent of each other, but contain coupled terms. Furthermore, the blade deflections arising from excitation of the tower modes vary with blade azimuth, so a step-by-step solution is required. The treatment which follows is limited to the fundamental blade and tower modes, but could be extended to encompass higher modes.

The equation of motion of the blade is given by Equation (5.62). The blade deflection for blade J may be written

$$x(r,\, t) = \mu(r).f_{\mathrm{J}}(t) + \mu_{\mathrm{TJ}}(r).f_{\mathrm{T}}(t) \tag{5.105}$$

where $\mu(r)$ is the first blade mode shape for a rigid tower and $\mu_{\mathrm{TJ}}(r)$ is the normalized rigid body deflection of blade J resulting from excitation of the tower first mode. Assuming the normalization is carried out with respect to hub deflection,

$$\mu_{\mathrm{TJ}}(r) = 1 + \frac{r}{L}\cos\psi_{\mathrm{J}} \tag{5.106}$$

where L is the depth below the hub of the intercept between the tangent to the top of the deflected tower and the undeflected tower axis, as illustrated in Figure 5.31. Substitution of Equation (5.105) into Equation (5.62) yields, with the aid of Equation (5.65):

$$\begin{aligned} &m(r)\mu(r)\ddot{f}_{\mathrm{J}}(t) + \hat{c}(r)\mu(r)\dot{f}_{\mathrm{J}}(t) + m(r)\omega^2\mu(r)f_{\mathrm{J}}(t) \\ &= q(r.t) - m(r)\mu_{\mathrm{TJ}}(r)\ddot{f}_{\mathrm{T}}(t) - \hat{c}(r)\mu_{\mathrm{TJ}}(r)\dot{f}_{\mathrm{T}}(t) \end{aligned} \tag{5.107}$$

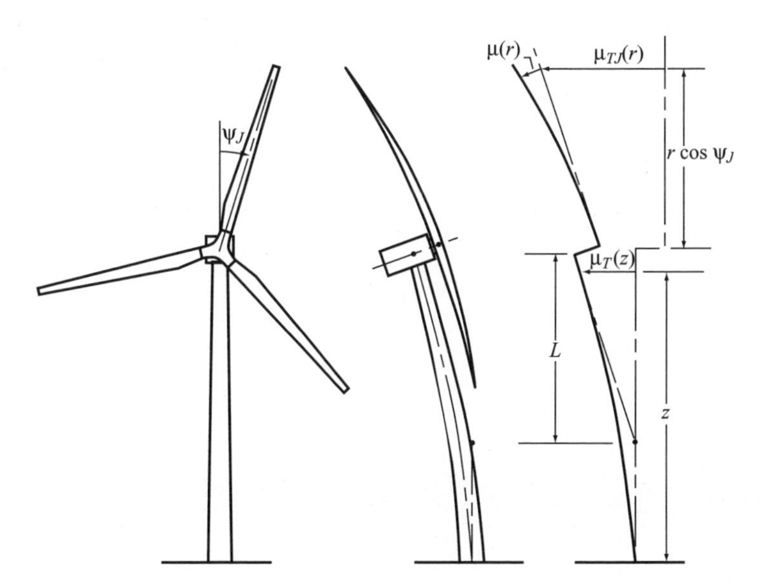

Figure 5.31 Fundamental Mode Shapes of Blade and Tower

where the coupled terms have been transferred to the right hand side. Multiplying through by $\mu(r)$ and integrating over the length of the blade gives:

$$m_1\ddot{f}_J(t) + c_1\dot{f}_J(t) + m_1\omega^2 f_J(t) = \int_0^R \mu(r)q(r,t)\,\mathrm{d}r - \int_0^R m(r)\mu(r)\mu_{TJ}(r)\,\mathrm{d}r.\ddot{f}_T(t)$$
$$- \int_0^R \hat{c}(r)\mu(r)\mu_{TJ}(r)\,\mathrm{d}r.\dot{f}_T(t) \tag{5.108}$$

By analogy with Equation (5.70), the equation of motion of the tower is

$$m_{T1}\ddot{f}_T(t) + c_{T1}\dot{f}_T(t) + m_{T1}\omega_T^2 f_T(t) = \int_0^H \mu_T(z)q(z,t)\,\mathrm{d}z \tag{5.109}$$

Here μ_T is the tower first mode shape and m_{T1} is the generalized mass of the tower, nacelle and rotor (including the contribution of rotor inertia), with respect to the first mode, given by

$$m_{T1} = \int_0^H m_T(z)\mu_T^2(z)\,\mathrm{d}z + m_N + m_R + I_R/L^2 \tag{5.110}$$

Here $m_T(z)$ is the mass per unit height of the tower, m_N and m_R are the nacelle and rotor masses, and I_R is the inertia of the rotor about the horizontal axis in its plane, which is constant over time for a three-bladed rotor. For a two-bladed, fixed-hub rotor it varies with rotor azimuth, and for a teetering rotor it is omitted altogether.

The major component of the loading on the tower, $q(z, t)$, is the load fed in at hub height, H, from the blades. The inertia forces on the blades due to rigid body motion associated with the tower first mode have been accounted for by including rotor mass and inertia in m_{T1}, and the corresponding damping forces can be accounted for in the calculation of the damping coefficient, c_{T1}. However, the aerodynamic loads on the blades and the inertia and damping forces associated with blade flexure – all of which are transmitted to the tower top – have to be included in the right-hand side of Equation (5.109) as

$$\mu_T(H).F + \left(\frac{\mathrm{d}\mu_T}{\mathrm{d}z}\right)_H .M = F + M/L \tag{5.111}$$

where

$$F = \sum_N q(r,t)\,\mathrm{d}r - \sum_N \int_0^R m_1(r)\mu(r)\,\mathrm{d}r.\ddot{f}_J(t) - \sum_N \int_0^R \hat{c}(r)\mu(r)\,\mathrm{d}r.\dot{f}_J(t) \tag{5.112}$$

and

$$M = \sum_N r\cos\psi_J q(r,t)\,\mathrm{d}r - \sum_N \int_0^R r\cos\psi_J m_1(r)\mu(r)\,\mathrm{d}r.\ddot{f}_J(t)$$
$$-\sum_N \int_0^R r\cos\psi_J \hat{c}(r)\mu(r)\,\mathrm{d}r.\dot{f}_J(t) \tag{5.113}$$

The suffix J refers to the Jth blade, and N in the summations is the total number of blades.

Hence

$$F + M/L = \sum_N \mu_{TJ} q(r,t)\,\mathrm{d}r - \sum_N \int_0^R m_1(r)\mu(r)\mu_{TJ}(r)\,\mathrm{d}r.\ddot{f}_J(t)$$
$$-\sum_N \int_0^R \hat{c}(r)\mu(r)\mu_{TJ}(r)\,\mathrm{d}r.\dot{f}_J(t)$$

and Equation (5.109) becomes

$$m_{T1}\ddot{f}_T(t) + c_{T1}\dot{f}_T(t) + m_{T1}\omega_T^2 f_T(t) = \sum_N \mu_{TJ} q(r,t)\,\mathrm{d}r - \sum_N \int_0^R m_1(r)\mu(r)\mu_{TJ}(r)\,\mathrm{d}r.\ddot{f}_J(t)$$
$$-\sum_N \int_0^R \hat{c}(r)\mu(r)\mu_{TJ}(r)\,\mathrm{d}r.\dot{f}_J(t) \tag{5.114}$$

omitting the term for loading on the tower itself.

Equations (5.108) and (5.114) provide $(N+1)$ simultaneous equations of motion with periodic coefficients μ_{TJ} corresponding to the $(N+1)$ degrees of freedom assumed. The procedure for the step-by-step dynamic analysis which is based on these equations may be summarized as follows:

(1) Substitute the displacements, velocities and aerodynamic loads at the beginning of the first time step into Equations (5.108) and (5.114), and solve for the initial accelerations.

(2) Formulate the *incremental* equations of motion for the time step, based on Equations (5.108) and (5.114), retaining the coupled terms on the right-hand side, i.e., as pseudo forces.

(3) Assume initially that the coupled terms are *constant* over the duration of the time step, so that they disappear from the incremental equations of motion altogether, rendering them uncoupled.

(4) Solve the uncoupled incremental equations of motion to obtain the increments of displacement and velocity over the time step. Adopting the linear accelera-

tion method (Section 5.8.5), the expressions for the displacement and velocity increments at the tip of blade J are as follows:

$$\Delta f_J = \frac{\Delta Q_J + m_1\left(\frac{6}{h}\dot{f}_{J0} + 3\ddot{f}_{J0}\right) + c_1\left(3\dot{f}_{J0} + \frac{h}{2}\ddot{f}_{J0}\right)}{m_1\omega^2 + \frac{3c_1}{h} + \frac{6m_1}{h^2}} \tag{5.115}$$

$$\Delta\dot{f}_J = \frac{3}{h}\Delta f - 3\dot{f}_0 - \frac{h}{2}\ddot{f}_0 \tag{5.116}$$

The derivation of these expressions parallels that for the absolute values of displacement and velocity at the end of the time step, given in Section 5.8.5. Similar expressions obtain for the displacement and velocity increments at the hub due to tower flexure.

(5) Solve Equations (5.108) and (5.114) for the accelerations at the *end* of the time step.

(6) Solve the incremental equations of motion again – this time including the changes in the coupled terms on the right hand side over the time step – to obtain revised increments of displacement and velocity over the time step.

(7) Repeat Step 5 and Step 6 until the increments of displacement and velocity converge.

(8) Repeat Steps 1–7 for the second and subsequent time steps.

If the analysis is being carried out to obtain the response to deterministic loads, advantage may be taken of the fact that the behaviour of each blade mirrors that of its neighbours with an appropriate phase difference. This means that the number of equations of motion can be reduced to two, and the analysis iterated over a number of revolutions until a steady-state response is achieved. For example, in the case of a machine with three blades, A, B and C the values of blade B and blade C tip velocities and accelerations, which are required on the right-hand side of Equation (5.114), would be equated to the corresponding values for blade A occurring $T/3$ and $2T/3$ earlier (T being the period of blade rotation).

Figure 5.32 shows the results from the application of the above procedure to the derivation of blade tip and hub displacements in response to tower shadow loading, considering only the blade and tower fundamental modes. The machine is three bladed and the parameters chosen are, as far as the rotor is concerned, generally the same as for the rigid tower example in Section 5.8.5 illustrated in Figure 5.25. The tower natural frequency is 1.16 Hz, and the tower damping ratio (which is dominated by the aerodynamic damping of the blades) is taken as 0.022.

It can be seen that the tower response is sinusoidal at blade passing frequency, which is the forcing frequency. The amplitude is only about one fiftieth of the maximum blade tip displacement of about 30 mm, reflecting the large generalized

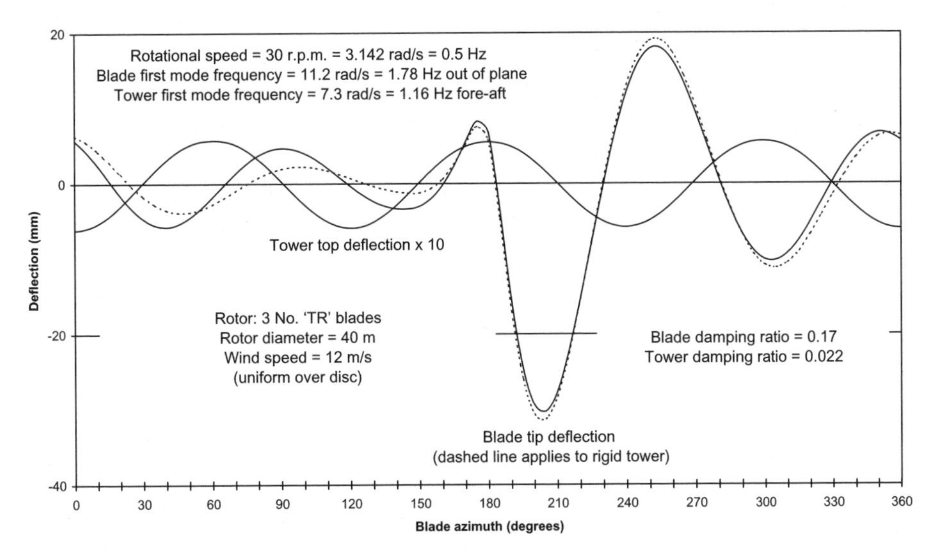

Figure 5.32 Tower Top and Blade Tip Deflections Resulting from Tower Shadow, Considering Fundamental Mode Responses Only

mass associated with the tower mode relative to that associated with the blade mode. The tower shadow effect causes the blade to accelerate rapidly upwind as it passes the tower, with the maximum deflection occurring at an azimuth of about 205°. Also plotted on Figure 5.32 is the deflection that would occur if the nacelle were fixed, and it is seen from the comparison that one effect of tower flexibility is to slightly reduce the peak deflection. However, a more significant effect of the tower motion is the maintenance of the amplitude of the subsequent blade oscillations at a higher level prior to the next tower passing.

The modal analysis method outlined above forms the basis for a number of codes for wind turbine dynamic analysis, such as the Garrad Hassan BLADED code (Bossanyi, 2000). Typically these codes encompass several blade modes, both out-of-plane and in-plane, and several tower modes, both fore–aft and side-to-side, together with drive train dynamics (see Section 5.8.10).

Rather than use modal analysis, the dynamic behaviour of coupled rotor/tower systems can also be investigated using finite elements. Standard finite-element dynamics packages are, however, inappropriate to the task, because they are only designed to model the displacements of structures with fixed geometry. Lobitz (1984) has pioneered the application of the finite-element method to the dynamic analysis of wind turbines with two-bladed, teetering rotors, and Garrad (1987) has extended it to three-bladed, fixed-hub machines. In both cases, equations of motion are developed in matrix form for the blade and tower displacement vectors and then amalgamated using a connecting matrix which is a function of blade azimuth and satisfies the compatibility and equilibrium requirements at the tower/rotor interface. Solution of the equations is carried out by a step-by-step procedure. The finite-element method is more demanding of computing power, so the modal analysis method is generally preferred.

5.8.10 Wind turbine dynamic analysis codes

A large modern turbine is a complex structure. Relatively sophisticated methods are required in order to predict the detailed performance and loading of a wind turbine. These methods should take into account:

- the aerodynamics of the rotating blade, including induced flows (i.e., the modification of the flow field caused by the turbine itself), three-dimensional flow effects, and dynamic stall effects when appropriate;
- dynamic analysis of the blades, drive train, generator and tower, including the modification of the aerodynamic forces due to the vibrational velocities of the structure;
- dynamic response of subsystems such as the yaw system and blade pitch control system;
- control algorithms used during normal operation, start-up and shut-down of the turbine;
- temporal and spatial variations of the wind field impinging on the turbine, including the three-dimensional structure of the turbulence itself.

Starting from a wind turbulence spectrum, it is possible to develop techniques in the frequency domain which account for many of these aspects, including rotational sampling of the turbulence by the blades, the response of the structure, and the control system. These techniques are set out in Sections 5.7.5, 5.8.6, 5.12.4 and elsewhere. However, although frequency domain methods are elegant and computationally efficient, they can only be applied to linear time-invariant systems, and therefore cannot deal with some important aspects of wind turbine behaviour, such as:

- stall hysteresis;
- non-linearities in subsystems such as bearing friction, pitch rate limits, and non-linear aspects of control algorithms;
- variable speed operation;
- start-up and shut-down.

As a result, time-domain methods are now used almost exclusively for wind turbine design calculations. The ready availability of computing power means that the greater computational efficiency of frequency domain methods is no longer such an important consideration.

A number of codes are available commercially for the calculation of wind turbine performance and loads using time-domain simulations. These simulations use numerical techniques to integrate the equations of motion over time, by subdividing the time into short timesteps (which may be of fixed or variable length), as described in Section 5.8.5. In this way, all the non-linearities and non-stationary aspects of the system, such as those listed above, can be dealt with to any desired level of accuracy. A useful comparative survey of such codes is given by Molenaar and Dijkstra (1999).

Two principal approaches to the modelling of structural dynamics are embodied in these software packages. Some use a full finite-element representation of the structure, which is broken down into small elements. The equations of motion are solved for each element, with boundary conditions matched at the interfaces between elements. An example of such a code is Adams-WT (Hansen, 1998), which consists of a general purpose finite-element code (Adams) interfaced to an aerodynamic module.

The other main approach is the modal analysis method as described in the preceding section, in which simple finite-element methods are used to predict just the first few modes of vibration of the structure as a whole, or of its main parts. The equations of motion for these modes, which include periodic coefficients, are then derived and solved with appropriate boundary conditions over each time step. This gives a much smaller set of equations (although the equations themselves may be more complex). The higher frequency modes of the system tend to contribute very little to the system dynamics and loads, and so the modal method generally gives a very good approximation to the performance of the structure. An example of a code based on this approach is *Bladed for Windows* (Bossanyi, 2000), which allows the most important rotor and tower modes to be calculated. These are then linked to the remaining system dynamics (drive train, control systems, etc.), and to an aerodynamics module similar to that of Adams-WT.

Both of these codes include a full three-dimensional, three-component model of the turbulent wind field computed using the Veers method (Veers, 1988) as described in Section 5.7.6. *Bladed for Windows* additionally has an offshore module, allowing stochastic wave loading and current loading on the tower to be modelled for an offshore turbine. As with the effect of aerodynamics, the effect of the vibrational velocities of the structure on the hydrodynamic forces is significant. This leads to considerable interactions between the wind and wave loading. Jamieson *et al.* (2000) have demonstrated that if wind and wave loading are treated in isolation from each other, an over-conservative design is likely to result.

The use of sophisticated calculation methods such as those described above are rapidly becoming mandatory for the certification of wind turbines, particularly at the larger sizes. A few illustrative examples of results obtained with *Bladed for Windows* are described below.

Figure 5.33 shows a *Bladed for Windows* simulation of the in- and out-of-plane bending moments at the root of one of the blades, during operation in steady, sheared wind. The in-plane moment is almost a sinusoidal function of azimuth, being dominated by the gravity loading due to the self-weight of the blade which, relative to the blade, changes direction once per revolution. The mean is offset from zero because of the mean positive aerodynamic torque developed by the blade. There is a slight distortion of the sinusoid, partly because of the variation of aerodynamic torque due to wind shear and the effect of tower shadow, and partly because of the in-plane vibration of the blade.

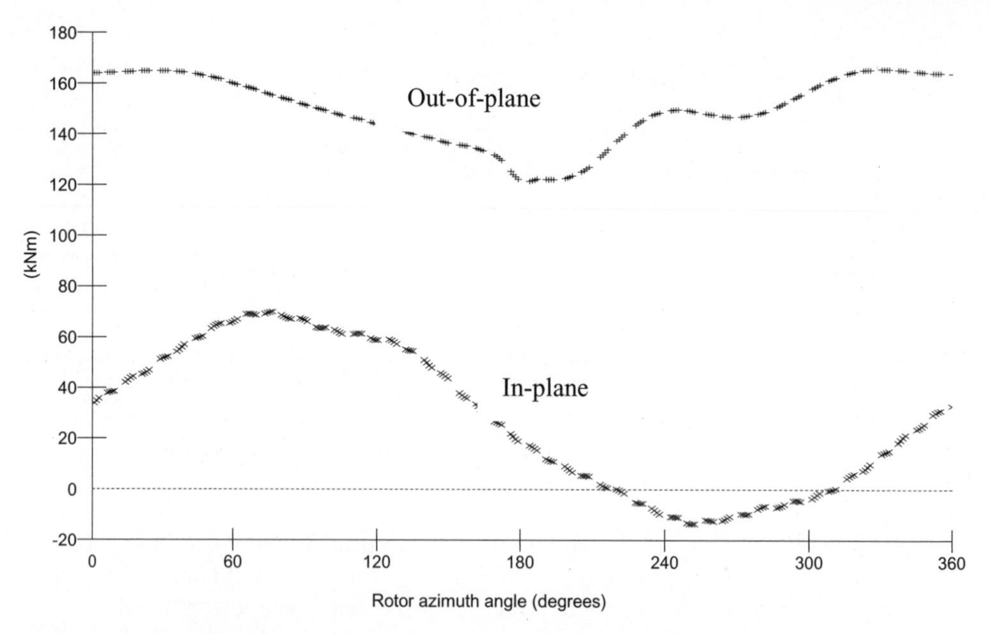

Figure 5.33 Blade Root Bending Moment in Steady Wind

The out-of-plane moment is always positive, the mean value being dominated by the aerodynamic thrust on the blade. There is a systematic variation with azimuth resulting from the wind shear, giving a lower load at 180° azimuth (bottom dead centre) than at 0°. A sharp dip at 180° is also visible, and this is the effect of the tower shadow (the reduction in wind speed in the vicinity of the tower). The blade out-of-plane vibrational dynamics contribute a significant higher-frequency variation.

In turbulent wind, the loads take on a much more random appearance, as shown in Figure 5.34. The out-of-plane load in particular is varying with wind speed and, as this is a pitch-controlled machine, with pitch angle. The in-plane load is more regular, as it is always dominated by the reversing gravity load.

Spectral analysis provides a useful means of understanding these variations. Figure 5.35 shows auto-spectra of the blade root out-of-plane bending moment and the hub thrust force. The out-of-plane bending moment is dominated by peaks at all multiples of the rotational frequency of 0.8 Hz. These are caused mainly by the rotational sampling of turbulence by the blade as it sweeps around, repeatedly passing through turbulent eddies. Wind shear and tower shadow also contribute to these peaks. A small peak due to the first out-of-plane mode of vibration at about 3.7 Hz is just visible. There is also a significant effect of the first tower fore–aft mode of vibration at about 0.4 Hz.

This tower effect is also visible in the spectrum of the hub thrust force. However, this force is the sum of the shear forces at the roots of the three blades. These forces are 120° out of phase with each other, with the result that the peak at the rotational frequency ($1P$) is eliminated, as are the peaks at multiples of this frequency such as

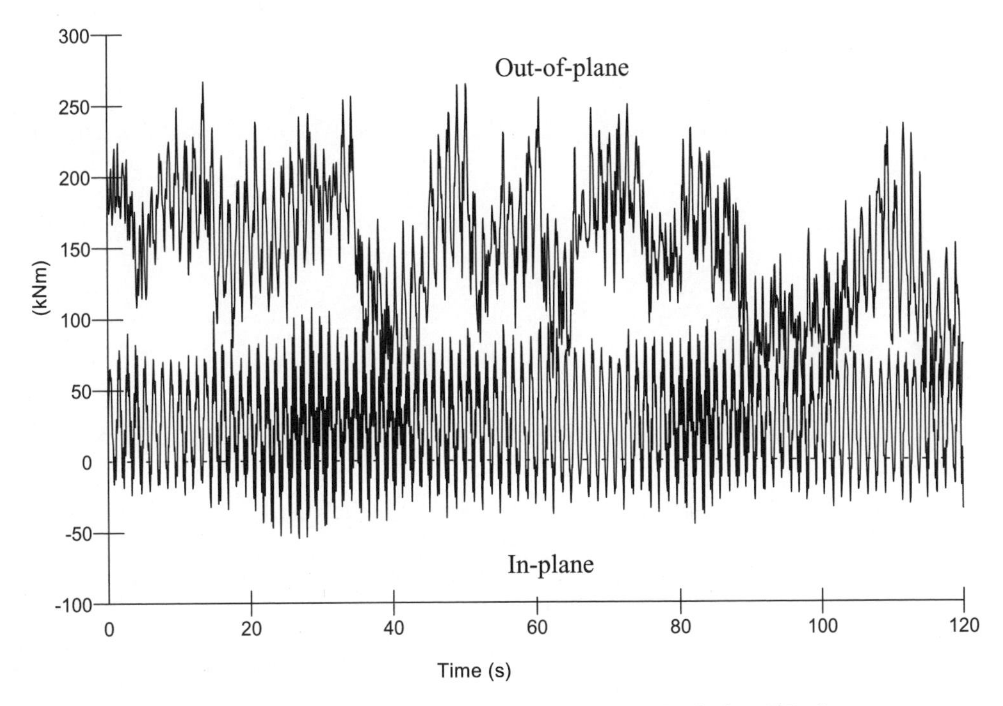

Figure 5.34 Blade Root Bending Moment in Turbulent Wind

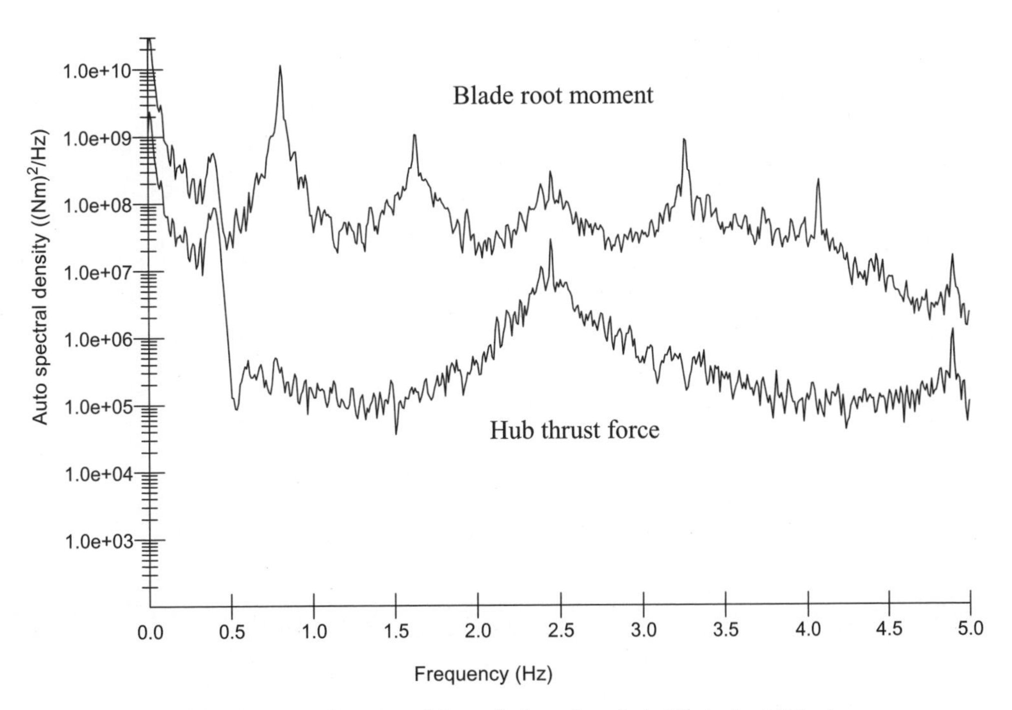

Figure 5.35 Spectra of Out-of-plane Loads in Turbulent Wind

$2P$, $4P$ etc. Only the peaks at multiples of $3P$ remain, since at these frequencies the three blades act in phase with each other.

This effect is even more significant in the in-plane load spectra (Figure 5.36). Of the blade load peaks at multiples of $1P$, only the relatively small peaks at $3P$ and $6P$ come through to the hub torque. The $1P$ peak in the blade load, which is dominated by gravity, is particularly large, but it is completely eliminated from the hub torque. The tower peak at 0.4 Hz is visible in both loads. A large blade load peak at the first in-plane blade vibrational mode at 4.4 Hz is also seen, but this is a mode which does not include any rotation at the hub, and consequently is not seen in the hub torque. Some higher frequency blade modes (not shown) will be coupled with hub rotation.

5.8.11 Aeroelastic stability

Aeroelastic instability can arise when the change in aerodynamic loads resulting from a blade displacement is such as to exacerbate the displacement rather than diminish it, as is normally the case. A theoretical example would be a teetering rotor operating in stalled flow, where the rate of change of lift coefficient with angle of attack is negative, so that the aerodynamic damping is negative likewise. In such circumstances, teeter excursions would be expected to grow until the limits of the negative damping band or of the teeter stops were reached. In practice, this phenomenon can be avoided if the blade is designed so that the blade root flapwise bending moment increases monotonically with wind speed over the full wind speed operational range (Armstrong and Hancock, 1991).

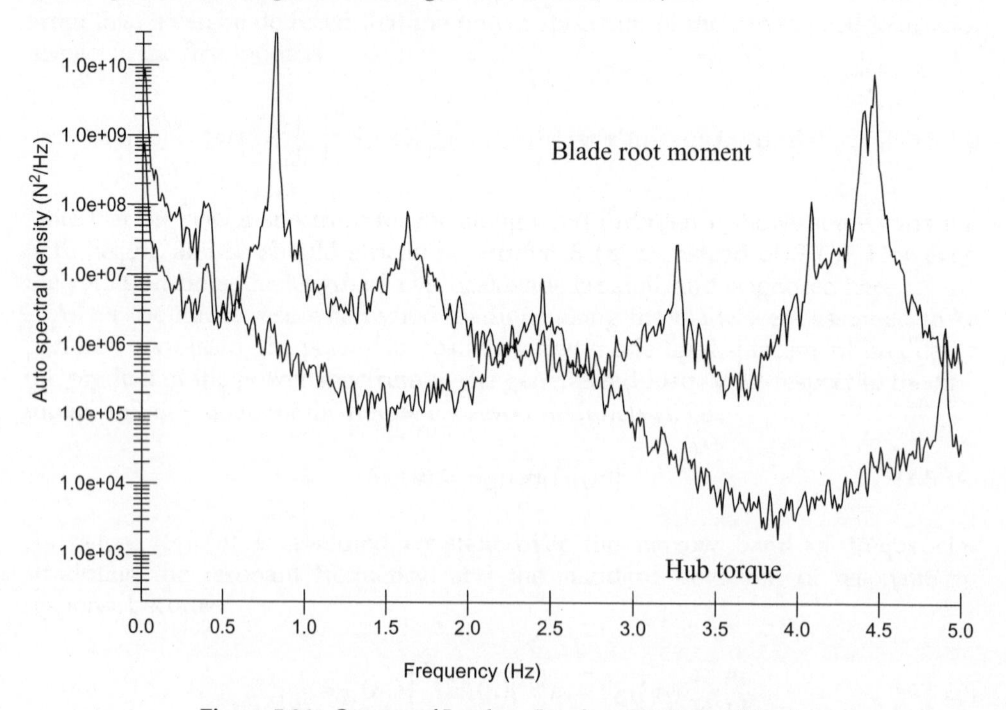

Figure 5.36 Spectra of In-plane Loads in Turbulent Wind

A real instance of incipient aeroelastic instability was the development of an edgewise blade resonance under stalled conditions on some larger three-bladed machines. A negative rate of change of lift coefficient with angle of attack is believed to have been the prime cause – see Section 7.1.9.

Another potential instance of aeroelastic instability is classical flutter, encountered in the design of helicopter rotors, in which the blade structure is such that out-of-plane flexure in the downwind direction results in blade twisting, causing an increase in the angle of attack. During the development of some of the early large machines, the dangers of aeroelastic instability were considered to be a real concern, and much analysis work was directed to demonstrating that individual turbine designs would not be susceptible to it. However, partly no doubt because of the high torsional rigidity of the closed cell hollow structure adopted for most wind turbine blades, aeroelastic instability has not yet been found to be critical in practice, and stability analyses are no longer regarded as an essential part of the design process. This may change, however, if designs become more flexible.

5.9 Blade Fatigue Stresses

5.9.1 Methodology for blade fatigue design

The verification of the adequacy of a blade design in fatigue requires knowledge of the fatigue loading cycles expected over the lifetime of the machine at different radii, derivation of the resultant stress cycles and calculation of the corresponding fatigue damage number in relation to known fatigue properties of the material. The procedure is less or more complicated, depending upon whether blade loading in one or two planes is taken into account. If bending about only the weaker principal axis is taken into account, considering only aerodynamic lift forces, the steps involved are as follows.

(1) Derive the individual fatigue load spectra for each mean wind speed and for each radius. This is a non-trivial task because, unless wind simulation is used, the information on the periodic and stochastic load components is available in different forms, i.e., as a time history and a power spectrum respectively. Sections 5.9.2 and 5.9.3 consider methods of addressing this difficulty.

(2) Synthesize the complete fatigue load spectrum at each radius from the separate load spectra for each mean wind speed, including start-ups and shutdowns (see Section 5.5.1).

(3) Convert the fatigue load cycles (expressed as bending moments) to fatigue stresses by dividing by the appropriate section modulus. (The section modulus with respect to a particular principal axis is defined as Second Moment of Area of the cross section about that axis divided by the distance of the point under consideration from the axis.)

(4) Sum the fatigue damage numbers, n_i/N_i, according to Miner's rule, for each moment range 'bin' in the fatigue load spectrum, according to the appropriate $S-N$ curve for the material. $S-N$ curves for different blade materials are considered in Section 7.1.6 and 7.1.7, together with the allowance necessary for mean stress.

Sections 5.9.2 and 5.9.3 are concerned with the first step of the sequence above. For a given mean wind speed, the periodic component of blade loading will be invariant over time, and the stochastic component will be stationary. As noted in Section 5.7.5, the stochastic component can be analysed either in the frequency domain (provided that a linear relationship between incident wind speed and blade loadings can be assumed) or in the time domain, i.e., by using wind simulation. Section 5.9.2 considers how the deterministic and stochastic components may be combined if the latter have been analysed in the frequency domain, while Section 5.9.3 looks in detail at the option of assessing fatigue damage completely in the frequency domain.

If the fatigue damage resulting from both in-plane and out-of-plane loading is to be computed, it is necessary to revise the ordering of the steps above, in order to derive the periodic and stochastic components of the stress variation for each point under consideration and for each mean wind speed. For a chosen point, the procedure becomes as shown below.

(A1) For a given mean wind speed, calculate the time histories of the bending moments about the principal axes resulting from the periodic load components over one blade rotation. The derivation of aerodynamic moments from blade element loads is illustrated in Figure 5.37.

(A2) Convert these bending moment time histories to stress time histories by dividing by the appropriate section modulus, and adding them together.

(B) For the same mean wind speed, convert the power spectrum of the stochastic bending moment component (which, because of the linearity assumption, arises from fluctuating lift only) to a power spectrum of stress at the chosen point.

(C) Calculate the fatigue damage resulting from the combined periodic and stochastic stress components, using the methods of Sections 5.9.2 and 5.9.3.

(D) Repeat the above steps for the other mean wind speeds.

(E) Add together the fatigue damages arising at each mean wind speed to obtain the total fatigue damage during normal running.

5.9.2 Combination of deterministic and stochastic components

Previous sections have shown how the deterministic (i.e., periodic) and stochastic components of blade bending moments can be characterized in terms of time

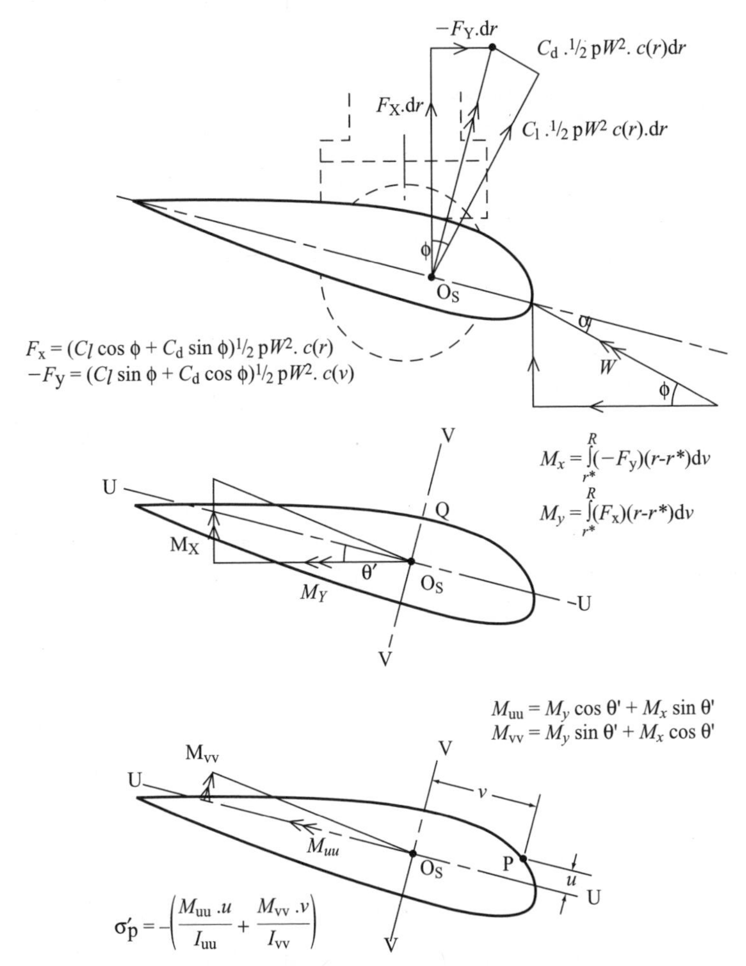

Figure 5.37 Derivation of Blade Bending Stresses at Radius r^* due to Aerodynamic Loads

histories and power spectra respectively. Unfortunately the spectral description of the stochastic loading is not in a suitable form to be combined with the time history of the periodic loading, but this difficulty can be resolved by one of two methods, as follows.

(1) The power spectrum of the stochastic component can be transformed into a time history by inverse Fourier transform, which can then be added directly to the time history of the periodic component. Applications of this method have been reported by Garrad and Hassan (1986) and Warren *et al.* (1988). With the subsequent development of wind simulation techniques, this method is no longer commonly used, because the use of transformations to generate time-histories of *wind speed* rather than of wind loading avoids the need to assume that wind speed and wind loading are linearly related when deriving the power spectrum of the stochastic load component.

(2) A probability density function for the load cycle ranges can be derived empirically, based on the spectral properties of the power spectrum of the stochastic and periodic components of loading combined.

The second approach is considered in the next Section.

5.9.3 Fatigue prediction in the frequency domain

The probability density function (p.d.f.) of peaks of a narrow band, Gaussian process are given by the well-known Rayleigh distribution. As each peak is associated with a trough of similar magnitude, the p.d.f. of cycle ranges is Rayleigh likewise.

Wind turbine blade loading cannot be considered as narrow band, despite the concentration of energy at the rotational frequency by 'gust slicing' (Section 5.7.5), and neither can it be considered as Gaussian because of the presence of periodic components. Dirlik (1985) produced an empirical p.d.f. of cycle ranges applicable to both wide and narrow band Gaussian processes, in terms of basic spectral properties determined from the power spectrum. This was done by considering 70 power spectra of various shapes, computing their rainflow cycle range distributions and fitting a general expression for the cycle range p.d.f. in terms of the first, second and fourth spectral moments. Dirlik's expression for the cycle range p.d.f. is:

$$p(S) = \frac{\frac{D_1}{Q}\mathrm{e}^{-Z/Q} + \frac{D_2 Z}{R^2}\mathrm{e}^{-(Z^2/2R^2)} + D_3 Z\mathrm{e}^{-(Z^2/2)}}{2\sqrt{m_o}} \tag{5.117}$$

where

$$Z = S/\sqrt{m_o},\ D_1 = \frac{2(x_m - \gamma^2)}{1+\gamma^2},\ D_2 = \frac{(1-\gamma-D_1+D_1^2)}{1-R},\ D_3 = 1 - D_1 - D_2$$

$$Q = \frac{1.25(\gamma - D_3 - D_2 R)}{D_1},\ R = \frac{\gamma - x_m - D_1^2}{(1-\gamma-D_1+D_1^2)},\ x_m = \frac{m_1}{m_0}\sqrt{\frac{m_2}{m_4}},\ \gamma = \frac{m_2}{\sqrt{m_0 m_4}},$$

$$m_i = \int_0^\infty n^i S_\sigma(n)\,\mathrm{d}n \qquad S_\sigma(n) \text{ is the power spectrum of stress,}$$

and S is the cycle stress range.

Although the Dirlik cycle range p.d.f. was not intended to apply to signals containing periodic components, several investigations (Hoskin *et al.* (1989), Morgan and Tindal (1990), Bishop *et al.* (1991)) have been carried out to determine its validity for wind turbine fatigue damage calculations, using monitored data for flapwise bending from the MS1 wind turbine on Orkney. Cycle range p.d.f.s were calculated from power spectra of monitored strains using the Dirlik formula and fatigue damage rates derived from these p.d.f.s compared with damage rates derived directly from the monitored signal by rainflow cycle counting. The ratio of damage calculated by the Dirlik method to damage calculated by the rainflow method ranged from 0.84 to 1.46, from 1.01 to 2.48 and from 0.73 to 2.34 in the three investigations listed above, using a S/N curve exponent of 5 in each case, as the blade structure was of steel. In view of the fact that the calculated damage rates vary as the fifth power of the stress ranges, these results indicate that the Dirlik method is capable of giving quite accurate results, despite the presence of the periodic components.

There are two main drawbacks to the application of the Dirlik formula to power spectra containing periodic components. First, the presence of large spikes in the spectra due to the periodic components renders them very different from the smooth distributions Dirlik originally considered, and second, information about the relative phases of the periodic components is lost when they are transformed to the frequency domain. Morgan and Tindal (1990) illustrate the effect of varying phase angles by a comparison of plots of $(\cos \omega t + 0.5 \cos 3\omega t)$ and $(\cos \omega t - 0.5 \cos 3\omega t)$ which is reproduced in Figure 5.38. For a material with a S/N curve exponent of 5, stresses conforming to the first time history would result in 5.25 times as much fatigue damage as stresses conforming to the second.

Bishop, Wang and Lack (1995) developed a modified form of the Dirlik formula to include a single periodic component, using a neural network approach to determine the different parameters in the formula from computer simulations.

Madsen *et al.* (1984) adopted a different approach to the problem of determining fatigue damage resulting from combined stochastic and periodic loading, involving the derivation of a single equivalent sinusoidal loading that would produce the same fatigue damage as the actual loading. The method applies a reduction factor, g, which is dependent on bandwidth, to account for the reduced cycle ranges implicit in a wide band as opposed to a narrow band process, and utilizes Rice's p.d.f. for the peak value of a single sinusoid combined with a narrow band stochastic process,

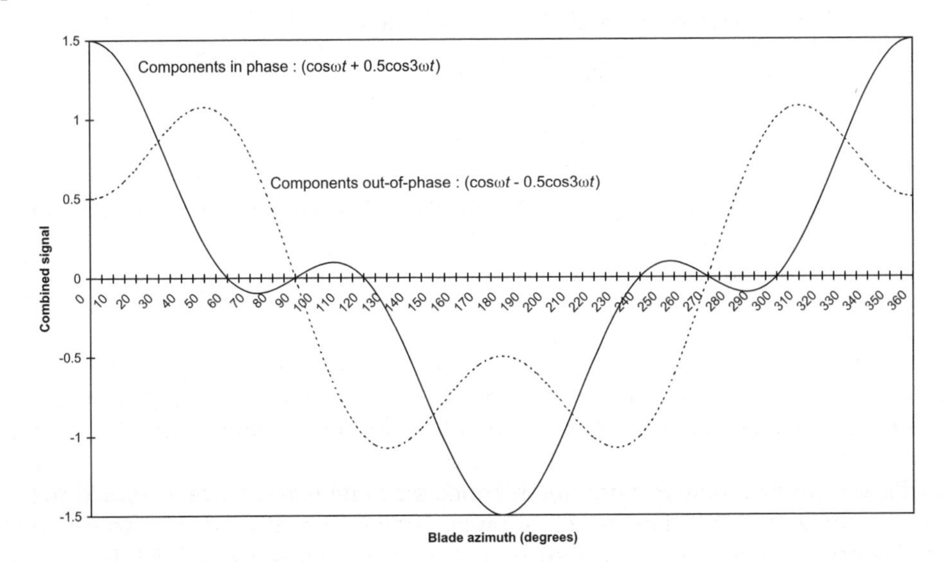

Figure 5.38 Effect of Variation of Phase Angle between Harmonics on Combined Signal

substituting half the maximum range of the periodic signal, including harmonics, for the amplitude of the sinusoid. A fuller summary is given in Hoskin, Warren and Draper (1989). They concluded, along with Morgan and Tindal (1990), that the Madsen method yielded slightly less accurate fatigue damage values than the Dirlik method for the MS1 monitored data for flapwise bending referred to above.

5.9.4 Wind simulation

Wind simulation, which was introduced in Section 5.7.6, has two significant advantages over the methods described above for fatigue damage evaluation. First, it can handle non-linear relationships between wind speed fluctuations and blade loadings in the calculation of stochastic loads, and second, it avoids the difficulty of deriving the fatigue stress ranges arising from combined periodic and stochastic load components. It is therefore currently the favoured method for detailed fatigue design. The procedure is essentially as follows.

(1) Generate a three-dimensional 'run of wind' for the chosen mean wind speed, with the desired shear profile and tower shadow correction.

(2) Perform a step-by-step dynamic analysis on the turbine operating in this wind field, to obtain in-plane and out-of-plane bending moment time histories at different radii.

(3) Convert these bending moment time histories to time histories of bending moments about the principal axes.

(4) Compute stress–time histories at chosen points on each cross section.

(5) Derive the number of cycles in each stress range 'bin' by Rainflow Cycle Counting (see Section 5.9.5 below).

(6) Scale up the cycle numbers in line with the predicted number of hours of operation at the chosen mean wind speed.

(7) Calculate corresponding fatigue damage numbers based on the applicable S/N curve.

(8) Repeat above steps for different mean wind speeds, and total the resulting fatigue damages at each point.

A computationally simpler alternative is to generate a one-dimensional 'run of wind' (in which only the longitudinal component of turbulence is modelled), and run a number of simulations at different, fixed yaw angles.

The duration of wind simulations is limited by available computer power, with a time history length of 600 s being frequently chosen. A consequence of this is that a single simulation will not provide an accurate picture of the infrequent high-stress range fatigue cycles, which can have a disproportionate effect on fatigue damage for materials with high m value such as those used for blades. However, this inaccuracy can be reduced (and quantified) by running several simulations with different random number seeds at each wind speed – see Thomsen (1998).

5.9.5 Fatigue cycle counting

As noted in Section 5.9.4, the dynamic analysis of turbine behaviour in a simulated wind field yields time histories of loads or stresses which then need to be processed to abstract details of the fatigue cycles. There are two established methods of fatigue cycle counting: the reservoir method and the rainflow method, both of which yield the same result.

In the reservoir method, the load or stress history (with time axis horizontal) is imagined as the cross section of a reservoir, which is successively drained from each low point, starting at the lowest and working up. Each draining operation then yields a load or stress cycle (see BS 5400 (1980) for a full description).

The rainflow method was first proposed by Matsuishi and Endo in 1968, and its title derives from the concept of water flowing down the 'rooves' formed when the time history is rotated so that the time axis is vertical. However, the following description not involving the rainflow analogy may be easier to understand.

The first step is to reduce the time history to a series of peaks and troughs, which are then termed extremes. Then each group of four successive extremes is examined in turn to determine whether the values of the two intermediate extremes lie between the values of the initial and final extremes. If so, the two intermediate extremes are counted as defining a stress cycle, which is then included in the cycle count, and the two intermediate extremes are deleted from the time history. The process is continued until the complete series of extremes forming the time history has been processed in this way. Then the sequence remaining will consist simply of a diverging and a converging part from which the final group of stress ranges can be extracted (see 'Fatigue Characteristics' in the IEA series of Recommended Practices for Wind Turbine Testing and Evaluation (1984) for a full description of the method and for details of algorithms that can be used for automating the process).

Although, in principle, the fatigue cycles obtained from, say, a 600 s time history could be listed individually, it is normal to reduce the volume of data by allocating individual cycles to a series of equal load or stress ranges known as 'bins' – e.g., 0–2, 2–4, 4–6 N/mm^2 etc. The fatigue spectrum is then presented in terms of the number of cycles falling into each 'bin'.

5.10 Hub and Low-speed Shaft Loading

5.10.1 Introduction

The loadings on the hub consist of the aerodynamic, gravity and inertia loadings on the blades and the equal and opposite (discounting hub self-weight) reaction from the shaft. For fixed hub machines, the loading on the shaft will include a significant moment arising from blade aerodynamic loads, but in the case of teetered two bladed rotors this moment will be virtually eliminated. In either case, however, the cantilevered low-speed shaft will experience large fluctuating moments due to rotor weight as it rotates. Figure 5.39 shows a low-speed shaft and front bearing in a factory prior to assembly.

The shaft moments due to out-of-plane loads on the blades can be expressed as moments about a pair of rotating axes, one perpendicular to blade 1 and the other parallel to it. In the case of a three-bladed rotor, these moments are respectively as follows:

$$M_{YS} = \Delta M_{Y1} - \tfrac{1}{2}(\Delta M_{Y2} + \Delta M_{Y3}) \qquad M_{ZS} = \frac{\sqrt{3}}{2}(\Delta M_{Y3} - \Delta M_{Y2}) \tag{5.118}$$

Here ΔM_{Y1}, ΔM_{Y2} and ΔM_{Y3} are the fluctuations of the blade out-of-plane moments about the hub centre (M_{Y1}, M_{Y2} and M_{Y3}) about the mean value (see Figure 5.40).

5.10.2 Deterministic aerodynamic loads

The deterministic aerodynamic loads on the rotor may be split up into a steady component, equal for each blade, and a periodic component, also equal for each

Figure 5.39 Low-speed Shaft and Front Bearing Before Assembly. The hub mounting flange at the right-hand end is bolted to a temporary support to allow the bearing to be threaded on the shaft. (Reproduced by permission of NEG Micon)

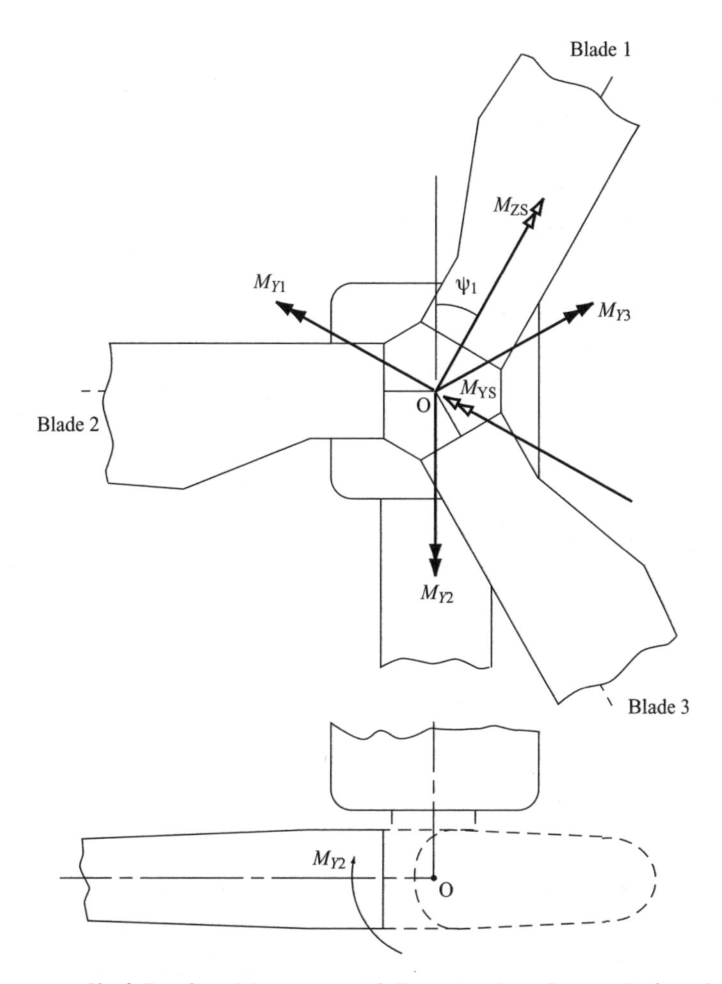

Figure 5.40 Shaft Bending Moments, with Rotating Axis System Referred to Blade 1

blade, but with differing phase angles. The blade root out-of-plane bending moments due to the first component will be in equilibrium, and will apply a 'dishing' moment to the hub which will result in tensile stresses in the front and compression stresses in the rear. These stresses will be uniaxial for a two bladed rotor, and biaxial for a three-bladed rotor.

The fluctuations in out-of-plane blade root bending moment due to wind shear, shaft tilt and yaw misalignment will often be approximately sinusoidal, with a frequency equal to the rotational frequency. Using Equations (5.118), it is easily shown that, for a sinusoidally varying blade root bending moment with amplitude M_o, the amplitude of the resulting shaft bending moment is $1.5M_o$ for a three-bladed machine and $2M_o$ for a rigid hub two-bladed machine.

In the case of wind shear conforming to a power law, the loading on a horizontal blade is always greater than the average of the loadings on blades pointing vertically upwards and downwards, so the loading departs significantly from sinusoidal. The shaft bending moment fluctuations due to wind shear with a 0.2

exponent are compared in Figure 5.41 for two- and three-bladed rigid hub machines operating at 30 r.p.m. in a hub-height wind speed of 12 m/s. The ratio of moment ranges is still close to 2:1.5.

5.10.3 Stochastic aerodynamic loads

The out-of-plane blade root bending moments arising from stochastic loads on the rotor will result in both a fluctuating hub 'dishing' moment (see above) and fluctuating shaft bending moments. For a two-bladed, rigid hub rotor, the shaft moment is equal to the difference between the two out-of-plane blade root bending moments, or teeter moment, the standard deviation of which is given by Equation (5.102). Similarly, the standard deviation of the mean of these two moments (i.e., the 'dishing' moment) is given by Equation (5.103).

The derivation of the standard deviation of the shaft moment for a three-bladed machine is at first sight more complicated, as the integration has to be carried out over three blades instead of two. However, if the shaft moment about an axis parallel to one of the blades, M_{ZS} (Figure 5.35), is chosen, the contribution of loading on that blade disappears, and the expression for the shaft moment standard deviation becomes:

$$\sigma_{\text{Mzs}}^2 = \left(\tfrac{1}{2}\Delta\Omega\frac{\mathrm{d}C_l}{\mathrm{d}\alpha}\right)^2 \int_{-R}^{R}\int_{-R}^{R} \kappa_u^{\mathrm{o}}(r_1, r_2, 0)c(r_1)c(r_2)\frac{\sqrt{3}}{2}r_1\frac{\sqrt{3}}{2}r_2|r_1||r_2|\,\mathrm{d}r_1\,\mathrm{d}r_2 \quad (5.119)$$

where the limits of the integrations refer to the other two blades. $\kappa_u^{\mathrm{o}}(r_1, r_2, 0)$ is given by Equation (5.51), with $\Omega\tau$ set equal to zero when r_1 and r_2 are radii to points on the same blade, and replaced by $2\pi/3$ when r_1 and r_2 are radii to points on different blades. Note that, compared with the two-bladed case, the cross correla-

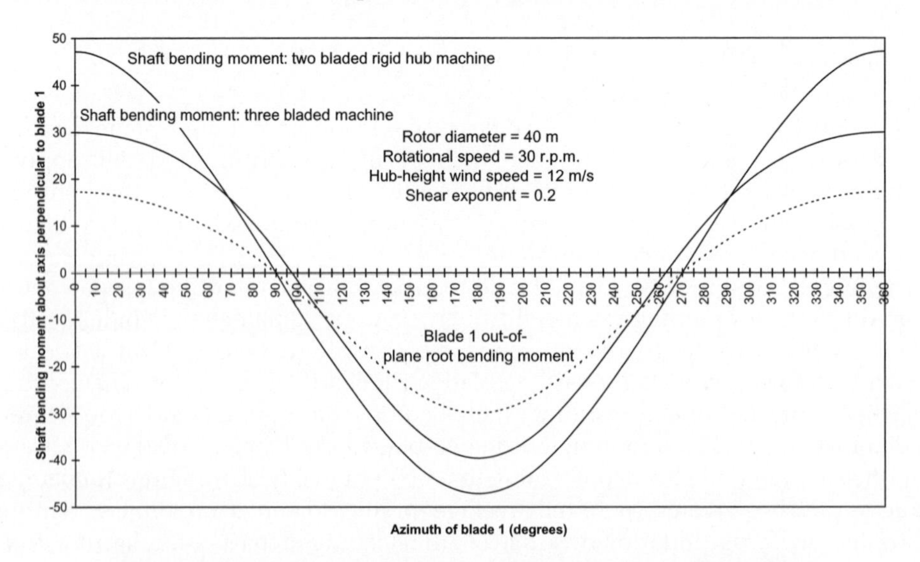

Figure 5.41 Shaft Bending Moment Fluctuation due to Wind Shear

tion function, $\kappa_u^o(r_1, r_2, 0)$ will be increased when r_1 and r_2 relate to *different* blades, because of the reduced separation between the two blade elements resulting from the 120° degree angle between the blades. Equation (5.119) can be rewritten in terms of the normalized cross correlation function, $\rho_u^o(r_1, r_2, 0) = \kappa_u^o(r_1, r_2, 0)/\sigma_u^2$, as follows:

$$\sigma_{Mzs}^2 = \sigma_u^2 \left(\tfrac{1}{2}\rho\Omega \frac{\mathrm{d}C_l}{\mathrm{d}\alpha}\right)^2 \int_{-R}^{R}\int_{-R}^{R} \rho_u^o(r_1, r_2, 0) c(r_1) c(r_2) \frac{\sqrt{3}}{2} r_1 \frac{\sqrt{3}}{2} r_2 |r_1||r_2| \,\mathrm{d}r_1\,\mathrm{d}r_2 \tag{5.119a}$$

In the case of 40 m diameter turbines with TR blades operating in wind with a turbulence length scale of 73.5 m, the standard deviation of shaft moment due to stochastic loading for a three-bladed machine is 82 percent of that for a two-bladed, fixed hub machine rotating at the same speed. This ratio would rise to $\sqrt{3}/2$ if the effect on the cross correlation function of the 120° blade spacing were ignored. It is worth noting that, for a three-bladed machine, the standard deviation of the shaft moment M_{YS} due to stochastic loading is the same as that of M_{ZS}.

By analogy with the derivation of the shaft moment above, the standard deviation of the hub 'dishing' moment for a three-bladed machine due to stochastic loading is given by:

$$\sigma_{Mh}^2 = \tfrac{1}{4}\sigma_u^2 \left(\tfrac{1}{2}\rho\Omega \frac{\mathrm{d}C_l}{\mathrm{d}\alpha}\right)^2 \int_{-R}^{R}\int_{-R}^{R} \rho_u^o(r_1, r_2, 0) c(r_1) c(r_2) \frac{\sqrt{3}}{2} r_1^2 \frac{\sqrt{3}}{2} r_2^2 \,\mathrm{d}r_1\,\mathrm{d}r_2 \tag{5.120}$$

where the integrations are carried out over two blades only, and the cross correlation function is modified as before to account for the 120° angle between the blades. It can be shown that

$$\tfrac{1}{4}\sigma_{MZS}^2 + \sigma_{Mh}^2 = \tfrac{3}{4}\sigma_{My1}^2 \tag{5.121}$$

5.10.4 Gravity loading

An important component of shaft loading is the cyclic cantilever bending moment due to rotor weight, which usually has a dominant effect on shaft fatigue design. As an illustration, a rotor consisting of three TR blades, each weighing 2 tonnes, and a 5 tonne hub cantilevered 0.85 m beyond the shaft main bearing, will produce a maximum shaft gravity moment of about 90 kNm. This compares with a shaft moment range due to wind shear of 70 kNm for a hub height wind of 12 m/s and a shear exponent of 0.2, and a shaft moment standard deviation of 50 kNm due to turbulence, taking a turbulence intensity of 20 percent and the same hub-height mean wind speed. Note that the shaft moment due to wind shear relieves that due to gravity, so it would be wise to adopt a smaller wind shear exponent for shaft fatigue calculations.

5.11 Nacelle Loading

5.11.1 Loadings from rotor

The previous section considered the moments applied to the shaft by the rotor hub using an axis system rotating with the shaft. In addition to these moments, the shaft also experiences an axial load due to rotor thrust, and radial forces arising from differential blade edgewise loadings and any out-of-balance centrifugal force.

In order to calculate loadings on the elements of the nacelle structure, it is first necessary to transform the shaft loads (or the constituent blade loads) defined in terms of the rotating axis system into nacelle loads expressed in terms of a fixed axis system. Here the conventional system in which the x-axis is downwind, the y-axis is horizontal to starboard and the z-axis is vertically upwards will be adopted (Figure C2). Thus the moments acting on the nacelle about the y- and z-axes as a result of the blade root out-of-plane bending moments are as follows for a three bladed machine with shaft tilt η:

$$M_{YN} = M_{Y1}\cos\psi + M_{Y2}\cos(\psi - 120°) + M_{Y3}\cos(\psi - 240°) \tag{5.122}$$

$$M_{ZN} = (M_{Y1}\sin\psi + M_{Y2}\sin(\psi - 120°) + M_{Y3}\sin(\psi - 240°))\cos\eta \tag{5.123}$$

where ψ is the azimuth of blade 1 (see Figure 5.42).

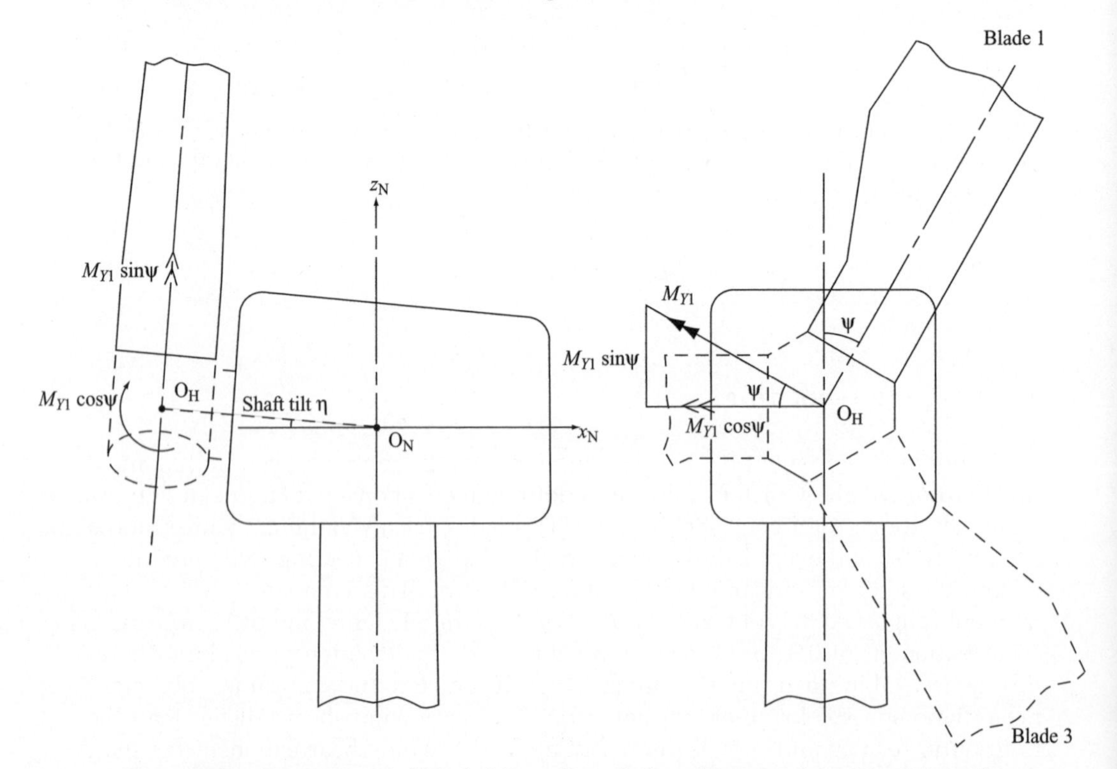

Figure 5.42 Components of Blade 1 Root Out-of-plane Moment about Fixed Axes Set

It is instructive to compare the moments acting on the nacelle due to deterministic loading for three-bladed and two-bladed machines. The fluctuations of out-of-plane root bending moment due to wind shear and yaw misalignment are approximately proportional to the cosine of blade azimuth for an unstalled blade. Substituting $M_{Y1} = M_0 \cos \psi$, $M_{Y2} = M_0 \cos(\psi - 2\pi/N)$ etc into Equations (5.122) and (5.123) yields $M_{YN} = 1.5M_0$ and $M_{ZN} = 0$ for a three-bladed machine, whereas the corresponding results for a rigid hub two-bladed machine are $M_{YN} = M_0(1 + \cos 2\psi)$ and $M_{ZN} = M_0 \sin 2\psi \cos \eta$. Thus the moments on the nacelle are constant for a three-bladed machine, but continually fluctuating with amplitude M_0 for a rigid hub two-bladed machine. Parallel results are obtained for $M_{Y1} = M_0 \sin \psi$ which approximates to the out-of-plane root bending moment due to shaft tilt – again for an unstalled blade. The full comparison is given in Table 5.6 below.

It is clear that the moments acting on the nacelle due to deterministic loading are much more benign for a three-bladed rotor than for a two-bladed rotor with rigid hub.

In the case of three-bladed machines, the standard deviation of shaft bending moment due to stochastic rotor loading is independent of the rotating axis chosen (Section 5.10.3), so the standard deviation of the resulting moment on the nacelle will take the same value about both the nacelle y- and z-axes.

5.11.2 Cladding loads

Except in the case of sideways wind loading, cladding loads are not usually of great significance. They may be calculated according to the rules given in standard wind loading codes. For sideways wind loading, a drag factor of 1.2 will generally be found to be conservative.

Table 5.6 Comparison Between Nacelle Moments due to Deterministic Loads for Two- and Three-bladed Machines

	Nacelle moments resulting from out-of-plane blade root bending moment fluctuations due to **wind shear and yaw misalignment** approximated by: $M_{Y1} = M_0 \cos \psi$, $M_{Y2} = M_0 \cos(\psi - 2\pi/N)$ etc		Nacelle moments resulting from out-of-plane blade root bending moment fluctuations due to **shaft tilt** approximated by: $M_{Y1} = M_0 \sin\psi$, $M_{Y2} = M_0 \sin(\psi - 2\pi/N)$ etc	
	Nacelle nodding moment, M_{YN}	Nacelle yaw moment, M_{ZN}	Nacelle nodding moment, M_{YN}	Nacelle yaw moment, M_{ZN}
Three-bladed machine	$1.5M_0$	Zero	Zero	$1.5M_0 \cos \eta$
Two-bladed, rigid hub machine	$M_0(1 + \cos 2\psi)$	$M_0 \sin 2\psi \cos \eta$	$M_0 \sin 2\psi$	$M_0(1 - \cos 2\psi)\cos \eta$

5.12 Tower Loading

5.12.1 Extreme loads

As noted in Section 5.4.1, it is customary to base the calculation of extreme loads on a non-operational turbine on the 50 year return 3 s gust. Several loading configurations may need to be considered, and the critical load case for the tower base will generally differ from that for the tower top. In addition, it is necessary to investigate the extreme operational load cases, as these can sometimes govern instead if the tip speed is high in relation to the design gust speed.

In the case of non-operational, stall-regulated machines, the critical case for the tower base occurs when the wind is blowing from the front and inducing maximum drag loading on the blades. By contrast, sideways wind loading to produce maximum lift on a blade pointing vertically upwards or rear wind loading on the rotor with one blade shielded by the tower will produce the maximum tower top bending moment.

One of the benefits of pitch-regulation of three-bladed machines is that blade feathering at shut-down considerably reduces non-operational rotor loading. The critical configuration as far as tower base bending moment is concerned is sideways wind loading, with two of the blades inclined at 30° to the vertical. The horizontal component of the loading on these blades is $\cos^3 30°$ of the loading on a vertical blade, so that the total rotor loading is only 43.3 percent ($= 100.\sqrt{3}/4\%$) of the maximum experienced by a stall-regulated machine.

The cases of sideways wind loading on a wind turbine referred to above can only arise if the yaw drive is disabled by grid loss for sufficient time for a 90° wind direction change to take place. In many areas, the level of grid security will be high enough for this possibility to be ruled out, so that sideways wind loading need not be considered if the yaw drive is programmed to remain operational in high winds. The critical non-operational load case for a three-bladed, pitch-regulated machine then occurs when the wind is from the front with a 15° to 20° yaw error. The rotor load is maximum when one of the blades is vertical, and is similar in magnitude to loading produced by a sideways wind. However, as the load results from blade lift rather than drag, it is at right angles to the loading on the tower, so the total moment at the tower base is significantly less.

Information on the drag factors appropriate for cylindrical and lattice towers is to be found in Eurocode 1, part 2–4 (1997), and in national codes such as BS 8100 (1986) or DS 410 (1983). The drag factor for a cylindrical tower is typically 0.6–0.7. Rotor loading is generally the dominating component of tower base moment for stall-regulated machines, but with pitch-regulated machines the contributions of tower loading and rotor loading are often of similar magnitude.

5.12.2 Dynamic response to extreme loads

Just as in the case of the single, stationary cantilevered blade considered in Section 5.6.3, the quasistatic bending moments in the tower calculated for the extreme gust speed will be augmented by inertial moments resulting from the excitation of resonant tower oscillations by turbulence. As before, it is convenient to express this augmentation in terms of a dynamic factor, Q_D, defined as the ratio of the peak moment over a 10 min period, including resonant excitation of the tower, to the peak quasistatic moment over the same period. Thus

$$M_{\text{Max}} = \tfrac{1}{2}\rho U_{e50}^2 H \oint C_f \left(\frac{z}{H}\right)^{1+2\alpha} \mathrm{d}A . Q_D \tag{5.124}$$

where U_{e50} is the 50 year return gust speed at hub height, z is height above ground, H is the hub height, C_f is the force factor (lift or drag) for the element under consideration, α is the shear exponent, taken as 0.11 in IEC 61400-1, and

$$Q_D = \frac{1 + g\left(2\frac{\sigma_u}{\overline{U}}\right)\sqrt{K_{SMB} + \frac{\pi^2}{2\delta} R_u(n_1) K_{Sx}(n_1) \lambda_{M1}^2}}{1 + g_0\left(2\frac{\sigma_u}{\overline{U}}\right)\sqrt{K_{SMB}}} \tag{5.17}$$

The integral sign $\oint$ signifies that the integral is to be undertaken over each blade and the tower.

The derivation of Equation (5.17) is explained in Section 5.6.3 and the Appendix in relation to a cantilevered blade. The essentially similar procedure for a tower supporting a rotor and nacelle is as follows.

(1) Calculate the resonant size reduction factor, $K_{Sx}(n_1)$, which reflects the effect of the lack of correlation of the wind fluctuations at the tower natural frequency along the blades and tower. Adopting an exponential expression for the normalized co-spectrum as before, Equation (A5.25) becomes:

$$K_{Sx}(n_1) = \frac{\oint\oint \exp[-C s n_1/\overline{U}] C_f^2 c(r) c(r') \mu_1(r) \mu_1(r') \, \mathrm{d}r \, \mathrm{d}r'}{\left(\oint C_f c(r) \mu_1(r) \, \mathrm{d}r\right)^2} \tag{5.125}$$

where the integral sign $\oint$ denotes integration over the blades and the tower, r and r' denote radius in the case of the blades and depth below the hub in the case of the tower, s denotes the separation between the elements $\mathrm{d}r$ and $\mathrm{d}r'$, C_f is the relevant force coefficient, $c(r)$ denotes chord in the case of the blades and diameter in the case of the tower, and $\mu_1(r)$ denotes the tower first mode shape.

This expression can be considerably simplified by setting $\mu_1(r)$ to unity for the rotor and ignoring the tower loading contribution entirely. This is not unreasonable, as only loading near the top of the tower is of significance, and this does not add much to the spatial extent of the loaded area.

(2) Calculate the damping logarithmic decrement, δ, for the tower first mode. The aerodynamic component is given by

$$\delta_a = 2\pi\xi_a = 2\pi \frac{c_{a1}}{2m_{T1}\omega_1} = \frac{\oint \hat{c}_a(r)\mu_1^2(r)\,dr}{2m_{T1}n_1} \tag{5.126}$$

where m_{T1} is the generalized mass of the tower, nacelle and rotor (including the contribution of rotor inertia) with respect to the first mode given by Equation (5.110), and n_1 is the tower natural frequency in Hz. For a stall-regulated machine facing the wind, the rotor contribution to aerodynamic damping is simply $\rho\overline{U}C_D A_R/2m_{T1}n_1$ where A_R is the rotor area.

(3) Calculate the standard deviation of resonant nacelle displacement according to

$$\frac{\sigma_{x1}}{\bar{x}_1} = 2\frac{\sigma_u}{\overline{U}}\frac{\pi}{\sqrt{2\delta}}\sqrt{R_u(n_1)}\sqrt{K_{Sx}(n_1)} \tag{5.6}$$

(4) Calculate the ratio λ_{M1}, which relates the ratio of the standard deviation of resonant tower base moment to the mean value to the corresponding ratio for nacelle displacement as follows:

$$\frac{\sigma_{M1}}{\overline{M}} = \lambda_{M1}\frac{\sigma_{x1}}{\bar{x}_1} \tag{5.7}$$

If $\mu_1(r)$ is set to unity for the rotor, λ_{M1} is given by:

$$\lambda_{M1} = \frac{\int_0^H m(z)\mu_1(z)z\,dz\left\{C_d A_R + \int_0^H C_f\left[\frac{U(z)}{\overline{U}}\right]^2 d(z)\mu_1(z)\,dz\right\}}{m_{T1}H\left\{C_d A_R + \int_0^H C_f\left[\frac{U(z)}{\overline{U}}\right]^2 d(z)\frac{z}{H}\,dz\right\}} \tag{5.127}$$

where z is the height up the tower measured from the base, and H is the hub height. If the loading on the tower is relatively small, this approximates to

$$\lambda_{M1} = \frac{\int_0^H m(z)\mu_1(z)z\,dz}{m_{T1}H} \tag{5.127a}$$

which is close to unity because the tower head mass dominates the integral.

(5) Calculate the size reduction factor for the root bending moment quasistatic or background response, K_{SMB}, which reflects the lack of correlation of the wind fluctuations along the blades and tower. K_{SMB} may be derived from a similar expression to that for the resonant size reduction factor given in Equation (5.125), but with the exponential function modified to $\exp[-s/0.3L_u^x]$.

(6) Calculate the peak factors for the combined (i.e., resonant plus quasistatic) and quasistatic responses in terms of the respective zero up-crossing frequencies. (In estimating the zero up-crossing frequency of the quasistatic response, the blade area should be replaced by the rotor area in Equation (A5.57).

(7) Substitute the parameter values derived in Steps (1)–(6) into Equation (5.17) to obtain the dynamic factor, Q_D.

The procedure is illustrated by the following example.

Example 5.3: Estimate the dynamic factor, Q_D, for the extreme tower base moment for a stationary 40 m diameter three-bladed stall-regulated turbine.

Data:
Hub height, $H = 35$ m
Tower first mode natural frequency, $n_1 = 1.16$ Hz
Generalized mass of tower, nacelle and rotor, $m_{\mathrm{T1}} = 28\,000$ kg
Rotor area, $A_{\mathrm{R}} = 23.9\ \mathrm{m}^2$
Tower top diameter = 2 m, base diameter = 3.5 m
50 year return 10 min mean wind speed at hub height, $\overline{U} = 50$ m/s
Roughness length, $z_0 = 0.01$ m
Turbulence intensity at hub height, $\sigma_u/\overline{U} = 1/\ln(H/z_0) = 0.1225$ (based on Eurocode 1, Part 2.4 (1997) for $z_0 = 0.01$)
Blade drag factor = 1.3
Tower drag factor = 0.6
Air density, $\rho = 1.225\ \mathrm{kg/m^3}$

The non-dimensional power spectral density of longitudinal wind turbulence, $R_u(n)$, is calculated at the tower first mode natural frequency according to the Kaimal power spectrum defined in Eurocode 1 (see Equation (A5.8) in Appendix) as 0.0425. The decay constant, C, in the exponential expression for the normalized co-spectrum in Equation (5.125) is taken as 9.2. The calculation of Q_{D} procedes as follows.

Size reduction factor for resonant response – Equation (5.125) with tower loading ignored and $\mu_1(r)$ taken as unity over the rotor	0.166
Aerodynamic damping logarithmic decrement, δ_a	
– rotor contribution, $\sigma \overline{U} C_{\mathrm{d}} A_{\mathrm{R}}/2m_{\mathrm{T1}} n_1$	0.079
– tower contribution, $\sigma \overline{U} C_{\mathrm{d}} \int_0^H d(z)\mu_1^2(z)\mathrm{d}z/2m_{\mathrm{T1}} n_1$	0.007
Structural damping logarithmic decrement, δ_s	0.02
Damping logarithmic decrement, total, $\delta_a + \delta_s$	0.106

Ratio of standard deviation of resonant tower base moment to mean value – Equations (5.6), (5.7) and (5.127)

$$\frac{\sigma_{M1}}{\overline{M}} = 2\frac{\sigma_u}{\overline{U}}\frac{\pi}{\sqrt{2\delta}}\sqrt{R_u(n_1)}\sqrt{K_{Sx}(n_1)}\lambda_{M1}$$

$$= 2 \times 0.1225 \times 6.823\sqrt{0.0425} \times \sqrt{0.166} \times 1.02 \qquad 0.1432$$

Size reduction factor for quasistatic or background response, K_{SMB} – see Step (5) above 0.837

Ratio of standard deviation of tower base moment quasistatic response to mean value

$$\frac{\sigma_{MB}}{\overline{M}} = 2\frac{\sigma_u}{\overline{U}}\sqrt{K_{SMB}} = 2 \times 0.1225 \times \sqrt{0.837} \qquad 0.2241$$

Ratio of standard deviation of total tower base moment response to mean value

$$\frac{\sigma_M}{\overline{M}} = \sqrt{\left(\frac{\sigma_{MB}}{\overline{M}}\right)^2 + \left(\frac{\sigma_{M1}}{\overline{M}}\right)^2} = \sqrt{0.2241^2 + 0.1432^2} \qquad 0.2660$$

Zero up-crossing frequency of quasistatic response, n_0 – see Step (6) above 0.31 Hz

Zero up-crossing frequency of total tower base moment response, ν – Equation (A5.54) 0.68 Hz

Peak factor, g, based on ν – Equation (A5.42) 3.63

Ratio of extreme tower base moment to mean value

$$\frac{M_{\max}}{\overline{M}} = 1 + g\left(\frac{\sigma_M}{\overline{M}}\right) = 1 + 3.63(0.2660) \qquad 1.966$$

Peak factor, g_0, based on n_0 – Equation (A5.42) 3.41

Ratio of quasistatic component of extreme tower base moment to mean value

$$= 1 + g_0\left(\frac{\sigma_M}{\overline{M}}\right) = 1 + 3.41(0.2241) \qquad 1.764$$

Dynamic factor, $Q_D = 1.966/1.764$ – Equation (5.17) 1.115

It is apparent from Equation (5.17) that a key parameter determining the dynamic factor is the damping value. If the rotor contribution to aerodynamic damping were not available, the damping in the above example would be reduced by a factor of four, which would increase the dynamic factor to 1.41. This is of relevance to non-operational pitch-regulated machines facing into the wind, because if the lift loading on a blade is near the maximum, the aerodynamic damping may be near zero or even negative.

5.12.3 Operational loads due to steady wind (deterministic component)

Tower fore-and-aft bending moments result from rotor thrust loading and rotor moments. The moments acting on the nacelle due to deterministic rotor loads have already been described in Section 5.11.1. Although the thrust loads on individual blades vary considerably with azimuth as a result of yaw misalignment, shaft tilt or wind shear, the fluctuations on different blades balance each other, so that the total rotor thrust shows negligible azimuthal variation as a result of these effects. For example, on two-bladed machines, a wind shear exponent of 0.2 results in a rotor thrust variation of about +/−1 percent.

Tower shadow loading results in a sinusoidal tower top displacement at blade passing frequency (see Figure 5.32). Figure 5.43 illustrates the variation of rotor thrust with wind speed for stall and pitch regulated 40 m diameter three-bladed machines.

5.12.4 Operational loads due to turbulence (stochastic component)

Analysis in the frequency domain

Except near the top of the tower, the dominant source of fore-aft stochastic tower bending moments is rotor thrust. The standard deviation of rotor thrust can be expressed in terms of the turbulence intensity and the cross correlation function between wind fluctuations at different points on the rotor, following the method used for deriving the standard deviation of stochastic blade root bending moment in Section 5.7.5. As before, a linear relation between the wind fluctuations and the resultant load fluctuations is assumed, so that the perturbation of loading per unit length of blade, q, at radius r is given by

$$q = \tfrac{1}{2}\rho\Omega rc\frac{\mathrm{d}C_l}{\mathrm{d}\alpha}u \tag{5.25}$$

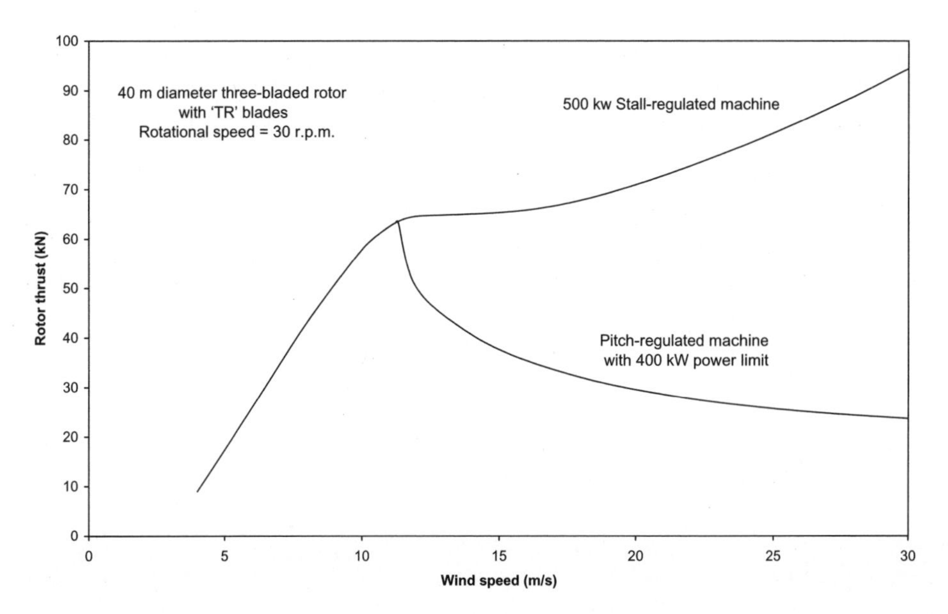

Figure 5.43 Rotor Thrust During Operation in Steady, Uniform Wind: Variation with Wind Speed for Similar Stall-regulated and Pitch Regulated Machines

and the perturbation of rotor thrust by

$$\Delta T = \left(\tfrac{1}{2}\rho\Omega\frac{\mathrm{d}C_l}{\mathrm{d}\alpha}\right)\oint uc(r)r\,\mathrm{d}r \tag{5.128}$$

where the integral sign $\oint$ signifies that the integration is carried out over the whole rotor. Hence the following expression for the variance of the rotor thrust is obtained:

$$\sigma_T^2 = \left(\tfrac{1}{2}\rho\Omega\frac{\mathrm{d}C_l}{\mathrm{d}\alpha}\right)^2 \sigma_u^2 \oint\oint \rho_u^{\mathrm{o}}(r_1, r_2, 0)c(r_1)c(r_2)r_1 r_2\,\mathrm{d}r_1 r_2 \tag{5.129}$$

where $\rho_u^{\mathrm{o}}(r_1, r_2, 0)$ is the normalized cross correlation function, $\kappa_u^{\mathrm{o}}(r_1, r_2, 0)/\sigma_u^2$ for points at radii r_1 and r_2 on the same or on different blades. $\kappa_u^{\mathrm{o}}(r_1, r_2, 0)$ is given by Equation (5.51), with $\Omega\tau$ replaced by the phase angle between the two blades on which r_1 and r_2 measured. For a three-bladed, 40 m diameter rotor and an integral length scale of 73.5 m, the reduction in the standard deviation of the stochastic rotor thrust fluctuations is about 20 percent due to the lack of correlation of the wind speed variations over the rotor. If the machine is rotating at 30 r.p.m. in an 8 m/s wind and the turbulence intensity is 20 percent, the rotor thrust standard deviation will be about 9 kN – i.e., 25 percent of the steady value

The derivation of the expression for the power spectrum of rotor thrust parallels that for the power spectrum of blade root bending moment (Section 5.7.5), yielding:

$$S_T(n) = \left(\tfrac{1}{2}\rho\Omega\frac{\mathrm{d}C_l}{\mathrm{d}\alpha}\right)^2 \oint\oint S_{u\,J,K}^{\mathrm{o}}(r_1, r_2, n)c(r_1)c(r_2)r_1 r_2\,\mathrm{d}r_1\,\mathrm{d}r_2 \tag{5.130}$$

where $S_{u\,J,K}^{0}(r_1, r_2, n)$ is the rotationally sampled cross spectrum for points at radii r_1 and r_2 on blades J and K respectively. Note that on a machine with three blades, A, B and C, $S_{u\,J,K}^{\mathrm{o}}(r_1, r_2, n)$ is complex when J and K are different, but $S_{u\,A,B}^{\mathrm{o}}(r_1, r_2, n)$ and $S_{u\,A,C}^{\mathrm{o}}(r_1, r_2, n)$ are complex conjugates, so the double integral in Equation (5.130) is still real. An example power spectrum of rotor thrust for a three bladed machine is shown in Figure 5.44. It can be seen that there is some concentration of energy at the blade passing frequency of 1.5 Hz due to gust slicing, but that the effect is not large. The concentration effect is significantly greater for two-bladed machines (see Figure 5.45). This shows the power spectrum of rotor thrust for a two-bladed machine with the same blade planform, but rotating 22.5 percent faster to give comparable performance.

In addition to thrust fluctuations, longitudinal turbulence will also cause rotor torque fluctuations and in-plane rotor loads due to differential loads on different blades, both of which will result in tower sideways bending moments. The expression for the in-plane component of aerodynamic lift per unit length, $-F_Y(r) = \frac{1}{2}\rho W^2 C_l c(r)\sin\phi$, can be differentiated with respect to the wind fluctuation as follows:

$$-\frac{\mathrm{d}F_Y}{\mathrm{d}u} = \tfrac{1}{2}\rho c(r)\frac{\mathrm{d}}{\mathrm{d}u}[W^2\sin\phi C_l] = \tfrac{1}{2}\rho c(r)\frac{\mathrm{d}}{\mathrm{d}u}[W\{U_\infty(1-a)+u\}C_l]$$

$$\cong \tfrac{1}{2}\rho c(r)W\left[C_l + \sin\phi\frac{\mathrm{d}C_l}{\mathrm{d}\alpha}\right]$$

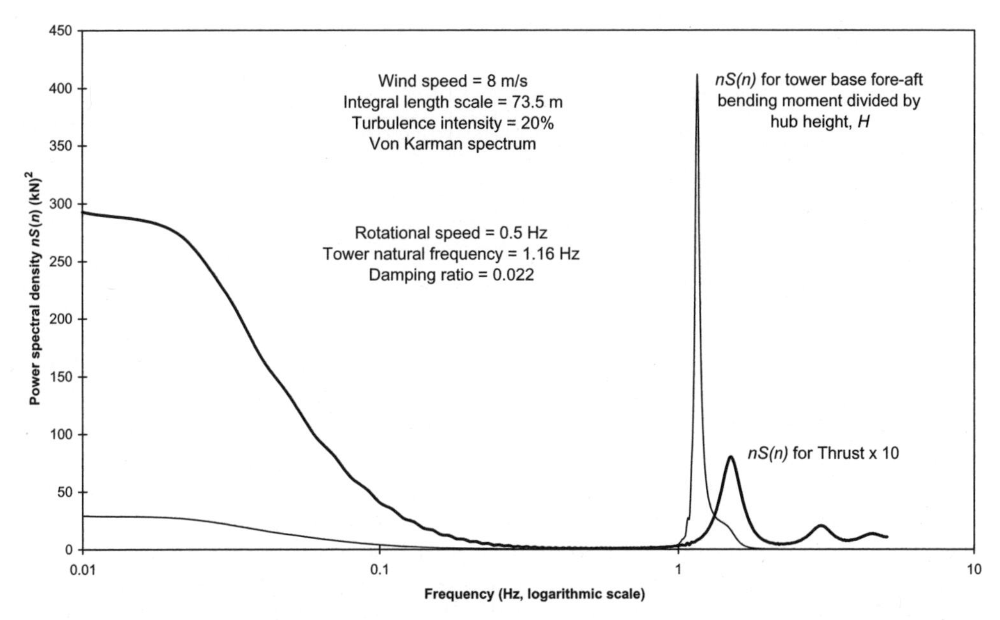

Figure 5.44 Power Spectra of Rotor Thrust and Resultant Tower Base Fore–aft Bending Moment for Three-bladed, 40 m Diameter Turbine

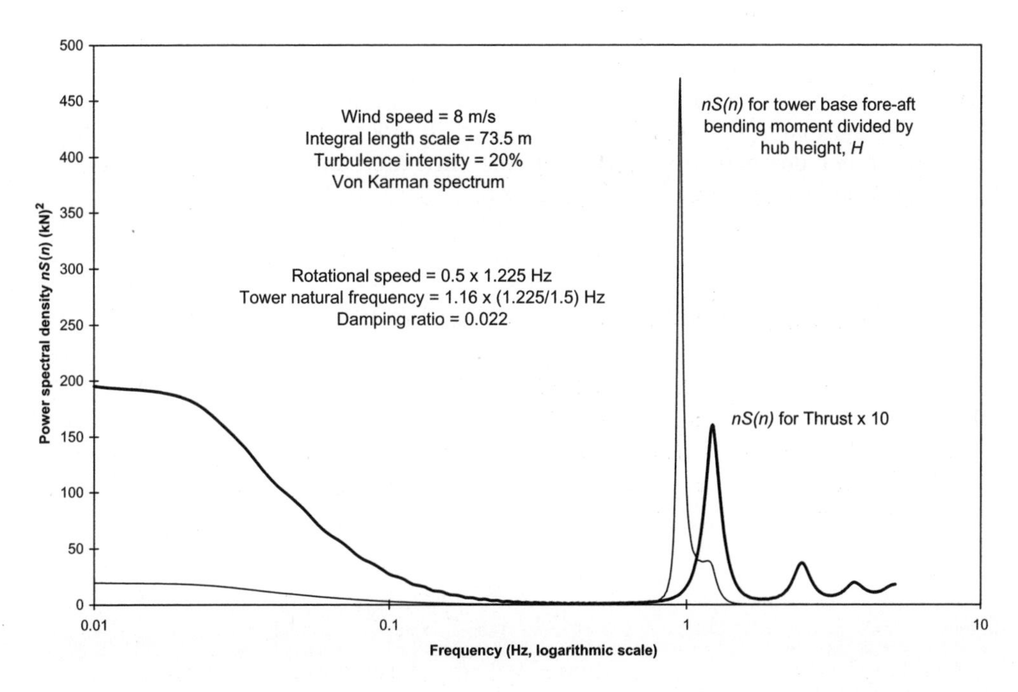

Figure 5.45 Power Spectra of Rotor Thrust and Resultant Tower Base Fore–aft Bending Moment for Two-bladed, 40 m Diameter Turbine

so, approximately,

$$-\frac{\mathrm{d}F_Y}{\mathrm{d}u} = \left(\tfrac{1}{2}\rho\Omega\frac{\mathrm{d}C_l}{\mathrm{d}\alpha}\right)c(r)r\left[\frac{C_l}{\mathrm{d}C_l/\mathrm{d}\alpha} + \sin\phi\right] \tag{5.131a}$$

Thus the standard deviation of rotor torque is approximately given by

$$\sigma_Q = \left(\tfrac{1}{2}\rho\Omega\frac{\mathrm{d}C_l}{\mathrm{d}\alpha}\right)\sigma_u\left\{\oint r^2 c(r)\left[\frac{C_l}{\mathrm{d}C_l/\mathrm{d}\alpha} + \sin\phi\right]\mathrm{d}r\right\} \tag{5.131b}$$

(which parallels Equation (5.26)) provided the relationship between blade loading and wind speed fluctuation remains linear and the turbulence length scale is large compared with rotor diameter. Equation (5.131b) can be used to derive an expression for the variance of the rotor torque in the same way as for rotor thrust above. At the top of the tower the stochastic M_X (i.e., side-to-side) moment due to rotor torque fluctuations is typically of the same order of magnitude as the stochastic M_Y (i.e., fore-aft) moment due to differential out-of-plane loads on the rotor, but at the tower base the dominant effect of rotor thrust loading means that the stochastic side-to-side moments are usually significantly less than the stochastic fore-aft moments before the excitation of tower resonance is taken into account.

Analysis in the time domain

As noted in Section 5.7.5, there are situations, such as operation in stalled flow, when the linear relationship between blade loading and wind speed fluctuations required for analysis in the frequency domain does not apply. In these cases, recourse must be made to analysis in the time domain using wind simulation techniques such as described in Section 5.7.6.

5.12.5 Dynamic response to operational loads

The power spectrum of rotor thrust will usually contain some energy at the tower natural frequency, leading to dynamic magnification of deflections, and hence of tower bending moments. The power spectrum of hub deflection, $S_{x1}(n)$, resulting from the excitation of the tower first fore-aft flexural mode, is related to the power spectrum of rotor thrust by

$$S_{x1}(n) = \frac{S_T(n)}{k_1^2}\frac{1}{[(1 - n^2/n_1^2)^2 + 4\xi_1^2 n^2/n_1^2]} \tag{5.132}$$

This relation is analagous to Equation (5.90), and derived in the same way.

The amplitude of tower base fore-aft moment at resonance in the first mode, M_{Y1}, can be derived from the corresponding amplitude of hub deflection, x_{H1}, as follows

$$M_{Y1} = \omega_1^2 x_{H1} \int_0^H m(z)\mu(z) z \, \mathrm{d}z = \omega_1^2 x_{H1} m_{T1} H \frac{\int_0^H m(z)\mu(z) z \, \mathrm{d}z}{H \int_0^H m(z)\mu^2(z) \, \mathrm{d}z} \tag{5.133}$$

The quotient on the right-hand side is close to unity because of the dominance of the tower head mass, so, substituting k_1 for $\omega_1^2 m_{T1}$, the equation reduces to $M_{Y1} = x_{H1} k_1 H$, which applies at any exciting frequency. Hence the power spectrum for the tower base fore-aft bending moment due to rotor thrust loading is given by

$$S_{My1}(n) = S_T(n) H^2 \frac{1}{[(1 - n^2/n_1^2)^2 + 4\xi^2 n^2/n_1^2]} \tag{5.134}$$

The aerodynamic damping is almost entirely provided by the rotor, the damping ratio for the first tower mode being approximately

$$\xi_{a1} = N \frac{\frac{1}{2}\rho\Omega \frac{\mathrm{d}C_l}{\mathrm{d}\alpha} \int_0^R r c(r) \, \mathrm{d}r}{2 m_{T1} \omega_1} \tag{5.135}$$

where N is the number of blades (see Section 5.8.4). The overall damping ratio is obtained by adding this to the structural damping ratio for the tower (see Table 5.4), and is generally low compared to the blade first mode damping because of the large tower head mass. The effect of a low damping ratio is illustrated by the power spectrum of fore-aft tower bending moment shown in Figure 5.44, which has a very high peak at the tower natural frequency of 1.16 Hz, despite this frequency being somewhat removed from the blade passing frequency of 1.5 Hz. The damping ratio is calculated as 0.022, consisting of 0.019 due to aerodynamic damping and 0.03 due to structural damping (for a welded steel tower).

In the example shown in Figure 5.44, the tower dynamic response increases the standard deviation of the tower base fore-aft bending moment by 15 percent. However, the effect of tower dynamic response results in a larger increase of 25 percent in the case of the two-bladed machine featured in Figure 5.45 despite the reduction in tower natural frequency in proportion to blade-passing frequency.

It is important to note that the rotor provides negligible aerodynamic damping in the side-to-side direction, so that effectively the only damping present is the structural damping. This means that, even though the side-to-side loadings are small in relation to the fore-aft loads, the side-to-side tower moment fluctuations can sometimes approach the fore-aft ones in magnitude.

5.12.6 Fatigue loads and stresses

The tower moments at height z are related to the hub-height loads as follows:

$$M_Y(z, t) = F_X(H, t).(H - z) + M_Y(H, t) \quad M_X(z, t) = -F_Y(H, t).(H - z) + M_X(H, t)$$

$$M_Z(z, t) = M_Z(H, t) \tag{5.136}$$

For three-bladed machines, the five hub-height fatigue loads are almost entirely stochastic, because the deterministic load component is either constant (for a given mean wind speed) or negligible, and it is instructive to consider how they relate to one another. Recognizing that the centre of any gust lying off the rotor centre will be located at a random azimuth, then it is clear that the rotor out-of-plane loads, i.e., the moment about the horizontal axis, $M_Y(H, t)$, the hub moment about the vertical axis, $M_Z(H, t)$, and the rotor thrust, $F_X(H, t)$, will all be statistically independent of each other. The same will apply to the rotor in-plane loads – the rotor torque, $M_X(H, t)$, and the sideways load, $F_Y(H, t)$. However, as the out-of-plane and in-plane loads on a blade element are both assumed to be proportional to the local wind speed fluctuation, u, it follows that the rotor torque fluctuations will be in phase with the rotor thrust fluctuations, and the rotor sideways load fluctuations will be in phase with the fluctuations of the hub moment about the horizontal axis, $M_Y(H, t)$.

The above relationships have implications for the combination of fatigue loads. Clearly the power spectrum of the fore-aft tower moment at height z, $S_{My}(z, n)$, can be obtained by simply adding the power spectrum of the hub moment about the horizontal axis to $(H - z)^2$ times the power spectrum of the rotor thrust. Similarly the power spectrum of the side-to-side tower moment at height z, $S_{Mx}(z, n)$, can be obtained by adding the power spectrum of the rotor torque to $(H - z)^2$ times the power spectrum of the rotor sideways load.

Having obtained power spectra for the M_X, M_Y and M_Z moments at height z, the corresponding fatigue load spectra can be derived with reasonable accuracy by means of the Dirlik method described in Section 5.9.3. As the tower stress ranges will be enhanced by tower resonance, the input power spectra should incorporate dynamic magnification, as outlined in Section 5.12.5.

Fatigue stress ranges due to bending about the two axes can easily be calculated separately from the $M_X(z)$ and $M_Y(z)$ fatigue spectra, but the stress ranges due to the two fatigue spectra combined cannot be calculated precisely because of lack of information about phase relationships. However, as noted above, the $M_X(H)$ component of the $M_X(z)$ fluctuations is in phase with the $F_X(H)$ component of the $M_Y(z)$ fluctuations, and the $F_Y(H)$ component of the $M_X(z)$ fluctuations is in phase with the $M_Y(H)$ component of the $M_Y(z)$ fluctuations so the stress ranges due to the $M_X(z)$ and $M_Y(z)$ fatigue spectra combined can be conservatively calculated as if they were in phase too. Theoretically this means pairing the largest $M_X(z)$ and $M_Y(z)$ loading cycles, the second largest, the third largest and so on, right through the fatigue spectra, and calculating the stress range resulting from each pairing. In practice, of course, the $M_X(z)$ and $M_Y(z)$ load cycles are distributed between two sets of equal size 'bins', so they have to be reallocated to bins in a two-dimensional matrix of descending load ranges, as shown in the grossly simplified example given in Tables 5.7 and 5.8 below:

Table 5.7 Example M_X and M_Y Fatigue Spectra

ΔM_Y (kNm)	No. of ΔM_Y cycles	ΔM_X (kNm)	No. of ΔM_X cycles
200–300	5	100–150	10
100–200	15	50–100	40
0–100	80	0–50	50

Table 5.8 Joint M_X and M_Y Cycle Distribution

	ΔM_Y(kNm)			
ΔM_X (kNm)	200–300	100–200	0–100	Total No. of M_X cycles
100–150	5	5		10
50–100		10	30	40
0–50			50	50
Total No. of M_Y cycles	5	15	80	

For a circular tower, the stress ranges would have to be computed at several points around the circumference in order to identify the location (with respect to the nacelle axis) where the fatigue damage was maximum.

A simpler but potentially cruder approach to the combination of the two fatigue spectra is to use the 'Damage Equivalent Load' method. This involves the calculation of constant amplitude fatigue loadings, $M_{X.Del}$ and $M_{Y.Del}$, of, say 10^7 cycles each, that would respectively produce the same fatigue damages as the M_X and M_Y spectra, using the S/N curve appropriate to the fatigue detail under consideration. If the M_X and M_Y fluctuations are treated as being in-phase as before, the combined 'Damage Equivalent Load' moment is $\sqrt{M_{X.Del}^2 + M_{Y.Del}^2}$.

References

American Society of Civil Engineers, (1993). *ASCE 7-93: Minimum design loads for buildings and other structures.*

Armstrong, J. R. C. and Hancock, M., (1991). 'Feasibility study of teetered, stall-regulated rotors' *ETSU Report No. WN 6022.*

Batchelor, G. K., (1953). *The theory of homogeneous turbulence.* Cambridge University Press, UK.

Bishop, N. W. M., Zhihua, H. and Sheratt, F., (1991). 'The analysis of non-gaussian loadings from wind turbine blades using frequency domain techniques.' *Proceedings of the BWEA Conference*, pp 317–323.

Bishop, N. W. M., Wang, R. and Lack, L., (1995). 'A frequency domain fatigue predictor for wind turbine blades including deterministic components.' *Proceedings of the BWEA Conference*, pp 53–58.

Bossanyi, E. A., (2000). 'Bladed for Windows theory and user manuals'. *282/BR/009 and 282/BR/010.* Garrad Hassan and Partners Ltd.

British Standard Institution, (1972). 'Code of basic data for the design of buildings'. *CP3 Chapter V, Part 2, Wind loads.*

British Standard Institution, (1980). *BS 5400: Part 10: 1980 Steel, concrete and composite bridges – Code of practice for fatigue.*

British Standard Institution, (1986). *BS 8100: Part 1: 1986 Lattice towers and masts – Code of practice for loading.'*

Clough, R. W. and Penzien, J., (1993). *Dynamics of structures.* McGraw Hill, New York, USA.

Danish Standards, (1992). *DS 472: Loads and Safety of Wind Turbine Construction.* (*First Edition*).

Danish Standards, (1983). *DS 410: Loads for the design of structures.* (*Third Edition*).

Davenport, A. G., (1964). 'Note on the distribution of the largest value of a random function with application to gust loading.' *Proc. Inst. Civ. Eng.*, **28**, 187–196.

Dirlik, T., (1985). *Application of computers in fatigue analysis.* Ph.D Thesis, University of Warwick, UK.

Dutch Standard, (1988). 'Safety regulations for wind generators'. *NEN 6096, Draft Standard, Second Edition.*

Eurocode, (1997). *1: Basis of design and actions on structures – Part 2.4: Actions on structures – Wind actions.*

Garrad, A. D. and Hassan, U., (1986). 'The dynamic response of wind turbines for fatigue life and extreme load predicition.' *Proceedings of the EWEA Conference*, pp 401–406.

Garrad, A. D., (1987). 'The use of finite-element methods for wind turbine dynamics.' *Proceedings of the BWEA Conference*, pp 79–83.

Germanischer Lloyd, (1993). 'Rules and Regulations: IV – Non-Marine Technology: Part 1 – Wind Energy: Regulation for the Certification of Wind Energy Conversion Systems.' (amended 1994 and 1998).

Hansen, A. C., (1998). *Users guide to the wind turbine dynamics computer programs YawDyn and AeroDyn for Adams, Version 11.0.* University of Utah, USA.

Hoskin, R. E., Warren, J. G. and Draper, J., (1989). 'Prediction of fatigue damage in wind turbine rotors.' *Proceedings of the EWEC*, pp 389–394.

International Electrotechnical Commission, (1997). *IEC 61400-1: Wind turbine generator systems – Part 1: Safety Requirements.* (*Second Edition*).

International Energy Agency, (1984). *International Recommended Practices for Wind Energy Conversion Systems Testing: 3. Fatigue Characteristics.*

Jamieson, P. and Hunter, C., (1985). 'Analysis of data from Howden 300 kW wind turbine on Burgar Hill Orkney.' *Proceedings of the BWEA Conference*, pp 253–258.

Jamieson, P., Camp, T. R. and Quarton, D. C., (2000). 'Wind turbine design for offshore.' *Proceedings of the Offshore Wind Energy in Mediterranean and European Seas, CEC/EWEA/IEA*, Sicily, pp 405–414.

Lobitz, D. W. A., (1984). 'NASTRAN based computer program for structural dynamic analysis of HAWTs.' *Proceedings of the EWEA Conference.*

Madsen, P. H. *et al.*, (1984). 'Dynamics and fatigue damage of wind turbine rotors during steady operation.' *Riso National Laboratory, R-512.* Riso National Laboratory, Roskilde, Denmark.

Matsuishi, M. and Endo, T., (1968). 'Fatigue of metals subject to varying stress.' Japanese Society for Mechanical Engineers.

Molenaar, D. P. and Dijkstra, S., (1999). 'State-of-the-art of wind turbine design codes: main features overview for cost-effective generation'. *Wind Engng.*, **23**, 5, 295–311.

Morgan, C. A. and Tindal, A. J., (1990). 'Further analysis of the Orkney MS-1 data.' *Proceedings of the BWEA Conference*, pp 325–330.

Petersen, J. T. *et al.*,. (1998). 'Prediction of dynamic loads and induced vibrations in stall'. *Report No. R-1045.* Riso National Laboratory, Roskilde, Denmark.

Putter, S. and Manor, H., (1978). 'Natural frequencies of radial rotating beams.' *J. Sound Vib.*, **56**, 2, pp 175–185.

Rasmussen, F., (1984). 'Aerodynamic performance of a new LM 17.2 m rotor.' *Riso National Laboratory Report No. M-2467.* Riso National Laboratory, Roskilde, Denmark.

Thomsen, K., (1998). 'The statistical variation of wind-turbine fatigue loads'. *Report No. R-1063.* Riso National Laboratory, Roskilde, Denmark.

Thomsen, K. and Madsen, P. H., (1997). 'Application of statistical methods to extreme loads for wind turbines.' *Proceedings of the EWEC*, pp 595–598.

Veers, P. S., (1988). 'Three-dimensional wind simulation.' *SAND88-0152, Sandia National Laboratory.*

Warren, J. G. *et al.*, (1988). 'Prediction of fatigue damage in wind turbine rotors.' *Proceedings of the BWEA Conference.*

Appendix: Dynamic Response of Stationary Blade in Turbulent Wind

A5.1 Introduction

As described in Chapter 2, the turbulent wind contains wind speed fluctuations over a wide range of frequencies, as described by the power spectrum. Although the bulk of the turbulent energy is normally at frequencies much lower than the blade first mode out-of-plane frequency, which is typically over 1 Hz, the fraction close to the first mode frequency will excite resonant blade oscillations. This appendix describes the method by which the resonant response may be determined. Working in the frequency domain, expressions for the standard deviations of both the tip displacement and root bending moment responses are derived, and then the method of deriving the peak value in a given period is described. Initially the wind is assumed to be perfectly correlated along the blade, but subsequently the treatment is extended to include the effect of spatial variation.

A5.2 Frequency Response Function

A5.2.1 Equation of motion

The dynamic response of a cantilever blade to the fluctuating aerodynamic loads upon it is most conveniently investigated by means of modal analysis, in which the the excitations of the various different natural modes of vibration are computed separately and the results superposed. Thus the deflection $x(r, t)$ at radius r is given by:

$$x(r, t) = \sum_{i=1}^{\infty} f_i(t)\mu_i(r)$$

Normally, in the case of a stationary blade, the first mode dominates and higher modes do not need to be considered. The equation of motion for the ith mode, which is derived in Section 5.8.1, is as follows:

$$m_i\ddot{f}_i(t) + c_i\dot{f}_i(t) + m_i\omega_i^2 f_i(t) = \int_0^R \mu_i(r)q(r, t)\,\mathrm{d}r \tag{A5.1}$$

where $q(r, t)$ is the applied loading, $f_i(t)$ is the tip displacement, $\mu_i(r)$ is the non-dimensional mode shape of the ith mode, normalized to give a tip displacement of unity, ω_i is the natural frequency in radians per second, m_i is the generalized mass, $\int_0^R m(r)\mu_i^2(r)\,\mathrm{d}r$, and c_i is the generalized damping, $\int_0^R \hat{c}(r)\mu_i^2(r)\,\mathrm{d}r$.

A5.2.2 Frequency response function

If $q(r, t)$ varies harmonically, with frequency ω and amplitude $q_0(r)$, then it can be shown that:

$$f_i(t) = \frac{1}{m_i} \frac{\int_0^R \mu_i(r) q_0(r)\,\mathrm{d}r}{\sqrt{(\omega_i^2 - \omega^2)^2 + (c_i/m_i)^2 \omega^2}} \cos(\omega t + \phi_i)$$

$$= \frac{1}{m_i \omega_i^2} \frac{\int_0^R \mu_i(r) q_0(r)\,\mathrm{d}r}{\sqrt{(1 - \omega^2/\omega_i^2)^2 + (c_i/m_i\omega_i^2)^2 \omega^2}} \cos(\omega t + \phi_i) \tag{A5.2}$$

Defining $k_i = m_i \omega_i^2$, and noting that the damping ratio $\xi_i = c_i/2m_i\omega_i$, this becomes:

$$f_i(t) = \frac{1}{k_i} \frac{\int_0^R \mu_i(r) q_0(r)\,\mathrm{d}r}{\sqrt{(1 - \omega^2/\omega_i^2)^2 + 4\xi_i^2 \omega^2/\omega_i^2}} \cos(\omega t + \phi_i) = A_i \cos(\omega t + \phi_i) \tag{A5.3}$$

The numerator $\int_0^R \mu_i(r) q_0(r)\,\mathrm{d}r$. is the amplitude of the equivalent loading at the tip of the cantilever that would result in the same tip displacement as the loading $q(r, t)$, and is known as the generalized load with respect to the ith mode, $Q_i(t)$. Thus the ratio between the tip displacement amplitude and the amplitude of the generalized load is

$$\frac{A_i}{\int_0^R \mu_i(r) q_0(r)\,\mathrm{d}r.} = \frac{1}{k_i \sqrt{(1 - \omega^2/\omega_i^2)^2 + 4\xi_i^2 \omega^2/\omega_i^2}}$$

$$= \frac{1}{k_i \sqrt{(1 - n^2/n_i^2)^2 + 4\xi_i^2 n^2/n_i^2}} = |H_i(n)| \tag{A5.4}$$

The ratio $|H_i(n)|$ is the modulus of the complex frequency response function, $H_i(n)$, and its square can be used to transform the power spectrum of the wind incident on the blade into the power spectrum of the ith mode tip displacement. Thus, in the case of the dominant first mode, the tip displacement in response to a harmonic generalized loading, $Q_1(t)$, of frequency n is given by

$$x_1(R, t) = f_1(t) = Q_1(t)|H_1(n)|$$

and the power spectrum of the first mode tip deflection is $S_{1x}(n) = S_{Q1}(n)|H_1(n)|^2$.

In what follows, the simplifying assumption is made initially that the wind is perfectly correlated along the blade.

A5.3 Resonant Displacement Response Ignoring Wind Variations along the Blade

A5.3.1 Linearization of wind loading

For a fluctuating wind speed $U(t) = \overline{U} + u(t)$, the wind load per unit length on the blade is $\frac{1}{2}C_f\rho U^2(t)c(r) = \frac{1}{2}C_f\rho[\overline{U}^2 + 2\overline{U}u(t) + u^2(t)]c(r)$, where C_f is the lift or drag coefficient, as appropriate, and $c(r)$ is the local blade chord dimension. In order to permit a linear treatment, the third term in the square brackets, which will normally be small compared to the first two, is ignored, so that the fluctuating load $q(r, t)$ becomes $C_f\rho\overline{U}u(t)c(r)$.

A5.3.2 First mode displacement response

Setting $q(r, t) = C_f\rho\overline{U}u(t)c(r)$, the first mode tip displacement response to a sinusoidal wind fluctuation of frequency n $(= \omega/2\pi)$ and amplitude $u_o(n)$ given by Equation (A5.3) becomes

$$f_1(t) = \int_0^R \mu_1(r)C_f\rho\overline{U}c(r)\,dr\; u_o(n)|H_1(n)|\cos(2\pi nt + \phi_1)$$

$$= C_f\rho\overline{U}\int_0^R \mu_1(r)c(r)\,dr\; u_o(n)|H_1(n)|\cos(2\pi nt + \phi_1) \tag{A5.5}$$

Hence power spectrum of first mode tip displacement is

$$S_{1x}(n) = \left[C_f\rho\overline{U}\int_0^R \mu_1(r)c(r)\,dr\right]^2 S_u(n)|H_1(n)|^2 \tag{A5.6}$$

where $S_u(n)$ is the power spectrum for the along wind turbulence. Thus the standard deviation of the first mode tip displacement is given by

$$\sigma_{1x}^2 = [C_f\rho\overline{U}\int_0^R \mu_1(r)c(r)\,dr]^2\int_0^\infty S_u(n)|H_1(n)|^2\,dn \tag{A5.7}$$

A5.3.3 Background and resonant response

Normally the bulk of the turbulent energy in the wind is at frequencies well below the frequency of the first out-of-plane blade mode. This is illustrated in Figure A5.1, where a typical power spectrum for wind turbulence is compared with the square, $|H_1(n)|^2$, of an example frequency response function for a 1 Hz resonant frequency. The power spectrum is that due to Kaimal (and adopted in Eurocode 1, 1997):

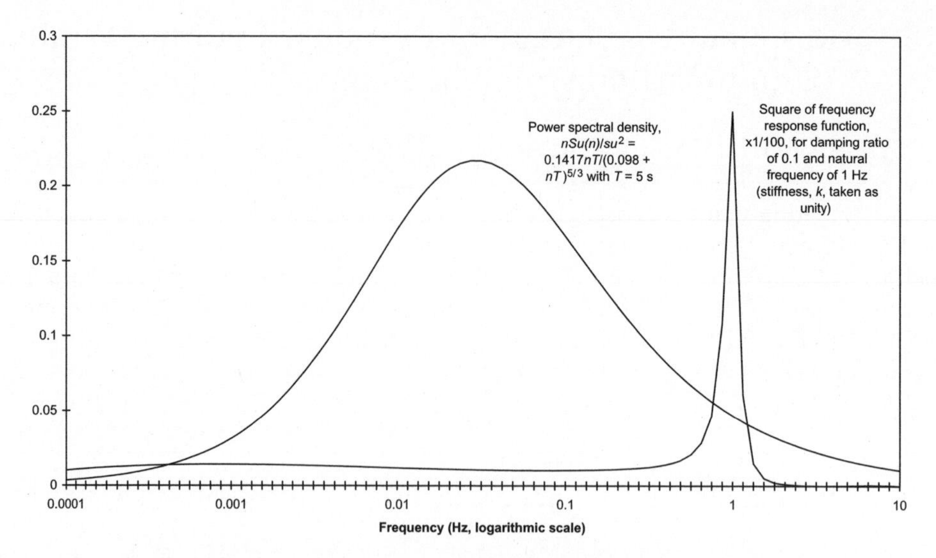

Figure A5.1 Power Spectrum and Frequency Response Function

$$nS_u(n) = \sigma_u^2 \frac{0.1417 n L_u^x / \overline{U}}{(0.098 + n L_u^x / \overline{U})^{\frac{5}{3}}} \tag{A5.8}$$

and is plotted as the non-dimensional power-spectral density function, $R_u(n) = nS_u(n)/\sigma_u^2$, against a logarithmic frequency scale. The time scale, $L_u^x/\overline{U}$, chosen is 5 s, based on a mean wind speed, $\overline{U}$, of 45 m/s and an integral length scale, L_u^x, of 225 m.

In view of the fact that the resonant response usually occurs over a narrow band of frequencies on the tail of the power spectrum, it is normal to treat it separately from the quasistatic response at lower frequencies, and to ignore the variation in $nS_u(n)$ on either side of the resonant frequency, n_1 (see, for example, Wyatt (1980)). The variance of total tip displacement then becomes:

$$\sigma_x^2 = \sigma_B^2 + \sigma_{x1}^2$$

in which the variance of the first mode resonant response, σ_{x1}, is given by

$$\sigma_{x1}^2 = \left[C_f \rho \overline{U} \int_0^R \mu_1(r) c(r)\, dr \right]^2 S_u(n_1) \int_0^\infty |H_1(n)|^2\, dn \tag{A5.9}$$

and the resonant response of higher modes, σ_{x2}^2, σ_{x3}^2 etc are ignored. The non-resonant response, σ_B, is termed the background response, and can be derived from simple static beam theory.

It has been shown by Newland (1984) that $\int_0^\infty |H_1(n)|^2\, dn$ reduces to $(\pi^2/2\delta)(n_1/k_1^2)$, where δ is the logarithmic decrement of damping. The logarithmic decrement, δ, is 2π times the damping ratio, ξ_1, defined as $\xi_1 = c_1/2m_1\omega_1$. Hence Equation (A5.9) becomes

$$\sigma_{x1}^2 = \left[C_f \rho \overline{U} \int_0^R \mu_1(r) c(r)\, dr \right]^2 S_u(n_1) \frac{\pi^2}{2\delta} \frac{n_1}{k_1^2} \tag{A5.10}$$

For comparison, the first mode component, $\bar{x}_1$, of the steady response is obtained by setting $\omega = 0$ and $q_0(r) = \frac{1}{2}\rho \overline{U}^2 C_f c(r)$ in Equation (A5.3), yielding

$$\bar{x}_1 = \tfrac{1}{2}\rho \overline{U}^2 C_f \frac{1}{k_1} \int_0^R \mu_1(r) c(r)\, dr \tag{A5.11}$$

Hence the ratio of the standard deviation of the first mode resonant response to the first mode component of the steady response is

$$\frac{\sigma_{x1}}{\bar{x}_1} = 2\frac{\sigma_u}{\overline{U}} \frac{\pi}{\sqrt{2\delta}} \sqrt{\frac{n_1 S_u(n_1)}{\sigma_u^2}} = 2\frac{\sigma_u}{\overline{U}} \frac{\pi}{\sqrt{2\delta}} \sqrt{R_u(n_1)} \tag{A5.12}$$

Note that towards the upper tail of the power spectrum of along wind turbulence, where n_1 is likely to be located, $\sqrt{R_u(n_1)}$ tends to $\sqrt{0.1417/(nL_u^x/\overline{U})^{\frac{2}{3}}}$.

A5.4 Effect of Across-wind Turbulence Distribution on Resonant Displacement Response

In the foregoing treatment, the wind was assumed to be perfectly correlated along the blade. The implications of removing this simplifying assumption will now be examined.

The fluctuating load on the blade, $q(r, t)$, becomes $C_f \rho \overline{U} u(u, r) c(r)$ per unit length, and the generalized fluctuating load with respect to the first mode becomes

$$Q_1(t) = \int_0^R \mu_1(r) q(r, t)\, dr = C_f \rho \overline{U} \int_0^R u(r, t) c(r) \mu_1(r)\, dr \tag{A5.13}$$

The standard deviation, σ_Q, of $Q(t)$ is given by

$$\sigma_{Q1}{}^2 = \frac{1}{T}\int_0^T Q_1^2(t)\, dt = (\rho \overline{U} C_f)^2 \frac{1}{T}\int_0^T \left[\int_0^R u(r, t) c(r) \mu_1(r)\, dr\right] \left[\int_0^R u(r', t) c(r') \mu_1(r')\, dr'\right] dt$$

$$= (\rho \overline{U} C_f)^2 \int_0^R \int_0^R \left[\frac{1}{T}\int_0^T u(r, t) u(r', t)\, dt\right] c(r) c(r') \mu_1(r) \mu_1(r')\, dr\, dr' \tag{A5.14}$$

Now the expression within the square brackets is the cross correlation function, $\kappa_u(r, r', \tau) = E\{u(r, t) u(r', t + \tau)\}$ with τ set equal to zero. The cross correlation function is related to the cross spectrum, $S_{uu}(r, r', n)$, as follows:

$$\kappa_u(r, r', \tau) = \tfrac{1}{2}\int_{-\infty}^{\infty} S_{uu}(r, r', n)\exp(i2\pi n\tau)\,\mathrm{d}n$$

giving

$$\kappa_u(r, r', 0) = \left[\frac{1}{T}\int_0^T u(r, t)u(r', t)\,\mathrm{d}t\right] = \int_0^{\infty} S_{uu}(r, r', n)\,\mathrm{d}n \text{ for } \tau = 0 \tag{A5.15}$$

Hence

$$\sigma_{Q1}{}^2 = (\rho\overline{U}C_f)^2\int_0^R\int_0^R\left[\int_0^{\infty} S_{uu}(r, r', n)\,\mathrm{d}n\right]c(r)c(r')\mu_1(r)\mu_1(r')\,\mathrm{d}r\,\mathrm{d}r' \tag{A5.16}$$

The normalized cross spectrum is defined as $S_{uu}^N(r, r', n) = S_{uu}(r, r', n)/S_u(n)$, and like $S_{uu}(r, r', n)$, is in general a complex quantity, because of phase differences between the wind speed fluctuations at different heights. As only in-phase wind speed fluctuations will affect the response, we consider only the real part of the normalized cross spectrum, known as the normalized co-spectrum, and denoted by $\psi_{uu}^N(r, r', n)$. Substituting in Equation (A5.16), we obtain:

$$\sigma_{Q1}{}^2 = (\rho\overline{U}C_f)^2\int_0^R\int_0^R\left[\int_0^{\infty} S_u(n)\psi_{uu}^N(r, r', n)\,\mathrm{d}n\right]c(r)c(r')\mu_1(r)\mu_1(r')\,\mathrm{d}r\,\mathrm{d}r' \tag{A5.17}$$

From this, it can be deduced that the power spectrum of the generalized load with respect to the first mode is

$$S_{Q1}(n) = (\rho\overline{U}C_f)^2\int_0^R\int_0^R S_u(n)\psi_{uu}^N(r, r', n)c(r)c(r')\mu_1(r)\mu_1(r')\,\mathrm{d}r\,\mathrm{d}r' \tag{A5.18}$$

Note that the power spectrum for the along wind turbulence shows some variation with height, and so should strictly be written $S_u(n, z)$ instead of $S_u(n)$. However, the variation along the length of a vertical blade is small, and is ignored here.

As for the initial case when wind loadings along the blade were assumed to be perfectly correlated, the power spectrum for first mode tip displacement is equal to the product of the power spectrum of the generalized load (with respect to the first mode) and the square of the frequency response function, i.e.,

$$S_{1x}(n) = S_{Q1}(n)|H_1(n)|^2 \tag{A5.19}$$

As before, $S_{Q1}(n)$ is assumed constant over the narrow band of frequencies straddling the resonant frequency, and the standard deviation of resonant tip response becomes:

$$\sigma_{x1}^2 = S_{Q1}(n_1)\int_0^{\infty}|H_1(n)|^2\,\mathrm{d}n = S_{Q1}(n_1)\frac{\pi^2}{2\delta}\frac{n_1}{k_1^2} \tag{A5.20}$$

A5.4.1 Formula for normalized co-spectrum

It remains to evaluate $S_{Q1}(n_1) = (\rho\overline{U}C_f)^2 S_u(n_1) \int_0^R \int_0^R \psi_{uu}^N(r, r', n)\; c(r)c(r')\mu_1(r)\mu_1(r')\,\mathrm{d}r\,\mathrm{d}r'$. The normalized co-spectrum, $\psi_{uu}^N(r, r', n)$, must decrease as the spacing $[r - r']$ between the two points considered increases, and intuitively it is to be expected that the decrease would be more rapid for the higher frequency components of wind fluctuation. On an empirical basis, Davenport (1962) has proposed an exponential expression for the normalized co-spectrum as follows:

$$\psi_{uu}^N(r, r', n) = \exp[-C|r - r'|n/\overline{U}] \tag{A5.21}$$

where C is a non-dimensional decay constant. Davenport noted that measurements by Cramer (1958) indicated values of C ranging from 7 in unstable conditions to 50 in stable conditions, but recommended the use of the lower figure as being the more conservative despite the likelihood of stable conditions in high winds. Dyrbye and Hansen (1997) quote Riso measurements reported by Mann (1994) which indicate a value of C of 9.4, and they recommend a value of 10 for use in design. A value of 9.2 is implicitly assumed in Eurocode 1 (1997).

There is an obvious inconsistency in the exponential expression for the normalized co-spectrum – when it is integrated up over the plane perpendicular to the wind direction, the result is positive instead of zero as it should be. This has led to the development of more complex expressions by Harris (1971) and Krenk (1995). However, the Davenport formulation will be used here, giving

$$\begin{aligned}\sigma_{x1}^2 &= S_{Q1}(n_1)\frac{\pi^2}{2\delta}\frac{n_1}{k_1^2}\\ &= (\rho\overline{U}C_f)^2 S_u(n_1)\int_0^R\int_0^R \exp[-C|r - r'|n_1/\overline{U}]c(r)c(r')\mu_1(r)\mu_1(r')\,\mathrm{d}r\,\mathrm{d}r'\left[\frac{\pi^2}{2\delta}\frac{n_1}{k_1^2}\right]\end{aligned} \tag{A5.22}$$

The resonant response can be expressed in terms of the first mode component, $\bar{x}_1$, of the steady response,

$$\tfrac{1}{2}\rho\overline{U}^2 C_f \frac{1}{k_1}\int_0^R \mu_1(r)c(r)\,\mathrm{d}r$$

from Equation (A5.11) giving

$$\frac{\sigma_{x1}^2}{\bar{x}_1^2} = 4\frac{\sigma_u^2}{\overline{U}^2}\frac{\pi^2}{2\delta}R_u(n_1)\frac{\displaystyle\int_0^R\int_0^R \exp[-C|r - r'|n_1/\overline{U}]c(r)c(r')\mu_1(r)\mu_1(r')\,\mathrm{d}r\,\mathrm{d}r'}{\left(\displaystyle\int_0^R c(r)\mu_1(r)\,\mathrm{d}r\right)^2} \tag{A5.23}$$

Hence

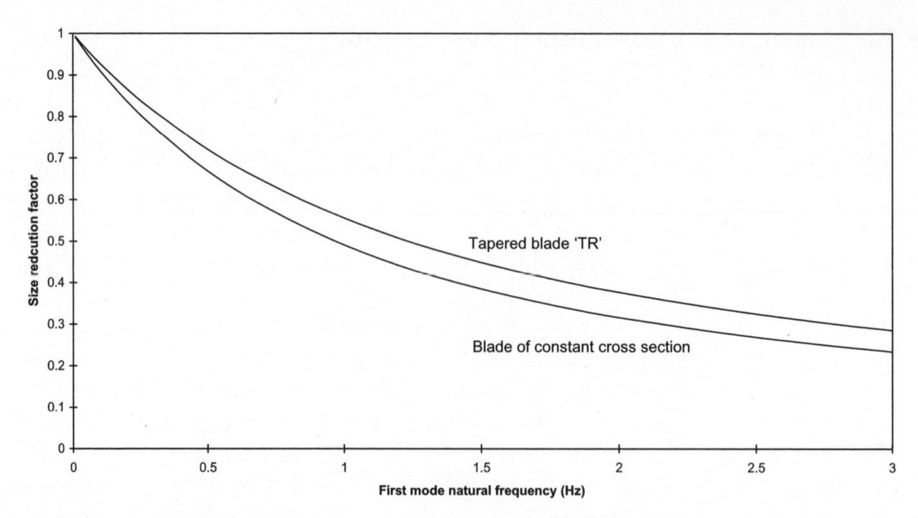

Figure A5.2 Size Reduction Factors for the First Mode Resonant Response due to Lack of Correlation of Wind Loading along the Blade-Variation with Frequency for 20 m Blade

$$\frac{\sigma_{x1}}{\bar{x}_1} = 2\frac{\sigma_u}{\overline{U}}\frac{\pi}{\sqrt{2\delta}}\sqrt{R_u(n_1)}\sqrt{K_{Sx}(n_1)} \tag{A5.24}$$

where

$$K_{Sx}(n_1) = \frac{\int_0^R \int_0^R \exp[-C|r-r'|n_1/\overline{U}]c(r)c(r')\mu_1(r)\mu_1(r')\,\mathrm{d}r\,\mathrm{d}r'}{\left(\int_0^R c(r)\mu_1(r)\,\mathrm{d}r\right)^2} \tag{A5.25}$$

is denoted the size reduction factor, which results from the lack of correlation of the wind along the blade. As an example, the size reduction factor, $K_{Sx}(n_1)$, is plotted out against frequency in Figure A5.2 for the case of a 20 m blade with chord $c(r) = 0.0961R - 0.06467r$ (Blade TR), assuming a decay constant C of 9.2, and a mean wind speed $\overline{U}$ of 45 m/s. The mode shape taken is the same as for the example in section 5.6.3 (see Figure 5.3). Also shown for comparison is the corresponding parameter for a uniform cantilever.

A5.5 Resonant Root Bending Moment

For design purposes it is the augmentation of blade bending moments due to dynamic effects that is of principal significance. The ratio of the standard deviation of the first mode resonant root bending moment to the steady root bending moment (allowing for the lack of correlation of wind fluctuations along the blade) is derived below.

Defining $M_1(t)$ as the fluctuating root bending moment due to wind excitation of the first mode, we have

$$M_1(t) = \int_0^R m(r)\ddot{x}_1(t, r) r \, \mathrm{d}r = \int_0^R m(r)\omega_1^2 x_1(t, r) r \, \mathrm{d}r = \omega_1^2 f_1(t) \int_0^R m(r)\mu_1(r) r \, \mathrm{d}r \tag{A5.26}$$

Hence the standard deviation of $M_1(t)$,

$$\sigma_{M1} = \omega_1^2 \sigma_{x1} \int_0^R m(r)\mu_1(r) r \, \mathrm{d}r \tag{A5.27}$$

The steady root bending moment,

$$\overline{M} = \int_0^R \tfrac{1}{2}\rho \overline{U}^2 C_f c(r) r \, \mathrm{d}r = \tfrac{1}{2}\rho \overline{U}^2 C_f \int_0^R c(r) r \, \mathrm{d}r \tag{A5.28}$$

Hence the ratio

$$\frac{\sigma_{M1}}{\overline{M}} = \frac{\omega_1^2 \sigma_{x1} \displaystyle\int_0^R m(r)\mu_1(r) r \, \mathrm{d}r}{\tfrac{1}{2}\rho \overline{U}^2 C_f \displaystyle\int_0^R c(r) r \, \mathrm{d}r} \tag{A5.29}$$

Substituting the expression for σ_{x1} from Equation (A5.22), we obtain

$$\frac{\sigma_{M1}}{\overline{M}} = \frac{\omega_1^2 \rho \overline{U} C_f \dfrac{\pi}{\sqrt{2\delta}} \dfrac{\sqrt{n_1 S_u(n_1)}}{k_1} \displaystyle\int_0^R m(r)\mu_1(r) r \, \mathrm{d}r \sqrt{\int_0^R \int_0^R \exp[-C|r - r'|n_1/\overline{U}] c(r) c(r') \mu_1(r) \mu_1(r') \, \mathrm{d}r \, \mathrm{d}r'}}{\tfrac{1}{2}\rho \overline{U}^2 C_f \displaystyle\int_0^R c(r) r \, \mathrm{d}r} \tag{A5.30}$$

Noting that $R_u(n) = nS_u(n)/\sigma_u^2$, and that $k_1 = m_1 \omega_1^2$, this simplifies to

$$\frac{\sigma_{M1}}{\overline{M}} = 2 \frac{\sigma_u}{\overline{U}} \frac{\pi}{\sqrt{2\delta}} \sqrt{R_u(n)} \frac{\displaystyle\int_0^R m(r)\mu_1(r) r \, \mathrm{d}r}{m_1 \displaystyle\int_0^R c(r) r \, \mathrm{d}r} \sqrt{\int_0^R \int_0^R \exp[-C|r - r'|\mathrm{n}_1/\overline{U}] c(r) c(r') \mu_1(r) \mu_1(r') \, \mathrm{d}r \, \mathrm{d}r'} \tag{A5.31}$$

where $m_1 = \int_0^R m(r)\mu_1^2(r)\,\mathrm{d}r$ is the generalized mass with respect to the first mode, and the exponential expression within the double integral allows for the lack of correlation of wind fluctuations along the blade. Substituting $(\int_0^R c(r)\mu_1(r)\,\mathrm{d}r).\sqrt{K_{Sx}(n_1)}$ for the square root of the double integral, using Equation A5.25, leads to

$$\frac{\sigma_{M1}}{\overline{M}} = 2\frac{\sigma_u}{\overline{U}}\frac{\pi}{\sqrt{2\delta}}\sqrt{R_u(n)}\frac{\int_0^R m(r)\mu_1(r)r\,\mathrm{d}r}{m_1\int_0^R c(r)r\,\mathrm{d}r}\left(\int_0^R c(r)\mu_1(r)\,\mathrm{d}r\right).\sqrt{K_{Sx}(n_1)} \tag{A5.32}$$

Defining the ratio of the integrals,

$$\frac{\int_0^R m(r)\mu_1(r)r\,\mathrm{d}r}{m_1\int_0^R c(r)r\,\mathrm{d}r}\left(\int_0^R c(r)\mu_1(r)\,\mathrm{d}r\right) = \frac{\dfrac{\sigma_{M1}}{\overline{M}}}{\dfrac{\sigma_{x1}}{\bar{x}_1}} \tag{A5.33}$$

as λ_{M1} we obtain

$$\frac{\sigma_{M1}}{\overline{M}} = 2\frac{\sigma_u}{\overline{U}}\frac{\pi}{\sqrt{2\delta}}\sqrt{R_u(n_1)}\sqrt{K_{Sx}(n_1)}\lambda_{M1} \tag{A5.34}$$

A5.6 Root Bending Moment Background Response

The root bending moment background response can be expressed in terms of the standard deviation of the root bending moment excluding resonant effects. If the wind is perfectly correlated along the blade, this is given by

$$\sigma_{MB} = C_f\rho\overline{U}\sigma_u\int_0^R c(r)r\,\mathrm{d}r \tag{A5.35}$$

However, if the lack of correlation of wind fluctuations along the blade is taken into account,

$$\sigma_{MB} = C_f\rho\overline{U}\sigma_u\sqrt{\int_0^R\int_0^R \rho_u(r-r')c(r)c(r')rr'\,\mathrm{d}r\,\mathrm{d}r'} \tag{A5.36}$$

where $\rho_u(r-r')$ is the normalized cross correlation function between simultaneous wind-speed fluctuations at two different blade radii, and is defined as

$$\rho_u(r-r') = \frac{1}{\sigma_u^2}E\{u(r,t)u(r',t+\tau)\} \tag{A5.37}$$

with τ set equal to zero. Measurements indicate that the normalized cross correlation function decays exponentially, so it can be expressed as

$$\rho_u(r - r') = \exp[-|r - r'|/L_u^r] \tag{A5.38}$$

where L_u^r is the integral length scale for the longitudinal turbulence component measured in the across wind direction along the blade, and is thus defined as $\int_0^\infty \rho_u(r - r)\mathrm{d}(r - r')$. As the integral length scale for longitudinal turbulence measured vertically in the across wind direction (L_u^z) is, if anything, less than that measured horizontally (L_u^y), it is conservative to treat it as being equal to that measured horizontally, with the result that L_u^r can be taken as equal to L_u^y also. Typically L_u^y is approximately equal to 30 percent of L_u^x, the integral length scale for longitudinal turbulence measured in the along wind direction. Observing that

$$\overline{M} = \tfrac{1}{2}\rho\overline{U}^2 C_\mathrm{f}\int_0^R c(r)r\,\mathrm{d}r$$

we can therefore write

$$\frac{\sigma_{MB}}{\overline{M}} = 2\frac{\sigma_u}{\overline{U}}\sqrt{K_{SMB}} \tag{A5.39}$$

where K_{SMB}, the size reduction factor for the root bending moment background response, is defined as

$$K_{SMB} = \frac{\int_0^R\int_0^R \exp[-|r - r'|/0.3\mathrm{L}_u^x]c(r)c(r')rr'\,\mathrm{d}r\,\mathrm{d}r'}{\left(\int_0^R c(r)r\,\mathrm{d}r\right)^2} \tag{A5.40}$$

For a blade with a uniform chord, the integral is straightforward, giving

$$K_{SMB} = 4\left[\frac{2}{3\phi} - \frac{1}{\phi^2} + \frac{2}{\phi^4} - \exp(-\phi)\left\{\frac{2}{\phi^3} + \frac{2}{\phi^4}\right\}\right] \quad \text{where } \phi = \frac{R}{0.3L_u^x} \tag{A5.41}$$

As an example, K_{SMB} comes to 0.927 for the case of $R = 20$ m and $L_u^x = 230$ m, indicating that the effect of the lack of correlation of the wind fluctuations is rather small.

For blades with a normal tapering chord, K_{SMB} can be evaluated numerically. In the case of a blade with a tip chord equal to 25 percent of the maximum chord, K_{SMB} is 0.924 for the same value of ϕ as before. It is seen that the taper has a negligible effect on the end result.

A5.7 Peak Response

One of the key parameters required in blade design is the extreme value of the out-of-plane bending moment. The 50 year return moment is defined as the expected maximum moment occurring during the mean wind averaging period when the mean takes the 50 year return value. Treating the moment as a Gaussian process, Davenport (1964) has shown that the expected value of the maximum departure from the mean is the standard deviation multiplied by the peak factor, g, where

$$g = \sqrt{2\ln(\nu T)} + \frac{0.5772}{\sqrt{2\ln(\nu T)}} \tag{A5.42}$$

In this formula, ν is the mean zero-upcrossing frequency of the root moment fluctuations, and T is the mean wind speed averaging period. The variance of the root bending moment is, in the same way as for the tip displacement, equal to the sum of the variances of the background and resonant root bending moment responses, i.e.,

$$\sigma_M^2 = \sigma_{MB}^2 + \sigma_{M1}^2 \tag{A5.43}$$

Hence, from Equations (A5.39) and (A5.34), we obtain

$$\sigma_M^2 = \sigma_{MB}^2 + \sigma_{M1}^2 = 4\overline{M}^2\frac{\sigma_u^2}{\overline{U}^2}\left[K_{SMB} + \frac{\pi^2}{2\delta}R_u(n_1)K_{Sx}(n_1)\lambda_{M1}^2\right] \tag{A5.44}$$

Thus

$$M_{\max} = \overline{M} + g\sigma_M = \overline{M}\left(1 + 2g\frac{\sigma_u}{\overline{U}}\sqrt{K_{SMB} + \frac{\pi^2}{2\delta}R_u(n_1)K_{Sx}(n_1)\lambda_{M1}^2}\right) \tag{A5.45}$$

The mean zero up-crossing frequency of the root moment fluctuations, ν, is defined as

$$\nu = \sqrt{\frac{\int_0^\infty n^2 S_M(n)\,\mathrm{d}n}{\int_0^\infty S_M(n)\,\mathrm{d}n}} \tag{A5.46}$$

where $S_M(n)$ is the power spectrum of the root moment fluctuations. If we separate the power spectrum of the background response from the first mode resonant response at frequency n_1, then the above expression can be written

$$\nu = \sqrt{\frac{\left(\int_0^\infty n^2 S_{MB}(n)\,\mathrm{d}n\right) + n_1^2\sigma_{M1}^2}{\sigma_{MB}^2 + \sigma_{M1}^2}} \tag{A5.47}$$

Now

$$S_{MB}(n) = (C_f\rho\overline{U})^2 S_u(n)\int_0^R\int_0^R \psi_{uu}^N(r,\, r',\, n)c(r)c(r')rr'\,\mathrm{d}r\,\mathrm{d}r' \tag{A5.48}$$

and

$$\overline{M} = C_f\tfrac{1}{2}\rho\overline{U}^2\int_0^R c(r)r\,\mathrm{d}r \tag{A5.49}$$

so

$$S_{MB}(n) = 4\frac{\overline{M}^2}{\overline{U}^2}\,\frac{S_u(n)\int_0^R\int_0^R \psi_{uu}^N(r,\, r',\, n)c(r)c(r')rr'\,\mathrm{d}r\,\mathrm{d}r'}{\left(\int_0^R c(r)r\,\mathrm{d}r\right)^2} \tag{A5.50}$$

Defining

$$K_{SMB}(n) = \frac{\int_0^R\int_0^R \psi_{uu}^N(r,\, r',\, n)c(r)c(r')rr'\,\mathrm{d}r\,\mathrm{d}r'}{\left(\int_0^R c(r)r\,\mathrm{d}r\right)^2} \tag{A5.51}$$

we obtain

$$S_{MB}(n) = 4\frac{\overline{M}^2}{\overline{U}^2}S_u(n)K_{SMB}(n) \tag{A5.52}$$

Substituting into Equation (A5.47) gives

$$\nu = \sqrt{\frac{4\dfrac{\overline{M}^2}{\overline{U}^2}\left(\int_0^\infty n^2 S_u(n)K_{SMB}(n)\mathrm{d}n\right) + n_1^2\sigma_{M1}^2}{\sigma_{MB}^2 + \sigma_{M1}^2}} \tag{A5.53}$$

Noting from Equation (A5.52) that $\sigma_{MB}^2 = (4\overline{M}^2/\overline{U}^2)\int_0^\infty S_u(n)K_{SMB}(n)\,\mathrm{d}n$, the expression for ν becomes

$$\nu = \sqrt{\frac{n_0^2\sigma_{MB}^2 + n_1^2\sigma_{M1}^2}{\sigma_{MB}^2 + \sigma_{M1}^2}} \tag{A5.54}$$

where

$$n_0 = \frac{\int_0^\infty n^2 S_u(n) K_{SMB}(n)\,\mathrm{d}n}{\int_0^\infty S_u(n) K_{SMB}(n)\,\mathrm{d}n} \tag{A5.55}$$

Substituting $\psi_{uu}^N(r, r', n) = \exp[-C(r - r')n/\overline{U}]$ into the expression for $K_{SMB}(n)$ in the numerator of Equation (A5.55) gives

$$\int_0^\infty n^2 S_u(n) K_{SMB}(n)\,\mathrm{d}n = \int_0^\infty n^2 S_u(n) \frac{\int_0^R \int_0^R \exp[-C(r - r')n/\overline{U}]c(r)c(r')rr'\,\mathrm{d}r\,\mathrm{d}r'}{\left(\int_0^R c(r)r\,\mathrm{d}r\right)^2}\,\mathrm{d}n \tag{A5.56}$$

For high frequencies, the double integral is, in the limit, inversely proportional to frequency, so the integrand $n^2 S_u(n) K_{SMB}(n)$ is proportional to $n^2 n^{-5/3} n^{-1} = n^{-2/3}$ and the integral does not converge. Consequently it is necessary to take account of the chordwise lack of correlation of wind fluctuation at high frequencies and, if this is done, it is found that, in the limit, the integrand is proportional to $n^{-5/3}$ for which the integral is finite. The evaluation of the integral $\int_0^\infty n^2 S_u(n) K_{SMB}(n)\,\mathrm{d}n$ taking chordwise lack of correlation into account is a formidable task, so the use of an approximate formula for the frequency, n_0, is preferable, especially as the influence of n_0 on the peak factor, g, is slight. One formula is given in Eurocode 1 (1997), but a simpler one due to Dyrbye and Hansen (1997) for a uniform cantilever is as follows:

$$n_0 = 0.3\frac{\overline{U}}{\sqrt{L_u^x \sqrt{Rc}}} \tag{A5.57}$$

Here R is the blade tip radius and c is the blade chord, assumed constant. For a tapering chord, the mean chord, $\bar{c}$, can be substituted.

A5.8 Bending Moments at Intermediate Blade Positions

A5.8.1 Background response

Denoting the standard deviation of the quasistatic or background bending moment fluctuations at radius r^* as $\sigma_{MB}(r^*)$, it is apparent that

$$\frac{\sigma_{MB}(r^*)}{\sigma_{MB}(0)} = \sqrt{\frac{K_{SMB}(r^*)}{K_{SMB}(0)}} \frac{\int_{r^*}^{R} c(r)[r - r^*]\mathrm{d}r}{\int_0^R c(r) r \,\mathrm{d}r} \tag{A5.58}$$

The ratio of the steady moment at radius r^* to that at the root is $\int_{r^*}^{R} c(r)[r - r^*]\,\mathrm{d}r / \int_0^R c(r) r\,\mathrm{d}r$, so the ratio of the standard deviation of the quasistatic fluctuations at radius r^* to the steady value there is

$$\frac{\sigma_{MB}(r^*)}{\overline{M}(r^*)} = \frac{\sigma_{MB}(r^*)}{\sigma_{MB}(0)} \frac{\sigma_{MB}(0)}{\overline{M}(0)} \frac{\overline{M}(0)}{\overline{M}(r^*)} = \frac{\sigma_{MB}(0)}{\overline{M}(0)} \sqrt{\frac{K_{SMB}(r^*)}{K_{SMB}(0)}} \tag{A5.59}$$

Generally, the square root will be close to unity, so $\sigma_{MB}(r^*)/\overline{M}(r^*)$ will be nearly constant.

A5.8.2 Resonant response

In Section A5.5 it was shown that the standard deviation of the first mode resonant root bending moment is equal to $\omega_1^2 \sigma_{x1} \int_0^R m(r)\mu_1(r) r\,\mathrm{d}r$ (Equation A5.27). The corresponding quantity at other radii can be derived similarly, giving

$$\sigma_{M1}(r^*) = \omega_1^2 \sigma_{x1} \int_{r^*}^{R} m(r)\mu_1(r)[r - r^*]\mathrm{d}r \tag{A5.60}$$

Hence the ratio of the standard deviation of the first mode resonant root bending moment at radius r^* to the steady value there is

$$\frac{\sigma_{M1}(r^*)}{\overline{M}(r^*)} = \frac{\sigma_{M1}(r^*)}{\sigma_{M1}(0)} \frac{\sigma_{M1}(0)}{\overline{M}(0)} \frac{\overline{M}(0)}{\overline{M}(r^*)} = \frac{\int_{r^*}^{R} m(r)\mu_1(r)[r - r^*]\,\mathrm{d}r}{\int_0^R m(r)\mu_1(r) r\,\mathrm{d}r} \frac{\int_0^R c(r) r\,\mathrm{d}r}{\int_{r^*}^{R} c(r)[r - r^*]\,\mathrm{d}r} \frac{\sigma_{M1}(0)}{\overline{M}(0)} \tag{A5.61}$$

References

Cramer, H. E., (1958). 'Use of power spectra and scales of turbulence in estimating wind loads.' *Second National Conference on Applied Meteororlogy*, Ann Arbor, Michigan, USA.

Davenport, A. G., (1962). 'The response of slender, line-like structures to a gusty wind.' *Proc. Inst. Civ. Eng.*, **23**, 389–408.

Davenport, A. G., (1964). 'Note on the distribution of the largest value of a random function with application to gust loading.' *Proc. Inst. Civ. Eng.*, **28**, 187–196.

Dyrbye, C., and Hansen, S. O., (1997). *Wind loads on structures.* John Wiley and Sons.

Eurocode 1, (1997). *Basis of design and actions on structures – Part 2.4: Actions on structures – Wind actions.*

Harris, R. I., (1971). The nature of the wind. *Proceedings of the CIRIA Conference*, pp 29–55.

Krenk, S., (1995). 'Wind field coherence and dynamic wind forces.' *Symposium on the advances in Non-linear Stochastic Mechanics*. Kluwer, Dordrecht, Germany.

Mann, J., (1994). 'The spatial structure of neutral atmospheric surface-layer turbulence.' *J. Ind. Aerodyn.*, **1**, 167–175.

Newland, D. E., (1984). *Random vibrations and spectral analysis.* Longman, UK.

Wyatt, T. A., (1980). 'The dynamic behaviour of structures subject to gust loading'. *Proceedings of the CIRIA Conference*, ''Wind engineering in the eighties'' pp 6-1-6-22.

6
Conceptual Design of Horizontal-Axis Turbines

6.1 Introduction

Within the general category of horizontal axis wind turbines for grid applications there exists a great variety of possible machine configurations, power control strategies and braking systems. This chapter looks at the different areas where design choices have to be made, and considers the advantages and disadvantages of the more conventional options in each case. Inevitably there are situations in which decisions in one area can impact on those in another, and some of these are noted.

Alongside these discrete design choices there are several fundamental design parameters, such as rotor diameter, machine rating and rotational speed, which also have to be established at the start of the design process. Continuous variables such as these lend themselves to mathematical optimization, as described in the opening sections of the chapter.

6.2 Rotor Diameter

The issue of what size of turbine produces energy at minimum cost has been fiercely debated for a long time. Protagonists of large machines cite economies of scale and the increase in wind speed with height in their favour. From the other camp, the 'square-cube law', whereby energy capture increases as the square of the diameter, whereas rotor mass (and therefore cost) increases as the cube, is advanced as an argument against.

In reality, both arguments are correct, and there is a trade-off between economies of scale and a variant of the 'square-cube law' which takes into account the wind shear effect. This trade-off can be examined with the help of simple cost modelling, which is considered next.

6.2.1 Cost modelling

The sensitivity of the cost of energy to changes in the values of parameters governing turbine design can be examined with the aid of a model of the way component costs vary in response. The normal procedure is to start with a baseline design, for which the costs of the various components are known. In a rigorous analysis, the chosen parameter is then assigned a different value and a fresh design developed, leading to revised component weights, based on which new component costs can be assigned.

In general, the cost of a component will not simply increase *pro rata* with its mass, but will contain elements that increase more slowly. An example is the tower surface protective coating , the cost of which increases approximately as the *square* of the tower height, if all dimensions are proportional to this height. If the design parameter variation considered is only about +/−50 percent, it is usually sufficiently accurate to represent the relationship between component cost and mass as a linear one with a fixed component:

$$C(x) = C_B\left(\mu\frac{m(x)}{m_B} + (1-\mu)\right) \tag{6.1}$$

where $C(x)$ and $m(x)$ are the cost and mass of the component respectively when the design parameter takes the value x, and C_B and m_B are the baseline values; μ is the proportion of the cost that varies with mass, which will obviously differ for different baseline machine sizes.

The choice of the value of μ inevitably requires considerable expertise as regards the way manufacturing costs vary with scale, which may be limited in the case of products at the early stage of development. In view of this, the effort of developing fresh designs for different design parameter values may well not be justified, so resort is often made to scaling ratios based on similarity relationships. This approach is adopted in the investigation of optimum machine size which follows.

6.2.2 Simplified cost model for machine size optimization—an illustration

The baseline machine design is taken as a 60 m diameter, 1.5 MW turbine, with the costs of the various components taken from Fuglsang and Thomsen (1998). These are given in Table 6.1 as a percentage of the total.

Machine designs for other diameters are obtained by scaling *all* dimensions of all components in the same proportion, except in the case of the gearbox, generator, grid connection and controller. Rotational speed is kept inversely proportional to rotor diameter to maintain constant tip speed, and hence constant tip speed ratio at a given wind speed. As a result, all machine designs reach rated power at the same wind speed, so that rated power is proportional to diameter squared. Consequently the low-speed shaft torque increases as diameter cubed, which is the basis for assuming the gearbox mass increases as the cube of rotor diameter, even though the gearbox ratio changes. Hence, if a blanket value of μ of 0.9 is adopted for simplicity,

Table 6.1 Component Costs Expressed as a Percentage of Total Machine Cost for a 1.5 MW, 60 metre diameter Wind Turbine on Land (from Fuglsang and Thomsen (1998))

Component	Cost as a percentage of total	Component	Cost as a percentage of total
Blades	18.3%	Controller	4.2%
Hub	2.5%	Tower	17.5%
Main shaft	4.2%	Brake system	1.7%
Gearbox	12.5%	Foundation	4.2%
Generator	7.5%	Assembly	2.1%
Nacelle	10.8%	Transport	2.0%
Yaw system	4.2%	Grid connection	8.3%
		TOTAL	100%

the cost of all components apart from generator, controller and the grid connection, for a machine of diameter D, is given by:

$$C_1(D) = 0.8C_T(60)\left(0.9\left(\frac{D}{60}\right)^3 + 0.1\right) \tag{6.2}$$

where $C_T(60)$ is the total cost of the baseline machine.

The rating of the generator and the grid connection is proportional only to the diameter squared. It is assumed that Equation (6.1) applies to the cost of these components, but with mass replaced by rating. Thus, if μ is taken as 0.9 once more, the cost of the generator and grid connection are given by:

$$C_2(D) = 0.158C_T(60)\left(0.9\left(\frac{D}{60}\right)^2 + 0.1\right) \tag{6.3}$$

The controller cost is assumed to be fixed. Hence the resulting turbine cost as a function of diameter is:

$$\begin{aligned} C_T(D) &= C_T(60)\left(0.8\left\{0.9\left(\frac{D}{60}\right)^3 + 0.1\right\} + 0.158\left\{0.9\left(\frac{D}{60}\right)^2 + 0.1\right\} + 0.042\right) \\ &= C_T(60)\left(0.72\left(\frac{D}{60}\right)^3 + 0.1422\left(\frac{D}{60}\right)^2 + 0.1378\right) \end{aligned} \tag{6.4}$$

As the tower height, along with all other dimensions, is assumed to increase in proportion to rotor diameter, the annual mean wind speed at hub height will increase with rotor diameter because of wind shear. The energy yield should thus be calculated taking this effect into account. The cost of energy (excluding operation and maintenance costs) can then be calculated in €/kWh/annum by dividing the turbine cost by the annual energy yield. The variation of energy cost with diameter, calculated according to the assumptions described above, is plotted in Figure 6.1 for two levels of wind shear, corresponding to roughness lengths, z_0, of 0.001 m and 0.05 m, the hub-height mean wind speed being scaled according to the relation

$$\overline{U}(z) \propto \ln(z/z_0) \tag{2.10}$$

(see Section 2.6.2). Also included is a plot for the case of zero wind shear. It is apparent that the level of wind shear has a noticeable effect on the optimum machine diameter, which varies from 44 m for zero wind shear to 52 m for the wind shear corresponding to a surface roughness length of 0.05 m, which is applicable to farmland with boundary hedges and occasional buildings. Strictly, the impact of the increased annual mean wind speed with hub height on the fatigue design of the rotor and other components should also be taken into account, which would reduce the optimum machine size slightly.

It should be emphasized that the optimum sizes derived above depend critically on the value of μ adopted. For example, if μ were taken as 0.8 instead of 0.9, the optimum diameter in the absence of wind shear would increase to 54 m, although the minimum cost of energy would alter by only 0.3 percent. A more sophisticated approach would allocate different values of μ to different components, as is done in Fuglsang and Thomsen (1998). Ideally these would be based on cost data on components of the same design but different sizes.

The cost model outlined above provides a straightforward means of investigating scale effects on machine economics for a chosen machine design. In practice, the use of different materials or different machine configurations may prove more economic at different machine sizes, and will yield a series of alternative cost *versus* diameter curves.

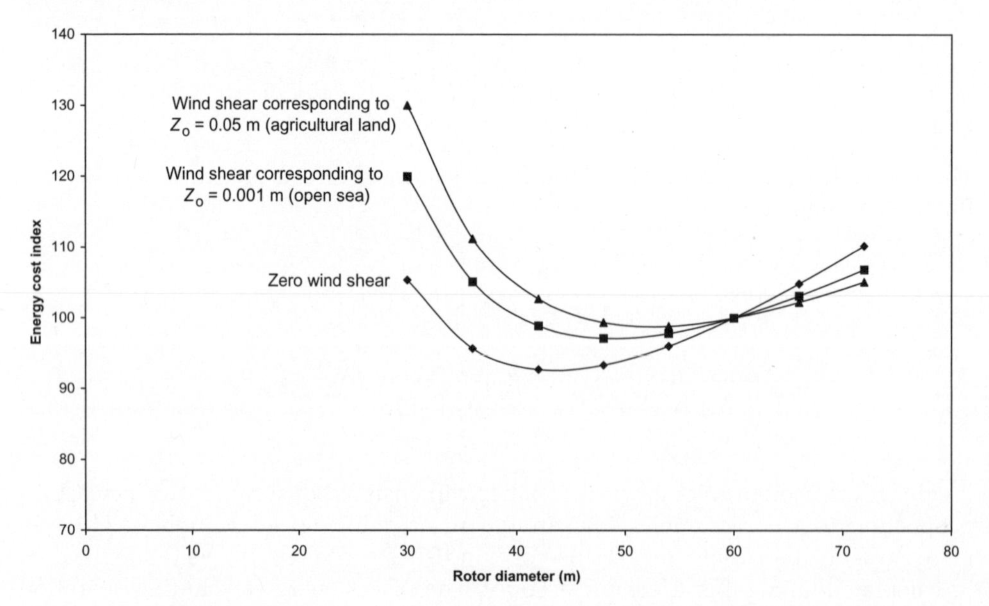

Figure 6.1 Variation of Optimum Turbine Size with Wind Shear (Assuming Constant Hub Height to Diameter Ratio)

6.3 Machine Rating

The machine rating determines the wind speed (known as rated wind speed) at which rated power is reached. If the rating is too high, the rated power will only be reached rarely, so the cost of the drive train and generator will not be justified by the energy yield. On the other hand, if the rating is reduced below the optimum then the cost of the rotor and its supporting structure will be excessive in relation to energy yield.

The investigation of the optimum relationship between rotor diameter and rated power can be carried out with the help of the cost modelling technique described in the previous section.

6.3.1 Simplified cost model for optimizing machine rating in relation to diameter

Assuming that the blade planform and twist distribution are fixed, the annual energy yield can be calculated for a number of rated wind speeds, for a given annual mean wind speed and Weibull shape factor. The turbine rotational speed is assumed to vary in proportion to the rated wind speed for simplicity. The aim of the optimization is to obtain the minimum cost of energy, which requires knowledge of how the costs of the various turbine components would be affected by the rating change. Although, in theory, this could only be rigorously derived from carrying out a series of detailed turbine designs, in practice it is possible to obtain a useful indication of cost trends by identifying the parameters driving the design of each component category and investigating their dependence on the rated wind speed. If the cost split between various components is known for a baseline machine, these cost trends can then be applied to it in order to determine the optimum rating. In this case the cost shares given in Table 6.1 for a 60 m diameter, 1.5 MW machine are used once again. The machine is assumed to be pitch regulated.

The parameters determining the design of the major components are set out first.

(1) *Blade weight*: the following assumptions are made:

- the blade planform is constant;
- the blade design is governed by out-of-plane bending moments in fatigue;
- the out-of-plane bending moment fluctuations are proportional to the product of the wind speed fluctuation and the rotational speed (see Equation (5.25) in Section 5.7.5);
- the rotational speed is proportional to rated wind speed as already stated;
- the blade skin thickness is proportional to the out-of-plane bending moment ranges.

Hence the blade skin thickness and therefore the blade weight are proportional to the rated wind speed.

(2) *Hub weight:* it is assumed that this is proportional to the blade out-of-plane bending moments in fatigue, as in the case of the blade itself.

(3) *Low speed shaft weight:* this is assumed to be governed by the shaft bending moment due to the cantilevered rotor and hub weights, which are taken as proportional to rated wind speed

(4) *Gearbox and brake:* gearbox and brake design are governed by the rated torque, P/Ω. The rated power is proportional to the cube of the rated wind speed, and the rotational speed is proportional to the rated wind speed, so the torque varies as the rated wind speed squared. The weights of the gearbox and brake are therefore taken to be proportional to the rated speed squared.

(5) *Generator:* generator design is governed by rated power. The weight of the generator is therefore assumed to be proportional to the cube of the rated wind speed.

(6) *Nacelle structure and yaw system:* it is assumed that the design of these are governed by the fluctuating moment on the nacelle due to differential blade out-of-plane root bending moments, which depend on blade out-of-plane bending moment fluctuations. The weights are therefore taken to be proportional to rated wind speed.

(7) *Tower weight:* the tower design may be governed either by fatigue loading during operation or by extreme loads with the turbine shutdown, so both possibilities will be considered. In the latter case the tower loading will be independent of rated wind speed, while in the former it will be mainly governed by rotor thrust fluctuations, which are assumed to be proportional to rated wind speed.

(8) *Foundation:* the foundation design is governed by extreme loads rather than by fatigue, so it is independent of rated speed.

(9) *Grid connection:* the weight of cables, switchgear and transformers are assumed to be proportional to rated power, and hence proportional to rated speed cubed.

(10) *Controller, assembly and transport:* these are assumed to be fixed.

The various components listed above are classified into different categories according to the way in which their weights vary with rated wind speed in Table 6.2. Also tabulated are the component costs as a percentage of the total for the baseline machine, together with the sum for each category.

As noted in Section 6.2.1, the relationship between the cost of a component and its mass can be approximated by a linear relationship of the form of Equation (6.1). As before, μ is assumed to take a value of 0.9 for all components. When the cost functions for all the components are added together, the following expression is

Table 6.2 Percentage Contribution of Different Components to Machine Cost from Table 6.1, Classified According to the Power Law Assumed to Define the Relationship Between the Component Mass and the Machine Rated Wind Speed

Components for which the weight/cost is independent of rated wind speed		Components for which the weight varies as rated wind speed		Components for which the weight varies as rated wind speed squared		Components for which the weight varies as rated wind speed cubed	
Component	Cost	Component	Cost	Component	Cost	Component	Cost
Foundation	4.2%	Blades	18.3%	Gearbox	12.5%	Generator	7.5%
Controller	4.2%	Hub	2.5%	Brake system	1.7%	Grid connection	8.3%
Assembly	2.1%	Main shaft	4.2%				
Transport	2.0%	Nacelle	10.8%				
		Yaw system	4.2%				
		Tower	17.5%				
Total	12.5%	Total	57.5%	Total	14.2%	Total	15.8%

obtained for machine cost as a function of the ratio of the rated wind speed to that of the baseline machine, $U_R U_{RB}$:

$$C_T = C_{TB}(0.125 + 0.575\{0.1 + 0.9(U_R/U_{RB})\} + 0.142\{0.1 + 0.9(U_R/U_{RB})^2\}$$
$$+ 0.158\{0.1 + 0.9(U_R/U_{RB})^3\})$$
$$= C_{TB}(0.2125 + 0.5175(U_R/U_{RB}) + 0.1278(U_R/U_{RB})^2 + 0.1422(U_R/U_{RB})^3) \quad (6.5)$$

A measure of the cost of energy is obtained by dividing the machine cost from Equation (6.1) by the annual energy yield, which is calculated for each rated wind speed by combining the corresponding power curve with the Weibull distribution of wind speeds. This exercise has been carried out for the 60 m diameter, 1.5 MW pitch-regulated baseline machine, assuming an annual mean wind speed of 7 m/s, and taking the rated wind speed of the baseline machine as 14.15 m/s. It is found that the optimum rated wind speed is 12.4 m/s or 1.77 times the annual mean, giving an optimum power rating of 1010 kW. The variation in cost of energy with rated power on either side of the optimum is very small, as can be seen from Figure 6.2, with a departure of 200 kW from the optimum producing an energy cost increase of only about 1 percent. If μ is assumed to take the lower value of 0.8, the optimum rated wind speed rises to 12.7 m/s, giving an optimum power rating of 1135 kW.

It was noted above that if the tower design is governed by extreme winds when the turbine is shut-down, then its cost is a fixed element in the total. The machine cost formula above then becomes:

$$C_T = C_{TB}(0.37 + 0.36(U_R/U_{RB}) + 0.1278(U_R/U_{RB})^2 + 0.1422(U_R/U_{RB})^3) \quad (6.6)$$

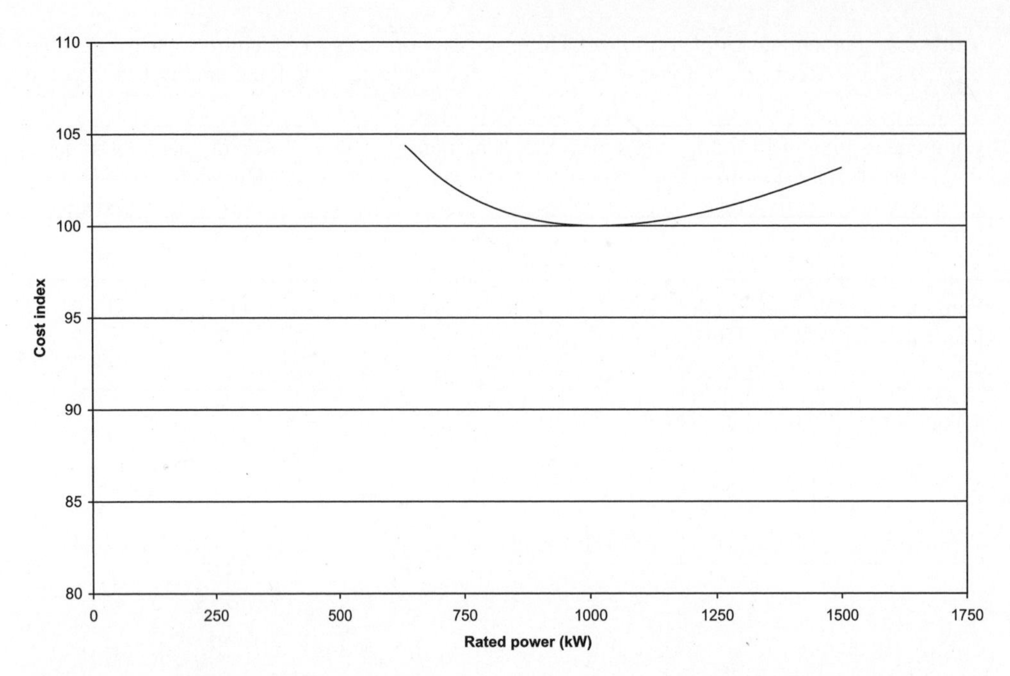

Figure 6.2 Variation in Cost of Energy with Rated Power for 60 m Diameter, Pitch-regulated Machine and Annual Wind Speed of 7 m/s, Based on Cost Model Defined by Equation (6.5)

which results in an increase in optimum rated power from 1012 kW to 1170 kW for an annual mean wind speed of 7 m/s for the pitch-regulated machine considered above.

6.3.2 Relationship between optimum rated wind speed and annual mean

The optimum power rating is of course heavily dependant on the annual mean wind speed, U_{ave}. The optimum rated wind speed, U_{Ro} for the above 60 m diameter pitch-regulated machine is given for a range of annual mean wind speeds in Table 6.3, from which it is apparent that the ratio U_{Ro}/U_{ave} is approximately constant.

A similar exercise can be carried out to determine the optimum rated power of a stall-regulated machine, and would yield similar results. However, because stall-regulated machines reach rated power at a substantially higher wind speed than pitch-regulated machines of the same rating (see Figure 6.4), the U_{Ro}/U_{ave} ratio for stall-regulated machines is typically over 2.

6.3.3 Specific power of production machines

It is instructive to investigate the relationship between rated power and swept area for production machines, and these quantities are plotted against each other in Figure 6.3 for 75 machines in production in 1996. Although different machines will

Table 6.3 Variation of Optimum Rated Wind Speed with Annual Mean for Pitch-regulated Machines

Annual mean wind speed, U_{ave}(m/s)	Optimum rated wind speed, U_{Ro}(m/s)	Ratio U_{Ro}/U_{ave}	Optimum rated power (kW)	Specific power, defined as rated power per unit swept area (kW/m^2)	Cost index, with cost of energy for a.m.w.s. of 7.5 m/s taken as100
7	12.4	1.77	1012	358	114
7.5	13.1	1.74	1187	420	100
8	13.7	1.72	1376	487	89
8.5	14.4	1.69	1579	558	80
9	15.0	1.67	1797	635	72

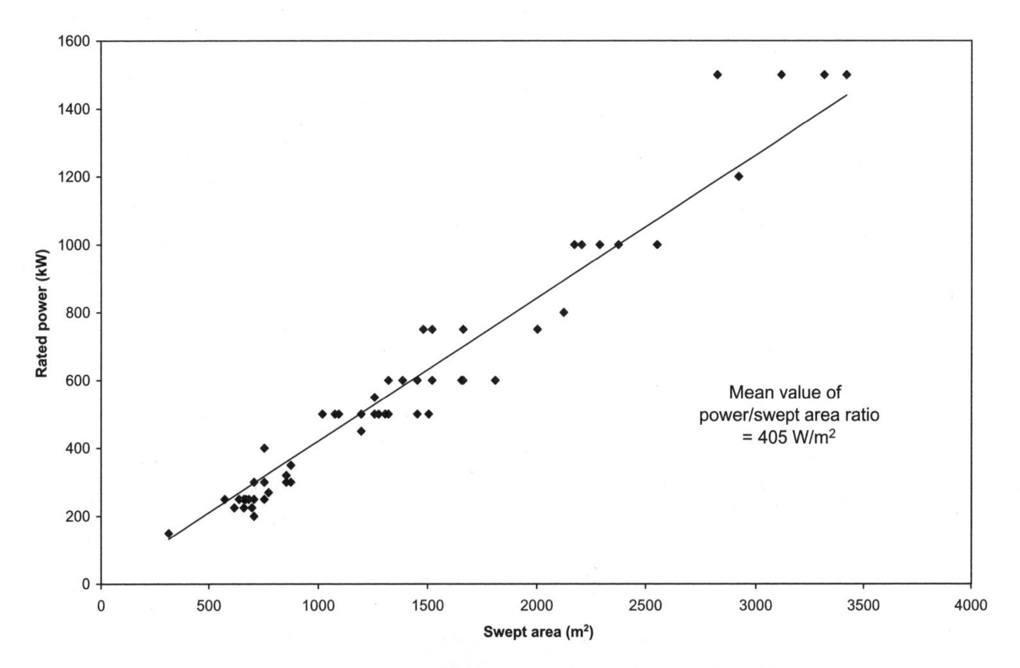

Figure 6.3 Rated Power *versus* Swept Area for Turbines in Production in 1997

have been designed for different annual mean wind speeds, the degree of scatter is not large, and a clear trend is apparent, with the line of best fit being close to a straight line passing through the origin. The mean specific power, defined as rated power divided by swept area, is 405 W/m^2 for the 75 machines – close to the optimum value in Table 6.3 for an annual wind speed of 7.5 m/s.

6.4 Rotational Speed

The aim of the wind turbine designer is the production of energy at minimum cost, subject to constraints imposed by environmental impact considerations. However, blade designs optimized for a number of different rotational speeds but the same rated power produce substantially the same energy yield, so the choice of rotational speed is based on machine cost rather than energy yield.

One of the key cost drivers is the rotor torque at rated power, as this is the main determinant of the drive train cost. For a given tip radius and machine rating, the rotor torque is inversely proportional to rotational speed, which argues for the adoption of a high rotational speed. However increasing the rotational speed has adverse effects on the rotor design, which are explored in the following sections.

6.4.1 Ideal relationship between rotational speed and solidity

Equation (3.67a) in Section 3.7.2 gives the chord distribution of a blade optimized to give maximum power at a particular tip speed ratio in terms of the lift coefficient, ignoring drag and tip loss:

$$\sigma_r \lambda C_l = \frac{8/9}{\sqrt{(1-\frac{1}{3})^2 + \lambda^2\mu^2\left[1 + \frac{2}{9\lambda^2\mu^2}\right]^2}} \tag{3.67a}$$

where λ is the tip speed ratio, σ_r is the solidity and $\mu = r/R$. Over the outboard half of the blade, which produces the bulk of the power, the local speed ratio, $\lambda\mu$, will normally be large enough to enable the denominator to be approximated as $\lambda\mu$, giving:

$$\sigma_r \lambda C_l = \frac{Nc(\mu)}{2\pi R}\lambda C_l = \frac{8}{9\lambda\mu} \tag{6.7}$$

where N is the number of blades. After rearrangement, this gives

$$c(\mu)\left(\frac{\Omega R}{U_\infty}\right)^2 = \frac{16\pi R}{9 C_l N}\frac{1}{\mu} \tag{6.8}$$

Hence it can be seen that, for a family of designs optimized for different rotational speeds at the same wind speed, the blade chord at a particular radius is inversely proportional to the square of the rotational speed, assuming that N and R are fixed and the lift coefficient is maintained at a constant value by altering the local blade pitch to maintain a constant angle of attack.

Note that Equation (6.8) does not apply if energy yield is optimized over the full range of operating wind speeds for a pitch-regulated machine. In this case, it has been demonstrated that the blade chord at a particular radius is approximately inversely proportional to rotational speed rather than to the square of it (Jamieson and Brown, 1992).

6.4.2 Influence of rotational speed on blade weight

The effect of rotational speed on blade weight can be explored with reference to the family of blade designs just described. As in Section 6.3.1, it is assumed that the blade design is governed by out-of-plane bending moments in fatigue and that the moment fluctuations are proportional to the product of the wind speed fluctuation, the rotational speed and the chord scaling factor (see Equation (5.25) in Section 5.7.5). By Equation (6.8) the chord scaling factor is inversely proportional to the square of the rotational speed, so the moment fluctuations simply vary inversely as the rotational speed.

The thickness to chord ratios at each radius are assumed to be unaffected by the chord scaling, so the blade section modulus for out-of-plane bending at a given radius is proportional to the product of the blade shell skin thickness, $w(r)$, and the square of the local chord. Thus

$$Z(r) \propto w(r)(c(r))^2 \propto w(r)/\Omega^4 \qquad (6.9)$$

In order to maintain the fatigue stress ranges at the same level, we require the blade section modulus, $Z(r)$, to vary as the moment fluctuations, which, as shown above, vary inversely as rotational speed. Thus

$$Z(r) \propto 1/\Omega \text{ so } w(r)/\Omega^4 \propto 1/\Omega \text{ and } w(r) \propto \Omega^3 \qquad (6.10)$$

Blade weight is proportional to the skin thickness times chord, and thus varies as rotational speed.

6.4.3 Optimum rotational speed

On the basis of the assumptions of Section 6.4.2 (which will by no means always apply), blade weight increases in proportion to rotational speed. However, the blade out-of-plane fatigue loads, which may govern the design of the nacelle structure and tower, vary *inversely* as the rotational speed. It is therefore likely that, as rotational speed is increased, there will be a trade-off between reducing costs of the drive train, nacelle structure and tower on the one hand and increasing rotor cost on the other, which will determine the optimum value.

6.4.4 Noise constraint on rotational speed

The aerodynamic noise generated by a wind turbine is approximately proportional to the fifth power of the tip speed. It is therefore highly desirable to restrict turbine rotational speed, especially when the wind speed – and therefore ambient noise levels – are low. Consequently manufacturers of turbines to be deployed at normal sites on land generally limit the tip speed to about 65 m/s. Experience suggests that this results in wind turbine noise levels on a par with ambient levels at a distance of 400 m, which is the normal minimum spacing between turbines and habitations.

6.4.5 Visual considerations

There is a consensus that turbines are more disturbing to look at the faster they rotate.

6.5 Number of Blades

6.5.1 Overview

European windmills traditionally had four sails, perhaps because pre-industrial techniques for attaching the sail stocks to the shaft lent themselves to a cruciform arrangement in which the stocks for opposite sails formed a continuous wooden beam. By contrast the vast majority of horizontal axis wind turbines manufactured today have either two or three-blades, although at least one manufacturer used to specialize in one-bladed machines. As the latter are relatively unusual, consideration of them will be restricted to Section 6.5.7, and the rest of Section 6.5 will concentrate on two- and three-bladed machines.

In comparing the relative merits of machines with differing numbers of blades, the following factors need to be considered:

- performance,
- loads,
- cost of rotor,
- impact on drive train cost,
- noise emission,
- visual appearance.

Some of these factors are strongly influenced by rotational speed and rotor solidity, and the ideal relationship between these parameters and the number of blades is considered in the next section. Section 6.5.3 investigates alternative two-bladed derivatives of a realistic three-bladed baseline design and compares their relative energy yields and notional costs. Section 6.5.4 reviews the differences in loading imposed by two- and three-bladed rotors on the supporting structure, and Section 6.5.5 considers the constraint on rotational speed imposed by noise emission. Visual appearance is considered briefly in Section 6.5.6.

6.5.2 Ideal relationship between number of blades, rotational speed and solidity

The effect of the number of blades on the blade chord and rotational speed of a machine optimized for a particular wind speed is given by Equation (6.8):

$$Nc(\mu)\left(\frac{\Omega R}{U_\infty}\right)^2 = \frac{16\pi R}{9C_l}\frac{1}{\mu}$$

Hence it can be seen that, if the number of blades is reduced from three to two, increasing the chord by 50 percent or the rotational speed by 22.5 percent are two of the options for preserving optimized operation at the selected wind speed. (It is assumed that the lift coefficient is maintained at a constant value by altering the local blade pitch to maintain a constant angle of attack.)

6.5.3 Some performance and cost comparisons

Clear-cut comparisons between two- and three-bladed machines are notoriously difficult because of the impossibility of establishing equivalent designs. Conceptually, the simplest option is to increase the chord by 50 percent at all radii and leave everything else – including rotational speed – unchanged. In the absence of tip loss, the induction factors, and hence the annual energy yield, remain the same, but when tip loss is included, the annual energy yield drops by about 3 percent. However, retention of the same rotor solidity largely negates one of the main benefits of reducing the number of blades, namely reduction in rotor cost, and so this option will not be pursued further. Instead it is proposed to take a realistic blade design for a three-bladed machine and look at the performance and cost implications of using the same blade on a two-bladed machine rotating at different speeds.

Performance comparisons are affected both by the power rating in relation to swept area (Section 6.3) and by the aerofoil data used. In this case a 40 m diameter stall-regulated three-bladed turbine with TR blades (see Example 5.1 in Section 5.6.3) operating at 30 r.p.m. is adopted as the baseline machine, and a power rating of 500 kW. is chosen, so that the specific power (398 W/m^2) is close to the norm. Empirical three-dimensional aerofoil data for a LM 19.0 blade is used (see Figure 5.9), with maximum lift coefficient increasing from blade tip to blade root, as this results in more accurate power curve predictions. The data are taken from Petersen *et al.* (1998). The blade twist distribution is set to give maximum energy yield at a site where the annual mean wind speed is 7 m/s, while limiting the maximum power to 500 kW. The design is thus somewhat different from the ideal design considered in the preceding section, which was optimized for a particular *wind speed* (see Figure 6.4 for the predicted power curve).

Two options for a corresponding 40 m diameter stall-regulated two-bladed design at a site with the same annual mean wind speed are examined and the notional energy costs compared with that for the baseline three-bladed machine. The costs of the two-bladed design options in relation to the baseline three-bladed machine are considered with reference to changes in the cost of the components, using the cost shares given in Table 6.1 and the methodology of Section 6.3.1.

As before, the blade weight is assumed to increase linearly with rotational speed, but the cost element for the blades at the baseline rotational speed is reduced by one third. The weights of the hub, shaft, nacelle and yaw system are also assumed

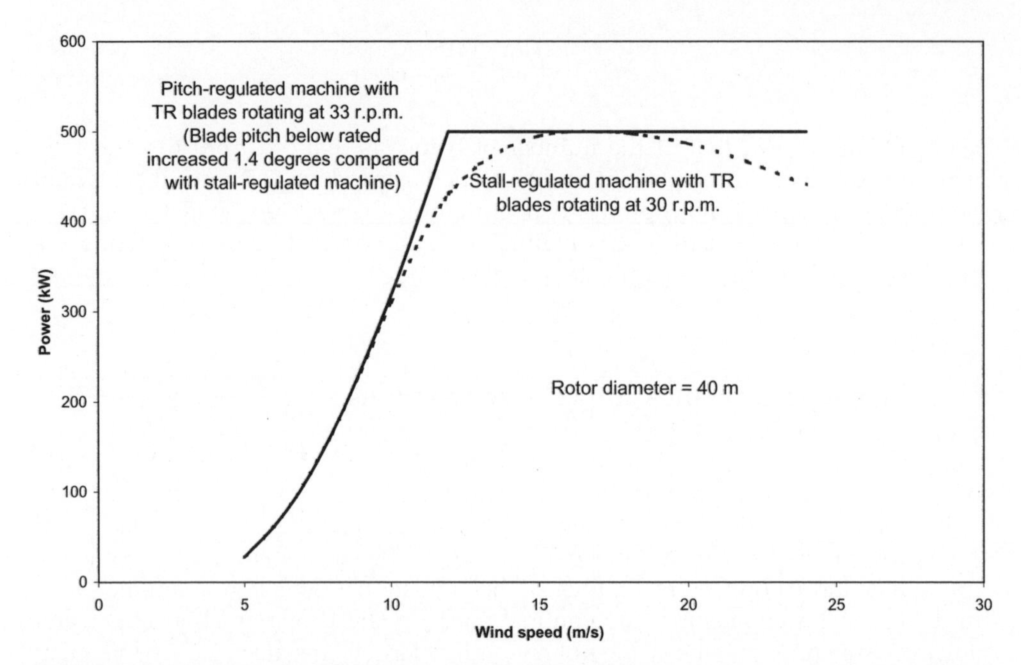

Figure 6.4 Comparison of Power Curves for 500 kW Stall-regulated and Pitch-regulated Machines with the Same Planform and Twist Distribution

to increase with rotational speed, but no account is taken of the increased loads on these components for a fixed-hub, two-bladed machine. Tower design is assumed to be governed by fatigue in the first instance, so tower weight is taken as proportional to rotational speed. The cyclic thrust loads on the rotor due to turbulence are virtually the same for two- and three-bladed machines rotating at the same speeds if the blade planforms are the same, so the tower cost element at the baseline rotational speed is left unchanged.

The weights of the gearbox and brake are taken to be proportional to the rated torque, P_R/Ω, while those of the generator and of the cables and equipment forming the grid connection are taken as proportional to rated power, P_R. The foundation cost element is reduced by a quarter for the two-bladed machine in recognition of the reduced extreme tower base overturning moment.

The various components are classified into different categories according to the way in which their weights vary with rotational speed and rated power in Table 6.4. Also tabulated are the two-bladed machine component costs as a percentage of the total for the baseline three-bladed machine, together with the sum for each category.

Adopting Equation (6.1) with $\mu = 0.9$ once more for the relationship between the cost of a component and its mass, the following expression is obtained for machine cost as a function of rotational speed and rated power:

$$C_T = C_{TB}(0.114 + 0.514\{0.1 + 0.9(\Omega/\Omega_B)\} + 0.142\{0.1 + 0.9(P_R/P_{RB})(\Omega_B/\Omega)\} + 0.158\{0.1 + 0.9(P_R/P_{RB})\}) \quad (6.11)$$

$$= C_{TB}(0.1954 + 0.4626(\Omega/\Omega_B) + 0.1278(P_R/P_{RB})(\Omega_B/\Omega_{RB}) + 0.1422(P_R/P_{RB}))$$

Table 6.4 Contribution of Different Components to the Cost of a Two-bladed Machine (Expressed as Percentages of Three-bladed Baseline Machine Cost) and Classified According to the Relationship Assumed Between the Component Mass and Rotational Speed/rated Torque/Rated Power

Components for which the weight/cost is independent of rated power or rotational speed		Components for which the weight varies as rotational speed, Ω		Components for which the weight varies as rated torque, P_R/Ω		Components for which the weight varies as rated power, P_R	
Component	Cost	Component	Cost	Component	Cost	Component	Cost
Foundation	3.1%	Blades	12.2%	Gearbox	12.5%	Generator	7.5%
Controller	4.2%	Hub	2.5%	Brake system	1.7%	Grid connection	8.3%
Assembly	2.1%	Main shaft	4.2%				
Transport	2.0%	Nacelle	10.8%				
		Yaw system	4.2%				
		Tower	17.5%				
Total	11.4%	Total	51.4%	Total	14.2%	Total	15.8%

Here, P_{RB} and Ω_B are the baseline values of rated power and rotational speed, 500 kW and 30 r.p.m. respectively. The two design options can now be examined.

(a) *Planform and rotational speed unchanged from baseline*
The maximum output power drops by almost exactly one third due to the reduction in the number of blades, but the reduction in energy yield is less severe at 19 percent. This is because, although the coefficient of performance (C_P) for the two-bladed machine is very nearly two thirds that of the three-bladed machine at the low tip speed ratio ($62.8/16 = 3.9$) corresponding to peak power output, the maximum value of C_P is almost as large as that of the three-bladed machine (see Figure 6.5, which compares the $C_P - \lambda$ curves).
The reduced number of blades and reduced rated power lead to an overall cost reduction of 16 percent (made up of 6 percent on the blades, 1 percent on the foundation and 9 percent on the gearbox, brake, generator and grid connection), leading to an increase in energy cost of 4 percent compared with the baseline three-bladed machine.

(b) *Chord distribution and diameter unchanged, but rotational speed increased by 18 percent and blade pitch adjusted to give 500 kW power rating*
In this two-bladed design variant, the rotational speed and blade pitch are chosen to maximize energy yield while restricting the rated power to that of the baseline design. The resultant annual energy yield is 4 percent less than for the three-bladed machine.
The option of increasing rotational speed is an attractive one as far as the drive train is concerned because, for a given machine rating, it results in a reduction in drive train torque and hence in gearbox cost. In this case, the

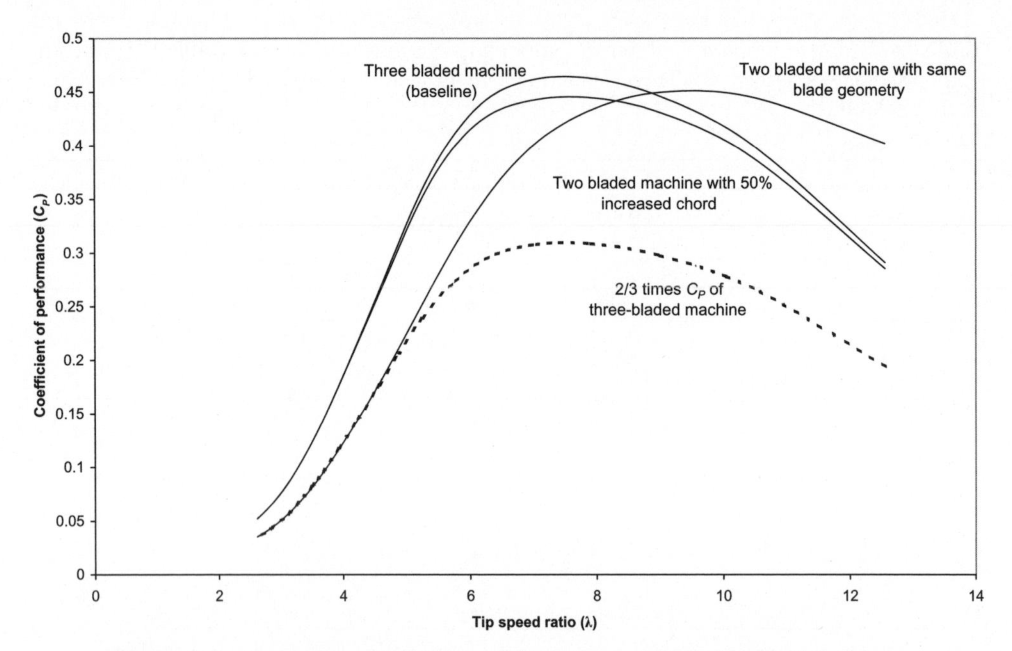

Figure 6.5 Comparison of C_P Curves for Two- and Three-bladed Machines

reduction in gearbox cost due to an 18 percent increase in rotational speed would yield a 2 percent reduction in total machine cost.

As the blade skin thickness is assumed to increase in proportion to rotational speed, the saving associated with eliminating one blade will be offset by an 18 percent increase in the weight of the remaining two, resulting in a 21 percent reduction in rotor cost, and a 4 percent reduction in overall cost. The cost savings on the blades, drive train and foundations are offset by cost increases on the hub, shaft, nacelle, yaw drive and tower due to increased rotational speed, resulting in an overall cost saving of only 1 percent. Hence the energy cost is 3 percent higher than for the baseline machine.

It is apparent that, with the tower design assumed dependent on fatigue loads, the two-bladed variants (a) and (b) considered above result in a small increase in the cost of energy relative to the three-bladed machine. However, if the tower design is governed by extreme loads rather than fatigue loads, the situation is reversed; see Table 6.5, in which it is assumed that the reduction of extreme load due to the reduced number of blades results in a 20 percent reduction in tower cost.

The results shown in Table 6.5 indicate that two-bladed, rigid-hub machines are unlikely to yield significant cost benefits *vis-à-vis* three-bladed machines, even if the tower design is determined by extreme loading. However, the results should be treated with caution, because the cost changes are based on a simplistic treatment of blade loadings, and of their knock-on effects on other components. (Loads on rigid-hub two-bladed machines are compared with those on three-bladed machines in more detail in the next section.)

The loadings on the nacelle of a two-bladed machine can be reduced significantly by the introduction of a teeter hinge between the rotor and the low-speed shaft,

Table 6.5 Comparison of Two-bladed Design Variants Utilizing the Same Blades with 40 m Diameter, 500 kW Three-bladed Baseline Machine

Variant	Rotational speed (r.p.m.)	Rated power (kW)	Annual energy yield (MWh)	Reduction in annual energy yield compared with baseline machine	Tower cost governed by	Reduction in overall machine cost	Increase/ reduction in cost of energy
(a)	30	331	1054	19%	Fatigue loading	16%	+4%
					Extreme loading	20%	−1%
(b)	35.4	500	1256	4%	Fatigue loading	1%	+3%
					Extreme loading	7%	−3%

with consequent potential cost benefits. The hinge eliminates the transfer of out-of-plane aerodynamic moments from the rotor to the low-speed shaft, resulting in large reductions in the operational loadings on the shaft, nacelle and yaw drive. The dependence of these loads on rotational speed is also largely removed, with the result that the optimum rotational speed for a two-bladed machine in energy cost terms is increased, approaching the value giving maximum energy yield.

Although teetering provides scope for significant cost savings on the shaft, nacelle and yaw drive (which account for nearly 20 percent of the baseline machine cost), these savings are offset by the additional costs associated with the teeter hinge and teeter restraint system.

6.5.4 Effect of number of blades on loads

Moment loadings on the low speed shaft and nacelle structure from three-bladed and rigid-hub two-bladed machines were examined in Sections 5.10 and 5.11, and are compared in Table 6.6 below for machines of the same diameter and rotational speed. The stochastic loading comparison is based on a turbulence length scale to rotor diameter ratio of 1.84.

It is seen that loadings from a rigid-hub two-bladed rotor are significantly larger than from a three-bladed rotor. However, in most two-bladed machine designs, the rotor is allowed to teeter instead of being rigidly mounted, with the result that aerodynamic moments on the shaft and nacelle structure quoted in Table 6.6 are eliminated, and the blade out-of-plane root bending moments are reduced. The benefits and drawbacks of teetering the rotor are examined in Section 6.6.

The rotor thrust variations at blade passing frequency due to stochastic loading, which are a dominant factor in tower fatigue design, are very similar for two- and three-bladed machines rotating at the same speed. However, two-bladed machines usually rotate faster than three-bladed machines of the same diameter, so the cyclic rotor thrust variations are higher.

Table 6.6 Comparison of Loads on Shaft and Nacelle for Three-bladed and Rigid-hub Two-bladed Machines

	Deterministic loading arising from wind shear and/or yaw misalignment, in terms of blade root out-of-plane bending moment amplitude, M_o		Stochastic loading
Location of moment loading	Three-bladed machine	Rigid-hub two-bladed machine	% increase for rigid-hub two-bladed machine compared with three-bladed machine
Shaft bending moment amplitude	1.5 M_o	2 M_o	22%
Nacelle nodding moment	1.5 M_o	$M_o(1 + \cos 2\psi)$	22%
Nacelle yaw moment	Zero	$M_o \sin 2\psi$	22%

6.5.5 Noise constraint on rotational speed

As noted in Section 6.5.3, there may be significant cost benefits to be gained from a two-bladed design with increased rotational speeds, because, in addition to the blade saving, the cost of the whole of the drive train is reduced because of the reduced torque. However, as noted in Section 6.4.2, it is normal to restrict tip speed to about 65 m/s in order to limit aerodynamic noise emission. At 62.8 m/s, the tip speed of the baseline machine discussed in Section 6.5.3 is within this limit, but the tip speed of option (b) of 74 m/s would be less likely to be acceptable, except at remote sites or offshore. This subject is considered further in Section 6.9.

6.5.6 Visual appearance

Although the assessment of visual appearance is essentially subjective, there is an emerging consensus that three-bladed machines are more restful to look at than two-bladed ones. One possible reason for this is that the apparent 'bulk' of a three-bladed machine changes only slightly over time, whereas a two-bladed machine appears to contract down to a one-dimensional line element, when the rotor is vertical, twice per revolution. A secondary factor is that two-bladed machines generally rotate faster, which an observer can also find more disturbing.

6.5.7 Single-bladed turbines

Apart from the saving in rotor cost itself, the single-bladed turbine concept is an attractive one because of the reduction in drive train cost realizable through increased rotational speed (Section 6.5.2). An obvious disadvantage is the resulting increased noise emission resulting from the faster rotation, but this would not be an issue offshore. Another consideration is the reduced yield due to increased tip loss. For example, a 40 m diameter machine consisting of a TR blade rotating at 48 r.p.m., with twist distribution reoptimized to give maximum energy yield, will achieve the same maximum power output as the baseline design, but provide 12 percent less energy.

The single blade must be counterweighted to eliminate torque fluctuations and any whirling tendency due to centrifugal loads. Furthermore, as a rigid hub would expose the nacelle to very large nodding and yawing moments in comparison with two- or three-bladed machines, it is customary to mount the rotor on a teeter hinge, so that the unbalanced aerodynamic out-of-plane moment can be resisted by a centrifugal couple, thereby reducing the hub moment. However, the teeter motion of the blade is significantly greater than that of a two-bladed machine, so it is normal to mount the rotor downwind. Morgan (1994) reports that particular difficulties have been encountered in predicting teeter excursions after grid loss and emergency stops, leading to excessive risk of teeter stop impacts.

6.6 Teetering

6.6.1 Load relief benefits

Two-bladed rotors are often mounted on a teeter hinge – with hinge axis perpendicular to the shaft axis, but not necessarily perpendicular to the longitudinal axis of the blades – in order to prevent differential blade root out-of-plane bending moments arising during operation. Instead, differential aerodynamic loads on the two-blades result in rotor angular acceleration about the teeter axis, with large teeter excursions being prevented by the restoring moment generated by centrifugal forces, as described in Section 5.8.8. However, when the machine is shut-down, the centrifugal restoring moment is absent, so differential blade loading will cause the rotor to teeter until it reaches the teeter end stops which need to be suitably buffered. Consequently the teeter hinge is unlikely to provide any amelioration of extreme blade root out-of-plane moments when the machine is shut-down.

The load relief afforded by the teeter hinge benefits the main structural elements in the load path to the ground in varying degrees, as outlined below:

(a) *Blade*. The main benefit is the elimination of the cyclic variations in out-of-plane bending moment due to yaw (Figure 5.10), shaft tilt, wind shear (Figure 5.11) and tower shadow (Figure 5.14). By contrast, there is only a small reduction in blade root out-of-plane bending moment due to stochastic loadings – see the example in Section 5.8.8, where an 11 percent reduction is quoted. Thus, teetering results in a large overall reduction in out-of-plane fatigue loading, although the significance of this will be tempered by the influence of the unaltered edgewise gravity moment.

(b) *Low-speed shaft*. Low-speed shaft design is governed by fatigue loading, which is normally dominated by the cyclic gravity moment due to the cantilevered rotor mass. On a rigid hub machine, the shaft moment 'Damage Equivalent Load' or DEL (defined in Section 5.12.6) due to deterministic and stochastic rotor out-of-plane loadings combined can be of similar magnitude, so the insertion of a teeter hinge can produce a substantial reduction in overall shaft moment DEL. It should be noted, however, that the cyclic shaft moment due to wind shear *relieves* that due to gravity on a rigid hub machine, so teetering is not beneficial in respect of this load component.

A rough estimate of the overall shaft moment DEL on a rigid-hub machine, excluding yaw error and tower shadow effects, can be obtained by taking the square root of the sum of the squares of the shaft moment DEL due to stochastic loads and that due to the combined cyclic loads due to gravity, wind shear and shaft tilt.

(c) *Nacelle structure*. The provision of a teeter hinge should eliminate nodding and yawing moments on the nacelle completely during operation, leaving only rotor torque, thrust and in-plane loadings. This will benefit the fatigue design of the nacelle structure considerably, but not the extreme load design, for the reasons already explained.

(d) *Yaw bearing and yaw drive*. Rigid-hub machines experience severe yaw moments due to both deterministic and stochastic loads, which were underestimated on many early designs. The introduction of a teeter hinge dramatically reduces yaw moments during operation by eliminating rotor out-of-plane moments on the hub, but yaw moments due to in-plane loads on the rotor still remain.

The relative magnitude of the yaw moments due to in-plane as opposed to out-of-plane loads on a rigid-hub rotor can be appreciated by comparing the effect of wind speed fluctuation, u, on the in-plane and out-of-plane loads on a blade element. Assuming that the blade is not stalled and that ϕ is small, the in-plane load per unit length is, from Equation (5.131a), given approximately by:

$$-F_Y = \left(\tfrac{1}{2}\rho\Omega\frac{dC_l}{d\alpha}\right)c(r)ru\left[\frac{C_l}{dC_l/d\alpha} + \sin\phi\right] \tag{6.12}$$

whereas the out-of-plane load per unit length is, from Equation (5.25) approximately

$$F_X = \left(\tfrac{1}{2}\rho\Omega\frac{dC_l}{d\alpha}\right)c(r)ru \tag{6.13}$$

Defining the distance between the hub centre and the tower centre-line as e, it is seen that the yaw moment due to the in-plane rotor load is

$$\frac{e}{r}\left[\frac{C_l}{dC_l/d\alpha} + \sin\phi\right]$$

times the yaw moment due to out-of-plane load. As e is typically about one tenth of the tip radius, it is seen that the yaw moments due to in-plane loads are at least an order of magnitude smaller than those due to out-of-plane moments, so that the introduction of the teeter hinge results in a very significant reduction.

(e) *Tower*. The fatigue loadings due to the M_Y moment and M_Z torque will clearly be significantly reduced at the top of the tower if the rotor is teetered, but the effect will be negligible towards the base where thrust loads dominate the moments.

6.6.2 Limitation of large excursions

Some limitation on teeter excursions has to be provided, if only to prevent collision between the blade and the tower. If the teeter hinge is located close to the axis of the blades, with the low-speed shaft passing through an aperture in the wall of the hub shell (see Figure 6.6), then the maximum teeter excursion is governed by the size of the aperture.

The teeter response to deterministic and stochastic loads is considered in Section 5.8.8. Although it is evident that a permitted teeter angle range of the order of $\pm 5°$ will accommodate the vast majority of teeter excursions during normal operation, it is usually impracticable to accommodate the largest that can occur. Hence, in order to minimize the occurrence of metal-to-metal impacts on the teeter end stops, buffers incorporating spring and/or damping elements normally have to be fitted. These also perform an important role in limiting the much larger teeter excursions that would otherwise arise during start-up and shut-down, when the centrifugal restoring moment is reduced.

6.6.3 Pitch–teeter coupling

As described in Section 5.8.8, the magnitude of teeter excursions can be reduced by coupling blade pitch to teeter angle, in order to generate an aerodynamic restoring moment proportional to the teeter angle. This can be done simply by setting the teeter hinge at an angle, known as the Delta 3 angle, to the perpendicular to the rotor axis. Alternatively, on pitch-controlled machines, pitch–teeter coupling can be introduced by actuating the blade pitch by the fore-aft motion of a rod passing through a hollow low-speed shaft (see Figure 6.6).

6.6.4 Teeter stability on stall-regulated machines

At first sight, it might be thought that the teeter motion of a stalled rotor would be unstable because of negative damping resulting from the negative slope of the C_l–α curve post-stall. However, two-dimensional aerodynamic theory is a poor predictor of post-stall behaviour, and it has proved possible to design teetered rotors that are stable in practice, such as the Gamma 60 (Falchetta *et al.*, 1996) and Nordic 1000

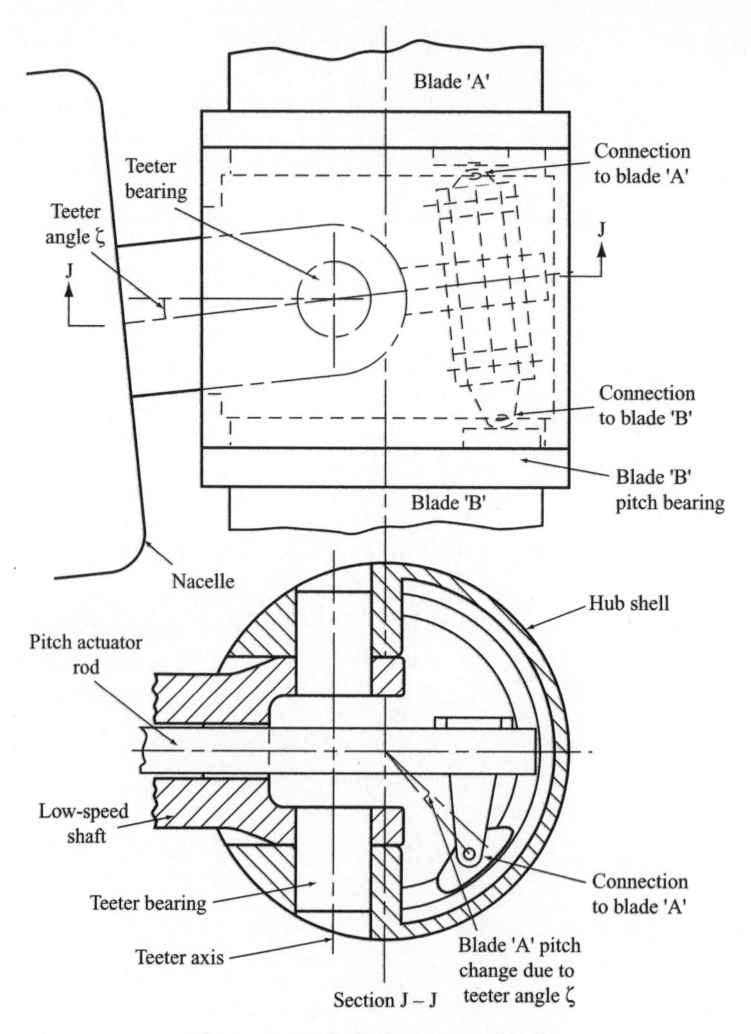

Figure 6.6 Pitch–Teeter Coupling

(Engstrom *et al.*, 1997). The concept is explored in detail in investigations by Armstrong and Hancock (1991) and Rawlinson-Smith (1994).

6.7 Power Control

6.7.1 Passive stall control

The simplest form of power control is passive stall control, which makes use of the post-stall reduction in lift coefficient and associated increase in drag coefficient to place a ceiling on output power as wind speed increases, without the need for any changes in blade geometry. The fixed-blade pitch is chosen so that the turbine reaches its maximum or rated power at the desired wind speed. Stall-regulated machines suffer from the disadvantage of uncertainties in aerodynamic behaviour post-stall which can result in inaccurate prediction of power levels and blade loadings at rated wind speed and above. These aspects are considered in greater detail in Section 4.2.2.

6.7.2 Active pitch control

Active pitch control achieves power limitation above rated wind speed by rotating all or part of each blade about its axis in the direction which reduces the angle of attack and hence the lift coefficient – a process known as blade feathering. The main benefits of active pitch control are increased energy capture, the aerodynamic braking facility it provides and the reduced extreme loads on the turbine when shut-down (see also Sections 4.2.5, 4.2.7 and 8.2.1).

The pitch change system has to act rapidly, i.e., to give pitch change rates of 5°/s or better in order to limit power excursions due to gusts enveloping the whole rotor to an acceptable value. However, it is not normally found practicable to smooth the cyclic power fluctuations at blade passing frequency due to blades successively slicing through a localized gust (Section 5.7.5) with the result that the large power swings of up to about 100 percent can sometimes occur.

The extra energy obtainable with pitch control is not all that large. A pitch-regulated machine with the same power rating as a stall-regulated machine, utilizing the same blades and rotating at the same speed will operate at a larger pitch angle below rated wind speed than the stall-regulated machine, in order to reduce the angle of attack and hence increase the power output at wind speeds approaching rated. If the 500 kW, 40 m diameter, 30 r.p.m. stall-regulated machine described in Section 6.5.3 is taken as baseline, a 500 kW, 30 r.p.m. pitch-regulated machine utilizing the same blades at optimum pitch would produce about 2 percent more energy. The optimum rotational speed is found to be about 33 r.p.m., which increases the energy gain to about 4 percent. The power curve of the 500 kW, 40 m diameter pitch controlled machine rotating at 33 r.p.m. is compared with the power curve of the corresponding stall-regulated machine utilizing the same blades, but rotating at 30 r.p.m. in Figure 6.4. Note that the knee in the power curve at rated speed will be more rounded in practice because the pitch control will not keep pace with the higher frequency components of turbulence.

Figure 6.7 shows a family of power curves for a range of positive pitch angles for the 500 kW, 40 m diameter pitch-controlled machine rotating at 33 r.p.m.. The intersections of these curves with the 500 kW abscissa define the relationship between steady wind speed and pitch angle required for power control. It is readily apparent from the power curve gradients at the intersection points that rapid changes of wind speed will result in large power swings when the mean wind speed is high.

The range of blade pitch angles required for power control is typically from 0° (often referred to as 'fine pitch'), at which the tip chord is in the plane of rotation or very close to it, and about 35°. However, for effective aerodynamic braking, the blades have to be pitched to 90° or full feather, when the tip chord is parallel to the rotor shaft with the leading edge into the wind.

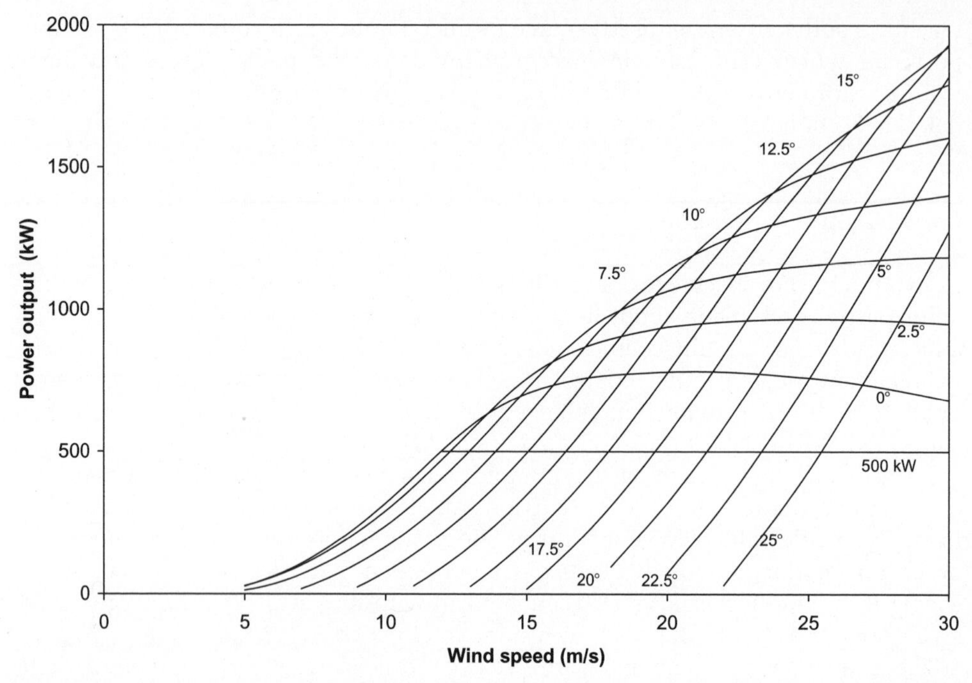

Figure 6.7 Power Curves for Different Pitch Angles: 40 m Diameter Rotor Rotating at 33 r.p.m.

A variety of pitch actuation systems have been adopted (see also Section 8.5). They are divided between those in which each blade has its own actuator and those in which a single actuator pitches all the blades. The former arrangement has the advantage that it provides two or three independent aerodynamic braking systems to control overspeed, and the disadvantage that it requires very precise control of pitch on each blade in order to avoid unacceptable pitch angle differences during normal operation. An advantage of the latter arrangement is that the pitch actuator, e.g. a hydraulic cylinder, can be located in the nacelle, producing fore-aft motion of the pitch linkages in the hub by means of a rod passing down the middle of a hollow low-speed shaft (see Figure 6.8). Alternatively, the axial position of the rod can be controlled by means of a ball-screw and ball-nut arrangement, in which the ball-nut is driven by a servomotor. Normally the ball-nut is driven at the same speed as the rotor, but when a change of pitch is required the ball-nut rotational speed is altered temporarily. This system is arranged to be fail-safe, so that should the servomotor or its control system fail, the servomotor is braked automatically and the ball-nut drives the blade pitch to feather.

Where hydraulic cylinders are used to pitch blades individually, they are mounted within the hub and each piston rod is usually connected directly to an attachment on the blade bearing (see Figure 6.9). The attachment point follows a circular path as the blade pitches, so the cylinder has to be allowed to pivot. The alternative solution of employing an electric motor to drive a pinion engaging with teeth on the inside of the blade bearing consequently appears rather neater (see Figure 6.10). Both systems require a hollow shaft to accommodate either hydraulic

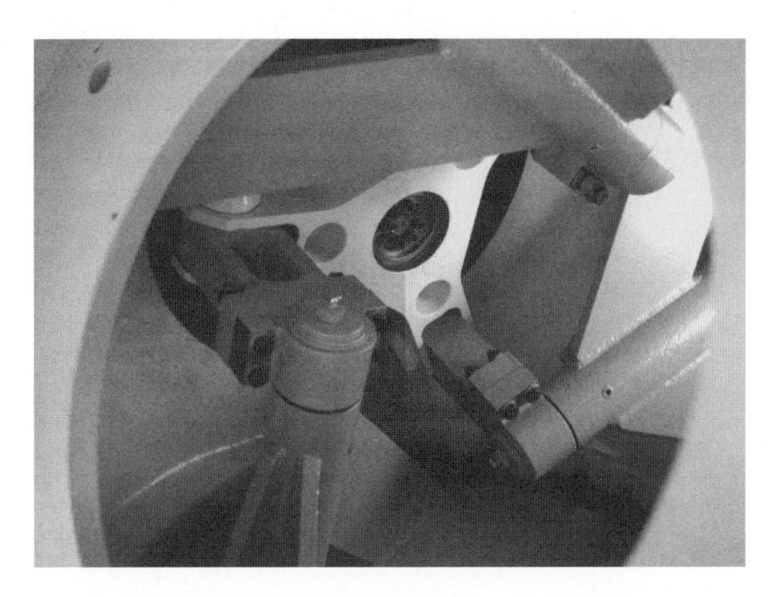

Figure 6.8 Pitch Linkage System Used in Conjuction with a Single Hydraulic Actuator Located in the Nacelle. (The central triangular 'spider' is connected to the actuator by a rod passing through the hollow low-speed shaft. Links from the spider drive the blade pitch *via* braced arms cantilevering into the hub from each blade. Each arm is parallel to its blade axis, but eccentric to it).

Figure 6.9 Blade Pitching System Using Separate Hydraulic Actuators for Each Blade. (Each actuator cylinder is supported on gimbal-type mountings bolted to the hub, and its piston applies a pitching torque to the blade via a cantilevered conical tube eccentric to the blade axis. The blade is attached to the outer ring of the pitch bearing).

Figure 6.10 Blade Pitching System Using a Separate Electric Motor for Each Blade. (A pinion, driven by the motor via a planetary gearbox, engages with gear teeth on the inside of the inner ring of the pitch bearing, to which the blade is bolted. The blade is not attached to the bearing in this photograph, so the fixing holes are visible).

hoses or power cables for pitch actuation together with signal cables for pitch angle sensing. In addition, appropriate slip rings are required at the rear end of the shaft.

Methods of providing back-up power supplies to ensure blade feathering in the event of grid-loss are considered in Section 8.5.

Although full-span pitch control is the option favoured by the overwhelming majority of manufacturers, power control can still be fully effective even if only the outer 15 percent of the blade is pitched. The principal benefits are that the duty of the pitch actuators is significantly reduced, and that the inboard portion of the blade remains in stall, significantly reducing the blade load fluctuations. On the other hand partial-span pitch control has several disadvantages as follows:

- the introduction of extra weight near the tip,
- the difficulty of physically accommodating the actuator within the blade profile,
- the high bending moments to be carried by the tip-blade shaft,
- the need to design the equipment for the high centrifugal loadings found at large radii,
- the difficulty of access for maintenance.

It should be apparent from the above brief survey of pitch actuation systems that the design of the hardware required for pitch-regulation is a significant task.

Moreover, as regards the controller, pitch-regulation introduces the need for fast response closed loop control, which is not required for the supervisory functions on a stall-regulated machine. Thus the benefits of pitch control have to be weighed carefully against all the additional costs involved, including the cost of maintenance.

Another factor that needs to be considered is fatigue loading. This increases significantly on full-span pitch-regulated machines, because the rate of change of lift coefficient with angle of attack remains at about 2π (see Equation (A3.9) in Appendix A3.1) instead of reducing to zero as the blade goes into stall, with the result that rapid changes in wind speed above rated will cause bigger thrust load changes.

Pitch system controller design is considered in detail in Chapter 8.

6.7.3 Passive pitch control

An attractive alternative to active control of blade pitch to limit power is to design the blade and/or its hub mounting to twist under the action of loads on the blades in order to achieve the desired pitch changes at higher wind speeds. Unfortunately, although the principle is easy to state, it is difficult to achieve it in practice, because the required variation in blade twist with wind speed generally does not match the corresponding variation in blade load. In the case of stand-alone wind turbines, the optimization of energy yield is not the key objective, so passive pitch control is sometimes adopted, but the concept has not been utilized as yet for many grid-connected machines.

Corbet and Morgan (1991) give a survey of how different types of blade loads might be utilized. Harnessing the centrifugal load is obviously promising in the case of variable speed machines, and this has been demonstrated using a screw cylinder and preloaded spring to passively control each tip blade, within the Dutch FLEXHAT programme. When the centrifugal load on the tip exceeds the preload, the tip blade is driven outwards against the spring and pitches (see Figure 6.13 for illustration of the concept).

Joose and Kraan (1997) have proposed replacing this mechanism by a maintenance-free 'Tentortube', which would twist under tension loading. This tube would be carbon-fibre reinforced with all the fibres set at an angle to the axis, so that centrifugal loading induced twist. It would be placed inside a hollow steel tip shaft, which would carry the aerodynamic loading on the tip blade.

6.7.4 Active stall control

Active stall control achieves power limitation above rated wind speed by pitching the blades initially *into* stall, i.e., in the opposite direction to that employed for active pitch control, and is thus sometimes known as negative pitch control. At higher wind speeds, however, it is usually necessary to pitch the blades back towards feather in order to maintain power output at rated.

A significant advantage of active stall control is that the blade remains essentially stalled above the rated wind speed, so that gust slicing (see Section 6.7.2) results in much smaller cyclic fluctuations in blade loads and power output. It is found that only small changes of pitch angle are required to maintain the power output at rated, so pitch rates do not need to be as large as for positive pitch control. Moreover, full aerodynamic braking requires pitch angles of only about −20°, so the travel of the pitch mechanism is very much reduced compared with positive pitch control.

Figure 6.11 compares schedules of pitch angle against wind speed for active stall control and active pitch control for the same blade. The active stall control schedule is derived from the intersection of the family of power curves for different negative pitch angles with the 500 kW abscissa in Figure 6.12, while the active pitch control schedule is derived from Figure 6.7.

The principal disadvantage of active stall control is the difficulty in predicting aerodynamic behaviour accurately in stalled flow conditions. Active stall control is considered further in Section 8.2.1.

6.7.5 Yaw control

As most horizontal-axis wind turbines employ a yaw drive mechanism to keep the turbine headed into the wind, the use of the same mechanism to yaw the turbine out of wind to limit power output is obviously an attractive one. However, there are two factors which militate against the rapid response of such a system to limit power: first, the large moment of inertia of the nacelle and rotor about the yaw axis,

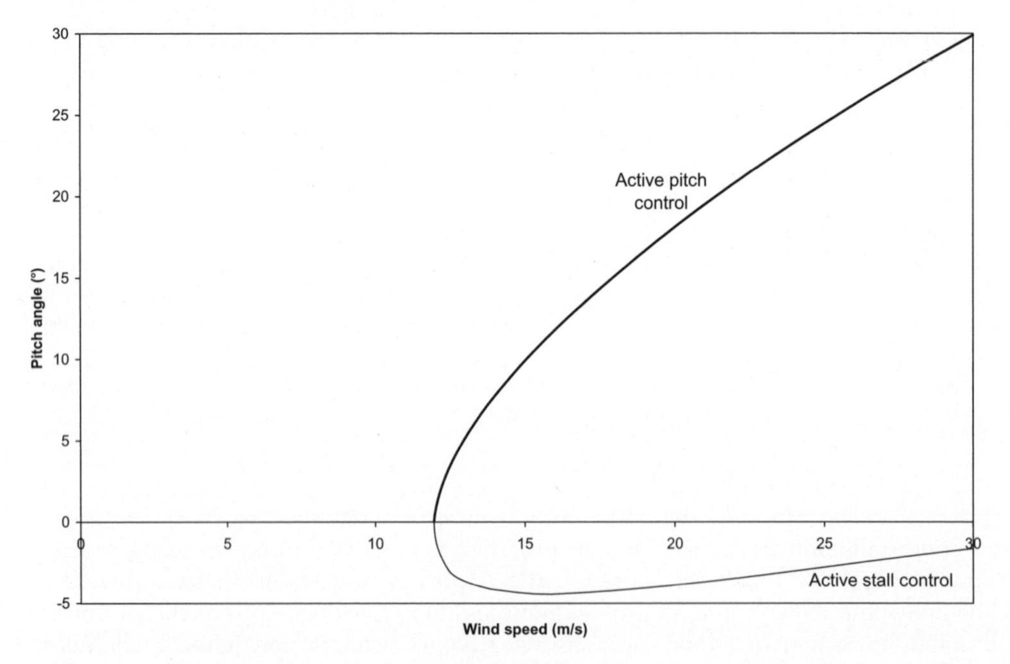

Figure 6.11 Specimen Pitch Angle Schedules for Active Pitch Control and Active Stall Control

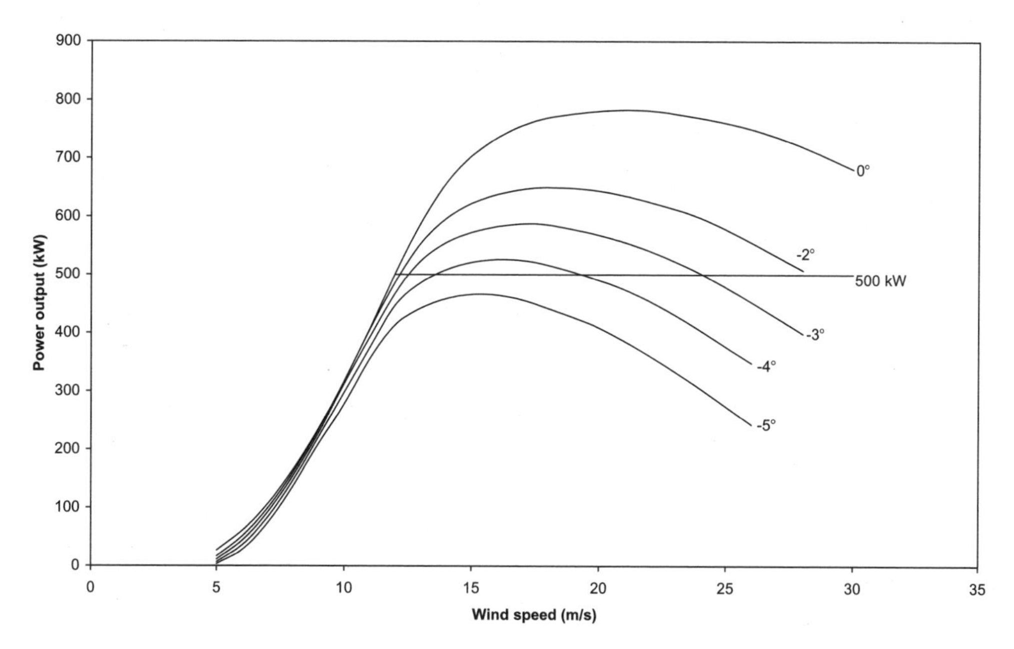

Figure 6.12 Power Curves for Different Negative Pitch Angles: 40 m Diameter Rotor Rotating at 33 r.p.m.

and second, the cosine relationship between the component of wind speed perpendicular to the rotor disc and the yaw angle. The latter factor means that, at small initial yaw angles, yaw changes of, say, 10° only bring about reductions in power of a few percent, whereas blade pitch changes of this magnitude can easily halve the power output. Thus active yaw control is only practicable for variable speed machines where the extra energy of a wind gust can be stored as rotor kinetic energy until the yaw drive has made the necessary yaw correction. This design philosophy has been exploited successfully in Italy on the 60 m diameter Gamma 60 prototype, which has an impressive maximum yaw rate of 8°/s (Coiante *et al.*, 1989).

6.8 Braking Systems

6.8.1 Independent braking systems—requirements of standards

DS 472 (Section 5.1.4) and the GL rules (Section 5.1.3) both require that a wind turbine shall have two independent braking systems. On the other hand, IEC 61400-1 (Section 5.1.2) does not explicitly require the provision of two braking systems (stating that the protection system shall include one or more systems capable of bringing the rotor to rest or to an idling state), but it does require the protection system to remain effective even after the failure of any non-safe-life protection system component.

IEC 61400-1 and the GL rules require that at least one of the braking systems should act on the rotor or low-speed shaft, but DS 472 goes further and requires that one should be an aerodynamic brake.

Normal practice is to provide both aerodynamic and mechanical braking. However, if independent aerodynamic braking systems are provided on each blade, and each has the capacity to decelerate the rotor after the worst-case grid loss, then the mechanical brake will not necessarily be designed to do this. The function of the mechanical brake in this case is solely to bring the rotor to rest, i.e., to park it, as aerodynamic braking is unable do this.

6.8.2 Aerodynamic brake options

Active pitch control

Blade pitching to feather (i.e., to align the blade chord with the wind direction) provides a highly effective means of aerodynamic braking. Blade pitch rates of 10° per second are generally found adequate, and this is of the same order as the pitch rate required for power control. The utilization of the blade pitch system for start-up and power control means that it is regularly exercised with the result that the existence of a dormant fault is highly unlikely.

In machines relying solely on blade pitching for emergency braking, independent actuation of each blade is required, together with fail-safe operation should power or hydraulic supplies passing through a hollow low-speed shaft from the nacelle be interrupted. In the case of hydraulic actuators, oil at pressure is commonly stored in accumulators in the hub for this purpose.

Pitching blade tips

Blade tips which pitch to feather have become the standard form of aerodynamic braking for stall-regulated turbines. Typically the tip blade is mounted on a tip shaft, as illustrated in Figure 6.13, and held in against centrifugal force during normal operation by a hydraulic cylinder. On release of the hydraulic pressure (which is triggered by the control system, or directly by an overspeed sensor), the tip blade flies outwards under the action of centrifugal force, pitching to feather simultaneously on the shaft screw. The length of the tip blade is commonly some 15 percent of the tip radius.

The ability of the control system to trigger blade tip activation is of crucial importance. On a number of early machine designs, the blade tips were centrifugally activated only, so there could be long periods without overspeed events when they did not operate. As a result there was a risk of seizure when operation was eventually required. With the now commonplace arrangement enabling the control system to activate the tip as well, the system can be routinely tested automatically. The penalty is that the low-speed shaft needs to be hollow to accommodate the feed to the hydraulic cylinder.

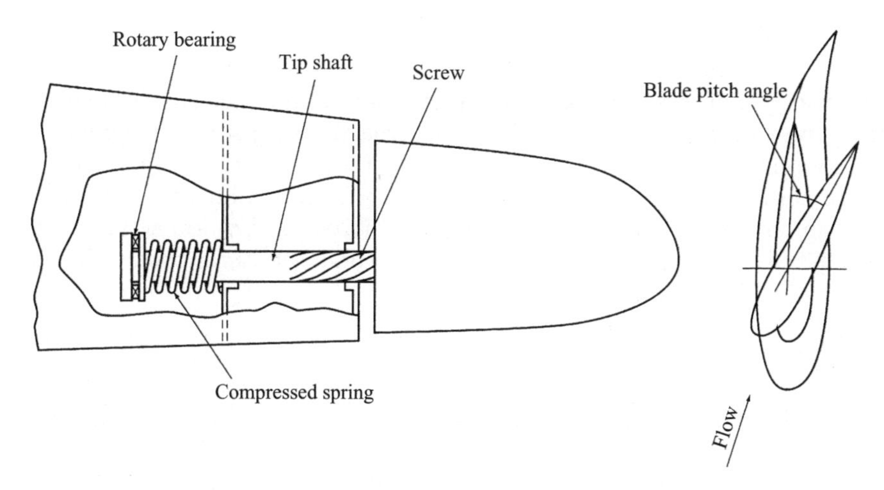

Figure 6.13 Passive Control of Tip Blade, Using Screw on Tip Shaft and Spring

Spoilers

Spoilers are hinged flaps, which conform to the aerofoil profile when retracted, and stick out at right angles to it when deployed. However, although such devices have been used in the past, they have to be of considerable length in order to decelerate the rotor adequately (Jamieson and Agius, 1990). Moreover, unless the design allows for their operation to be regularly tested, there is a risk that they will fail to deploy when actually needed.

Other devices

Various other devices have been suggested, such as

- ailerons,
- the sliding leading edge device, or SLEDGE, in which a length of leading edge at the tip slides radially outwards,
- the flying leading edge device, or FLEDGE, in which the whole leading edge together with an adjacent section of the camber face is pitched towards feather.

Jamieson and Agius (1990) and Armstrong and Hancock (1991) give useful surveys of these and other aerodynamic braking devices, and note that the SLEDGE device, which utilizes only 2 percent or 3 percent of the blade area, is highly effective aerodynamically. Derrick (1992) examines the capabilities of the SLEDGE and FLEDGE devices for both braking and power control in more detail. Despite their promise, these devices have not yet found commercial application.

6.8.3 Mechanical brake options

As noted in Section 6.8.1, the duty of the mechanical brake need only be that of a parking brake on machines where the aerodynamic brakes can be actuated independently. However, on pitch-regulated machines where blade position is controlled by a single actuator, full independent braking capability has to be provided by the mechanical brake. It is worth noting that several manufacturers of stall-regulated machines fitted with independent tip brakes ensure that the mechanical brake can stop the rotor unassisted. This may be to satisfy requirements in certain countries that two independent braking systems of a different type are provided.

A wind turbine brake typically consists of a steel brake disc acted on by one or more brake callipers. The disc can be mounted on either the rotor shaft (known as the low-speed shaft) or on the shaft between the gearbox and the generator (known as the high-speed shaft). The latter option is much the more common because the braking torque is reduced in inverse proportion to the shaft speeds, but it carries with it the significant disadvantage that the braking torques are experienced by the gear train. This can increase the gearbox torque rating required by as much as *ca* 50 percent, depending on the frequency of brake application (see Section 7.4.5). Another consideration is that the material quality of brake discs mounted on the high-speed shaft is more critical, because of the magnitude of the centrifugal stresses developed.

The brake callipers are almost always arranged so that the brakes are spring applied and hydraulically retracted, i.e., fail-safe.

Aerodynamic braking is much more benign than mechanical braking as far as loading of the blade structure and drive train is concerned, so it is always used in preference for normal shut-downs.

6.8.4 Parking *versus* idling

Although a mechanical parking brake is essential for bringing the rotor to rest for maintenance purposes, many manufacturers allow their machines to idle in low winds and some do so during high wind shut-downs. The idling strategy has two clear advantages: it reduces the frequency of imposition of braking loads on the gear train, and gives the impression to members of the public that the turbine is operating even when it is not generating. On the other hand, gearbox and bearing lubrication must be maintained throughout.

6.9 Fixed Speed, Two-speed or Variable-speed Operation

Wind turbine rotors develop their peak efficiency at one particular tip speed ratio (see Figure 3.15), so fixed speed machines operate sub-optimally, except at the wind speed corresponding to this tip speed ratio. Energy capture can clearly be increased by varying the rotational speed in proportion to the wind speed so that the turbine is always running at optimum tip speed ratio, or alternatively a slightly reduced improvement can be obtained by running the turbine at one of two fixed speeds so that the tip speed ratio is closer to the optimum than with a single fixed speed.

Noise considerations are often of greater significance than energy capture in the decision to opt for non-fixed speed operation. As noted in Section 6.4.4, the aerodynamic noise generated by a wind turbine is approximately proportional to the fifth power of the tip speed. Both variable speed and two-speed operation allow the rotational speed to be substantially reduced in low winds, thus reducing turbine aerodynamic noise dramatically when it could otherwise be objectionable because of low ambient noise.

6.9.1 Two-speed operation

At one time, two-speed operation was relatively expensive to implement because separate generators were required for each speed of turbine rotation. Either generators of differing numbers of poles were connected to gearbox output shafts rotating at the same speed, or generators with the same number of poles were connected to output shafts rotating at differing speeds. The rating of the generator for low-speed operation would normally be much less than the turbine rating.

The development of induction generators with two sets of independent windings has allowed the number of poles within a single generator to be varied by connecting them together in different ways. Standard generators of this type are now available which can be switched between four- and six-pole operation, giving a speed ratio of 1.5. Given correct selection of the two operating speeds, this ratio produces close to the optimum increase of energy capture.

It is important to note that, in the case of stall-regulated machines, only limited energy gains are to be had by converting from fixed-speed to two-speed operation for a given rated power, because the maximum rotational speed of the two-speed machine is restricted to the rotational speed of the fixed speed machine in order to limit the power to the required value. Energy gains are only of the order of 2 or 3 percent, but, nevertheless, two-speed operation is often deemed worthwhile on stall regulated machines because of noise considerations.

In the case of pitch-regulated machines, the energy gain obtainable by moving to two-speed operation depends on the power rating and rotational speed of the baseline fixed speed machine. Where these parameters have been chosen to be close to the optimum, in relation to rotor diameter and rotor chord respectively, an energy gain of only about 3 percent is attainable. However, energy gains of up to 10 percent are possible when the baseline design is sub-optimal.

Some disadvantages of two-speed operation are:

- additional generator cost,
- extra switchgear is required, which is subjected to a demanding duty in terms of frequency of operation,
- control of turbine speed is required at each speed change,
- energy is lost while the generator is disconnected during each speed change.

6.9.2 Variable-speed operation

By interposing a frequency converter between the generator and the network, it is possible to decouple the rotational speed from the network frequency. As well as allowing the rotor speed to vary, this also allows the generator air-gap torque to be controlled.

Variable-speed operation has a number of advantages:

- below rated wind speed, the rotor speed can be made to vary with wind speed to maintain peak aerodynamic efficiency;

- the reduced rotor speed in low winds results in a significant reduction in aerodynamically-generated acoustic noise – noise is especially important in low winds, where ambient wind noise is less effective at masking the turbine noise;

- the rotor can act as a flywheel, smoothing out aerodynamic torque fluctuations before they enter the drive train – this is particularly important at the blade passing frequency;

- direct control of the air-gap torque allows gearbox torque variations above the mean rated level to be kept very small;

- both active and reactive power can be controlled, so that unity power factor can be maintained – it is even possible to use a variable speed wind farm as a source of reactive power to compensate for the poor power factor of other consumers on the network; variable speed turbines will also produce a much lower level of electrical flicker.

In practice, losses in the frequency converter may amount to several per cent, counteracting the increased aerodynamic efficiency below rated. In terms of energy capture, there is often little to choose between a two-speed and a variable-speed machine. The load reduction possibilities, however, mean that most large MW-scale turbines now use variable speed in some form. Variations in aerodynamic torque at blade passing frequency are particularly significant in larger turbines because of the size of the rotor compared to the lateral and vertical length scales of turbulence.

Clearly there is a significant cost associated with the variable-speed drive or frequency converter, which must be weighed against the advantages. Other drawbacks include increased complexity, although there is no particular evidence of reduced availability due to power converter problems, and the generation of electrical noise and harmonics by the inverter system. Modern PWM inverters produce much lower levels of undesirable harmonics than earlier devices because of the high switching frequencies which can be achieved. Electrical noise can be a problem for control signals within the turbine if insufficient care is taken to shield cables. Fibre optic transmission is increasingly being used, and this is not affected.

There are two principal methods of achieving variable speed operation: 'broad range' variable speed, in which the generator stator is connected to the network *via* the frequency converter, and 'narrow range' variable speed where both the generator stator and rotor are connected to the network, the stator directly, and the rotor through slip rings and a frequency converter.

Broad range variable speed allows the speed to vary under load from zero to the full rated speed, but all the power output has to pass through the frequency converter. Narrow range variable speed allows a much cheaper frequency converter, since only a fraction of the power passes through it, but the speed can only vary by a more limited amount, typically ±30–50% either side of synchronous speed. In practice, this is enough to achieve almost all the advantages of variable-speed operation. This is the most commonly used approach. A disadvantage is the maintenance requirements of the slip rings.

6.9.3 Variable-slip operation

Variable slip represents a compromise between fixed- and variable-speed operation (Bossanyi and Gamble 1991, Pedersen 1995). The variable-slip generator is essentially an induction generator with a variable resistor in series with the rotor circuit, controlled by a high-frequency semiconductor switch. Below rated, this acts just like a conventional fixed-speed induction generator. Above rated, however, control of the resistance effectively allows the air-gap torque to be controlled and the slip speed to vary, so the behaviour is then similar to a variable-speed system. A speed range of about 10 percent is typical.

This is cheaper than a variable-speed system, and gives some of the advantages, in particular the control of torque in the drive train and the smoothing of aerodynamic torque variations above rated. It does not offer increased aerodynamic efficiency below rated (although it does not suffer from frequency converter losses), and it does not allow any control of the power factor. Electrical flicker will, however, be reduced above rated.

Slip rings can be avoided by mounting the variable resistors and control circuitry on the generator rotor. An advantage of mounting these externally *via* slip rings is that it is then easier to dissipate the extra heat which is generated above rated, and which may otherwise be a limiting factor at large sizes.

6.9.4 Other approaches to variable-speed operation

There are other possible approaches to variable-speed operation, although none of these has found significant commercial application. They include:

- use of a differential gearbox, with the third shaft controlled by a variable speed electric motor/generator (Law, Doubt and Cooper, 1984, Burton, Mill and Simpson, 1990) or by a hydraulic pump/motor (Henderson *et al.*, 1990);

- mechanical continuously-variable transmission systems such as have been developed for automotive applications.

6.10 Type of Generator

Fixed-speed wind turbines differ from almost all conventional generating plant by using induction rather than synchronous generators. This choice is driven by the requirement for significant damping in the drive train due to the cyclic variations in the torque developed by the aerodynamic rotor.

Both synchronous and induction generators have similar winding arrangements on the stator which, when connected to the three-phase network voltage, produce a fixed-speed, rotating magnetic field. However, the rotors of the two machines are quite different (Hindmarsh, 1984, McPherson, 1990). A synchronous machine has magnets mounted on its rotor and the rotor magnetic field then locks into that produced by the stator leading to operation at synchronous speed. For power generation applications, electromagnets are used on the rotor excited by an externally applied direct current. Although the rotor operates at the same speed as the stator magnetic field it leads the stator field by an angle depending on the applied torque. In contrast, the rotor of a conventional induction machine has a 'squirrel cage' winding into which currents are induced as the rotor bars cut the magnetic field produced by the stator. Hence, an induction generator can only develop torque at a rotational speed slightly greater than that of the stator field. This 'slip speed' is proportional to the applied torque.

Therefore, to a first approximation, the behaviour of a synchronous machine may be considered to be analogous to a torsional spring. The torque is proportional to the angle between the rotor and the stator field. This angle is known as the load or power angle. In contrast, an induction generator can be thought of as a torsional damper where the torque is proportional to the difference in speed between the rotor and the stator field (the slip speed). This is illustrated in simple schematic form in Figure 6.14. It may be seen that if the simple model of a fixed-speed wind turbine, equipped with a synchronous generator, is excited by the cyclic torque from the wind-turbine rotor then there is no damping in the drive train to control the torsional oscillations. It is a simple two-spring, two-mass system. In contrast, with an induction generator, the connection of the generator to the network is represented by a torsional damper. The main cyclic torque of the wind turbine rotor

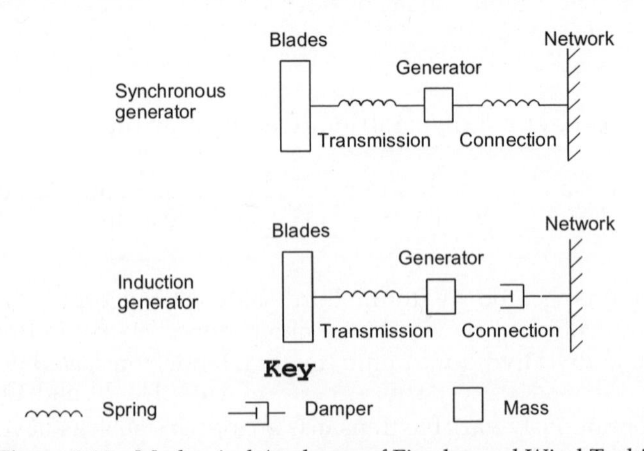

Figure 6.14 Mechanical Analogue of Fixed-speed Wind Turbines

will be at blade passing frequency and it is an unfortunate coincidence that this often matches quite closely the natural frequency of oscillation of a small synchronous generator connected to an electrical network.

In practice, synchronous generators are often fitted with additional cage damper windings but it is not practical to provide the degree of damping required for wind turbine applications. Also at higher ratings (above, say, 1 MW) second-order effects tend to reduce the damping available from induction generators (Saad-Saoud and Jenkins, 1999). However, the basic principle remains that the damping provided by induction generators is necessary for fixed-speed wind turbines.

In contrast, the generators of variable-speed wind turbines are not connected directly to the network but are de-coupled through solid-state frequency converters. Hence synchronous generators may be used.

6.10.1 Historical attempts to use synchronous generators

Induction generators are much less useful than synchronous generators for large-scale power generation.

- The damping action results in higher energy losses in the rotor than with synchronous generators. It is then, of course, necessary to arrange for the removal of the heat dissipated in the rotor.
- All the reactive power necessary to energize the magnetic circuits must be supplied from the network (or by local capacitors). If local capacitors are used then there is the danger of self-excitation, when connection to the network is lost.
- There is no direct control over the terminal voltage or reactive power flow.
- Induction generators do not produce sustained fault current for three-phase faults on the network.
- They suffer from problems of voltage instability. This was not an important issue with limited wind generation but with very large wind farms is becoming of concern.

Hence, in the early development of wind turbines considerable efforts were made to use synchronous generators. These involved a number of innovative solutions to the provision of damping. For example, both Westinghouse in the USA and Howden in the UK used fluid couplings in the drive train to provide damping. The Wind Energy Group in the UK mounted a 250 kW synchronous generator using a spring-damper system and connected a 3 MW synchronous generator through a sophisticated variable-speed mechanical gearbox (Law, Doubt and Cooper, 1984). However, these and other similar approaches using synchronous generators on large prototype wind turbines are now of historical interest only.

6.10.2 Direct-drive generators

There is considerable interest in the application of generators driven directly by the wind-turbine rotor without a speed increasing gearbox and a number of manufacturers offer such wind turbines. However, the power output of any rotating electrical machine may be generally described by (Laithwaite and Freris, 1980):

$$P = KD^2Ln$$

where D is the rotor diameter, L is the length, n is the rotational speed, and K is a constant.

Thus it may be seen that if the rotational speed is reduced then it is necessary either to lengthen the generator in proportion or to increase the diameter. It is cheaper to increase the diameter as this raises the power by the square rather than linearly. Thus, direct-drive generators for wind turbines tend to have rather large diameters but with limited length (Figure 6.20).

Induction generators require a rather small radial distance between the surface of the rotor and the stator (known as the air-gap). This is necessary to ensure an adequate air-gap magnetic flux density as all the excitation is provided from the stator. In contrast, synchronous generators have excitation systems on the rotor and so can operate with larger air-gaps. It is difficult to manufacture large diameter electrical machines with small air gaps for mechanical and thermal reasons. Hence direct-drive wind turbines use synchronous generators (either with permanent magnet excitation or, more usually, with a wound rotor and electromagnets providing the field). The use of a synchronous generator, in turn, leads to the requirement for solid-state frequency conversion equipment to de-couple the generator from the network and permit variable-speed operation.

6.11 Drive-train Mounting Arrangement Options

6.11.1 Low-speed shaft mounting

The functions of the low-speed shaft are the transmission of drive torque from the rotor hub to the gearbox, and the transfer of all other rotor loadings to the nacelle structure. Traditionally the mounting of the low-speed shaft on fore and aft bearings has allowed these two functions to be catered for separately; the gearbox is hung on the rear end of the shaft projecting beyond the rear bearing and the drive torque is resisted by a torque arm. The front bearing is positioned as close as possible to the shaft/hub flange connection, in order to minimize the gravity moment due to the cantilevered rotor mass, which usually governs shaft fatigue design. The spacing between the two bearings will normally be greater than that between front bearing and rotor hub in order to moderate the bearing loads due to shaft moment (see Figure 6.15 for an illustration of a typical arrangement).

The opposite approach is to make the gearbox an integral part of the load path between the low-speed shaft and tower top i.e., an 'integrated gearbox'. The fore

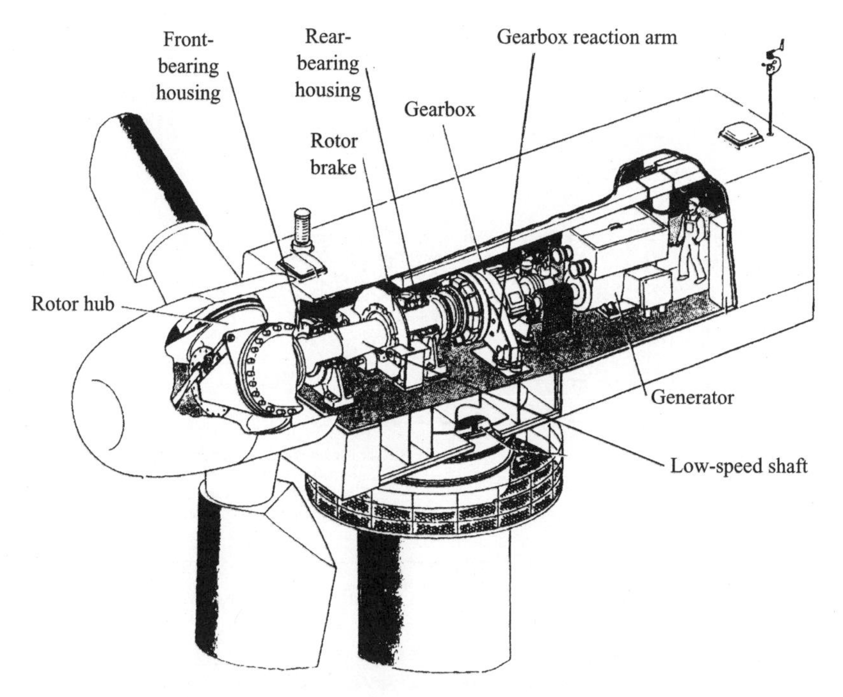

Figure 6.15 View of Nacelle Showing Traditional Drive Shaft Arrangement

and aft low-speed shaft bearings are absorbed within the gearbox, which moves to the front of the nacelle in order to minimize the rotor cantilever distance, and the gearbox casing then transmits the loads to the nacelle bedplate (Figure 6.21). Clearly this approach requires a much more robust gearbox casing, which must not merely resist the rotor loads, but do so without deflecting sufficiently to impair its functioning. Moreover its fore-aft length has to be increased in order to moderate the bearing loads due to shaft moment. The benefits lie in the reduced extent of the bedplate and the elimination of separate bearings requiring separate provision for lubrication, but a significant disadvantage is that gearbox replacement requires the removal of the rotor.

A configuration which is becoming increasingly popular is one intermediate between the two extremes described above, in which only the rear low-speed shaft bearing is absorbed into the gearbox. The gearbox is usually set well back from the front bearing in order to reduce the rear bearing loads, and is rigidly fixed to supporting pedestals positioned on either side of the nacelle. Typical arrangements are shown in Figure 6.16, which shows a cross section through the nacelle of the Nordex N-60 turbine, and in Figure 6.17. Note that the shaft tapers down in diameter towards the rear reflecting the reducing bending moment. The advantage of this arrangement is that the gearbox casing is not called upon to carry any moments due to cantilevered rotor mass or rotor out-of-plane loadings.

Figures 6.18 and 6.19 are aerial views of the nacelle of a NEG-Micon 1.5 MW machine with a similar drive train arrangement, after installation of the low-speed shaft.

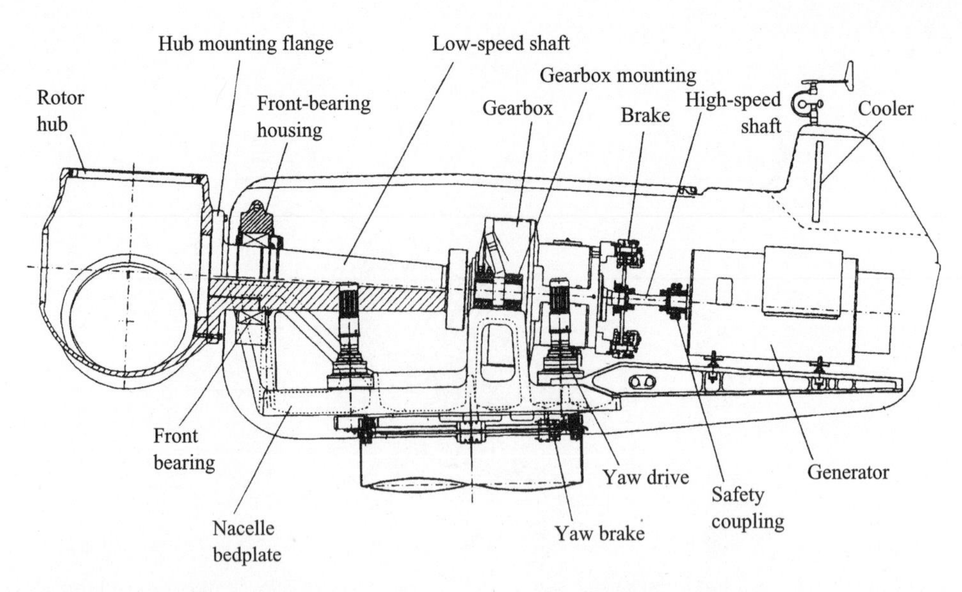

Figure 6.16 Nacelle Arrangement for the Nordex N60 Turbine (Reproduced by permission of Nordex)

Figure 6.17 Drive Train Side View. (From left to right the components visible through the cut-out in the nacelle wall are: (1) low-speed shaft front bearing, (2) low-speed shaft, (3) gearbox mountings, (4) gearbox, (5) high-speed shaft with brake, (6) generator. The fabricated nacelle bedplate is also visible).

In the case of wind turbines with direct-drive generators, the low-speed shaft arrangement is dramatically different. The low-speed shaft, which now connects the rotor hub to the rotor of the generator, is hollow, so that it can be mounted on a concentric fixed shaft cantilevered out from the nacelle bedplate (see Figure 6.20).

Figure 6.18 Turbine Assembly in the Air (1). (View of nacelle of 1.5 MW NEG Micon turbine after installation of low-speed shaft (front) and gearbox. The ring of bolt holes in the low-speed shaft flange for hub mountings are clearly visible). (Reproduced by permission of NEG Micon)

6.11.2 High-speed shaft and generator mounting

The generator is normally mounted to the rear of the gearbox on an extension of the nacelle bedplate and the connecting drive shaft – the 'high speed shaft' – is fitted with flexible couplings at each end, to cater for small misalignments between the generator and gearbox.

The generator axis is normally offset from the low-speed shaft axis. This is because, except in the case of machines fitted with a mechanical brake acting on the rotor, access is required to the rear end of the low-speed shaft for actuation of aerodynamic braking. Usually the generator is either offset to one side of the nacelle, which introduces asymmetry into the nacelle bedplate, or it is offset vertically upwards, which requires a vertical step in the bedplate.

A much more compact arrangement can be obtained by bolting the generator rigidly onto the rear of the gearbox *via* an adaptor tube (see Figure 6.21). The surfaces of the mating interfaces have to be carefully machined to ensure shaft alignment, and suitable access has to be provided to the coupling between the generator and gearbox output shafts. Despite the neatness of this layout, it has only been adopted by one or two manufacturers.

One consequence of locating the generator in the nacelle is that power cables running down the tower are required to twist as the nacelle yaws. On some large machines, the problems associated with the twisting of heavy cables have been avoided by mounting the generator vertically in the top of the tower, and driving

Figure 6.19 Turbine Assembly in the Air (2). (View of low-speed shaft and front bearing after installation on 1.5 MW NEG Micon turbine). (Reproduced by permission of NEG Micon)

the high-speed shaft *via* a bevel gear. An alternative solution to the problem of heavy twisting cables, however, is to leave the generator in the nacelle and to transform to a higher voltage there as well.

6.12 Drive-train Compliance

The rotational dynamics of the drive train can have a major effect on loading. The effect is very different in fixed- and variable-speed turbines, but in each case the consequence of ignoring drive-train dynamics at the design stage can be very severe.

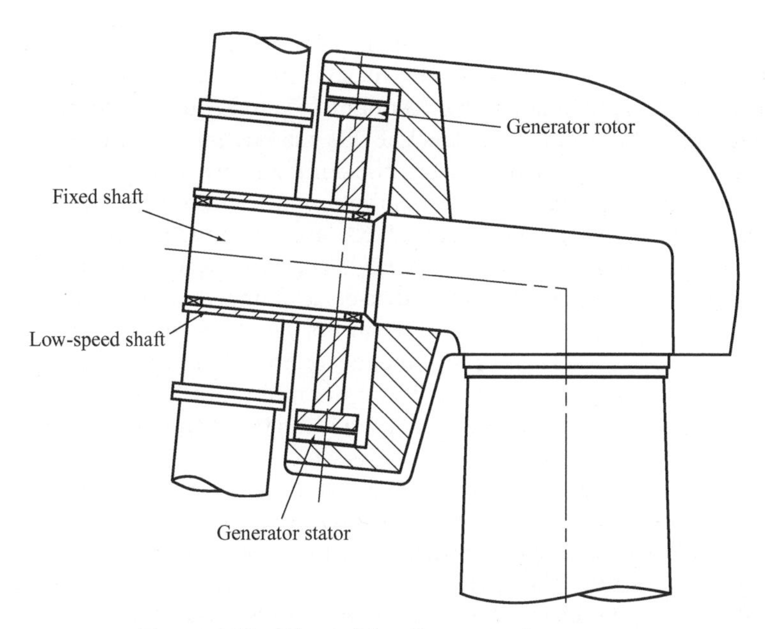

Figure 6.20 Direct-drive Generator Arrangement

Figure 6.21 Integrated Gearbox on the Zond Z-750 turbine. (The gearbox is mounted on a circular nacelle bedplate, with the hub to the left and generator at the rear. An electrically driven yaw drive can be seen beneath the generator).

In the variable-speed case, the dynamics may be quite simple: the drive train may be modelled as a rotor and a generator inertia, separated by a torsional spring. Typically the natural frequency of this resonant system is quite high, of the order of 3–4 Hz. However, this mode is subject to very little damping, especially above rated where the generator torque is held constant. (Below rated the torque will be varied as a function of rotational speed, thus providing a small amount of damping.) There is very little aerodynamic damping from the rotor, and this mode of vibration can potentially generate very large gearbox torque oscillations. Chapter 8 explains how the control system can be used to damp this mode by appropriate control of the generator torque, but it is important to ensure that the resonant frequency does not coincide with a significant forcing frequency such as 6P, which can make it very difficult to achieve sufficient damping through the control system.

In the fixed-speed case, the directly-coupled induction generator provides a lot of damping since the air-gap torque increases steeply with generator speed. The torque–slip curve completely changes the dynamics compared to the variable-speed case, resulting in a much lower first mode frequency, typically closer to 1 Hz. The damping factor is strongly dependent on the generator slip: a generator with 0.5% rated slip can give a peak dynamic magnification of perhaps 2 to 5 at the resonant frequency, whereas with 2 percent slip the peak magnification may be no more than 1 to 1.5. The position of the peak with respect to blade-passing frequency is critical: if the blade-passing frequency is close to the peak, very large gearbox torque and electrical power oscillations will occur at this frequency. Pitch control may further exacerbate these.

With a two-bladed turbine the blade-passing frequency tends to be closer to the resonant peak; with a three-bladed turbine the blade-passing frequency is typically higher, where the dynamic magnification is much lower. Nevertheless, even for three-bladed turbines it is not uncommon for power and torque oscillations at the blade passing frequency to be as large as ±50–100 percent during pitch controlled operation in high winds.

The use of a high-slip generator greatly improves the situation, but there are two main drawbacks. First, each 1 percent of slip corresponds to 1 percent of extra losses, which significantly reduces the energy yield below rated wind speed. Second, these extra losses equate with heat dissipation in the generator, making it more difficult to keep the generator cool, especially in large machines.

An alternative to high generator slip which has occasionally been used is a fluid coupling between the gearbox and the generator. This is also a device which generates a torque proportional to slip speed, and it suffers from the same draw-backs as a high-slip generator.

Another technique which has sometimes been used is to reduce the resonant frequency by introducing additional torsional flexibility into the drive train. This can be done by means of a quill shaft, a flexible low-speed coupling, or flexible mounts for the gearbox or even for the whole bedplate. The frequency reduction is, however, accompanied by a further loss of damping, and it may therefore be necessary to incorporate additional mechanical damping with the torsional flex-ibility, which is not always easy to engineer. Torsional flexibility in the high-speed shaft is not usually practical because of the large angular movement required to achieve the necessary flexibility: half a revolution may be necessary, compared to just one or two degrees at the low-speed shaft. An interesting variant (Leithead and Rogers, 1995) is to mount the generator on flexible mounts. This system can be tuned to absorb energy at the blade-passing frequency through an additional mode of vibration of the generator casing against its mountings. This mode also affects the generator slip speed (the difference between rotor and casing speeds) and is therefore damped by the slip curve. Nevertheless, generator casing displacements would still need to be of the order of 10–15°, which is still not easy to engineer.

6.13 Rotor Position with Respect to Tower

6.13.1 Upwind configuration

The upwind configuration is the one most commonly chosen. The principal advantage is that the tower shadow effect is much less for the same blade–tower spacing, reducing both dynamic loads on the blade and rhythmic noise effects. Set against this is the need to take great care to avoid the risk of blade–tower strikes with upwind machines, requiring accurate prediction of blade deflections under turbulent wind loading.

The clearance between the undeflected blade and the tower can be increased by tilting the low-speed shaft upwards or by increasing the rotor overhang. It is desirable to keep the rotor overhang small in order to minimize low-speed shaft and nacelle bedplate bending moments, so the low-speed shaft is normally tilted upwards by 5° or 6° to provide the necessary blade–tower clearance, at the cost of a very small reduction in power output.

6.13.2 Downwind configuration

The wind velocity deficit behind a wind-turbine tower is much greater than that in front of it, to the extent that Powles (1983) has reported a turbulent region with essentially no forward velocity extending up to four tower diameters downstream of an octagonal tower. Beyond this distance, recovery is relatively rapid, with the deficit reduced to about 25 percent at seven tower diameters downstream.

In addition to the mean wind-speed velocity deficit behind the tower, vortex shedding results in additional wind-speed fluctuations over and above those already present due to turbulence. The two effects combine to present a harsh environment to the blades immediately behind the tower. The blades are subjected to a large negative impulsive load each time they pass the tower, which contributes significantly to blade fatigue damage, and the audible tower 'thump' that results is liable to be unwelcome. Designers usually mitigate both effects by positioning the rotor plane well clear of the tower, but this inevitably increases nacelle costs somewhat.

An important benefit of the downwind configuration is that it allows the use of very flexible blades without the risk of tower strike. Such blades benefit by being less severely unloaded by the tower shadow, because wind loading deflects them further from the tower in the first place.

6.14 Tower Stiffness

A key consideration in wind turbine design is the avoidance of resonant tower oscillations excited by rotor thrust fluctuations at rotational or blade-passing frequency. The damping ratio may be only 2–3 percent for tower fore-aft oscillations and an order of magnitude less for side-to-side motion, so unacceptably large stresses and deflections could develop if the blade-passing frequency and tower natural frequency were to coincide. Rotational frequency is less of a concern, because cyclic loadings at this frequency only arise if there are geometrical differences between blades.

Wind-turbine towers are customarily categorized according to the relationship between the tower natural frequency and the exciting frequencies. Towers with a natural frequency greater than the blade-passing frequency are said to be stiff, while those with a natural frequency lying between rotational frequency and blade-passing frequency are said to be soft. If the natural frequency is less than rotational frequency, the tower is described as soft–soft.

If the tower is designed to meet strength requirements and no more, its frequency category is primarily determined by the ratio of tower height to turbine diameter, with the higher ratios producing the softer towers. The principal benefits of stiff towers are modest – they allow the turbine to run up to speed without passing through resonance, and tend to radiate less sound. However, since stiff towers usually require the provision of extra material not otherwise required for strength, soft towers are generally preferred.

6.15 Personnel Safety and Access Issues

An integral part of wind-turbine design is the inclusion of the necessary safety provisions for operation and maintenance staff. Minimum requirements include the following:

- ladder access to the nacelle – this needs to be fitted with a fall-arrest device, unless ladder runs are short and protected by intermediate landings; careful attention needs to be paid to the route between the tower top and nacelle to avoid hazards arising from sudden yawing movements;

- an alternative means of egress from the nacelle, for use in case of fire in the tower – this can take the form of an inertia-reel device, enabling personnel to lower themselves through a hatch in the nacelle floor;

- locking devices for immobilizing the rotor and the yawing mechanism – rotor brakes and yaw brakes are not considered sufficient, because of the risk of accidental release and the occasional need to deactivate them for maintenance purposes; the rotor locking device should act on the low-speed shaft, so that its effectiveness is not dependent on the integrity of the gearbox – typically the device consists of a pin mounted in a fixed housing, which can be engaged in a hole in a shaft-mounted disc;

- guards to shield any rotating parts within the nacelle;
- suitable fixtures for the attachment of safety harnesses for personnel working outside the nacelle.

The designer needs to assess the requirement for all-weather access to the nacelle at an early stage. Lattice towers afford no protection from the weather when climbing, so the number of days on which access for maintenance is possible will be restricted. Similar restrictions will arise if the nacelle cover has to be opened to the elements in order to provide space for personnel to enter.

Consideration also needs to be given to the means of raising and lowering tools and spares. If the interior of the tower is interrupted by intermediate platforms, these operations have to be performed outside, with consequent weather limitations.

Standard rules for electrical safety apply to all electrical equipment. However, particular care must be taken with the routing of electrical cables between tower and nacelle, in order to avoid potential damage due to chafing when they twist. If the power transformer is located in the tower base or nacelle instead of in a separate enclosure at ground level, it should be partitioned off to minimize the fire risk to personnel.

References

Armstrong, J. R. C. and Hancock, M., (1991). 'Feasibility study of teetered, stall-regulated rotors'. *ETSU Report No. WN 6022.*

Bossanyi, E. A. and Gamble, C. R., (1991). 'Investigation of torque control using a variable slip induction generator' *ETSU WN-6018*, Energy Technology Support Unit, Harwell, UK.

Burton, A. L., Mill, P. W. and Simpson, P. B., (1990). 'LS1 post-synchronization commissioning'. *Proceedings of the 12th BWEA Conference*, pp 183–193. Mechanical Engineering Publications, Bury St Edmunds, UK.

Coiante, D. *et al.*, (1989). 'Gamma 60 1.5 MW wind turbine generator'. *Proceedings of the European Wind Energy Conference*, pp 1027–1032.

Corbet, D. C. and Morgan, C. A., (1991). 'Passive control of horizontal axis wind turbines'. *Proceedings of the 13th BWEA Conference*, pp 131–136. Mechanical Engineering Publications, Bury St Edmunds, UK.

Derrick, A., (1992). 'Aerodynamic characteristics of novel tip brakes and control devices for HAWTs'. *Proceedings of the 14th BWEA Conference*, pp 73–78. Mechanical Engineering Publications, Bury St Edmunds, UK.

Engstrom, S. *et al.*, (1997). 'Evaluation of the Nordic 1000 Prototype'. *Proceedings of the European Wind Energy Conference*, pp 213–216.

Falchetta, M. *et al.*, (1996). 'Structural behaviour of the Gamma 60 prototype'. *Proceedings of the European Union Wind Energy Conference*, pp 268–271.

Fuglsang, P. and Thomsen, K., (1998). 'Cost optimization of wind turbines for large-scale offshore wind farms'. *Riso National Laboratory Report No. R-1000*. Riso National Laboratory, Roskilde, Denmark.

Hindmarsh, J., (1984). *Electrical machines and their applications*. Butterworth Heinemann, UK.

Henderson, G. M., *et al.*, (1990). 'Synchronous wind power generation by means of a torque-limiting gearbox'. *Proceedings of the 12th BWEA Conference*, pp 41–46. Mechanical Engineering Publications, Bury St Edmunds, UK.

Jamieson, P. and Agius, P., (1990). 'A comparison of aerodynamic devices for control and overspeed protection of HAWTs'. *Proceedings of the 12th BWEA Conference*, pp 205–213. Mechanical Engineering Publications, Bury St Edmunds, UK.

Jamieson, P. and Brown, C. J., (1992). 'The optimization of stall-regulated rotor design'. *Proceedings of the 14th BWEA Conference*, pp 79–84, Mechanical Engineering Publications, Bury St Edmunds, UK.

Joose, P. A. and Kraan, I., (1997). 'Development of a tentortube for blade tip mechanisms'. *Proceedings of the European Wind Energy Conference*, pp 638–641.

Laithwaite, E. R. and Freris, L. L., (1980). *Electric energy: its generation, transmission and use.* McGraw-Hill, Maidenhead, UK.

Law, H., Doubt, H. A. and Cooper, B. J., (1984). 'Power control systems for the Orkney wind-turbine generators'. *GEC Engineering No 2.*

Leithead, W. E. and Rogers, M. C., (1995). 'Improving damping by a simple modification to the drive train'. *Proceedings of the 17th BWEA Conference*, pp 273–278. Mechanical Engineering Publications, Bury St Edmunds, UK.

McPherson, G., (1990). *An introduction to electrical machines and transformers. Second Edition.* John Wiley and Sons, New York, US.

Pedersen, T. K., (1995). 'Semi-variable speed – a compromise?'. *Proceedings of the Wind Energy Conversion, 17th BWEA Conference*, pp 249–260. Mechanical Engineering Publications, Bury St Edmunds, UK.

Morgan, C., (1994). 'The prospects for single-bladed horizontal axis wind turbines'. *ETSU Report No. W/45/00232/REP*. Energy Technology Support Unint, Harwell, UK.

Petersen, T. P., *et al.*, (1998). 'Prediction of dynamic loads and induced vibrations in stall'. *Riso National Laboratory Report No. R-1045*. Riso National Laboratory, Roskilde, Denmark.

Powles, S. J. R., (1983). 'The effects of tower shadow on the dynamics of horizontal axis wind turbines'. *Wind Engng*, **7**, 1, 26–42.

Rawlinson-Smith, R. I., (1994). 'Investigation of the teeter stability of stalled rotors'. *ETSU Report No. W.43/00256/REP*. Energy Technology Support Unit, Harwell, UK.

Saad-Saoud, Z. and Jenkins, N., (1999). 'Models for predicting flicker induced by large wind turbines'. *IEEE Transactions on Energy Conversion*, **14**, 3, 743–748.

7
Component Design

7.1 Blades

7.1.1 Introduction

A successful blade design must satisfy a wide range of objectives, some of which are in conflict. These objectives can be summarized as follows:

(1) maximize annual energy yield for the specified wind speed distribution;

(2) limit maximum power output (in the case of stall regulated machines);

(3) resist extreme and fatigue loads;

(4) restrict tip deflections to avoid blade/tower collisions (in the case of upwind machines);

(5) avoid resonances;

(6) minimize weight and cost.

The design process can be divided into two stages: the aerodynamic design, in which objectives (1) and (2) are satisfied, and the structural design. The aerodynamic design addresses the selection of the optimum geometry of the blade *external* surface – normally simply referred to as the *blade geometry* – which is defined by the aerofoil family and the chord, twist and thickness distributions. The structural design consists of blade material selection and the determination of a structural cross section or *spar* within the external envelope that meets objectives (4) to (6). Inevitably there is interaction between the two stages, as the blade thickness needs to be large enough to accommodate a spar which is structurally efficient.

The focus of Section 7.1 is on blade structural design. After a brief consideration of the aerodynamic design in Section 7.1.2, practical constraints on the optimum design are noted in Section 7.1.3 and forms of blade structure surveyed in Section 7.1.4. An overview of the properties of some potential blade materials is given in Section 7.1.5 and the properties of glass-fibre reinforced plastic (GFRP) and laminated wood are considered in more detail in Sections 7.1.6 and 7.1.7. Governing load cases are considered in Sections 7.1.8 with reference to both stall- and pitch-regulated machines. Subsequent sections touch upon blade resonance, panel buckling design and blade root fixings.

7.1.2 Aerodynamic design

The aerodynamic design encompasses the selection of aerofoil family and optimization of the chord and twist distributions. The variation of thickness to chord ratio along the blade also has to be considered, but this ratio is usually set at the minimum value permitted by structural design considerations, as this minimizes drag losses.

The process for optimizing the blade design of machines operating at a fixed tip speed ratio is described in Section 3.7.2, where analytical expressions for the blade geometry parameter,

$$\sigma_r \lambda C_l = \frac{Nc(\mu)}{2\pi R} \lambda C_l$$

and the local inflow angle, ϕ, are derived as a function of the local tip speed ratio, $\lambda\mu = \lambda r/R$. (Equations (3.67a) and (3.68a)). If $\lambda\mu \gg 1$, the expressions can be approximated by

$$\sigma_r \lambda C_l = \frac{Nc(\mu)}{2\pi R} \lambda C_l = \frac{8}{9\lambda\mu} \text{ and } \phi = \frac{2}{3\lambda\mu} \tag{7.1}$$

If it is decided to maintain the angle of attack, α, and hence the lift coefficient, C_l, constant along the blade, then these relations translate to

$$c(\mu) = \frac{16\pi R}{9C_l N\lambda^2} \frac{1}{\mu} \text{ and } \beta = \frac{2}{3\lambda\mu} - \alpha \tag{7.2}$$

so that both the chord and twist are inversely proportional to radius.

In the case of machines operating at constant rotational speed, and hence at varying tip speed ratio, no parallel analytical solution for the optimum blade geometry exists. Instead resort must be made to numerical methods based on blade element – momentum theory, for example using Equations (3.51b) and (3.52a) in Section 3.8.6.

For pitch-regulated machines, the annual energy capture attributed to the annular ring swept out by each blade element is determined for the chosen wind speed distribution, and the variation of energy capture with blade chord and twist at each 'blade station' computed. In this way the values of blade chord and twist at each 'blade station' yielding maximum energy capture are identified.

For stall-regulated machines, the method is similar, but the total annual energy capture has to be maximized within the constraint of limiting the maximum total power output to the machine rating. The results of such an investigation are reported by Fuglsang and Madsen (1995).

7.1.3 Practical modifications to optimum design

The result of the optimization described in the previous section is typically a blade geometry in which both blade chord and blade twist vary approximately inversely with radius, as illustrated in Figure 3.19. However, because the inboard section of the blade makes only a small contribution to total power output (Figure 3.30), the aerofoil section is generally not continued inboard of about 15 percent radius in practice, and the chord at this radius is substantially reduced, to perhaps half the theoretical optimum. It is then often found expedient to taper the chord uniformly over the active length of the blade, with the tip chord and chord taper set so that the chord distribution approximates closely to the optimum over the outboard half of the blade (Figure 3.20).

The blade root area is normally circular in cross section in order to match up with the pitch bearing in the case of pitchable blades, or to allow pitch angle adjustment at the bolted flange (to compensate for non-standard air density) in the case of stall-regulated blades. The transition from the root section to the aerofoil section outboard of 15 percent radius should be a smooth one for structural reasons, with the result that the latter section will have a high thickness to chord ratio of up to about 50 percent.

7.1.4 Form of blade structure

A hollow shell corresponding to the defined blade envelope clearly provides a simple, efficient structure to resist flexural and torsional loads and some blade manufacturers adopt this form of construction (see Figure 7.1). However, in the case of small and medium size machines, where the out-of-plane loads dominate, there is greater benefit in concentrating skin material in the forward half of the blade, where the blade thickness is a maximum, so that it acts more efficiently in resisting out-of-plane bending moments (see Figures 7.2 and 7.3). The weakened areas of the

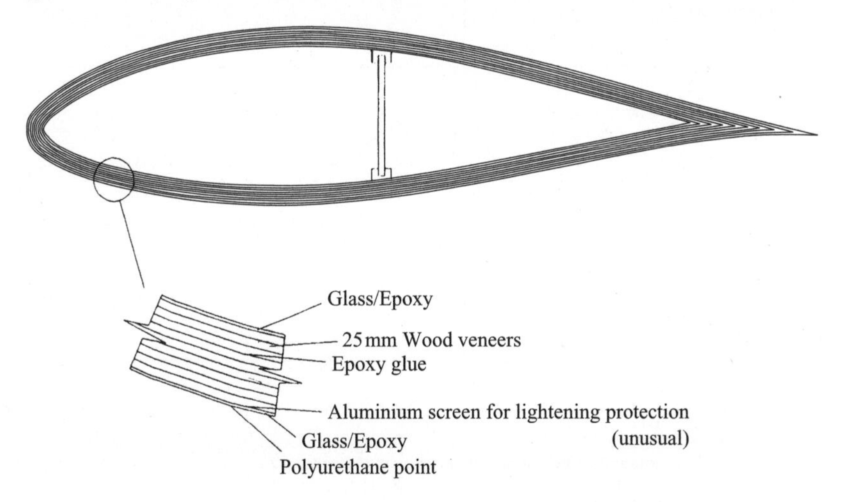

Figure 7.1 Wood/Epoxy Blade Construction Utilizing Full Blade Shell (Reproduced from Corbet (1991) by permission of the DT1 Renewable Energy R&D Programme)

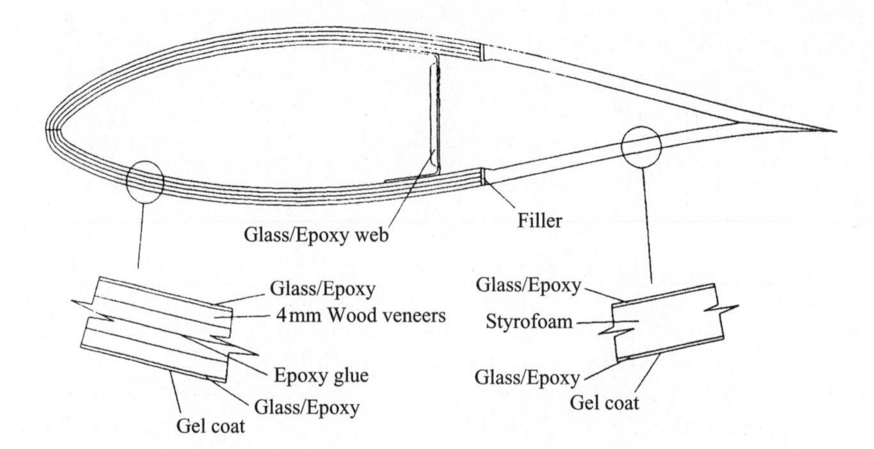

Figure 7.2 Wood/Epoxy Blade Construction Utilizing Forward Half of Blade Shell (Reproduced from Corbet (1991) by permission of the DT1 Renewable Energy R&D Programme)

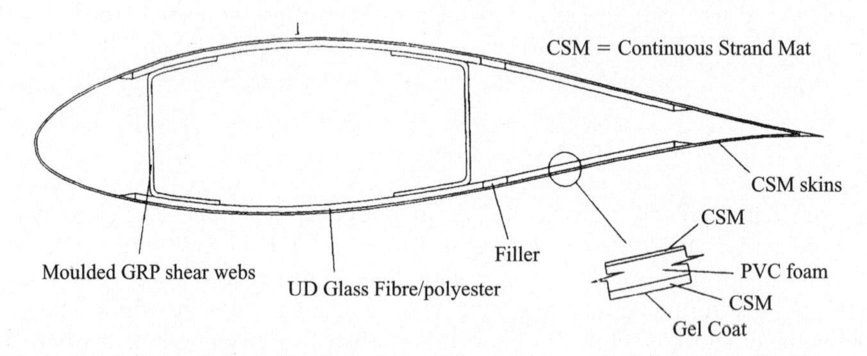

Figure 7.3 Glass-fibre Blade Construction Using Blade Skins in Forward Portion of Blade Cross Section and Linking Shear Webs. (Reproduced from Corbet (1991), by permission of the DT1 Renewable Energy, R&D Programme)

shell towards the trailing edge are then typically stiffened by means of sandwich construction utilizing a PVC foam filling.

The hollow shell structure defined by the aerofoil section is not very efficient at resisting out-of-plane shear loads, so these are catered for by the inclusion of one or more shear webs oriented perpendicular to the blade chord. If the load-bearing structure is limited to a compact closed hollow section spar, consisting of two shear webs and the skin sections between them (see Figure 7.4), then a GFRP blade lends itself to semi-automatic lay-up on a rotating mandrel which can be withdrawn after curing.

7.1.5 Blade materials and properties

The ideal material for blade construction will combine the necessary structural properties – namely high strength to weight ratio, fatigue life and stiffness – with low cost and the ability to be formed into the desired aerofoil shape.

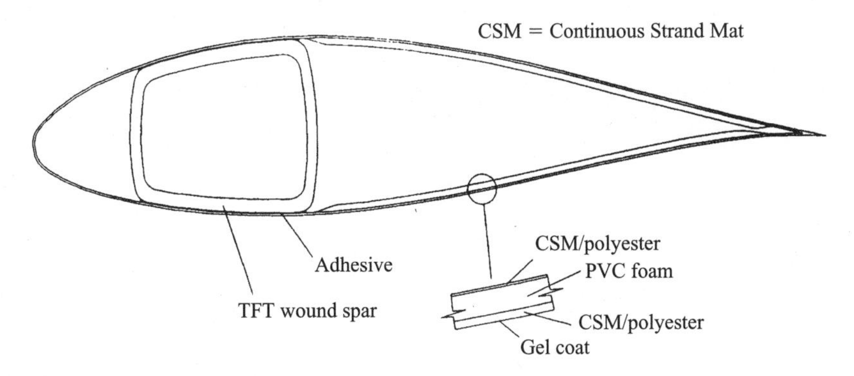

Figure 7.4 Glass-fibre Blade Construction Using Compact Spar Wound with Transverse Filament Tape (TFT) on Mandrel. (Reproduced from Corbet (1991), by permission of the DTI Renewable Energy R&D Programme)

Table 7.1 lists the structural properties of the materials in general use for blade manufacture and those of some other candidate materials. For comparative purposes, values are also presented of:

- compressive strength-to-weight ratio,
- fatigue strength as a percentage of compressive strength,
- stiffness-to-weight ratio,
- a panel stability parameter, $E/(UCS)^2$.

It is evident that glass- and carbon-fibre composites (GFRP and CFRP) have a substantially higher compressive strength-to-weight ratio compared with the other materials. However, this apparent advantage is not as decisive as it appears, for two reasons. First of all, the fibres of some of the plies making up the laminated blade shell have to be aligned off-axis (typically at $\pm 45°$) to resist shear loads, giving reduced strengths in the axial direction. Secondly, the relatively low Young's modulus of these composites means that resistance to buckling of the thin skins governs the design rather than simple compression yielding. The likelihood that buckling will govern is inversely related to the panel stability parameter, $E/(UCS)^2$, given in the last column of the table, so that materials with high values, such as wood composites will be least sensitive to buckling. As a result wood composite blades are generally lighter than equivalent glass-fibre composite blades. Design against buckling is considered in Section 7.1.10.

It should be noted that the low strength of wood laminate compared with other materials renders it unsuitable for blades with slender chords operating at high tip speed, where the flapwise bending moments during operation are inevitably high in relation to blade thickness. For example, Jamieson and Brown (1992) have shown that, in the case of a family of stall-regulated machines, the blade stress is highly sensitive to rotational speed, increasing as the fourth power, if the skin thickness to chord ratio is maintained constant. Although stresses can be reduced by increasing the skin thickness, this represents a less and less efficient use of the additional

Table 7.1 Structural Properties of Materials for Wind-turbine Blades

Material (NB: UD denotes unidirectional fibres – i.e., all fibres running longitudinally)	Ultimate tensile strength (UTS) (MPa)	Ultimate compressive strength (UCS) (MPa)	Specific gravity (s.g.)	Compressive strength to weight ratio UCS/s.g.	Mean fatigue strength at 10^7 cycles (amplitude) (MPa)	Mean fatique strength as percentage of UCS	Young's Modulus, E (GPa)	Stiffness to weight ratio E/s.g. (GPa)	Panel stability parameter $E/(\mathrm{UCS})^2$ $(\mathrm{MPa})^{-1}$
	(Mean for composites, minimum for metals)								
1 Glass/polyester *ply* with 50% fibre volume fraction and UD lay-up	860–900 [1] [2]	~720 [1]	1.85	390	140 [3]	19%	38 [2]	20.5	0.07
2 Glass/epoxy *ply* with 50% fibre volume fraction and UD lay-up	Properties are generally very close to those for GRP given above								
3 Glass/polyester *laminate* with 50% fibre volume fraction and 80% of fibres running longitudinally	690–720	~580	1.85	310	120	21%	33.5	18	0.1

4	Carbon fibre/epoxy *ply* with 60% fibre volume fraction and UD lay-up	1830 [4]	1100 [4]	1.58	700	350 [5]	32%	142 [4]	90	0.12
5	Khaya ivorensis/epoxy laminate	82 [6]	50 [6]	0.55	90	15 [7]	30%	10 [8]	18	4
6	Birch/epoxy laminate	117 [9]	81 [10]	0.67	121	16.5 [7]	20%	15 [10]	22.5	2.3
7	High Yield Steel (Grade Fe 510)	510	510	7.85	65	50 [11]	10%	210	27	0.81
8	Weldable aluminium alloy AA6082 (formerly H30)	295 [12]	295 [12]	2.71	109	17 [13]	6%	69 [12]	25.5	0.79

Sources:
[1] Mayer (1996) Figure 2.4.
[2] Barbero (1998) Table 1.1.
[3] Mayer (1996) Fig. 14.4.
[4] Carbon fibres exhibit a wide range of properties; figures given here are for one example only, taken from [2].
[5] Based on S–N curve index of $m = 14$, taken from GL rules.
[6] Bonfield and Ansell (1991) Moisture content = 10%.
[7] Based on S–N curve index of $m = 13.4$ for scarf-jointed wood laminates, taken from Hancock and Bond (1995).
[8] Bonfield *et al.* (1992).
[9] Mayer (1996) Table 7.3.
[10] Hancock (Personal Communication). Moisture content = 10%.
[11] Mean value for butt-welded joints with weld profile ground smooth (Class C), taken from BS 5400, Part 10 (1980).
[12] CP 118: 1969 'Code of practice for structural aluminium'.
[13] Mean value estimated from mean minus two standard deviations value for ground butt welded joint with shallow thickness transition, Detail Cat 221, in IIW 'Fatigue design of welded joints and components'.

material beyond a skin thickness to chord ratio of 3–4 percent, especially in the outboard part of the blade, where the blade thickness to chord ratio is low.

Fatigue performance is conveniently measured by mean fatigue strength at 10^7 cycles, as a percentage of ultimate compressive strength. Clearly, carbon-fibre and khaya/epoxy perform best with a value of about 30 percent. The low value for welded steel (10 percent), combined with steel's low strength-to-weight ratio, renders it uncompetitive for large diameter machines where gravity fatigue loading becomes important, although it was chosen for some of the early prototype megawatt scale machines when the fatigue properties of composite materials were less well understood.

The stiffness-to-weight ratio determines blade natural frequency. Apart from CFRP, the values in the table are all in a relatively small range (18–27 GPa), indicating that material choice will generally only have a marginal effect on dynamic behaviour.

From the above brief survey, it is apparent that the material with the best all-round structural properties is carbon-fibre composite. However, it has not found common use because it is an order of magnitude more costly than other materials. Instead, the most popular material is glass/polyester, followed by glass/epoxy and wood/epoxy.

Steel is the cheapest material in the raw state, and can be formed into tapering, curved panels following the aerofoil profile, except in the sharply curved region near the leading edge. However, it is much harder to introduce a twist into such panels, and this consideration, together with the poor fatigue properties, means that steel is rarely used. By contrast, glass- and carbon-fibre composites lend themselves to wet lay-up in half-moulds profiled to give the correct aerofoil shape, planform and twist. Laminated wood composite blades are built up in a similar way, but the veneer thickness has to be restricted to enable the veneers to flex to the required curvature during lay-up.

In the following paragraphs, the properties of the materials in most common use for blade manufacture are considered in more detail.

7.1.6 Properties of glass/polyester and glass/epoxy composites

As noted in Table 7.1, the properties of glass/polyester and glass/epoxy plies with the same fibre volume fraction and lay-up are generally very similar, i.e., the influence of the matrix is slight. They will therefore be treated as the same material in the discussion that follows, except in relation to fatigue, where some differences have been noted. The glass used in blade construction is E-glass, which has good structural properties in relation to its cost.

The plate elements forming the spar of a GFRP blade are normally laminates consisting of several plies, with fibres in different orientations to resist the design loads. Within a ply (typically 0.5–1.0 mm in thickness), the fibres may all be arranged in the same direction, i.e. UD or unidirectional or they may run in two directions at right angles in a wide variety of woven or non-woven fabrics.

Although the strength and stiffness properties of the fibres and matrix are well-defined, only some of the properties of a ply can be derived from them using simple rules. Thus, for a ply reinforced by UD fibres, the longitudinal stiffness modulus, E_1, can be derived accurately from the rule of mixtures formula

$$E_1 = E_f V_f + E_m(1 - V_f) \tag{7.3}$$

where E_f is the fibre modulus (72.3 GPa for E-Glass), E_m is the matrix modulus (in the range 2.7–3.4 GPa) and V_f is the fibre volume fraction. On the other hand, the inverse form of this formula, e.g.,

$$\frac{1}{E_2} = \frac{(1 - V_f)}{E_m} + \frac{V_f}{E_f} \tag{7.4}$$

significantly underestimates the transverse modulus, E_2, and the in-plane shear modulus, G_{12}. More accurate formulae based on more sophisticated models are given in Barbero (1998).

The longitudinal tensile strength of a ply reinforced by UD fibres, σ_{1t}, can be estimated from:

$$\sigma_{1t} = \sigma_{fu}\left[V_f + \frac{E_m}{E_f}(1 - V_f)\right] \tag{7.5}$$

where σ_{fu} is the ultimate tensile strength of the fibres. However, the tensile strengths of E-glass single fibres (3.45 GPa) cannot be realized in a composite, where fibre strength reductions of up to 50 percent have been measured. Accordingly, a value of σ_{fu} of 1750 MPa should be used in Equation (7.5).

The longitudinal compressive strength of a ply reinforced by UD fibres is always significantly less than the tensile strength because of microbuckling of the fibres, which is governed by the shear strength of the matrix and the degree of fibre misalignment. A strength reduction of at least 15 percent should be allowed for, assuming minimum fibre misalignment.

Clearly, longitudinal stiffness and strength are both limited by the fibre volume fraction obtainable. For hand lay-up, fibre volume contents of 30–40 percent are typical, but the use of 'vacuum bagging', in which trapped air and excess volatile compounds, such as residual solvent, are extracted, consolidates the composite and allows a volume fraction of 50 percent or more to be achieved. The use of 'pre-pregs', which are unidirectional fibres or woven fabric pre-impregnated with partially cured epoxy resin, results in similar increased fibre volume fractions.

Fatigue properties

When expressed in terms of stress, the fatigue properties of composite laminates extend over a wide range, depending on fibre volume fraction and the number of plies with fibres in the longitudinal direction. However, data from constant stress amplitude fatigue test results become much more intelligible if stress ranges are converted into initial strain ranges, allowing the fatigue properties of composites with different lay-ups to be compared. (The Young's modulus of a composite reduces over time during a fatigue test – hence the need to specify that the strain range is measured at the start of the test.)

The fatigue behaviour of composites depends on both the stress range *and* the mean stress level, which can both be described in terms of the maximum stress, σ_{max}, and the ratio of minimum to maximum stress, R. It is convenient initially to consider fatigue behaviour under reverse loading, i.e., with $R = -1$, for which the mean stress is zero, and then relate behaviour at other R ratios to it.

The constant amplitude fatigue behaviour of glass-fibre composites can best be characterized by a linear relationship between the logarithm of the number of cycles and the logarithm of the stress or strain amplitude, viz:

$$\varepsilon = \varepsilon_0 N^{-1/m} \text{ or } N = K\varepsilon^{-m} \text{ where } K = (\varepsilon_0)^m \text{ or } \log N = \log K - m \log \varepsilon \tag{7.6}$$

Echtermeyer, Hayman and Ronold (1996) carried out a regression analysis on a total of 111 constant amplitude, reverse loading fatigue test results for 10 different laminates tested at DnV, assuming that they all conformed to the same ε–N curve, and obtained values for ε_0, $\log K$ and m of 2.84 percent, 3.552 and 7.838 respectively, with a standard deviation of $\log N$ of 0.437. The DnV regression line is compared with another derived from 19 tests on a 0°/+45°, −45° laminate at ECN, giving $\varepsilon_0 = 2.34$ percent, $\log K = 3.775$ and $m = 10.204$, in Figure 7.5. The researchers did not constrain the regression lines to pass through the strain value at either UTS or UCS at $\log N = 0$ (*ca* 2.4 percent and 2.0 percent); had they done so the DnV line would have had a shallower slope, i.e., a larger value of m. After comparison with regressions on other fatigue test datasets, they concluded that the DnV line provided a reasonable basis for initial design.

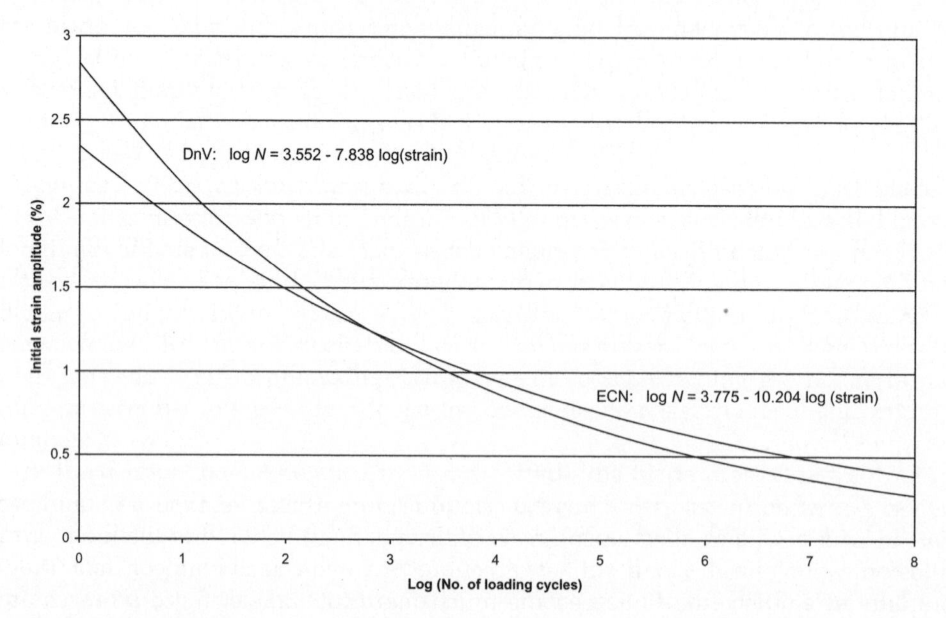

Figure 7.5 Strain-life Regression Lines Fitted to Results of Constant Amplitude, Reverse Loading Fatigue Tests on GFRP Composites

In the GL rules, the ε–N curve for design takes the same form as Equation (7.6) for the case of reverse loading, but ε_0 is set equal to the strain at ultimate tensile load and partial materials safety factors are included. GL specifies different values of the index, m, for composites with polyester and epoxy matrixes – 9 and 10 respectively – although opinion is divided as to whether the difference is justified.

Constant amplitude tests at other R ratios generally show reducing fatigue lives as the mean stress increases above zero – whether in tension or in compression. It is customary to represent the results on a constant life diagram (also known as a Goodman diagram), in which the stress range to failure is plotted against mean stress for different fatigue lives. Regression analyses can be carried out on families of test results at different R ratios to give a series of ε–N relations in the form of Equation (7.6) which can be used to plot the constant life diagram. Such an exercise was carried out on the Dutch 'FACT' database of fatigue tests on composites for wind turbines (Joose and van Delft, 1996), and some results of this work have been reproduced in Figure 7.6 (dashed lines).

In the preparation of design rules, it is common practice to make the simplifying assumption that the strain amplitude reduces linearly with increasing mean strain for a given fatigue life, reaching zero at a mean strain corresponding to either the ultimate tensile or compressive strength. Such linear constant life lines are shown in Figure 7.6 as heavy lines. Constant life lines for design are obtained from characteristic material properties divided by appropriate partial safety factors, as opposed to the characterisitc properties used in Figure 7.6. Thus the design strain amplitude when the mean stress is compressive becomes:

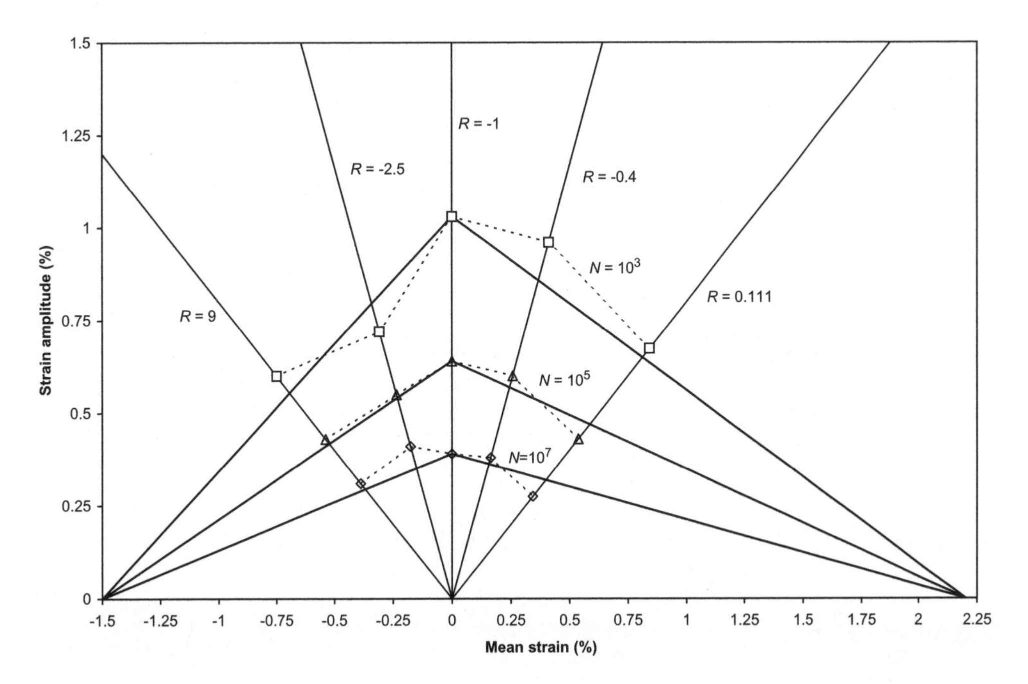

Figure 7.6 Constant Life Diagram Showing Variation in Fatigue Strain Amplitude with Mean Strain for Lives of 10^3, 10^5 and 10^7 Cycles for GFRP Composites

$$\varepsilon_d(\overline{\sigma}, N) = \varepsilon_{0d} N^{-1/m} \left(1 - \frac{\overline{\sigma}}{\sigma_{cd}}\right) \tag{7.7}$$

where $\varepsilon_{0d} = \varepsilon_{0k}/\gamma_{mf}$, $\sigma_{cd} = \sigma_{ck}/\gamma_{mu}$, ε_0 is the value of ε given by the ε–N curve when $\log N$ is zero, $\overline{\sigma}$ is the mean stress for the loading cycles under consideration, and σ_{cd} is the design compressive stress. γ_{mf} is the partial safety factor for fatigue strength, γ_{mu} is the partial safety factor for ultimate strength, and the suffices d and k signify design and characteristic values respectively.

Equation (7.7), together with its equivalent for mean tensile loading, can be used to calculate the permissible number of load cycles, N_i, for each strain range in the fatigue loading spectrum for the point in the blade cross section under examination – incorporating the appropriate partial safety factor for the consequences of failure. These are then combined with the predicted number of cycles for each strain range, n_i, to yield Miner's damage sum, $\sum_i (n_i/N_i)$, which is normally required to be less than unity.

There is inevitably a degree of uncertainty as regards the accuracy of Miner's rule in predicting the fatigue damage due to variable amplitude loading from constant amplitude test data. In order to investigate this, fatigue test programmes have been carried out using the WISPER (Wind SPEctrum Reference) and WISPERX variable amplitude fatigue load spectra, which have been devised to be representative of those experienced by wind turbine blades. (WISPERX is a modification of WISPER in which the large number of small cycles, accounting for approximately 90 percent of the total, are omitted to reduce test durations.) For each test specimen, the WISPER (or WISPERX) load sequence is scaled to give a chosen maximum stress level and applied repeatedly until the specimen fails.

Van Delft, de Winkel and Joose (1996) analysed the results of a series of tests carried out at ECN and Delft Technical University on a 0°, ±45° laminate and found that, for a maximum stress of about 150 MPa, the actual fatigue lives of specimens subjected to repetitions of the WISPER or WISPERX load sequences were about 100 times less than predicted for these sequences on the basis of constant amplitude, reverse loading test data and Miner's rule, with the effect of mean stress allowed for using the linear relation described above. The $R = -1$ test data led to an S–N curve given by $N = (\sigma/\sigma_{tu})^{-10}$, where σ is the amplitude of the stress cycles and σ_{tu} is the ultimate tensile strength, so the number of cycles to failure for constant amplitude loading at other R values was taken as $N = (\sigma/\sigma_{tu}(1 - \overline{\sigma}/\sigma_{tu}))^{-10}$ for a tensile mean and $N = (\sigma/\sigma_{tu}(1 - \overline{\sigma}/\sigma_{cu}))^{-10}$ for a compressive mean in calculating the Miner's damage sum. The difference in fatigue lives at the stated maximum stress level quoted above translates to an approximate ratio of 1:1.5 between actual and predicted maximum stress levels of the WISPER sequence to cause failure over the design fatigue life, which would clearly use up a substantial proportion of the safety factors used in design. However, other investigators working with different laminates have found reasonable agreement between measured and predicted fatigue lives under WISPER loading.

Material safety factors

Limit-state design requires that the characteristic strength of a material be divided by a partial safety factor for material strength. In the case of GFRP, this factor needs to take account of degradation of the material over time, as well as the material's inherent variability.

The GL rules lay down that the material safety factor for calculating the design strength of GFRP under extreme loads is to be calculated as the product of

- a basic factor, γ_{MO}, of 1.35
- a factor, $C_{2a} = 1.5$, to account for the influence of ageing,
- a factor, $C_{3a} = 1.1$ to account for strength reduction at higher temperatures,
- a factor, $C_{4a} = 1.2$ for hand lay-up laminates or 1.1 where manufacture is partially automated,
- a factor, $C_{5a} = 1.1$ if the laminate is not post cured.

These rules result in a material safety factor in the range 2.45–2.94. In the case of fatigue loads, the ageing factor of 1.5 is omitted and the factor accounting for lay-up is replaced by one taking account of the type of fibre reinforcement.

7.1.7 Properties of wood laminates

Although laminated wood/epoxy is classed as a composite, it is markedly different in form from GFRP. Individual plies are made up of large sheets of wood veneer (Figure 7.7) instead of a multiplicity of fibres laid up in a matrix, and the epoxy behaves as an adhesive rather than a matrix, bonding the sheets together at the longitudinal and transverse joints and bonding each ply to its neighbour. Thus the fibre volume fraction is close to 100 percent and the anisotropic properties of the wood laminate derive principally from the anisotropic properties of the wood itself.

Wood strength properties are much greater in the direction parallel to the grain, so all the veneers are orientated with the grain parallel to the blade axis, in order to resist blade-bending loads efficiently. However, the veneers cannot be produced in lengths much greater than 2.5 m, so transverse joints have to be included, which introduce lines of weakness not normally found in GFRP blades. The effect is minimized by staggering the joints, and by using scarf joints in preference to butt joints.

The epoxy adhesive has a secondary function of sealing the veneers against moisture ingress; additional moisture protection is provided by a layer of glass/epoxy on both the external and internal surfaces. It is important to maintain moisture content at a low level, because veneer strength decreases about 6 percent for every 1 percent rise in moisture content.

A comparison of some of the properties of wood laminates used, or considered for use, in wind turbine blades is given in Table 7.2. Khaya ivorensis, an African mahoghany, and Douglas fir used to be the main species used for blade manufacture in the UK and US respectively, but environmental pressures have led to the phasing out of Khaya in favour of European species such as poplar and birch.

The table gives tensile strengths of unjointed specimens. Bonfield *et al.* (1992) report the results of tests on jointed specimens, which showed a significant

Figure 7.7 Blade Production. View of Veneer Lay-up in Mould to Make One Blade Skin. The Blade is Completed by Glueing Face and Camber Skins Together. (Reproduced by permission of NEG-Micon)

Table 7.2 Properties of Unjointed Wood/Epoxy Laminates

Species	Specific gravity	Mean tensile strength along the grain (MPa)	Mean compression strength along the grain (MPa)	Young's Modulus along the grain (GPa)	Shear strength (MPa)
Khaya ivorensis	0.55	82	50	10	9.5
Poplar	0.45	63	52	10	9
Baltic pine	0.55	105	40	16	
Birch	0.67	117	81	15	16
Beech	0.72	103	69	10	16
Douglas fir	0.58	100	61	15	12

reduction in tensile strength to 50 MPa for butt jointed Khaya. Scarf jointed Khaya specimens, with a 1:6 length to thickness ratio, performed much better, achieving a tensile strength of 75 MPa. In all cases the joints in the different veneers making up the laminate were staggered.

An important consideration for design is the variability of strength properties, particularly as wood is an inherently variable material. Strength tends to increase with density, and density varies according to the growing conditions of the tree and the part of the tree from which the wood is taken. Such variability can be reduced by careful grading and the rejection of damaged veneers before laminating. Bonfield and Ansell (1991) report compression tests on 32 carefully selected Khaya samples which yielded the compression strength of 50 MPa given in the table with a standard deviation of only 3 MPa. It should be noted that the lack of annual growth rings in equatorially grown wood may reduce the degree of scatter. Wood strengths perpendicular to the grain are typically much less than those along the grain – for example, the compressive strength of transversely loaded Khaya is only 12.6 MPa.

Fatigue properties

The fatigue properties of wood laminates have been the subject of a sustained programme of work at Bath University, starting with Khaya and then extending to other species (Bonfield *et al.* (1992)). A useful summary of this work appears in Bond and Ansell (1998). The general conclusion is that wood performs very well in fatigue with a shallow $S-N$ curve slope, and that fatigue strengths at high cycles do not vary greatly between species.

If the $S-N$ curve for constant amplitude, reverse loading ($R = -1$) fatigue is normalized with respect to the ultimate compressive strength, σ_{cu} – i.e., $\sigma = \sigma_{cu}N^{-1/m}$, then the results of tests on unjointed Khaya indicate a value of the index m of about 20. However, the value of m reduces to about 16 for scarf-jointed khaya, poplar and beech, and to about 13 for butt-jointed specimens. Hancock and Bond (1995) have proposed the use of an index of 13.4 for design purposes for scarf-jointed wood laminates in general.

Testing at other R ($= \sigma_{min}/\sigma_{max}$) ratios allows constant life diagrams to be plotted – see, for example, the diagram for scarf-jointed poplar in Figure 7.8, taken from

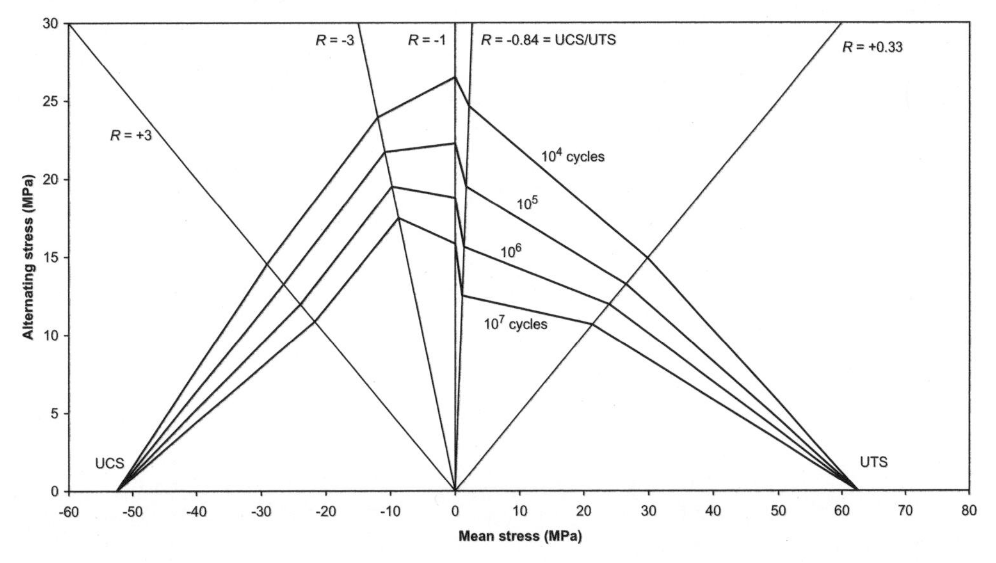

Figure 7.8 Constant Life Diagram for Scarf-Jointed Poplar Derived from 50 percent Median Regression Lines on $S-N$ Fatigue Test Data

Bond and Ansell (1998). Note the relatively low stress ranges at $R = -0.84$ (= UCS/UTS), which may be due to simultaneous occurrence of compressive and tensile damage. Despite this, the simplification of the constant life diagram to a series of straight lines between the $R = -1$ stress range for each fatigue life and either the UTS or UCS is reasonably accurate.

Material safety factors

The material safety factor applied when timber is used in building construction is normally high, e.g., about 3–4. However, there are a number of reasons for adopting a much lower value in blade design:

(1) laminated construction is used, so any defects are very localized;

(2) the moisture content is carefully controlled during manufacture, and the blade skin is then very effectively sealed against further moisture ingress;

(3) creep effects are negligible as the gravity loads change direction because of blade rotation and the wind loads are temporary in nature.

Accordingly a partial safety factor of only about 1.5 is normal for design against extreme loads.

7.1.8 Governing load cases

Extreme loading during operation: stall-regulated machines

As described in Chapter 5, wind turbine design codes specify a number of load cases consisting of various combinations of defined wind speed and direction changes – some of them involving external or machine faults – which are an attempt to define an envelope of the worst loadings to be expected in practice. It is instructive to take one such code, IEC 61400-1, and compare the blade loadings arising from the different load cases for a particular design. The WTG chosen is a 40 m diameter, 500 kW stall-regulated Class II machine fitted with TR blades (see Figure 5.2(a)) and operating at a single rotational speed of 30 rpm. The rated wind speed, U_r, and cut out speed, U_O, are 16 m/s and 25 m/s respectively. The shaft tilt with respect to the horizontal is taken as 5° so, allowing for a 8° inclination of the flow to the horizontal as specified in the code, the maximum shaft tilt with respect to the flow is 13°. Category A turbulence characteristics are assumed.

The table below summarizes the conditions applicable to the power production load cases (apart from those involving a machine fault) and compares the resulting peak out-of-plane blade bending moment at 60 percent radius, ignoring dynamic effects. In the cases involving a change of wind direction, the maximum yaw error is simply taken as equal to the maximum direction change i.e., ignoring any action taken by the control system to correct it.

Table 7.3 Wind Conditions for Power Production Load Cases for Class IIA 40 m Diameter WTG with 16 m/s Rated Speed and Calculated Maximum Out-of-plane Bending Moments at 12 m Radius Ignoring Dynamics, for Stall-regulated Machine

Load case	Description	Hub-height wind speed (m/s)			Wind direction change = yaw error	Maximum shaft tilt with respect to flow	Maximum inclination of flow to shaft axis	Critical blade azimuth for maximum bending moment	Maximum out-of-plane bending moment at 12 m radius (kNm)
		Initial	Gust	Max					
1.1	Normal turbulence model – extreme			40.4	27°	13°	30°	~25°	130
(1.2)	(Normal turbulence model – fatigue)								
1.3	Extreme coherent gust with direction change	16	15	31	45°	13°	46.8°	~15°	118
1.4	Electrical fault (not considered)								
1.5	Grid loss with 1 year extreme operating gust	25	11.6	36.6	–	13°	13°	90°	115
1.6	50 year extreme operating gust	25	15.5	40.5	–	13°	13°	90°	108
1.7	Extreme wind shear	25	–	25	–	13°	13°	90°	90
1.8	50 year extreme direction change	25	–	25	47.8	13°	49.5°	~15°	108
1.9	Extreme coherent gust	16	15	31	–	13°	13°	90°	88

The blade loadings are calculated using empirical three-dimensional aerofoil data taken from Petersen *et al.* (1998) – see Figure 5.9. This displays a gentler stall than typical two-dimensional data, so there is no significant reduction in blade out-of-plane bending moment as the blade goes into stall. Above about 20 m/s, the out-of-plane bending moment begins to increase progressively once again as drag begins to become significant. The predicted variation of blade 12 m radius out-of-plane bending moment with wind speed is plotted out for a 0.2 shear exponent and a range of yaw angles on Figure 7.9, with the yaw direction defined as positive when the lateral component of air flow with respect to the rotor disc is in the same direction as the blade movement at zero azimuth (i.e., at 12 o'clock). The effect of this increase in relative velocity outweighs that of the reduction of angle of attack at wind speeds beyond stall, so the bending moment at 0° azimuth is increased by negative yaw. Maximum moments occur at negative yaw angles and 0° azimuth rather than at positive yaw angles and 180° azimuth, because wind shear augments the wind speed in the former case. Also plotted is the variation of bending moment with wind speed for a 13° shaft tilt with respect to flow and 90° azimuth, which is the critical configuration for load cases not involving a change in wind direction.

Considering the deterministic load cases 1.3 and 1.5 to 1.9 initially, it is interesting to note that the maximum out-of-plane bending moments lie within a relatively close range in four of the six cases. It should be pointed out that, in the grid-loss case, the bending moment depends on the rotor acceleration after loss of load, which is largely determined by rotor inertia and the time delay to tip brake deployment. The bending moment quoted is a notional one, based on a generous 1.5 s time delay until full deployment and an inertia value calculated for a fibreglass rotor. With lighter rotors,

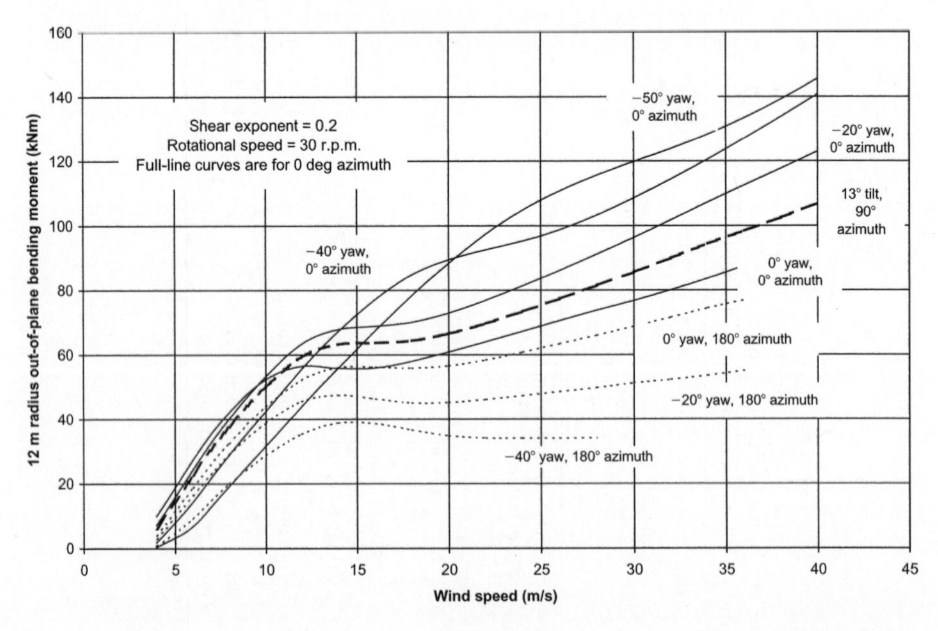

Figure 7.9 Variation of 12 m Radius Out-of-plane Bending Moment with Wind Speed at Various Yaw Angles for an Example 40 m Diameter Stall-regulated Machine with TR Blades

the grid-loss case is liable to become the critical one as regards out-of-plane bending moments, unless the tip deployment time can be reduced sufficiently.

The 'Normal turbulence model' specified for load case 1.1 includes random variations of both the longitudinal wind velocity component about the 10 min mean and the lateral component about a zero mean. Both components are taken to be Gaussian distributed, with the standard deviation (m/s) of the longitudinal component of turbulence given by

$$\sigma_1 = 0.18(\tfrac{1}{3}15 + \tfrac{2}{3}U_{\text{hub}}) \tag{7.8}$$

for a Class A site and the standard component of the lateral component of turbulence taken as $0.8\sigma_1$. The design load case is then the set of conditions existing over at least one rotation cycle that produces the extreme value of the loading under evaluation.

Utilizing the probability distribution of the longitudinal wind velocity component about the 10 min mean in combination with a Weibull distribution of 10 min mean wind speeds on the one hand, and the probability distribution of the lateral wind velocity component on the other, it is possible to establish, for wind speed 'bins' corresponding to different values of the longitudinal wind speed component, the maximum yaw angle that would occur for at least one rotation cycle over the machine design life. For each case, the out-of-plane bending moment at 12 m radius is calculated (allowing for shaft tilt), enabling the critical case to be identified. It is seen from Table 7.1 that the combination of 40.4 m/s wind speed and 27° yaw angle is the most severe, producing a moment of 130 kNm, which is 10 percent larger than the moment produced by the worst of the deterministic load cases (load case 1.3). Note that the 40.4 m/s wind speed here is the resultant of a longitudinal velocity component, assumed parallel to the shaft axis in plan, of 36 m/s and a lateral component of 18.3 m/s. Thus the yaw error arises from an additional lateral component which increases the total wind speed, in contrast to load cases 1.3 and 1.8 where the wind merely changes direction.

The derivation of the extreme 12 m radius out-of-plane bending moment for the 'Normal turbulence model' load case described above is conservative on three counts, because no allowance is made for the following:

- lack of correlation of the wind over the outer 40 percent of the blade,
- limitation on maximum wind speed seen during operation by high wind cut-out,
- limitation on maximum yaw angle by yaw control.

The alleviation of extreme loadings by high wind cut-out and yaw control depend on the averaging times applied to the wind speed and direction signals by the control system.

Extreme loading during operation: pitch-regulated machines

The characterization of extreme operational loadings on pitch-regulated machines is inevitably more complicated than for stall-regulated machines, although at the same time it should be more accurate because of the avoidance of uncertainties associated with stall. It is instructive to focus comparisons on the blade bending moment about the weak axis at 60 percent radius once again. This time it is referred to as the flapwise bending moment rather than the out-of-plane (of rotation) moment because of blade pitching.

Figure 7.10 presents the variation of 60 percent radius flapwise bending moment with short-term mean wind speed at several yaw angles for a 500 kW, 40 m diameter pitch-regulated machine fitted with TR blades and rotating at 33 r.p.m. The rated speed, V_R, is 12 m/s and other parameters, including the wind shear exponent, are the same as in the stall-regulated example above. The figure only shows the bending moments resulting from slow variations in wind speed, i.e., those which can be followed by the pitch control system, so moments arising from faster wind speed fluctuations must be added to obtain the total. The curves are very different in shape from those obtained for the stall-regulated machine. The 12 m radius flapwise bending moment reaches a peak at rated wind speed, and then drops off sharply, becoming negative by about 24 m/s in the case of zero yaw and zero wind shear. This is because the blade has pitched to such an extent that the outboard section of the blade is providing a braking torque to counteract the increased torque from the inboard section. At high wind speeds and yaw angles, large negative bending moments are developed, which approach the magnitude of the peak positive moment at rated speed. Note that the bending moment *reduces* with negative yaw angle at zero azimuth, instead of increasing as it does for stall-regulated operation. This is because blade pitching renders angle of attack, which is reduced under these conditions, more critical than relative velocity. Plots of the

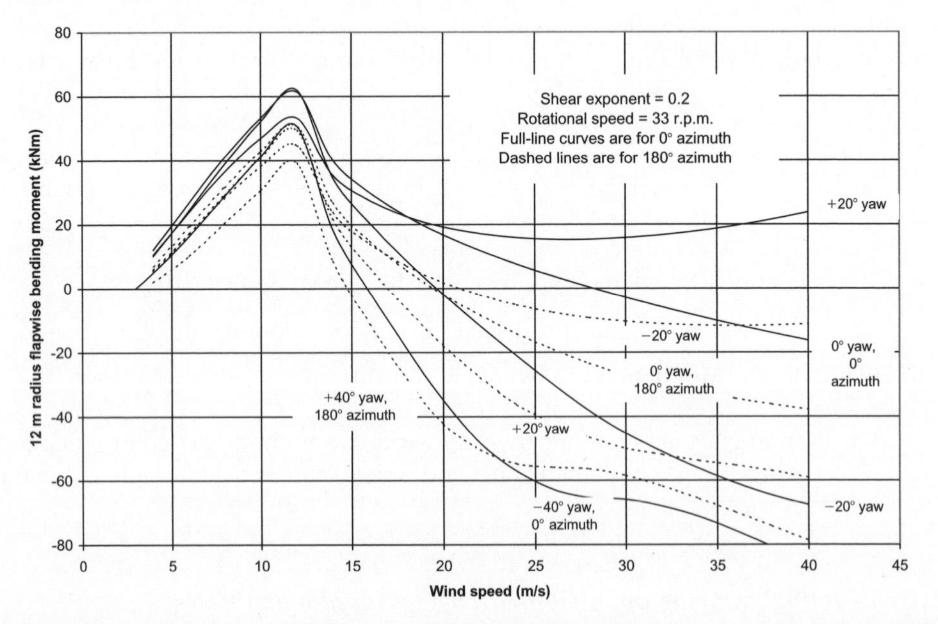

Figure 7.10 Variation of 12 m Radius Flapwise Bending Moment with Short-term Mean Wind Speed at Various Yaw Angles for an Example 40 m Diameter Pitch-regulated Machine with TR Blades

variation of flapwise bending moment with short-term mean wind speed at inboard blade cross sections are essentially similar to those in Figure 7.10, because moments are dominated by loadings on the outboards portion of the blade.

To the extent that the pitch control system can keep pace with the wind speed transients, the curves in Figure 7.10 can be used to provide an approximate indication of the extreme bending moments arising from some of the IEC 61400-1 deterministic load cases. Leaving aside the grid-loss case (where the outcome is largely determined by rotor inertia and emergency pitching rate), it is seen that the extreme moments are only about one half of the maximum value for the stall-regulated machine.

The spectrum of the longitudinal wind speed fluctuations will contain significant energy at frequencies above the level at which the pitch control system can respond, and these have to be considered in the analysis of the 'Normal turbulence model' load case. Figure 7.11 illustrates the perturbations in 12 m radius flapwise bending moment for the above machine, as a result of such high frequency wind speed fluctuations, with respect to a sharp rise above rated wind speed (12 m/s) and sharp falls below steady winds of 24, 28 and 32 m/s. In the first case considered, the yaw angle is $-20°$ and the azimuth $0°$, as this configuration yields the largest positive bending moment at rated wind speed. In the second case, the yaw angle is $-40°$, which exceeds the maximum value predicted over the design life, and provides an upper bound on the largest negative moment at short-term mean wind speeds around the cut-out value.

It is apparent from Figure 7.11 that rapid wind speed increases above rated wind speed will produce a significant increase in bending moment, but rapid reductions in wind speed below, say, a 24 m/s steady value will not, as in this case the blade goes into negative stall. Over the machine lifetime, the maximum increase in wind

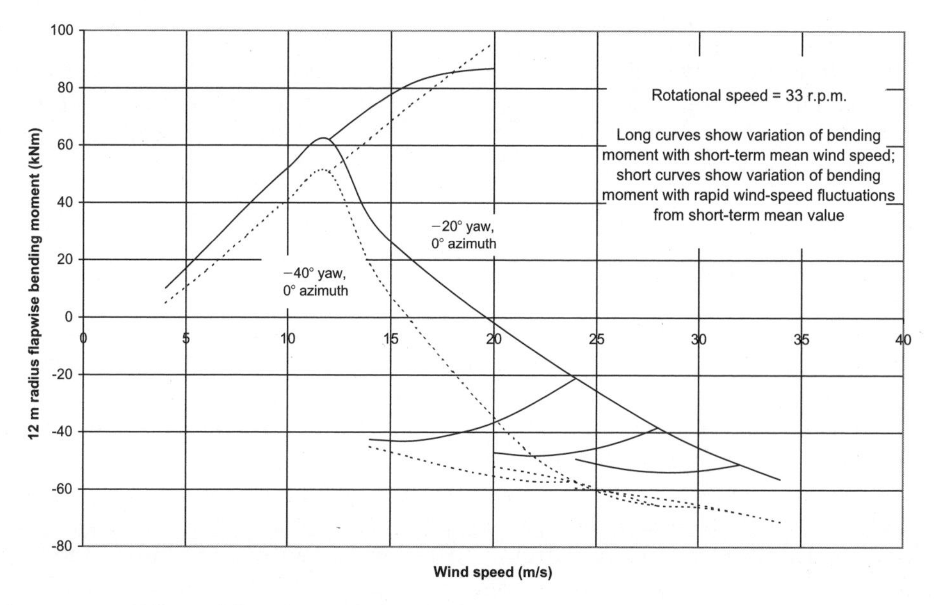

Figure 7.11 Effect of Rapid Wind-Speed Fluctuations on 12 m Radius Flapwise Bending Moment for an Example 40 m Diameter Pitch-regulated Machine with TR Blades

speed above rated that does not produce a blade pitch response can be estimated using

$$u_{\max} = \sigma_{\mathrm{u}} \sqrt{\frac{\int_{\Omega/2}^{\infty} S_u(n)}{\int_0^{\infty} S_u(n)}} \left[\sqrt{2\ln(\Omega T)} + \frac{\gamma}{\sqrt{2\ln(\Omega T)}} \right]$$

$$= (\sigma_u)_{n>\Omega/2} \left[\sqrt{2\ln(\Omega T)} + \frac{0.5772}{\sqrt{2\ln(\Omega T)}} \right] \tag{7.9}$$

where $(\sigma_u)_{n>\Omega/2}$ is the standard deviation of wind speed fluctuations above the pitch response cut-off frequency (assumed to be half the rotational frequency) and T is the total period of operation in the wind speed band centred on the rated speed at the yaw angle under consideration. For a 12 m/s rated wind speed, $\sigma_u =$ 2.34 m/s from Equation (7.8) and $(\sigma_u)_{n>\Omega/2} = 2.34(0.4)^{0.5} = 1.48$ m/s. Taking a wind speed band of 2 m/s and yaw angles between −20° and −40°, the expression in square brackets (i.e., the peak factor) comes to 5.5, so that the lifetime extreme value of the wind speed increase without pitch response is about 8 m/s. If the wind speed fluctuations over the outer 8 m of blade are treated as perfectly correlated, this results in a maximum value of 12 m radius flapwise bending moment of 96 kNm (see Figure 7.11), which is over 50 percent greater than that occurring in a steady 12 m/s wind. Thus the extreme flapwise bending moment during operation occurs at winds around rated rather than around the upper cut-out speed – a phenomenon which is a normal feature of pitch-regulated machines. Also, the extreme flapwise bending moment is less than for the similarly rated stall-regulated machine considered above.

The relative criticality of the load cases corresponding to extreme turbulent wind speed fluctuations and the occurrence of the 50 year extreme operating gust during operation at rated wind speed, which are exemplified by IEC load cases 1.1 and 1.6 respectively, will be determined by pitch control system performance.

Extreme loading at standstill

The derivation of stationary blade loads is described in Section 5.6. Figure 7.12 shows the out-of-plane bending moment distribution for a TR blade under the action of a 60 m/s wind, corresponding to the 50 year return extreme 3 s gust specified for a Class II wind turbine in IEC 61400-1. A uniform lift coefficient of 1.5 is assumed. Two curves are shown: the lower curve is the quasistatic bending moment while the upper one incorporates dynamic magnification due to excitation of resonant oscillations (see Section 5.6.3).

For comparison purposes, the extreme operational out-of-plane bending moment (excluding dynamics) for the example stall-regulated machine (load case 1.1, Table 7.1) is also plotted. It is clear that the operational load case is critical over most of the blade length, though only by a small margin. The non-operational bending

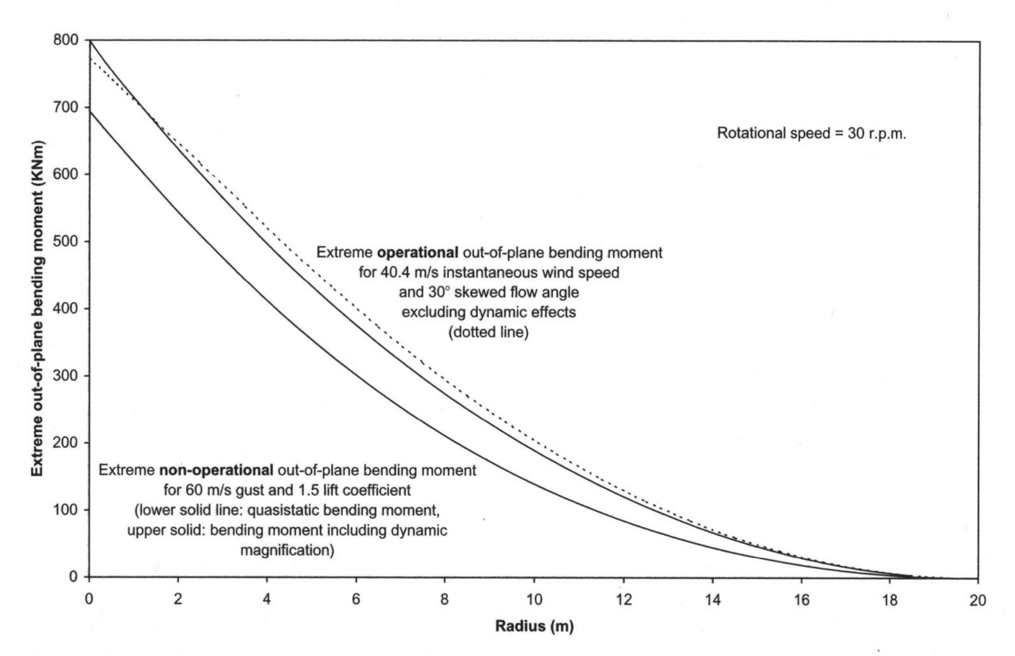

Figure 7.12 Comparison of Extreme Operational and Non-operational Out-of-plane Bending Moment Distributions for a 40 m Diameter Stall-regulated Machine with TR Blades

moment distribution is more highly curved than the operational one, so the former is more likely to govern at the root and the latter to govern outboard.

Extreme non-operational loads tend to govern blade out-of-plane bending on pitch-regulated machines, as extreme operational out-of-plane bending moments are generally significantly less than on stall-regulated machines.

Fatigue loading

The importance of fatigue loading relative to extreme loading is very much a function of material properties. As the vast majority of blades are manufactured from composite materials with similar fatigue properties, discussion in this sub-section will be based on these.

As set out in Sections 7.1.6 and 7.1.7, composite materials are characterized by a very shallow S–N curve i.e., the reciprocal index m in the relation $\sigma \propto N^{-1/m}$ for constant amplitude, reversed loading ($R = -1$) is typically 10 or more. As a result, fatigue damage can be dominated by the small number of high range stress cycles associated with unusual wind conditions, rather than by the routine medium range cycles.

The other significant property of composite materials is the increase in fatigue damage with mean stress level, which is usually accounted for by scaling up the stress amplitude entered in the $R = -1$ S–N curve formulation by the factor $1/(1 - \overline{\sigma}/\sigma_d)$, where σ_d is the design strength in compression for a compression mean or in tension for a tension mean. This increases the relative importance of stress cycles with a high mean.

Behaviour of stall-regulated machines in fatigue

For stall-regulated machines, the highest out-of-plane bending moment ranges and means normally occur at high wind speeds and yaw angles. This is illustrated in Figure 7.9, which shows the variation in this moment with wind speed and yaw angle at 60 percent radius for a 40 metre diameter machine, based on the three-dimensional data referred to at the start of Section 7.1.8 above. Note that above rated wind speed, the bending moment plots level off, so that a given departure of the lateral wind component from the zero mean, sustained over half a revolution, results in a larger bending moment fluctuation than a change in the longitudinal component of twice this magnitude. For example, if the mean wind speed is 24 m/s, a lateral component of 6 m/s (corresponding to a yaw angle of 14°) causes a bending moment variation of 20 kNm when the blade rotates from 0° to 180° azimuth, compared to a variation of 17 kNm as a result of a ±6 m/s fluctuation in longitudinal wind speed (which, in any case, could only occur after many blade rotations).

Similar comments apply to vertical wind speed fluctuations, but here there is a built-in initial tilt angle between the air flow and the shaft axis because of shaft angle tilt and updraft. Thus bending moment plots derived from three-dimensional wind simulations above rated are dominated by fluctuations at blade-passing frequency which bloom and decay as the angle between the air flow and the shaft axis rises and falls. Superimposed on these are lower frequency fluctuations caused by changes in the longitudinal wind speed.

Clearly high wind/high yaw cycles will be a major source of fatigue damage, although the contribution of cycles at wind speeds below stall may also be important, because of the more rapid variation of moment with wind speed there, and the much increased number of cycles.

Thomsen (1998) has investigated for blade root out-of-plane bending on a 1.5 MW, 64 m diameter three-bladed machine, taking a constant turbulence intensity of 15 percent and a S–N curve index of 12. The results, including allowance for mean stress, are plotted in Figure 7.13 (dotted), and indicate that the damage is concentrated at wind speeds of 20 m/s and above. The figure also shows the effect of adopting a steeper S–N curve (with $m = 10$) and the IEC Class A turbulence distribution (with increasing intensities as mean wind speed decreases). In each case, the relative damage contribution at high wind speeds is reduced, but the switch to the IEC turbulence distribution causes the more significant change.

It should be noted that the relative contributions of different wind speeds to lifetime fatigue damage are also dependent on the shape of the bending moment/wind speed characteristics. Thus for the machine with the bending moment/wind speed characteristics at 60 percent radius presented in Figure 7.9, the peak damage occurs at 10 m/s, if the IEC Class A turbulence intensity distribution is assumed (see Figure 7.15).

Behaviour of pitch-regulated machines in fatigue

For pitch-regulated machines, the highest flapwise bending moment ranges occur at high wind speeds and yaw angles, but the largest mean values occur around

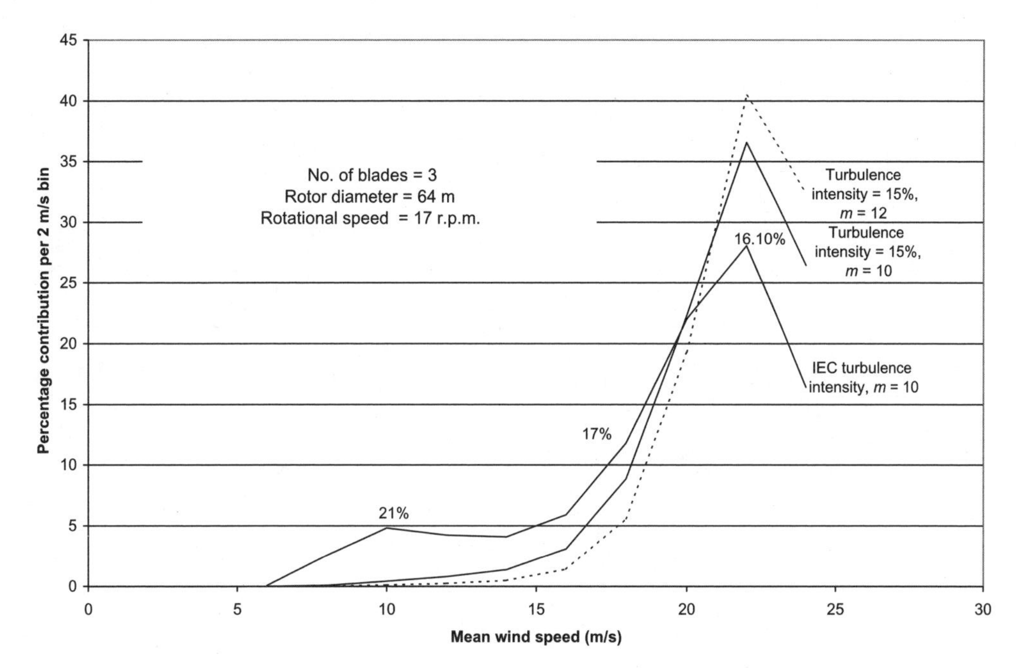

Figure 7.13 Relative Contribution to Life-time Fatigue Damage for Different Wind Speeds for a 1.5 MW Stall-regulated Machine, Including Effect of Mean Load, after Thomsen (1998)

rated wind speed. Moreover, blade pitching results in a rapid fall-off in bending moment with short-term mean wind speed just above rated. This behaviour is illustrated in Figure 7.10, which shows the variation in flapwise moment with short-term mean wind speed and yaw angle at 60 percent radius for a 40 m diameter machine. It transpires that the combination of the steep bending moment/ short-term wind speed characteristic, high mean bending moment and large number of loading cycles just above rated wind speed results in more fatigue damage at this wind speed than at higher wind speeds, where the increasing bending moment fluctuations due to yaw offset are mitigated by reducing mean loads and numbers of cycles.

The nature of the bending moment fluctuations at a mean wind speed just above rated is shown on Figure 7.14, which is a time history obtained from a three-dimensional wind speed simulation, for the machine with the bending moment/ short-term mean wind speed characteristics presented in Figure 7.10, (with the response to high frequency wind speed fluctuations allowed for separately). As with the case of a stall-regulated machine operating at high wind speed discussed above, there are considerable bending moment fluctuations at the rotational speed, but this time they are largely due to spatial variations in longitudinal wind speed across the disc (i.e., 'gust slicing') rather than due to yaw or tilt offset. In addition, there are large low frequency bending moment fluctuations as a result of short-term mean wind speed changes – indeed, inspection of the bending moment and short-term mean wind speed plots reveals an inverse relationship between the two.

The fatigue damage in flapwise bending at 12 m radius arising from operation of the above machine at different mean wind speeds ignoring dynamics is plotted out

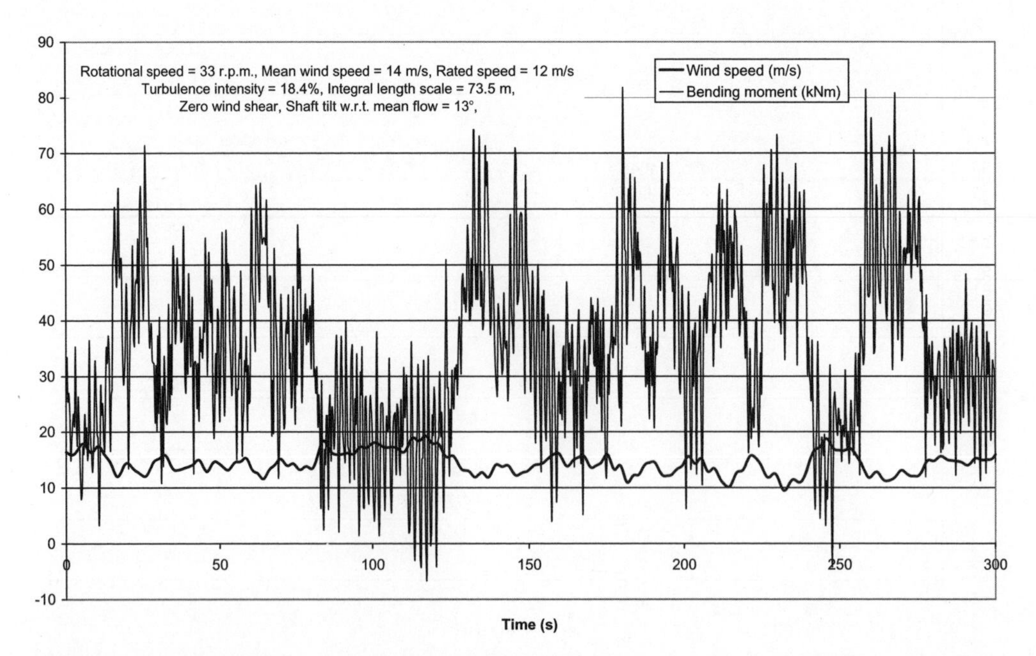

Figure 7.14 Time History of Flapwise Bending Moment at 12 m Radius and Short-term Mean Wind Speed for 40 m Diameter Pitch-regulated Machine with Three TR Blades, Based on Three-dimensional Wind Simulation with 14 m/s Mean

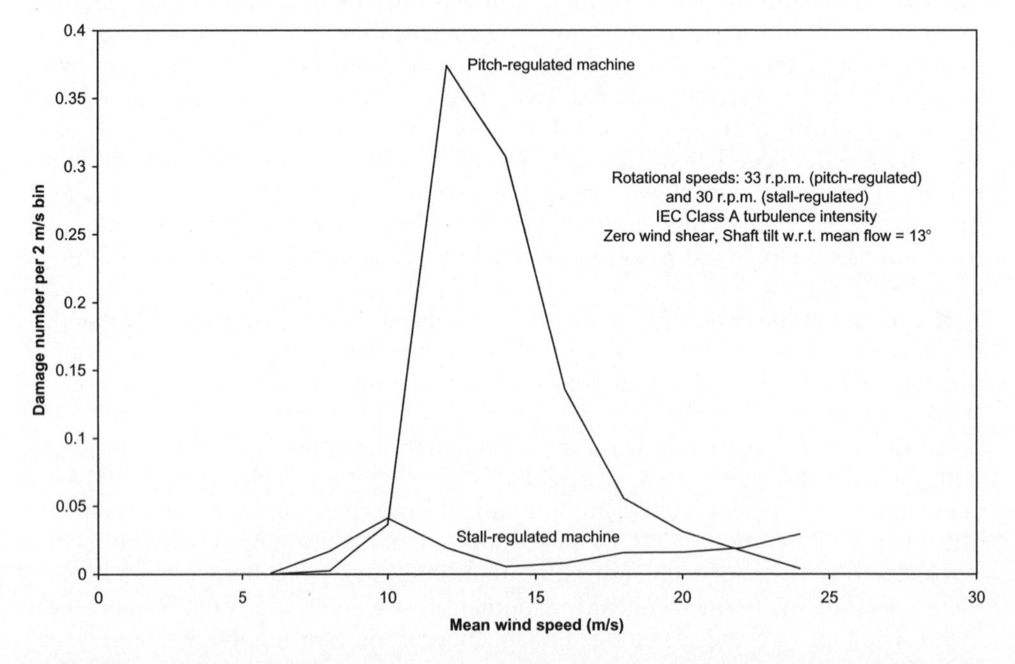

Figure 7.15 Variation of Blade Fatigue Damage in Flapwise Bending at 12 m Radius with Mean Wind Speed, for Similar 40 m Diameter Pitch- and Stall-regulated Machines, Ignoring Dynamics

on Figure 7.15, and is compared with that for a similar stall-regulated machine having the same section modulus. The cross section is designed to resist the extreme bending moment for the stall-regulated machine of 130 kNm. In both cases the $S-N$ curve index is taken as 10, and the IEC Class A turbulence intensity assumed. It is apparent that the pitch-regulated machine fatigue damage is concentrated around rated speed, and that the total damage is an order of magnitude greater than the total for the stall-regulated machine. As the 12 m radius section modulus to resist the extreme flapwise moment is likely to be less for the pitch-regulated machine, fatigue loading is likely to be more critical than indicated by the comparison in the figure.

Factors affecting fatigue criticality

The relative criticality of fatigue and extreme loading is determined by the material properties and safety factors adopted, as well as by the loadings themselves. As an aid to comparison, the fatigue loading can be described in terms of the notional one cycle equivalent load, $\sigma_{eq(n=1)}$, which is defined as the stress range of the single reverse loading cycle that would cause the same total fatigue damage as the actual fatigue loading on the basis of the design $S-N$ curve. Then fatigue is critical if

$$\frac{\sigma_{eq(n=1)}}{2\sigma_{0d}} > \gamma_L \frac{\sigma_{ext}}{\sigma_{cd}} \tag{7.10}$$

where σ_{0d} is the stress *amplitude* given by the reverse loading fatigue design curve at $N = 1$, σ_{ext} is the stress resulting from the extreme loading case, γ_L is the load factor and σ_{cd} is the design compression stress (which is assumed not to be governed by buckling considerations). The condition may be rewritten in terms of characteristic stress values as follows:

$$\frac{\sigma_{eq(n=1)}}{\sigma_{ext}} > 2\gamma_L \frac{\gamma_{mu}}{\gamma_{mf}} \frac{\sigma_{0k}}{\sigma_{ck}} \tag{7.11}$$

or as

$$\frac{\sigma_{eq(n=1)}}{\sigma_{ext}} > 2.7 \frac{\gamma_{mu}}{\gamma_{mf}} \frac{\sigma_{0k}}{\sigma_{ck}} \quad \text{with } \gamma_L \text{ set to } 1.35.$$

As is implicit from the survey of GFRP and wood laminate properties in Sections 7.1.6 and 7.1.7, the value of σ_{0k}/σ_{ck} can vary between about 1.0 and 1.4. If the derivations of the characteristic ultimate and fatigue strengths are statistically similar, the IEC rules indicate that the respective partial materials safety factors should be taken as equal, whereas, as noted in Section 7.1.6, the GL rules for GFRP imply a value of 1.5 for the γ_{mu}/γ_{mf} ratio. Thus in principal the parameter

$$2\gamma_L \frac{\gamma_{mu}}{\gamma_{mf}} \frac{\sigma_{0k}}{\sigma_{ck}}$$

governing fatigue criticality can take a wide range of values of between 2.7 and about 6.

In deriving the fatigue damage plots in Figure 7.15, a mid-range value of 4 has been adopted, resulting in total damages of 0.96 and 0.17 for the pitch and stall regulated machines respectively. However, if, the minimum value of 2.7 were adopted, corresponding to $\sigma_{0k} = \sigma_{ck}$ and $\gamma_{mu} = \gamma_{mf}$, the fatigue damages would rise by a factor of 50 for $m = 10$.

The other important material property governing the criticality of fatigue loading is, of course, the slope index of the log–log S–N curve, m, which affects the value of the notional one cycle equivalent load, $\sigma_{eq(n=1)}$. With the high values applicable to wood laminates, fatigue is much less likely to govern.

Other sources of variability

There are a number of other sources of variability in fatigue damage calculations, apart from uncertainty about the material properties themselves, some of which are detailed below.

(1) IEC 61400-1 provides two alternative stochastic turbulence models – those due to Von Karman and Kaimal. The Von Karman model is isotropic, whereas in the Kaimal model, which is more realistic in this respect, the standard deviations of lateral and vertical turbulences are 80 percent and 50 percent of the longitudinal turbulence respectively. In the case of stall-regulated machines, where wind misalignment at high wind speeds is often the main source of fatigue damage, the choice of turbulence model could clearly have a decisive effect.

(2) When the fatigue assessment is based on simulations of limited duration (typically 300 to 600 s), the damage is often dominated by a few extreme cycles, which are subject to significant statistical variation from one simulation to another. Accordingly several simulations at a given mean wind speed are necessary to obtain an accurate result (Thomsen, 1998).

(3) In allowing for the reduction in fatigue strength due to mean stress. For example, in Equation (7.7), the mean stress can either be calculated over each stress range obtained by rainflow cycle counting or over the length of the simulation.

Fatigue due to gravity loading

In-plane fatigue loads arise from gravity loading and fluctuations in the in-plane aerodynamic loadings, but gravity loadings dominate for machines large enough to be grid connected, rendering the loading calculation relatively straightforward.

In order to compare the approximately constant amplitude in-plane fatigue loading with the spectrum of out-of-plane fatigue loading, it is convenient to express both as equivalent loads at a specified number of cycles, n_{eq}. Often the 1 Hz equivalent load is used, in which case the number of cycles, n_{eq}, is equal to the number of seconds in the machine lifetime during which the machine operates. For machines of 40 m diameter and above, the fatigue equivalent load for in-plane bending at the blade root is typically greater than that for out-of-plane bending.

Over most of the blade length, the chord dimension is much larger than the blade thickness, so the section modulus for edgewise bending will generally exceed that for flapwise bending. However, for blades attached to the hub or pitch-bearing by a circular ring of bolts, which is the normal arrangement, the blade structure adjacent to the root is a cylindrical shell, which will have the same section modulus about both axes if the wall thickness is uniform. As a consequence, the blade root is the first area that should be checked for in-plane fatigue loading.

The procedure can be illustrated for a 20 m tip radius TR blade in GFRP, designed for an extreme static root moment of 750 kNm at 0.5 m radius. Taking the gravity moment at the root as 124 kNm, and assuming 2.6×10^8 revolutions over the machine lifetime, the notional one cycle in-plane bending equivalent load range, $M_{eq(n=1)}$, becomes $248(2.6 \times 10^8)^{0.1} = 1720$ kNm and $\sigma_{eq(n=1)}/\sigma_{ext} = 1720/750 = 2.3$. This is less than the minimum value of 2.7 of the right-hand expression in the inequality (7.11), so in-plane fatigue loading does not govern. In practice, the cylinder wall thickness would have to be increased significantly to prevent buckling, rendering fatigue less critical still at this diameter.

It may be concluded from the above example that in-plane fatigue loadings are not a significant consideration in the design of stall-regulated blades constructed in GFRP or wood laminates, until much larger diameters are reached. They cannot be ignored entirely, because of blade twist and because they add to the fatigue stress ranges due to flapwise bending at points on the cross section away from the neutral axis for edgewise bending. In-plane fatigue loadings are of more significance for pitch-regulated machines because gravity loadings will contribute increasingly to flapwise loadings as the pitch angle increases.

Tip deflection

Under extreme operating conditions, tip deflections of up to about 10 percent of blade radius can occur, so care is needed to avoid the risk of blade/tower collisions in the case of upwind machines. GL specify that the quasi-static tip deflection under the extreme unfactored operational loading is not to exceed 50 percent of the clearance without blade deflection, which implies a safety factor of 2. IEC, on the other hand, require no blade/tower contact when the extreme loads are multiplied by the combined partial safety factors for loads and the blade material.

It is instructive to compare the tip deflections for similar blades designed in different materials. If the skin thickness distributions are chosen so that the design compression strength of each material is fully mobilized under the extreme load case, then the tip deflection will be proportional to the design compression strength to Young's Modulus ratio, σ_{cd}/E, of the blade material. These ratios are compared for different materials in Table 7.4.

It is clear from the table that a GFRP blade will be about twice as flexible as blades in the other materials (apart from steel), provided that the spar is stocky enough for buckling not to govern the design, e.g., as in Figure 7.4. In the case of thin-walled cross sections, however, such as that in Figure 7.3, the GFRP compres-

Table 7.4 Design Strength-to-stiffness Ratios for Different Wind Turbine Blade Materials

Material	Ultimate compression strength, σ_{cu} (MPa)	Partial safety factor for material strength, γ_{mu}	Compression design strength, σ_{cd} (MPa)	Young's Modulus, E (GPa)	Strength-to-stiffness ratio, $(\sigma_{cd}/E) \times 10^3$
Glass/polyester *laminate* with 50% fibre volume fraction and 80% UD	580	2.94	197 (ignoring buckling)	32.5	6.1
Carbon fibre/epoxy *ply* with 60% fibre volume fraction and UD lay-up	1100	2.94	374	142	2.6
Khaya/epoxy laminate	50	1.5	33	10	3.3
Birch/epoxy laminate	81	1.5	54	15	3.6
	Yield strength, σ_y	γ_{my}			
High-yield steel (Grade Fe 510)	355	1.1	323	210	1.54
Weldable aluminium alloy AA6082	240	1.1	218	69	3.2

sive design stress has to be reduced significantly to guard against buckling, with the result that blade flexibility is reduced. For example, in the case of the 19.5 m TR blade, the compressive design stress is reduced to about 90 MPa, resulting in a tip deflection of about 2 m under extreme loading, and a strength-to-stiffness ratio less than that for wood/epoxy. In this connection, it is noteworthy that Hancock, Sonderby and Schubert (1997) records a 130 percent proof load test on a 31.2 m birch/epoxy blade which resulted in a tip deflection of 3.4 m.

7.1.9 Blade resonance

One of the most important objectives of blade design is the avoidance of resonant oscillations, which, in a mild form, exacerbate fatigue damage and in an extreme form can lead to rapid failure. The excitation of blade resonance can be minimized by maximizing the damping and ensuring that the blade flapwise and edgewise natural frequencies are well separated from the exciting frequencies – i.e. the rotational frequency and its harmonics, particularly the blade passing frequency – and from the frequencies of other vibration modes with which there is an identifiable risk of coupled oscillations.

Vibrations in stall

On stall-regulated machines, the lift curve slope $dC_l/d\alpha$, goes negative when a section of the blade goes into stall, resulting in local negative aerodynamic damping of blade motion in the lift direction. If the overall aerodynamic damping for a particular mode shape is negative, and exceeds the modal structural damping in magnitude, then divergent oscillations can develop from any initial disturbance, regardless of the relationship between the mode natural frequency and exciting frequencies. The first mode in each direction is most susceptible to such behaviour because the structural damping increases with frequency while the aerodynamic damping diminishes. If conditions favouring first-mode oscillations are to be avoided, the factors affecting the aerodynamic damping of both edgewise and flapwise oscillations need to be understood, so these are explored below.

Consider a turbine operating in steady conditions in a perpendicular airflow. If a blade cross section at radius r experiences out-of-plane and in-plane perturbations with velocities $\dot{x}$ in the downwind direction and $\dot{y}$ in the direction opposite to that of blade rotation (assumed clockwise), then the relative velocity triangle is as in Figure 7.16(a). The lift and drag forces per unit length on a blade element, L and D, can be resolved into out-of-plane and in-plane forces F_X and F_Y (see Figure 7.16(b)), leading to Equations (5.18) and (5.19). Ignoring the small rotational induction factor, which is very small, these may be rewritten as:

$$F_Y = W[-C_l(U_\infty(1-a) - \dot{x}) + C_d(\Omega r - \dot{y})]\tfrac{1}{2}\rho c \tag{7.12}$$

$$F_X = W[C_l(\Omega r - \dot{y}) + C_d(U_\infty(1-a) - \dot{x})]\tfrac{1}{2}\rho c \tag{7.13}$$

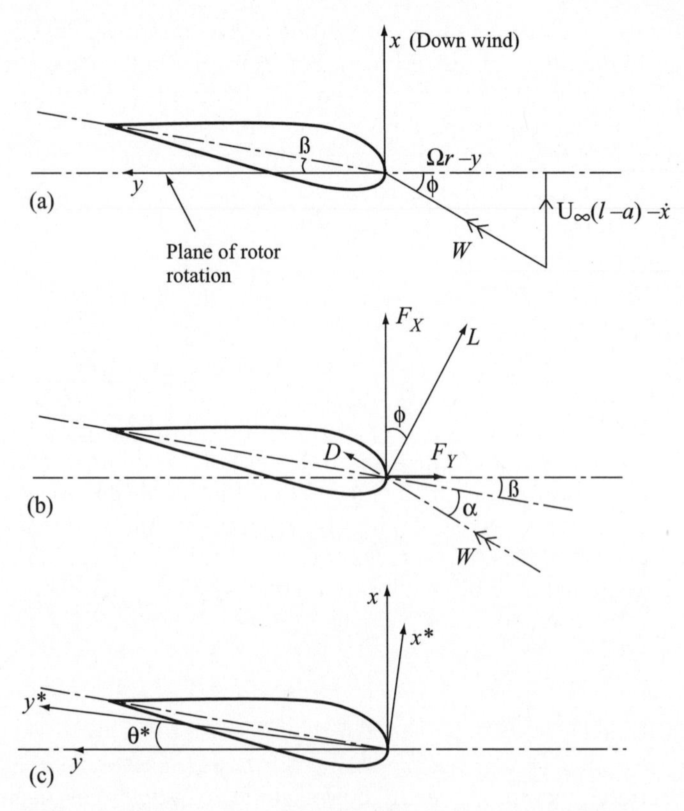

Figure 7.16 (a) Velocity Diagram for Vibrating Blade (Looking Towards Hub); (b) Out-of-plane and In-plane Components of Lift and Drag Forces; (c) Directions of Vibration, x^* and y^*

Here U_∞ is the free stream wind speed and $U_\infty(1 - a)$ the reduced wind speed at the rotor plane as usual. The damping coefficients per unit length for vibrations in the in-plane and out-of-plane directions are then given by

$$\hat{c}_Y(r) = -\frac{\partial F_Y}{\partial \dot{y}} \tag{7.14a}$$

$$\hat{c}_x(r) = -\frac{\partial F_X}{\partial \dot{x}} \tag{7.14b}$$

Analagous 'cross' coefficients relating the in-plane force to the out-of-plane velocity and *vice versa* can also be defined as:

$$\hat{c}_{YX}(r) = -\frac{\partial F_Y}{\partial \dot{x}} \tag{7.15a}$$

$$\hat{c}_{XY}(r) = -\frac{\partial F_X}{\partial \dot{y}} \tag{7.15b}$$

Substituting V for $U_\infty(1-a)$ for brevity, the in-plane damping coefficient is derived as follows:

$$\hat{c}_Y(r) = -\frac{\partial F_Y}{\partial \dot{y}} = -\tfrac{1}{2}\rho c\left\{\frac{\partial W}{\partial \dot{y}}[-C_l V + C_d \Omega r] + W\left[-\frac{\partial C_l}{\partial \dot{y}}V + \frac{\partial C_d}{\partial \dot{y}}\Omega r - C_d\right]\right\} \quad (7.16)$$

Noting that

$$\frac{\partial W}{\partial \dot{y}} = -\frac{\Omega r}{W}$$

and

$$\frac{\partial C_l}{\partial \dot{y}} = \frac{\partial C_l}{\partial \alpha}\frac{\partial \alpha}{\partial \dot{y}} = \frac{\partial C_l}{\partial \alpha}\frac{\partial \phi}{\partial \dot{y}} = \frac{\partial C_l}{\partial \alpha}\frac{V}{W^2}$$

this equation becomes:

$$\hat{c}_Y(r) = \tfrac{1}{2}\rho c\frac{\Omega r}{W}\left\{-VC_l + \frac{V^2}{\Omega r}\frac{\partial C_l}{\partial \alpha} + \frac{2\Omega^2 r^2 + V^2}{\Omega r}C_d - V\frac{\partial C_d}{\partial \alpha}\right\} \quad (7.17)$$

The 'cross' coefficients and the out-of-plane damping coefficient and are derived by a similar procedure:

$$\hat{c}_{YX}(r) = \tfrac{1}{2}\rho c\frac{\Omega r}{W}\left\{-\frac{\Omega^2 r^2 + 2V^2}{\Omega r}C_l - V\frac{\partial C_l}{\partial \alpha} + VC_d + \Omega r\frac{\partial C_d}{\partial \alpha}\right\} \quad (7.18)$$

$$\hat{c}_{XY}(r) = \tfrac{1}{2}\rho c\frac{\Omega r}{W}\left\{+\frac{2\Omega^2 r^2 + V^2}{\Omega r}C_l - V\frac{\partial C_l}{\partial \alpha} + VC_d - \frac{V^2}{\Omega r}\frac{\partial C_d}{\partial \alpha}\right\} \quad (7.19)$$

$$\hat{c}_x(r) = \tfrac{1}{2}\rho c\frac{\Omega r}{W}\left\{+VC_l + \Omega r\frac{\partial C_l}{\partial \alpha} + \frac{\Omega^2 r^2 + 2V^2}{\Omega r}C_d + V\frac{\partial C_d}{\partial \alpha}\right\} \quad (7.20)$$

It is apparent from inspection of the expressions for the two damping coefficients, $\hat{c}_Y$ and $\hat{c}_X$, that the choice of an aerofoil with a gentler stall – i.e., with a smaller lift curve slope after stall onset – will increase the damping coefficient in both cases. Note that the modal damping coefficient is dominated by the damping per unit length over the outboard part of the blade, so it is important to select an aerofoil with a gentle stall in this area only.

The choice of aerofoil also affects performance, so there is merit in expressing the damping coefficients in terms of the power output in order to investigate possible trade-offs between them. It transpires that the damping and 'cross' coefficients per unit length can be formulated quite simply in terms of the power output per unit length of blade, $P'(r, V) = \Omega r(-F_Y)$, and the blade thrust per unit length, F_X, as follows:

$$\hat{c}_Y = -\frac{2}{\Omega^2 r^2}P' + \frac{V}{\Omega^2 r^2}\frac{\partial P'}{\partial V} = \frac{1}{\Omega^2 r^2}\left(-2P' + V\frac{\partial P'}{\partial V}\right) \tag{7.21}$$

$$\hat{c}_{XY} = -\frac{\partial F_Y}{\partial \dot{x}} = \frac{\partial F_Y}{\partial V} = \frac{1}{\Omega r}\frac{\partial}{\partial V}(\Omega r F_Y) = -\frac{1}{\Omega r}\frac{\partial P'}{\partial V} \tag{7.22}$$

$$\hat{c}_{XY} = \frac{1}{\Omega r}\left(2F_X - V\frac{\partial F_X}{\partial V}\right) \tag{7.23}$$

$$\hat{c}_X = -\frac{\partial F_X}{\partial \dot{x}} = \frac{\partial F_X}{\partial V} \tag{7.24}$$

Equations (7.21) and (7.23) are derived from the equations

$$\Omega r \hat{c}_Y + V\hat{c}_{YX} = 2F_Y = -2P'/\Omega r$$

and

$$\Omega r \hat{c}_{XY} + V\hat{c}_X = 2F_X$$

which may be verified using Equations (7.17) to (7.20).

From Equation (7.21) it is clear that the damping coefficient in the in-plane direction, $\hat{c}_Y$, will always be negative when $2(P'/V)$ exceeds $\partial P'/\partial V$, and that a negative power curve slope should be avoided if the size of the negative damping is to be kept small.

Effect of blade twist

In the discussion so far, damping of vibrations in the out-of-plane and in-plane directions only has been considered. In practice, blade twist will result in the flapwise and edgewise vibrations taking place in directions rotated from the out-of-plane and in-plane directions in the same sense as the blade twist, but by a lesser amount (see Section 5.8.1). If we define x^*- and y^*-axes in the directions of the flapwise and edgewise displacements, each making an angle of θ^* to the x- and y-axes respectively, as shown in Figure 7.16(c), then the edgewise damping coefficient per unit length is given by:

$$\hat{c}_{Y^*} = \hat{c}_Y \cos^2\theta^* - (\hat{c}_{YX} + \hat{c}_{XY})\sin\theta^* \cos\theta^* + \hat{c}_X \sin^2\theta^* \tag{7.25}$$

Substitution of Equations (7.21)–(7.24) in Equation (7.25) yields:

$$\hat{c}_{Y^*} = \cos^2\theta^*\left[\frac{1}{\Omega^2 r^2}\left(-2P' + V\frac{\partial P'}{\partial V}\right)\right] + \cos\theta^* \sin\theta^*\left[\frac{1}{\Omega r}\left(-\frac{\partial P'}{\partial V} + 2F_X - V\frac{\partial F_X}{\partial V}\right)\right]$$
$$+ \sin^2\theta^*\left(\frac{\partial F_X}{\partial V}\right) \tag{7.26}$$

This expression also gives the flapwise damping coefficient per unit length if θ^* is replaced by $\theta^* + 90°$ throughout.

The variation of the damping coefficient $\hat{c}_{Y^*}$ per unit length at 14 m radius with vibration direction, θ^*, at three different wind speeds is illustrated in Figure 7.17 for a specimen aerofoil section on a 20.5 m tip radius blade rotating at 29 r.p.m. The data are taken from Petersen *et al.* (1998), and do not include allowance for the axial induction factor. It can be seen that negative damping is worst at 20 m/s, and that negative edgewise damping is ameliorated by increasing θ^* at the expense of increasing negative flapwise damping.

Although a plot of the local damping coefficient at *ca* 70 percent radius can provide a useful indication of trends, the best guide to the likelihood of divergent oscillations is provided by the modal damping coefficient for the mode under consideration. This is obtained by multiplying the right-hand side of Equation (7.26) by the square of the local modal amplitude and integrating over the length of the blade.

If comparison of the first mode edgewise and flapwise modal damping coefficients shows there is a benefit to be gained from altering the direction of vibration, small changes can be made by redistributing material within the blade cross section. Alternatively the blade pitch could be altered in conjunction with a compensatory change in aerofoil camber so that the aerodynamic properties for any given inflow angle are unchanged.

The prediction of edgewise vibrations in stall is examined in detail by Petersen *et al.* (1998), whose work provides the basis of the introductory survey given here.

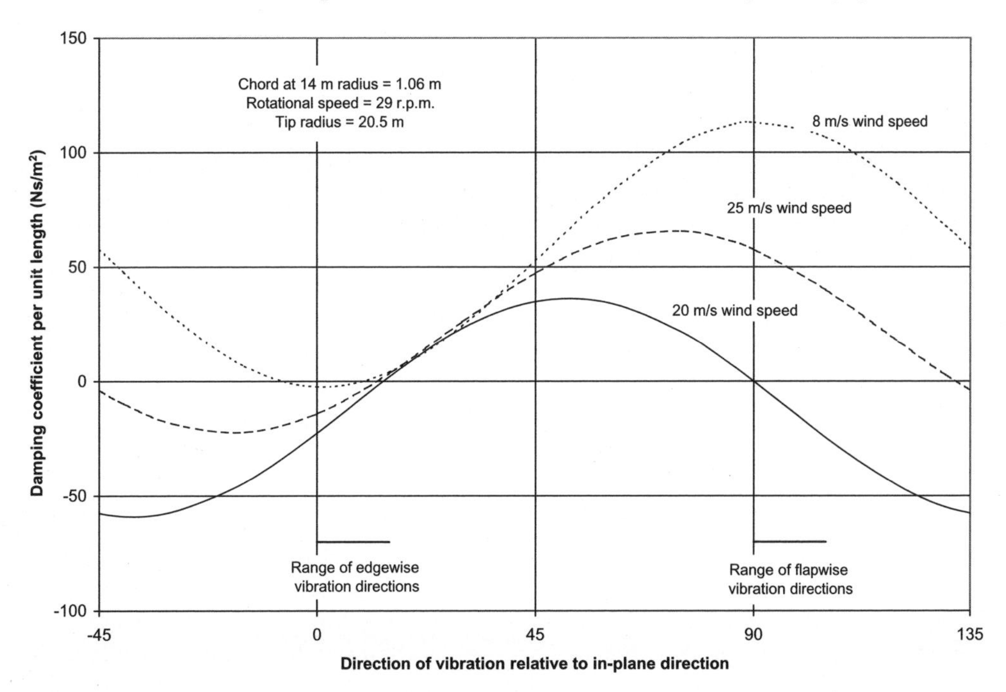

Figure 7.17 Variation in Damping Coefficient at 14 m Radius with Vibration Direction for Example Aerofoil

They concluded that the fundamental cause of edgewise blade oscillations that had been observed on some stall-regulated machines of 40 m diameter and over was negative aerodynamic damping, but found that the use of dynamic stall models improved the level of agreement with measurements.

Coupling of edgewise blade mode and rotor whirl modes

A further important finding was that, on one machine subject to stall-induced vibrations which was investigated in detail, there was coupling between the blade first edgewise mode and one of the second rotor whirl modes. The rotor whirl modes arise from the combination of simultaneous nodding and yawing oscillations of the rotor shaft, which occur at the same frequency during operation due to gyroscopic effects. As a result, the rotor hub traces out a circular or elliptical path, running either in the same direction as rotor rotation or in reverse, which explains the existence of two first and second modes.

The explanation for the coupling was as follows. When a pair of blades vibrate in the edgewise direction in anti-phase, they impart a sinusoidally varying in-plane force to the rotor hub even though their edgewise root bending moments cancel out. The direction of this oscillating force rotates with the rotor, so it has horizontal and vertical components of the form $\sin(\omega_1 t + \eta).\sin\Omega t$ and $\sin(\omega_1 t + \eta).\cos\Omega t$, where ω_1 is the frequency of the blade first edgewise mod, and Ω is the speed of rotor rotation. With respect to stationary axes the in-plane loads on the hub therefore act at two frequencies – namely $\omega_1 + \Omega$ and $\omega_1 - \Omega$. In the case of the machine investigated by Petersen *et al.*, the upper frequency of $2.9 + 0.5 = 3.4$ Hz coincided with the backward second rotor whirl mode, allowing interaction between this mode and the blade first edgewise mode.

Simulations were carried out on an aeroelastic model of the turbine at various wind speeds and satisfactory agreement obtained between simulated and measured behaviour. In particular, the simulation at 23.2 m/s predicted the build up of large blade root edgewise moment oscillations at the first mode frequency, as observed on the real machine at this wind speed. Significantly, when the latter simulation was repeated with the rotor shaft stiffness increased sufficiently to increase the backward second rotor whirl mode frequency to 3.6 Hz, the predicted blade root edgewise moment oscillations were negligible by comparison.

Mechanical damping

An alternative strategy for preventing damaging edgewise vibrations is the incorporation of a tuned mass damper inside the blade towards the tip. The performance of such a damper on a 22 m tip radius blade is reported by Anderson, Heerkes and Yemm (1998). It was found that the fitting of a damper tuned to the first mode edgewise frequency, and weighing only 0.4 percent of the total blade weight, effectively suppressed the edgewise vibrations which had previously been observed during high wind speed operation.

7.1.10 Design against buckling

The stress at which a slender plate element without imperfections buckles under compression loading is known as the critical buckling stress. The derivation of the critical buckling stresses for thin-walled curved panels bounded by stiffeners, which typically form the blade load-bearing structure, is relatively straightforward when the panel material is isotropic and solutions are provided in Timoshenko and Gere (1961). These do not apply to composite materials such as the GFRP and wood laminates commonly used in blade manufacture, however, as these are anisotropic, but solutions can be derived for a symmetric laminate using the energy method, as outlined below.

Consider a long cylindrical panel of length L and radius r, supported along two generators and subtending an angle ψ at the cylinder axis, which is axially loaded in compression. If it deflects to form n half-waves around the circumference between supports and m half-waves along its length, then its out-of-plane deflection can be written as:

$$w = C \sin \frac{n\pi\theta}{\psi} \sin \frac{m\pi x}{L} \tag{7.27}$$

where θ and x are the co-ordinates of the deflected point with respect to one of the long edges and one end respectively. In the absence of in-plane direct strains in the plate, this out-of-plane deflected profile will result in circumferential deflections

$$v_0 = \frac{C\psi}{n\pi} \cos \frac{n\pi\theta}{\psi} \sin \frac{m\pi x}{L} \tag{7.28}$$

These deflections will result in in-plane shear stresses, which reach a maximum at the corners of each rectangular buckled panel. In practice, additional in-plane deflections occur to moderate these shear stresses, as follows:

$$u = A \sin \frac{n\pi\theta}{\psi} \cos \frac{m\pi x}{L} \quad \text{in the axial direction}$$

$$v = B \cos \frac{n\pi\theta}{\psi} \sin \frac{m\pi x}{L} \quad \text{in the circumferential direction} \tag{7.29}$$

The in-plane strain energy is calculated as

$$U_2 = \tfrac{1}{2}h \iint (\sigma_1\varepsilon_1 + \sigma_2\varepsilon_2 + \tau\gamma)\, r\, \mathrm{d}\theta\, \mathrm{d}x \tag{7.30}$$

with the suffices 1 and 2 denoting the axial and circumferential directions respectively, so that

$$\varepsilon_1 = \frac{\partial u}{\partial x}, \ \varepsilon_2 = \frac{\partial v}{r\partial\theta}, \ \gamma = \frac{\partial u}{r\partial\theta} + \frac{\partial(v_0 + v)}{\partial x} \tag{7.31}$$

Substituting $\sigma_1 = E_x(\varepsilon_1 + v_y\varepsilon_2)/(1 - v_x v_y)$, $\sigma_2 = E_y(\varepsilon_2 + v_x\varepsilon_1)/(1 - v_x v_y)$ and $\tau = G_{xy}\gamma$, where E_x, E_y and G_{xy} are the longitudinal, transverse and shear moduli of the laminate respectively (obtained by averaging the corresponding moduli of the individual plies) and v_x and v_y are the effective Poisson's ratios, the in-plane strain energy becomes:

$$U_2 = \frac{h}{2(1 - v_x v_y)} \iint [E_x\varepsilon_2^2 + E_y\varepsilon_1^2 + 2E_x v_y \varepsilon_1\varepsilon_2 + (1 - v_x v_y)\gamma^2 G_{xy}] r \, d\theta \, dx \qquad (7.32)$$

Substituting the expressions for ε_1, ε_2, and γ from Equation (7.31) and integrating over the width of the panel, ψr, and the length of one half wave, L/m, we obtain

$$U_2 = \frac{E_x h}{1 - v_x v_y} \psi r \frac{L}{m} \left(\frac{m\pi}{L}\right)^2 \frac{C^2}{8} \left[\alpha^2 + \beta^2 \frac{E_y}{E_x}\left(\frac{n}{\lambda}\right)^2 + 2v_y\alpha\beta\left(\frac{n}{\lambda}\right) \right.$$
$$\left. + (1 - v_x v_y)\frac{G_{xy}}{E_1}\left\{\alpha\left(\frac{n}{\lambda}\right) + \beta + \frac{\psi}{n\pi}\right\}^2\right] \qquad (7.33)$$

where $\lambda = m\psi r/L$ and the ratios $\alpha = A/C$ and $\beta = B/C$ are yet to be determined.

The expression for the strain energy of curvature is derived as follows. Replacing the angular coordinate θ by the linear coordinate y ($= r\theta$), the bending energy absorbed in an area $dx.dy$ is:

$$dU_b = -\tfrac{1}{2}\left(M_x \frac{\partial^2 w}{\partial x^2} + M_y \frac{\partial^2 w}{\partial y^2}\right) dx.dy$$

where

$$M_x = -D_x \frac{\partial^2 w}{\partial x^2} - D_{xy}\frac{\partial^2 w}{\partial y^2}$$

and

$$M_y = -D_y \frac{\partial^2 w}{\partial y^2} - D_{xy}\frac{\partial^2 w}{\partial x^2}$$

for a specially orthotropic laminate, i.e. one in which the reinforcement in each layer is either oriented at 0° or 90°, or is bi-directional with the same amount of fibres at $+\theta°$ and $-\theta°$. D_x and D_y are the flexural rigidities of the laminate when flat, for bending about the y-axis and x-axis respectively, and D_{xy} is the 'cross flexural rigidity' i.e. the moment per unit width about one axis generated by unit curvature about the other. Hence

$$dU_b = \tfrac{1}{2}\left(D_x\left(\frac{\partial^2 w}{\partial x^2}\right)^2 + 2D_{xy}\frac{\partial^2 w}{\partial x^2}\frac{\partial^2 w}{\partial y^2} + D_y\left(\frac{\partial^2 w}{\partial y^2}\right)^2\right)\mathrm{d}x\,\mathrm{d}y \tag{7.34}$$

The twisting energy absorbed in an area $\mathrm{d}x\,\mathrm{d}y$ is:

$$dU_t = \tfrac{1}{2}(M_{xy} + M_{yx})\frac{\partial^2 w}{\partial x \partial y}\mathrm{d}x\,\mathrm{d}y$$

where

$$M_{xy} = 2\left[\int_{-h/2}^{h/2} G_{xy}(z).z^2\,\mathrm{d}z\right]\frac{\partial^2 w}{\partial x \partial y}$$

in which z is the distance measured from the mid-plane of the laminate, $G_{xy}(z)$ is the in-plane shear modulus at that distance and h is the laminate thickness. Denoting the torsional rigidity,

$$\left[\int_{-h/2}^{h/2} G_{xy}(z).z^2\,\mathrm{d}z\right]$$

by D_T, then

$$dU_t = \tfrac{1}{2}.4D_\mathrm{T}\left(\frac{\partial^2 w}{\partial x \partial y}\right)^2\mathrm{d}x\,\mathrm{d}y \tag{7.35}$$

The total strain energy of curvature over the width of the panel and the length of one half wave is found by substituting the out-of-plane deflection given by Equation (7.27) in Equations (7.34) and (7.35) and integrating over this area, which gives:

$$U_1 = U_b + U_t = \frac{C^2}{8}\frac{\psi r L}{m}D_x\left(\frac{m\pi}{L}\right)^4\left[1 + \left(\frac{n}{\lambda}\right)^4\frac{D_y}{D_x} + \left(\frac{n}{\lambda}\right)^2\left\{2\frac{D_{xy}}{D_x} + 4\frac{D_T}{D_x}\right\}\right] \tag{7.36}$$

The energy absorbed by the panel during buckling as a result of in-plane strains and out-of-plane curvature is equal to the work done by the critical axial load as the panel shortens. The shortening of the panel over one half wave length is given by

$$\int_0^{L/m}\tfrac{1}{2}\left(\frac{\partial w}{\partial x}\right)^2\mathrm{d}x = \frac{\pi^2}{4}C^2\frac{m}{L}\sin^2\frac{n\pi\theta}{\psi} \tag{7.37}$$

so the work done by the axial force of N_x per unit width over the panel width is

$$T_1 = \frac{\pi^2}{8}C^2\frac{m}{L}\psi r N_x \tag{7.38}$$

The equality $T_1 = U_1 + U_2$ yields the critical value of the axial force as follows:

$$(N_x)_{cr} = D_x\left(\frac{m\pi}{L}\right)^2\left[1+\left(\frac{n}{\lambda}\right)^4\frac{D_y}{D_x}+\left(\frac{n}{\lambda}\right)^2\left\{2\frac{D_{xy}}{D_x}+4\frac{D_T}{D_x}\right\}\right]$$
$$+\frac{E_x h}{1-v_x v_y}\left[\alpha^2+\beta^2\frac{E_y}{E_x}\left(\frac{n}{\lambda}\right)^2+2\alpha\beta v_y\frac{n}{\lambda}+(1-v_x v_y)\frac{G_{xy}}{E_x}\left\{\alpha\frac{n}{\lambda}+\beta+\frac{\psi}{n\pi}\right\}^2\right] \tag{7.39}$$

Noting that

$$\frac{m\pi}{L}=\frac{m\psi r}{nL}\frac{n\pi}{\psi r}=\frac{\lambda}{n}\frac{n\pi}{\psi r}$$

this equation becomes

$$(\sigma_x)_{cr} = \frac{D_x}{h}\left(\frac{\lambda}{n}\frac{n\pi}{\psi r}\right)^2\left[1+\left(\frac{n}{\lambda}\right)^4\frac{D_y}{D_x}+\left(\frac{n}{\lambda}\right)^2\left\{2\frac{D_{xy}}{D_x}+4\frac{D_T}{D_x}\right\}\right]$$
$$+\frac{E_x}{1-v_x v_y}\left[\alpha^2+\beta^2\frac{E_y}{E_x}\left(\frac{n}{\lambda}\right)^2+2\alpha\beta v_y\frac{n}{\lambda}+(1-v_x v_y)\frac{G_{xy}}{E_x}\left\{\alpha\frac{n}{\lambda}+\beta+\frac{\psi}{n\pi}\right\}^2\right] \tag{7.40}$$

The right-hand side of Equation (7.40) contains four unknowns, the number of transverse half waves, n, the ratio of longitudinal to transverse half wave length, n/λ, and the factors α and β. Assuming that there is only one transverse half wave, as is normally the case, the expression is minimized with respect to α, and β for each value of n/λ, and then with respect to n/λ to obtain the critical stress.

The results of this exercise are illustrated for a particular curved laminate panel in Figure 7.18. The radius of curvature, r, of 1150 mm and thickness, h, of 15 mm are chosen to be representative of the values likely to obtain at 70 percent radius on a blade with 20 m tip radius. The laminate has 80 percent of its plies with fibres oriented axially and 20 percent with fibres at ±45° to resist shear loads. In each case, the fibre volume fraction is 50 percent. The ±45° plies are concentrated about the laminate mid-plane so that they do not detract significantly from the longitudinal flexural rigidity. Thus, if the longitudinal modulus of the UD plies is denoted by E_1, and the Poisson's ratios by v_{12} and v_{21}, the longitudinal flexural rigidity is given approximately by:

$$D_x = \frac{E_1 h^3}{12(1-v_{12}v_{21})} \tag{7.41}$$

The other ply and laminate properties required for evaluation of the critical stress are detailed on the figure. Note that in the derivation of the laminate in-plane stiffness properties it is necessary to transform the in-plane stiffness properties of

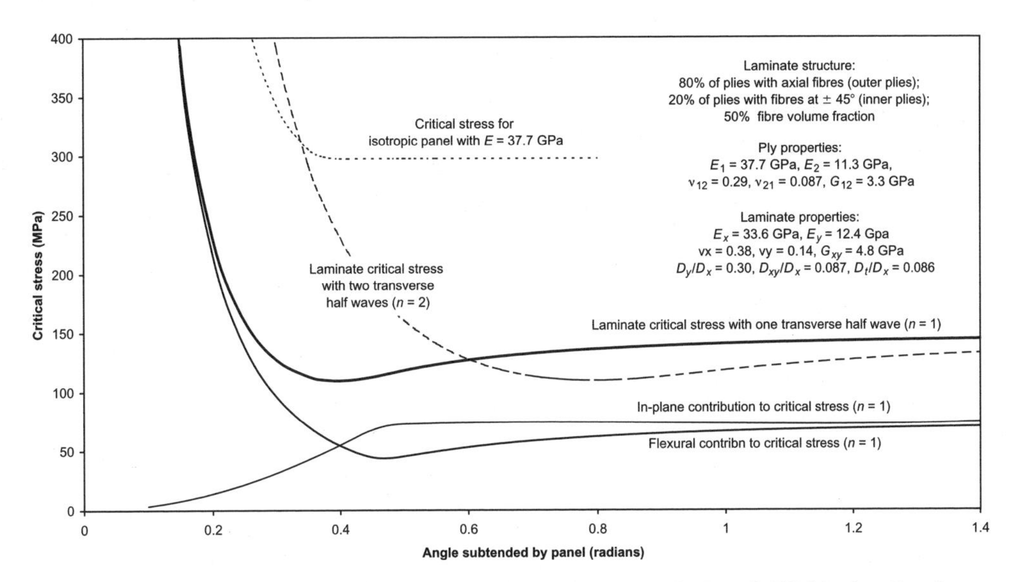

Figure 7.18 Variation of Axial Critical Buckling Stress with Panel Width for Specimen Curved Anisotropic Panel with Radius 1150 mm and Thickness 15 mm

the ±45° plies obtained initially in relation to the ply axes (which are parallel to the fibre directions) to the set of properties in relation to the global x- and y-axes of the laminate as a whole (see, for example, Barbero (1998) for the requisite formulae).

The heavy curve in Figure 7.18 shows the variation in axial critical stress with panel width (in terms of subtended angle) when the buckled shape has only a single half wave in the transverse direction and the fine lines below show the separate in-plane and flexural contributions. The minimum stress of 110 MPa occurs when the angle subtended by the panel is about 20°, but there is only a gradual increase in critical stress as the angle increases above this. When the subtended angle exceeds about 35°, buckling with two half waves in the transverse direction takes over as the critical mode – see dashed line. Also shown for comparison is the critical buckling stress variation for an isotropic plate with Young's modulus equal to the longitudinal modulus of the UD plies of the laminate – see dotted line. The minimum critical stress in this case is 298 MPa, about 2.7 times as big.

7.1.11 Blade root fixings

The fixing of the blade root to the hub is one of the most critical areas of blade design, because the order of magnitude difference between the relative stiffnesses of the steel hub and the blade material – usually GFRP or wood – militates against a smooth load transfer. The connection is usually made by steel bolts, which can either be embedded in the blade material in the axial direction or aligned radially to pass through the blade skin, but in either case stress concentrations are inevitable.

Figure 7.19 illustrates four different types of blade root fixings in section. The blade structure is usually a cylindrical shell at the root, in which case the stud or bolt fixings are arranged in a circle. Figure 7.19(a) shows the carrot connector,

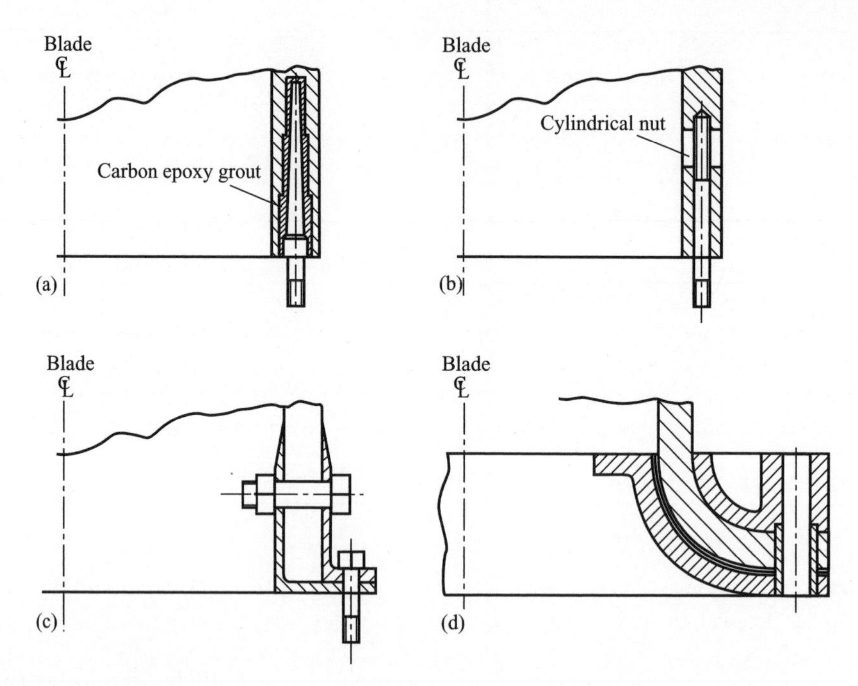

Figure 7.19 (a) Carrot Connector; (b) T-bolt Connector; (c) Pin-hole Flange; (d) Trumpet Flange

which is the standard fixing for laminated wood blades. The connector consists of a tapered portion carbon-epoxy grouted into a stepped hole drilled into the end of the blade, together with a projecting threaded stud for attachment to the hub or pitch bearing. Connectors are either machined from high strength steel or cast from spheroidal graphite iron (SGI). They are normally preloaded to reduce fatigue loading. A similar connector, in which the embedded portion is cylindrical rather than tapered, is in common use on GFRP blades.

Figures 7.19(b)–(d) show three further fixing arrangements used on GFRP blades. The T-bolt connector, shown in Figure 7.19(b), consists of a steel stud inserted into a longitudinal hole in the blade skin, which engages with a cylindrical nut held in a transverse hole. The stud is preloaded to reduce fatigue loading. The 'pin-hole flange' arrangement in Figure 7.19(c) uses the same method of load transfer between the GRP and the steel – i.e., bearing on a transverse rod – but the interface does not lend itself to preloading. Moreover the bolts attaching the flange to the hub are eccentric to the blade skin, so the flange has to resist the resultant bending moment as well.

In the trumpet flange detailed in Figure 7.19(d), the blade root is splayed out in the form of a trumpet mouth and clamped between inner and outer flanges by the ring of bolts which attach the flange to the hub. These bolts also pass through the GFRP skin to provide positive anchorage. Again the flange has to resist bending moments arising from the eccentricity of the fixing bolts to the blade skin where it emerges from the flange. The pin-hole and trumpet flange arrangements are rarely used for larger blades.

The stress distributions calculated for blade root fixings are subject to significant levels of uncertainty, so it is normal to conduct both static and fatigue tests on them to verify the suitability of the design. Static pull-out failures of carrot connectors occur as a result of shearing of the wood surrounding the grout, but fatigue failures can also occur in the connector itself or the grout. However, SGI studs subjected to $R = 0.1$ fatigue loading at over 60 percent of the UTS have survived for approximately 10^6 cycles.

Mayer (1996) records the results of fatigue tests on the other blade root fixings featured in Figure 7.19, but in no case did failure occur as a result of fatigue of the GFRP in the region of the root fixing. In the case of the T-bolt fixing arrangement, failure occurred in the studs rather than in the GRP. The pin-hole flange specimens developed fatigue cracks in the GFRP in areas remote from the root fixings and the trumpet flange specimens developed cracks in the flanges themselves.

7.2 Pitch Bearings

On pitch-regulated machines a bearing similar to a crane slewing ring is interposed between each blade and the hub to allow the blade to be rotated or 'pitched' about its axis. A typical arrangement is as shown in Figure 7.20, in which the inner and outer rings of the bearing are bolted to the blade and hub respectively. The different types of bearings available can be classified according to the rolling elements used and their arrangement, in order of increasing moment capacity:

(a) single-row roller bearings, with alternate rollers inclined at $+45°$ and $-45°$ to the plane of the bearing;

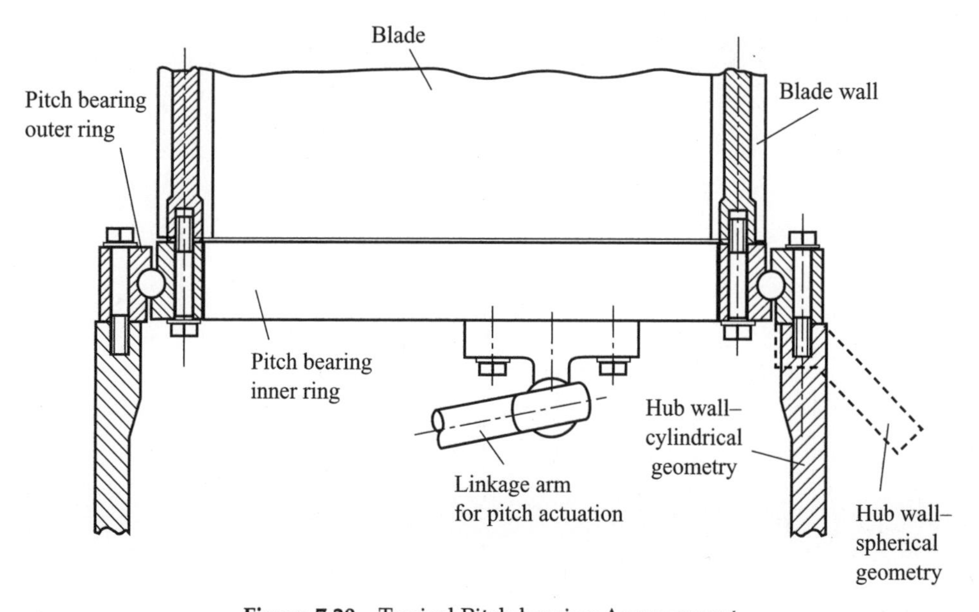

Figure 7.20 Typical Pitch-bearing Arrangement

(b) single-row ball bearings,

(c) double-row ball bearings;

(d) three-row roller bearings.

These are shown in cross section in Figure 7.21. The single-row ball bearing slewing rings are normally designed to transmit axial loads in both directions and are therefore known as four point contact bearings. Low contact stresses are achieved by making the radii on each side of the grooves only slightly larger than that of the balls.

At low wind speeds, the cyclic in-plane bending moment at the blade root due to gravity is of similar magnitude to the out-of-plane moment due to blade thrust, so bearing loads will alternate in direction over portions of the bearing circumference. Accordingly it is desirable to avoid the risk of play by preloading the bearing. This can be achieved relatively easily on bearings in which one of the rings is split on a plane normal to the axis, such as types (c) and (d), but is more difficult when both rings are solid. In this case it is necessary to force the rolling elements into the races one by one during manufacture.

The bearing selected for a particular application needs to have sufficient moment capacity to both resist the extreme blade root bending moments and provide adequate fatigue life. Manufacturers' catalogues typically specify both the extreme moment capacity and the steady moment loading that will give a life of, say, 30 000 bearing revolutions, so the wind turbine designer's chief task is to convert the

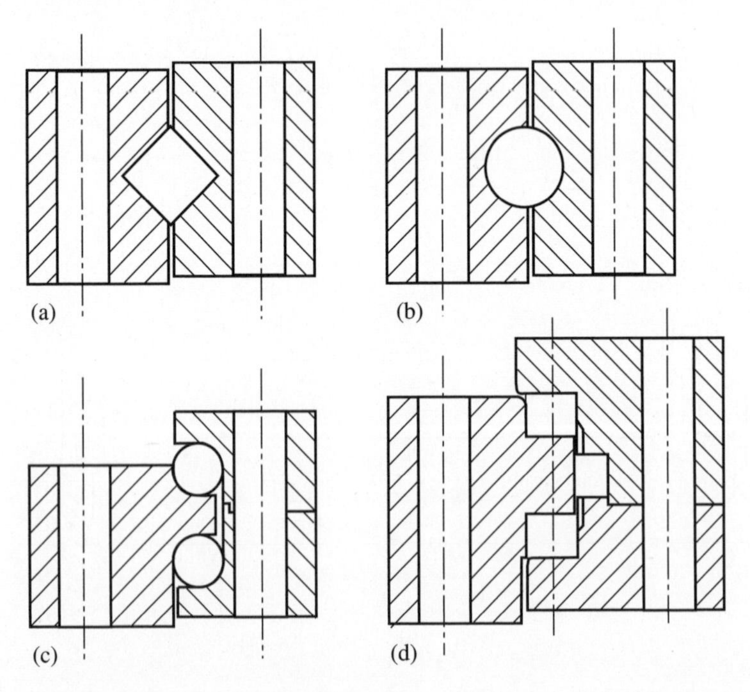

Figure 7.21 (a) Single-row Crossed Roller Bearings; (b) Single-row Ball Bearings; (c) Double-row Ball Bearings; (d) Three-row Roller Bearings

anticipated pitch bearing duty into the equivalent constant loading at the appropriate number of revolutions. If the rolling elements are ball bearings, the bearing life is inversely proportional to the cube of the bearing loading, so the equivalent loading at N revolutions of the pitch bearing can be calculated according to the formula:

$$M_{\mathrm{eqt}} = \left[\frac{\sum_i n_i M_i^3}{N} \right]^{1/3} \tag{7.42}$$

where n_i is the total pitch bearing movement anticipated over the design life at moment loading M_i, expressed as a number of revolutions. In the case of roller bearings, the index of the S–N curve is 10/3 instead of 3, so the above formula should be modified accordingly. As the blade root out-of-plane moment drops as the wind speed increases above rated, the fatigue damage will be concentrated at wind speeds near rated.

The total pitch bearing movement over a period of operation at a particular wind speed is a function of the turbulence intensity and the pitch control algorithm, and is best predicted by means of a wind simulation. The mean blade pitching rate during operation above rated wind speed is found to be of the order of 1 degree/s, assuming the pitch system only responds to wind speed fluctuations at a frequency less than the speed of rotation.

The performance of slewing ring bearings such as those employed as pitch bearings is critically dependent on the extent of bearing distortion under load, so manufacturers normally specify a maximum axial deflection and tilt of the bolted contact surfaces. For example, the limiting values given by Rothe-Erde for a single-row ball bearing slewing ring with a 1000 mm track diameter are 0.6 mm and 0.17° respectively. Local tilting of the bearing rings could clearly be minimized if the blade wall, bearing track and hub wall were all positioned in the same plane. However, this would necessitate the provision of flanges, so the simpler arrangement shown is Figure 7.20, in which the fixing bolts are inserted centrally into the blade and hub walls, is generally preferred. The designer must then ensure that the blade and hub structures are of sufficient stiffness to limit the bearing distortion due to the eccentric loading to acceptable values.

It is standard practice to preload the bearing fixing bolts in order to minimize bolt fatigue loading. Grade 10.9 bolts are commonly used so that the preload can be maximized.

7.3 Rotor Hub

The relatively complex three-dimensional geometry of rotor hubs favours the use of casting in their manufacture, with spheroidal graphite iron being the material generally chosen.

Two distinct shapes of hub for three-bladed machines can be identified: tri-cylindrical or spherical. The former consists of three cylindrical shells concentric with the blade axes, which flare into each other where they meet, while the latter consists simply of a spherical shell with cut-outs at the three-blade mounting positions. Diagrams of both types are shown in Figure 7.22, while an actual spherical hub is illustrated in Figure 7.23. The structural action of the hub in resisting three loadings is discussed in the following paragraphs.

(1) *Symmetric rotor thrust loading*. The blade root bending moments due to symmetric rotor thrust loading put the front of the hub in bi-axial tension near the rotor axis and the rear in bi-axial compression, while the thrust itself generates out-of-plane bending stresses in the hub shell adjacent to the low speed shaft flange connection. The load paths are easy to visualize in this case.

(2) *Thrust loading on a single blade*. This generates out-of-plane bending stresses in the hub shell at the rear, and in-plane tensile stresses around a curved load path between the upwind side of the blade bearing and the portion of the low-speed shaft flange connection remote from the blade (see dashed line in Figure 7.22(b). The resultant lateral loads will result in out-of-plane bending.

(3) *Blade gravity moments*. On the tri-cylindrical hub, equal and opposite blade gravity moments are communicated via the cylindrical shells to areas near the rotor axis at front and rear where they cancel each other out. It is less straightforward to visualize the corresponding load paths on the spherical hub, as out-of-plane bending is likely to be mobilized.

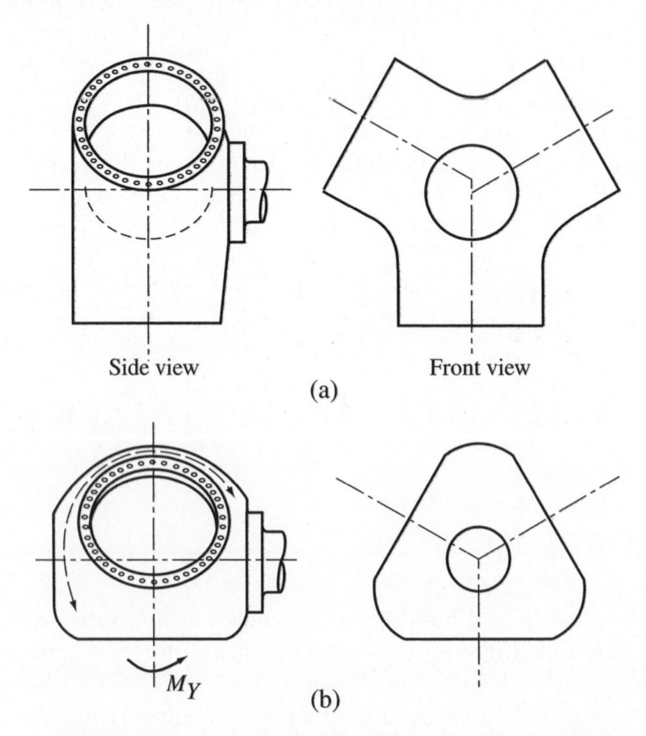

Figure 7.22 (a) Tri-cylindrical Hub; (b) Spherical Hub

Figure 7.23 Rotor Hub. View of Spherical-shaped Rotor Hub for the 1.5 MW NEG Micon Turbine Awaiting Installation. The Hub and Spinner are Temporarily Oriented with the Rotor Shaft Axis Vertical. The Turbine is Stall-regulated, so Slotted Blade Fixing Holes are Provided to Allow for Fine Adjustment of Blade Pitch to Suit the Site. (Reproduced by permission of NEG-Micon)

The complexity of the stress states arising from the latter two types of loading renders finite-element analysis of rotor hubs more or less mandatory. At the most, six load cases need to be analysed, corresponding to the separate application of moments about the three axes and forces along the three axes at a single hub/blade interface. Then the distribution of hub stresses due to combinations of loadings on different blades can be obtained by superposition. Similarly the fluctuation of hub stresses over time can be derived by inputting the time histories of the blade loads obtained from a wind simulation.

The critical stresses for hub design are the in-plane stresses at the inner or outer surface, where they reach a maximum because of shell bending. For any one location on the hub, these are defined by three quantities at each surface: the in-plane direct stresses in two directions at right angles, and the in-plane shear stress. In general, these stresses will not vary in-phase with each other over time, so the principal stress directions will change, complicating the fatigue assessment.

There is, as yet, no generally recognized procedure for calculating the fatigue damage accumulation due to multi-axial stress fluctuations, although the following methods have been used, despite their acknowledged imperfections. They all cater for one or more series of repeated stress cycles rather than the random stress fluctuations resulting from turbulent loading.

(1) *Maximum shear method*. Here the fatigue evaluation is based on the maximum shear stress ranges, calculated from either the $(\sigma_1 - \sigma_2)/2$, $\sigma_1/2$ or $\sigma_2/2$ time histories. The effect of mean stress is allowed for using the Goodman relationship:

$$\frac{\tau_a}{S_{SN}} + \frac{\tau_m}{S_{Su}} = \frac{1}{\gamma} \tag{7.43}$$

where τ_a is the alternating shear stress, τ_m is the mean shear stress, S_{SN} is the alternating shear stress for N loading cycles from the material S–N curve, S_{Su} is the ultimate shear strength, and γ is the safety factor.

Having used Equation (7.43) to determine S_{SN}, the permitted number of cycles for this loading range can be derived from the S–N curve, enabling the corresponding fatigue damage to be calculated.

(2) *ASME Boiler and pressure vessel code method*. This is similar to the maximum shear method, but the shear stress ranges are based on notional principal stresses calculated from the *changes* in the values of σ_x, σ_y, σ_z, τ_{xy}, τ_{yz} and τ_{zx} from datum values occurring at one of the extremes of the stress cycle. Mean stress effects are not included.

(3) *Distortion energy method*. In this method, the fatigue evaluation is based on the fluctuations of the effective or Von Mises stress. In the case of the hub shell, the stress perpendicular to the hub surface (and hence the third principal stress) is zero, so the effective stress is given by:

$$\sigma' = \sqrt{\frac{(\sigma_1 - \sigma_2)^2 + \sigma_1^2 + \sigma_2^2}{2}} \tag{7.44}$$

As the effective stress is based on the distortion energy, it is a scalar quantity, so it needs to be assigned a sign corresponding to that of the dominant principal stress. The effect of mean stress is allowed for in the same way as for the maximum shear method, except that the stresses in Equation (7.43) are now direct stresses instead of shear stresses. S–N curves for spheroidal graphite iron are given in Hück (1983).

7.4 Gearbox

7.4.1 Introduction

The function of the gearbox is to step up the speed of rotor rotation to a value suitable for standard induction generators, which, in the case of fixed-speed machines or two-speed machines operating at the higher speed, is usually 1500 r.p.m. plus the requisite slip. For machines rated between 300 kW and 2000 kW, with upper rotational speeds between 48 and 17 r.p.m., overall gear ratios of between about 1:31 and 1:88 are therefore required. Normally these large step-ups are achieved by three separate stages with ratios of between 1:3 and 1:5 each.

The design of industrial fixed ratio gearboxes is a large subject in itself and well beyond the scope of the present work. However, it is important to recognize that the use of such gearboxes in wind turbines is a special application, because of the unusual environment and load characteristics, and the sections which follow focus on these aspects. Sections 7.4.2 to 7.4.6 consider variable loading, including drive train dynamics and the impact of emergency braking loads, and examine how gear fatigue design is adapted to take account of it. The relative benefits of parallel and epicyclic shaft arrangements are discussed in Section 7.4.7, while subsequent sections deal with noise reduction measures, and lubrication and cooling. A useful reference is the American Gear Manufacturers Association Information Sheet (1996) which covers the special requirements of wind turbine gearboxes in some detail.

7.4.2 Variable loads during operation

The torque level in a wind turbine gearbox will vary between zero and rated torque according to the wind speed, with excursions above rated on fixed-speed pitch-regulated machines due to slow pitch response. The short-term torque fluctuations will be subject to dynamic magnification to the extent that they excite drive train resonances (see Section 7.4.3 below). In addition there will be occasional much larger torques of short duration due to braking events, unless the brake is fitted to

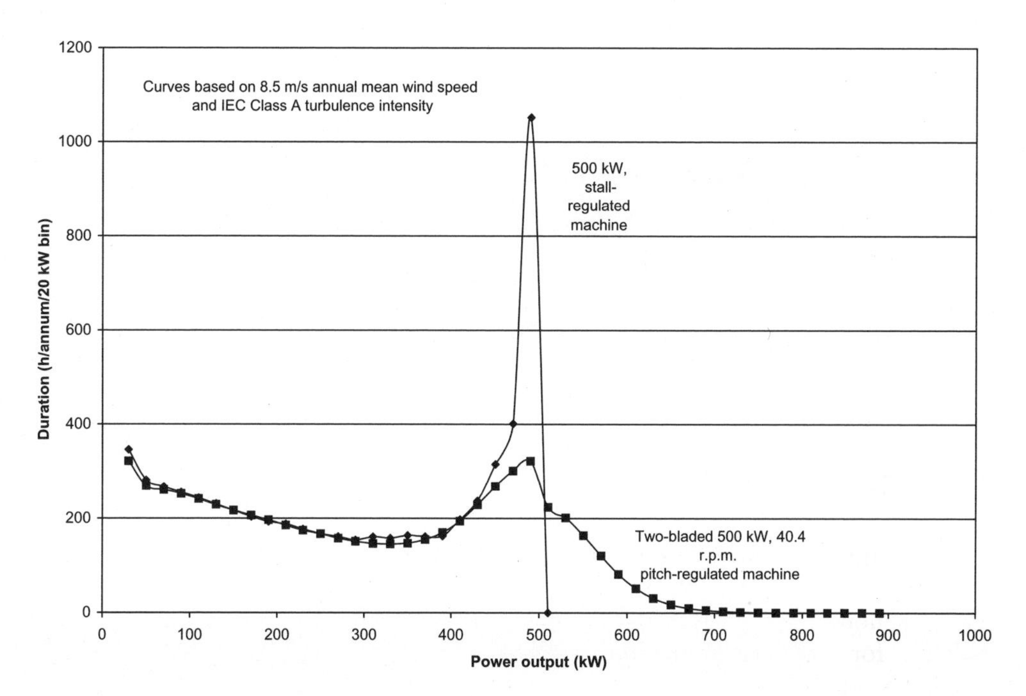

Figure 7.24 Load Duration Curves for 500 kW, Two-bladed Pitch-regulated and 500 kW, Stall-regulated Fixed-speed Machines

the low-speed shaft. Figure 7.24 shows example load–duration curves (excluding dynamic effects and braking) for two 500 kW, fixed-speed machines – one stall- and the other pitch-regulated. The curve for the former is calculated by simply combining the power curve with the distribution of instantaneous wind speeds, which is obtained by superposing the turbulent variations about each mean wind speed on the Weibull distribution of hourly means. Excursions above rated power are not included.

In the case of a pitch-regulated machine, the pitch control system is not normally designed to respond to wind speed fluctuations at blade-passing frequency or above, as this would impose excessive loads on the control mechanism. Thus there is no attenuation of the significant power fluctuations that occur at blade-passing frequency due to turbulence, which are illustrated for the example two-bladed, 500 kW machine operating in a 20 m/s mean wind with 16.5 percent turbulence intensity in Figure 7.25.

The load duration curve for a fixed-speed pitch-regulated machine can be derived approximately from the distribution of instantaneous wind speeds below rated wind speed, and the distribution of short-term mean wind speeds (i.e. those to which the pitch system can respond) above. The former can be combined with the power curve to give the power distribution due to instantaneous winds below rated directly, while the winds above rated are assumed to produce Gaussian spreads of power outputs about the rated value, with the standard deviation depending on the short-term mean wind. The standard deviation of power fluctuations when the pitch control system is operational can be related to that portion of the wind fluctuations above the pitch control system cut off frequency:

$$\sigma_P^2 = \frac{1}{N^2}\sum_j \sum_k \left[\int_{\Omega}^{\infty} S_u^o(r_j, r_k, n)\,\mathrm{d}n\right]\left(\frac{\mathrm{d}p}{\mathrm{d}u}\right)_j \left(\frac{\mathrm{d}p}{\mathrm{d}u}\right)_k \tag{7.45}$$

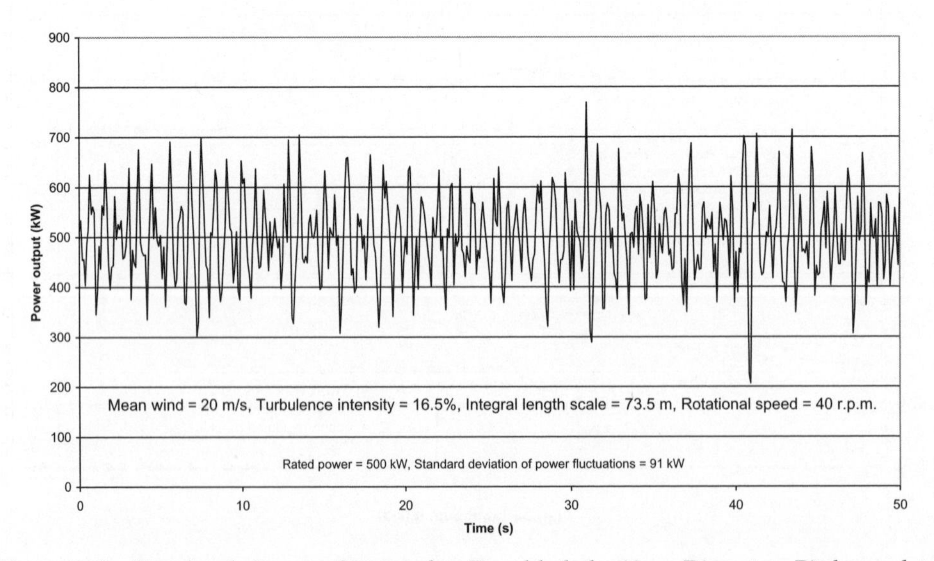

Figure 7.25 Simulated Power Output for Two-bladed, 40 m Diameter Pitch-regulated Machine Operating in Above-rated Wind Speed

where $S_u^o(r_i, r_k, n)$ is the rotationally sampled cross spectrum of the wind speed fluctuations at a pair of points, j and k, on the rotor (see Section 5.7.5) and $(dp/du)_j$ is the rate of change with wind speed of the power generated by the blade elements at r_j on all N blades if the pitch does not change. The summations are carried out over the whole rotor, and give $\sigma_P = 0.213(dP/du)\sigma_u = 91$ kW for the example two-bladed machine operating at 40.4 r.p.m. in a 20 m/s mean wind with 16.5 percent turbulence intensity. Here dP/du is the rate of change of turbine power with wind if the pitch does not change. The standard deviation of the power fluctuations for a three-bladed machine of similar size would be about one third less.

7.4.3 Drive-train dynamics

All wind turbines experience aerodynamic torque fluctuations at blade-passing frequency and multiples thereof because of the 'gust slicing' phenomenon, and these fluctuations will inevitably interact with the dynamics of the drive train, modifying the torques transmitted. The resulting drive train torque fluctuations can be assessed by dynamic analysis of a drive train model consisting of the following elements connected in series:

- a body with rotational inertia and damping (representing the turbine rotor),
- a torsional spring (representing the gearbox),
- a body with rotational inertia (representing the generator rotor),
- a torsional damper (modelling the resistance produced by slip on an the induction generator), and
- a body of infinite rotational inertia rotating at constant speed (the mechanical equivalent of the electrical grid).

The inertias, spring stiffness and damping must all be referred to the same shaft.

7.4.4 Braking loads

Most turbines have the mechanical brake located on the high-speed shaft, with the result that braking loads are transmitted through the gearbox. If, as is usually the case, the mechanical brake is one of the two independent braking systems required, then it must be capable of decelerating the rotor to a standstill from an overspeed, e.g., after a grid loss. This typically requires a torque of about three times rated torque.

The mechanical brake is only required to act alone during emergency shut-downs, which are comparatively rare. During normal shut-downs the rotor is decelerated to a much lower speed by aerodynamic braking, so the duration of mechanical braking is much less, but the braking torque is the same, unless there is provision for two different braking torque levels.

Figure 7.26 is a typical record of low-speed shaft torque during a normal shut-

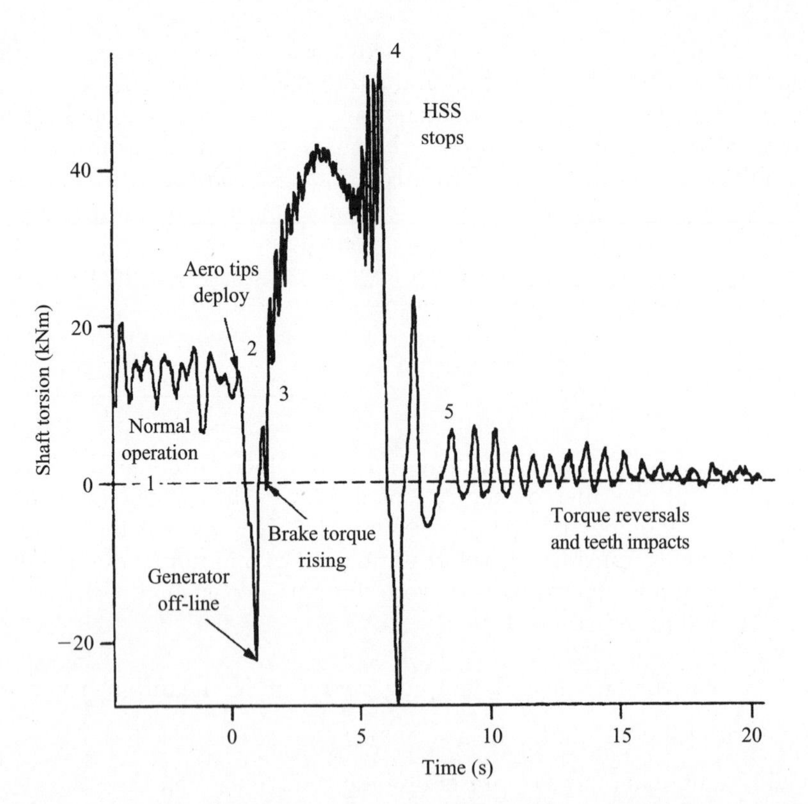

Figure 7.26 Low-speed Shaft Torque During Braking at Normal Shut-down. Extracted from AGMA/AWEA 921–A97, Recommended practices for design and specification of gearboxes for wind-turbine generator systems, with permission of the publisher, the American Gear Manufacurers Association, 1500 King Street, Suite 201, Alexandria, Virginia 22314, USA.

down, in which the mechanical brake is applied as soon as the generator has been taken off-line. It is apparent that the braking torque is far from constant, taking about 2 s to reach its first maximum and then falling off slightly before reaching a higher maximum just before the high-speed shaft stops. Following this, there are significant torque oscillations due to the release of wind-up in the drive train. These result in torque reversals accompanied by tooth impacts and take some time to decay.

Although braking loads are infrequent and of short duration, their magnitude means that they can have a decisive effect on fatigue damage. The AGMA/AWEA document (1996) recommends that the time histories of braking and other transient events are simulated with the aid of a dynamic model of the drive train for input into both the gear extreme load design calculations and the fatigue load spectrum.

7.4.5 Effect of variable loading on fatigue design of gear teeth

Gear teeth must be designed in fatigue to achieve both acceptable contact stresses on the flanks and acceptable bending stresses at the roots. In non-wind turbine applications, gearboxes typically operate at rated torque throughout their lives, so the gear strengths are traditionally modified by 'life factors' which are derived from the material S–N curves on the basis of the predicted number of tooth load cycles for the gear in question. The British code for gear design, BS436 (British Standards Institution, 1986), recognizes an endurance limit for both contact stress and bending stress, so that the life factors are unity when the number of tooth load cycles exceeds 10^9 and 3×10^6 respectively, but increase for lesser numbers of cycles.

The Hertzian compression stress between a pair of spur gear teeth in contact at the pitch point (i.e., at the point on the line joining the gear centres) is given by

$$\sigma_C = \sqrt{\frac{F_t}{bd_1}\frac{E}{\pi(1-v^2)}\frac{u+1}{u}\frac{1}{\sin\alpha\cos\alpha}} \tag{7.46}$$

where F_t is the force between the gear teeth at right angles to the line joining the gear centres, b is the gear face width, d_1 is the pinion pitch diameter, u is the gear ratio (greater than unity), and α is the pressure angle, i.e., the angle at which the force acts between the gears – usually 20°–25°.

Note that the contact stress increases only as the square root of the force between the teeth because the area in contact increases with the force as well.

The maximum bending stress at the tooth root is given by

$$\sigma_B = \frac{F_t h}{\frac{1}{6}bt^2}K_S \tag{7.47}$$

where h is the maximum height of single tooth contact above the critical root section, t is the tooth thickness at the critical root section, K_S is a factor to allow for stress concentration at the root.

For gearing operating at rated torque only, the designer needs to show that the resultant bending stress multiplied by an appropriate safety factor is less than the endurance limit multiplied by the life factor, Y_N, and a number of stress modifying factors, as follows

$$\sigma_B.\gamma \leqslant \sigma_{B\,\mathrm{lim}}.Y_N.Y_R Y_X \ldots\ldots\ldots \tag{7.48}$$

A similar calculation is required in relation to the contact stress.

Given the predicted turbine load spectrum (Section 7.4.2), which should include dynamic effects (see Section 7.4.3), it is then necessary to establish the required design torque at the endurance limit. Normally this is done by invoking Miner's rule and determining the infinite life torque for which the design torque spectrum yields unity fatigue damage in conjunction with the prescribed S–N curve. Y_N in Equation (7.48) can then be set to unity, as the life factor has been accounted for in the derivation of the required infinite life torque.

Figure 7.27 shows specimen torque–endurance curves laid down by BS 436 for case-hardened gears for tooth bending and tooth contact stress (with no pitting allowed) plotted in terms of the torque at the endurance limit. Hence in each case the design infinite life torque, T_∞, is calculated according to:

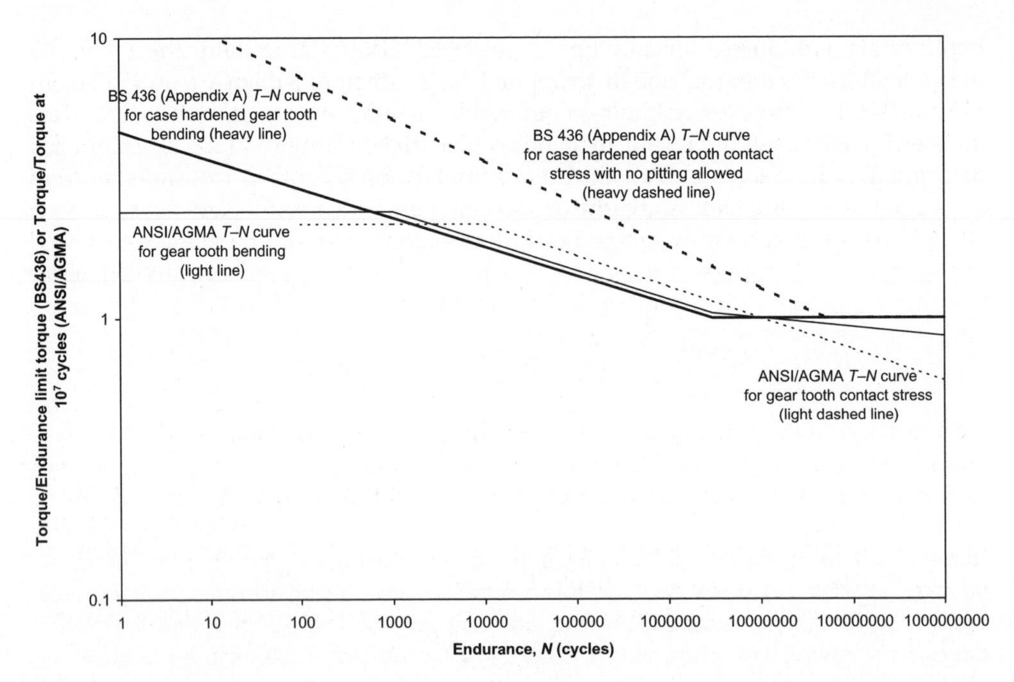

Figure 7.27 Specimen Torque-endurance Curves for Gear Tooth Design under Variable Amplitude Loading

$$T_\infty = \left[\sum_i \left(\frac{N_i}{N_\infty} T_i^m \right) \right]^{1/m} \tag{7.49}$$

where N_i is the number of cycles at torque level T_i, and torques less than T_∞ are omitted from the summation. The number of cycles at the lower knee of the torque–endurance curve, N_∞, is always 3×10^6 cycles for tooth bending but is generally higher for contact stress, varying according to the material. Note that the slope index, m, of the torque–endurance curve for contact stress is half that of the contact stress–endurance curve because contact stress only increases as the square root of torque (Equation 7.46).

Leaving braking loads out of consideration to begin with, the design infinite life torque will be equal to the rated torque if there are no power fluctuations above rated, because the number of gear tooth loading cycles at rated torque will be well above N_∞. For example, in the case of the 500 kW stall-regulated machine featured in Figure 7.24, the teeth on the critical pinion driven by the 30 r.p.m. low-speed shaft will experience $3 \times 30 \times 60 \times 1050 \times 20 = 1.13 \times 10^8$ load cycles at rated torque over 20 years, assuming a first stage gear ratio of 3. On the other hand, for the 500 kW, two-bladed pitch-regulated machine, the power fluctuations above rated detailed in the Figure 7.24 load–duration curve result in a design infinite life torque for the first stage pinion tooth bending stress of 1.36 times the rated torque, with most of the damage coming from torques just above this value. (The first stage gear ratio is assumed to be three as before and the turbine rotational speed is taken as 40.4 r.p.m.) The design infinite life torque for tooth contact stress is only 1.17 ×

rated torque – significantly less than for bending, as expected from comparison of the BS 436 torque–endurance curves in Figure 7.27.

Figure 7.27 also shows specimen torque–endurance curves derived from S–N curves in the American National Standards Institute/AGMA 2001-C95 plotted in terms of the torque at 10^7 cycles. The torque–endurance curve for tooth bending stress, which is based on a middle of the range Brinell Hardness value of 250 HB, closely parallels the selected BS 436 curve, except that the curve continues with a very shallow slope beyond 3×10^6 cycles instead of displaying an endurance limit. The design torques at 10^7 cycles for tooth bending for the example 500 kW machines featured in Figure 7.24 are similar to the design infinite life torques obtained using the BS 436 torque–endurance curves.

The ANSI/AGMA (1995) torque–endurance curves for tooth contact stress are significantly more conservative than the selected BS 436 curve. This is particularly so in the case of the ANSI curve selected, which is the one recommended for wind turbine applications, in view of the elimination of the lower knee. The absence of the lower knee increases the design torque at 10^7 cycles for tooth contact to 1.4 times the rated torque for the stall regulated machine, but the figure for the pitch regulated machine is only about 10 percent higher.

From the above discussion, the general conclusion can be drawn that tooth bending fatigue usually governs the increased gearbox rating required to take care of load excursions above rated.

The effect of braking loads on the design infinite life torque according to BS 436 can be illustrated with respect to the example machines discussed in Section 7.4.2. Although the mechanical brake must be capable of decelerating an overspeeding rotor unassisted, a shut-down under these conditions will be a very rare event. Accordingly the typical emergency shut-down considered for fatigue design purposes is deceleration from normal rotational speed under the action of mechanical and aerodynamic braking combined, with an assumed stopping time of 3 s. An emergency shut-down frequency of 20 per annum is assumed. Normal shut-downs are assumed to occur on average twice a day, with a stopping time of 1.5 s, because of the reduced rotational speed at which mechanical braking is initiated for parking. In each case the braking torque is assumed to remain constant at three times rated torque throughout the brake application for simplicity. Based on these assumptions, the percentage increases in design infinite life torque for gear tooth bending in fatigue, due to the inclusion of braking loads in the load spectrum, are shown in Table 7.5 for emergency braking alone on the one hand and normal plus emergency shut-downs on the other.

Also shown in the table are the percentage increases in the ANSI/AGMA design life for gear tooth bending at 10^7 cycles due to the inclusion of braking loads. It is seen that the inclusion of emergency braking loads alone makes very little difference to design torques in the case of the pitch-regulated machine, but is significant in the case of the stall-regulated machine. The addition of braking loads at normal shut-downs incurs a much greater penalty in both cases because of the large number of stops involved, indicating that provision for brake application at reduced torque on these occasions would probably be worthwhile. Note that the larger percentage increases in design torques due to braking indicated by BS 436 are a consequence of the assumption that there is an endurance limit.

Table 7.5 Illustrative Increases in Design Torques for Gear Tooth Bending due to Inclusion of Braking Loads in Fatigue Load Spectrum, According to BS 436 and ANSI/AGMA Rules

	500 kW Stall-regulated machine		500 kW Two-bladed pitch-regulated machine	
	Percentage increase in BS 436 design infinite life torque for tooth bending	Percentage increase in ANSI/AGMA 250 HB design torque at 10^7 cycles for tooth bending	Percentage increase in BS 436 design infinite life torque for tooth bending	Percentage increase in ANSI/AGMA 250 HB design torque at 10^7 cycles for tooth bending
Emergency braking at 3 × FLT	30%	16%	4%	3%
Emergency plus normal braking, each at 3 × FLT	65%	47%	25%	21%

7.4.6 Effect of variable loading on fatigue design of bearings and shafts

Bearing lives are approximately inversely proportional to the cube of the bearing loading. Applying Miner's rule, the equivalent steady bearing loading over the gearbox design life can thus be calculated from the load duration spectrum according to the formula

$$F_{\text{eqt}} = \left[\frac{\sum_i N_i F_i^3}{\sum_i N_i} \right]^{1/3} \tag{7.50}$$

where N_i is the number of revolutions at bearing load level F_i. Gravity often dominates the loading on the low-speed shaft bearings, but on the other shafts the bearing loads result from drive torque only, so the bearing load duration spectrum can be scaled directly from the torque duration spectrum. Note that the $S{-}N$ curve for bearings is much steeper than those for gear tooth design, so that occasional large braking loads will be of less significance.

The nature of the fatigue loading of intermediate shafts is essentially different from that of gear teeth, as the former is governed by the torque *fluctuations* as opposed to the absolute torque magnitude. Consequently the fatigue load spectrum for shaft design should be derived from rainflow cycle counts on simulated torque time histories rather than on the load duration curve used for gear tooth design.

7.4.7 Gear arrangements

Parallel axis gears may be arranged in one of two ways in each gear stage. The simplest arrangement within a stage consists of two external gears meshing with each other and is commonly referred to as 'parallel shaft'. The alternative 'epicyclic' arrangement consists of a ring of planet gears mounted on a planet carrier and meshing with a sun gear on the inside and an annulus gear on the outside. The sun and planets are external gears and the annulus is an internal gear as its teeth are on the inside. Usually either the annulus or planet carrier are held fixed, but the gear ratio is larger if the annulus is fixed.

The epicyclic arrangement allows the load to be shared out between the planets, reducing the load at any one gear interface. Consequently the gears and gearbox can be made smaller and lighter, at the cost of increased complexity. The scope for material savings are greatest in the input stages of the gear train, so it is common to use the epicyclic arrangement for the first two stages and the parallel shaft arrangement for the output stage. A further advantage of epicyclic gearboxes is greater efficiency as a result of the reduced sliding that takes place between the annulus and planet teeth.

The derivation of the optimum gear ratio in a series of parallel shaft stages is fairly straightforward and is described below. Equation (7.47) for tooth bending stress can be modified as follows

$$\sigma_B = \frac{F_t h}{\frac{1}{6}bt^2} K_S = F_t \frac{6(h/m)}{bm(t/m)^2} K_S = F_t \frac{6z_1(h/m)}{bd_1(t/m)^2} K_S \tag{7.47a}$$

where m is the module, defined as d_1/z_1 for spur gears and z_1 is the number of pinion teeth. If the ratios h/m and t/m are treated as constants, then the bending stress is proportional to the number of teeth for a given size of gear. Hence the design of the gears is governed by contact stress because, in principle, the bending stress can always be reduced by reducing the number of pinion teeth. Thus, based on Equation (7.46), the permitted tangential force, F_t, is proportional to $bd_1u/(u+1)$ so that the permitted low-speed shaft torque, $T_{LSS} = F_t d_2/2$ is given by

$$T_{LSS} \propto d_2 b d_1 u/(u+1) = bd_2^2/(u+1) \tag{7.51}$$

Hence the volumes of the low-speed shaft gear wheel and the meshing pinion can be expressed as $V_2 = kT_{LSS}(u+1)$, where k is a constant, and $V_1 = V_2/u^2$ respectively. These can be used to derive an expression for the volume of gears in a drive train with an infinite number of stages each with the same ratio. It is found that the total gear volume is a minimum for a gear stage ratio of 2.9, but increases by only 10 percent when the ratio drops to 2.1 or rises to 4.3.

The gear teeth of parallel shaft gear stages are only loaded in one direction, so the permitted alternating bending stress in fatigue, σ_{alt}, is modified to account for the non-zero mean value in accordance with the Goodman relation:

$$\frac{\sigma_{\text{alt}}}{\sigma_{\text{lim}}} = 1 - \frac{\overline{\sigma}}{\sigma_{\text{ult}}} \tag{7.52}$$

where σ_{lim} is the permitted alternating bending stress with zero mean, $\overline{\sigma}$ is the mean bending stress and σ_{ult} is the ultimate tensile strength. Setting $\overline{\sigma} = \sigma_{\text{alt}}/2$ results in:

$$\sigma_{\text{alt}} = \frac{\sigma_{\text{lim}}\sigma_{\text{ult}}}{(\sigma_{\text{ult}} + \sigma_{\text{lim}})} \tag{7.53}$$

If the $\sigma_{\text{lim}}/\sigma_{\text{ult}}$ ratio is 0.2, then $\sigma_{\text{alt}} = 0.833\sigma_{\text{lim}}$ and the permitted peak bending stress at the endurance limit is $1.667\sigma_{\text{lim}}$. In epicyclic gearboxes, by contrast, the gear teeth on the planet wheels are loaded in both directions, so the permitted peak bending stress at the endurance limit is only $0.5\sigma_{\text{lim}}$. As the number of teeth on the smallest gear cannot be reduced indefinitely, this means that tooth bending is more likely to govern in the case of epicyclic gearing.

The minimum total gear volume for an infinite series of epicyclic gear stages with fixed annuli is obtained for a gear stage ratio of two, which implies that the radius of the sun gear is the same as that of the annulus gear and that there are an infinite number of planets! This is not realistic, and the annulus radius is in practice typically double the sun radius, giving a gear ratio of three. It is instructive to compare the volume of gears for an epicyclic and parallel gear stages with this ratio, assuming that tooth bending stress governs in each case.

For the parallel stage, it can be shown using Equation (7.47a) that the volume of the pinion is

$$\frac{\pi}{4}bd_1^2 = k_{\text{B}}F_t d_1 z_1/\sigma_{\text{alt}} = k_{\text{B}}\frac{F_t d_2 z_1}{\sigma_{\text{alt}}u} = k_{\text{B}}\frac{2T_{\text{LSS}}z_1}{0.833\sigma_{\text{lim}}u} \tag{7.54}$$

where k_{B} is a constant. This gives a volume for gear wheel and pinion of $2.4k_{\text{B}}T_{\text{LSS}}z_1(1 + 1/u^2)u/\sigma_{\text{lim}} = 8k_{\text{B}}T_{\text{LSS}}z_1/\sigma_{\text{lim}}$ for $u = 3$. For the epicyclic stage, the volume of the planet, which is assumed to have the same number of teeth as the pinion of the parallel stage, i.e., the minimum permissible, is

$$\frac{\pi}{4}bd_{Pl}^2 = k_{\text{B}}F_t d_{\text{PL}} z_1/(0.5\sigma_{\text{lim}}) \tag{7.55}$$

If the low-speed shaft drives the planet carrier and the N planets are spaced at 1.15 diameters, then the low speed shaft torque is

$$T_{\text{LSS}} = F_t N(r_{\text{A}} + r_{\text{S}})$$

where

$$N = \frac{\pi(r_{\text{A}} + r_{\text{S}})}{1.15(r_{\text{A}} - r_{\text{S}})}$$

r_A is the annulus radius and r_S is the sun radius. Hence, putting $a = r_A/r_S$, the volume of a planet is

$$k_B T_{LSS} \frac{1.15(a-1)}{\pi(a+1)^2 r_S} \frac{2d_{PL} z_1}{\sigma_{lim}}$$

and the volume of the sun is $4/(a-1)^2$ times as big. The total volume of planets and sun becomes

$$V = k_B T_{LSS} \frac{1}{a+1} \frac{2d_{PL} z_1}{r_S \sigma_{lim}} \left(1 + \frac{4}{(a-1)^2 N}\right) \tag{7.56}$$

Substituting $a = 2$, we obtain

$$d_{PL} = r_s \text{ and } N = \frac{3\pi}{1.15} = 8.195$$

which is rounded down to 8, giving

$$V = k_B T_{LSS} \frac{2z_1}{3\sigma_{lim}} \left(1 + \frac{4}{8}\right) = k_B T_{LSS} z_1 / \sigma_{lim}$$

Hence the volume of the sun and planets of the epicyclic stage is only one eighth of the volume of the gearwheel and pinion of the equivalent parallel stage, assuming the designs are governed by gear tooth bending stress. If contact stress were to govern, the relative volume of the epicyclic stage would be even less.

The dramatic materials savings obtainable with epicyclic gearboxes depend on equal sharing of loads between the planets. Although this is theoretically achievable through accuracy of manufacture, it is in practice desirable to introduce some flexibility in the planet mountings to take up any planet position errors – for example by supporting the planets on slender pins cantilevered out from the planet carrier. Note that the fatigue design of such pins is, like the design of intermediate shafts, governed by torque *fluctuations* rather than by torque absolute magnitude.

7.4.8 Gearbox noise

The main source of gearbox noise arises from the meshing of individual teeth. Loaded teeth deflect slightly, so that if no tooth profile correction is made, unloaded teeth are misaligned when they come into contact, resulting in a series of impacts at the meshing frequency. It is therefore standard practice to adjust the tooth profile – usually by removing material from the tip area of both gears, referred to as 'tip relief' – to bring the unloaded teeth back into alignment at the rated gear loading. In the case of wind turbines, the gear loading is variable, so it is necessary to select the load level at which the tip relief provides the correct compensation. If the tip relief load level is too high, there will be excessive loss of tooth contact near the tips at low powers, while if it set too low the noise level at rated power will be too high.

However, if gearbox noise is expected to be more intrusive at low wind speeds, when it is less likely to be masked by aerodynamic noise, then a low compensation load level should be selected.

Helical gears are usually quieter than spur gears (with teeth parallel to the gear axis) because the width of the tooth comes into mesh over a finite time interval rather than all at once. Moreover, the peak tooth deflections of helical gears are less than those of spur gears because there are always at least two teeth in contact rather than one, and because the varying bending moment across the tooth width means that the less heavily loaded portions of the tooth can provide restraint to the part that is most heavily loaded. As a result, the tooth misalignments due to insufficient/excessive tip relief at a particular load level will be reduced.

Epicyclic gears are normally quieter than parallel shaft gears because the reduced gear size results in lower pitch line velocities. However, this benefit is lost if spur gears are used rather than helical gears, in order to avoid problems with planet alignment. One way of maintaining the alignment of helical planet gears is to provide thrust collars on the sun and annulus.

As the annulus of an epicyclic gear stage is often fixed, it would be convenient to integrate it with the gearbox casing. However, this would enable annulus gear meshing noise to be radiated directly from the casing, so it is preferable to make the annulus a separate element, supported on resilient mountings. Similarly, resilient gearbox mountings should be used to attenuate the transmission of gearbox noise to the nacelle structure and tower.

The noise produced by gear tooth meshing can reach the environment outside the wind turbine by a variety of routes:

- through the shaft directly to the blades, which may radiate efficiently,
- through the resilient mounts of the gearbox to the support structure and thereby to the tower, which can radiate efficiently under some circumstances,
- through the resilient mounts of the gearbox to the support structure and thereby to the nacelle structure, which can also radiate,
- through the casing wall to the nacelle air and then *via* air intake and exhaust ducts,
- through the casing wall to the nacelle air and then *via* the nacelle structure.

All these paths are modally dense and it is virtually impossible to design out a selected frequency. If noise is a problem then the options are to reduce the source sound level, perhaps by improving the tip relief as described above, or to modify the major path to reduce transmission. Identification of the major path is not straightforward, but one way of doing so is to use Statistical Energy Analysis (SEA), which combines a theoretical model with extensive field measurements. The path may not be simple, as non-linearity in the system can make one path the predominant one at low wind speeds and another path critical at higher wind speeds. Treatment of a radiating path can involve damping treatment such as shear layer damping or even just sand or bitumen layers added to the tower wall, for instance. In some cases the treatment can have more than one effect. When blades are the major source of radiation and damping material is added inside the blades then this material can act as a stiffening material as well as a damping mechanism. Sometimes it is useful to add tuned absorbers to parts of the structure to damp out one particular frequency. An alternative use of such tuned absorbers is to design them to raise the impedance at the tuned frequency so that the offending vibration does not pass that point on the structure.

7.4.9 Integrated gearboxes

As noted in Section 6.11.1, the cases of integrated gearboxes must be very robust, in order to transmit the rotor loads to the nacelle structure without experiencing deflections which would impair the proper functioning of the gears. In view of the complex shape of the casing, stress distributions due to each load vector usually have to be determined using finite-element (FE) analysis – these can then be superposed in line with the different extreme load combinations. The fatigue analysis will require the superposition of stress histories resulting from simultaneous time histories of rotor thrust, yaw moment and tilt moment derived from simulations at different wind speeds.

7.4.10 Lubrication and cooling

The function of the lubrication system is to maintain an oil film on gear teeth and the rolling elements of bearings, in order to minimize surface pitting and wear (abrasion, adhesion and scuffing). Varying levels of the elastohydrodynamic lubrication provided by the oil film can be identified, depending on oil film thickness. These range from full hydrodynamic lubrication, which exists when the metal surfaces are separated by a relatively thick oil film, to boundary lubrication when the asperities of the metal surfaces may be separated by lubricant films only a few molecular dimensions in thickness. Scuffing, which is a severe form of adhesive wear involving localized welding and particle transfer from one gear to the other, can occur under boundary lubrication conditions, which are promoted by high loading and low pitch line velocity and oil viscosity.

Two alternative methods of lubrication are available: splash lubrication and pressure fed. In the former, the low-speed gear dips into an oil bath and the oil thrown up against the inside of the casing is channelled down to the bearings. In the latter, oil is circulated by a shaft driven pump, filtered and delivered under pressure to the gears and bearings. The advantage of splash lubrication is its simplicity and hence reliability, but pressure fed lubrication is usually preferred for the following reasons:

- oil can be positively directed to the locations where it is required by jets,
- wear particles are removed by filtration,

- the churning of oil in the bath, which can result in a net efficiency loss, is avoided,
- the oil circulation system enables heat to be removed much more effectively from the gearbox by passing the oil through a cooler mounted outside the nacelle,
- it allows for intermittent lubrication when the machine is shut-down if a standby electric pump is incorporated.

With a pressure fed system, it is normal practice to fit temperature and pressure switches downstream of the filter to trip the machine for excessive temperature or insufficient pressure.

Guidance on the selection of lubricant, which has to take into account the ambient temperatures at the site in question, is given in the AGMA/AWEA (1996) document. Sump heaters may be needed to enable oil to be circulated when the turbine starts up at low temperatures.

7.4.11 Gearbox efficiency

Gearbox efficiency can vary between about 95 percent and 98 percent, depending on the relative number of epicyclic and parallel shaft stages and on the type of lubrication.

7.5 Generator

7.5.1 Induction generators

The induction generators commonly used on fixed-speed wind turbines are very similar to conventional industrial induction motors. In principle the only differences between an induction machine operating as a generator and as a motor are the direction of power flow in the connecting wires, whether torque is applied to or taken from the shaft and if the rotor speed is slightly above or below synchronous. The size of the market for induction motors is very large and so, in many cases, an induction generator design will be based on the same stator and rotor laminations as a range of induction motors in order to take advantage of high manufacturing volumes. Some detailed design modifications, e.g., changes in rotor bar material, may be made by the machine manufacturers to reflect the different operating regime of wind turbine generator, particularly the need for high efficiency at part load, but the principles of operation are those of conventional induction motors.

The synchronous speed, which is determined by the number of magnetic poles will be in the range of 1500 r.p.m. (4 pole), 1000 r.p.m. (6 pole) or 750 r.p.m. (8 pole) for connection to a 50 Hz network. For commercial reasons it is common to use a voltage of only 690 V even for large generators and in some very large wind turbines the resulting high currents have led to the decision to locate the turbine transformer in the nacelle. The physical protection of the generator is arranged to avoid the ingress of moisture, i.e. a totally enclosed design, and in some wind turbines liquid cooling is used to reduce air-borne noise. A high slip at rated power output is often requested by the wind turbine designer as this increases the damping in the wind turbine drive train but at the expense of losses in the rotor.

Figure 7.28 shows the conventional equivalent circuit of an induction machine that may be used to analyse its steady-state behaviour (Hindmarsh, 1984, and McPherson, 1990). The slip (s) is the difference between the angular velocity of the stator and rotor:

$$s = \frac{\omega_s - \omega_r}{\omega_s} \tag{7.57}$$

and so for motor operation it is positive and for generator operation negative.

Figure 7.29 shows how the active power varies with slip for a 1 MW induction machine. A convention has been chosen with the current flowing into the circuit and so the normal operating region for a generator is between O and A. At 1 MW generation (A) the slip is around −0.8 percent with the rotor rotating faster than the stator field. It may be seen that the maximum power that may be generated before the peak of the curve is reached is only some 1.3 MW. This is because the generator

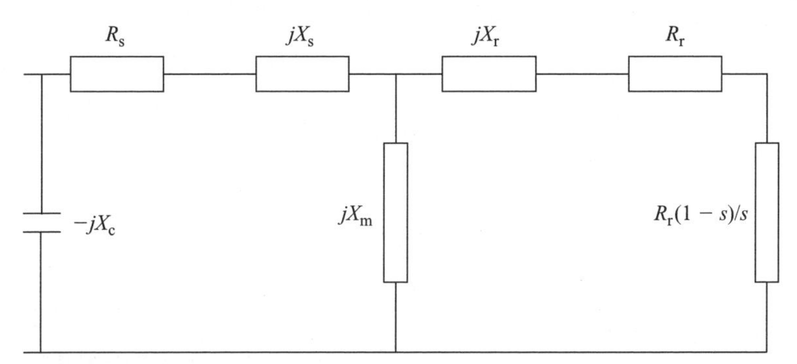

Figure 7.28 Equivalent Circuit of an Induction Machine with Power Factor Correction Capacitors

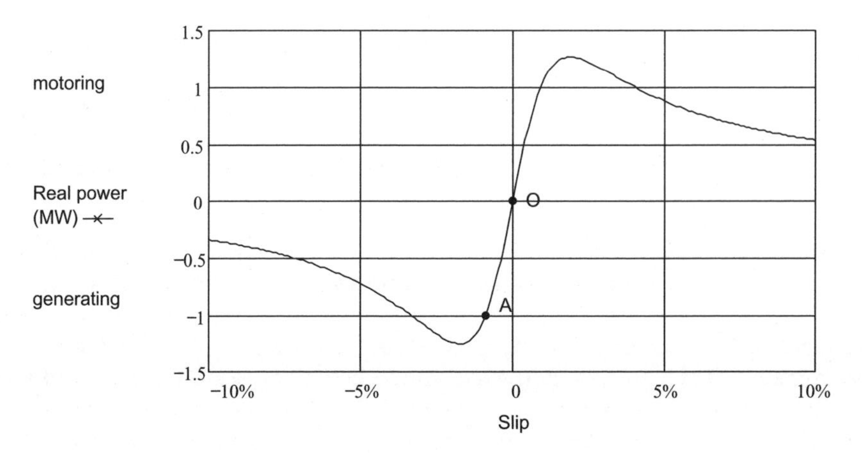

Figure 7.29 Variation of Active Power with Slip of a 1 MW Induction Machine

has been represented as connected to a distribution network with a low short-circuit level (and hence a high impedance) and the impedance of the local turbine transformer is included in the calculation. This source impedance acts to lower the maximum power which may be exported before the peak of the curve is reached, and instability occurs. It also leads to asymmetry in the torque–slip curve.

Figure 7.30 shows how the reactive power drawn by the generator varies with slip. The normal generating operating region is again shown as O–A. At point A (1 MW output) the generator (with its fixed local power factor correction capacitors) draws some 600 kVAr. It may be seen that the reactive power requirement increases very rapidly if the output power rises above its rated value.

Figures 7.29 and 7.30 may be combined to give the conventional 'circle diagram' representation of an induction machine shown in Figure 7.31. Again the normal generating region is shown as O–A. The fixed power factor correction capacitors act to reduce the requirement for reactive power and so translate the circle diagram on the y-axis towards the origin.

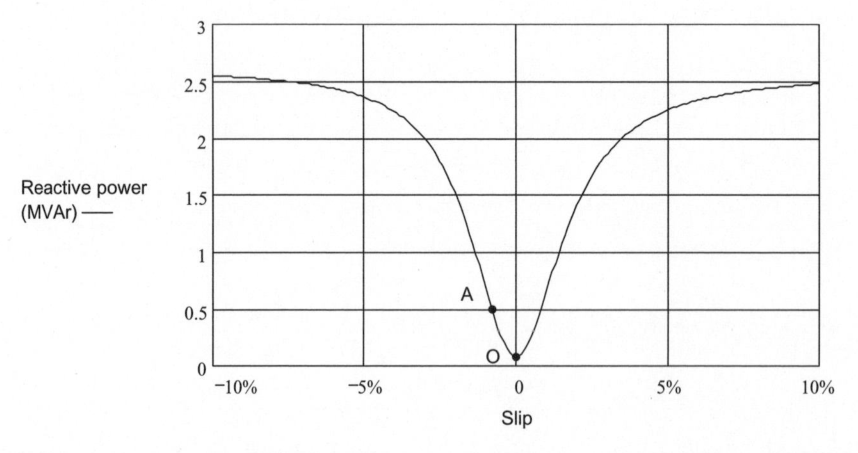

Figure 7.30 Variation of Reactive Power with Slip for a 1 MW Induction Machine

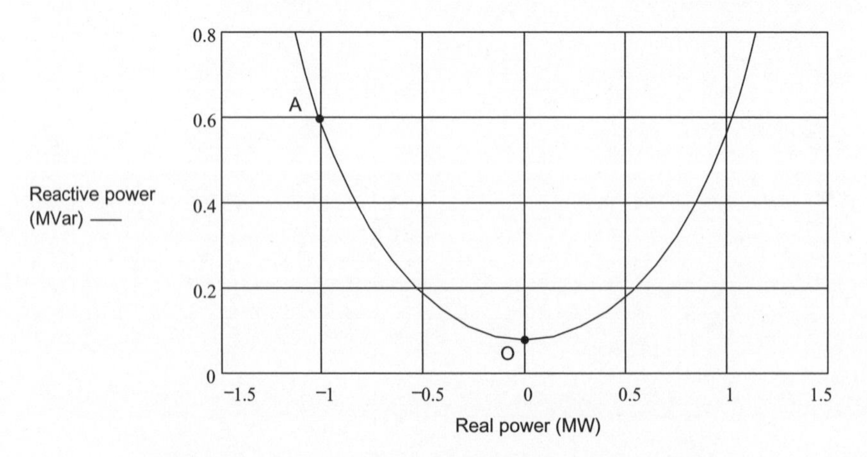

Figure 7.31 Circle Diagram of 1 MW Induction Machine

The equations used to describe the steady-state performance of induction generators are given in any standard undergraduate textbook (e.g. Hindmarsh, 1984, and McPherson, 1990). Dynamic analysis is more complex but is dealt with by Krause (1986).

7.5.2 Variable-speed generators

There are two fundamental approaches to electrical variable-speed operation. Either all the output power of the wind turbine may be passed through the frequency converter to give a broad range of variable speed operation or a restricted speed range may be achieved by converting only a fraction of the output power.

Figure 7.32 shows in schematic form how a broad range, variable-speed generation system may be configured. Early broad range variable-speed wind turbines used a diode rectifier bridge in the generator converter and a naturally commutated, thyristor, current source, converter on the network side (Freris, 1990). However, naturally commutated thyristor converters consume reactive power and generate considerable characteristic harmonic currents. On weak distribution systems it is difficult to provide suitable filtering and power factor correction for this type of equipment. Hence modern practice is to use two voltage source converters (Heier, 1998) with either a synchronous or induction generator. Each converter consists of a Graetz Bridge (as shown in Figure 10.17) with Insulated Gate Bipolar Transistors (IGBTs) as the switching elements. The bridges are switched rapidly (typically between 2–6 kHz) with some form of Pulse Width Modulation to produce a close approximation to a sine wave. The generator converter rectifies all the power to DC, which is then inverted by the network converter. Operation of this type of voltage source converter is described in Mohan, Undeland and Williams (1995).

Control strategies vary but one approach is to control the generator converter to maintain the DC link voltage at a constant value and then use the network converter to control the power flowing out of the system and hence the torque on the generator (Jones and Smith, 1993). A power bandwidth of 200–500 radians/s is quoted in this paper indicating the very fast control possible with such equipment with an overall efficiency of 92.1 percent consisting of 95.9 percent for the generator and 96 percent for the power electronics. The network side converter may be

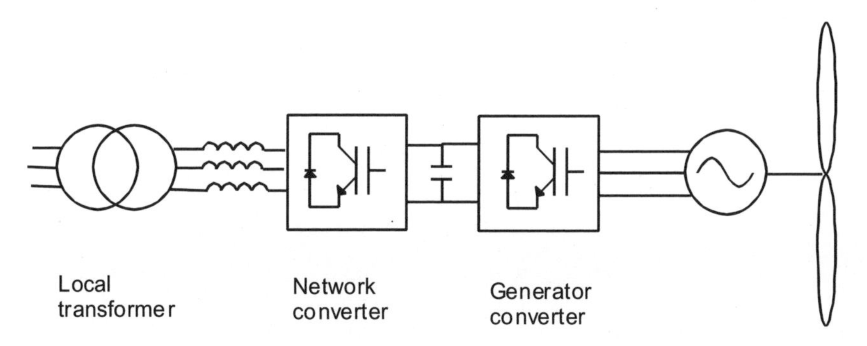

Figure 7.32 Broad Range Variable Speed Generation

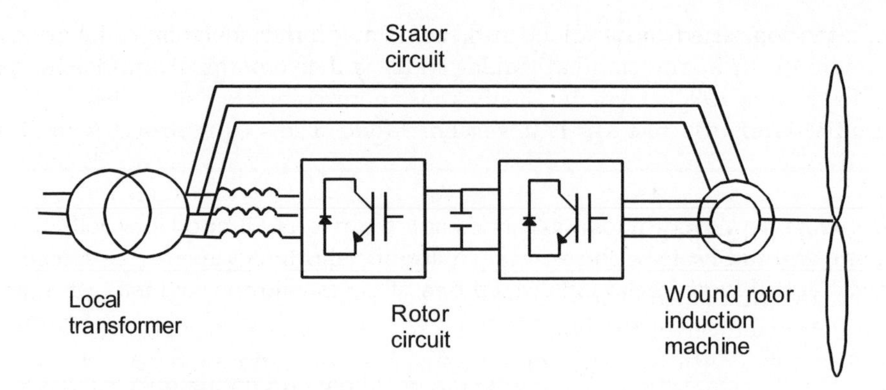

Figure 7.33 Narrow Range Variable Speed Generation

arranged to operate at any power factor within the rating of the equipment with very low harmonic distortion.

Figure 7.33 shows how narrow band variable speed control may be implemented. A wound rotor induction generator is used and control is possible over all four quadrants of speed and torque. The generator stator is connected directly to the network with a frequency converter in the rotor circuit. This 'doubly fed' concept was used in early large prototype wind turbines, e.g., the 3 MW Growian constructed in Germany in the early 1980s (Warneke, 1984) and the Boeing MOD 5B in the USA at the same time. At that time, cyclo-converters were used to change the frequency of the rotor circuit but modern practice is to use two voltage source converters similar to the broad range variable-speed designs but having a lower power rating. Again, control techniques vary but one approach is to use vector control techniques on the machine side converter to adjust torque and the excitation of the generator independently. The network side bridge maintains the DC link voltage and supplies additional reactive power to maintain the power factor of the wind turbine (Pena, Clare and Asher, 1996). Doubly-fed induction generators are becoming increasingly common in large wind turbines where the benefits of limited variable-speed operation are required but at the reduced cost of controlling only a fraction of the output power. A possible future development is the use of the brushless doubly-fed generator where rather than use slip rings the rotor is excited by a second controlled stator winding. This approach eliminates the requirement for slip-rings and brushes but is not yet used commercially.

7.6 Mechanical Brake

7.6.1 Brake duty

As indicated in Section 6.8.3, a mechanical brake can be called on to fulfil a variety of roles, according to the braking philosophy adopted for the machine in question. The minimum requirement is for the mechanical brake to act as a parking brake, so that the machine can be stopped for maintenance purposes. The brake will also be used to bring the rotor to a standstill during high wind shut-downs for the majority of machine designs, and during low-speed shut-downs as well in some cases. Aerodynamic braking is used to decelerate the rotor initially, so the mechanical brake torque can be quite low. However, IEC 61400-1 requires that the mechanical brake be capable of bringing the rotor to a complete stop from a hazardous idling state in any wind speed less than the 1 year return period 3 s gust (see Table 5.1).

If the mechanical brake is required to arrest the rotor in the event of a complete failure of the aerodynamic braking system, then there are two deployment options to consider. Either the mechanical brake can be actuated when an overspeed resulting from the failure of the aerodynamic system is detected, or actuated simultaneously with the aerodynamic brake as part of the standard emergency shut-down procedure. The advantage of the former strategy is that the mechanical brake will rarely, if ever, have to be deployed in this way, so that some pad or even disc damage can be tolerated when deployment actually occurs. In addition, fatigue loading of the gearbox will be reduced if the brake is mounted on the high-speed shaft. On the other hand, if the mechanical brake is actuated before significant overspeed has developed, then the aerodynamic torque to be overcome by the mechanical brake in the event of aerodynamic braking failure will be less.

The most severe emergency braking case will arise following a grid loss during generation in winds above rated. In the case of pitch-regulated machines, the maximum overspeed will occur after grid loss at rated wind speed because the rate of change of aerodynamic torque with rotational speed decreases and soon becomes negative at higher wind speeds. Conversely, if the pitch mechanism should jam, the braking duty becomes more severe at wind speeds at or above cut-out, because much higher aerodynamic torques are developed as the rotor slows down and the angle of attack increases. For stall-regulated machines the critical wind speed is generally at an intermediate value between rated and cut-out.

7.6.2 Factors governing brake design

The braking torque provided by callipers gripping a brake disc (Figure 7.34) is simply the product of twice the calliper force, the coefficient of friction (typically 0.4), the number of callipers and the effective pad radius. Callipers providing clamping forces of up to 500 KN are available. However, the brake design is also limited by:

- centrifugal stresses in the disc,
- pad rubbing speed,
- power dissipation per unit area of pad, and
- disc temperature rise.

The nature of these constraints is described below.

The critical stress generated by centrifugal stresses is in the tangential direction at

Figure 7.34 High-speed Shaft Brake Disc and Calliper (Reproduced by permission of NEG-Micon)

the inner radius of the brake disc, but it is governed principally by the disc rim speed according to the following formula:

$$\sigma_\theta(a) = \frac{3+\nu}{4}\rho\omega^2 b^2\left(1 + \frac{1-\nu}{3+\nu}\frac{a^2}{b^2}\right) \tag{7.58}$$

where a and b are the inner and outer disc radii respectively and ω is the disc rotational speed. One brake manufacturer, Twiflex, quotes a maximum safe disc rim speed of around 90 m/s for their discs manufactured in spheroidal graphite cast iron.

Brake pads are generally made from sintered metal or a cheaper, resin-based material. The former can accept rubbing speeds of up to 100 m/s, but some manufacturers quote permitted rubbing speeds for the latter of only about 30 m/s. However, Wilson (1990) reports satisfactory performance of resin-based pads at a rubbing speed of up to 105 m/s if the power dissipation rate per unit area, Q, is kept low enough. The criterion, ascribed to Ferodo, is that $Q = \mu PV \leqslant 11.6\ \text{MW/m}^2$, where μ is the coefficient of friction, P is the brake-pad pressure in KN/m^2 and V is the rubbing speed in m/s. This requires the pad pressure to be reduced to 275 KN/m^2, assuming a friction coefficient of 0.4.

During braking the kinetic energy of the rotor and drive train together with the additional energy fed in by the aerodynamic torque are dissipated in the brake disc and pads as heat, resulting in rapid initial temperature rise near the surface of the brake disc. The rate of energy dissipation is equal to the product of the braking torque and the disc rotational speed, so in the latter stages of braking the rate of energy dissipation cannot sustain the high surface temperatures and they begin to fall again.

The coefficient of friction for pads of resin-based materials is sensibly constant at a level of about 0.4 at temperatures up to 250°C, but begins to drop thereafter, reaching 0.25 at 400°C. Although in theory the brake can be designed to reach the latter temperature, in practice the varying torque complicates the calculations and leaves little margin of error against a runaway loss of brake torque. Accordingly 300°C is often taken as the upper temperature limit for resin-based pads.

Sintered metal pads have a constant coefficient of friction of about 0.4 up to a temperature of at least 400°C, but manufacturers indicate that the material can perform satisfactorily at temperatures up to 600°C on a routine basis, or up to 850° intermittently. Wilson (1990) reports a reduced friction coefficient of 0.33 at 750°C. Such temperatures cannot be realized in practice because the temperature of the disc itself is limited to 600°C in the case of spheroidal graphite cast iron or to a much smaller value in the case of steel (*op. cit.*).

Clearly the use of the more expensive sintered brake pads allows the brake disc to absorb much more energy. However, the sintered metal is a much more effective conductor of heat than resin-based material, so it is often necessary to incorporate heat insulation into the calliper design to prevent overheating of the oil in the hydraulic cylinder. A method of calculating brake-disc temperature rise is given in the next section.

7.6.3 Calculation of brake disc temperature rise

The build up in temperature across the width of a brake disc over the duration of the stop can be calculated quite easily if a number of assumptions are made. First, the heat generated is assumed to be fed into the disc at a uniform intensity over the areas swept out by the brake pads as the disc rotates. This is a reasonable approximation for a high-speed shaft-mounted brake and for a low-speed shaft-mounted brake with several callipers until rotation has almost ceased, but the energy input by this stage is much lower. Within the disc heat flow is assumed to be perpendicular to the disc faces only, i.e., radial flows are ignored.

Consider a brake-disc slice at a distance x from the nearest braking surface, of thickness Δx and cross-sectional area A. The rate of heat flow away from the nearest braking surface entering the slice is $\dot{Q} = -kA(\mathrm{d}\theta/\mathrm{d}x)$ (where θ is the temperature and k the thermal conductivity) and the rate of heat flow leaving it on the far side is $\dot{Q} + (\mathrm{d}\dot{Q}/\mathrm{d}x)$. The temperature rise of an element of thickness Δx over a time interval Δt is given by

$$\Delta\theta A \Delta x \rho C_p = \Delta Q = -\frac{\mathrm{d}\dot{Q}}{\mathrm{d}x}\Delta x \Delta t = kA\frac{\mathrm{d}^2\theta}{\mathrm{d}x^2}\Delta x \Delta t$$

where ρ is the density and C_p is the specific heat, so that

$$\frac{d\theta}{dt} = \frac{k}{\rho C_p}\frac{d^2\theta}{dx^2} \tag{7.59}$$

Adopting a finite-element approach, Equation (7.59) can be written

$$\theta(x, t + \Delta t) = \theta(x, t) + \frac{k}{\rho C_p}\frac{\Delta t}{(\Delta x)^2}[\theta(x + \Delta x, t) + \theta(x - \Delta x, t) - 2\theta(x, t)] \tag{7.60}$$

Substituting values of $k = 36$ W/m per °K, $C_p = 502$ J/kg per °K and $\rho = 7085$ kg/m^3 for Grade 450 spheroidal graphite cast iron yields a value for the thermal diffusivity $\alpha = k/(\rho C_p)$ of 1.01×10^{-5} m^2/s. If the time increment, Δt, is selected at 0.025 s and the element thickness is taken as 1.005 mm, then Equation (7.60) simplifies to

$$\theta(x, t + \Delta t) = 0.25[\theta(x + \Delta x, t) + \theta(x - \Delta x, t) + 2\theta(x, t)] \tag{7.61}$$

This equation can be used to calculate the temperature distribution across the brake disc, starting with a uniform distribution and imposing suitable increments at the braking surfaces at the boundaries. The behaviour at the boundaries is simpler to follow through if they are treated as planes of symmetry like the disc mid-plane, with imagined discs flanking the real one. The temperature increment at the boundary at each time step, which is added to that calculated from Equation (7.61), is given by

$$\Delta\theta_0 = \frac{2T\omega(t)\Delta t}{\rho C_p S} \tag{7.62}$$

where T is the braking torque (assumed constant), $\omega(t)$ is the disc rotational speed at time t, and S is the area swept out by the brake pad (or pads) on one side of the disc. For a disc diameter D and pad width w, S is $\pi(D - w)w$. The factor 2 is required because heat is assumed to flow into the imagined disc as well as into the real one. Hence the initial temperature build up can be calculated as illustrated in Table 7.6, taking an arbitrary value of $\Delta\theta_0$ of 40°C. (The gradual reduction in $\Delta\theta_0$ over time due to deceleration is ignored here for simplicity.)

The brake-disc surface temperature rise is found to be a minimum when the ratio of the braking torque to the maximum aerodynamic torque is about 1.6. As the ratio is reduced below this value, the extended stopping time results in more energy being abstracted from the wind, so temperatures begin to rise rapidly. On the other hand, the maximum brake temperature is relatively insensitive to increases in the ratio above 1.6. The variation in maximum brake-disc surface temperature with braking torque is illustrated for the emergency braking of a stall-regulated machine following an overspeed in Figure 7.35, where the continuous line gives the surface temperature rise calculated by the finite-element method outlined above. It transpires that the maximum temperature rise can be estimated quite accurately by the following empirical formula

Table 7.6 Illustrative Example of Calculation of Brake Disc Temperature Rise Using Finite Element Model

Time step	Time (s)	Element Distance from braking surface (mm)	0 0	1 1.0	2 2.0	3 3.0	4 4.0	5 5.0
1		Initial temperature	0	0	0	0	0	0
		Boundary temperature increment	40					
	0.025	Temperature at end of time step	20	10	0	0	0	0
2		Boundary temperature increment	40					
		Sum	60	10	0	0	0	0
	0.05	Temperature at end of time step	35	20	2.5	0	0	0
3		Boundary temperature increment	40					
		Sum	75	20	2.5	0	0	0
	0.075	Temperature at end of time step	47.5	29.4	6.3	0.6	0	0
4		Boundary temperature increment	40					
		Sum	87.5	29.4	6.3	0.6	0	0
	0.1	Temperature at end of time step	58.5	38.2	10.6	1.9	0.1	0

$$\theta_{\max} - \theta_0 = \frac{E}{\sqrt{t}} \frac{1}{64\,600 w(D-w)} = \frac{E}{\sqrt{t}} \frac{2\pi}{64\,600 S} \qquad (7.63)$$

where E is the total energy dissipated in Joules, t is the duration of the stop in seconds and S is the area of the disc surfaces swept by the brake pads. The temperature derived using this formula is plotted as a dotted line in Figure 7.35 for comparison.

7.6.4 High-speed shaft brake design

A key parameter to be chosen in brake design is the design braking torque. The coefficient of friction can vary substantially above and below the design value due to such factors as bedding in of the brake pads and contamination, so the design braking torque calculated on the nominal friction value must be increased by a suitable materials factor. Germanischer Lloyd specify a materials factor of 1.2 for the coefficient of friction, and add in another factor of 1.1 for possible loss of calliper spring force. If these factors are adopted, the minimum design braking moment is 1.78 times the maximum aerodynamic torque, after including the aerodynamic load factor of 1.35. A small additional margin of, say, 5 percent should be added to

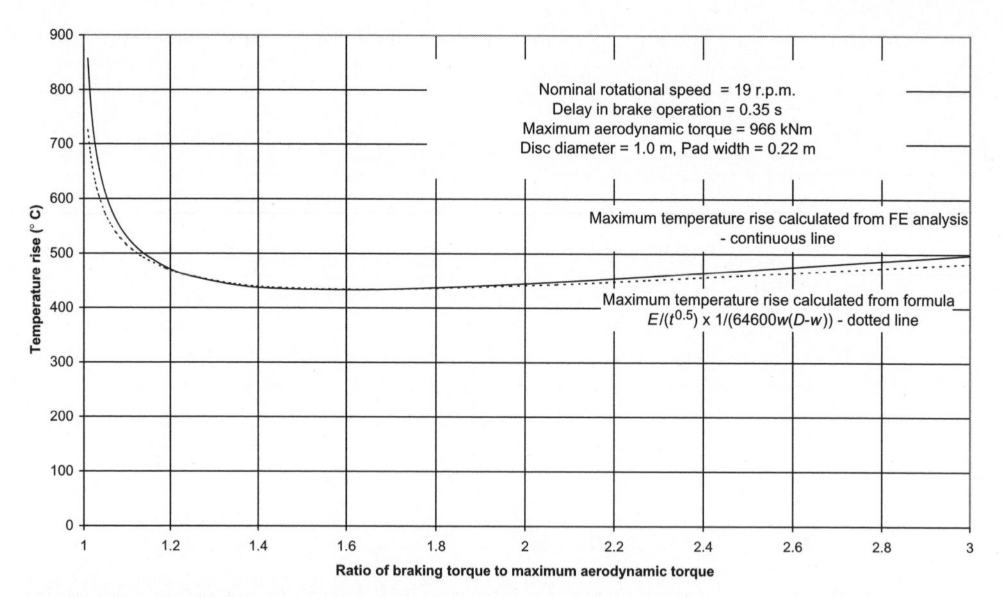

Figure 7.35 Brake Disc Surface Maximum Temperature Rise for Emergency Braking of 60 m Diameter 1.3 MW Stall-regulated Turbine from 10 percent Overspeed in 20 m/s Wind with HSS Brake Acting Alone

ensure that the rotor is still brought to rest without a very large temperature rise should the 1.78 safety factor be completely eroded.

The procedure to be followed for the design of a brake on the high-speed shaft (HSS) can conveniently be illustrated by an example.

Example 7.1: Design a HSS brake for a 60 m diameter, 1.3 MW stall-regulated machine capable of shutting the machine down in a 20 m/s wind from a 10 percent overspeed occurring after a grid loss, with or without assistance from the aerodynamic braking system. The nominal LSS and HSS rotational speeds are 19 r.p.m. and 1500 r.p.m. respectively, ignoring generator slip. Assume that the brake application delay time is 0.35 s, and that the inertia of the turbine rotor, drive train, brake disc and generator rotor – all referred to the low-speed shaft – totals 2873 Tm^2.

(a) *Derivation of the brake design torque*: The peak aerodynamic torque occurs when the maximum rotational speed is reached just prior to brake application. The first step is to determine the relationship between rotational speed and aerodynamic torque for the stated wind speed of 20 m/s. From this the acceleration of the rotor and build-up of aerodynamic torque during the 0.35 s delay before the brake comes on can be determined. The speed increase in this case is 1 r.p.m., giving a maximum rotor speed of $19 \times 1.1 + 1 = 21.9$ r.p.m. and peak aerodynamic torque of 966 kNm. Hence the brake design torque is $966 \times 1.78 \times 1.05 = 1800$ kNm referred to the low-speed shaft, or $1800 \times 19/1500 = 22.8$ kNm at the brake.

(b) *Brake-disc diameter selection*: The maximum rotor speed corresponds to a high-speed shaft speed of $21.9 \times (1500/19) = 1729$ r.p.m. = 181 rad/s, so the maximum permissible brake-disc radius as regards centrifugal stresses is about $90/181 = 0.497$ m. It is advisable to choose the largest permitted size in order to minimize temperature rise, so 1.0 m diameter is selected in this case. The pad rubbing speed will be quite acceptable if sintered pads are used.

(c) *Selection of number and size of brake pads*: The total brake-pad area is governed by the need to keep the maximum power dissipation per unit pad area below 11.6 MW/m^2. The power dissipation is equal to the product of the braking torque and the rotational speed, so it is at a maximum at the onset of braking, i.e. $22.8 \times 181 = 4128$ kW, giving a required total area of the brake pads of $4128/11\,600 = 0.356$ m^2. This area can be provided by four callipers fitted with 0.22×0.22 m pads, giving 0.387 m^2 in all.

(d) *Maximum brake-disc temperature check*: The variation in disc surface temperature over the duration of the stop can be calculated using the finite-element method outlined in the preceding section. The resulting variation in this case is plotted in Figure 7.36. The surface temperature reaches a maximum of 440°C, just after halfway through the stop, which lasts 4.7 s from the time the brake comes on. This temperature is well below the limit for sintered pads.

(e) *Calliper force:* The braking friction force required is 58.5 kN, calculated from the torque divided by the effective pad radius of 0.39 m. Hence the required calliper force is $58.5/(8 \times 0.4) = 17.3$ kN which is rather low for a calliper sized for a 0.22×0.22 m brake pad.

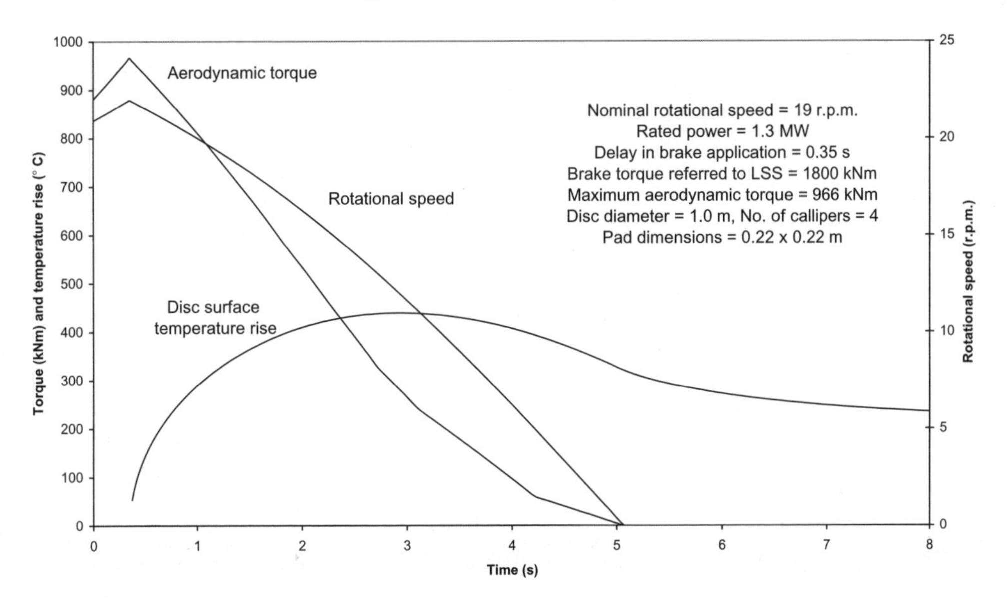

Figure 7.36 Emergency Braking of Stall-regulated 60 m Diameter Turbine from 10 percent Overspeed in 20 m/s Wind with HSS Mechanical Brake Acting Alone

The design process outlined above results in an excessive number of lightly-loaded callipers, because of the limitation on power dissipation per unit area. If the relative infrequency of emergency braking events allowed this limitation to be relaxed, then a more economic solution would result.

7.6.5 Two level braking

During normal as opposed to emergency shut-downs, the rotor is decelerated to a much lower speed by aerodynamic braking before the brake is applied, so the brake torque required is much reduced. In view of the benefit of reduced loads on the braking system, and on the gearbox in particular, some manufacturers arrange for a reduced braking torque for normal shut-downs. This is achieved on the usual 'spring applied, hydraulically released' brake callipers by allowing oil to discharge from the hydraulic cylinder *via* a pressure relief valve when the brake is applied, so that the hydraulic pressure drops to a reduced level. After the rotor has come to rest, the remaining hydraulic pressure can be released, so that the brake torque rises to the full level.

7.6.6 Low-speed shaft brake design

The procedure for designing a low-speed shaft disc brake is much simpler than that for the high-speed shaft brake, because the limits on disc-rim speed, pad-rubbing speed, power dissipation per unit area and temperature rise do not influence the design, which is solely torque driven. The large braking torque required means that a brake placed on the low-speed shaft will be much bulkier than one with the same duty placed on the low-speed shaft. For example the design LSS braking torque of 1800 kNm from the example above would require a 1.8 m diameter disc fitted with seven callipers.

A study by Corbet, Brown and Jamieson (1993), which investigated a range of machine diameters, concluded that the brake cost would double or treble if the brake were placed on the low-speed shaft rather than on the high-speed shaft. However, when the extra gearbox costs associated with a high-speed brake were taken into account, the cost advantage of the high-speed shaft brake disappeared.

7.7 Nacelle Bedplate

The functions of the nacelle bedplate are to transfer the rotor loadings to the yaw bearing and to provide mountings for the gearbox and generator. Normally it is a separate entity, although in machines with an integrated gearbox, the gearbox casing and the nacelle bedplate could, in principle, be a single unit. The bedplate can either be a welded fabrication consisting of longitudinal and transverse beam members or a casting sculpted to fit the desired load paths more precisely. One fairly common arrangement is a casting in the form of an inverted frustum which supports the low-speed shaft main bearing at the front and the port and starboard gearbox supports towards the rear, with the generator mounted on a fabricated platform projecting to the rear and attached to the main casting by bolts.

Although conventional methods of analysis can be used to design the bedplate for extreme loads, the complicated shape renders a finite-element analysis essential for calculating the stress concentration effects needed for fatigue design. Fatigue analysis is complicated by the need to take into account up to six rotor load components. However, given stress distributions for each load component obtained by separate FE analyses, the stress-time history at any point can be obtained by combining appropriately scaled load component time histories previously obtained from a load case simulation.

7.8 Yaw Drive

The yaw drive is the name given to the mechanism used to rotate the nacelle with respect to the tower on its slewing bearing, in order to keep the turbine facing into the wind and to unwind the power and other cables when they become excessively twisted. It usually consists of an electric or hydraulic motor mounted on the nacelle, which drives a pinion mounted on a vertical shaft *via* a reducing gearbox. The pinion engages with gear teeth on the fixed slewing ring bolted to the tower, as shown in Figure 7.37. These gear teeth can either be on the inside or the outside of the tower, depending on the bearing arrangement, but they are generally located on the outside on smaller machines so that the gear does not present a safety hazard in the restricted space available for personnel access.

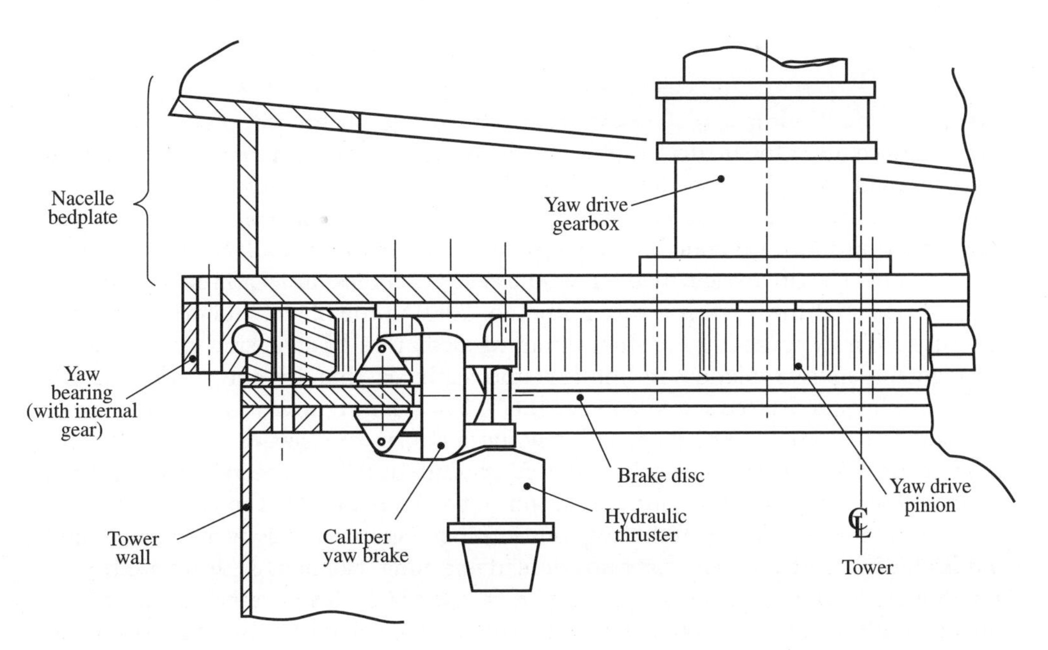

Figure 7.37 Typical Arrangement of Yaw Bearing, Yaw Drive and Yaw Brake

The yaw moments on rigid hub machines arise from differential loading on the blades, which may be broken down into deterministic and stochastic components. On a three-bladed machine, the dominant cyclic yaw loading is at $3P$, but it is generated by $2P$ blade loading, as is demonstrated below. Defining blade out-of-plane root bending moments containing harmonics of the rotational frequency, Ω, as follows

$$M_{Yj} = \sum_n a_n \sin\left(n\left\{\omega t + \frac{2\pi(j-1)}{3}\right\} + \phi_n\right) \tag{7.64}$$

Hence the yaw moment from all three blades is given by

$$M_{ZT} = \sin\omega t \sum_n a_n \sin(n\omega t + \phi_n) + \sin\left(\omega t + \frac{2\pi}{3}\right)\sum_n a_n \sin\left(n\left\{\omega t + \frac{2\pi}{3}\right\}\phi_n\right)$$
$$+ \sin\left(\omega t + \frac{2\pi}{3}\right)\sum_n a_n \sin\left(n\left\{\omega t - \frac{2\pi}{3}\right\} + \phi_n\right) \tag{7.65}$$

i.e.

$$M_{ZT} = \sum_n a_n\left[\sin\omega t \sin(n\omega t + \phi_n)\left\{1 - \cos\frac{2\pi n}{3}\right\} + \sqrt{3}\cos\omega t\cos(n\omega t + \phi_n)\sin\frac{2\pi n}{3}\right] \tag{7.66}$$

For the first four harmonics this gives

$$M_{ZT} = 1.5\{a_1\cos\phi_1 - a_2\cos(3\omega t + \phi_2) + a_4\cos(3\omega t + \phi_4\} \tag{7.67}$$

Thus it is seen that the blade out-of-plane bending harmonics at $2P$ and $4P$ produce yaw moment at $3P$, while those at $1P$ and $3P$ produce steady and zero yaw moments respectively. The main sources of blade out-of-plane loading at $2P$ are tower shadow and turbulence.

As turbine size increases, the turbine diameter becomes larger in relation to gust dimensions, and the scope for differential loading on the blades due to turbulence increases. The expression for the standard deviation of the stochastic yawing moment on a three-bladed machine is the same as that for the shaft moment standard deviation – see Equation (5.119).

Anderson *et al.* (1993) investigated yaw moments on two sizes of Howden three-bladed turbines (33 m dia, 330 kW and 55 m dia, 1 MW) and concluded that the major source of cyclic yaw loading is stochastic at $3P$. Yaw error, on the other hand, was not found to make a significant contribution. Given that yaw error results in a blade out-of-plane load fluctuation at blade-passing frequency, this result is in accordance with Equation (7.67). Several different strategies have been evolved for dealing with the large cyclic yaw moments that arise on rigid hub machines due to turbulence, as follows.

(1) *Fixed yaw* A yaw brake is provided in the form of one or more callipers acting on an annular brake disc and is designed to prevent unwanted yaw motion under all circumstances (see Figure 7.37). This can require six callipers on a 60 m diameter machine. During yawing, the yaw motors drive against the brake callipers, which are partly released, so that the motion is smooth.

(2) *Friction damped yaw* Yaw motion is damped by friction in one of three different ways. In the first, the nacelle is supported on friction pads resting on a horizontal annular surface on the top of the tower. The yaw drive has to work against the friction pads, which also allow slippage under extreme yaw loads. This system was employed on the 500 kW Vestas V39 and the 3 MW WEG LS1.

   In the second, the nacelle is mounted on a conventional rolling-element slewing bearing, and the friction is provided by a permanently applied brake, using the same configuration as for fixed yaw. Optionally, the pressure on the brake pads can be increased when the machine is shut-down for high winds.

   In the third, the nacelle is supported on a three-row roller-type slewing bearing (see Figure 7.21(d), but with the rollers replaced by pads of elastomer composite to generate friction.

(3) *Soft yaw* This is hydraulically-damped fixed yaw. The oil lines to each side of the hydraulic yaw motor are each connected to an accumulator *via* a choke valve, allowing limited damped motion to and fro to alleviate sudden yaw loads. This system is used on the 300 kW WEG MS3, which has a two-bladed, teetered rotor, but experiences significant yaw loads when teeter impacts occur.

(4) *Damped free yaw* A hydraulic yaw motor is used as before, but the oil lines to each side of the motor are connected together in a loop *via* a check valve, rather than being connected to a hydraulic power pack. This arrangement prevents sudden yaw movements in response to gusts, but depends on yaw stability over the full range of wind speeds. Unfortunately, yaw stability in high winds is rare.

(5) *Controlled free yaw* This is the same as damped free yaw, except that provision is made for yaw corrections when necessary. This strategy was adopted successfully on several Windmaster machines, including the two-bladed, fixed-hub 750 kW machine.

Friction damped yaw is the strategy most commonly adopted.

7.9 Tower

7.9.1 Introduction

The vast majority of wind turbine towers are constructed from steel. Concrete towers are a perfectly practicable alternative but, except at the smaller sizes, they require the transfer of a substantial element of work from the factory to the turbine site, which has not normally proved economic. Accordingly, this section concentrates on the two types of steel towers – tubular and lattice. The restrictions on first-mode natural frequency are considered first.

7.9.2 Constraints on first-mode natural frequency

As noted in Section 6.14, it is important to avoid the excitation of resonant tower oscillations by rotor thrust fluctuations at blade-passing frequency or, to a lesser extent, at rotational frequency. Dynamic magnification impacts directly on fatigue loads, so the further the first-mode tower natural frequency is from the exciting frequencies, the better. Unfortunately, it is generally the case that the natural frequency of a tower designed to be of adequate strength for extreme loads is of the same order of magnitude as the blade-passing frequency.

In the case of machines operating at one of two fixed speeds, the latitude available for the selection of the tower natural frequency is more restricted. Figure 7.38 shows the variation of dynamic magnification factor with tower natural frequency for excitation at upper and lower blade-passing and rotational frequencies for a three-bladed machine with a 3:2 ratio between the upper and lower speeds. The curves are plotted for a damping ratio of zero, but the difference if the curves were plotted for a realistic damping ratio of about 2 percent would be imperceptible. The figure also shows the tower natural frequency bands available if the dynamic magnification ratio were to be limited to 4 for all four sources of excitation. It is apparent that the minimum dynamic magnification ratio obtainable with a tower natural frequency between the upper and lower blade-passing frequencies is 2.6, for a tower natural frequency of 0.85 times the upper blade-passing frequency. However, in view of the fact that the rotor thrust load

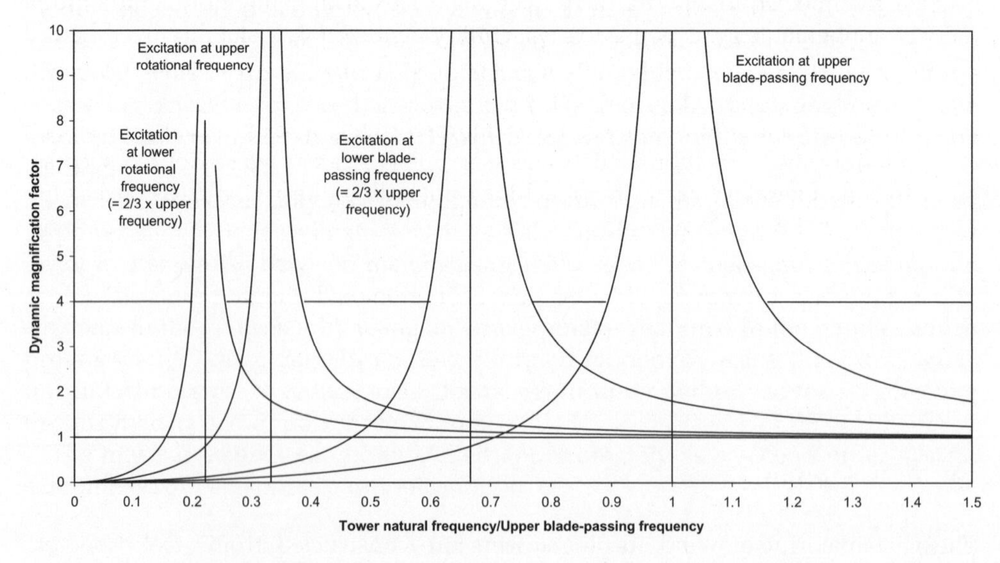

Figure 7.38 Variation of Dynamic Magnification Factor with Tower Natural Frequency for a Two-speed, Three-bladed Machine

fluctuations will be significantly smaller during operation at the lower rotational speed, it would be advisable to select a somewhat lower tower natural frequency than this to minimize overall fatigue damage.

Once a satisfactory tower design – in terms of strength and natural frequency– has been evolved for a given turbine, it is a straightforward matter to scale up the machine to larger rotor sizes, provided all the tower dimensions are scaled similarly, the hub-height wind speed is unchanged, and the tip speed is maintained constant. It can be shown that in these circumstances the tower natural frequency varies inversely with rotor diameter, as does the rotational speed of the rotor, so that the dynamic magnification factors are unchanged. Similarly, tower stresses due to extreme wind loading are the same as before.

The situation is less straightforward if the tower height is to be varied for a particular turbine. Assuming, as before, that the extreme hub-height wind speed remains the same, and that the wind loading on the tower is negligible compared with the wind loading on the rotor, then the tower base overturning moment is simply proportional to hub height H. Constant stresses can be maintained at the tower base by scaling all cross section dimensions up in proportion to the cube root of the hub height. If the same scaling is maintained all the way up the tower, then the tower natural frequency will vary as $\sqrt{I_B/H^3} = \sqrt{H^{4/3}/H^3} = 1/H^{5/6}$, neglecting tower mass, where I_B is the second moment of area of the tower base cross section. Thus doubling the tower height would result in a 44 percent reduction in natural frequency. Alternatively, if the tower base overturning moment were assumed to vary as $H^{1.5}$ to allow for the effect of wind shear on hub-height wind speed and the contribution of wind loading on the tower, then constant tower base stresses could be maintained by scaling the cross section dimensions up by $\sqrt{H}$. On this basis, tower natural frequency would vary as $1/\sqrt{H}$. The practical consequences of 'tuning' the tower natural frequency are discussed with respect to tubular towers in the next section.

7.9.3 Steel tubular towers

In the absence of buckling, a waisted conical shell, with a semi angle of 45° below the critical zone for tip clearance, would be the most efficient structure for transferring a horizontal rotor thrust acting in any direction to ground level. However, apart from the practicalities of transport and erection, instability of thin-walled shells in compression precludes such a design solution, and the steel tubular towers in common use have a very modest taper. It can be noted in passing that the manufacture of gently tapering towers has only been made possible by the development of increasingly sophisticated rolling techniques, and that early tubular towers were constructed from a series of cylindrical tubes of decreasing diameter with short 'adaptor' sections welded between them.

A tapered tower is generally fabricated from a series of pairs of plates rolled into half frusta and joined by two vertical welds. The height of each frustum so formed is limited to 2 or 3 m by the capacity of the rolling equipment. Care has to be taken in the execution of the horizontal welds to minimize local distortion, which weakens the tower under compression loading.

Assuming that a tower design with a uniform taper is to be adopted, the key design parameters to establish are the diameter and wall thickness at the tower base. The tower top diameter, on the other hand, is governed by the size of the yaw bearing, and wall thicknesses at intermediate heights can generally be interpolated between the tower base value and a sensible minimum at the top of about one hundredth of the local radius.

The main considerations determining the tower dimensions at the base are buckling of the shell wall in compression, strength under fatigue loading and stiffness requirements for 'tuning' the natural frequency. These are dealt with in separate sub-sections below.

As machines get larger, another important consideration is the maximum tower base diameter that can be accommodated on the highway when tower sections are transported overland. In the flat terrain of North Germany and Denmark, this limit is generally 4.0–4.2 m, but elsewhere it will often be less.

Design against buckling

Given perfect geometry, the strength of a cylindrical steel tube in axial compression is the lesser of the yield strength and the elastic critical buckling stress, given by

$$\sigma_{cr} = 0.605Et/r \tag{7.68}$$

where r is the cylinder radius and t is the wall thickness. Yield strength governs for r/t less than $0.605E/f_y$, which equates to 506 for mild steel, with $f_y = 245$ MPa. However, the presence of imperfections, particularly those introduced by welding, means that the tower-wall compression resistance is significantly reduced, even at the relatively low tower-wall radius to thickness ratios normally adopted. There is quite a wide disparity between the provisions of different national codes, with some making an explicit link between compression resistance and tolerances on imperfections and others not. The recommendations produced by the European Convention for Constructional Steelwork (ECCS, 1988) contain relatively straightforward empirical rules for the design of thin-walled cylinders in compression, which are based on sets of experimental results from several sources. These will eventually be superseded by the provisions of Part 1–6 of Eurocode 3 'Supplementary rules for shell structures', but in the meantime the ECCS rules relating to cylinders subject to bending loads are set out here.

The first step is to calculate a critical stress reduction coefficient for axial loading, α_0, from which a parallel coefficient for bending loading, α_B, is derived:

$$\alpha_0 = \frac{0.83}{\sqrt{1 + 0.01r/t}} \text{ for } r/t < 212, \quad \alpha_0 = \frac{0.70}{\sqrt{0.1 + 0.01r/t}} \text{ for } r/t > 212 \tag{7.69}$$

and

$$\alpha_B = 0.1887 + 0.8113\alpha_0 \tag{7.70}$$

These values apply if the out-of-plane deviations, w, of the cylinder measured with either

(a) a rod of length $L = 4\sqrt{rt}$ placed vertically, away from welds, or

(b) a circular template of the same length placed horizontally, away from welds, or

(c) a rod of length $L = 25t$ placed vertically across horizontal welds,

do not exceed 1 percent of the respective rod or template length. The value of α_B is halved if the maximum value of the imperfection ratio, w/L, is 2 percent, and may be interpolated for intermediate values of w/L.

The buckling strength, σ_u, is then given in terms of the yield strength and the elastic critical buckling stress (Equation (7.68)) as follows

$$\sigma_u = f_y\left[1 - 0.4123\left(\frac{f_y}{\alpha_B\sigma_{cr}}\right)^{0.6}\right] \text{ for } \alpha_B\sigma_{cr} > f_y/2 \text{ and } \sigma_u = 0.75\alpha_B\sigma_{cr} \text{ for } \alpha_B\sigma_{cr} < f_y/2 \tag{7.71}$$

The buckling strength of a mild-steel cylinder in bending as a fraction of the yield strength is plotted against the radius to wall-thickness ratio for imperfection ratios (w/L) of 1 percent and 2 percent on Figure 7.39.

The effect of the choice of tower base diameter on total tower weight is best illustrated by reference to a concrete example. Consider the design of a 50 m hub

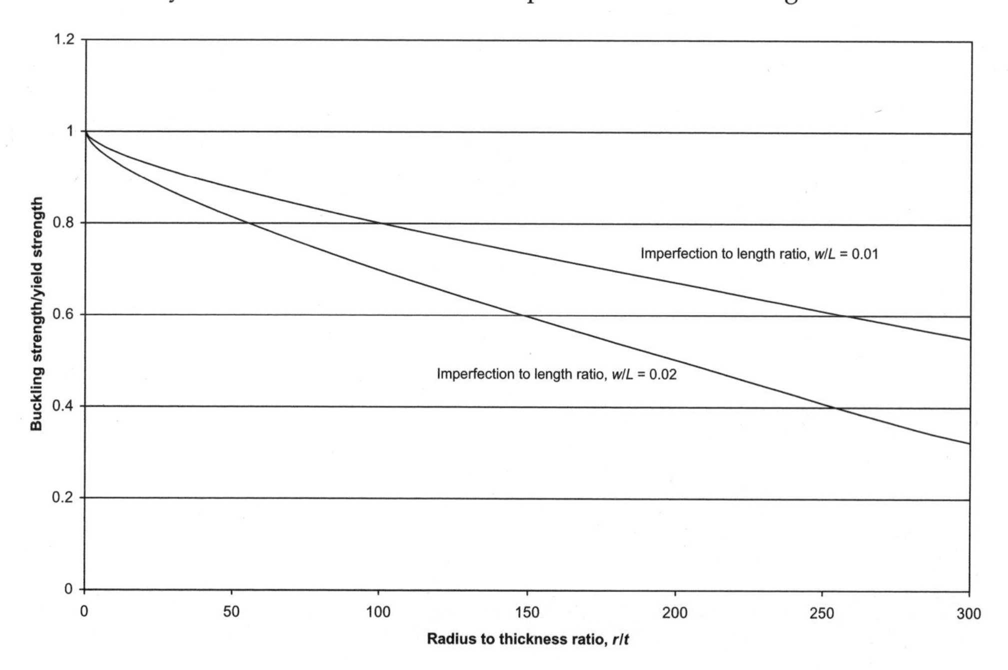

Figure 7.39 Buckling Strength in Bending of Thin-walled Mild-steel Cylinder for Varying Radius to Thickness Ratio and Different Levels of Imperfection Based on ECCS Rules

height tower in mild steel for a 60 m diameter, three-bladed, stall-regulated turbine at a site with a 60 m/s extreme wind speed. The tower base wall thickness required to resist the overturning moment produced by this wind speed has been calculated for a range of tower base diameters with the aid of Equation (7.71) and plotted on Figure 7.40. Corresponding tower weights have also been plotted, based on a tower top diameter and wall thickness of 2.25 m and 11 mm respectively and assuming an idealized linear wall-thickness variation between tower top and tower base. It can be seen that the tower weight reaches a minimum value at about 4.5 m diameter, indicating that beyond this point the reduction in cross-sectional area for constant section modulus is offset by the effects of the reducing buckling strength and the increasing wind loading on the tower itself. The weight penalty resulting from restricting the tower base diameter to 4.0 m for transport purposes would, in this case, be negligible.

Fatigue design

Clear rules for the fatigue design of steel-welded structures are given in Eurocode 3, where a family of S–N curves is defined for different weld details. On a log–log plot these curves in fact consist of two straight lines, with slopes of 1/5 and 1/3 for numbers of cycles above and below 5×10^6 respectively. In addition, there is a cut-off limit at $N = 10^8$ cycles, so that stress cycles with a stress range smaller than that defined at 10^8 cycles are deemed not to cause any fatigue damage at all.

Excluding the tower doorway (which is considered later) the critical weld details on a steel tubular tower are likely to be at welded attachments for intermediate

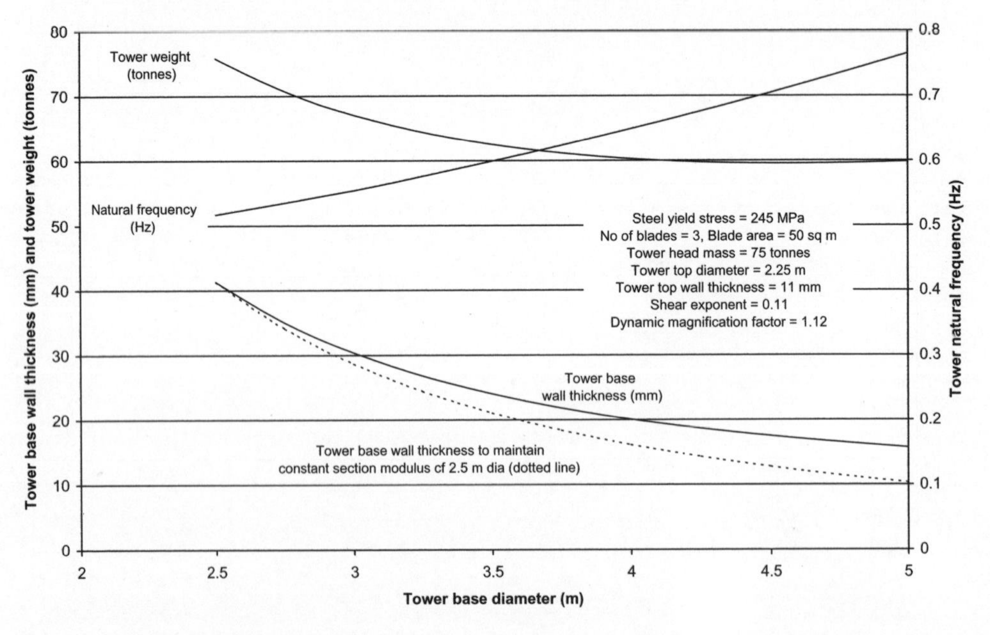

Figure 7.40 Variation in Tower Base Wall Thickness with Diameter Required for Support of 60 m Diameter Stall-regulated Wind Turbine at 50 m Height in 70 m/s Extreme Wind

platform and cable support members and the horizontal welds to the tower base flange and intermediate bolted flanges. Assuming a full penetration butt weld is provided, the detail category number for the horizontal welds is 71 (where the number 71 indicates the stress range applicable at 2×10^6 cycles in MPa). The detail category number for longitudinal welded attachments reduces as the length of the attachment increases, but if the attachment length can be restricted to 100 mm, the detail category number of 71 applies here as well. The S–N curve for this detail category is shown in Figure 7.41.

Eurocode 3 lays down different partial safety factors for fatigue strength, γ_{Mf}, according to the consequences of failure and the ease of inspection. If the local failure of a component does not lead rapidly to the failure of the structure, then it is termed 'fail-safe', and $\gamma_{Mf} = 1.0$ provided the joint detail is accessible for inspection. On the other hand, $\gamma_{Mf} = 1.25$ for an accessible non 'fail-safe' component. The factors increase to 1.15 and 1.35 respectively for joint details that are difficult to access. In a welded tubular structure, there is no barrier to the propagation of a fatigue crack that has reached a critical length, so the welds have to be considered non-fail-safe. Given that the tower welds are visible and accessible with relative ease from inside the tower, a partial safety factor for fatigue strength of 1.25 would appear reasonable.

The derivation of fatigue load spectra and the combination of stress ranges due to M_X and M_Y load spectra are discussed in Section 5.12.6.

Relative criticality of extreme and fatigue loads

The relative criticality of buckling failure (under extreme loads) and fatigue loads depends on a variety of factors. However, fatigue is more likely to be critical on pitch-regulated machines than on stall-regulated ones, because of the increased

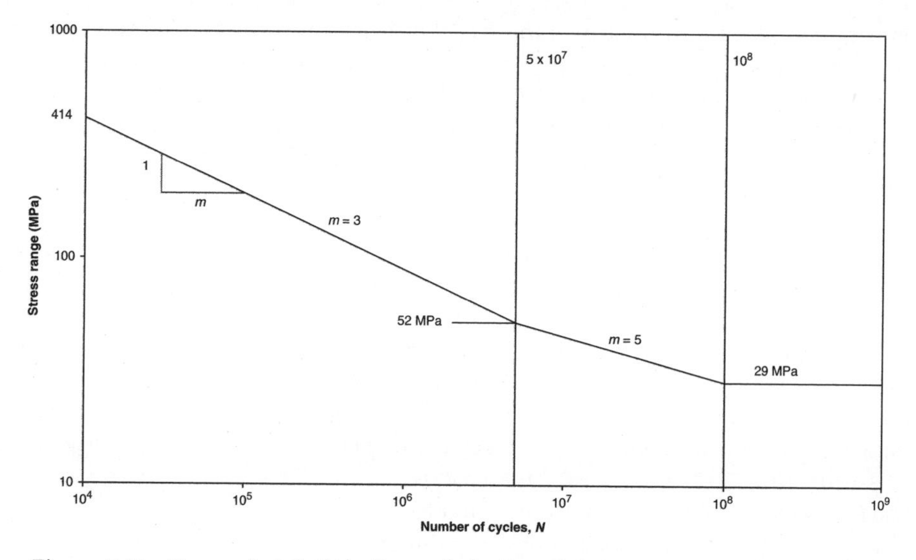

Figure 7.41 Eurocode 3 Fatigue Strength for Detail Category 71 (Butt-welded Joint)

rotor thrust fluctuations above rated and the reduced extreme loading at standstill. Fatigue is also more likely to be critical at low wind-speed sites, because the percentage reduction in extreme loads is less than the percentage reduction in fatigue equivalent load.

Tuning of tower natural frequency

Considerable scope exists, at least in theory, for adjusting the tower natural frequency to a suitable value by varying the base diameter, while maintaining the necessary strength against extreme and fatigue loading. The effect on natural frequency of varying tower base diameter by a factor of two, for a case where extreme loading governs, is illustrated for a 60 m diameter stall-regulated machine at 50 m hub-height in Figure 7.40. The frequency increases from 0.517 Hz for a 2.5 m base diameter to 0.765 Hz for a 5.0 m diameter. Now the rotational speed of a 60 m diameter turbine to yield a 60 m/s tip speed is about 19 r.p.m. If we assume that the machine is two speed, with a lower rotational speed of $19 \times 2/3 =$ 12.67 r.p.m., then the lower blade passing frequency will be 0.633 Hz, right in the middle of the available tower natural frequency range. Adopting a $+15\%/-15\%$ frequency exclusion zone, the tower natural frequency is required to be less than 0.538 Hz or more than 0.728 Hz. However, a frequency of 0.728 Hz would require a diameter of about 4.7 m (without making the tower wall thicker than necessary for the strength requirement), which is likely to be ruled out by transport considerations. Thus the only strength limited design option is one with a base diameter of 2.75 m, with a weight penalty of about 10 T compared with the 60 T optimum design, giving a natural frequency of about 0.535 Hz. Alternatively a 4 m base diameter could be chosen and the wall thickness increased by 37% to 27.5 mm to give a frequency of about 0.728 Hz. However, the weight penalty in this case is over 15 T.

The above case study illustrates the fact that it is not always economic to satisfy the natural frequency requirements for a particular combination of turbine and hub height. In these circumstances it may well be preferable to change the hub height. For example, a hub height of 55 m would work much better for the case described, with a tower base diameter of 3.5 m yielding a natural frequency of 0.535 Hz and a tower weight of 74 T.

Joints between tower sections

Towers are normally fabricated in several sections for transport reasons, so joints are required. Welding on site is an expensive operation, so bolted joints are almost always used, although sleeved joints, in which each tapered tower section is threaded over the one beneath and forced into place by jacking, have been used successfully.

The structurally most effective joint is made with friction grip bolted splice plates oriented vertically and sandwiching the walls of the abutting tower sections between them. Provided the grip force is adequate, the joint will not slip even under the extreme load, with the result that the bolts are not subject to fatigue loads. Unfortunately, apart from the effect of splice plates on the external appearance, there are practical difficulties of joint assembly, because bolting requires the provision of some form of personnel access on the outside of the tower. Nevertheless splice plates are used on some towers.

The most popular bolted arrangement is the internal flanged joint as illustrated in Figure 7.42. The flanges are butt welded to the ends of the mating sections, with the flange outer edge flush with the tower wall. Alternatively the flange may be formed with a stub section of tower wall already attached. Such flanges, which are termed weld neck flanges, provide a smoother transition from wall to flange (as illustrated in the lower half of Figure 7.42) and result in a higher butt-weld fatigue category.

After assembly, each bolt is torqued or tensioned to induce a preload between the flanges in order to minimize in-service bolt fatigue stresses. The bolt should be initially sized to resist the prying force induced by the extreme tower-wall tensile stresses – taking the fulcrum adjacent to the flange inner edge – and then checked for fatigue.

The fatigue calculation for the bolts in a flanged joint depends on the relationship between the bolt load and tower wall stress, which only remains linear while contact is maintained over the full flange width. The VDI Guideline (Verein

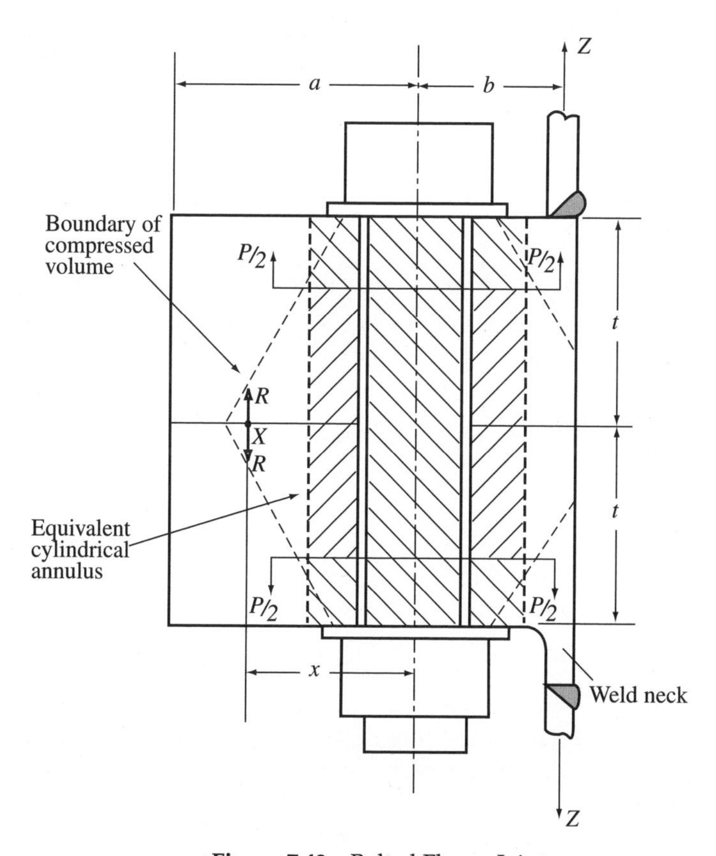

Figure 7.42 Bolted Flange Joint

Deutscher Ingenieure, 1986), VDI 2230, gives a method for calculating the bolt load increment as a proportion of the load increment in the 'tributary' width of tower wall under these conditions. The axial loading on the flanged joint and the effect of the moment due to the eccentricity of loading are considered separately. The axial load is assumed to be shared between the bolt and the preloaded flanges in proportion to the stiffnesses of the load paths, which, in the case of the flanges, is based on a reduced cross-sectional area related to the volume compressed by the preload according to

$$A_{\text{ers}} = \frac{\pi}{4}\left(d_w^2 - d_h^2\right) + \frac{\pi}{8} d_w (D_A - d_w)\left[(x+1)^2 - 1\right] \text{ where } x = \sqrt[3]{\frac{l_k d_w}{D_A^2}} \tag{7.72}$$

and d_w is the washer face diameter on the bolt head and nut, d_h is the bolt hole diameter, l_k is the clamping length between bolt head and nut, and D_A is twice the distance from the bolt centreline to the nearest flange edge, or the bolt spacing, whichever is the less.

The Guideline recognizes that the effective plane of introduction of the external load will not necessarily be immediately under the bolt head or nut, but may lie nearer the flange mid-plane, giving the load paths distinguished by different cross hatching in Figure 7.42. Stresses due to the eccentricity of the tower wall load to the flange contact area are dealt with by ordinary bending theory applied to the whole contact area.

The VDI 2230 method outlined above no longer applies once a gap has opened up between the flanges at the outer edge. For larger fluctuations in the externally applied load, Z, the fulcrum model can be used, although it is inevitably conservative at low loads. The axial load, P, applied to the bolt/flange combination is calculated on the basis that a fulcrum exists at X, a distance x from the bolt, so that $P = Z(1 + b/x)$, and the load share between the bolt and the compressed volume of flange is calculated according to the relative stiffnesses as before.

In Figure 7.43, the two linear relationships between bolt load increment and externally applied load are compared with experimental results for a particular test specimen with a single flange bolt. It is assumed that the planes of introduction of the load on the bolt/flange combination are immediately under the bolt head and nut in each case. Line OA shows the VDI 2230 model, with the point A representing the limit of its validity. Line OB shows the fulcrum model, with B representing the point at which the preload between the flanges at the position of the bolts disappears. Thereafter, the bolt load varies as $Z(1 + b/x)$, i.e. along line BC for $x/a = 0.7$. It may be noted from Figure 7.43 that a value of x/a of 0.8 results in better agreement with the test results at high loads, but these are not of interest for design purposes.

Schmidt and Neuper (1997) have proposed a more sophisticated model identified as 'Model C', which combines aspects of the two models already described and gives a bolt load characteristic consisting of the three straight lines OA, AB and BC (see Figure 7.43). Clearly this agrees much better with the experimental results, but it adds to the complexity of the fatigue load calculation.

Uniformity of bolt loading around the tower clearly depends on the accuracy of the mating flange surfaces. Schmidt, Winterstetter and Kramer (1999) have investi-

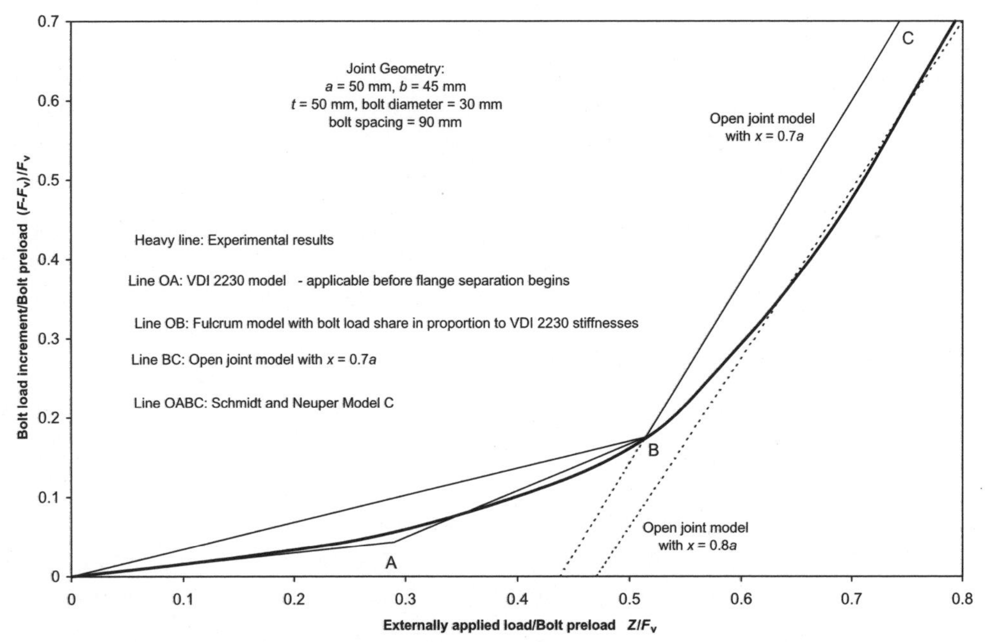

Figure 7.43 Flange Joint Bolt Load Variation with Externally Applied Load – Experimental Results and Engineering Models Compared

gated the effects of various imperfections using a finite-element model and made tentative suggestions regarding permitted tolerance levels.

Tower tie-down

The tower is normally fitted with a base flange, which can either be attached to the foundation by screwed rods cast into the concrete or bolted to an embedded tower stub. This sub section is concerned with the former arrangement.

The screwed rods are normally anchored in some way at their base, and their capacity to resist overturning moment is determined by the pull-out resistance of the semi-circle of bolts on the upwind side. As this is governed by the concrete shear strength, the rods have to be anchored quite deep into the concrete, so that their length is typically similar to the tower base radius.

The fatigue loads in the tie-down rods can be considerably reduced by pre-tensioning. The share of tower-wall uplift loads taken by the rods can be based on an estimate of the relative stiffnesses of the rod and the loaded volume of the concrete, assuming a dispersion angle of about 30° in the radial direction. The screwed rods should be sheathed, so that the pre-tension is applied over the full length.

Tower doorways

A doorway is required for access at or near the tower base, and additional doorways are sometimes required for a transformer in the tower base or for maintenance access to the blade tip mechanism. Typically they have vertical sides with semi-circular ends at top and bottom. Vertical stiffeners have to be provided as standard down each side to compensate for the missing section of wall and to resist compression buckling, but attention has to be paid to the weld detail at the stiffener ends, where stress concentration due to the opening is likely to be an additional factor.

The weld detail at the stiffener end can be eliminated by reinforcing the inside edge of the doorway with a continuous flange all the way round. The detail category of the flange to tower wall butt weld under transverse loading is then 71, but there is no stress concentration factor to contend with at the top and bottom of the doorway.

7.9.4 Steel lattice towers

Steel lattice towers are usually assembled from angle sections, with bolting used for attaching the bracing members to the legs and splicing the leg sections together. Typically the towers are square in plan with four legs, facilitating the attachment of the bracing members.

One of the advantages of lattice towers is that material savings can be obtained by splaying the legs widely apart at the base, without jeopardizing stability or posing transport problems. The latitude for doing this higher up is limited by tip clearance considerations, so waisted tower designs are common. A more elegant tower design results if the legs are rolled to a gentle concave curve, however.

The loads in the legs (or 'chords') result from the tower bending moments, while the loads in the bracing (or 'web') members result from a combination of tower shear and torsional loads. In each case member buckling under extreme loads has to be considered, and fatigue loading at the joints. Two devices are employed to improve member stability: the web members are arranged as pairs of intersecting diagonals rather than adopting a single triangulated system, so that the tension diagonal can stabilize the compression diagonal at each intersection, and the web/chord intersection points on either side of each chord member are staggered vertically to reduce the spacing of chord supports restraining flexure about the minor axis. Note that care with detailing is needed at the waist, if present, in order to ensure adequate lateral restraint for the chords at the change of direction.

Fatigue loading of bolts is avoided by the use of friction grip bolts. Accordingly galvanizing is normally used for corrosion protection rather than painting, in order to achieve an adequate coefficient of friction.

7.10 Foundations

The design of wind-turbine foundations is largely driven by the tower base overturning moment under extreme wind conditions. A variety of slab, multi-pile and mono-pile solutions have been adopted for tubular towers, and these are discussed in turn below.

7.10.1 Slab foundations

Slab foundations are chosen when competent material exists within a few metres of the surface. The overturning moment is resisted by an eccentric reaction to the weight of the turbine, tower, foundation and overburden (allowing for buoyancy, if the water table can rise above the base of the slab). The eccentricity of the reaction, and hence the magnitude of the restoring moment, is limited by the load carrying capacity of the sub-strata, which determines the width of the area at the edge of the slab required to carry the gravity loads. Brinch Hansen (1970) provides straightforward rules for calculating the slab bearing capacity under these conditions, based on the simplifying assumption of uniform loading over the loaded area. However, if the substrata behave elastically, tilting of the slab foundation is likely to result in a linear distribution of bearing stress over the loaded area, so an alternative approach is to base the design on the maximum rather than the average value. The GL rules additionally require that positive bearing stress exists over the whole width of the foundation when the turbine is operating, which limits the maximum overturning moment under these conditions to $WB/6$, where W is the gravity load and B is the slab width. This requirement can add significantly to the required foundation size.

Four alternative slab foundation arrangements are shown in Figure 7.44. Figure 7.44(a) shows a slab of uniform thickness, with its upper surface just above ground level, which is chosen when bedrock is near the ground surface. The main reinforcement consists of top and bottom mats to resist slab bending and the slab is made thick enough for shear reinforcement not to be required. The second variant shown in Figure 7.44(b) is a slab surmounted by a pedestal. This is used when the

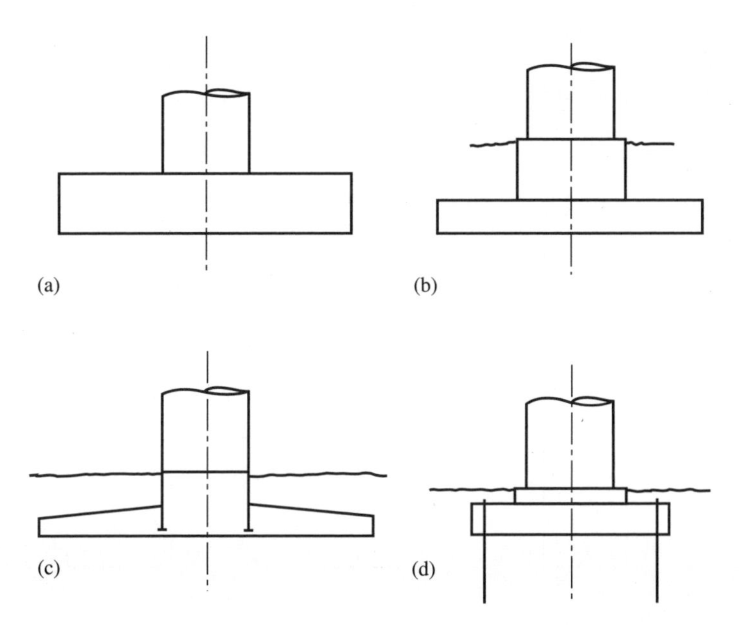

Figure 7.44 (a) Plain Slab; (b) Stub and Pedestal; (c) Stub Tower Embedded in Tapered Slab; (d) Slab Held Down by Rock Anchors

bedrock is at a greater depth than the slab thickness required to resist the slab bending moments and shear loads. The gravity load on the substrata is increased by virtue of the overburden, so the overall slab plan dimensions can be reduced somewhat.

The third variant, shown in Figure 7.44(c) is similar to the second, but embodies two possible modifications which can be applied independently: replacement of the pedestal by a stub tower embedded in the slab, and introduction of a tapering slab depth. The stub tower has to be perforated near the top of the slab to allow radial top face reinforcement to pass through it, and reinforcement to resist punching shear loads from the tower stub bottom flange must be incorporated. Tapering the slab depth has the merit of saving material, but is difficult to execute.

Rock anchors eliminate the need to add weight to a gravity foundation for counterbalance purposes, and thus enable the foundation size to be significantly reduced, provided bearing capacities are sufficiently high (see Figure 7.44(d)). Specialist contractors are needed for rock anchor installation, so they only find occasional use.

The ideal shape of gravity foundation in plan is a circle, but in view of the complications of providing circular formwork, an octagonal shape is usually chosen instead. Sometimes slabs are square in plan to simplify the shuttering and reinforcement further.

7.10.2 Multi-pile foundations

In weaker ground, a piled foundation often makes more efficient use of materials than a slab. Figure 7.45(a) illustrates a foundation consisting of a pile cap resting on eight cylindrical piles arranged in a circle. Overturning is resisted by both pile vertical and lateral loads, the latter being generated by moments applied to the head of each pile. Consequently the reinforcement must be arranged to provide full

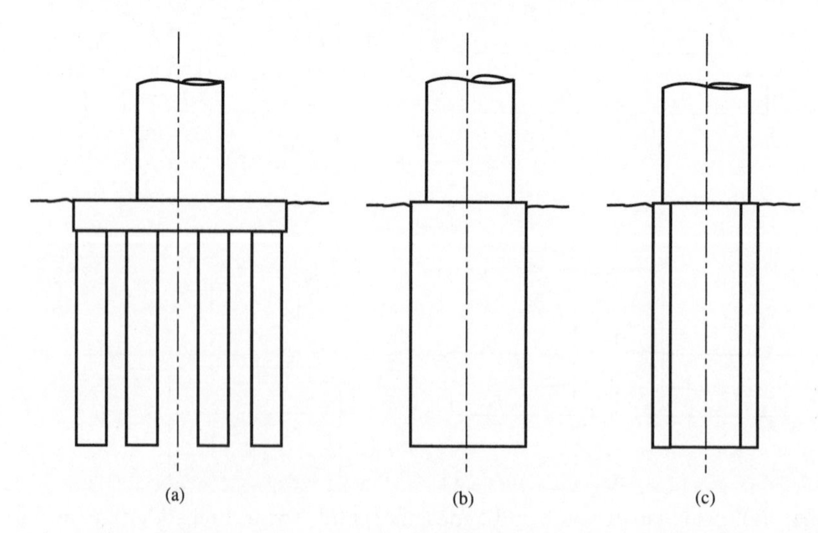

Figure 7.45 (a) Pile Group and Cap; (b) Solid Mono-pile; (c) Hollow Mono-pile

moment continuity between the piles and the pile cap. Holes for the piles can be auger drilled and the piles cast *in situ* after the positioning of the reinforcement cage.

7.10.3 Concrete mono-pile foundations

A concrete mono-pile foundation consists of a single large diameter concrete cylinder, which resists overturning by mobilizing soil lateral loads alone (see Figure 7.45(b)). These lateral loads can be calculated conservatively for sand by using either simple Rankine theory for passive pressures on retaining walls, which ignores soil/wall friction, or Coulomb theory, which includes it. However, in the case of a mono-pile, friction on the sides of the soil wedge notionally displaced when the pile begins to tilt provides further resistance, and this is accounted for in the solution due to Brinch Hansen (1961).

This type of foundation is an attractive option when the water table is low and the soil properties enable a deep hole to be excavated from above without the sides caving in. However, while simple, the concept is relatively expensive in terms of materials. The hollow cylinder variant illustrated in Figure 7.45(c) uses materials much less extravagantly by replacing the concrete in the body of the cylinder, which has no structural role to play, with fill.

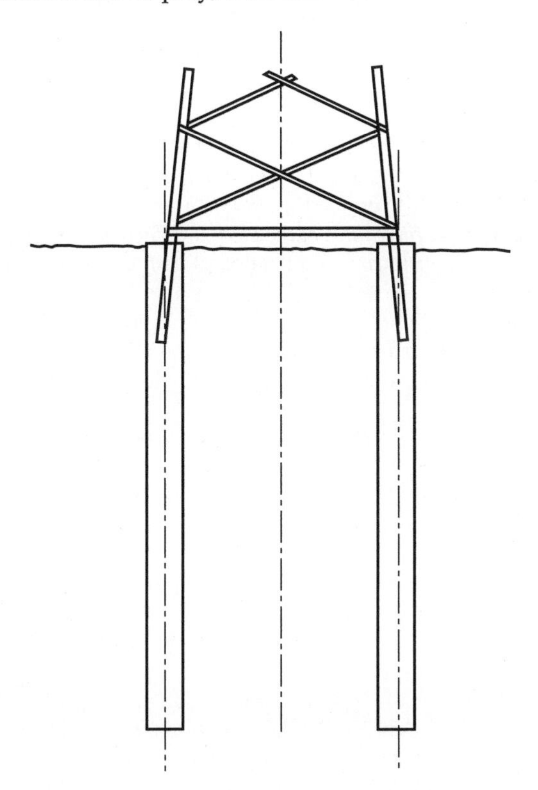

Figure 7.46 Piled Foundation for Steel Lattice Tower

7.10.4 Foundations for steel lattice towers

The legs of steel lattice towers are relatively widely spaced, and lend themselves to separate foundations. Bored cast *in situ* piles are commonly used (see Figure 7.46). The mechanism for resisting overturning is simply uplift and downthrust on the piles, but the piles must also be designed for the bending moments induced by the horizontal shear load. Pile uplift is resisted by friction on the surface of the piles, which depends on both the soil/pile friction angle and the lateral soil pressure. Considerable uncertainty surrounds the magnitude of these quantities, so Eurocode 7 recommends the use of pile testing to establish pile capacity.

The angle sections forming the base of the tower legs are cast in place when the concrete for the piles is poured. A framework is assembled in advance, incorporating the leg base sections, so that the legs can be set at the correct spacing and inclination before concreting.

References

American Gear Manufacturers Association/American Wind Energy Association, (1996). *'Recommended practices for design and specification of gearboxes for wind turbine generator systems'. AGMA/AWEA 921-A97.*

American National Standards Institute/American Gear Manufacturers Association, (1995). *'Fundamental rating factors and calculation methods for involute spur and helical gear teeth'. ANSI/AGMA 2001-C95.*

Anderson, C. G. *et al.*, (1993). 'Yaw system loads of HAWTS'. *ETSU Report No W/42/00195/REP.* Energy Technology Support Unit, Harwell, UK.

Anderson, C. G., Heerkes, H. and Yemm, R., (1998). 'Prevention of edgewise vibration on large stall-regulated blades'. *Proceedings of the BWEA Conference*, pp 95–102.

Barbero, E. J., (1998). *Introduction to composite materials design.* Taylor and Francis, Philadelphia, USA.

Bond, I. P. and Ansell, M. P., (1998). 'Fatigue properties of jointed wood composites. Part I: Statistical analysis, fatigue master curves and constant life diagrams'. *J. Mat. Sci.*, **33**, 2751–2762, and 'Part II: Life prediction: analysis for variable amplitude loading'. *J. Mat. Sci.*, **33**, 4121–4129.

Bonfield, P. W. and Ansell, M. P., (1991). 'Fatigue properties of wood in tension, compression and shear'. *J. Mat. Sci.*, **26**, 4765–4773.

Bonfield, P. W. *et al.*, (1992). 'Fatigue testing of wood composites for aerogenerator blades. Part VII: Alternative wood species and joints'. *Proceedings of the BWEA Conference*, pp 243–249.

Brinch Hansen, J., (1961). 'The ultimate resistance of rigid piles against transverse forces'. *Danish Geotechnical Institute Report No. 12.*

Brinch Hansen, J., (1970). 'A revised and extended formula for bearing capacity'. *Danish Geotechnical Institute Bulletin No. 28.*

British Standards Institution, (1986). 'BS436: Spur and helical gears – Part 3 Method for calculation of contact and root bending stress limitations for metallic involute gears'.

Corbet, D. C., (1991). 'Investigation of materials and manufacturing methods for wind turbine blades'. *ETSU Report No. W/44/00261.* Energy Technology Support Unit, Harwell, UK.

Corbet, D. C., Brown, C. and Jamieson, P., (1993). 'The selection and cost of brakes for horizontal axis stall-regulated wind turbines'. *ETSU Report No. WN 6065*. Energy Technology Support Unit, Harwell, UK.

Echtermeyer, A. T., Hayman, E. and Ronold, K. O., (1996). 'Comparison of fatigue curves for glass composite laminates.' *Design of Composite Structures against Fatigue,* (Ed. R. M. Mayer). Mechanical Engineering Publications, Bury St Edmunds, UK.

European Convention for Constructional Steelwork, (1988). *Recommendations on buckling of shells.*

Freris, L. L. (ed) (1990). *Wind energy conversion systems,* Prentice Hall.

Fuglsang, P. L. and Madsen, H. A., (1995). 'A design study of a 1 MW stall-regulated rotor'. *Riso National Laboratory Report No. R-799.* Riso National Laboratory, Roskilde, Denmark.

Hancock, M. and Bond, I. P., (1995). 'The new generation of wood composite wind turbine rotor blades – design and verification'. *Proceedings of the BWEA Conference,* pp 47–52.

Hancock, M. Sonderby, O. and Schubert, M. A., (1997). 'Design, development and testing of a 31 m wood composite stall regulated blade for serial production'. *Proceedings of the EWE Conference,* pp 206–212.

Heier, S., (1998). *Grid integration of wind energy conversion systems.* John Wiley & Sons, Chichester, UK.

Hindmarsh, J., (1984). *Electrical machines and their application.* Pergamon Press, Oxford, UK.

Hück, M., (1983). 'Calculation of S/N curves for Steel, Cast Steel and Cast Iron – Synthetic S/N curves'. *Verein Deutsher Eisenhüttenleute Report No. ABF 11.* Verlag Stahleisen, Düsseldorf, Germany.

Jamieson, P. and Brown, C. J., (1992). 'The optimization of stall-regulated rotor design'. *Proceedings of the BWEA Conference,* pp 79–84.

Jones, R. and Smith, G. A., (1993). 'High quality mains power from variable-speed wind turbines'. *IEE Conference, Renewable Energy – Clean Power 2001.*

Joose, P. A. and van Delft, D. R. V., (1996). 'Has fatigue become a wearisome subject? – Overview of 12 years' of materials research in the Netherlands'. *Proceedings of the EUWEC Conference,* pp 902–906.

Krause, P. C., (1986). *Analysis of electric machinery.* McGraw-Hill, New York, USA.

Mayer, R. M., (1996). *Design of composite structures against fatigue.* Mechanical Engineering Publications, Bury St Edmunds, UK.

McPherson, G., (1990). *An introduction to electrical machines and transformers.* John Wiley & Sons, New York, USA.

Mohan, N., Undeland, T. M. and Williams, W. P., (1995). *Power electronics, converters application and design, Second Edition.* John Wiley & Sons, New York, USA.

Petersen, J. T. *et al.*, (1998). 'Prediction of dynamic loads and induced vibrations in stall'. *Riso National Laboratory Report No. R-1045.* Riso National Laboratory, Roskilde, Denmark.

Pena, R., Clare, J. C. and Asher, G. M., (1996). 'Doubly fed induction generator using back-to-back PWM converters and its application to variable speed wind-energy generators'. *IEE Proceedings Electric Power Applications,* **143**, pp 231–241.

Schmidt, H. and Neuper, M., (1997). 'Zum elastostatischen Tragverhalten exzentrisch gezogener L-Stöße mit vorgespannten Scrauben' ('On the elastostatic behaviour of an eccentrically tensioned L-joint with prestressed bolts'), *Stahlbau,* **66**, 163–168.

Schmidt, H., Winterstetter, T. A. and Kramer, M., (1999). 'Non-linear elastic behaviour of imperfect, eccentrically tensioned L-flange ring joints with prestressed bolts as basis for fatigue design'. *Proceedings of the European Conference On Computational Mechanics.*

Thomsen, K., (1998). 'The statistical variation of wind turbine fatigue loads.' *Riso National Laboratory Report No. R-1063.* Riso National Laboratory, Roskilde, Denmark.

Timoshenko, S. P. and Gere, J. M., (1961). *Theory of Elastic Stability, Second Edition.* McGraw-Hill, New York, USA.

Van Delft, D. R. V., de Winkel, G. D. and Joose, P. A., (1996). 'Fatigue behaviour of fibreglass wind turbine blade material under variable amplitude loading'. *Proceedings of the EUWEC Conference*, pp 914–918.

Verein Deutscher Ingenieure, (1986/1988). 'VDI 2230 Part 1: Systematic calculation of high duty bolted joints – Joints with one cylindrical bolt'.

Warneke, O., (1984). 'Use of a double-fed induction machine in the Growian large wind energy converter'. *Siemens Power Engineering*, **VI**, 1, pp 56–59.

Wilson, R. A., (1990). 'Implementation and optimization of mechanical brakes and safety systems'. *Proceedings of a DEn/BWEA Workshop on 'Mechanical systems for wind turbines'*.

8
The Controller

In the most general sense, the wind turbine control system consists of a number of sensors, a number of actuators, and a system consisting of hardware and software which processes the input signals from the sensors and generates output signals for the actuators. The sensors might include, for example:

- an anemometer,
- a wind vane,
- at least one rotor speed sensor,
- an electrical power sensor,
- a pitch position sensor,
- various limit switches,
- vibration sensors,
- temperature and oil level indicators,
- hydraulic pressure sensors,
- operator switches, push buttons, etc.

The actuators might include a hydraulic or electric pitch actuator, sometimes a generator torque controller, generator contactors, switches for activating shaft brakes, yaw motors, etc.

The system that processes the inputs to generate outputs usually consists of a computer or microprocessor-based controller which carries out the normal control functions needed to operate the turbine, supplemented by a highly reliable hard-wired safety system. The safety system must be capable of overriding the normal controller in order to bring the turbine to a safe state if a serious problem occurs.

8.1 Functions of the Wind-turbine Controller

8.1.1 Supervisory control

Supervisory control can be considered as the means whereby the turbine is brought from one operational state to another. The operational states might, for example, include:

- stand-by, when the turbine is available to run if external conditions permit,
- start-up,
- power production,
- shutdown, and
- stopped with fault.

It is possible to envisage other states, or it may be useful to further subdivide some of these states. As well as deciding when to initiate a switch from one state to another, the supervisory controller will carry out the sequence control required. As an example, the sequence control for start-up of a fixed-speed pitch-regulated wind turbine might consist of the following steps:

- power-up the pitch actuator;
- release the shaft brake;
- ramp the pitch position demand at a fixed rate to some starting pitch;
- wait until the rotor speed exceeds a certain small value;
- engage the closed loop pitch control of speed;
- ramp the speed demand up to synchronous speed;
- wait until the speed has been close to the target speed for a specified time;
- close the generator contactors;
- engage the closed loop pitch control of power; and
- ramp the power demand up to the rated level.

The supervisory controller must check that each stage is successfully completed before moving on to the next. If any stage is not completed within a certain time, or if any faults are detected, the supervisory controller should change to shut-down mode.

8.1.2 Closed-loop control

The closed-loop controller is usually a software-based system that automatically adjusts the operational state of the turbine in order to keep it on some pre-defined operating curve or characteristic. Some examples of such control loops are:

- control of blade pitch in order to regulate the power output of the turbine to the rated level in above-rated wind speeds;
- control of blade pitch in order to follow a predetermined speed ramp during start-up or shut-down of the turbine;
- control of generator torque in order to regulate the rotational speed of a variable-speed turbine;
- control of yaw motors in order to minimize the yaw tracking error.

Some of these control loops may require very fast responses in order to prevent the turbine wandering far from its correct operating curve. Such controllers may need to be designed very carefully if good performance is to be achieved without detrimental effects on other aspects of the turbine's operation. Others, such as yaw control, are typically rather slow acting, and careful design is then much less critical.

This chapter examines the main issues behind closed-loop controller design, and presents some of the techniques that can be used to effect a successful design.

8.1.3 The safety system

It is helpful to consider the safety system as quite distinct from the main or 'normal' control system of the turbine. Its function is to bring the turbine to a safe condition in the event of a serious or potentially serious problem. This usually means bringing the turbine to rest with the brakes applied.

The normal wind-turbine supervisory controller should be capable of starting and stopping the turbine safely in all foreseeable 'normal' conditions, including extreme winds, loss of the electrical network, and most fault conditions which are detected by the controller. The safety system acts as a back-up to the main control system, and takes over if the main system appears to be failing to do this. It may also be activated by an operator-controlled emergency stop button.

Thus the safety system must be independent from the main control system as far as possible, and must be designed to be fail-safe and highly reliable. Rather than utilizing any form of computer or micro-processor based logic, the safety system would normally consist of a hard-wired fail-safe circuit linking a number of normally open relay contacts that are held closed when all is healthy. Then if any one of those contacts is lost, the safety system trips, causing the appropriate fail-safe actions to operate. This might include disconnecting all electrical systems from the supply, allowing fail-safe pitching to the feather position, and allowing the spring-applied shaft brake to come on.

The safety system might, for example, be tripped by any one of the following:

- rotor overspeed, i.e., reaching the hardware overspeed limit – this is set higher than the software overspeed limit which would cause the normal supervisory controller to initiate a shut-down (see Figure 8.1 for typical arrangement of rotor speed sensing equipment on low-speed shaft);

- vibration sensor trip, which might indicate that a major structural failure has occurred;
- controller watchdog timer expired: the controller should have a watchdog timer which it resets every controller timestep – if it is not reset within this time, this indicates that the controller is faulty and the safety system should shut down the turbine;
- emergency stop button pressed by an operator;
- other faults indicating that the main controller might not be able to control the turbine.

Figure 8.1 Low-speed Shaft Speed Sensing System. (Three proximity sensors mounted on a bracket attached to the front of the (integrated) gearbox register the passage of the teeth on the shaft circumference, and provide independent speed signals for the control and safety systems. The flange onto which the hub is bolted is immediately to the left of the teeth).

8.2 Closed-loop Control: Issues and Objectives

8.2.1 Pitch control (see also Sections 4.2.4 and 6.7.3)

Pitch control is the most common means of controlling the aerodynamic power generated by the turbine rotor. Pitch control also has a major effect on all the aerodynamic loads generated by the rotor.

Below rated wind speed, the turbine should simply be trying to produce as much power as possible, so there is generally no need to vary the pitch angle. The aerodynamic loads below rated wind speed are generally lower than above rated, so again there is no need to modulate these using pitch control. However, for fixed-speed turbines, the optimum pitch angle for aerodynamic efficiency varies slightly with wind speed. Therefore, on some turbines, the pitch angle is varied slowly by a few degrees below rated in response to a heavily averaged anemometer or power output signal.

Above rated wind speed, pitch control provides a very effective means of regulating the aerodynamic power and loads produced by the rotor so that design limits are not exceeded. In order to achieve good regulation, however, the pitch control needs to respond very rapidly to changing conditions. This highly active control action needs very careful design as it interacts strongly with the turbine dynamics.

One of the strongest interactions is with the tower dynamics. As the blades pitch to regulate the aerodynamic torque, the aerodynamic thrust on the rotor also changes substantially, and this feeds into the tower vibration. As the wind increases, the pitch angle increases to maintain constant torque, but the rotor thrust decreases. This allows the downwind tower deflection to decrease, and as the tower top moves upwind the relative wind speed seen by the rotor increases. The aerodynamic torque increases further, causing more pitch action. Clearly if the pitch-controller gain is too high this positive feedback can result in instability. It is therefore vital to take the tower dynamics into account when designing a pitch controller.

Below rated wind speed, the pitch setting should be at its optimum value to give maximum power. It follows that when the wind speed rises above rated, either an increase or a decrease in pitch angle will result in a reduction in torque. An increase in pitch angle, defined as turning the leading edge into wind, reduces the torque by decreasing the angle of attack and hence the lift. This is known as pitching towards feather. A decrease in pitch, i.e., turning the leading edge downwind, reduces the torque by increasing the angle of attack towards stall, where the lift starts to decrease and the drag increases. This is known as pitching towards stall.

Although pitching towards feather is the more common strategy, some turbines pitch towards stall. This is commonly known as active stall or assisted stall (see Section 6.7.4). Pitching to feather requires much more dynamic pitch activity than pitching to stall: once a large part of the blade is in stall, very small pitch movements suffice to control the torque. Pitching to stall results in significantly greater thrust loads because of the increased drag. On the other hand, the thrust is much more constant once the blade is stalled, so thrust-driven fatigue loads may well be smaller.

A further problem with pitching to stall is that the lift curve slope at the start of the stalled region is negative, i.e., the lift coefficient decreases with increasing angle of attack. This results in negative aerodynamic damping, which can result in instability of the blade bending modes, both in-plane and out of plane. This can be a problem also with fixed pitch stall-regulated turbines.

Most pitch controlled turbines use full-span pitch control, in which the pitch bearing is close to the hub. It is also possible, though not common, to achieve aerodynamic control by pitching only the blade tips, or by using ailerons, flaps, air-jets or other devices to modify the aerodynamic properties. These strategies will result in most of the blade being stalled in high winds. If only the blade tips are pitched, it may be difficult to fit a suitable actuator into the outboard portion of the blade; accessibility for maintenance is also difficult.

8.2.2 Stall control

Many turbines are stall-regulated, which means that the blades are designed to stall in high winds without any pitch action being required. This means that pitch actuators are not required, although some means of aerodynamic braking is likely to be required, if only for emergencies (see Section 6.8.2).

In order to achieve stall-regulation at reasonable wind speeds, the turbine must operate closer to stall than its pitch-regulated counterpart, resulting in lower aerodynamic efficiency below rated. This disadvantage may be mitigated in a variable-speed turbine, when the rotor speed can be varied below rated in order to maintain peak power coefficient.

In order for the turbine to stall rather than accelerate in high winds, the rotor speed must be restrained. In a fixed speed turbine the rotor speed is restrained by the generator, which is governed by the network frequency, as long as the torque remains below the pull-out torque. In a variable speed turbine, the speed is maintained by ensuring that the generator torque is varied to match the aerodynamic torque. A variable-speed turbine offers the possibility to slow the rotor down in high winds in order to bring it into stall. This means that the turbine can operate further from the stall point in low winds, resulting in higher aerodynamic efficiency. However, this strategy means that when a gust hits the turbine, the load torque not only has to rise to match the wind torque but also has to increase further in order to slow the rotor down into stall. This removes one of the main advantages of variable-speed operation, namely that it allows very smooth control of torque and power above rated.

8.2.3 Generator torque control (see also Sections 6.9 and 7.5)

The torque developed by a fixed-speed (i.e., directly-connected) induction generator is determined purely by the slip speed. As the aerodynamic torque varies, the rotor speed varies by a very small amount such that the generator torque changes to match the aerodynamic torque. The generator torque cannot therefore be actively controlled.

However, if a frequency converter is interposed between the generator and the network, the generator speed will be able to vary. The frequency converter can be actively controlled to maintain constant generator torque or power output above rated wind speed. Below rated, the torque can be controlled to any desired value, for example with the aim of varying the rotor speed to maintain maximum aerodynamic efficiency.

There are several means of achieving variable-speed operation. One is to connect the generator stator to the network through a frequency converter, which must then be rated for the full power output of the turbine. Alternative arrangements include a wound rotor induction generator with its stator connected directly to the network, and with its rotor connected to the network through slip rings and a frequency converter. This means that the frequency converter need only be rated to handle a fraction of the total power, although the larger this fraction, the larger the achievable speed range will be.

A special case is the variable slip induction generator, where active control of the resistance in series with the rotor windings allows the torque/speed relationship to be modified. By means of closed-loop control based on measured currents, it is possible to maintain constant torque above rated, effectively allowing variable-speed operation in this region. Below rated it behaves just like a normal induction generator (Bossanyi and Gamble, 1991, Pedersen, 1995).

8.2.4 Yaw control

Turbines whether upwind or downwind, are generally stable in yaw (Section 3.10) in the sense that if the nacelle is free to yaw, the turbine will naturally remain pointing into the wind. However, it may not point exactly into wind, in which case some active control of the nacelle angle may be needed to maximize the energy capture. Since a yaw drive is usually required anyway, e.g. for start-up and for unwinding the pendant cable, it may as well be used for active yaw tracking. Free yaw has the advantage that it does not generate any yaw moments at the yaw bearing. However, it is usually necessary to have at least some yaw damping, in which case there will be a yaw moment at the bearing.

In practice, most turbines do use active yaw control. A yaw error signal from the nacelle-mounted wind vane is then used to calculate a demand signal for the yaw actuator. Frequently the demand signal will simply be a command to yaw at a slow fixed rate in one or the other direction. The yaw vane signal must be heavily averaged, especially for upwind turbines where the vane is behind the rotor. Because of the slow response of the yaw control system, a simple dead-band controller is often sufficient. The yaw motor is switched on when the averaged yaw error exceeds a certain value, and switched off again after a certain time or when the nacelle has moved through a certain angle.

More complex control algorithms are sometimes used, but the control is always slow-acting, and does not demand any special design considerations. One exception is the case of active yaw control to regulate aerodynamic power in high winds, as used on the variable speed Gamma 60 turbine referred to in Section 6.7.5. This clearly requires very rapid yaw rates, and results in large yaw loads and gyroscopic and asymmetric aerodynamic loads on the rotor. This method of power regulation would be too slow for a fixed-speed turbine, and even on the Gamma 60 the speed excursions during above-rated operation were quite large.

8.2.5 Influence of the controller on loads

As well as regulating the turbine power in high winds, and perhaps optimizing it in low winds, it is clear that the action of the control system can have a major impact on the loads experienced by the turbine. The design of the controller must take into account the effect on loads, and at least ensure that excessive loads will not result from the control action. It is possible to go further than this, and explicitly design the controller with the reduction of certain loads as an additional objective.

The reduction of certain loads is clearly compatible with the primary objective of limiting power in high winds. For example, the limitation of power output is clearly compatible with reduction of gearbox torque. In other cases, however, there may be a conflict in which case the controller design is bound to be a compromise involving a trade-off between competing goals. For example, there is a clear trade-off between good control of power output and pitch actuator loads. The more actuator activity can be tolerated, the better the power control can be.

The interaction between pitch control and tower vibration referred to in Section 8.2.1 is another important example, since the amount of tower vibration has a major effect on tower base loads. Blade, hub and other structural loads will also be influenced by pitch control activity. Generator torque control can have a major impact on gearbox loads, as described below.

8.2.6 Defining controller objectives

The primary objective of the closed-loop controller can usually be stated fairly simply. For example, the primary objective of the pitch controller may be to limit power or rotor speed in high winds. There may be more than one 'primary' objective, as in the case where the pitch or torque controller is also used to optimize energy capture in low winds.

However, since the controller can also have a major effect on structural loads and vibrations, it is vital to consider these when designing the control algorithm. Thus a fuller description of the pitch controller objectives might be:

- to regulate aerodynamic torque in above-rated wind speeds;
- to minimize peaks in gearbox torque;
- to avoid excessive pitch activity;
- to minimize tower base loads as far as possible by controlling tower vibration, and
- to avoid exacerbating hub and blade root loads.

Clearly some of these objectives conflict with others, so the control design process will inevitably involve some degree of trade-off or optimization. In order to do this, it is necessary to be able to quantify the different objectives. It is usually almost impossible to do this with any precision, because the various loads may affect not only the costs of different components (sometimes in complex ways) but also their reliability. Even the trade-off between energy capture and component cost is not straightforward, as it will depend on the wind regime, the discount rate, and knowledge of future prices for the sale of electricity. Therefore, some degree of judgement will always be required in arriving at an acceptable controller design.

8.2.7 PI and PID controllers

A brief general description is given here of PI and PID controllers, since they will be referred to a number of times in the subsequent sections. The proportional and integral (PI) controller is an algorithm which is very widely used for controlling all kinds of equipment and processes. The control action is calculated as the sum of two terms, one proportional to the control error, which is the difference between the desired and actual values of the quantity to be controlled, and one proportional to the integral of the control error. The integral term ensures that in the steady state the control error tends to zero – if it did not, the integral term would make the control action continue to increase. The proportional term makes the algorithm more responsive to rapid changes in the quantity being controlled.

A differential term is often added, which gives a contribution to the control action proportional to the rate of change of the control error. This is then known as a PID controller. In terms of the Laplace operator s, which can usefully be thought of as a differentiation operator, the PID controller from measured signal x to control signal y can be written as follows:

$$y = \left(K_\mathrm{p} + \frac{K_\mathrm{i}}{s} + \frac{K_\mathrm{d} s}{1 + sT_\mathrm{d}}\right) x \tag{8.1}$$

where K_P, K_i and K_d are the proportional, integral and derivative gains respectively. The denominator of the differential term is essentially a low-pass filter, and is needed to ensure that the gain of the algorithm does not increase indefinitely with frequency, which would make the algorithm very sensitive to signal noise. Setting $K_\mathrm{d} = 0$ results in a PI controller.

It is often the case that the control action is subject to limits. For example, if the control action represents the blade pitch used to control power above rated, then when the power drops below rated the pitch will be limited to the fine pitch setting, and will not be allowed to drop further. In this situation the integral term of the PI or PID controller will grow more and more negative as the power remains below rated. Then when the wind speed rises again and the power rises above rated, the integral term will start to grow again towards zero, but until it gets close to zero it will more than compensate for the proportional and derivative terms. Therefore the pitch may remain 'stuck' at fine pitch for a considerable time, depending on how long the power has been below rated, until the integral term has come back close to zero. This is known as integrator wind-up, and clearly it must be prevented. This is done in effect by disabling the integrator when the pitch is on the limit. This is known as 'integrator desaturation', which is described more fully in Section 8.6.

The design of PI and PID controllers, including the choice of gains, is described in more detail in Section 8.4.

8.3 Closed-loop Control: General Techniques

This section outlines the principles behind many of the types of closed-loop controllers to be found in wind-turbines. Mathematical methods for designing the closed-loop algorithms are covered in Section 8.4.

8.3.1 Control of fixed-speed, pitch-regulated turbines

A fixed-speed pitch-regulated turbine usually means a turbine that has an induction generator connected directly to the AC network, and which therefore rotates at a nearly constant speed. As the wind speed varies, the power produced will vary roughly as the cube of the wind speed. At rated wind speed, the electrical power generated becomes equal to the rating of the turbine, and the blades are then pitched in order to reduce the aerodynamic efficiency of the rotor and limit the power to the rated value. The usual strategy is to pitch the blades in response to the power error, defined as the difference between the rated power and the actual power being generated, as measured by a power transducer. The primary objective is then to devise a dynamic pitch control algorithm that minimizes the power error, although as explained above, this may not be the only objective.

The main elements of the control loop are shown in Figure 8.2. A PI or PID algorithm is often used for the controller.

When the power falls below rated, the pitch demand saturates at the fine pitch limit, maximizing the aerodynamic efficiency of the rotor. Since the optimum pitch angle depends on the tip speed ratio, it is possible to increase energy capture below rated by a small percentage if the fine pitch limit is varied in response to the wind speed. The measured power itself is the best available measure of wind speed over

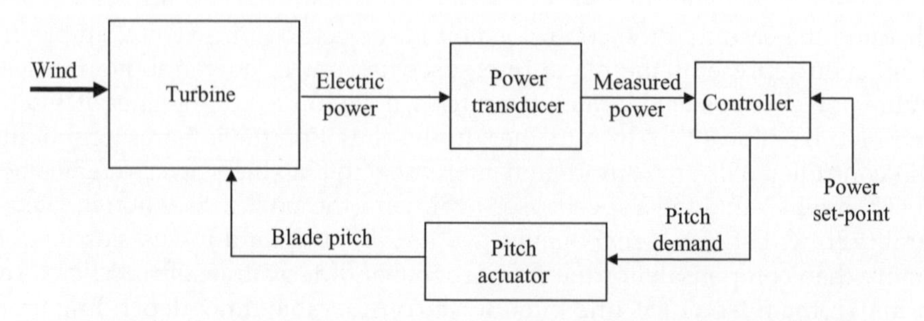

Figure 8.2 Main Control Loop for a Fixed-speed Pitch-regulated Turbine

the whole turbine. However, the fine pitch limit should be varied relatively slowly compared to the control-loop dynamics. Satisfactory performance can be obtained by changing the fine pitch limit in response to a moving average of the measured power. The moving average time constant should be significantly longer than the blade-passing frequency to avoid unnecessary pitch activity below rated.

8.3.2 Control of variable-speed pitch-regulated turbines

With a variable-speed generator, it becomes possible to control the load torque at the generator directly, so that the speed of the turbine rotor can be allowed to vary between certain limits. An often-quoted advantage of variable-speed operation is that below rated wind speed, the rotor speed can be adjusted in proportion to the wind speed so that the optimum tip speed ratio is maintained. At this tip speed ratio the power coefficient, C_P, is a maximum, which means that the aerodynamic power captured by the rotor is maximized. This is often used to suggest that a variable-speed turbine can capture much more energy than a fixed-speed turbine of the same diameter. In practice, however, it may not be possible to realize as much gain as this simple argument would suggest.

Maximum aerodynamic efficiency is achieved at the optimum tip speed ratio $\lambda = \lambda_{opt}$, at which the power coefficient C_P has its maximum value $C_{P(max)}$. Since the rotor speed Ω is then proportional to wind speed U, the power increases with U^3 and Ω^3, and the torque with U^2 and Ω^2. The aerodynamic torque is given by

$$Q_a = \tfrac{1}{2}\rho A C_Q U^2 R = \tfrac{1}{2}\rho \pi R^3 \frac{C_P}{\lambda} U^2 \tag{8.2}$$

Since $U = \Omega R/\lambda$ we have

$$Q_a = \tfrac{1}{2}\rho \pi R^5 \frac{C_P}{\lambda^2} \Omega^2 \tag{8.3}$$

In the steady state therefore, the optimum tip speed ratio can be maintained by setting the load torque at the generator, Q_g, to balance the aerodynamic torque, i.e.,

$$Q_g = \frac{1}{2}\frac{\pi \rho R^5 C_P}{\lambda^3 G^3}\omega_g^2 - Q_L \tag{8.4}$$

Here Q_L represents the mechanical torque loss in the drive train (which may itself be a function of rotational speed and torque), referred to the high-speed shaft. The generator speed is $\omega_g = G\Omega$, where G is the gearbox ratio.

This torque–speed relationship is shown schematically in Figure 8.3 as line BC. Although it represents the steady-state solution for optimum C_P, it can also be used dynamically to control generator torque demand as a function of measured generator speed. In many cases, this is a very benign and satisfactory way of controlling generator torque below rated wind speed.

For tracking peak C_P below rated in a variable-speed turbine, the quadratic

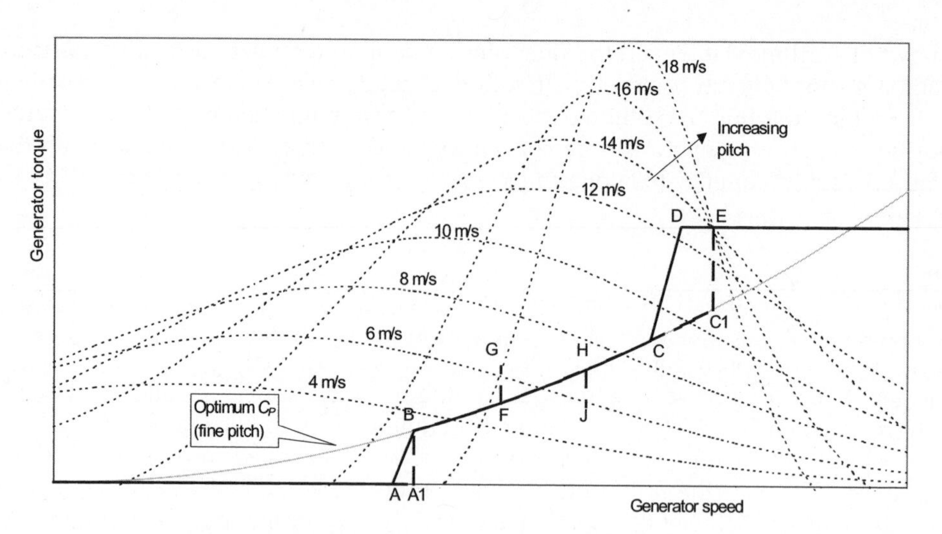

Figure 8.3 Schematic Torque–Speed Curve for a Variable-speed Pitch-regulated Turbine

algorithm of Equation (8.4) works well and gives smooth, stable control. However, in turbulent winds, the large rotor inertia prevents it from changing speed fast enough to follow the wind, so rather than staying on the peak of the C_P curve it will constantly fall off either side, resulting in a lower mean C_P. This problem is clearly worse for heavy rotors, and also if the C_P–λ curve has a sharp peak. Thus in optimizing a blade design for variable-speed operation, it is not only important to try to maximize the peak C_P, but also to ensure that the C_P–λ curve is reasonably flat-topped.

It is possible to manipulate the generator torque to cause the rotor speed to change faster when required, so staying closer to the peak of the C_P curve. One way to do this is to modify the torque demand by a term proportional to rotor acceleration (Bossanyi, 1994):

$$Q_g = \frac{1}{2}\frac{\pi\rho R^5 C_P}{\lambda^3 G^3}\omega_g^2 - Q_L - B\dot{\omega}_g \tag{8.5}$$

For a stiff drive train, and ignoring frequency converter dynamics, the torque balance gives:

$$I\dot{\Omega} = Q_a - GQ_g \tag{8.6}$$

where I is the total inertia (of rotor, drive train and generator) and Ω is the rotational speed of the rotor. Hence

$$(I - G^2 B)\dot{\Omega} = Q_a - \frac{1}{2}\frac{\pi\rho R^5 C_P}{\lambda^3 G^2}\omega_g^2 + GQ_L \tag{8.7}$$

The effective inertia is reduced from I to $I - G^2B$, allowing the rotor speed to respond more rapidly to changes in wind speed.

Another possible method is to use available measurements to make an estimate of the wind speed, calculate the rotor speed required for optimum C_P, and then use the generator torque to achieve that speed as rapidly as possible. The aerodynamic torque can be expressed as

$$Q_a = \tfrac{1}{2}\rho A C_Q R U^2 = \tfrac{1}{2}\rho\pi R^5 \Omega^2 C_Q/\lambda^2 \tag{8.8}$$

where R is the turbine radius, Ω the rotational speed, and C_Q the torque coefficient. If drive train torsional flexibility is ignored, a simple estimator for the aerodynamic torque is

$$Q_a^* = GQ_g + I\dot{\Omega} = GQ_g + IG\dot{\omega}_g \tag{8.9}$$

where I is the total inertia. A more sophisticated estimator could take into account drive train torsion, etc. From this it is possible to estimate the value of the function $F(\lambda) = C_Q(\lambda)/\lambda^2$ as

$$F^*(\lambda) = \frac{Q_a^*}{\tfrac{1}{2}\rho\pi R^5 (G\omega_g)^2} \tag{8.10}$$

Knowing the function $F(\lambda)$ from steady-state aerodynamic analysis, one can then deduce the current estimated tip speed ratio λ^*. The desired generator speed for optimum tip speed ratio can then be calculated as

$$\omega_d = \omega_g \hat{\lambda}/\lambda^* \tag{8.11}$$

where $\hat{\lambda}$ is the optimum tip speed ratio to be tracked. A simple PI controller can then be used, acting on the speed error $\omega_g - \omega_d$, to calculate a generator torque demand which will track ω_d. The higher the gain of PI controller, the better the C_P tracking, but at the expense of larger power variations. Simulations for a particular turbine showed that a below-rated energy gain of almost 1 percent could be achieved, with large but not unacceptable power variations.

Holley, Rock and Chaney (1999) demonstrated similar results with a more sophisticated scheme, and also showed that a perfect C_P tracker would capture 3 percent more energy below rated by demanding huge power swings of plus and minus three to four times rated power, which is totally unacceptable.

As turbine diameters increase in relation to the lateral and vertical length scales of turbulence, it becomes more difficult to achieve peak C_P anyway because of the non-uniformity of the wind speed over the rotor. Thus if one part of a blade is at its optimum angle of attack at some instant, other parts will not be.

In most cases, it is actually not practical to maintain peak C_P from cut-in all the way to rated wind speed. Although some variable-speed systems can operate all the way down to zero rotational speed, this is not the case with limited range variable-speed systems based on wound rotor induction generators, which are popular especially on larger turbines. These systems only need a power converter rated to handle a fraction of the turbine power, which is a major cost saving. This means that in low wind speeds, just above cut-in, it may be necessary to operate at an essentially constant rotational speed, with the tip speed ratio above the optimum value.

At the other end of the range, it is usual to limit the rotational speed to some level, usually determined by aerodynamic noise constraints, which is reached at a wind speed which is still some way below rated. It is then cost-effective to increase the torque demand further, at essentially constant rotational speed, until rated power is reached. Figure 8.3 illustrates some typical torque–speed trajectories, which are explained in more detail below. Turbines designed for noise-insensitive sites may be designed to operate along the optimum-C_P trajectory all the way until rated power is reached. The higher rotational speed implies lower torque and in-plane loads, but higher out-of-plane loads, for the same rated power. This strategy might be of interest for offshore wind-turbines.

8.3.3 Pitch control for variable-speed turbines

Once the rated torque has been reached, no further increase in load torque can occur, so the turbine will start to speed up. Pitch control is then used to regulate the rotor speed, with the load torque held constant. A PI or PID controller is often satisfactory for this application. In some situations it may be useful to include notch filters on the speed error to prevent excessive pitch action at, for example, the blade-passing frequency or the drive train resonant frequency (see Section 8.3.5).

Rather than maintain a constant torque demand while the pitch control is regulating the rotational speed, it is possible to vary the torque demand in inverse proportion to the measured speed in order to keep the power output, rather than the torque, at a constant level. Provided the pitch controller is able to maintain the speed close to the set point, there will be little difference between these two approaches. The reduction of load torque with increasing speed has a slight destabilizing effect on the pitch controller, but this is often not serious, and provided the gearbox torque and rotor speed variations are not greatly affected, the constant power approach is attractive from the perspective of power quality and flicker.

8.3.4 Switching between torque and pitch control

In practice, acoustic noise, loads or other design constraints usually mean that the maximum allowable rotor speed is reached at a relatively low wind speed. As the wind speed increases further, it is desirable to increase the torque and power without any further speed increase, in order to capture more energy from the wind. The simplest strategy is to implement a torque–speed ramp: line CD in Figure 8.3 Once rated power or torque is reached, pitch control is used to maintain the rotor speed at its rated value. In order to prevent the torque and pitch controllers from interfering with one another, the speed set-point for the pitch controller is set a little higher, at point E in Figure 8.3. If the speed set-point were at D then there would constantly be power dips in above-rated winds, whenever the speed fell transiently below D. Furthermore the pitch controller would act below rated, as the pitch and torque controllers would both be trying to control the speed.

It would be an improvement if the torque–speed trajectory A–B–C–D–E in Figure 8.3 could be changed to A1–B–C1–E. The turbine would then stay close to optimum C_P over a wider range of wind speeds, giving slightly higher energy capture for the same maximum operating speed (Bossanyi, 1994). The vertical sections A1–B and C1–E can be achieved by using a PI controller for the torque demand, in response to the generator speed error with the set point at A1 or C1. Transitions between constant speed and optimum C_P operation are conveniently handled by using the optimum-C_P curve as the upper torque limit of the PI controller when operating at A1, or the lower limit when at C1. The set point flips between A1 and C1 when the measured speed crosses the mid-point between A1 and C1. Despite this step change in set point the transition is completely smooth because the controller will be saturated on the optimum-C_P limit curve both before and after the transition.

This logic can easily be extended to implement 'speed exclusion zones', to avoid speeds at which blade-passing frequency would excite, for example, the tower resonance, by introducing additional speed set points and some logic for switching between them – see lines FG, HJ in Figure 8.3. When the torque demand exceeds G for a certain time, the set point ramps smoothly from F to H. Then if it falls below J, the set point ramps back again.

Another advantage of PI control of the torque is that the 'compliance' of the system can be controlled. Controlling to a steep ramp (CD in Figure 8.3) can be quite harsh in that the torque demand will be varying rapidly up and down the slope. A PI controller, on the other hand, can be tuned to achieve a desired level of 'softness'. With high gain, the speed will be tightly controlled to the set point, requiring large torque variations. Lower gains will result in more benign torque variations, while the speed is allowed to vary more around the set point.

In order to use point C1 as the speed set point for both the torque and the pitch controllers, it is necessary to decouple the two. One technique is to arrange some switching logic which ensures that only one of the control loops is active at any one time. Thus below rated the torque controller is active and the pitch demand is fixed at fine pitch, while above rated the pitch controller is active and the torque demand is fixed at the rated value. This can be done with fairly simple logic, although there will always be occasions when the controller is caught briefly in the 'wrong' mode. For example, if the wind is just below rated but rising rapidly, it might be useful to start pitching the blades a little before the torque demand reaches rated. If the pitch does not start moving until the torque reaches rated, it then has to move some way before it starts to control the acceleration, and a small overspeed may result.

A more satisfactory approach is to run both control loops together, but to couple them together with terms which drive one or the other loop into saturation when far above or below the rated wind speed. Thus most of the time only one of the controllers is active, but they can be made to interfere constructively when close to the rated point.

A useful method is to include a torque error term in the pitch PID in addition to the speed error. Above rated, since the torque demand saturates at rated, the torque error will be zero, but below rated it will be negative. An integral term will bias the pitch demand towards fine pitch, preventing the pitch controller from acting in low winds, while a proportional term may help to start the pitch moving a little before the torque reaches rated if the wind speed is rising rapidly.

It is also necessary to prevent the torque demand from dropping when operating well above rated wind speed. Here a useful strategy is a 'ratchet' which prevents the torque demand from falling while the pitch is not at fine. This can also smooth over brief lulls in the wind around rated, using the rotor kinetic energy to avoid transient power drops.

8.3.5 Control of tower vibration

For both fixed and variable-speed machines the influence of the pitch controller on tower vibration and loading, described in Section 8.2.1, is one of the major constraints on the design of the control algorithm. The first tower fore-aft vibrational mode is essentially very lightly damped, exhibiting a strong resonant response which can be maintained at quite a high level even by a small amount of excitation which is naturally present in the wind. The strength of the response depends critically on the small amount of damping which is present, mostly aerodynamic damping from the rotor. The pitch control action modifies the effective damping of that mode. In designing the pitch controller, it is therefore important to avoid further reducing the already small level of damping, and if possible to increase it.

The design of control algorithms is covered in Section 8.4. This includes the choice of PID gains, as well as the addition of further terms to the controller which modify the overall dynamics in such a way as to help increase the tower damping. The use of modern control methods such as optimal state feedback is also discussed. This technique can help to achieve a suitable compromise between the competing objectives of speed or power control (achieved by regulating the in-plane loading) and tower vibration control (which depends on modifying the out-of-plane loading).

There is, however, only a certain amount of information in the measured speed or power signal. State estimators such as Kalman filters (Section 8.4.5) can be used to try to distinguish between the effects of wind speed changes and tower motion on the measured signal. However, it is also possible to enhance the information available to the controller by using an accelerometer mounted in the nacelle, which provides a very direct measure of tower fore-aft motion. By using this extra signal, it is in fact possible to reduce tower loads significantly without adversely affecting the quality of speed or power regulation.

The tower dynamics can be modelled approximately as a second-order system exhibiting damped simple harmonic motion, i.e.,

$$M\ddot{x} + D\dot{x} + Kx = F + \Delta F \tag{8.12}$$

where x is tower displacement and F is the applied force, which in this case is predominantly the rotor thrust. ΔF is the additional thrust caused by pitch action. We can equate M with the tower modal mass and K with the modal stiffness, such that the tower frequency is $\sqrt{K/M}$ rad/s. The damping term D is small. The effective damping can clearly be increased if ΔF is proportional to $-\dot{x}$. Clearly it is easier to measure acceleration than velocity, so the tower acceleration would have to be integrated to provide a measure of $\dot{x}$. A suitable gain for ΔF can be estimated from the partial derivative from pitch to thrust, $\partial F/\partial\beta$ where β is the pitch angle, in order to achieve any particular additional damping D_P:

$$\begin{aligned} \delta F &= \frac{\partial F}{\partial \beta}\delta\beta = -D_P\dot{x} \\ \delta\beta &= \frac{-D_P}{\partial F/\partial\beta}\dot{x} \end{aligned} \tag{8.13}$$

It may sometimes be necessary to place a notch filter in series with this feedback term to prevent unwanted feedback from other components of tower acceleration, for example at blade-passing frequency. Figure 8.4 shows the results of a simulation with and without such an acceleration feedback term, in combination with a PID controller to control rotational speed. The simulations were driven with a realistic three-dimensional turbulent wind input. The speed control was hardly affected, and although there is a significant increase in pitch actuator activity, the pitch rates required are far from excessive. Clearly this technique is capable of increasing the tower damping substantially, almost eliminating the resonant response and dramatically reducing tower base loads. However, it requires an accelerometer which is reliable enough to be included in the contol loop and will not contribute significantly to turbine down-time. There has been no significant commercial application of this technique to date.

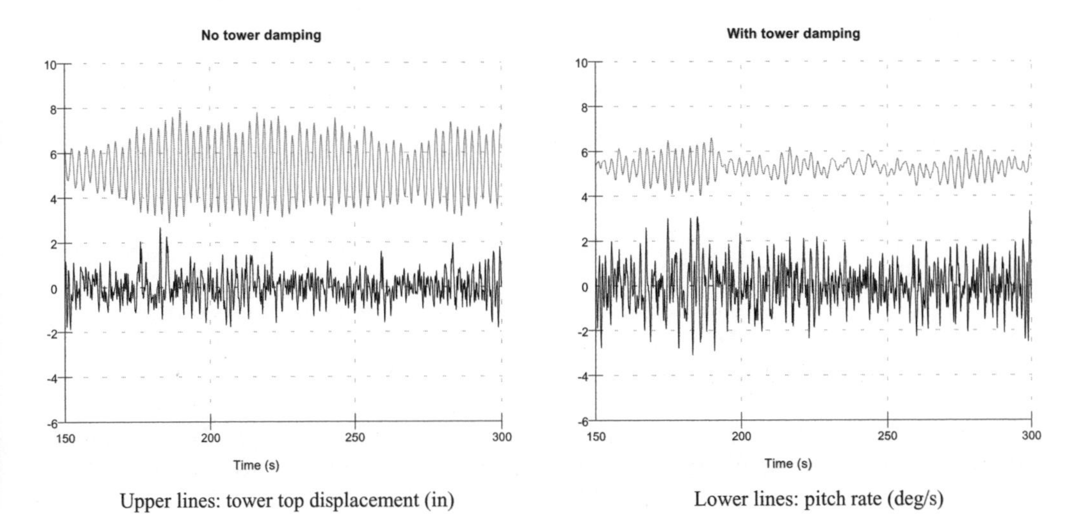

Figure 8.4 Use of a Tower Accelerometer to Help Control Tower Vibration

8.3.6 Control of drive train torsional vibration

A typical drive train can be considered to consist of a large rotor inertia and a (smaller) high-speed shaft inertia (mainly the generator and brake disc), separated by a torsional spring which represents twisting of shafts and couplings, bending of gear teeth and deflection of any soft mountings. Sometimes it is important to consider also the coupling of the torsional mode of vibration with the first rotor in-plane collective mode, in which case the drive train can be approximated by three inertias and two torsional springs. In some cases the coupling to the second tower side-to-side mode, which has a lot of rotation at the tower top, is also important.

In a fixed-speed turbine, the induction generator slip curve (Section 7.5) essentially acts like a strong damper, with the torque increasing rapidly with speed (see Figure 6.11). Therefore the torsional mode of the drive train is well damped and generally does not cause a problem. In a variable-speed turbine operating at constant generator torque, however, there is very little damping for this mode, since the torque no longer varies with generator speed. The very low damping can lead to large torque oscillations at the gearbox, effectively negating one of the principal advantages of variable-speed operation, the ability to control the torque.

Although it may be possible to provide some damping mechanically, for example by means of appropriately designed rubber mounts or couplings, there is a cost associated with this. Another solution, which has been successfully adopted on many variable-speed turbines, is to modify the generator torque control to provide some damping. Instead of demanding a constant generator torque above rated, (or a torque varying slightly in inverse proportion to speed in the case of the constant power algorithm described in Section 8.3.3), a small ripple at the drive train frequency is added on to this basic torque demand, with the phase adjusted to counteract the effect of the resonance and effectively increase the damping. A high-pass or band-pass filter of the form

$$G\frac{2\zeta\omega s(1+s\tau)}{s^2+2\zeta\omega s+\omega^2} \tag{8.14}$$

(where G is a gain) acting on the measured generator speed can be used to generate this additional ripple. The frequency ω must be close to the resonant frequency which is to be damped. The time constant τ can sometimes be used to compensate for time lags in the system. A root locus plot (Section 8.4) is very useful for tuning the filter parameters.

Although a very effective filter can be made by tuning it to give a frequency response with a very broad peak (large ζ), this may be detrimental to the overall performance in that low-frequency variations in torque and power are then introduced. Even with a narrow peak, there can be sufficient response at multiples of blade-passing frequency such as $3P$ or $6P$ to disturb the system, in which case a notch filter (Section 8.4) can be cascaded with the filter of Equation (8.14). Of course if the resonant frequency nearly coincides with, say, $6P$ then the resonance will be very difficult to control because it will be strongly excited.

Figure 8.5 shows some simulation results for a variable-speed turbine operating in simulated three-dimensional turbulence. A large drive train resonance can be

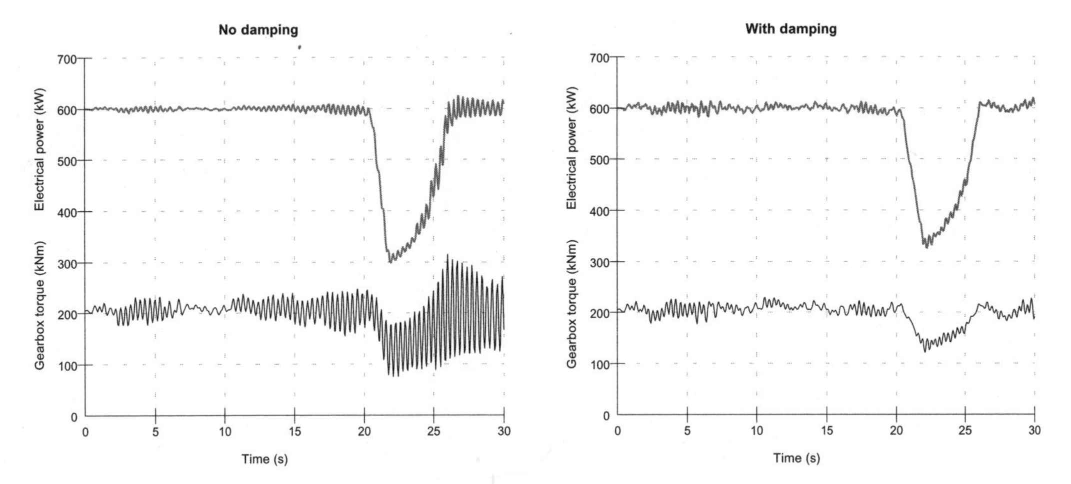

Figure 8.5 Effect of a Drive-Train Damping Filter

seen to be building up. Although the power and generator torque are smooth, the gearbox would be very badly affected. The effect of introducing a damping filter as described above is also shown. It almost completely damps out the resonance without increasing the electrical power variations. This is because the torque ripple needed to damp the resonance is actually very small, because the amount of excitation is small.

8.3.7 Variable-speed stall regulation

Figure 8.6 shows two power curves for the same rotor, one running as a 600 kW fixed-speed pitch-regulated turbine and one adjusted to run as a fixed-speed stall-

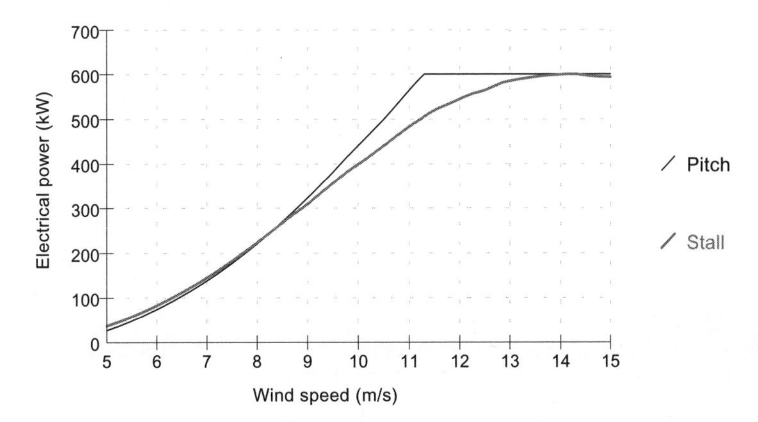

Figure 8.6 Comparison of Pitch and Stall Control

regulated turbine with the same rating. The rotational speed of the stall-regulated turbine has been reduced in order to limit the power to the same rated level. Therefore, although the stall-regulated turbine generates slightly more energy at very low wind speeds, as the blades approach stall above 8 m/s there is a large loss of output compared to the pitch-regulated machine. (In practice of course, if the turbine was designed to operate in stall, the blade design, solidity and rotor speed could be reoptimized, reducing this difference.)

By making use of variable speed, it is quite possible to correct this loss of energy by operating either turbine at the optimum tip speed ratio up to rated, or until the maximum r.p.m. is reached. At rated power, it is then possible to reduce the speed of the rotor to bring it into stall, although this has rarely been done to date on commercial machines. This can be done by closed-loop control of the generator torque in response to power error, allowing the turbine to follow exactly the same power curve as the pitch-regulated turbine. Thus the variable-speed stall-regulated turbine can achieve the same energy output as the variable-speed pitch-regulated turbine, but without the need for an active pitch mechanism. As explained in Section 8.2.2, however, significant torque and power transients will result from this strategy. The smooth torque and power, which are one of the main advantages of variable-speed systems, will therefore not be realized.

One simple and effective control algorithm for this case is illustrated in Figure 8.7. It consists of two nested loops, an outer power loop which demands a generator speed, and an inner speed loop which demands a generator torque. As in Section 8.3.5, a PI controller can be used for the inner loop. This is the same controller as for sections A1-B and C1-E of Figure 8.3, making it particularly easy to arrange the transition between control modes at the rated point since the inner loop is always active. A PI controller also works well for the outer loop.

8.3.8 Control of variable-slip turbines

The operating envelope for a variable-slip generator is shown in Figure 8.8. Below rated, the generator acts just like a conventional induction machine, with the torque related to the slip speed according to the slip curve AB. Once point B is reached, a resistor in series with the rotor circuit, previously short-circuited by a semiconductor switch, is progressively brought into play by switching the semiconductor

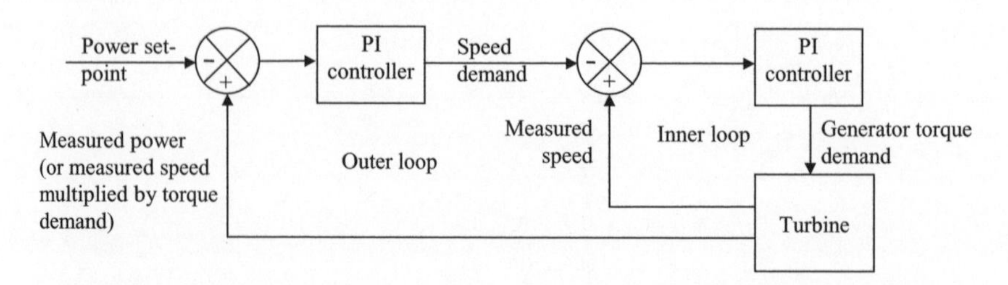

Figure 8.7 A Simple Control Algorithm for Variable-speed Stall-regulation

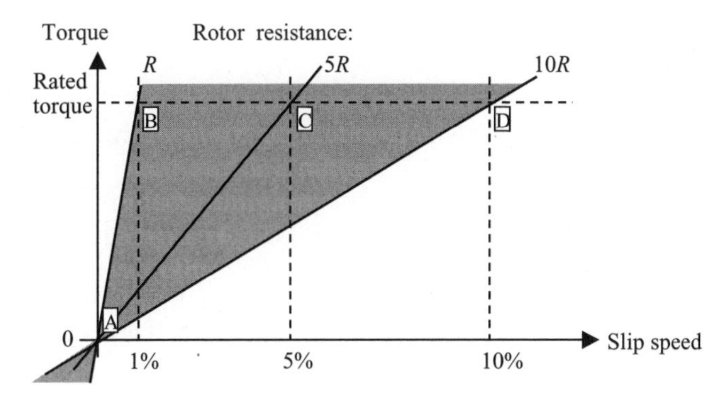

Figure 8.8 Operating Envelope for a Variable-slip Generator

switch on and off at several kHz, and varying the mark-space ratio to change the average resistance. As the average resistance increases, the generator slip curve changes so that its slope varies inversely with the total resistance of the rotor circuit. Figure 8.8 shows a typical example in which the rotor resistance can increase by a factor 10, changing the slip curve from AB to AD. By controlling the resistance therefore, the generator can operate anywhere within the shaded region. The resistance is usually varied by a closed-loop algorithm which seeks to regulate the torque to any desired value. For example, this might be a PI algorithm with torque error input and the mark-space ratio of the switch as output.

In practice, it is usual to keep the torque demand at the rated value. Then the generator will simply act as a conventional induction generator following the slip curve AB until rated torque is reached, at which point it will accelerate along the constant torque line BCD just like a variable-speed system. If the speed increases beyond D the torque will be forced to increase again. Pitch control is used to regulate the speed to a chosen set-point such as point C. The higher the speed C, the higher the mechanical power input for the same output power. Thus the power dissipated in the rotor circuit corresponds exactly to the slip. Therefore C should be chosen as low as possible to minimize the cooling requirements (as well as turbine loads which increase with speed). However, if C is too close to B then the torque will occasionally dip down the slope AB as the speed varies around the set point, causing power dips even when operating well above rated wind speed. How small the interval between B and C can be made depends on the rotor inertia and the responsiveness of the pitch control algorithm. As for a variable-speed system, the latter can be a PI or PID algorithm. It is possible to change the rate limits of the PID to force the pitch towards fine at maximum rate if the speed gets too close to B, or to feather at maximum rate if it gets too close to D.

As with a variable-speed system, it may be desirable to modify the torque demand as in Section 8.3.6 to control drive train torsional vibrations. However, in order to do this, it is necessary to be able to update the torque demand at relatively high frequency, at least five and preferably 10 times the drive train frequency which is typically of the order of 3–5 Hz.

8.3.9 Individual pitch control

Although individual or cyclic pitch control has been proposed many times, it has yet to find a place in commercial wind turbines. Asymmetrical loadings across the rotor are responsible for a significant contribution to fatigue loads, and in principle it should be possible to reduce these loads by controlling the pitch of each blade separately according to the conditions experienced by each blade. This may become particularly important for large wind turbines. However, in order to achieve any useful benefit, there must be some measurement available which can distinguish between the different blades, so that the controller can generate appropriate pitch demand signals for each.

The simplest measure which could be used is simply the rotor azimuth angle. Although in a turbulent wind the wind speed variations across the rotor are not particularly dependent on azimuth, there are some effects (wind shear, tower shadow, upflow and shaft tilt) which cause a systematic azimuth-dependent variation in the aerodynamic conditions at a point on the blade. In principle, the pitch of each blade could modified as a function of azimuth in order to reduce the loading variations caused by these effects, as long as the effects are constant. In practice, wind shear and upflow may vary significantly according to environmental conditions, although it may be possible to correlate these with wind direction. If the nacelle wind vane signal is also used, yaw misalignment can be added to the list.

In practice, however, it is very difficult to achieve any real gains in this way because of the dominance of stochastic variations due to turbulence. Also, especially for large rotors, the appropriate change in pitch will be different at different points along the blade, so the 'optimum' effect can never be achieved.

If it were possible to have some instantaneous measure of the asymmetrical loads, it might be possible to reduce them using individual pitch control. The use of strain gauges in the blade roots has been investigated by Caselitz *et al.*, (1997), although the unreliability of most strain gauges may make such an approach undesirable. Nevertheless, if a suitably reliable load measurement were available, the potential exists to reduce blade, hub, yaw bearing and tower fatigue loads significantly.

Taking the difference between the measured out-of-plane load on each blade and the mean of the signals from all three blades, an addition to the pitch demand for that blade can be calculated. Simulation studies have demonstrated that a reduction in fatigue loads on certain components of 10–20 percent is achievable with a relatively simple strategy of this sort, although the pitch rates required to achieve this were rather large, typically ± 10 deg/s or more on an almost continuous basis. As turbines grow larger, the pitch rates required will diminish, since the pitch action required is essentially at $1P$, the rotational frequency, and this frequency will decrease as rotor diameter increases.

This type of load reduction may also benefit from more advanced control design strategies such as the LQG technique described below, which may be able to achieve a better trade-off between load reduction and pitch activity.

Since there are significant possibilities for load reduction on large turbines using this approach, there may be a good case for research into more reliable sensors for measuring the blade root loads, or the resulting small deflections. Accelerometers in the blade tips are a possibility, but the difficulty of access for maintenance is a significant drawback.

Another theoretical possibility for individual pitch control is for actually generating yawing loads in response to measured yaw misalignment, in order to keep the turbine pointing into the wind without the use of a yaw motor. However, it is unlikely that the yaw motor can be dispensed with completely, as it will probably be needed to yaw the nacelle while the rotor is not turning, at least for cable unwinds, etc. This potential application for cyclic pitch control remains rather speculative.

The use of the azimuth signal to pitch each blade as it approaches bottom dead centre in order to reduce the risk of it striking the tower has been suggested, but to rely on this to avoid such a catastrophic event seems risky.

8.4 Closed-loop Control: Analytical Design Methods

Clearly the choice of controller gains is crucial to the performance of the controller. With too little overall gain, the turbine will wander around the set point, while too much gain can make the system completely unstable. Inappropriate combinations of gains can cause structural responses to become excited. This section outlines some of the techniques which have been found to be useful in designing closed-loop control algorithms for wind turbines, such as the gains of a PI or PID controller for example. Clearly it is only appropriate here to give some useful hints and pointers. There are many standard texts on control theory and controller design methods, to which the reader should refer for more detailed information, for example D'Azzo and Houpis (1981), Anderson and Moore (1979), and Astrom and Wittenmark (1990).

8.4.1 Classical design methods

A linearized model of the turbine dynamics is an essential starting point for controller design. This allows various techniques to be used for rapidly evaluating the performance and stability of the control algorithm. Detailed non-linear simulations using a three-dimensional turbulent wind input should then be used to verify the design before it is implemented on the real turbine.

For a variable-speed turbine below rated wind speed, a PI speed controller using demanded torque can be quite slow and gentle, and the linearized model can be very simple. It must include the rotational dynamics of the drive train, but other dynamics are not usually important. For pitch control, however, the aerodynamics of the rotor and some of the structural dynamics can be critical. The linearized model for pitch controller design should contain at least the following dynamics:

- rotor and generator rotation,
- tower fore-aft vibration,

- power or speed transducer response,
- pitch actuator response.

The generator characteristics are also necessary for fixed-speed systems, and drive train torsion is particularly important for variable-speed turbines. In all cases a linearized description of the aerodynamics of the rotor is required, for example as a set of partial derivatives of torque and thrust with respect to pitch angle, wind speed and rotor speed. The thrust is important as it affects the tower dynamics, which couple strongly with pitch control.

A typical linear model is shown in Figure 8.9. With such a linear model, it is then possible to vary the gains and other parameters, and then rapidly carry out a number of tests which help to evaluate the performance of the controller with those gain settings. Some of these tests are open-loop tests, which means they are applied to the open-loop system obtained by breaking the feedback loop, for example at the symbol **X** in Figure 8.9. Other tests are carried out on the closed-loop system. Before describing some of these tests, some basic theory on open and closed-loop dynamics is outlined.

Figure 8.10 shows a simplified general model in which the turbine (i.e., from pitch actuator to transducer in Figure 8.9) is represented by the 'plant model' with transfer function $G(s)$, and the control algorithm is represented by the controller transfer function $kC(s)$, where s is the laplace variable and k an overall controller gain.

Now the open-loop system can be represented by the transfer function $H(s) = kC(s)G(s)$. If the input to the transfer function is denoted x and the output is

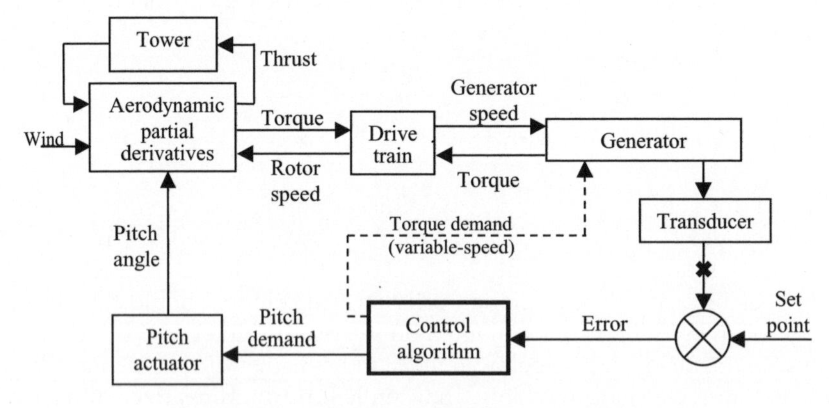

Figure 8.9 Typical Linearized Turbine Model

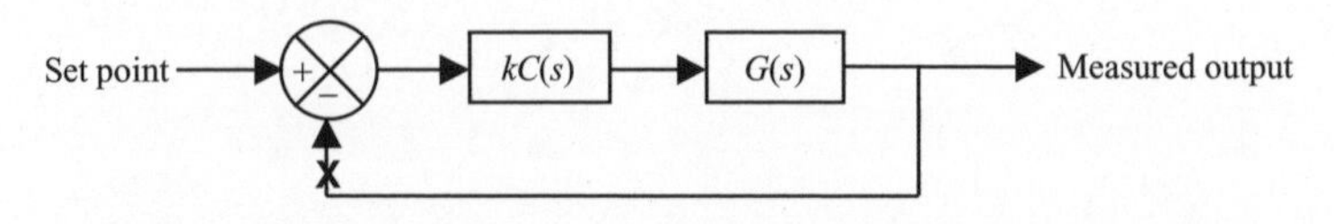

Figure 8.10 Simplified General Model of Plant and Controller

y, then $Y(s) = kH(s)X(s)$, where $X(s)$ and $Y(s)$ are the Laplace transforms of x and y. When the loop is closed at **X** the closed-loop dynamics can be derived as

$$Y'(s) = kH(s)(X(s) - Y'(s)) \tag{8.15}$$

where $Y'(s)$ is the Laplace transform of the closed-loop output. This can be rewritten as

$$Y'(s) = \frac{kH(s)}{1 + kH(s)} X(s) \tag{8.16}$$

In other words if the open-loop system is $H(s)$, the closed-loop system will have dynamics represented by $H'(s) = kH(s)/(1 + kH(s))$.

Now a linear transfer function can be expressed as the ratio of two polynomials in s. Thus for the open-loop system, $A(s)Y(s) = B(s)X(s)$, and so $H(s) = B(s)/A(s)$, where $A(s)$ and $B(s)$ are polynomials in s. The roots of the polynomial $A(s)$ give important information about the system response. Consider, for example, a first-order system

$$\tau \dot{y}_1 = x - y_1 \tag{8.17}$$

representing a first-order lagged response of y_1 with respect to x. This system can be represented by the transfer function

$$H(s) = \frac{B(s)}{A(s)} \tag{8.18}$$

where $B(s) = 1$ and $A(s) = 1 + \tau s$.

The single root of $A(s)$ is given by $\sigma = -1/\tau$, while Equation (8.17) has solutions of the form $y = a + be^{\sigma t}$, with $\sigma = -1/\tau$ again. These solutions are stable if τ is positive, in other words if the root of $A(s)$ is negative. A second-order system will have solutions of the form $y = a + be^{\sigma_1 t} + ce^{\sigma_2 t}$, where once again σ_1 and σ_2 are the roots of the second-order polynomial which forms the denominator of the transfer function. Now σ_1 and σ_2 may be real numbers or they may form a complex conjugate pair $\sigma \pm j\omega$. The solutions are stable if σ_1 and σ_2 are both negative, or if σ is negative. In general, it can be stated that a linear system is stable if all the roots of the denominator polynomial have negative real parts. These roots are known as the *poles* of the system, and they represent values of the Laplace variable which make the transfer function infinite. The roots of the numerator polynomial are known as the *zeros* of the system, since transfer function is zero at these points.

Now let us rewrite Equation (8.16) in terms of the polynomials A and B:

$$Y'(s) = \frac{kB(s)}{A(s) + kB(s)} X(s) \tag{8.19}$$

Clearly when the gain k is small, the closed-loop transfer function tends towards the open-loop transfer function kB/A. However, when the gain is large the poles will tend towards the roots of B. In other words as the gain increases from zero to infinity, the poles of the closed-loop system move from the open-loop poles and end up at the open-loop zeros. They move along complicated trajectories in the complex plane. A plot of these trajectories is known as a root locus plot, and is very useful for helping to select the feedback gain k. The gain is selected such that all the closed-loop poles are in the left half-plane, making the system stable, and preferably as well-damped as possible. The damping factor for a pole pair at $\sigma \pm j\omega = re^{j\theta}$ is given by $-\cos(\theta) = -\sigma/r$, as shown in Figure 8.11.

Figure 8.12 shows an example of a root locus plot for a variable-speed pitch controller. As the gain increases, the closed-loop poles (+) move from the open-loop poles (x), corresponding to zero feedback gain, to the open-loop zeros (O). (Actually there are usually more poles than zeros; the 'missing' zeros can be considered to be equally spaced around a circle of infinite radius.) In this example, the gain has been chosen to maximize the damping of the lightly-damped tower poles (B). Any further increase in gain would exacerbate tower vibration, eventually leading to

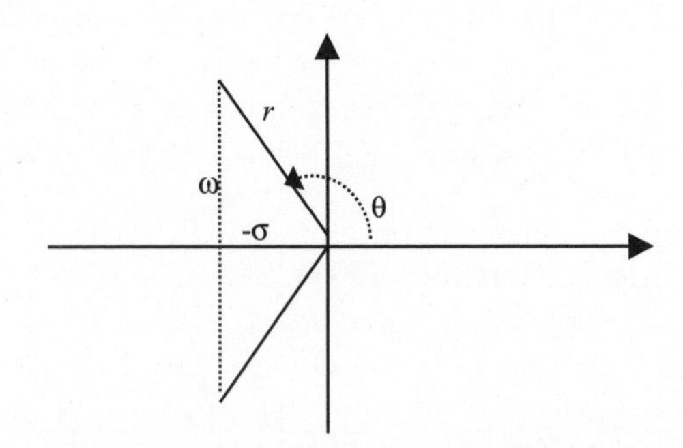

Figure 8.11 Damping Ratio for a Complex Pole Pair

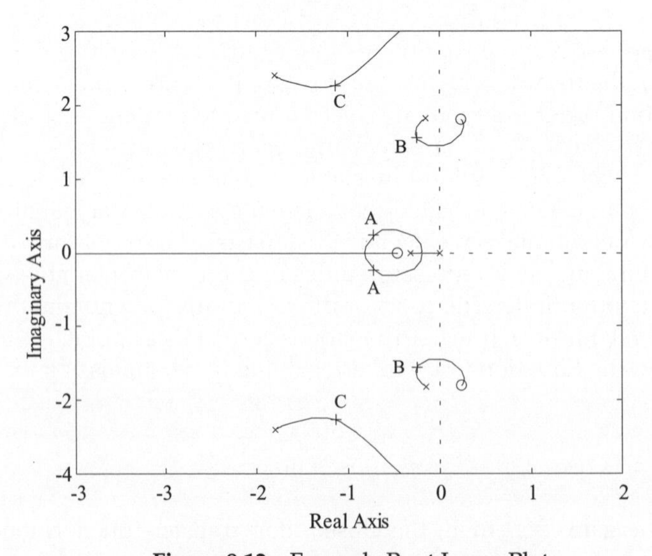

Figure 8.12 Example Root Locus Plot

instability as the poles cross the imaginary axis. At the chosen gain, the controller poles (A) are well-damped. The poles at (C) result from the pitch actuator dynamics. They remain sufficiently well-damped, although again, excessive gain would drive them to instability.

Although a root locus plot is useful for helping to select the overall controller gain, this can only be done once the other parameters defining the controller have been fixed. A PI controller (Equation (8.1) with $K_d = 0$) is characterized by only two parameters, K_P and K_i. It can be re-written as

$$y = K_p\left(1 + \frac{1}{sT_i}\right)x \tag{8.20}$$

where $T_i = K_P/K_i$ is known as the integral time constant. The root locus plot can be used to select K_P once T_i has been defined, but the shape of the loci will change with different T_i. However, it is relatively straightforward to iterate on the value of T_i, using the root locus plot each time to select K_P, until a suitable overall performance is achieved, using criteria such as those listed below. In the case of PID and more complex controllers, where more than two parameters must be selected, other ways must be found to select the parameters, although it is always possible to use a root locus plot for the final choice of the overall gain.

The choice of parameters will usually be an iterative process, often using a certain amount of trial and error, and on each iteration the performance of the resulting controller must be assessed. Useful measures of performance include the following.

- Gain and phase margins which are calculated from the open-loop frequency response, and give an indication of how close the system is to instability. If the margins are too narrow, the system may tend to become unstable. The system will be unstable if the open-loop system displays a 180° phase lag with unity gain. The phase margin represents the difference between the actual phase lag and 180° at the point where the open-loop gain crosses unity. A phase margin of at least 45° is usually recommended, although there is no firm rule. Similarly, the gain margin represents the amount by which the open-loop gain is less than unity where the open-loop phase lag crosses 180°. A gain margin of at least a few decibels is recommended.

- The cross-over frequency, which is the frequency at which the open-loop gain crosses unity, gives a useful measure of the responsiveness of the controller.

- The positions of the closed-loop poles of the system indicate how well various resonances will be damped.

- Closed-loop step responses, for example the response of the system to a step change in wind speed, give a useful indication of the effectiveness of the controller. For example, in tuning a pitch controller, the rotor speed and power excursions should return rapidly and smoothly to zero, the tower vibration should damp out reasonably fast, and the pitch angle should change smoothly to its new value, without too much overshoot and without too much oscillation.

- Frequency responses of the closed-loop system also give some very useful indications. For example, in the case of pitch controller:
  - the frequency response from wind speed to rotor speed or electrical power should die away at low frequencies, as the low frequency wind variations are controlled away;
  - the frequency response from wind speed to pitch angle must die away at high frequencies, and must not be too great at critical disturbance frequencies such as the blade-passing frequency, or the drive train resonant frequency in variable-speed systems;
  - the frequency response from wind speed to tower velocity should not have too large a peak at the tower resonant frequency,

  and so forth.

With experience it is possible, by examining measures such as these, to converge rapidly on a controller tuning which will work well in practice.

8.4.2 Gain scheduling for pitch controllers

Close to rated, since the fine pitch angle is selected to maximize power, it follows that the sensitivity of aerodynamic torque to pitch angle is very small. Thus a much larger controller gain is required here than at higher wind speeds, where a small change in pitch can have a large effect on torque. Frequently the torque sensitivity changes almost linearly with pitch angle, and so can be compensated for by varying the overall gain of the controller linearly in inverse proportion to the pitch angle. Such a modification of gain with operating point is termed a 'gain schedule'. However, the sensitivity of thrust to pitch angle varies in a different way, and because of its effect on tower dynamics, which couples strongly with the pitch controller, it may be necessary to modify the gain schedule further to ensure good performance in all winds.

It is therfore important to generate linearized models of the system corresponding to several different operating points between rated and cut-out wind speed, and to choose a gain schedule which ensures that the above performance measures are satisfactory over the whole range.

8.4.3 Adding more terms to the controller

It is often possible to improve the performance of a basic PI or PID controller by adding extra terms to modify the behaviour in a particular frequency range. For example, a pitch control algorithm may be found to cause a large amount of pitch actuator activity at a relatively high frequency, which is of little benefit in controlling the turbine and may be quite counter-productive. This may occur if some dynamic mode was not taken into account in the linearized model which was used to design the turbine. An example of this is the drive train torsional resonance in a variable-speed turbine, which can feed through to the measured generator speed and hence to the pitch control, causing high-frequency pitch activity which is of no benefit. Another likely cause is the pitch response to a major external forcing frequency, such as the blade-passing frequency. While a low pass filter in series with the controller will certainly reduce high-frequency response, the resulting phase shift at lower frequencies may significantly impair the overall performance of the controller. A better 'cure' for excessive activity at some well-defined frequency is to include a notch filter in series with the controller. A simple second-order notch filter tuned to filter out a particular frequency of ω rad/s has a transfer function

$$\frac{s^2+\omega^2}{s^2+2\zeta\omega s+\omega^2} \tag{8.21}$$

where the 'damping' parameter ζ represents the width or 'strength' of the notch filter. This should be increased until the filtering effect is sufficient at the target frequency, without too much detriment to the control performance at lower frequencies.

Another useful filter is the phase advance or phase lag filter,

$$\frac{(s+\omega_1)}{(s+\omega_2)} \tag{8.22}$$

which increases the open-loop phase lag between frequencies ω_1 and ω_2 ($\omega_1 < \omega_2$), or decreases it if $\omega_1 > \omega_2$. Phase advance can sometimes be useful for improving the stability margins. Open-loop gain and phase plots can therefore be useful for helping to select ω_1 and ω_2. A PID controller can be rewritten as a PI controller in series with a phase advance (or phase lag) filter.

A general second-order filter of the form

$$\frac{s^2+2\zeta_1\omega_1 s+\omega_1^2}{s^2+2\zeta_2\omega_2 s+\omega_2^2} \tag{8.23}$$

can sometimes be useful for modifying the frequency response in a particular area. With $\omega_1=\omega_2$ and $\zeta_1=0$ this is just a notch filter, as described above. With $\zeta_1 > \zeta_2$ the filter has a bandpass effect, which can be used to increase control action at a particular frequency.

A root locus plot is often useful for investigating the effect of such filters. With experience, the effect on the loci of placing the filter poles and zeros in particular ways can be anticipated. Such techniques can help to see how, for example, a pair of lightly-damped poles due to a structural resonance can be dragged further away from the imaginary axis, so as to increase the damping.

8.4.4 Other extensions to classical controllers

Other extensions to classical controllers have sometimes been used in order to further improve the performance in particular ways, for example the use of non-linear gains, and variable or asymmetrical limits.

Non-linear gains are sometimes used to penalize large peaks or excursions in controlled variables. For example, the gain of a PI pitch controller can be increased as the power or speed error increases. A simple way to do this is to add to the input signal to the PI controller a term proportional to the square or cube of the error (remembering to adjust the sign if the square is used). This technique should be used with caution, however, as too much non-linearity will drive the system towards instability, in much the same way as if the linear gain is too high. This technique requires a trial-and-error approach since it is very difficult to analyse the closed-loop behaviour of non-linear systems using standard methods. Adding the non-linear term only when the power or speed is above the set-point will help to reduce peaks, but will also cause a reduction in the mean power or speed, similar to a reduction in set-point.

Asymmetrical pitch rate limits can also be used to reduce peaks. By allowing the blades to pitch faster towards feather than towards fine, power or speed peaks will be reduced. Once again the mean level will also be reduced by introducing this asymmetry. However, this technique is somewhat more 'comfortable' than the use of non-linear gains, in that it remains a linear system constrained by limits.

There is often a desire to reduce the set-point in high winds, to reduce the infrequent but highly-damaging loads experienced in those conditions at the expense of a small loss of output. It is straightforward to reduce the set-point as a function of wind speed (the pitch angle is usually used as a measure of the rotor-averaged wind speed, as for gain scheduling). However, the most damaging loads occur during high turbulence, and so it would be better to reduce the set-point in high winds only when the turbulence is also high. Rather than actually reducing the set-point, asymmetrical rate limits provide a simple but effective means of achieving this effect, since the rate limits will only 'bite' when the turbulence is high.

A further extension of this technique is to modify the rate limits dynamically, even to the extent of changing the sign of a rate limit in order to force the pitch in one direction during certain conditions such as large power or speed excursions. A useful application of this is in the control of variable slip systems, where it is important to keep the speed above the minimum slip point (point B in Figure 8.8). If the speed falls below this point, it then ceases to vary much as it is constrained by the minimum slip curve, and so the proportional term in the PI controller ceases to respond. Modifying the rate limits as a function of speed error as in Figure 8.13 is a useful technique to prevent this happening.

8.4.5 Optimal feedback methods

The controller design methods described above are based on classical design techniques, and often result in relatively simple PI or PID algorithms together with

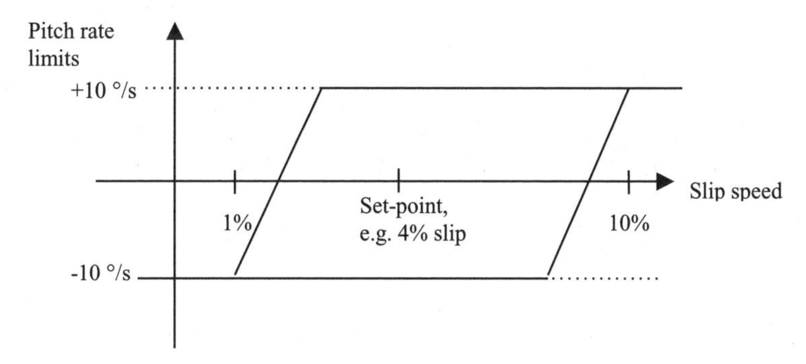

Figure 8.13 Pitch Rate Limit Modification for a Variable-slip Controller

various filters in series or in parallel, such as phase shift, notch or bandpass filters, and sometimes using additional sensor inputs. These methods can be used to design fairly complex high-order controllers, but only with a considerable amount of experience on the part of the designer.

There is, however, a huge body of theory (and practice, although to a lesser extent) relating to more advanced controller design methods, some of which have been investigated to some extent in the context of wind-turbine control, for example:

- self-tuning controllers,
- LQG/optimal feedback and H_∞ control methods,
- fuzzy logic controllers,
- neural network methods.

Self-tuning controllers (Clarke and Gawthrop, 1975) are generally fixed-order controllers defined by a set of coefficients, which are based on an empirical linear model of the system. The model is used to make predictions of the sensor measurements, and the prediction errors are used to update the coefficients of the model and the feedback law.

If the system dynamics are known, then some very similar mathematical theory can be used, but applied in a different way. Rather than fitting an empirical model, a linearized physical model is used to predict sensor outputs, and the prediction errors are used to update estimates of the system state variables. These variables may include rotational speeds, torques, deflections, etc. as well as the actual wind speed, and so their values can be used to calculate appropriate control actions even though those particular variables are not actually measured.

Observers

A subset of the known dynamics may be used to make estimates of a particular variable: for example, some controllers use a wind-speed observer to estimate the wind speed seen by the rotor from the measured power and/or rotational speed and the pitch angle. The estimated wind speed can then be used to define the appropriate desired pitch angle.

State estimators

Alternatively, using a full model of the dynamics, a Kalman filter can be used to estimate all the system states from the prediction errors (Bossanyi, 1987). This technique can explicitly use knowledge of the variance of any stochastic contributions to the dynamics, as well as noise on the measured signals, in a mathematically optimum way to generate the best estimates of the states. This relies on an assumption of Gaussian characteristics for the stochastic inputs. Thus it is possible explicitly to take account of the stochastic nature of the wind input by formulating a wind model driven by a Gaussian input. This can even be extended to include blade-passing effects.

The Kalman filter can readily take account of more than one sensor input in generating its 'optimal' state estimates. Thus it is ideal for making use of, for example, an accelerometer measuring tower fore-aft motion as well as the normal power or speed transducer. It would be straightforward to add other sensors, if available, to improve the state estimates further.

Optimal feedback

Knowing the state estimates, it is then possible to define a cost function, which is a function of the system states and control actions. The controller objective can then be defined mathematically: the objective is to minimize the selected cost function. If the cost function is defined as a quadratic function of the states and control actions (which is actually a rather convenient formulation), then it is relatively straightforward to calculate the 'optimal' feedback law. This is defined as the feedback law which generates control signals as a linear combination of the states such that the cost function will be minimized. Since a *L*inear model is required, with a *Q*uadratic cost function and *G*aussian disturbances, this is known as an LQG controller.

This cost function approach means that the trade-off between a number of partially competing objectives is explicitly defined, by selecting suitable weights for the terms of the cost function. This makes such a method ideal for a controller which attempts to reduce loads as well as achieving its primary function of regulating power or speed. Although it is not practical to calculate the weightings in the cost function rigorously, they can be adjusted in a very intuitive way. This approach is also readily configured for multiple inputs and outputs, so for example, as well as using generator speed and tower acceleration inputs, it can in principle simultaneously produce the pitch demand and torque demand outputs which will minimize the cost function.

Figure 8.14 illustrates the structure of the LQG controller, showing the state estimator and the optimal state feedback. For implementation, the entire controller can be reduced to a set of difference equations connecting the measured outputs (y)

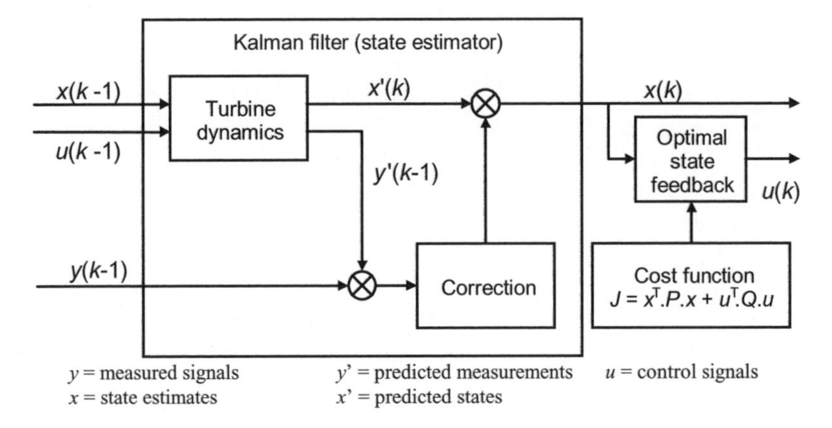

Figure 8.14 Structure of the LQG Controller

to the new control signals (u). This means that once the design is completed, the algorithm is easy to implement and does not require massive processing power.

The linearized dynamics of the system are expressed in discrete state-space form:

$$x'(k) = Ax(k-1) + Bu(k-1)$$

The Kalman gain L is calculated taking into account the stochastic disturbances affecting the system, and allows the state estimates to be improved by comparing the predicted sensor outputs y' to the actual outputs y:

$$x(k) = x'(k) + L(y(k-1) - y'(k-1))$$

where

$$y'(k-1) = Cx(k-1) + Du(k-1)$$

The optimal state feedback gain K generates the controls actions

$$u(k) = -Kx(k)$$

where K is calculated such that the quadratic cost function J is minimized. The cost function is

$$J = x^{\mathrm{T}}Px + u^{\mathrm{T}}Qu$$

(actually the integral, or the mean value over time, or the expected value of this quantity). P and Q are the state and control weighting matrices. It is usually more useful to define the cost function in terms of other quantities, v, which can be considered as extra (often un-measured) outputs of the system:

$$v = C_v\mathrm{x} + D_v u$$

Hence the cost function is

$$J = v^T Rv + u^T Su = x^T C_v^T RC_v x + u^T D_v^T RD_v u + u^T Su$$

so that $P = C_v^T RC_v$ and $Q = D_v^T RD_v + S$.

Another possibility is to generate optimal control signals directly as a function of the sensor outputs. This is known as optimal output feedback (Steinbuch, 1989). However, the mathematical solution of this problem is based on *necessary* conditions for optimality which are not always *sufficient* for optimality. Therefore the solutions generated can be, and in practice often are, non-optimal and potentially very far from optimal. This variation is therefore rather problematic.

As turbines become larger and the requirements placed on the controller become more demanding, advanced control methods such as LQG are likely to become increasingly used, although as yet there are few known examples of the practical application of these techniques in commercial wind-turbines. However, this approach was used to design a controller for a 300 kW fixed-speed two-bladed teetered turbine in the UK in 1992. After testing on a prototype turbine in the field, this controller was shown to give significant reductions in pitch activity and power excursions compared to the original PI controller, and it was subsequently adopted for the production machine and successfully used on over 70 turbines (Bossanyi, 2000).

LQG controllers are not necessarily robust, which means that they can be sensitive to errors in the turbine model. A similar approach is the H_∞ controller, in which uncertainties in the turbine and wind models can explicitly be taken into account. Such a controller was evaluated in the field on a 400 kW fixed-speed pitch-regulated turbine by Knudsen, Andersen and Töffner-Clausen (1997), who reported a reduction in pitch activity and some potential for reduced fatigue loads compared to a PI controller.

8.4.6 Other methods

Rule-based or 'fuzzy logic' controllers are useful when the system dynamics are not well known or when they contain important non-linearities. Control actions are calculated by weighting the outcomes of a set of rules applied to the measured signals. Although there has been some work on fuzzy controllers for wind-turbines, there is no clear evidence of benefits. In practice, quite a good knowledge of the system dynamics is usually available, and the dynamics can reasonably be linearized at each operating point, so there is no clear motivation for such an approach.

The same could be said of controllers based on neural networks. These are learning algorithms, which are 'trained' to generate suitable control actions using a particular set of conditions, and then allowed to use their learnt behaviour as a general control algorithm. While this is potentially a powerful technique, it is difficult to be sure that such a controller will generate acceptable control actions in all circumstances.

Nevertheless, there may be some potential for such methods where significant non-linearities or non-stationary dynamics are involved. These might be in the turbine itself (stall hysteresis might be one example), in the driving disturbances (the wind characteristics are not stationary), or in the controller objectives. For example, the controller objectives might change around rated wind speed, or non-linear effects such as fatigue damage might be included in the cost function.

8.5 Pitch Actuators (See also Section 6.7.2)

An important part of the control system of a pitch-controlled turbine is the pitch actuation system. Both hydraulic and electric actuators are commonly used, each type having its own particular advantages and disadvantages which should be considered at the design stage.

Smaller machines often have a single pitch actuator to control all the blades simultaneously, although there is an increasing trend to use individual pitch actuators for each blade on larger turbines. This has the advantage that it is then possible to dispense with the large and expensive shaft brake which would otherwise be needed. This is because of the requirement for a turbine to have at least two independent braking systems capable of bringing the turbine from full load to a safe state in the event of a failure. Provided the individual pitch actuators can be made independently fail-safe, and as long as the aerodynamic braking torque is always sufficient to slow the rotor down to a safe speed even if one pitch actuator has failed at the working pitch angle, then multiple actuators may be considered to be independent braking systems for this purpose. There may still be a need for a parking brake, at least for the use of maintenance crews, but this may then be fairly small. It must at least be capable of bringing the rotor to a complete stop from a low speed, not necessarily in high or extreme wind speeds, for long enough to allow a rotor lock to be inserted.

A collective pitch actuation system commonly consists of an electric or hydraulic actuator in the nacelle, driving a push-rod which passes through the centre of the gearbox and hollow main shaft. The push-rod is attached to the pitchable blade roots through mechanical linkages in the hub. The actuator in the nacelle is often a simple hydraulic cylinder and piston. A charged hydraulic accumulator ensures that the blades can always be feathered even if the hydraulic pump loses power. An alternative arrangement is to use an electric servo motor to drive a ballnut which engages with a ballscrew on the push-rod. Since the push-rod turns with the rotor, loss of power to the motor causes the ballscrew to wind the pitch to feather, giving fail-safe pitch action. This requires a fail-safe brake on servo motor to ensure that the ballnut stops turning if power is lost.

Individual pitch control requires separate actuators in the hub for each blade. Therefore there must be some means of transmitting power to the rotating hub to drive the actuators. This can be achieved by means of slip rings in the case of electric actuators, or a rotary hydraulic joint for hydraulic actuators. A rotary transformer can also be used to transmit electrical power to the hub without the inconvenience of slip rings, which require maintenance.

The need to ensure a backup power supply on the hub to enable the blades to pitch even in the event of power loss can be a source of problems. A hydraulic system needs an accumulator for each blade, while electric actuators usually have battery packs in the hub for this purpose. Such battery packs are large, heavy and expensive, and alternative methods such as the use of hub-mounted generators, which can always generate power as long as the hub is turning, have been proposed. If a battery is used, the actuator motors must either be DC motors or (more commonly) AC motors with a frequency converter, with the batteries on the DC link. Since this will form part of the safety system, the reliability of the inverter between the DC link and the pitch motor is important. A hub-mounted generator would produce either DC or variable frequency AC, and once again the reliability of the connection to the pitch motor is important. Since the pitch actuators have to be independently fail-safe, separate battery packs or generators and frequency converters etc. must be provided for each blade.

The friction in the pitch bearing is often a significant factor in the design of the pitch actuation system. The bearing friction depends on the loading applied to the bearing, and the large overturning moment acting on the bearing can lead to very high levels of friction.

A hydraulic actuator would usually be controlled by means of a proportional valve controlling the flow of oil to the actuator cylinder. The valve opening, and hence the oil flow rate, would be set in proportion to the required pitch rate. The demanded pitch rate may come directly from the turbine controller, or it might come from a pitch-position feedback loop. In this case the turbine controller generates a pitch-position demand. This is compared to the measured pitch position, and the pitch-position error is turned into a pitch-rate demand through a fast PI or PID control loop, implemented either digitally or by means of a simple analogue circuit.

In the case of an electric actuator, the motor controller usually requires a torque demand signal. This may be derived from a speed controller, which uses a fast PI or PID controller acting on speed error to generate a torque demand. Once again the speed demand may come directly from the turbine controller or from a position feedback loop.

Simpler actuators are sometimes used, particularly when a fast pitch response is not so important, as in the case of a turbine which is controlled by pitching to stall rather than to feather. In this case an actuator which merely pitches at a fixed rate in either direction may be adequate.

8.6 Control System Implementation

Previous sections have explained some of the techniques whereby control algorithms can be designed. The system and controller dynamics have been described in continuous time in terms of the Laplace operator, s. While it is possible to implement a continuous-time controller, for example using analogue circuitry, the use of digital controllers is now almost universal. The greater flexibility of digital systems is a factor here: simply by making software changes, the control logic can be changed completely.

A consequence of using digital control is that the control actions are calculated and updated on a discrete time step, rather than in continuous time. Control algorithms designed in continuous time must therefore be converted to discrete time for implementation in a digital controller. It is also possible to design controllers in discrete time, if the linearized model of the turbine is first discretized.

The following sections briefly describe some of the practical issues involved in implementing a control algorithm in a real digital controller. Once again, the reader is referred to standard control theory texts for more detailed treatments.

8.6.1 Discretization

Supposing a control algorithm has been designed in continuous time as a transfer function (such as Equation (8.1) for a PID controller, for example), it must be discretized before it can be implemented in a digital controller. Discretized transfer functions are usually represented in terms of the delay operator, z, where $z^{-k}x$ represents the value of x sampled k timesteps ago. As a simple example, a moving average or lag filter from x to y is often implemented as

$$y_k = Fy_{k-1} + (1 - F)x_k$$

This is a difference equation which can readily be implemented in code in a discrete controller. In terms of the delay operator, it can be written as

$$(1 - Fz^{-1})y = (1 - F)x$$

or alternatively as a transfer function consisting of a ratio of polynomials in z^{-1}:

$$y = \frac{(1 - F)}{(1 - Fz^{-1})}x$$

Now the Laplace operator can be considered as a differentiation operator, and so as a simple approximation, it might be possible to convert a continuous transfer function into discrete form by replacing s by $(1 - z^{-1})/T$, where T is the timestep.

In fact by simple algebraic manipulation, it is straightforward to show that with this substitution, the above discrete transfer function is in fact equivalent to the continuous transfer function representation of a first-order lag with time constant τ, namely

$$y = \frac{1}{1 + s\tau}x$$

with the factor F being given by $\tau/(T + \tau)$.

Clearly any discretized equation can only be an approximation to the continuous-time behaviour. There are other discretization methods, and the so-called bilinear or 'Tustin' approximation often works better in practice. In this case the Laplace operator is replaced by

$$\frac{2}{T}\frac{(1 - z^{-1})}{(1 + z^{-1})}$$

Discretization results in a phase shift compared to the continuous-time process. This phase shift increases with frequency. If the algorithm performance is particularly sensitive to the phase shift at a certain frequency, then the discretization can be 'pre-warped' to this frequency. Pre-warping modifies the phase shift so that the phase of the discrete transfer function is correct at the chosen frequency, but deviates at lower and higher frequencies. An example of a situation where pre-warping may be important is in the case of a drive train resonance damper in a variable-speed turbine (Section 8.3.6). The resonant frequency which is being targeted is usually fairly high, typically around 4 Hz, and unless the controller timestep is very short the phase lag caused by discretization may significantly affect the performance of the damping algorithm.

The approximation for s used for discretization with pre-warping about a frequency ω is

$$\frac{\omega}{\tan(\omega T/2)}\frac{(1 - z^{-1})}{(1 + z^{-1})}$$

8.6.2 Integrator desaturation

Controllers containing integral terms, such as PI or PID controllers, experience a particular problem known as integrator wind-up when the control action saturates at a limiting value. A common example is in pitch control, where the pitch angle is limited to the fine pitch position when the wind is below rated. For example, a PI power controller for a fixed-speed turbine can be represented as:

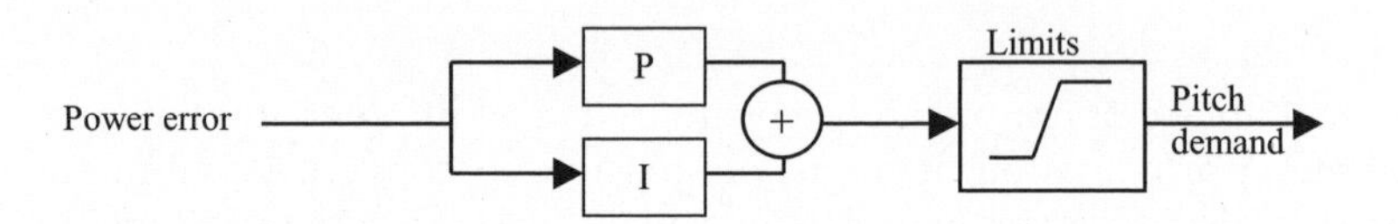

Above rated, the power error will be zero on average because of the integral term. Below rated, the pitch saturates at the fine pitch position, and the power error will remain negative. The integral of the error will therefore grow more and more negative, and only the application of the limits prevents the actual pitch demand from doing the same. However, when the wind speed reaches rated again and the power error becomes positive, it will take a long time before the integrated power error climbs back up to zero and starts to demand a positive pitch angle. To prevent this problem of integrator wind-up, the integral term must be prevented from integrating when the pitch is at the limit. This is easily implemented by separating out the integrator, $I(z)$, from the rest of the controller, $R(z)$. $R(z)$ generates a change in demanded pitch angle, and $I(z)$ then integrates this by adding it to the previous pitch demand *after* the limits have been applied.

As an example, a PI controller (Equation 8.20) discretized using the bilinear approximation would be:

$$y = K_p[(T/2T_i + 1) + (T/2T_i - 1)z^{-1}]\frac{1}{[1 - z^{-1}]}x = R(z)I(z)x$$

To avoid integrator wind-up, this can be implemented as follows:

$\Delta y_k = K_P[(T/2T_i + 1)x_k + (T/2T_i - 1)x_{k-1}]$ (implementation of $R(z)$)
$y_k^* = y_{k-1} + \Delta y_k$ (integrator $I(z)$ using previous *limited* output)
$y_k = \lim(y_k^*)$ (application of limits)

Higher-order controllers may need more sophisticated desaturation techniques.

References

Anderson, B. D. O. and More, J. B., (1979). *Optimal filtering*. Prentice Hall, New York, USA.

Astrom, K. J. and Wittenmark, B., (1990). *Computer controlled systems*. Prentice-Hall, New York, USA.

Bossanyi, E. A., (1987). 'Adaptive pitch control for a 250 kW wind-turbine'. *Proceedings of the Ninth BWEA Conference*. Mechanical Engineering Publications, Bury St Edmunds, UK.

Bossanyi, E. A. and Gamble, C. R., (1991). 'Investigation of torque control using a variable slip induction generator'. *ETSU WN-6018*, Energy Technology Support Unit, Harwell, UK.

Bossanyi, E. A., (1994). 'Electrical aspects of variable-speed operation of horizontal axis wind turbine generators'. *ETSU W/33/00221/REP*, Energy Technology Support Unit, Harwell, UK.

Bossanyi, E. A., (2000). 'Developments in closed-loop controller design for wind-turbines'. *Proceedings of the ASME Wind Energy Symposium*, AIAA/ASME.

Caselitz, P. *et al.*, (1997). 'Reduction of fatigue loads on wind-energy converters by advanced control methods'. *Proceedings of the European Wind Energy Conference*, pp. 555–558.

Clarke. D. and Gawthrop, P., (1975) 'Self-tuning controller'. *Proc. IEE*, **122**, 9, 929–934.

D'Azzo, J. J. and Houpis, C. H., (1981). *Linear control system analysis and design*. McGraw-Hill, New York, USA.

Holley, W., Rock, S. and Chaney, K., (1999). 'Control of variable speed wind turbines below rated wind speed'. *Proceedings of the Third ASME/JSME Conference (FEDSM99-5295–16)*.

Knudsen, T., Andersen, P. and Töffner-Clausen, S., (1997). 'Comparing PI and robust pitch controllers on a 400 kW wind turbine by full-scale tests'. *Proceedings of the European Wind Energy Conference*, pp. 546–550.

Pedersen, T. K., (1995). Semi-variable speed – a compromise? *Proceedings of the Wind Energy Conversion 17th British Wind Energy Association Conference*, pp. 249–260. Mechanical Engineering Publications, Bury St Edmunds, UK.

Steinbuch, M., (1989). *Dynamic modelling and robust control of a wind energy conversion system*. Ph.D. Thesis, University of Delft, Holland.

9

Wind Turbine Installations and Wind Farms

For any wind turbine installation, there are certain additional activities (e.g., construction of foundations and access roads, electrical connections, site erection, as well as project development and management) that must be undertaken. For flat onshore sites, which might be found typically in Denmark or North Germany, the total investment cost is approximately 1.3 times the ex-works turbine cost (EUREC Agency, 1996). In the UK, where sites are often located in more remote, upland areas the balance-of-plant costs (i.e., all costs other than the wind turbines) tend to be higher and a more typical breakdown is shown in Table 9.1. Commercial developers of wind farms will often prefer larger projects as, in that way, the fixed-costs, particularly electrical network connection and project development and management costs, may be spread over a bigger investment. A further encouragement for large projects is that the fixed costs of arranging project finance are high. However, there are individuals, community groups and commercial organizations who develop smaller wind farms and projects of this type are common in Denmark and Germany. The advantage of community involvement in the project is, of course, that planning permission is more readily obtained if it is seen that there is tangible benefit to local people.

9.1 Project Development

The development of a wind farm follows a broadly similar process to that of any other power generation project, but with the particular requirements that the wind

Table 9.1 Typical Breakdown of Costs for a 10 MW Wind Farm

Element of wind farm	% of total cost
Wind turbines	65
Civil works (including foundations)	13
Wind farm electrical infrastructure	8
Electrical network connection	6
Project development and management costs	8

turbines must be located in high wind speed sites to maximize energy production and their size makes visual effects a particularly important aspect of the environmental impact. Guidance on the development of wind energy schemes has been issued by the British Wind Energy Association (BWEA, 1994) in their *Best Practice Guidelines for Wind Energy Development*. A similar document is published by the European Wind Energy Association (EWEA). The three main elements of the development of a wind-farm project are identified as: (1) technical and commercial issues, (2) environmental considerations, and (3) dialogue and consultation. Perhaps surprisingly to many engineers and technologists the technical and commercial considerations are often the more straightforward and project success or failure hinges critically on environmental considerations and the success of the dialogue and consultation process with local inhabitants and planning authorities.

Wind-farm development may be divided into a number of phases:

- initial site selection,
- project feasibility assessment (including obtaining a power purchase contract for the wind farm output)
- preparation and submission of the planning application,
- construction,
- operation, and
- decommissioning and land reinstatement.

9.1.1 Initial site selection

Initially a desk-based study is carried out to locate a suitable site and to confirm it as a potential candidate for the location of a wind farm. It may be recalled (see section 4.5) that the mean power production for a wind turbine (assuming 100 percent availability) is given by:

$$E = T\int P(U)f(U)\,\mathrm{d}U \tag{9.1}$$

where $P(U)$ is the power curve of the wind turbine, $f(U)$ is the probability density function (PDF) of the wind speed, and T is the time period.

The power curve is available from the potential turbine supplier(s) while an initial estimate of the PDF of the wind speed may be obtained from a wind atlas (European Wind Atlas, 1989). The PDF is generally based on a Weibull distribution and takes account of regional climatology, roughness of the surrounding terrain, local obstacles and topology. PDFs are calculated for 12 30° sectors and integrated with the power curve usually using numerical evaluation techniques. At this stage of the project development only an approximate indication of the wind-farm output is required in order to confirm the potential of the site(s). However, it may be useful to supplement the wind atlas with some form of initial computer modelling.

In the UK a wind-speed database exists based on the NOABL airflow model (ETSU, 1999). The model represents the effect of topography on wind speed and provides estimates of annual mean wind speeds with a resolution of 1-km square at either 10 m, 25 m or 45 m above ground level. It does not take account of local thermally driven wind or local effects. These can have a considerable effect on wind speed and so on-site measurements are required for an accurate assessment.

If measured site wind speed data is available then the energy yield of a wind turbine can be estimated as shown in Figure 9.1. by combining the binned wind speed distribution with the power curve:

$$Energy = \sum_{i=1}^{i=n} H(U_i)P(U_i) \tag{9.2}$$

where $H(U_i)$ is the number of hours in wind speed bin U_i, $P(U_i)$ is the power output at that wind speed and there are n wind speed bins.

In addition to assessing the wind resource it is also necessary to confirm that road access is available, or can be developed at reasonable cost, for transporting the turbines and other equipment. Blades of large wind turbines can be up to 40 m in length and so clearly can pose difficulties for transport on minor roads. For a large wind farm, the heaviest piece of equipment is likely to be the main transformer if a substation is located at the site.

The local electricity utility should be able to provide information on the amount of generation which the distribution network can accept although for a first approximation it may be useful to consider rules-of-thumb as indicated in Table 10.4. Such rules-of-thumb give approximate information only but do serve to highlight those sites where the electrical connection will require long extensions of the power network with associated high cost and environmental impact.

The initial technical assessment will be accompanied by a review of the main environmental considerations. The most important constraints include special consideration of National Parks or other areas designated as being of particular

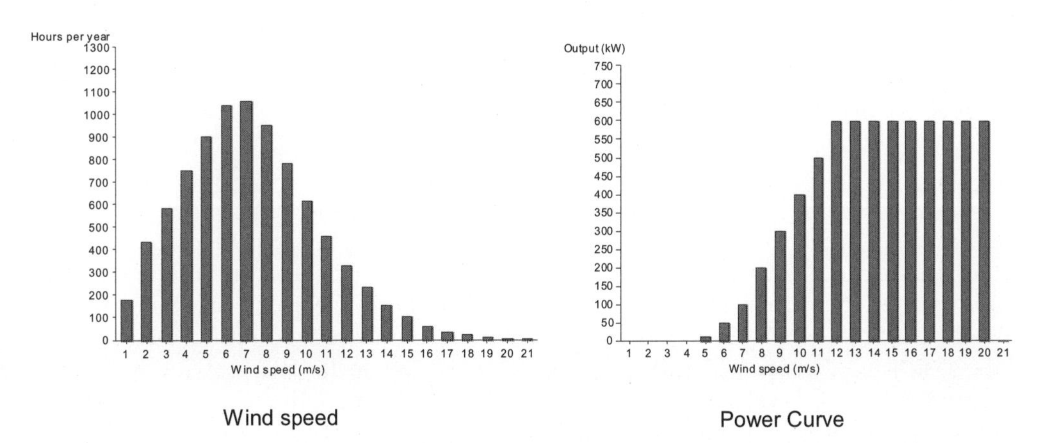

Figure 9.1 Annual Energy Calculation of a Wind Turbine

amenity value and ensuring that no turbine is located so close to domestic dwellings that a nuisance will be caused (e.g., by noise, visual domination or light shadow flicker). A preliminary assessment of visual effects is also required considering the visibility of the wind farm particularly from important public viewpoints. If, within the wind-farm perimeter, there are areas of particular ecological value, due to flora or fauna, then these need to be avoided as well as any locations of particular archaeological or historical interest. Communication systems (e.g., microwave links, TV, radar or radio) may be adversely effected by wind turbines and this needs to be considered at an early stage.

In parallel with the technical and environmental assessments it is normal to open discussion with the local civic or planning authorities to identify and agree the major potential issues which will need to be addressed in more detail if the project development is to continue.

9.1.2 Project feasibility assessment

Once a potential site has been identified then more detailed, and expensive, investigations are required in order to confirm the feasibility of the project. The wind farm energy output, and hence the financial viability of the scheme, will be very sensitive to the wind speed seen by the turbines over the life of the project. Hence it is not generally considered acceptable in complex terrain to rely on the estimates of wind speed made during the initial site selection but to use the measure–correlate–predict (MCP) technique to establish a prediction of the long-term wind resource (Derrick, 1993, Mortimer, 1994).

9.1.3 The Measure–correlate–predict technique

The MCP approach is based on taking a series of measurements of wind speed at the wind farm site and correlating them with simultaneous wind speed measurements made at a meteorological station. The averaging period of the site-measured data is chosen to be the same as that of the meteorological station data. In its simplest implementation, linear regression is used to establish a relationship between the measured site wind speed and the long-term meteorological wind speed data of the form:

$$U_{\text{site}} = a + bU_{\text{long-term}} \tag{9.1.3}$$

Coefficients are calculated for the 12 30° directional sectors and the correction for the site applied to the long-term data record of the meteorological station. This allows the long-term wind speed record held by the meteorological station to be used to estimate what the wind speed at the wind-farm site would have been over the last, say, 20 years. It is then assumed that the site long-term wind speed is represented by this estimate which is used as a prediction of the wind speed during the life of the project. Thus, MCP requires the installation of a mast at the wind farm site on which are mounted anemometers (usually cup anemometers) and a wind vane. If possible one anemometer is mounted at the hub height of the proposed wind turbines with others lower to allow wind shear to be measured. Measurements are made over at least a 6 month period (although clearly the more data obtained the more confidence there will be in the result) and correlated with the measurements made concurrently at the meteorological station.

There are a number of difficulties of using MCP (Landberg and Mortensen, 1993) including:

- with modern wind turbines, high site meteorological masts are necessary and these may themselves require planning permission;
- there may not be a suitable meteorological station nearby (within say 50–100 km) or with a similar exposure and wind climate;
- the data obtained from the meteorological station may not always be of good quality and may include gaps so that it may be time consuming to ensure that it is properly correlated with the site data;
- it is based on the assumption that the previous long-term record provides a good estimate of the wind resource over the lifetime of the wind farm.

Conventional MCP techniques assume that the wind direction distribution at the site is the same as that of the meteorological station. Recent investigations (Addison *et al.*, 2000) suggest that this assumption is source of significant error and a correlation technique based on artificial neural networks is proposed. Using this approach an improvement in predictive accuracy of 5–12 percent was obtained although it should be emphasized that these advanced techniques are not common industrial practice.

9.1.4 Micrositing

The conventional MCP technique is now well established and specially designed site data loggers, temporary meteorological masts and software programs for data processing are commercially available. The estimate of the long-term wind speeds obtained from MCP may then be used together with a wind-farm design package (e.g., WindFarmer, Windfarm) to investigate the performance of a number of potential turbine layouts. These sophisticated programs take the wind distribution data and combine them with topographic wind-speed variation and the effect of the wakes of the other wind turbines to generate the energy yield of any particular turbine layout. Two models commonly used to calculate the effect of topology and surface roughness of the site and surrounding area are the Wind Atlas Analysis and Application Program (WA^{s}P) and MS-Micro/3 (the latest in a series of versions of the MS3DJH/3R model) (Walmsley, Salmon and Taylor, 1982, Walmsley, Taylor and Keith, 1986). Other constraints such as turbine separation, terrain slope, wind turbine noise and land ownership boundaries may also be applied. Optimization techniques are then used to optimize the layout of the turbines for maximum energy yield of the site taking into account local wind speeds and wake effects. The packages also have visualization facilities to generate zones of visual impact, views of the wind farm either as wire frames (see Section 9.2.3) or as photomontages. Figure 9.2 shows an energy map for a prospective wind farm site created with wind farm design software. The areas on the top of the hills show the highest energy density.

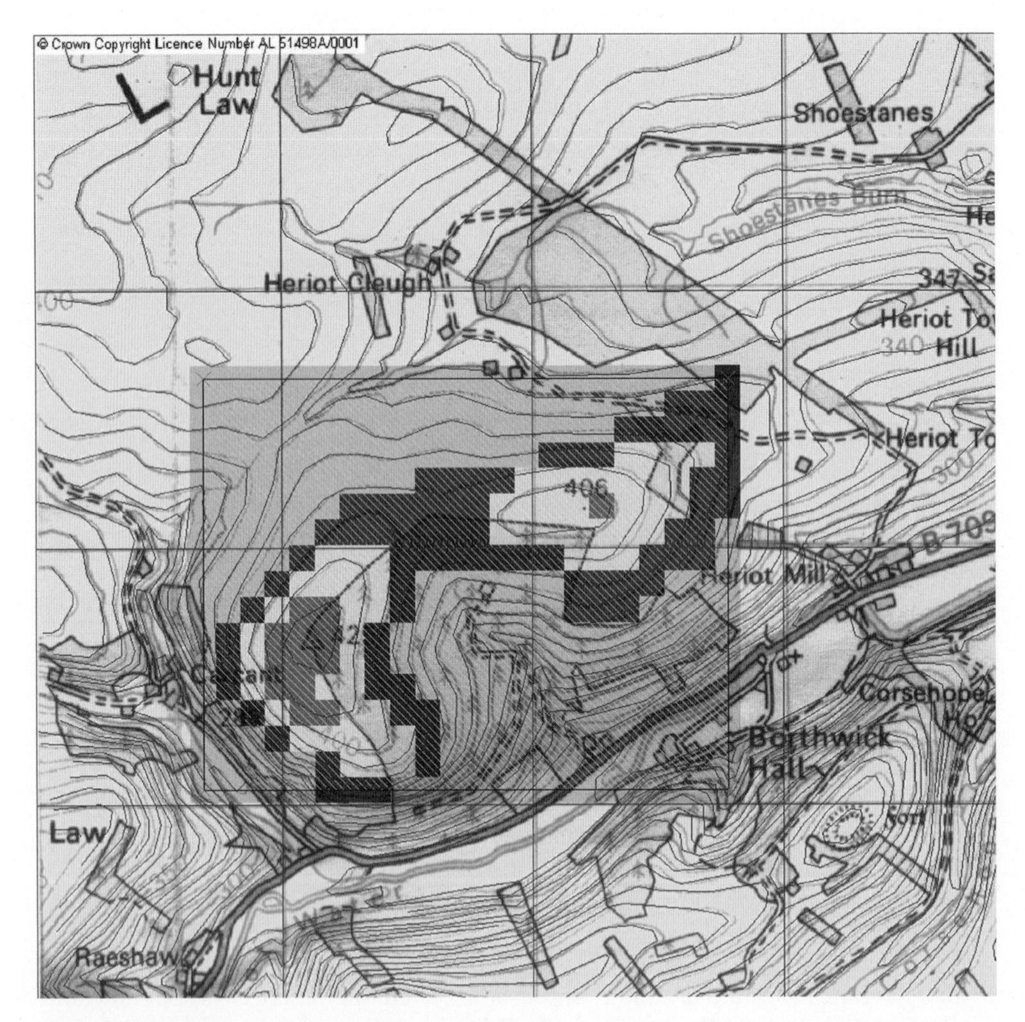

Figure 9.2 Example of Energy Map for a Prospective Wind farm Site Created with Wind farm Design Software

0-1127 Wm^{-2} 1127-1277 Wm^{-2} 1277-1427 Wm^{-2} 1427-1502 Wm^{-2}

(Courtesy of Garrad Hassan *www.garradhassan.co.uk*) Based upon the Ordnance Survey mapping on behalf of the Controller of Her Majesty's Stationary Office © Crown Copyright. MC 100014737.

9.1.5 Site investigations

At the same time as wind speed data are being collected more detailed investigations of the proposed site may also be undertaken. These include a careful assessment of existing land use and how best the wind farm may be integrated with, for example, agricultural operations. The ground conditions at the site also need to be investigated to ensure that the turbine foundations, access roads and construction areas can be provided at reasonable cost. Local ground conditions may influence the position of turbines in order to reduce foundation costs. It may also be important to undertake a hydrological study to determine whether spring water supplies are taken from the wind farm site and if the proposed foundations or cable trenches will cause disruption of the ground-water flow. More detailed investigations of the site access requirements will include assessment of bend radii, width, gradient and any weight restrictions on approach roads. Discussions are also likely to continue with the local electricity utility concerning the connection to the distribution network and the export of the wind farm power.

The planning application will require the preparation of an Environmental Statement and the scope of this is generally agreed, in writing, with the civic authorities during the Project Feasibility Assessment.

9.1.6 Public consultation

Prior to the erection of the site anemometer masts the wind farm developer may wish to initiate some form of informal public consultation. This is likely to involve local community organizations, environmental societies and wildlife trusts. It may also be appropriate to keep local politicians informed. Of course, the erection of meteorological masts does not necessarily imply that the wind farm will be constructed but, as they are highly visible structures, careful consultation is required to ensure that unnecessary public concern is allayed.

9.1.7 Preparation and submission of the planning application

Wind farms are recognized as having significant environmental impact and it is usual for an Environmental Statement to be required as a major part of the application for planning permission. The preparation of the Environmental Statement is an expensive and time-consuming undertaking and usually requires the assistance of various specialists. The purpose of a wind farm Environmental Statement may be summarized:

- to describe the physical characteristics of the wind turbines and their land-use requirements,
- to establish the environmental character of the proposed site and the surrounding area,

- to predict the environmental impact of the wind farm,
- to describe measures which will be taken to mitigate any adverse impact, and
- to explain the need for the wind farm and provide details to allow the planning authority and general public to make a decision on the application.

Topics covered in the Environmental Statement will typically include the following (BWEA, 1994).

Policy framework. The application is placed in the context of national and regional policy.

Site selection. The choice of the particular site that has been selected is justified.

Designated areas. The potential impact of the wind farm on any designated areas (e.g., National Parks) is evaluated.

Visual and landscape assessment. This is generally the most important consideration and is certainly the most open to subjective judgement. Hence it is usual to employ a professional consultancy to prepare the assessment. The main techniques which will be used include: zones of visual influence (ZVI) to indicate where the wind farm will be visible from, wireframe analysis which show the location of the turbines from particular views, and photomontage production which are computer generated images overlaid on a photograph of the site.

Noise assessment. After visual impact, noise is likely to be the next most important topic. Hence predications of the sound produced by the proposed development are required with special attention being paid to the nearest dwellings in each direction. It may be necessary to establish the background noise at the dwellings by a series of measurements so that realistic assessments can be made after the wind farm is in operation.

Ecological assessment. The impact of the wind farm, including its construction, on the local flora and fauna needs to be considered. This may well require site surveys at a particular season of the year.

Archaeological and historical assessment. This is an extension of the investigation undertaken during the site selection.

Hydrological assessment. Depending on the site, it may be necessary to evaluate the impact of the project on water courses and supplies.

Interference with telecommunication systems. Although wind turbines do cause some interference with television transmission this is normally only a local effect and can usually be remedied at modest cost. Any interference with major point-to-point communication facilities (e.g., microwave systems) or airfield radar is likely to be a much more significant issue.

Aircraft safety. The proximity to airfields or military training areas needs to be considered carefully.

Safety. An assessment is required of the safety of the site including the structural integrity of the turbines. Particular local issues may include highway safety and shadow flicker.

Traffic management and construction. The Environmental Statement addresses all phases of the project and so both the access tracks and the increase in vehicle movements on the public roads need to be considered.

Electrical connection. There may be significant environmental impact associated with the electrical connection (e.g., the construction of a substation and new circuit). Although this may be dealt with formally as a separate planning application it still needs to be considered, particularly as any requirement to place underground any long, high-voltage circuits will be very expensive.

Economic effects on the local economy, global environmental benefits. It is common to emphasize the benefit that the wind farm will bring both to the local economy and to reduction in gaseous emissions.

Decommissioning. The assessment should also include proposals for the decommissioning of the wind farm and the removal of the turbines at the end of the project. Decommissioning measures are likely to involve the removal of all equipment which is above ground and restoration of the surface of all areas.

Mitigating measures. It is obvious that the wind farm will have an impact on the local environment and so this section details the steps that are proposed to mitigate and adverse effects. This is likely to emphasize the attempts that have been made to minimize visual intrusion and control noise.

Non-technical summary. Finally a non-technical summary is required and this may be widely distributed to local residents.

9.2 Visual and Landscape Assessment

Modern wind turbines are large structures, sometimes over 100 m to the blade tip, and must be sited in exposed locations of high mean wind speed to operate effectively. The individual turbines must be spaced at at least 3–5 rotor diameters and hence large wind farms are extensive. There is generally a trade-off between energy yield and visibility of wind turbines and so visual and landscape assessment is a key aspect of the Environmental Statement required for any wind-farm development. The assessment of landscape and visual effects of a wind farm is normally undertaken by a professional consultant specializing in such work, who will attempt to quantify the impacts, but it is generally recognized that a degree of subjective interpretation is required for such work.

The assessment process is, of course, iterative and will influence the design and layout of the wind farm. However, it may be divided roughly into a number of areas:

- landscape character assessment (including landscape policy and designation),

- design and mitigation,
- assessment of impacts (including visibility and viewpoint analysis),
- shadow flicker.

The landscapes within which wind farms are constructed vary widely and the turbine layout is chosen to take account of the landscape character as well as wind conditions. Thus in Figure 9.3 the turbines are located along the field boundaries while linear arrangements are shown in Figures 9.4 to 9.6 either along ridges or the coast line.

It is also essential to consider the sociological aspects of the development. An individual's perception of a wind-farm development will be determined not only by the physical parameters (e.g., wind turbine size, number, colour, etc.) but also by his/her opinion of wind energy as part of the energy supply (Taylor and Rand, 1991).

9.2.1 Landscape character assessment

The fundamental step in minimizing the visual impact of a wind farm is to identify an appropriate site and ensure that the proposed development is in harmony with the location. Many exposed upland areas are likely to be of high amenity and have been designated as areas of significant landscape value or even as National Parks.

Figure 9.3 Windfarm of Six 660 kW Turbines in Flat Terrain (Reproduced by permission of Cumbria Wind Farms Ltd, Paul Carter)

Figure 9.4 Windfarm of 600 kW, 43 m diameter Wind Turbines, Tarifa, Spain (Reproduced by permission of NEG MICON; www.neg-micon.dk)

Figure 9.5 Windfarm of 7 600 kW Wind Turbines Located Along Coast (Reproduced by permission of Powergen Renewables/Wind Prospect Ltd www.windprospect.com)

Figure 9.6 Cameron Ridge in California USA (Reproduced by permission of Renewable Energy Systems Ltd www.res-ltd.com)

Development of wind farms within a National Park is likely to be extremely difficult and it is also important to recognize that any view of the site from inside a specially designated area will be considered to be important by the planning authorities.

Once a potential site has been identified, a baseline study is undertaken to establish the landscape and visual context of the area. This will include a description of the existing landscape based on fieldwork observations and up-to-date maps. It is necessary to describe the landform, landcover and land-use patterns. The landscape character will be described and its quality evaluated together with the location of potentially sensitive viewpoints.

Stanton (1994) suggests some characteristics of the wind farm visual appearance which are considered to be desirable. The development should be simple, logical and avoiding visual confusion. Although one landscape type is no more appropriate for a wind farm development than another this author considers that their suitability for different types of development varies greatly. For example, flat agricultural land is considered suitable for either a small number of wind turbines or large wind farms of similar regularly spaced machines while coastal areas are considered appropriate for large numbers of wind turbines but the development should relate to the linear quality of the coastal space. Such views are, of course, open to debate but they do illustrate that it is essential to consider the character of the landscape before a wind farm development is proposed.

Cumulative effects are also considered to be important and the impact of more than one wind farm being visible in an area is a major consideration. Local features such as buildings and hedges, which restrict views, may mean that a landscape has the capability to accommodate additional wind turbines but detailed investigation is required to establish this.

9.2.2 Design and mitigation

Wind turbine designers have recognized for some years that the overall form of the structure of a large wind turbine has to be pleasing and this aspect of industrial design is considered early in the development of a new machine. It is now generally recognized that for aesthetic reasons three-bladed turbines are preferred (see Section 6.5.6). Two-bladed rotors sometimes give the illusion of varying speed of rotation, which can be disconcerting. In addition, for a similar swept area, a two-bladed rotor will operate faster than one with three blades. There is considerable evidence (Taylor and Rand, 1991, Gipe, 1995) that a slower speed of rotation is more relaxing to the eye. This effect works to the advantage of the large modern wind turbines, which operate typically at 30–35 r.p.m. There is some speculation that, in the future, very large wind turbines designed specifically for remote offshore installation may revert to two-bladed rotors to realize the engineering benefits of this arrangement.

There would appear to be a difference between European and North American practice concerning the acceptability of lattice steel towers. There are clearly engineering benefits associated with lattice towers particularly in terms of the costs of the foundations (see Section 7.10.4), although the upper, tapered section of the tower must be such as to provide adequate clearance for the blades. From a distance and in certain light conditions lattice towers can disappear leaving only the rotors visible and this effect is generally thought to be undesirable. In Europe, solid tubular towers are generally preferred. Tower heights can vary greatly in response to both wind conditions and planning constraints. In Northern Germany, very high towers (>60 m) are used in order to maximize energy generation. Conversely, in western parts of the UK, which have high mean wind speeds, tower heights are reduced in order to minimize the area from which the wind farm is visible.

The appearance of the turbines can be influenced to some extent by their colour. In the upland sites of the UK, the turbines will generally be seen against the sky and so an off-white or mid-grey tone is generally considered to be appropriate. Where the wind farm is seen against other backgrounds a colour to blend in with the ground conditions may be more suitable. There is general agreement that the outer gel-coat of the blades should be a matt or semi-matt finish to minimize reflections. In the USA there have been a number of wind farms that used a variety of wind turbines sometimes of quite different design (i.e., with differing numbers of blades, direction of rotation, tower types etc). It is unlikely that such developments would be acceptable in highly populated parts of Europe.

The layout and design of the wind farm is important in determining if planning consent is to be received. The preliminary wind-farm layout will be determined by engineering considerations (e.g., local wind speeds, turbine separation, noise, geotechnical considerations, etc.) and will then be modified to take account of visual and landscape impacts. On open, flat land the turbines are often arranged in a regular layout in order to provide a simple and logical visual image with maximum power output. Alternatively on hill sites, or where there are significant hedges or field boundaries, the turbine layout is arranged around these features. When viewed from up to 1–2 km, wind turbines are considered to dominate the field of vision and so it is desirable that views within this distance can be minimized (e.g., by moving turbines to take make use of local screening features or by tree planting). When turbines are seen one behind another there is an increase in visual confusion and this 'stacking' effect is considered to be undesirable. Some viewpoints are likely to be particularly important and it may be appropriate to arrange the turbines so that these views of the wind farm are as clear and uncluttered as possible.

In addition to the wind turbines, there are a number of associated ancillary structures required. With smaller wind turbines, the local transformers may be located adjacent to the towers, often in an enclosure to provide protection against the weather and vandalism. However, the towers of many large turbines are wide enough to accommodate the transformers inside and this obviously reduces visual clutter on the wind farm. There is often a requirement for a main wind farm substation and some form of local control building. Engineering considerations would indicate that this substation should be located in the middle of the wind farm but, in order to reduce visual impact, it is often located some distance away where it gives minimal visual intrusion. Within European wind farms all power collection circuits are underground and it may also be appropriate, if rather expensive, to use underground cable to make the final connection to the local utility system. Roads are required within the wind farm for construction and it is an occasional requirement of the planning consent that they are re-vegetated after commissioning. This of course can lead to very considerable expense if the road has to be reinstated to allow a large crane to be used for maintenance or repair.

9.2.3 Assessment of impact

A major part of the Environmental Statement is the assessment of visual impact. Two main techniques are used: (1) visibility analysis using zones of visual impact (ZVI), and (2) viewpoint analysis using wire frames and photomontages.

Zones of visual impact show those areas of the surrounding country, usually up to 10–20 km radius, from which a wind turbine, or any part of a wind turbine, in a wind farm is visible. The ZVI is generated using computer methods based on a digital terrain model and shows how the local topology will influence the visibility of the wind farm. Usually ZVI techniques ignore local landscape features such as screening from trees and buildings. Also, weather conditions are not considered and clear visibility is assumed. Figure 9.7 shows an example of a ZVI generated using commercially available wind farm design software. The cumulative impact of a number of wind farms may be calculated in a similar manner.

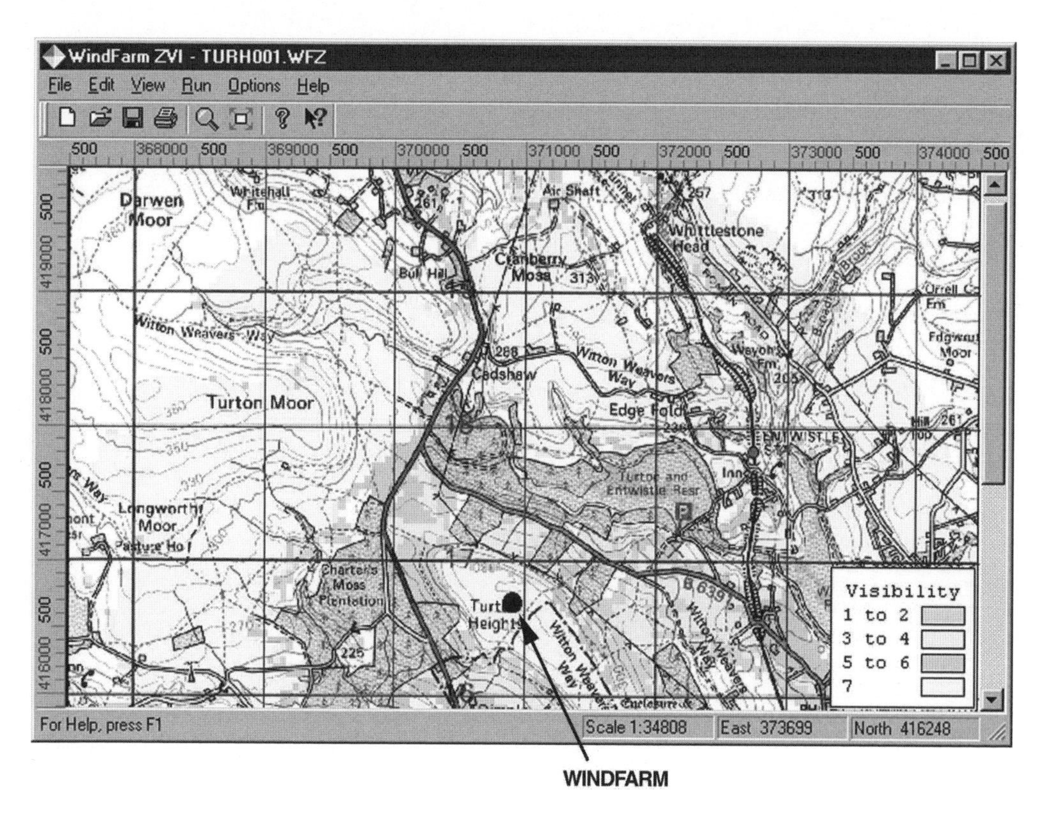

Figure 9.7 Example of Zone of Visual Impact of a Wind Farm (Courtesy of ReSoft www.resoft.co.uk) Based upon the Ordnance Survey Mapping on behalf of The Controller of Her Majesty's Stationary Office © Crown Copyright MC 100014737

Viewpoint analysis is based on selecting a number of important locations from which the wind farm is visible and applying professional judgement using quantitative criteria to assess the visual impact. The viewpoints are selected in consultation with the civic planning authorities and for a large wind farm up to 20 locations may be chosen. Although approaches vary, the assessment may involve consideration of three aspects: (1) the sensitivity of the landscape, (2) the sensitivity of the viewpoint, and (3) the magnitude of the change of view. Thus, for example, the landscape within a National Park will be of 'high' sensitivity while a landscape with existing discordant features such as old quarry workings may have a 'low' sensitivity. Similarly a viewpoint where the land use is residential or has high recreational value may have a 'high' sensitivity, while a viewpoint that is used only for indoor employment (e.g., a local industrial estate), might be considered to have a 'low' sensitivity. The magnitude of the impact can be described in a similar manner depending, for example, on the number of turbines visible, the distance to the wind farm and the prominence of the development. The overall significance of the impact is then assessed, again using quantitative terminology (such as substantial, moderate, slight, negligible, etc.), by combining these factors. Where a substantial impact is identified, acceptability will depend on whether it is considered that the wind farm will have a detrimental effect on the landscape quality.

Figure 9.8 shows a typical wireframe image generated from a viewpoint and Figure 9.9 shows a photomontage. Both of these images were generated using wind farm design software. Wireframe representations, provide an accurate impression of turbine position and scale while photomontages are the normal tool used to give an overall impression of the visual effect of a wind farm. Videomontages have been proposed in order to give an impression of the movement of turbine blades but this tool is not yet in regular use.

Figure 9.8 Example of Wire Frame View (Courtesy of ReSoft www.resoft.co.uk)

Figure 9.9 Example of Photomontage (Courtesy of ReSoft www.resoft.co.uk)

9.2.4 Shadow flicker

Shadow flicker is the term used to describe the stroboscopic effect of the shadows cast by rotating blades of wind turbines when the sun is behind them. The shadow can create a disturbance to people inside buildings exposed to such light passing through a narrow window. Although considered to be an important issue in Europe, and recognized in the operation of traditional windmills (Verkuijlen and Westra, 1984) shadow flicker has not generally been recognized as significant in the USA (Gipe, 1995).

The frequencies that can cause disturbance are between 2.5–20 Hz. The effect on humans is similar to that caused by changes in intensity of an incandescent electric light due to variations in network voltage from a wind turbine (see Section 10.5.1). In the case of shadow flicker the main concern is variations in light at frequencies of 2.5–3 Hz which have been shown to cause anomalous EEG (electroencephalogram) reactions in some sufferers from epilepsy. Higher frequencies (15–20 Hz) may even lead to epileptic convulsions. Of the general population, some 10 percent of all adults and 15–30 percent of children are disturbed to some extent by light variations at these frequencies (Verkuijlen and Westra, 1984).

Large modern three-bladed wind turbines will rotate at under 35 r.p.m. giving blade-passing frequencies of less than 1.75 Hz, which is below the critical frequency of 2.5 Hz. A minimum spacing from the nearest turbines to a dwelling of 10 rotor diameters is recommended to reduce the duration of any nuisance due to light flicker (Taylor and Rand, 1991). However, a spacing of this magnitude is likely to be required in any event by noise constraints and to avoid visual domination.

9.2.5 Sociological aspects

There are a number of computer-based tools available for quantifying visual effects and landscape architects and planners have developed techniques to place quantitative measures on visual impact using professional judgement. However, public attitudes, which ultimately determine whether a wind farm may be constructed, are influenced by many more complex factors. Public attitudes to wind farms have been studied on a number of occasions (e.g., ETSU, 1993, 1994) and Gipe (1995) discusses this subject in considerable detail. In general, the large majority of people approve of wind farms after they have been constructed although a significant minority remains opposed to them. In particular, there is the difficult issue that some local residents consider they are paying a high cost for a benefit, either financial or environmental, which accrues to others. The financial benefits may be shared with the community in a number of ways including by the development of co-operative or community-owned wind farms, while the environmental issue has to be addressed by consultation. Also, it is suggested that stationary wind turbines are less acceptable than those rotating and so maintaining high availability with a low cut-in wind speed is likely to improve public perception.

9.3 Noise

Noise from wind turbines is often perceived as one of the more significant environmental impacts (Wagner, Bareis and Guidati, 1996). During the early development of wind energy, in the 1980s, some turbines were rather noisy and this led to justified complaints from those living close to them. However, since then, there has been very considerable development both in techniques for reducing noise from wind turbines and in predicting the noise nuisance a wind farm will create.

The UK document Planning Policy Guidance Note (Department of the Environment, 1993) suggests that:

'A planning application for any wind-farm development could usefully be accompanied by the following information regarding details of the proposed turbine(s) and predicted noise levels:

- predicted noise levels at specific properties closest to the wind farm over the most critical range of wind speeds;
- measured background noise levels at the properties and wind speeds outlined above;
- a scale map showing the proposed wind turbine(s), the prevailing wind conditions, nearby existing developments;
- results of independent measurements of noise emission from the proposed wind turbine including the sound power and narrow-band frequency spectrum; in the case of a prototype turbine where no measurements are available, predictions should be made by comparison with similar machines.'

9.3.1 Terminology and basic concepts

Two distinctly different measures are used to describe wind turbine noise. These are the *sound power level* L_W of the source (i.e., the wind turbine) and the *sound pressure level* L_P at a location. Because of the response of the human ear, a logarithmic scale is used based on reference levels that correspond to the limit of hearing. The units of both L_P and L_W are the decibel (dB).

A noise source is described in terms of its *sound power level*, L_W:

$$L_W = 10 \log_{10}\left(\frac{W}{W_0}\right) \tag{9.4}$$

where W is the total sound power level emitted from the source (in Watts) and W_0 is a reference value of 10^{-12} W.

The *sound pressure level* L_P is defined as

$$L_P = 10 \log_{10}\left(\frac{P^2}{P_0^2}\right) \tag{9.5}$$

where P is the RMS value of the sound pressure and P_0 is a reference value of 2×10^{-5} Pa.

By simple algebraic manipulation it may be seen that the addition of n sound pressure levels (expressed in dB) is carried out as shown:

$$L_{Pn} = 10 \log_{10} \sum_{j=1}^{j=n} 10^{L_{P(j)}/10} \tag{9.6}$$

Thus, adding two sound pressure levels of the same magnitude results in an increase of 3 dB. Table 9.2 gives an indication of the typical range of sound pressure levels.

The human ear is capable of detecting sounds between 20 Hz and 20 kHz and spectral analysis is typically undertaken over this range. A narrow-band spectrum, with a defined bandwidth of measurement, gives the fullest information of the signal and may be used to detect particular tones. However, it is conventional to use octave and 1/3-octave bands for broadband analysis. The upper frequency of an octave band is twice that of the lower frequency while for the 1/3-octave band the upper frequency is $\sqrt[3]{2}$ times the lower frequency.

It is common to weight the measurements to reflect the response of the human ear with frequency. This is done by applying the so-called A-weighted filter. Measurements made with this filter are referred to as dBA or dB(A). Table 9.3 shows the centre frequencies of the octave bands together with the A-weighting in dB. It may be seen that frequencies below 250 Hz and above 16 kHz are heavily attenuated.

An equivalent sound pressure level $L_{\mathrm{eq},T}$ is the value of a continuous steady sound that, within the specified time interval (T) has the same mean square sound pressure level as the sound under consideration which varies with time (International Energy Agency, 1994)

$$L_{\mathrm{eq},T} = 10 \log_{10} \left(\frac{1}{T} \int_0^T \frac{P^2}{P_0^2} \, \mathrm{d}t \right) \tag{9.7}$$

A similar calculation may be undertaken using A–weighted values to give $L_{\mathrm{Aeq},T}$

Table 9.2 Examples of Sound Pressure Levels

Example	Sound pressure level (dB(A))
Threshold of hearing	0
Rural night time background	20–40
Busy general office	60
Inside factory	80–100
Jet aircraft at 100 m	120
Wind farm at 350 m	35–45

Table 9.3 Centre Frequency of Octave Bands and A-weighting

Octave band centre frequency (Hz)	A-weighting (dB)
16	−56.7
31.5	−39.4
63	−26.2
125	−16.1
250	−8.6
500	−3.2
1000	0.0
2000	1.2
4000	1.0
8000	−1.1
16000	−6.6

$$L_{\mathrm{Aeq},T} = 10\log_{10}\left(\frac{1}{T}\int_0^T \frac{P_{\mathrm{A}}^2}{P_0^2}\,\mathrm{d}t\right) \tag{9.8}$$

where $L_{\mathrm{Aeq},T}$ is the equivalent continuous A-weighted sound pressure level determined over time T and P_{A} is the instantaneous A-weighted sound pressure.

The exceedance level, L_{A90}, is defined as the A-weighted sound pressure level which is exceeded for 90 percent of the time. Some planning authorities prefer the use of the L_{A90} sound pressure level, particularly for measurements of background noise, as the L_{eq} measurement may be heavily influenced by short-term effects such as passing aircraft or traffic. A wind farm is a fairly constant source of noise and so its contribution to the L_{A90} sound pressure level (measured over a 10 min period) is likely to be some 1.5–2.5 dB(A) less than to the 10 min L_{Aeq} value (ETSU, 1997a).

The sound intensity I is the power transmitted through a unit area, A,

$$I = \frac{W}{A} \tag{9.9}$$

and far from the source of sound with a uniform flux

$$I = \frac{P^2}{Z_0} \tag{9.10}$$

where P is the RMS value of the sound pressure level, and Z_0 is the characteristic acoustic impedance.

Choosing a suitable value for $I_{\mathrm{ref}}(10^{-12}\ \mathrm{W/m^2})$, the sound pressure level may be expressed in terms of sound intensity as

$$L_{\mathrm{P}} = 10\log_{10}\left(\frac{I}{I_{\mathrm{ref}}}\right) \tag{9.11}$$

For spherical spreading

$$I = \frac{W}{4\pi r^2} \tag{9.12}$$

where r is the distance from the source.

Hence, under conditions of ideal spherical spreading

$$L_P = 10\log_{10}\left(\frac{W}{4\pi r^2 10^{-12}}\right) = L_W - 10\log_{10}(4\pi r^2) \tag{9.13}$$

and similarly, if hemispherical spreading is assumed

$$L_P = 10\log_{10}\left(\frac{W}{2\pi r^2 10^{-12}}\right) = L_W - 10\log_{10}(2\pi r^2) \tag{9.14}$$

Sound pressure level, from a point source, decays with distance according to the inverse square law under both spherical and hemi-spherical spreading assumptions. Hence for each doubling of distance the sound pressure level is reduced by 6 dB.

For a line source of noise the sound pressure level is given by

$$L_P = 10\log_{10}\left(\frac{L_{wl}}{2\pi r 10^{-12}}\right) = L_{wl} - 10\log_{10}(2\pi r) \tag{9.15}$$

(L_{wl} is the sound power level per unit length of the source.) Hence for cylindrical spreading the decay is only proportional to distance, resulting in a 3 dB reduction of sound pressure level for each doubling of distance perpendicular to the line source.

9.3.2 Wind-turbine noise

Noise from wind turbines is partly mechanical, and partly aerodynamic.

Mechanical noise is generated mainly from the rotating machinery in the nacelle particularly the gearbox and generator although there may also be contributions from cooling fans, auxiliary equipment (such as pumps and compressors) and the yaw system. Mechanical noise is often at an identifiable frequency or tone (e.g., caused by the meshing frequency of a stage of the gearbox). Noise containing discrete tones is more likely to lead to complaints and so attracts a 5dB penalty in many noise standards. Mechanical noise may be air-borne (e.g., the cooling fan of an air-cooled generator) or transmitted through the structure (e.g., gearbox meshing which is transmitted through the gearbox casing, the nacelle bed-plate, the blades and the tower).

Pinder (1992) is quoted by Wagner, Bareis and Guidati (1996) as giving the values of sound power level as shown in Table 9.4 for a 2 MW experimental wind turbine. It may be seen that the gearbox is the dominant noise source through structure-borne transmission.

Techniques to reduce the mechanical noise generated from wind turbines include careful design and machining of the gearbox, use of anti-vibration mountings and couplings to limit structure-borne noise, acoustic damping of the nacelle, and liquid cooling of the generator (see Section 7.4.8).

Aerodynamic noise is due to a number of causes (ETSU, 1997a):

Table 9.4 Sound Power Levels of Mechanical Noise of a 2 MW Experimental Wind Turbine (after Wagner, Bareis and Guidati, 1996)

Element	Sound power level (dB(A))	Air-borne or structure-borne
Gearbox	97.2	Structure-borne
Gearbox	84.2	Air-borne
Generator	87.2	Air-borne
Hub (from gearbox)	89.2	Structure-borne
Blades (from gearbox)	91.2	Structure-borne
Tower (from gearbox)	71.2	Structure-borne
Auxiliaries	76.2	Air-borne

- low-frequency noise,
- inflow turbulence noise,
- airfoil self noise.

Low-frequency noise is caused by changes in the wind speed experienced by the blades due to the presence of the tower or wind shear. Although this effect is very pronounced with downwind turbines it is also significant with upwind machines. The spectrum of the noise is dominated by blade-passing frequency (typically up to 3 Hz) and its harmonics (typically up to 150 Hz). It may be seen from Table 9.3 that the A-weighted filter heavily attenuates these frequencies and so this source does not make a major contribution to audible noise. However, it was reported that experimental, downwind turbines, constructed in the 1980s excited low-frequency vibrations in adjacent buildings. For upwind turbines increasing the clearance between the blades and tower will reduce this effect. Hubbard and Shephard (Spera, 1994) show interesting experimental evidence of low-frequency noise from a large 80 m diameter downwind turbine and provide a comparison with a 90 m diameter upwind machine.

Inflow turbulence creates broadband noise as the blades interact with the eddies caused by atmospheric turbulence. It generates frequencies up to 1000 Hz which are perceived by an observer as a 'swishing' noise. Turbulent inflow noise is considered to be influenced by the blade velocity, airfoil section and turbulence intensity. Wagner, Bareis and Guidati (1996) describe this phenomenon in some detail, remarking that it is not yet fully understood, and quote the results of one field experiment where the sound pressure level actually decreased with increasing turbulence intensity.

Airfoil self-noise is generated by the airfoil itself, even in steady, turbulent-free flow. It is typically broadband although imperfections in the blade surface may generate tonal components. The main types of airfoil self-noise include the following.

Trailing edge noise. This is perceived as a broadband swishing sound with frequencies in the range 750–2000 Hz. It is due to interaction of the turbulent boundary layer with the trailing edge of the blade. Trailing edge noise is a major source of higher frequency noise on wind turbines.

Tip noise. The literature is not clear as to whether three-dimensional tip effects are a major contributor to wind turbine noise. However, the majority of the blade noise, as well as the turbine power, emanates from the outer 25 percent of the blade and so there has been very considerable investigation of novel blade tips to reduce noise. Imperfections in the blade shape due to tip-brakes or other control surfaces are another potential source of noise.

Stall effects. Blade stall causes unsteady flow around the airfoil which can give rise to broadband sound radiation.

Blunt trailing edge noise. A blunt trailing edge can give rise to vortex shedding and tonal noise. It can be avoided by sharpening the trailing edge but this has implications for manufacturing and erection.

Surface imperfections. Surface imperfections such as those caused by damage during erections or due to lightning strikes can be a significant source of tonal noise.

The obvious approach to reducing aerodynamic noise is to lower the rotational speed of the rotor, although this may result in some loss of energy. The ability to reduce noise in low wind-speed conditions is a major benefit of variable-speed or two-speed wind turbines. An alternative technique would be to reduce the angle of attack of the blade although again with a potential loss of energy.

Stiesdal and Kristensen (1993) provide a comprehensive description of noise control methods applied to a 300 kW stall-regulated wind turbine. The gearbox was identified as an important source of tonal, mechanical noise. Control measures included detailed modifications to the design and manufacture of the gears and an additional insulating covering over the gearbox housing. Air-borne mechanical noise was minimized by total enclosure of the nacelle and careful design of sound baffles in the ventilation openings. Structure-borne noise was reduced by rubber mounting both the gearbox and generator and using a rubber coupling on the high-speed shaft. The long low-speed shaft (2 m) reduced the drive train vibrations transmitted to the rotor. Tip noise was minimized by the use of a 'tip torpedo' to control the tip-vortex while trailing-edge noise was controlled by specifying a 1–2 mm trailing edge thickness. Stall noise was reduced by using a turbulator strip on the leading edge of the blades near the tip that initiated stall in a controlled manner. Overall these measures resulted in a reduction of the sound power level of 3–4 dB and the elimination of significant tones. The most important noise control methods were considered to be the tip modification, the controlled stall and the gearbox improvements. The measured sound power level of the improved wind turbine was 96 dB(A) at a wind speed of 8 m/s (measured at 10 m height). This was an impressive improvement as typical sound power levels of a modern 600 kW wind turbine may be up to 100 dB(A) at a wind speed of 8 m/s (measured at 10 m height) increasing by 0.5–1 dB(A) per m/s.

9.3.3 Measurement, prediction and assessment of wind-farm noise

The sound power level of a wind turbine is normally determined by field experiments. These were originally specified in the IEA Recommended Practice (International Energy Agency, 1994) but are now described in an international standard (BS, 1999, IEC, 1998). Outdoor experiments are necessary because of the large size of modern wind turbines and the necessity to determine their noise performance during operation. The sound power level cannot be measured directly but is found from a series of measurements of sound pressure levels made around the turbine at various wind speeds from which the background sound pressure levels have been deducted. The method provides the apparent A-weighted sound power level at a wind speed of 8 m/s, its relationship with wind speed and the directivity of the noise source for a single wind turbine. It does not distinguish between aerodynamic and mechanical noise although there is a method proposed to determine if tones are 'prominent'. In addition to A-weighted sound pressure levels, octave or 1/3-octave and narrow-band spectra are measured.

The measurements are taken at a distance, R_0, from the base of the tower:

$$R_0 = H + \frac{D}{2} \tag{9.16}$$

where H is the hub height and D is the diameter of the rotor. This distance is a compromise to allow an adequate distance from the source but with minimum influence of the terrain, atmospheric conditions or wind-induced noise. The microphones are located on boards at ground level so that the effect of ground interference on tones may be evaluated. Four microphone positions are used as shown in Figure 9.10.

Simultaneous A-weighted sound pressure level measurements (more than 30 measurements each of 1–2 min duration) are taken with wind speed. All wind speeds are corrected to a reference height of 10 m with a terrain roughness length of $z_0 = 0.5$ m. The preferred method of determining wind speed, when the turbine is operating, is from the electrical power output of the turbine and the power curve. The main sound pressure level measurement is that of the downwind position while the other three microphones are used for determining directivity. Measurements are taken with and without the turbine operating over a range of at least 4 m/s which includes the 8 m/s reference. The sound pressure levels, with and without the turbine in operation, are then plotted against wind speed and linear regression used to find the 8 m/s values. The method of bins may be used to group the data. The sound pressure level of the turbine alone at the reference conditions, $LA_{eq,ref}$, is then calculated using

$$L_{Aeq,ref} = 10 \log_{10}(10^{L_{S+N}/10} - 10^{L_N/10}) \tag{9.17}$$

where L_{S+N} is the sound pressure level of the wind turbine and the background sound at 8 m/s, and L_N is the sound pressure level of the background with the wind turbine parked at 8 m/s.

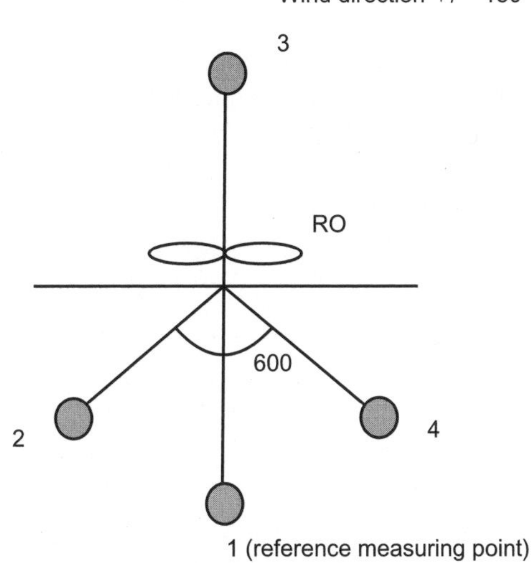

Figure 9.10 Recommended Pattern for Measuring Points from IEA Recommended Practices for Wind Turbine Testing (after IEA, 1994, and BS EN 61400–11)

The apparent A-weighted sound power level of the turbine is then calculated from

$$L_{\mathrm{WA,ref}} = L_{\mathrm{Aeq,ref}} + 10\log_{10}(4\pi R_i^2) - 6 \tag{9.18}$$

where R_i is the slant distance from the microphone to the wind turbine hub.

It may be seen that the calculation of sound power level assumes spherical radiation of the noise from the hub of the turbine. The subtraction of 6 dB is to determine the free field sound pressure level from the measurements and to correct for the approximate pressure doubling that occurs with the microphone located on a reflecting board at ground level.

The directivity, *DI*, is the difference between the A-weighted sound pressure level at measurement points $i = 2$–4 and the reference point 1, downstream of the turbine. It is calculated from

$$DI_i = L_{\mathrm{Aeq},i} - L_{\mathrm{Aeq},1} + 20\log_{10}\frac{R_i}{R_1} \tag{9.19}$$

One of the Appendices of the IEA document (International Energy Agency, 1994) provides a method to estimate the sound pressure level of a single turbine or group of turbines at a distance R provided that they are located in flat and open terrain. The calculation is based on hemispherical spreading (Equation 9.20));

$$L_P(R) = L_W - 10\log_{10}(2\pi R^2) - \Delta L_a \tag{9.20}$$

The correction ΔL_a is for atmospheric absorption and can be calculated from $\Delta L_a = R\alpha$ where α is a coefficient for sound absorption in each octave band and R is the distance to the turbine hub. An alternative approach is to use a similar equation to (9.20) but with α as 0.005 dB/m (as suggested in the Danish Statutory Order on Noise from Windmills, 1991) and L_W specified as a single, broadband sound power level.

If there are several wind turbines which influence the sound pressure level the individual contributions are calculated separately and summed using

$$L_{1+2+..} = 10\log_{10}(10^{L_1/10} + 10^{L_2/10} + \ldots) \tag{9.21}$$

Interestingly, the IEA Appendix concludes with the remark that in tests on small and medium sized wind turbines (55–300 kW), the noise levels predicted in this way were in reasonable agreement with measured noise levels (deviations of A-weighted sound pressure levels generally within +/−2 dB) but that serious differences were found when the prediction was based on more detailed prediction methods.

The IEA Method (1994) for determining sound pressure levels at a point is based on hemispherical spreading with a correction for atmospheric absorption. The assumption of hemispherical spreading gives a reduction of 6 dB per doubling of distance. Under some conditions, particularly downwind, this may be an optimistic assumption and a reduction of 3 dB per doubling of distance is more realistic. The IEA Method also ignores any effects of meteorological gradients (Wagner, Bareis and Guidati, 1996). Under normal conditions, air temperature decreases with height and so the sound speed will decrease with increasing height and cause the path of the sound to curve upwards. However, under conditions of temperature inversion, e.g., as might prevail on cold winter nights, the temperature increases with height causing the sound to curve downwards. Wind speed will have a similar effect. In the downwind direction the sound will be bent downwards while a shadow zone is formed upwind. The effect of the upwind shadow zone is more pronounced at higher frequencies.

A comprehensive study of noise propagation is reported in ETSU/W/13/00385/REP (ETSU, 2000). The results of this study, which included field experiments, support the concerns over the use of very complex models voiced by the IEA and the study concluded that '... significant and consistent correlation only exists between the sound pressure level and vector wind speed'. Hence the study proposes that a rather straightforward spherical propagation model should be assumed with additional terms included to account for directivity, air absorption, and in some cases special topographical features between the source and the converter.

A further IEA recommendation proposes how sound pressure level measurements should be taken at dwellings and other potentially sensitive locations (International Energy Agency, 1997). It is complimentary to International Energy Agency (1994) which provided guidance on the measurement of the source power of a wind turbine.

Permitted noise levels vary widely from country to country and even within countries according to local planing conditions. International practice was reviewed by the UK 'Working Group on Noise from Wind Turbines', (ETSU, 1997a). In Germany, the Netherlands and Denmark the limits are expressed in terms of a maximum permitted value of the sound pressure level at differing locations.

In contrast, the proposal of the UK Working Group is to base the permitted noise level of a wind farm on an increase of 5 dB(A) of the $L_{A90,10min}$ sound pressure level above background noise. The 5 dB(A) limit was selected as being a reasonable compromise between protecting the internal and external environment while not unduly restricting the development of wind energy itself which has other environmental benefits. In addition, it is suggested that a limit of less than 5 dB(A) would be difficult to monitor. BS4142 (British Standard, 1997), which is a general standard for industrial noise and may not be directly applicable to wind farms, states that a difference of +10 dB or more indicates that complaints are likely while a difference of around +5 dB is of marginal significance.

Applying this 5 dB(A) margin above, very quiet rural backgrounds would be too restrictive and so ... the UK Working Group also propose a lower fixed limit of 35–40 dB(A) during daytime and 43 dB(A) during night-time. The selection of the daytime limit of either 35 or 40 dB(A) is made by considering:

- the number of dwellings in the neighbourhood of the wind farm,
- the effect of noise limits on the number of kWh generated, and
- the duration and level of exposure.

The night-time lower limit of 43 dB(A) is derived from a 35 dB(A) sleep disturbance criterion, an allowance of 10 dB(A) for attenuation through an open window and with 2 dB subtracted for the use of $L_{A90,10min}$ rather than $L_{Aeq,10min}$. Examples of the noise criteria are shown in Figures 9.11 and 9.12. There is also a penalty for audible tones which rises to a maximum of 5 dB.

Table 9.5 Noise Limits for Sound Pressure Levels L_{Aeq} in Different European Countries (after Gipe, 1995). Note the Definitions of Location Vary from Country to Country; Further Details may be Found in ETSU (1997a)

Country	Commercial	Mixed	Residential	Rural
Germany				
Day	65	60	55	50
Night	50	45	40	35
Netherlands				
Day		50	45	40
Night		40	35	30
Denmark			40	45

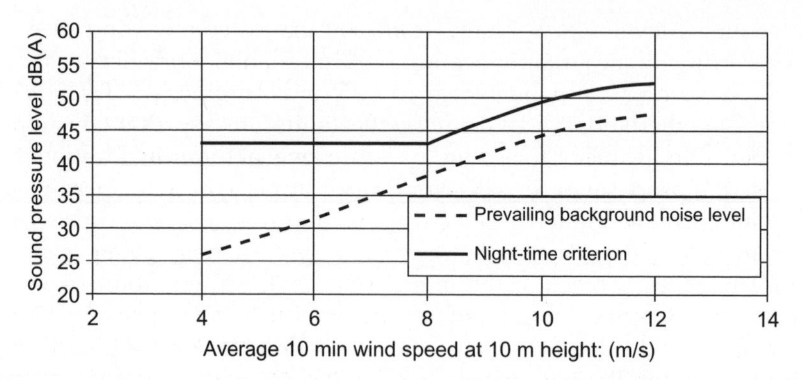

Figure 9.11 Example of Noise Criterion Proposed by the UK Working Group on Noise from Wind Turbines-Night-time Criterion (ETSU, 1997a) Reproduced by permission of ETSU on behalf of DTI

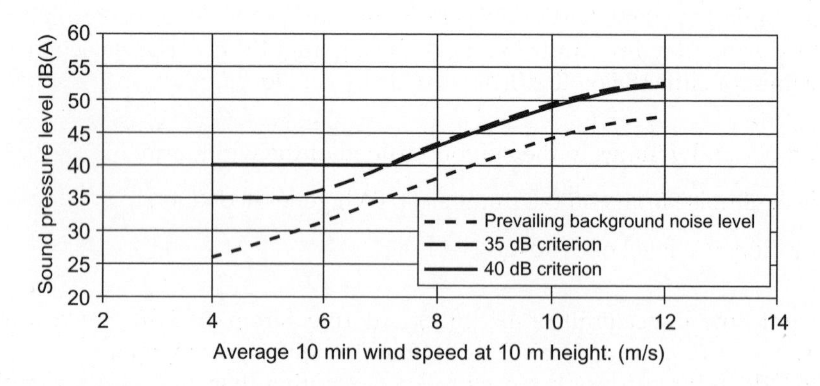

Figure 9.12 Example of Noise Criterion Proposed by the UK Working Group on Noise from Wind Turbines-Daytime Criterion (ETSU, 1997a) Reproduced by permission of ETSU on behalf of DTI

9.4 Electromagnetic Interference

Wind turbines have the potential to interfere with electromagnetic signals that form part of a wide range of modern communication systems and so their siting requires careful assessment in respect of electromagnetic interference (EMI). In particular, wind energy developments often compete with radio systems for hilltops and other open sites that offer high energy outputs from wind farms and good propagation paths for communication signals. The types of system that may be affected by EMI, and their frequency of operation, include VHF radio systems (30–300 MHz), UHF Television broadcasts (300 MHz–3 GHz) and microwave links (1–30 GHz). The interaction of wind turbines with defence and civilian radar used for air traffic control has also been the subject of investigation (ETSU, 1995).

Hall (1992) reported that tests on a fixed-speed 400 kW wind turbine showed that no radio transmissions attributable to a wind turbine could be detected at 100 m.

The electrical generator and associated control gear and electronics can produce radio frequency emissions but these may be minimized by appropriate suppression and screening at the generator. Rather than behave as an aerial the tubular tower had a substantial screening effect on all emissions. If the nacelle is metallic this will also screen the emissions from the nacelle itself. Although additional precautions may be necessary with the power electronic converters of variable-speed wind turbines, electromagnetic emission from wind turbines is not a common problem.

Scattering is, however, an important electromagnetic interference mechanism associated with wind turbines. An object exposed to an electromagnetic wave disperses incident energy in all directions and it is this spatial distribution that is referred to as scattering.

The problem of the interaction of wind turbines and radio communication systems is complex as the scattering mechanisms due to wind turbines are not readily characterized and the signal may be modulated by the rotation of the blades. There is also a large, and increasing, number of types of radio systems with quite different requirements for their effective operation. ETSU (1997b) indicates that it is expected that the electromagnetic properties of wind turbine rotors will be influenced by:

- rotor diameter and rotational speed,
- rotor surface area, planform and blade orientation including yaw angle,
- hub height,
- structural blade materials and surface finish,
- hub construction,
- surface contamination (including rain and ice), and
- internal metallic components including lightning protection.

The range of potential problems experienced by UK wind farms was investigated by sending questionnaires to developers of 99 wind farms in 1996 (ETSU, 1997b). There was one questionnaire for each operating or proposed wind farm. Of the 46 responses received, 26 projects had encountered potential problems. A summary of the replies is given in Table 9.6. The majority of problems were associated with potential interference with local TV reception or with TV rebroadcast links (RBL). In the UK, where a potential problem with TV reception is identified from a proposed wind farm, the broadcasting authorities conduct an investigation and may then lodge an objection to the project receiving planning permission. In most of the cases reported in the questionnaires, the objection was overcome by the developer providing a guarantee that any loss of signal would be made good at the expense of the wind-farm project if complaints were received. Three wind farms reported actual local TV interference. In two cases this was rectified by modifications to domestic aerials while in the other case a self-help relay transmitter was installed and the domestic aerials realigned.

A smaller number of potential problems were reported with VHF radio and microwave links. One project is recorded as having a broad range of problems including:

Table 9.6 Summary of Replies to Questionnaires Enquiring about Potential EMI Problems on UK Wind-farm Projects (ETSU, 1997) (Reproduced by permission of ETSU on behalf of DTI)

Signal giving potential problem	Number of projects
None	20
Local TV Reception	15
TV rebroadcast link	11
Microwave link	5
Local radio	2
Civil radar	1
Defence radar	3

- two communications masts on the site area,
- microwave links from one mast,
- local VHF radio using both masts, including the fire brigade, and
- potential interference to local TV reception and RBL.

However, after discussions with the appropriate agencies and on the basis of detailed calculations it was possible to construct a wind farm layout that did not lead to objections on EMI grounds.

ETSU (1997b) includes information of the practices of various radio authorities in determining if a wind farm is likely to give rise to EMI problems. For microwave fixed links it is important that there is a clear line of sight between the transmitter and receiver and that a proportion of the first Fresnel zone must be free from obstructions. The first Fresnel zone is an ellipsoid region of space that makes the major contribution to the signal received and within which component parts of the signal will be in phase. At least 60 percent of the first Fresnel zone must be free from obstruction in order to ensure 'free space' propagation conditions. The radius of the first Fresnel zone (R_F) is given by

$$R_F = \sqrt{\frac{\lambda d_1 d_2}{d_1 + d_2}}$$

where d_1 and d_2 are the distance from the two microwave terminals to the point of reference, and λ is the wavelength (Figure 9.13).

In practice it is usually considered necessary to be completely outside the first Fresnel zone of the link with an additional exclusion zone of either 200 m or 500 m in order to avoid unwanted reflections. These additional allowances for reflections are rather conservative rules-of-thumb and may be reduced following detailed studies.

For UHF television relay links, if the wind farm is outside 60° of the relative direction of the receive antenna no major problems are anticipated. If the site is

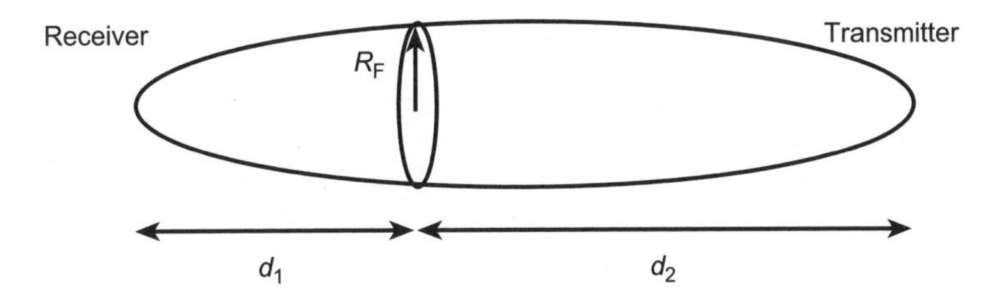

Figure 9.13 Illustration of First Fresnel Zone (Fresnel Ellipsoid)

within 15°–60° then some problems may be anticipated but may be overcome with a customized antenna. If the site is within 15° major problems may be anticipated. Similar considerations apply to domestic TV reception. No problems are anticipated if the wind farm is outside 60° of the relative direction of the domestic receive antenna, a good quality antenna being required for 20°–60° and a new source required if the wind farm is within 20°. It is emphasized by the radio authorities that each wind farm proposal needs to be considered individually.

9.4.1 Modelling and prediction of EMI from wind turbines

There are two fundamental interference mechanisms for EMI from wind turbines, back-scattering and forward-scattering (Moglia, Trusszi and Orsenigo, 1996). These are shown in Figure 9.14. Forward-scattering occurs when the wind turbine is located between the transmitter and receiver. The interference mechanism is one of scatter or refraction of the signal by the wind turbine and, for TV signals, it causes fading of the picture at the rotational speed of the blades. Back-scattering occurs when the turbine is located behind the receiver. This results in a time delay between the wanted signal and the reflected interference and gives rise to ghost or double images on a TV screen.

Moglia, Trusszi and Orsenigo (1996) (using the earlier work by van Kats and van

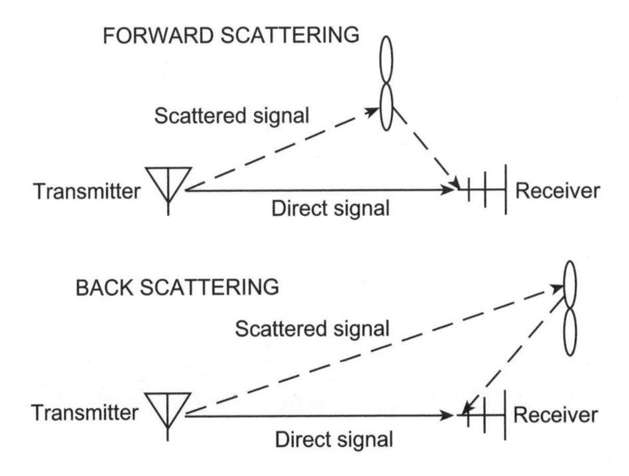

Figure 9.14 Interference Mechanisms of Wind Turbines with Radio Systems

Rees, 1989) provide an analysis of the electromagnetic interference caused by a wind turbine.

The useful carried signal received, C, is given by

$$C = P_T - A_{TR} + G_{TR} \tag{9.22}$$

where P_T is the transmitter power (dB), A_{TR} is the attenuation between the transmitter and receiver (dB), and G_{TR} is the receiver antenna gain in the direction of the required signal (dB). The interfering signal, I, is given by

$$I = P_T - A_{TW} + 10\log_{10}\left(\frac{4\pi\sigma}{\lambda^2}\right) - A_{WR} + G_{WR} \tag{9.23}$$

where A_{TW} is the attenuation between the transmitter and wind turbine (dB), A_{WR} is the attenuation between the wind turbine and receiver (dB), G_{WR} is the receiver antenna gain in the direction of the reflected signal (dB), $10\log_{10}(4\pi\sigma/\lambda^2)$ is the contribution to scattering of the wind turbine (dB), σ is the radar cross section (m^2). This may be understood as the effective area of the wind turbine. It is a function of the wind-turbine geometry and its dielectric properties together with the signal wavelength. λ is the wavelength of the signal (m).

It may be seen that the ratio of useful signal to interference is:

$$\left(\frac{C}{I}\right) = A_{TW} - 10\log_{10}\left(\frac{4\pi\sigma}{\lambda^2}\right) + A_{WR} - A_{TR} + G_{TR} - G_{WR} \tag{9.24}$$

Assume the distance between the transmitter and receiver is much greater than the distance from the wind turbine to the receiver denoted as r, then $A_{TW} = A_{TR}$. Assume the free space loss is $A_{WR} = 20\log_{10}(4\pi r/\lambda)$ and define the antenna discrimination factor as $\Delta G = G_{TR} - G_{WR}$.

Then (C/I) reduces to

$$\left(\frac{C}{I}\right) = 10\log_{10} 4\pi + 20\log_{10} r - 10\log_{10}\sigma + \Delta G \tag{9.25}$$

Thus the ratio of the useful carrier signal to interference may be improved by:

- increasing the distance from the turbine to the receiver, r,
- reducing the radar cross section, σ,
- improving the discrimination factor of the antenna, ΔG.

The carrier to interference ratio (C/I) defines the quality of a radio link. For example, a fixed microwave link may have a (C/I) requirement of 50–70 dB while for a mobile radio service the requirement may be only 15–30 dB (ETSU, 1997b).

Hence Equation (9.25) may be rearranged to define a 'forbidden zone' within which a wind turbine may not be located if an adequate carrier to interference ratio is to be maintained.

$$20\log_{10} r = \left(\frac{C}{I_{\text{required}}}\right) + 10\log_{10}\sigma - \Delta G - 11 \tag{9.26}$$

It may be seen that the 'forbidden zone' is critically dependent on the radar cross section, σ. Determination of the radar cross section of a wind turbine is not straightforward and a number of approaches are described in the literature. Van Kats and van Rees (1989) undertook a comprehensive series of site measurements on a 45 m diameter wind turbine. They estimated a radar cross section in the back-scatter region of 24 dBm^2 and a worst case value in the forward-scatter region of 46.5 dBm^2 (both values expressed as $10\log_{10}\sigma$).

Where measured results are not available, simple predictions may be made based on approximating the turbine blades to elementary shapes (Moglia, Trusszi and Orsenigo, 1996). For example, for a metallic cylinder the radar cross section is given by

$$\sigma = \frac{2\pi a L^2}{\lambda} \tag{9.27}$$

where a is the radius of the cylinder (m), L is the length of the cylinder (m), and λ is the signal wavelength (m), while for a rectangular metallic plate

$$\sigma = \frac{4\pi l^2 L^2}{\lambda} \tag{9.28}$$

where l is the width of the plate (m).

It is suggested that, in the back-scatter region, reflection is only caused by the metallic parts of the blades and so only these dimensions are used in the simple formulae. However, in the forward-scattering region the entire blades make a contribution although this is reduced because of the blade material (GRP) and shape. Hence, Moglia, Trusszi and Orsenigo (1996) apply the simple formulae but with a -5 dB correction for blade material and a -10 dB correction for blade shape (when using the rectangular approximation).

Hall (1992) quotes a radar cross-section model used by a number of researchers as:

$$10\log_{10}\sigma = 20\log_{10}\left[\frac{AX(\alpha)}{\lambda}\right] + 11 - C\ (\text{dBm}^2) \tag{9.29}$$

where A is the area of one blade (m^2), $X(\alpha)$ is a function describing the amplitude of the scattered signal in direction α, λ is the signal wavelength (m), C is a calibration constant, which may be assumed to be 15 dB.

Dabis and Chignell, in ETSU (1997b), dispute the use of these simple approaches to the determination of the radar cross section and suggest that more accurate computation is necessary. Their rather complex model is based on a physical optics formulation and assumes a conducting flat plate representation of the blades. It ignores any effects due to non-metallic blade materials and complex blade shapes. Although Dabis and Chignell provide interesting illustrative results of their method they consider significant further work is required to develop the technique further and to validate it against measured field data.

The simple approaches of Moglia, Trusszi and Orsenigo (1996) and van Kats and van Rees (1989) allow the calculation of interference regions or forbidden zones as indicated in Figure 9.15. The back-scattering region (for a given required (C/I) value) has a much smaller radius than the forward-scattering region. This is because the radar cross section is much smaller (only the metallic parts near the root of the blades were considered) and also it is possible to take advantage of the directivity of the receiving antenna (ΔG).

Equation (9.26) defines only the radii of the two regions and it is necessary to determine the angle over which the two regions extend. Van Kats and van Rees (1989) remark that some uncertainty exists as to where the back-scatter region changes into the forward-scatter region. Sengupta and Senior in Spera (1994) suggest that 80 percent of the region around the turbine should be in the back-scatter zone with the remaining 20 percent in the forward-scatter region. However, Moglia, Trusszi and Orsenigo (1996) use the formula

$$\Phi_{\text{border}} = \pi - \frac{\lambda}{L} \tag{9.30}$$

where λ is the wavelength and L the blade element radius. The back-scatter region extends between $0 < |\Phi| < \Phi_{\text{border}}$, while the forward-scatter region extends between $\Phi_{\text{border}} < |\Phi| < \pi$.

For the 45 m diameter turbine studied by van Kats and van Rees (1989), back-scatter region radii of 100 m (C/I) contour of 27 dB and 200 m (C/I) contour of 33 dB were determined. The forward scattering-region was much larger with radii of 1.3 km (C/I) contour of 27 dB and 2.7 km (C/I) contour of 33 dB. They suggest that a (C/I) value of 33 dB should result in no visible TV interference. Moglia, Trusszi and Orsenigo (1996) applied their method to medium-sized wind turbines (33 m and 34 m diameter) to give a back-scatter radius of approximately 80 m and a forward-scatter radius of 450 m for a 46 dB (C/I) contour. Their calculations were supported by measured site results, which gave reasonable agreement with the predictions. In these two examples back-scattering is unlikely to be a significant problem as other constraints (e.g., noise and visual effects) will ensure that the any dwelling is outside the 'forbidden zone'.

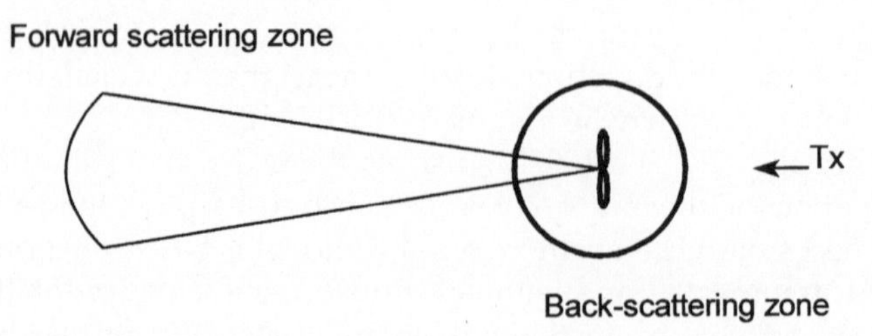

Figure 9.15 Example of simple calculation of interference regions of a wind turbine

In summary, it may be concluded that accurate analytical determination of the way wind turbines may interfere with radio communication systems remains difficult. In particular, techniques to determine the effects of irregular terrain and the details of the blade shape and materials have yet to be developed and validated. The approaches that rely on simplified assumptions to estimate the radar cross section can only be considered approximate. However, they do provide a useful qualitative understanding of the problem. In practice a dialogue with the local telecommunications authorities is required in order to determine if a wind turbine or wind farm development is likely to cause electromagnetic interference.

9.5 Ecological Assessment

Wind farms will often be constructed in areas of ecological importance and the Environmental Statement will include a comprehensive assessment of the local ecology, its conservation importance, the impact of the wind farm (during construction and operation) and mitigation measures. A study of the site hydrology is also likely to be included because of its importance for the ecology. It is suggested by English Nature (1994) that when considering the ecological impact of renewable energy schemes the following categories of effects should be considered:

- immediate damage to wildlife habitats during construction,
- direct effects on individual species during operation,
- longer-term changes to wildlife habitats as a consequence of construction or because of changed land use management practices.

Thus, the scope of the ecological assessment is likely to include:

- a full botanical survey including identification and mapping of plant species on the site,
- a desk and field survey of existing bird and non-avian fauna,
- an assessment of how the site hydrological conditions relate to the ecology,
- evaluation of the conservation importance of the ecology of the site,
- assessment of the potential impact of the wind farm,
- proposed mitigation measures, including which part(s) of the site should be avoided.

Gipe (1995) suggests that the main impact on plants and wildlife (excluding birds) is from road building and the disturbance of habitat. Direct loss of habitat due to a wind-farm construction is small (approximately 3 percent) and this can be reduced further by reinstating roads and construction areas once the wind farm is built. However, during construction there is likely to be considerable disturbance including frequent traffic of heavy vehicles and it may be necessary to timetable major work to avoid sensitive periods (e.g. breeding seasons).

9.5.1 Impact on birds

The impact of wind energy developments on birds was particularly controversial in the 1990s because of concern over raptors colliding with wind turbines in the USA and in the south of Spain at Tarifa. Elsewhere in Europe the main concern was not collisions of birds with turbines but issues associated with disturbance and habitat loss (Colson, 1995, ETSU, 1996). Since then, there has been very considerable research into the impact of wind farms on birds and it would appear that by careful assessment and monitoring, before construction and during operation of the wind farm, the environmental impact on birds may be managed effectively.

Lowther, in ETSU (1996), suggests that the presence of a wind farm may affect bird life in one or more of the following ways:

- collision,
- disturbance,
- habitat loss.

However, Lloyd (ETSU, 1996) suggests that, because the removal of habitat is comparatively small in a wind farm development, the main issues are the effect on bird behaviour either from construction or operation and direct mortality from birds colliding with wind turbines.

Gipe (1995) discusses US experience and provides Table 9.7 to give the estimated number of birds killed in Northern California. Colson quotes an estimated range of bird mortalities from wind-energy development in the USA prior to 1992 in the range of 0–0.117 birds/turbine year.

Table 9.7 Estimated Number of Birds Killed by Wind Turbines in Northern California (Gipe, 1995, Sources: for Altamont, Orloff and Flannery, California Energy Commission, 1992; for Solano, Howell and Noone, Solano County, 1992)

			Solano			
	Altamont raptors		All birds		Raptors	
	Low	High	Low	High	Low	High
Birds/year	164	403	17	44	11	2
Turbines	6800	6800	600	600	600	60
Birds/turbine year	0.024	0.059	0.029	0.074	0.018	0.04
MW	700	700	60	60	60	6
Birds/MW year	0.23	0.58	0.29	0.74	0.18	0.4

The wind farms of the Altamont pass in the early 1990s are not typical of European projects or even more recent developments in the USA. The turbines were very numerous, some 7000 turbines in 200 km^2, rather small in size (typically 100 kW), often on lattice steel towers and with close spacing along the rows facing the prevailing winds. Lloyd (ETSU, 1996) suggests that a number of factors may have led to the high collision rate among raptors. The Altamont turbines are often located on low rises and ridges in the pass to exploit local acceleration of the wind speed. A large number of the collisions appear to be at the end of the turbine rows where the birds may be attempting to fly around the densely packed turbines. In both Tarifa and Altamont there are few trees and it is suggested that some species used the lattice steel towers as perches and even for nesting.

The bird strike rate for early UK wind farms is quoted by Lowther (ETSU, 1996) and is shown in Table 9.8.

The three experimental turbines on Burgar Hill including a 3 MW 60 m diameter prototype were adjacent to the habitats of a number of bird populations of national significance, including 4 percent of the UK breeding population of the hen harrier. Studies were undertaken over a 9 year period and during that time four mortalities (three black-headed gulls and one peregrine) were noted as possibly being the result of turbine strike.

The Blyth Harbour wind farm [Still *et al.*, 1994] consists of nine 300 kW wind turbines positioned along the breakwater of a harbour which has been designated a Site of Special Scientific Interest. As it has the highest density of birds of any UK wind-farm site (110 varieties identified with more than 1100 bird movements per day) it was the subject of a monitoring programme. The main species that inhabit the area are: cormorant, eider, purple sandpiper and three types of gull. There was particular concern over the purple sandpiper, which winters in the harbour, and so special measures were taken to improve its roosting habitat by providing additional shelter. Over a 3 year period, 31 collision victims were identified. The mortality by collision was mainly in eider and among the gulls. It was calculated not to have a significant adverse impact on the local populations. In common with studies undertaken on Dutch and Danish coastal wind farms it appeared that most species had adapted to the wind turbine structures. With respect to disturbance and habitat loss, the study indicated that there was no significant long-term impact. The purple sandpiper was not adversely affected by the wind farm and, although temporary

Table 9.8 Bird Strikes at Wind Farms in the UK (Lowther in ETSU, 1996) (Reproduced by permission of ETSU on behalf of DTI)

Wind farm	Number of turbines	Bird strikes/ turbine year
Burgar Hill, Orkney	3	0.15
Haverigg, Cumbria	5	0
Blyth Harbour, Northumberland	9	1.34
Bryn Titli, Powys	22	0
Cold Northcott, Cornwall	22	0
Mynydd y Cemmaes, Powys	24	0.04

displacement of cormorant occurred during construction, the population returned to its precious haunts once construction was completed.

The Bryn Titli wind farm is adjacent to a Site of Special Scientific Interest, that holds important, breeding communities of: buzzard, peregrine, red grouse, snipe, curlew and raven. The wind farm was the subject of a bird impact study which showed no statistically significant impact on breeding birds and a bird strike study undertaken in 1994/95 indicated that it is unlikely that there were any collision mortalities during that time.

Clausager (ETSU, 1996) has undertaken an extensive review of both the American and European literature of the impact of wind turbines on birds. He concludes that, of the mainly coastal locations studied, the risk of death by collision with wind turbine rotors is minor and creates no immediate concern of an impact on the population level of common species. Drawing on 16 studies, and using a multiplier of 2.2 for birds not found, he estimates the highest number of mortalities due to collisions to be 6–7 birds/turbine year. With approximately 3500 wind turbines in Denmark this leads to maximum number of birds dying from collision as 20 000–25 000. This number is compared with at least one million birds being killed from traffic in Denmark each year. He dismisses the direct loss of habitat as being small and of minor importance but draws attention to the issues of changes in the area due to the wind farm construction particularly draining of low-lying areas. He also notes that some species of birds temporarily staying in an area may be adversely effected and that an impact has been recorded within a zone of 250–800 m, particularly for geese and waders. The need for further studies, particularly for offshore wind farms and developments on uplands, where only a limited number of investigations has been reported, is emphasized.

Both Lloyd and Colson propose mitigation measures that may be taken to protect important bird species while allowing wind-energy development to continue. These include the following.

- Baseline studies should be undertaken at every wind farm site to determine which species are present and how the birds use the site. This should be a mandatory part of the Environmental Statement for all wind turbines.

- Known bird migration corridors and areas of high bird concentrations should be avoided unless site specific investigation indicates otherwise. Where there are significant migration routes the turbines should be arranged to leave suitable gaps, (e.g., by leaving large spaces between groups of wind turbines).

- Microhabitats, including nesting and roosting sites, of rare/sensitive species should be avoided by turbines and auxiliary structures. (It may be noted that meteorological masts as well as wind turbines can pose a hazard for birds).

- Particular care is necessary during construction and it is proposed that access for contractors should be limited to avoid general disturbance over the entire site. If possible, construction should take place outside the breeding season. If this is not possible then construction should begin before the breeding season to avoid displacing nesting birds.

- Tubular turbine towers are preferred to lattice structures. Consideration should be given to using unguyed meteorological masts.

- Fewer large turbines are preferred to larger numbers of small turbines. Larger turbines with lower rotational speeds are probably more readily visible to birds then smaller machines.

- Within the wind farm the electrical power collection system should be underground.

- Turbines should be laid out so that adequate space is available to allow the birds to fly through them without encountering severe wake interaction. A minimum spacing of 120 m between rotor tips is tentatively suggested as having led to minimum collision mortalities on UK wind farms. Turbines should be set back from ridges and avoid saddles and folds which are used by birds to traverse uplands.

It is interesting to note that a 30 MW wind-farm development in Scotland was only able to proceed after an extensive study of raptors and the conversion of 450 ha of coniferous forest to heather dominated moorland and the exclusion of sheep from a further 230 ha (Madders and Walker, 1999). It is anticipated that this extension of the moorland habitat will increase the amount of prey away from the wind farm and so reduce the risk of collisions by golden eagles and other raptors.

9.6 Finance

9.6.1 Project appraisal

When evaluating power projects it has been conventional for many years to use techniques based on discounted cash flow (DCF) analysis (Khatib, 1997). This is based on the recognition that most people and organizations have a *time preference for money* and that they would rather receive money today not next year and would rather pay out money next year rather than today. The use of DCF analysis, with a high discount rate, tends to favour projects with short construction times, low capital costs and high operating costs. Thus, it makes renewable energy schemes, particularly offshore wind farms, rather difficult to justify, as a large initial capital investment is required with an income stream stretching into the future.

The mechanics of DCF analysis are simple and the calculation may be implemented easily on a spreadsheet and the main functions are often included in commercially available packages. Given a discount rate of r, the value of a sum in n years time is given by:

$$V_n = V_\mathrm{p}(1+r)^n \tag{9.31}$$

and so the present value of a sum received or paid in the future is given by:

$$V_{\mathrm{p}} = \frac{V_n}{(1+r)^n} \tag{9.32}$$

where V_n is the value of a sum in year n, and V_p is the present value of the sum.

Hence for a payment stream lasting m of years

$$V_{\mathrm{p}} = \sum_{n=1}^{n=m} \frac{V_n}{(1+r)^n} \tag{9.33}$$

This is a geometric series which, for equal payments (A), sums to

$$V_{\mathrm{p}} = \frac{1-(1+r)^{-n}}{r} A \tag{9.34}$$

This allows the calculation of the present value of any sum of money which is either paid or received in the future. The net present value (NPV) is simply the summation of all the present values of future income and expenditures. Using identical discounting techniques, it is possible to calculate other financial indicators:

- the benefit/cost ratio is the present value of all benefits divided by the present value of all costs,
- the internal rate of return (IRR) is the value of the discount rate that gives a net present value of zero.

These should not be confused with indicators not based on discounted values such as:

- payback period, expressed in years, which is the capital cost of the project divided by the annual average return;
- return on investment, which is the average annual return divided by the capital cost, expressed in percent.

For initial project appraisal some form of discounted cash flow is normally required and the non-discounted indicators are not used. The discount rate chosen reflects the value the lender places on the money and the risks that are anticipated in the project. In economic terms, the discount rate is an indication of the opportunity cost of capital to the organization. Opportunity cost is the return on the next best investment and so is the rate below which it is not worthwhile to invest in the project.

In the UK at present, discount rates of 8–12 percent would not be uncommon for commercial power projects reflecting the value placed on capital and the perceived level of risks. Some years ago (when the electricity supply industry was state-owned) the British Government discount rate required for power projects was 5 percent which was then raised to 8 percent. A number of continental European Governments still fund wind farm projects with rates of 4–5 percent.

In simple DCF analysis, general inflation is usually ignored and the discount rate is the so-called 'real' rate but it is possible to build in differential inflation if it is thought that some costs or income will rise at different rates over the life of the project. The financial model may be as sophisticated as required to reflect the concerns of the lender. It may, for example, include: (1) variations in electricity prices, (2) tax, (3) payments during construction, (4) residual value of the wind farm, and (5) operation and maintenance costs that vary over the life of the project.

There are generally considerable uncertainties with any financial modelling and it is often helpful to investigate the importance of the various assumptions and estimates in the model using a spider diagram. This is constructed by varying each of the important parameters one at a time from a base case. The slope of the resulting curves then indicates the sensitivity of the outcome to that particular parameter. The simple example of discounted cash flow analysis below is used to illustrate the concepts.

Example 9.1

Consider a 10 MW wind-farm project.

Capital cost £6 000 000
Annual capacity factor 28 percent
Annual operation and maintenance cost (O&M) £100 000
Sale price of electricity 0.04 £/kWh
Discount rate 10 percent
Project lifetime 15 years
The annual capacity factor is defined as the energy generated during the year (MWh) divided by wind farm rated power (MW) multiplied by the number of hours in the year. The determination of the capacity factor will, of course, be based on the best available information of wind speeds and turbine performance.
The annual energy yield is simply: $0.28 \times 10 \times 8760 = 24\,528$ MWh
The gross annual income is: $24\,528 \times 10^3 \times 0.04 =$ £981 120
Table 9.9 shows the simple DCF analysis while Figure 9.16 is the spider diagram to examine the sensitivity of the calculations.
The example serves to illustrate a number of points.

1. The net present value (NPV) is positive and so the project is suitable for further consideration.

2. With a 10 percent discount rate the discount factor in year 15 is 0.239. This indicates that the present value of the income (and expenditure) stream in that year is only 24 percent of its cash value.

3. The project is very sensitive to increase in capital costs or reductions in capacity factor. Any reduction in capacity factor could be caused either by low availability of the turbines or by low wind speeds. Steps need to be taken to control these

Table 9.9 Simple Discounted Cash Flow of a 10 MW Wind Farm

Year	Expenditure	Income	Net cash flow	Discount factor	Discounted cash flow
0	6 000 000		−6 000 000	1.000	−6 000 000
1	100 000	981 120	881 120	0.909	801 018
2	100 000	981 120	881 120	0.826	728 198
3	100 000	981 120	881 120	0.751	661 998
4	100 000	981 120	881 120	0.683	601 817
5	100 000	981 120	881 120	0.621	547 106
6	100 000	981 120	881 120	0.564	497 369
7	100 000	981 120	881 120	0.513	452 154
8	100 000	981 120	881 120	0.467	411 049
9	100 000	981 120	881 120	0.424	373 681
10	100 000	981 120	881 120	0.386	339 710
11	100 000	981 120	881 120	0.350	308 827
12	100 000	981 120	881 120	0.319	280 752
13	100 000	981 120	881 120	0.290	255 229
14	100 000	981 120	881 120	0.263	232 026
15	100 000	981 120	881 120	0.239	210 933
				NPV	490 936

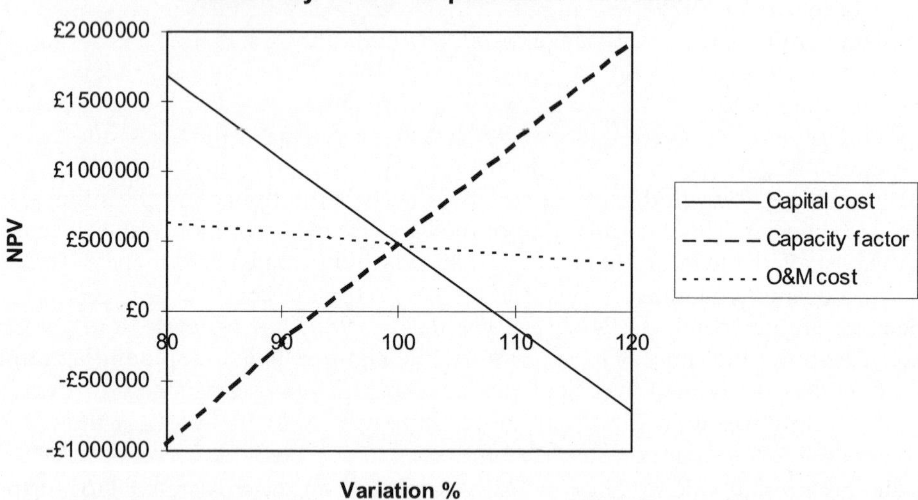

Figure 9.16 Sensitivity of Financial Model to Variations in Parameters

risks by obtaining a firm price for the wind farm, with guarantees on the availability of the turbines, and by ensuring that the projections of the wind speed at the site are based on the best data available.

4. The project is insensitive to operation and maintenance costs at the level included in the model.

5. In this simple model the costs have been discounted to the commissioning date (year zero). The complete payment for the wind farm is assumed to be made at commissioning and the income and O&M flows are assumed to be at the end of each years operation. If more precision is required then the financial model can be constructed to consider each quarter and the timing of the flows adjusted accordingly. Alternatively the annual income and expenditure may be taken to be in the middle of each year.

6. From trial and error calculations, the IRR is found to be 11.5 percent.

DCF modelling suffers from two major limitations. It does not reflect the external costs of power production (e.g., gaseous emissions from thermal plants or loss of visual amenity) and it does not represent cash flow. The lack of recognition of external costs is of major importance when governments or international lending agencies are considering alternative investments. Cash flow is a prime consideration for private developers of wind farm projects.

9.6.2 Project finance

Large companies (e.g., power utilities or major energy companies) may choose to develop small wind-farm projects using their own capital resources. The costs of the project are met from the general equity raised by the company with the liabilities of the project secured against the main corporate assets. The cost of borrowing, which influences the required discount rate for the project, depends on the financial strength of the company and it is possible to avoid the considerable expense involved in raising external finance. Hence this approach may allow cash-rich companies to develop projects at low cost.

Larger projects, and those developed by entrepreneurs, will tend to be based on project financing with a loan obtained from a bank or other financial institution. This has the advantage of reducing the requirement for capital from the developer but the loan repayment will have the first call on the income of the project. Project financing is also likely to offer tax advantages. If the loan is secured only on the project cash flow itself, this is referred to as 'limited recourse' financing. Alternatively, if the developer has a large parent company willing and able to provide suitable guarantees then the loan may be secured against the assets of the parent company which then appears as a liability on its balance sheet. Clearly the lenders of the debt will prefer to guarantee their loan against the assets of a large stable company rather than the project alone and this will be reflected in the repayment terms that are available. The debt/equity ratio can be as high as 80 percent but all risks must be identified and manageable if a suitable financial arrangement is to be made. If a limited recourse loan is made before a project is constructed then interest rates of 10–20 percent would be typical while these would be significantly reduced if the loan is re-financed after the project is commissioned and many of the construction risks are no longer significant.

Figure 9.17 shows a typical commercial structure of a wind-farm project. This is very similar to any other power project developed by independent power

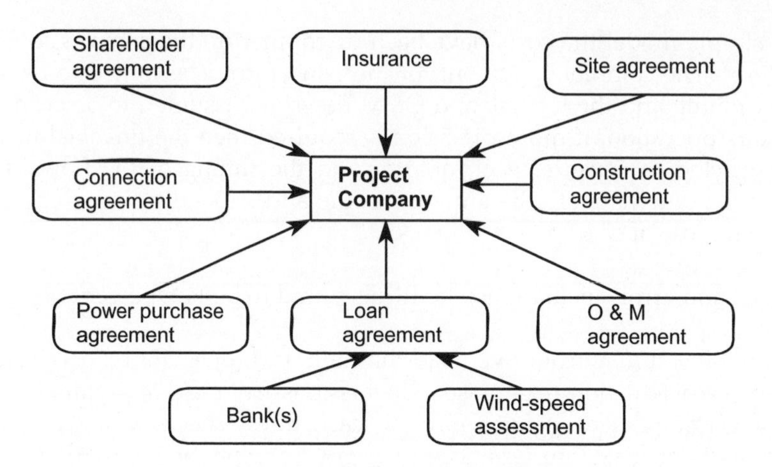

Figure 9.17 Typical Commercial Structure of a Wind-Farm Project

producers (IPPs) where the Project Company is merely the vehicle by which a large number of agreements are made. In order to limit risk, construction of the project does not start until all the agreements are in place and so-called 'financial close' is achieved.

The *power purchase agreement* is to sell the output electrical energy of the wind farm. To reduce risk, this should be at a defined price for the duration of the project. Under the UK Non Fossil Fuel Obligation or the German 'Electricity Feed-in-Law' this was reasonably easy to achieve but with the continuing liberalization of markets for electrical energy throughout the world it will become increasingly expensive to avoid exposure to varying energy prices.

The *loan agreement* is for the bank(s) to provide the debt finance for the project. An accurate and verifiable assessment of the wind resource is an essential prerequisite for this agreement although there is also likely to be a 'due diligence' investigation of the whole project to ensure all major risks are addressed.

The *construction agreement* is to purchase the wind turbines and construct the wind farm. To reduce risk this may be done on a 'turn-key' basis with the wind-turbine manufacturer taking responsibility for the entire wind farm.

The *O & M agreement* is to operate and maintain the wind farm for the life of the project.

The *site agreement* is to define the relationship with the landowners and to ensure access to the site and the wind resource for the duration of the project.

The *connection agreement* is to allow the wind farm to be connected to the electrical power system and so export its output. In a deregulated power system this is separate from the power purchase agreement.

The *shareholder agreement* is between the owners of the Project Company to define their rights and obligations.

It may be seen that a large number of agreements are required and it will typically take a year from project inception to financial close. During this time the developer is operating at risk, expending considerable resources, and so full development of this administrative/financial phase of the project is only commenced once feasibility has been demonstrated.

9.6.3 Support mechanisms for wind energy

Historically, electrical energy from wind turbines was not competitive in commercial markets with other forms of generation, particularly the use of a combined cycle gas turbine (CCGT) plant burning natural gas. Hence, in order to take account of external costs, and to meet commitments to reduce CO_2 emissions, various support mechanisms have been used by governments to encourage the development of wind power as well as other forms of renewable energy. These support mechanisms, together with the markets for electrical energy, are subject to very rapid change but the main principles are described.

Perhaps the most obvious approach to support wind power is to require fixed premium tariffs to be paid for all power generated by renewable sources. This was the basis for the 'Public Utilities Requirement to Purchase Act (PURPA) introduced in the USA in 1978 but abandoned in the late 1980s and the German 'Electricity Feed-in-Law'. A similar approach was adopted in Spain and Denmark for a period. The PURPA legislation was based on 'avoided cost', i.e., the marginal cost of generation from an alternative plant was paid. This varied from state to state but in California in the early 1980s the avoided cost was calculated to be US$0.06/kWh. This rate coupled with an inflation factor and tax incentives led to some 2000 MW of wind turbine capacity being installed between 1981 and 1989. However, development slowed dramatically as gas fired CCGT plant reduced the avoided cost. The German 'Electricity Feed-in-Law' required the utility to purchase wind generated power at 90 percent of the average price paid by all customers. As the bulk wholesale price of electrical energy is typically a third of the retail price it may be seen that this was a most attractive arrangement for wind farm operators. The cost of the 'Feed-in-Law' was borne by the utilities who made efforts to have the law rescinded. Denmark also operated a similar system with the utilities paying 70–85% of the retail price for wind-generated power. This rate, together with tax incentives, encouraged individuals and co-operatives to buy wind turbines and some 75% of turbines in Denmark are owned in this way.

The general effect of this type of premium tariffs is to provide very considerable stimulus to the development of wind power. However, this form of support may not apply strong downward pressure on wind farm costs. Hence, the price paid per kWh in Germany in 2000 was approximately double that available under the last round of the Non Fossil Fuel Obligation in the UK which was based on an auction of renewable contracts. However, it should be noted that the UK sites generally had higher wind speeds than many of those in Germany. It is difficult to reconcile the concept of fixed premium tariffs with that of a deregulated market for electrical energy and so this form of support may not be acceptable to European Union competition authorities in the long term.

Both the UK and Ireland have used a process based on competitive auctions of fixed-term, fixed-price power purchase contracts. This type of support mechanism has the effect of providing very considerable downward pressure on costs and also offers the developer stability as the price is guaranteed for the long term (in the UK for 15–20 years with an additional allowance for general inflation). The major disadvantages are that: (1) the auctions are held intermittently and so there is no continuous market for the turbine manufacturers; (2) very considerable costs are incurred in bidding which may be wasted if the bid is unsuccessful; and (3) there is little penalty on not constructing the projects and so some bids may be unrealistically low. The UK Non Fossil Fuel Obligation (NFFO) was remarkable in reducing the price for wind energy by two-thirds from 1989 (9p/kWh) to 1999 (2.9p/kWh) although the 1989 contracts were for only 9 years while the 1999 contracts were for 20 years. However, only limited capacity was actually built. The NFFO has now been abandoned and has been replaced by an obligation on electricity suppliers to provide a percentage of their energy from renewable sources and a system of tradable 'Green Certificates'.

The use of some form of 'Green Certificates' together with an obligation on users or suppliers to ensure that a fraction of their electrical energy comes from renewable sources is presently seen as the most likely future development for support mechanisms. There need be no physical connection between the 'Green Certificate' and the electrical energy and the 'Certificates' may be traded in one market and the energy (kWhs) in another. The advantages of this type of arrangement are that: (1) it is compatible with a competitive market for electrical energy; (2) it is likely to exert a downward pressure on costs; (3) it may develop a large market attractive to major companies; and (4) the full range of market instruments (e.g., futures markets) may be developed.

If Government support mechanisms are to be based on some form of 'Green Certificates' then the wind farm operator is still left with the problem of selling the energy. In markets for electrical energy, the wholesale price varies considerably throughout the day and year and so the wind farm operator is likely to be exposed to volatile prices. After privatization of the electricity supply industry in the UK a pool-type trading arrangement was developed whereby any generator might sell into the pool at the system marginal price of generation. Additional contracts made outside the pool, the so-called 'contracts for differences' were used by many conventional generators to manage the uncertainty of the variation of the pool price but the availability of the pool marginal price always provided a reference price for power. However, the pool is now being superseded by a system of bilateral energy contracts which will place a very high premium on controllability and predictability of generating plant. The New Electricity Trading Arrangements (NETA) will comprise three elements: (1) long-term bilateral contracts between generators, large customers and those supply companies serving smaller customers; (2) a short-term market, 24–3.5 hours ahead of delivery, between generators, customers and suppliers; (3) a balancing mechanism, 3.5 h ahead of delivery, operated by the National Grid Company (the Transmission Operator). Failure to match contracted demand with supply will probably result in higher costs in the short-term market or balancing market. This is likely to provide considerable difficulties for wind farm operators due to the intermittent and unpredictable nature their output.

Some consumers are prepared to pay a premium for electrical energy provided from renewable sources. This 'Green Pricing' is being developed in Germany and the UK but the extent of the market demand it is not yet clear. There are also difficulties with accrediting which generating plants are 'green' and suitable for such a scheme.

In parts of the USA 'net metering' is allowed for some small wind turbines. This is the use of a single electricity meter which operates backwards to reduce the measured kWh at times of export from the wind turbine and its local load, typically a house or farm. In economic terms this approach can be criticized as it ignores the variation with time in the value (and price) of electrical energy and also that a large element of the cost of electrical energy delivered to the customer is associated with the transmission and distribution facilities and their management.

References

Addison, J. F. D. *et al.*, (2000). 'A neural network version of the Measure–Correlate–Predict algorithm for estimating wind energy yield'. *International Congress and Exhibition on Condition Monitoring and Diagnostic Management*, Texas, USA.

British Standard, (1999). Wind turbine generator systems – Part 11: Acoustic noise measurement techniques'. *BS EN61400–11:1999*. BSI.

British Standard, (1997). 'Method for rating industrial noise affecting mixed residential and industrial areas'. *BS4142: 1997*. BSI.

British Wind Energy Association, (1994). 'Best practice guidelines for wind-energy development'. British Wind Energy Association, www.bwea.com.

Colson, E. W., (1995). 'Avian interactions with wind energy facilities: a summary'. *Proceedings of the American Wind Energy Association Conference, Windpower '95*, pp 77–86.

Danish Statutory Order on Noise from Windmills, (1991). 'Bekendtgørelse om stoj fra windmoller' (in Danish).

Department of the Environment, (1993). 'Planning Policy guidance note 22: Renewable energy, annex on wind energy'. *PPG22*, Department of the Environment, Welsh Office.

Derrick, A., (1993). 'Development of the Measure–Correlate–Predict strategy for site assessment'. *Proceedings of the European Wind Energy Association Conference*, pp 681–685.

English Nature, (1994). 'Nature conservation guidelines for renewable energy projects'. English Nature, Peterborough, UK.

ETSU, (1993). 'Attitudes to wind power – a survey of opinion in Cornwall and Devon'. *ETSU Report W/13/00354/038/REP*. ETSU, Harwell, UK.

ETSU, (1994). 'Cemmaes wind farm sociological impact study – final report'. *ETSU Report W/13/00300/REP*. ETSU, Harwell, UK.

ETSU, (1995). 'Wind farm radar study'. *ETSU Report W/32/00228/49/REP*. ETSU, Harwell, UK.

ETSU, (1996). 'Birds and wind turbines: can they co-exist'. *ETSU-N-113*. ETSU, Harwell, UK.

ETSU, (1997a). 'The assessment and rating of noise from wind farms'. *Final Report of the Working Group on Noise from Wind Turbines, ETSU-R-97*. ETSU, Harwell, UK.

ETSU, (1997b). 'Investigation of the interactions between wind turbines and radio systems aimed at establishing co-siting guidelines, Phase 1: Introduction and modelling of wind turbine scatter'. *ETSU Report W/13/00477/REP*. ETSU, Harwell, UK.

ETSU, (1999). 'UK Onshore wind energy resource', *ETSU Report R-99*, ETSU, Harwell, UK.

ETSU, (2000). 'A critical appraisal of wind-farm noise propagation', *ETSU Report W/13/00385/REP*. ETSU, Harwell, UK.

EUREC Agency, (1996). *'The future for renewable energy, prospects and directions'*. James and James, London, UK.

European Wind Atlas, (1989). Riso National Laboratory, Roskilde, Denmark, and www.windpower.dk.

EWEA. 'European best practice guidelines for wind-energy development'. European Wind Energy Association, www.ewea.org.

Gipe, P., (1995). *Wind energy comes of age*. John Wiley and Sons, New York, US.

Hall, H. H. (1992). 'The assessment and avoidance of electromagnetic interference due to windfarms'. *Wind Engng* **16**, 6, 326–337.

International Energy Agency, (1994). *Recommended practice for wind turbine testing 4. Acoustics – Measurements of noise emission from wind turbines, Third Edition.*

International Energy Agency, (1997). *Recommended practice for wind turbine testing 10. Measurements of noise emission from wind turbines at noise receptor locations, First Edition* IEC61400-11 1998.

International Energy Agency, (1998). *Wind turbine generator systems Part II: Acoustic noise measurement techniques.* IEC 61400-11: 1998.

Khatib, H., (1997). *Financial and economic evaluation of projects in the electricity supply industry.* IEE Power Series, London, UK.

Landberg, L. and Mortensen, N. G., (1993). 'A comparison of physical and statistical methods for estimating the wind resource at a site'. *Proceedings of the 15th British Wind Energy Association Conference* (ed. K. F. Pitcher). Mechanical Engineering Publications, Bury St Edmunds, UK pp 119–125.

Madders, M. and Walker, D. G., (1999). 'Solutions to raptor–wind farm interactions'. *Proceedings of the 21st British Wind Energy Association Conference* (Ed. P. Hinson). Mechanical Engineering Publications, Bury St Edmunds, UK, pp 191–195.

Moglia, A., Trusszi, G. and Orsenigo, L., (1996). 'Evaluation methods for the electromagnetic interferences due to wind farms'. *Proceedings of the Conference on Integration of Wind Power Plants in the Environment and Electric Systems, Paper 4.6.*

Mortimer, A. A., (1994). 'A new correlation/prediction method for potential wind farm sites'. *Proceedings of the 16th British Wind Energy Association Conference* (Ed. G. Elliot). Mechanical Engineering Publications, Bury St Edmunds, UK, pp 349–352.

Pinder, J. N., (1992). 'Mechanical noise from wind turbines'. *Wind Engng,* **16**, 3, 158–168.

Spera, D. A., (1994). *Wind turbine technology: fundamental concepts of wind turbine engineering.* ASME Press, New York, US.

Stanton, C., 'The visual impact and design of wind farms in the landscape'. *Proceedings of the 16th British Wind Energy Association Conference.* Mechanical Engineering Publications, Bury St Edmunds, UK, pp 249–255.

Stiesdal, H. and Kristensen, E., (1993). 'Noise control on the BONUS 300 kW wind turbine'. *Proceedings of the 15th British Wind Energy Association Conference.* Mechanical Engineering Publications, Bury St Edmunds, UK, pp 335–340.

Still *et al.*, (1994). 'The birds of Blyth Harbour'. *Proceedings of the British Wind Energy Association Conference.* Mechanical Engineering Publications, Bury St Edmunds, UK, pp 241–248.

Taylor, D. and Rand, M., (1991). 'Planning for wind energy in Dyfed'. *EERU 065.* Energy and Environment Research Unit, Open University, UK.

Van Kats, P. J. and van Rees, J., (1989). 'Large wind turbines: a source of interference for TV broadcast reception'. *Wind energy and the environment* (Ed. Swift-Hook, D. T.). IEE Energy Series, London, UK, pp 95–104.

Verkuijlen, E. and Westra, C. A., (1984). 'Shadow hindrance by wind turbines'. *Proceedings of the European Wind Energy Conference.* pp 356–361.

Wagner, S., Bareis, R. and Guidati, G., (1996). *Wind turbine noise.* Springer-Verlag, Berlin, Germany.

Walmsley, J. L., Salmon, J. R. and Taylor, P. A., (1982). 'On the application of a model of boundary-layer flow over low hills to real terrain'. *Boundary-layer Meteorology,* **23**, 17–46.

Walmsley, J. L., Taylor, P. A. and Keith, T., (1986). 'A simple model of neutrally stratified boundary-layer flow over complex terrain with surface roughness modulations – MS3DJH/3R'. *Boundary-layer Meteorology,* **36**, 157–186.

WA^{s}P – *www.wasp.dk* and Chapter 8 of *European Wind Atlas.* Riso National Laboratory, Roskilde, Denmark, www.windpower.dk.

10
Electrical Systems

10.1 Power-collection Systems

Figure 10.1 shows a schematic of a typical fixed-speed wind turbine electrical system. The main power circuit is from the generator, *via* three flexible pendant cables to a moulded case circuit breaker (MCCB). The MCCB is fitted with an overcurrent protection trip with an instantaneous setting against faults and a delayed (thermal) function which operates at a lower current level and is intended to detect overload of the generator. This arrangement follows normal practice for the control of large motors. An anti-parallel thyristor soft-start unit, normally with a bypass contactor, is used to reduce the current inrush when the generator is energized. There are a number of additional circuits including those for the power factor correction capacitors (PFC) and the auxiliary supplies. The power factor correction capacitors are switched in stages to provide a greater degree of control of reactive power and also to limit capacitive switching currents. Small reactors may be connected in series with the capacitors to reduce the inrush current.

Most of the auxiliaries use alternating current but there may also be a requirement to provide direct current for the wind turbine controller. The auxiliary circuits are protected by a smaller MCCB and fuses rated at lower currents. Surge diverters are fitted to protect the internal electrical system from overvoltages being transmitted from the site electrical network and another set of surge diverters may be connected at the generator terminals in the nacelle. The normal scope of supply of

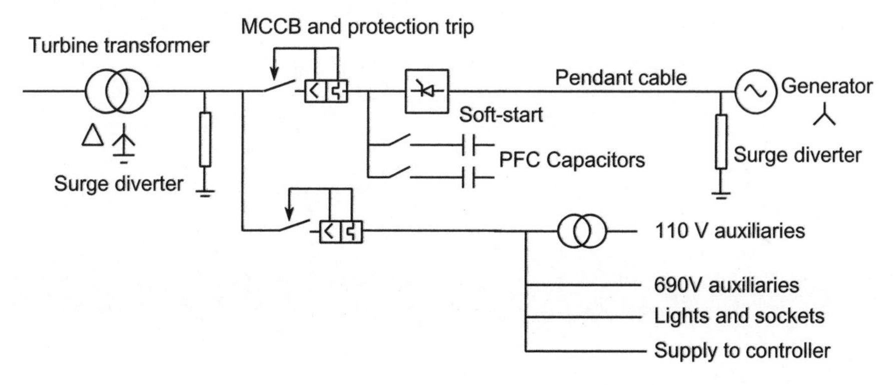

Figure 10.1 Electrical Schematic of a Fixed-speed Wind Turbine

the wind turbine manufacturer will end at the low voltage connection to the turbine transformer.

The voltage level chosen for the main power circuit from the generator to the tower base is usually less than 1000 V and is selected to be one of the internationally standard voltages, in Europe typically 690 V between phases. This voltage is surprisingly low for a large electrical machine and leads to rather high currents. For example a 600 kW wind turbine operating at 690 V requires a current of over 500 A. However, it is found to be convenient and cost-effective to restrict all voltages within the wind turbine to less than 1000 V as, in many countries, the safety regulations become very much more severe for voltages above this value and special precautions are required including the provision of dedicated equipment to earth the circuits before work on them is permitted. Further reasons for this low voltage include a wider choice of switchgear and flexible cables for the pendant and higher production volumes and hence lower costs of the generators.

However, this low generator voltage leads to a requirement for a transformer located in the tower or immediately adjacent to it (e.g., Figure 10.2). In some early wind farms a number of small wind turbines were connected to one transformer (Figure 10.3) but as turbine ratings have increased with consequently higher currents this approach would lead to excessive voltage variations and electrical losses in the low-voltage cables. It is only with rather small wind turbines located close together that the connection of multiple turbines to one transformer is likely to be cost-effective.

Figure 10.2 Transformer and Switchgear of 1.5 MW Wind Turbine, MV Switchgear on right, LV Switchgear on left (Reproduced by permission of NEG MICON; www.neg-micon.dk)

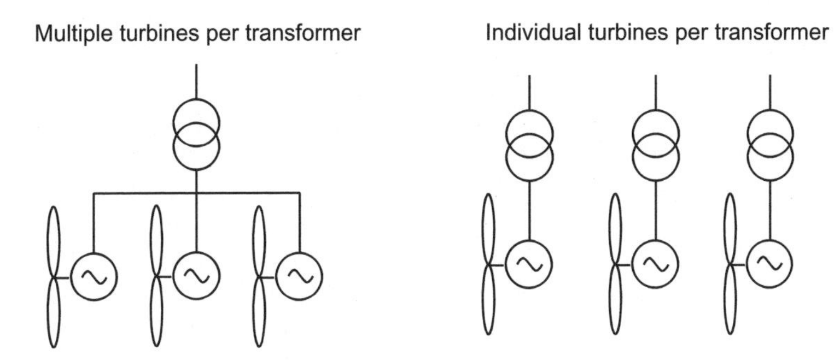

Figure 10.3 Grouping of Turbines to Transformers

The choice of medium-voltage (MV) level for the wind-farm power-collection system is usually determined by the practice of the local distribution utility. In this way cable and switchgear is readily available. Thus, in the UK the choice is generally between 11 kV and 33 kV while in continental Europe wind farm collection circuits are likely to be either 10 kV, 15 kV or 20 kV.

Local utility practice also influences the choice of the neutral earthing arrangement on the MV circuits. 11 kV circuits in the UK tend to have their neutral point either solidly earthed or connected to ground through a low-value resistor (typically 6.35 Ω to allow 1000 A to flow into a phase/ground fault). Solid neutral earthing is cheaper as it requires no extra equipment but can lead to high earth fault currents which may cause damage and high step or touch voltages (ANSI/IEEE, 1986). In contrast, on some continental European wind farms the neutral point of the medium voltage system is left unearthed. This allows operation of the wind farm to continue with a single phase-earth fault but with the voltages of the healthy phases raised by a factor of $\sqrt{3}$.

The design of the power collection is very similar to that of any MV power network but with one or two particular aspects. It is generally not necessary to provide any redundancy in the circuits to take account of MV equipment failures. Both operating experience and reliability calculations have shown that the power collection system of a wind farm is so much more reliable than the individual wind turbines that it is generally not cost-effective to provide any duplicate circuits. This also applies to the single circuit or transformer linking the wind farm to the utility network where duplication cannot usually be justified. If part of the wind power-collection circuit does fail then the only lost revenue is that of the energy output of that part of the wind farm. This is easily estimated and is generally modest. In contrast, if a public distribution utility circuit fails then the consequence is that the customers do not have access to electrical power. The commercial value of this loss of supply is more difficult to quantify but is generally considered to be several orders of magnitude higher than the price of the energy. Thus, public distribution circuits often have duplicate supplies while wind farm power-collection networks tend to consist of simple radial circuits with limited switchgear for isolation and switching. The wind turbine transformers are connected directly to the radial circuits.

In Europe almost all power-collection circuits within wind farms are based on underground cable networks. This is both for reasons of visual amenity and for safety as large cranes are required for the erection of the turbines. In other parts of the world (e.g., India and the USA) overhead MV lines are sometimes used within the wind farm to reduce costs. If cables are used then their low series inductive-reactance leads to small variations of voltage with current and so the circuits are dimensioned on considerations of electrical losses. Any wind farm project which reaches the stage of requiring a detailed design of the electrical system should also have a good estimate of the wind farm output. Thus it is straightforward to use these data to calculate the losses in the electrical equipment at various wind turbine output powers using a load-flow program, sum these over the life of the project and using discounted cash flow techniques choose the optimum cable size and transformer rating. In countries where wind-generated electricity attracts a high price then it is likely to be cost-effective to install cables and transformers with a thermal rating considerably in excess of that required at full wind-farm output in order to reduce losses. It may also be worthwhile to consider 'low-loss' transformer designs.

10.2 Earthing (Grounding) of Wind Farms

All electrical plant require a connection to the general mass of earth in order to:

- minimize shock hazards to personnel and animals,
- establish a low-impedance path for earth-fault currents and hence satisfactory operation of protection,
- improve protection from lightning and retain voltages within reasonable limits, and
- prevent large potential differences being established which are potentially hazardous to both personnel and equipment.

In the UK this subject is referred to as 'earthing' while in the US it is called 'grounding'. The terms may be considered to be synonymous.

Because earthing has obvious safety implications there is considerable guidance available in both national and international standards for conventional electrical plant. For earthing of AC substations, the US standard (ANSI/IEEE, 1986) is widely applied and Charlton has written an informative guide (Copper Development Association, 1997) which is mainly focused on UK practice.

Wind farms, however, have rather unusual requirements for earthing. They are often very extensive stretching over several kilometres, subject to frequent lightning strikes because of the height of modern wind turbines, and are often on high-resistivity ground being located on the tops of hills. Thus, normal earthing practice tends not to be easily applicable and special consideration is required. The IEEE Recommended Practice (1991), which is no longer current, recommended 'that the entire wind farm installation have a continuous metallic ground system connecting all equipment. This should include, but not be limited to, the substation, transformers, towers, wind-turbine generators and electronic equipment.' This practice is generally followed with bare conductor being laid in the power-collection cable trenches to provide both bonding of all parts of the wind farm as well as a long horizontal electrode to reduce the impedance of the earthing system.

A wind farm earthing system is required to operate effectively for both power frequency 50/60 Hz currents and lightning surges which have rise-times typically of less than 10 μs. Although it is conventional to use the same physical earthing network for power frequency currents and lightning surges the response of the earthing system to the high-frequency components of the lighting current is completely different to that at 50 Hz.

The performance of a wind-farm earthing system may be understood qualitatively by considering Figure 10.4.

At each wind turbine a local earth is provided by placing a ring of conductor around the foundation at a depth of about 1 m (sometimes known as a counterpoise earth) and by driving vertical rods into the ground. It is common to bond the steel reinforcing of the wind-turbine foundation into this local earthing network. The purpose of this local earth is to provide equipotential bonding against the effects of both lightning and power frequency fault currents and to provide one element of the overall wind-farm earthing system. Regulations differ for the value of this local earth but in the UK it is necessary to achieve a resistance to earth of less than 10 Ω (British Standard, 1999) before the connection of any power or SCADA cables. These local turbines' earths are shown as R_{turbine} in Figure 10.4. As the turbine-earthing network consists of a ring of only, say, 15 m in diameter and local driven rods it may be considered as purely resistive.

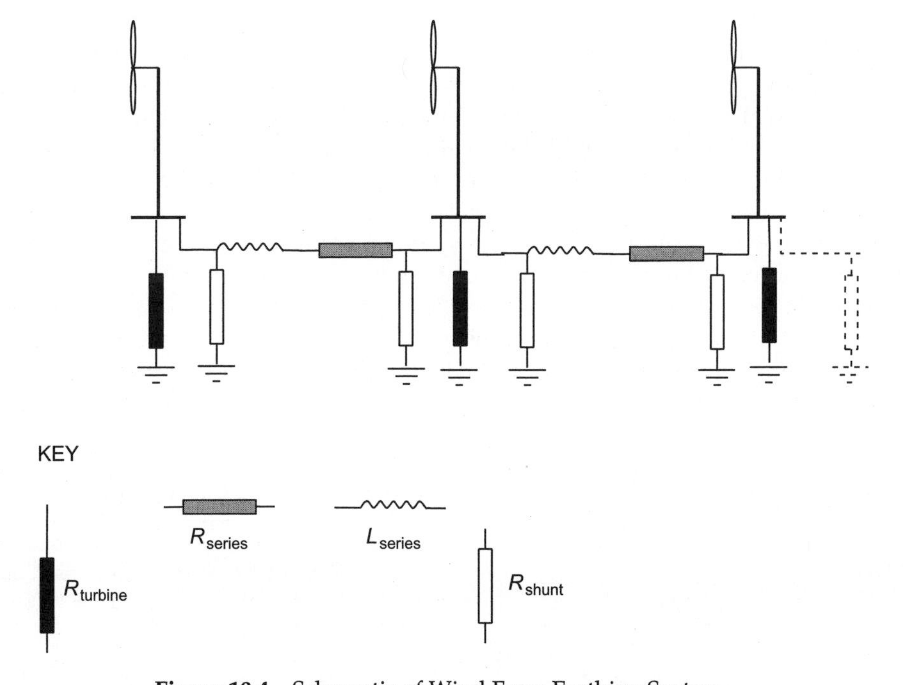

Figure 10.4 Schematic of Wind Farm Earthing System

However, the long horizontal electrodes linking one turbine to the next have a more complex behaviour (similar to a transmission line) and are represented in Figure 10.4 as a π equivalent circuit. Thus the resistance to ground is indicated by R_{shunt} while the series impedance is the combination of R_{series} and L_{series}. R_{series} comes about simply from the resistance of the earth wire while L_{series} is the self-inductance of the earth circuit. On the long earthing networks found on large wind farms this series impedance cannot be ignored. It may be seen immediately that for the high-frequency components of a lightning strike on a wind turbine the series inductance acts effectively to reduce the earthing network to only the local turbine earth. Even with 50 Hz fault current the series impedance leads to significantly higher earthing impedances than would be expected with geographically small earthing systems where series impedance may be neglected. In this discussion shunt capacitance has been ignored although this may become significant at very high frequencies.

This behaviour of wind-farm earthing systems has been confirmed by site measurements and Table 10.1 shows the measured values for the earth impedance at the main substations of two UK wind farms.

It may be seen that even at 50 Hz the impedance to earth consists of almost equal resistive and inductive parts (i.e., the X/R ratio of the circuit to earth is almost unity). This has important implications both for the design and testing of wind-farm earthing systems. It is not possible to use conventional calculation techniques which were designed for small, fully resistive, earth networks and it is necessary to take into account the effect of the long conductors. Sophisticated computer programs to do this are available and these may be used to consider both 50 Hz and lightning surge behaviour. Similarly, simple testing methods such as those developed by Tagg (1964) are only intended to measure resistance to earth and will give optimistic results on large wind farms. The only effective measurement method to determine the 50 Hz earth impedance of a large wind farm is the so-called current injection test. In this test a current (typically 10–20 A) is injected into the earth electrodes at the wind farm and the potential rise measured with respect to a 'true' earth. However, the return path of the injected current must be remote from the wind farm (usually some 5–10 km away) to ensure that it does not effect measurements at the wind farm. Hence it is conventional to use the de-energized connection circuit of the wind farm to the main power network as the path of the test current. The rise in potential of the wind-farm earthing systems is measured against a remote earth which classically is transferred to the site over a metallic telephone circuit. Ideally the route of the telephone circuit should be orthogonal to the power circuit to avoid induced effects. It may be seen that tests of this type are extremely

Table 10.1 Measured Earth Impedance at Substations of Two Wind Farms

Capacity of wind farm (MW)	Number of turbines	Route length of horizontal earth conductor (km)	Earth impedance at main substation (Ω at 50 Hz)
7.2	24	6.7	$0.89 + j0.92$
33.6	56	17.7	$0.46 + j0.51$

difficult and expensive to undertake as they involve taking the wind farm out of commission and arranging a temporary earth at the remote end of the current injection circuit. Also it is becoming increasingly difficult to use a telephone circuit to provide the remote earth reference as telecommunication companies increasingly use non-metallic media. However, it is very questionable whether simpler tests give useful results on large earthing networks with significant series inductance.

10.3 Lightning Protection

It is now recognized that lightning is a significant potential hazard to wind turbines and that appropriate protection measures need to be taken. Some years ago, it was thought that as the blades of wind turbines are made from non-conducting material (i.e., glass reinforced plastic (GRP) or wood-epoxy) then it was not necessary to provide explicit protection for these types of blades provided they did not include metallic elements for the operation of devices such as tip brakes. However, there is now a large body of site experience to show that lightning will attach to blades made from these materials and can cause catastrophic damage if suitable protection systems have not been fitted. Of course, if carbon fibre (which is conducting) is used to reinforce blades then additional precautions are necessary. In response to damage experienced on sites a number of new standards and recommendations for the lightning protection of wind turbines have been developed (Cotton *et al.*, 2000, International Energy Agency, 1997, IEC, 2000a).

Lightning is a complex natural phenomenon often consisting of a series of discharges of current. The term 'lightning flash' is used to describe the sequence of discharges which use the same ionized path and may last up to 1 s. The individual components of a flash are called 'strokes'.

Lightning flashes are usually divided into four main categories:

- downward inception, negative and positive polarity, and
- upward inception, negative and positive and polarity.

Generally flashes which start with a stepped leader from the thunder cloud and transfer negative charge to earth (downward inception – negative polarity) are the most common. Downward negative flashes typically consist of a high-amplitude burst of current lasting for a few microseconds followed by continuing current of several hundred amps. Then, following the extinction of the initial current transfer between cloud and earth there may be a number of restrikes. However, in some coastal areas there may be a majority of positive flashes of downward inception which consist of a single long duration strike with high charge transfer and specific energy content. Upward inception is normally associated with very high objects (e.g., communication towers) and the lightning has very different characteristics. Although the maximum peak current of this form of lightning is low (some 15 kA) the charge transfer can be very high and hence has significant potential for damage. The top of wind turbine blades can now be over 100 m above ground and so there is growing concern over the effect of upward negative lightning flashes. Upward positive flashes are rare.

Table 10.2 shows the parameters normally used to characterize lightning and some aspects of their potential for damage in wind turbines. The peak current of a single lightning stroke is over 200 kA but the median value is only approximately 30 kA. Corresponding values for charge transfer are 400 C (peak), 5 C (median) and specific energy 20 MJ/Ω (peak) and 55 kJ/Ω (median). The very large range of these parameters implies that the initial step in any consideration of lightning protection of a wind farm or wind turbine is to undertake a risk assessment (IEC, 2000a). The risk assessment will include consideration of the location of the turbines as the frequency and intensity of lightning varies considerably with geography and topology.

Table 10.3 shows some historical data of the frequency of lightning damage (Cotton *et al.*, 2000). Although this must be interpreted with care, as both wind-turbine design and lightning-protection systems were evolving rapidly during this period, the data do indicate the scale of the problem. A large number of the faults shown in Table 10.3 were due to indirect strokes effecting the wind turbine and wind farm control systems. Generally the number of faults was dominated by those to the control and electrical systems while blade damage gave the highest repair costs and loss of turbine availability, and hence reduction in wind farm revenue.

Figure 10.5 shows the techniques commonly used to protect blades against lightning. The main distinction is whether a limited number of receptors are used to intercept the lightning (Types A and B) or if an attempt is made to protect the entire blade (Types C and D). Type A shows how, using one or two receptors at the tip, the steel control wire of the tip brake may be used as the down conductor. In a blade without a moveable tip (Type B) then an additional conductor is installed. Type C has the down conductor located on the leading and trailing edges although there are considerable practical difficulties in firmly attaching a suitable conducting path to the leading edge. Type D shows the use of a conducting mesh on each side

Table 10.2 Effects of Various Aspects of Lighting on a Wind Turbine

Parameter	Effect on wind turbine
Peak current (A)	Heating of conductors, shock effects, electromagnetic forces
Specific Energy (J/Ω)	Heating of conductors, shock effects
Rate of rise of current (A/s)	Induced voltages on wiring, flashovers, shock effects
Long duration charge transfer (C)	Damage at arc attachments point or other arc sites (e.g., bearing damage)

Table 10.3 Wind-Turbine Lightning Damage Frequency (Cotton *et al.*, 2000)

Country	Period	Turbine years	Lightning faults	Faults per 100 turbine years
Germany	1991 to 1998	9204	738	8.0
Denmark	1991 to 1998	22 000	851	3.9
Sweden	1992 to 1998	1487	86	5.8

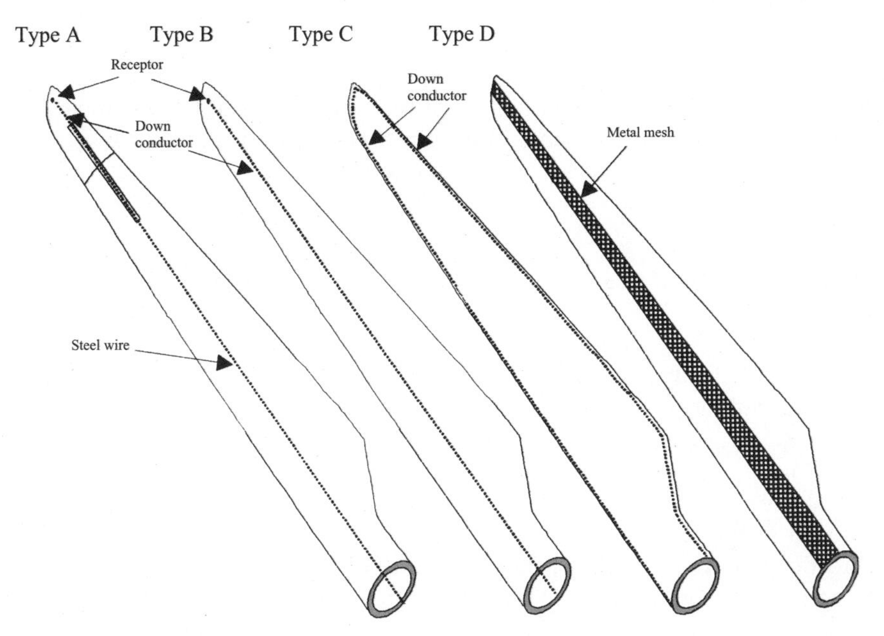

Figure 10.5 Blade Lightning-protection Methods (Reproduced from Cotton *et al.*, 2000, by Permission of Dr R. Sorenson, DEFU)

of the blade. The main mechanism of damage is when the lightning current forms an arc in air inside the blade. The pressure shock wave caused by the arc may explode the blade or, less dramatically, cause cracks in the blade structure. Thus, for effective protection it is essential that the lightning attaches directly to the protection system and then is conducted safely down the length of the blade in a metallic conductor of adequate cross section.

Service experience would seem to indicate that the use of a tip receptor works effectively for blades of up to 20 m in length. There are limited data available for very large wind turbines but it would appear to be questionable if a receptor only at the tip will give good protection for very long blades.

Once the lightning current has been conducted to the root of the blade there remains the problem of ensuring its safe passage to the outside of the tower and hence to earth. This is not straightforward as it is necessary for the current to pass across the pitch, shaft and yaw bearings while not damaging the generator and sensitive control equipment in the nacelle. The present understanding of how wind-turbine bearings may be damaged by the passage of lightning current is summarized in IEC (2000a). Generally large, heavily loaded, bearings are unlikely to be catastrophically damaged although there may be a reduction in their service life. There is, so far, no effective means of shunting lightning current around a large bearing as the bearing itself is the lowest inductance path and so the preferred route for the high-frequency current.

The control and electrical systems are protected against lightning by dividing the turbine into zones depending on whether direct attachment of lightning is possible and the magnitude of the current and hence electromagnetic field expected in each zone (IEC, 2000a). Within each zone components are protected to withstand the anticipated effects of lightning. The principal damage mechanisms to electrical and control systems are direct conduction and magnetic coupling while the main protective measures are good bonding, effective shielding and the use of appropriate surge protection devices at the zone boundaries.

10.4 Embedded (Dispersed) Wind Generation

The wind is a diffuse source of energy with wind farms and individual turbines often distributed over wide geographical areas, and so the public electricity distribution networks, which were originally constructed to supply customer loads, are usually used to collect the electrical energy. Thus wind generation is said to be *embedded* in the distribution network or the generation is described as being *dispersed*. The terms embedded generation and dispersed generation can be considered to be synonymous. Conventional distribution systems were designed for a unidirectional flow of power from the high-voltage transmission network to the customers. Significant generation was not considered in the initial design of the distribution networks and alters the way they operate. Connection of embedded generation to distribution networks has important consequences both for the wind-farm developer and the operator of the distribution network and is the subject of continuing interest (CIRED, 1999, CIGRE, 1998, Jenkins, 1995, 1996).

10.4.1 The electric power system

Figure 10.6 is a diagrammatic representation of a typical modern electric power system. The electrical power is generated by large central generating sets and is then fed into an interconnected high-voltage transmission system. The generating units may be fossil-fuel, nuclear or hydro sets and will have capacities of up to

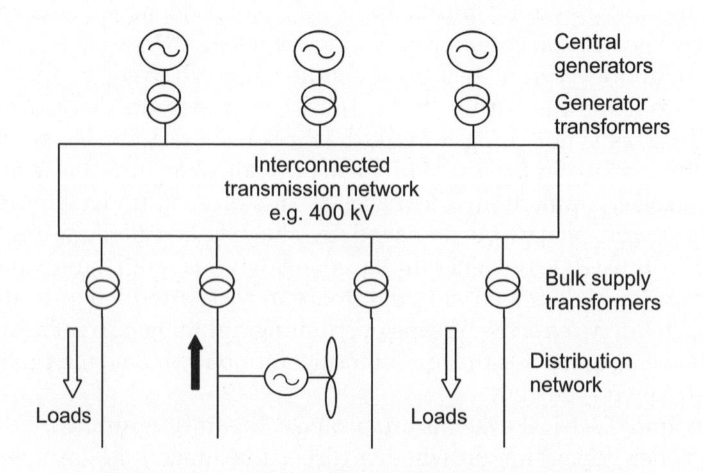

Figure 10.6 A Typical Large Utility Power System

1000 MW. The generation voltage is rather low (typically around 20 kV) to reduce the insulation requirements of the machine windings and so each generator has its own transformer to increase the voltage to that of the transmission system. The transmission network is interconnected, or meshed, and so there can be many paths for the electrical power to flow from the generator to the bulk supply transformers.

The bulk supply transformers are used to extract power from the transmission network and to provide it to the distribution networks at lower voltages. Practice varies from country to country but primary distribution voltages can be as high as 150 kV. Distribution networks are normally operated radially with a single path between the bulk supply transformers and the loads. In urban areas with high loads the distribution networks use large cables and transformers and so have a high capacity. However, in rural areas the customer load is often small and so the distribution circuits may have only a limited capability to transport power while maintaining the voltage within the required limits. Most wind farms are connected to rural, overhead distribution lines. The design of these circuits tends to be limited by consideration of voltage drop rather than thermal constraints and this severely limits their ability to accept wind generation.

10.4.2 Embedded generation

In the early days of electricity supply, each town or city had its own generating station supplying the local load. Thus all generation was local and embedded into the distribution networks. This arrangement suffered from two major problems: (1) the generating sets were rather small and hence of low efficiency, and (2) each station had to keep an additional generator running in case of breakdowns. Then in the 1930s it was found to be technically possible and cost-effective to interconnect these individual power stations so that larger, more efficient generating sets could be used and the requirement for reserve generation could be reduced. Over time, the size and voltage of the interconnected grid networks increased and so now voltages of 400 kV and even 765 kV are common and a single grid connects most of continental western Europe. There have also been significant advances in the technologies of central generating plant and, for example, large modern combined cycle gas turbine (CCGT) generating units have efficiencies approaching 60 percent.

However, over the last 10 years there has been a resurgence in interest in embedded generation stimulated by the need to reduce gaseous emissions from generating plant. Three major themes may be identified: (1) increased development of combined heat and power (CHP), (2) exploitation of diffuse renewable energy sources, and (3) deregulation of the electricity supply industry and the separation of generation and supply of energy from the operation of transmission and distribution networks.

CHP plants operate at very high overall thermal efficiencies and so reduce the total fossil-fuel consumed and hence the gaseous emissions. In addition to the existing commercially available technologies there are major development efforts in very small CHP generators for domestic use and also in the application of fuel cells.

Primary distribution circuits can accept power injections of up to 100 MW and so most on-shore wind generation is likely to be embedded in a distribution system.

There is also considerable interest in much smaller-scale renewable energy schemes, e.g., individual wind turbines or dispersed wind farms.

Finally, since the mid-1990s many countries have deregulated their electricity supply systems and separated the production and sale of electrical energy from the ownership and operation of the transmission and distribution networks. This has opened new commercial opportunities for embedded generation as access to the distribution network has become easier to obtain and the various costs and benefits of the electricity supply system have become more transparent.

10.4.3 Electrical distribution networks

The conventional function of an electrical distribution network is to transport electrical energy from a transmission system to customers' loads. This is to be done with minimal electrical losses and with the quality of the electrical power maintained. The voltage drop is directly proportional to the current while the series loss in an electrical circuit is proportional to the square of the current. Therefore the currents must be kept low which, for constant power transmitted, implies that the network voltage level must be high. However, high-voltage plant (e.g., lines, cables and transformers) are expensive due to the cost of insulation and so the selection of appropriate distribution network voltage levels is an economic choice.

Figure 10.7 is a schematic representation of a typical UK distribution system although most countries will have similar networks. Power is extracted from the

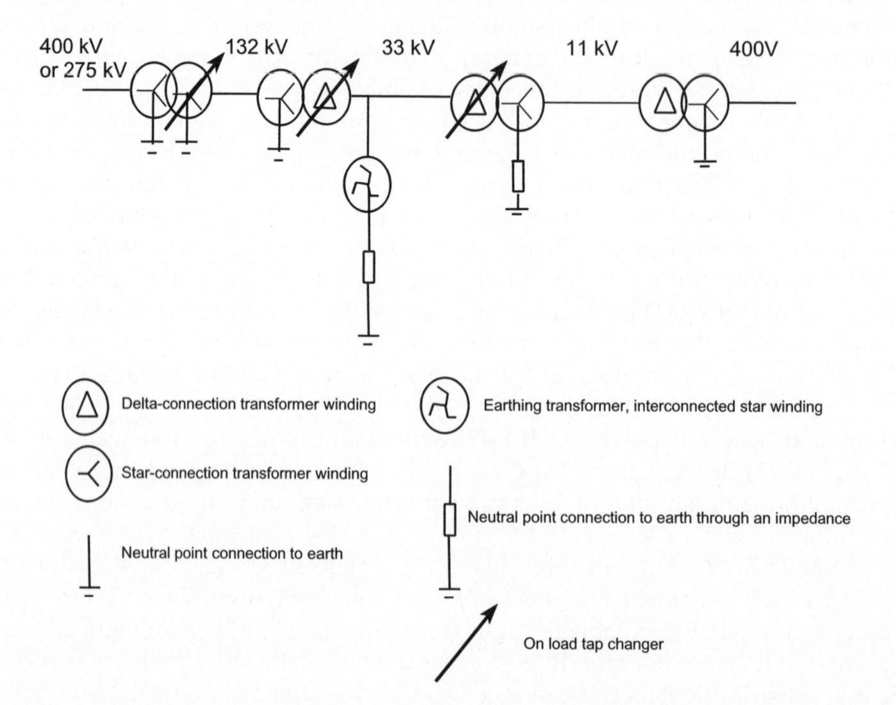

Figure 10.7 Typical UK Distribution Voltage Transformation and Earthing Arrangements

interconnected transmission grid and then transformed down to the primary distribution voltage (132 kV in this case). The electrical energy is then transported *via* a series of underground-cable and overhead-line circuits to the customers. Most customers typically receive electrical energy at 400 V (for a three-phase connection) or 230 V (for a single-phase connection). As the power required to be transmitted in a distribution circuit reduces the circuit voltage level is lowered by transformers.

The majority of distribution circuits are three phase although in some rural areas single-phase circuits are used. Only balanced three-phase networks are suitable for the connection of large wind generators. Three-phase transformer windings may be connected either in star or in delta and the winding arrangement chosen varies from region to region and generally follows historical practice. One advantage of the star connection is that the system neutral point is directly accessible and so can easily be grounded or earthed. Where a delta winding connection is used an accessible star point may be created by an earthing transformer. UK practice is to earth the neutral points of each voltage level at one point only, although in some continental European countries distribution networks are operated with the neutral point isolated.

Current passing through a circuit leads to a change in voltage and this is compensated for by altering the ratio (or taps) of the transformers. 11 kV/400 V transformers have fixed taps which can only be changed manually when no current flows. However, higher-voltage transformers have *on-load tap changers* that can be operated automatically when load current is passing. The simplest control strategy is to use an *automatic voltage controller* (AVC) to maintain the lower voltage terminals of the transformer close to a set-point (Figure 10.8). The AVC operates by measuring the voltage on the busbar, comparing it with a set-point value and then issuing an instruction to the on-load tap changer to alter the ratio of the transformer. Control systems of this type are unaffected by the presence of wind generators on the network. Even if the power flow through the transformer is reversed and power flows from the lower-voltage to the higher-voltage network, the control system will work satisfactorily. This is because the network source impedance is much less than the equivalent impedance of the wind generators and so the network voltage is controlled predominantly by the tap changer. Some designs of tap changer have a reduced current rating if power flows from the low-voltage side to the high-voltage side of the transformer. However, it is rare for embedded wind generation to require the full reverse power flow capacity of a transformer on the distribution network.

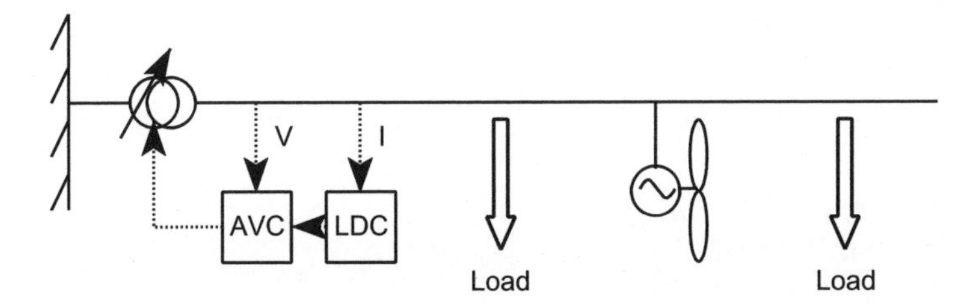

Figure 10.8 Voltage Control of a Distribution Circuit (after Lakervi and Holmes, 1995); AVC - Automatic Voltage Control, LDC - Line Drop Compensation

In some distribution networks, more sophisticated control techniques are used to control the voltage at a remote point along the circuit (Lakervi and Holmes, 1995). This is achieved by adding to the local voltage input a signal proportional to the voltage drop on the network. This is referred to as *line-drop compensation* (LDC). The LDC element measures the current flowing in the circuit and applies this to a simple model of the distribution circuit (i.e., the resistance and inductance to the point at which the voltage is to be controlled). The resulting voltage is then subtracted from the local voltage measurement. In practice, if wind generators are connected to the network then the current flow in the circuit will be changed, the effective impedance to the point at which the voltage is to be controlled will alter, and so the LDC technique will no longer work correctly. LDC is sometimes known as positive-reactance compounding and a further technique, negative-reactance compounding, may also not function correctly with embedded generation connected to the circuits. Further discussion of the effect of embedded generation on transformer voltage controllers is given by Thomson (2000).

The voltage change permitted in 11 kV circuits is small (typically +/−1 percent or 2 percent) as any voltage variation is passed directly through the fixed tap transformers to the customers' supply. In contrast, 33 kV and 132 kV circuits are allowed to operate over a wider voltage range (up to +/−6 percent) as the automatic on-load tap changes can compensate for variations in network voltage. Hence, it is common to find that only very limited capacities of wind generation can be connected at 11 kV as the high impedance of the circuits and relatively large currents flowing from the generators result in unacceptable variations in voltage. Determining the size of a wind farm which may be connected to a particular point in the distribution network requires a series of calculations based on the specific project data. However, Table 10.4 gives some indication of the maximum capacities which experience has indicated may be connected. Table 10.4 assumes that the wind farms are made up of a number of turbines and so the connection assessment is driven by voltage rise effects and not by power quality issues due to individual large machines.

The electrical 'strength' of a circuit is described by its *short-circuit* or *fault level*. The short-circuit level is the product of the pre-fault voltage and the current which would flow if a three-phase symmetrical fault were to occur. Clearly, this combination of current and voltage cannot occur simultaneously but the fault level (expressed in MVA) is a useful parameter which gives an immediate understanding of the capacity of the circuit to deliver fault current and resist voltage variations. In

Table 10.4 Indication of Possible Connection of Wind Farms

Location of connection	Maximum capacity of wind farm (MW)
Out on 11 kV network	1–2
At 11 kV busbars	8–10
Out on 33 kV network	12–15
At 33 kV busbar	25–30
On 132 kV network	30–60

the 'per-unit system' the fault level is merely the inverse of the magnitude of the source impedance (Weedy and Cory, 1998).

10.4.4 The per-unit system

The per-unit system is simply a technique used by electrical power engineers to simplify calculation by expressing all values as a ratio:

$$\frac{\text{actual value (in any unit)}}{\text{base or reference value (in the same unit)}}$$

Its advantages include: (1) a reduction in the appearance of $\sqrt{3}$ in the calculations, (2) that similar per-unit values apply to systems of various sizes, and (3) by careful choice of voltage bases the solution of networks containing several transformers is facilitated.

For simple calculations, such as the those required for assessment of wind farms on simple distribution circuits, all that is required is for these steps to be followed.

(1) Assume an arbitrary base VA (e.g., 10 MVA)

(2) Select voltage bases of each voltage level of the network (e.g., 33 kV and 11 kV). These voltage bases should be related by the nominal turns ratios of the transformers.

(3) Calculate the appropriate real and reactive power flows at the generator terminals as a per-unit value (i.e., for a power flow of 5 MW with the base VA chosen as 10 MVA, the per-unit value is $P = 5/10$ or 0.5 per unit, and similarly for a reactive power flow of 1 MVAr the per-unit value is $Q = 1/10$ or 0.1 per unit).

(4) If necessary, transform the impedances of the circuit from ohmic to per-unit values using a base impedance of $Z_{\text{base}} = V_{\text{base}}^2/VA_{\text{base}}$.

(5) The base current at any particular voltage level may be calculated from: $I_{\text{base}} = VA_{\text{base}}\sqrt{3} \times V_{\text{base}}$.

The per-unit system is used in Example 10.1.

10.4.5 Power flows, slow-voltage variations and network losses

If the output from an embedded wind turbine generator is absorbed locally by an adjacent load then the impact on the distribution network voltage and losses is likely to be beneficial. However, if it is necessary to transport the power through the distribution network then increased losses may occur and slow-voltage variations may become excessive.

If the wind generator operates at unity power factor (i.e., reactive power $Q = 0$), then the voltage rise in a lightly-loaded radial circuit (Figure 10.9), is given approximately by:

$$\Delta V = V_1 - V_0 = PR/V_0 \tag{10.1}$$

Operating the generator at a leading power factor (absorbing reactive power) acts to reduce the voltage rise but at the expense of increased network losses. In this case the voltage rise is given by:

$$\Delta V = V_1 - V_0 = (PR - XQ)/V_0 \tag{10.2}$$

The source impedance of an overhead 11 kV distribution circuit may, typically, have a ratio of inductive reactance to resistance (X/R ratio) of 2. An uncompensated induction generator at rated output, typically, has a power factor of 0.89 leading, i.e., $P = -2Q$. Thus, under these conditions, there is no apparent voltage rise in the circuit at full power. However, the real power loss (W) in the circuit is given approximately by:

$$W = (P^2 + Q^2)R/V_0^2 \tag{10.3}$$

The reactive power drawn by the generator acts to limit the voltage rise but higher real power losses are incurred in the circuit.

Equations (10.1)–(10.3) for voltage rise and circuit losses are approximate only and do not apply to heavily-loaded circuits. A simple but precise calculation for voltage rise in any radial circuit may be carried out using an iterative technique as follows. (Complex quantities are indicated by bold type.)

At the generator busbar of Figure 10.9 the *apparent power* (sometimes known as the *complex power*) $\mathbf{S}_1$ is given by:

$$\mathbf{S}_1 = P - jQ \tag{10.4}$$

(for a generator operating at lagging power factor, exporting VArs, $\mathbf{S}_1$ would be given by $P + jQ$).

By definition $S = \mathbf{VI}^*$, where * indicates the complex conjugate. Therefore the current flowing in the circuit is given by:

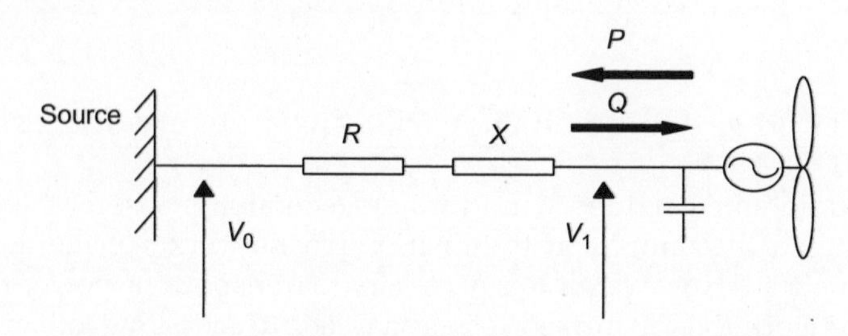

Figure 10.9 Fixed-speed Wind Turbine on a Radial Circuit

$$I = S_1^*/V_1^* = (P + jQ)/V_1^* \tag{10.5}$$

The voltage rise in the circuit is given by $\boldsymbol{IZ}$ and so:

$$V_1 = V_0 + IZ = V_0 + (R + jX)(P + jQ)/V_1^* \tag{10.6}$$

It is common for the network voltage V_0 to be defined and the generator busbar voltage V_1 to be required. V_1 can be obtained using the simple iterative expression:

$$V_1^{(n+1)} = V_0 + (R + jX)(P + jQ)/V_1^{*(n)} \tag{10.7}$$

where n is the iteration number.

This is a very simple form of the conventional Gauss–Seidel load flow algorithm (Weedy and Cory, 1998). Once the calculation converges an accurate solution is obtained. More complex load flow calculations may be carried out using commercially available programs. These include models of transformers, with their tap changers, and can solve large interconnected circuits in only a few iterations using more advanced algorithms.

Example 10.1 Calculation of Voltage Rise in a Radial Circuit (Figure 10.10)

Consider a 5 MW wind farm operating at a leading power factor of 0.98. The network voltage (V_0) is $(1 + j0)$ per unit and the circuit impedance (Z) is $(0.05 + j0.1)$ per unit on a 10 MVA base.

A power factor of 0.98 leading implies a reactive power draw of 1.01 MVAr.

Thus, following Equation (10.7), the calculation becomes

$$V_1^{(n+1)} = 1 + (0.05 + j0.1)(0.5 + j0.101)/V_1^{*(n)}$$

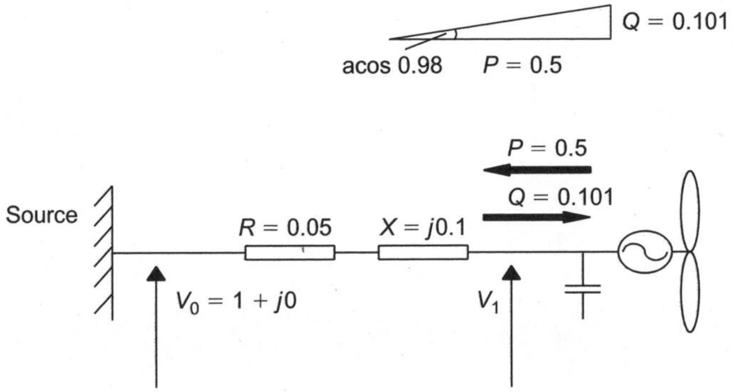

Figure 10.10 Example of Calculation of Voltage Rise on a Radial Circuit (All Values in Per-unit)

For the 1st iteration ($n = 0$) assume $V_1^{*(0)} = 1 + j0$; then $V_1^{(1)}$ may be calculated to be

$$V_1^{(1)} = 1.0149 + j0.0551$$

For the 2nd iteration ($n = 1$)

$$V_1^{*(1)} = 1.0149 - j0.0551$$

then

$$V_1^{(2)} = 1.0117 + j0.0549$$

For the 3rd iteration ($n = 2$)

$$V_1^{*(2)} = 1.0117 - j0.0.0549$$

then

$$V_1^{(3)} = 1.0117 + j0.0551$$

and the procedure has converged.

Therefore $V_1 = 1.013$ per unit at an angle of 3^0, i.e., the voltage at the generator terminals is 1.3 percent above that at the source. The angle between the two voltage vectors is small (3^0).

The approximate calculation ($V_1 = V_0 + PR - XQ$) indicates a voltage V_1 of 1.015 per unit (i.e., a rise of 1.5 percent).

The current (I) in the circuit may be calculated from:

$$I = S_1^*/V_1^* = (0.5 + j0.101)/(1.0117 - j0.0551)$$

$$= 0.4873 + j0.1264 \text{ per unit}$$

$$|I| = 0.503 \text{ per unit}$$

With a connection voltage of 33 kV the base current is given by

$$I_{\text{base}} = VA_{\text{base}}/\sqrt{3} \times V_{\text{base}} = 5 \times 10^6/1.732 \times 33 \times 10^3 = 87.5 \text{ A}$$

and so the magnitude of the current flowing in the 33 kV circuit is 44 A.

The real power loss in the circuit (W) is $W = I^2R = 0.0127$ per unit, or 127 kW.

The symmetrical short-circuit level at the generator, before connection of the generators is simply:

$$S'' = 1/|Z| = 1/(0.05^2 + 0.1^2)^{1/2}$$

$S'' = 8.94$ per unit or 89.4 MVA.

Once the induction generators of a wind farm are connected then they will make a contribution to the short-circuit current seen by a circuit breaker as it closes on to a fault. This is typically some five times the rating of the generator, e.g., some 25 MVA in this case. However, for a three-phase symmetrical fault the fault current contribution from an induction machine decays rapidly and will make only a small contribution to the fault current which must be interrupted by an opening circuit breaker. Detailed guidance on the calculation of fault currents, including the contribution from induction generators is given in IEC (1998).

10.4.6 Connection of embedded wind generation

Distribution utilities have an obligation to operate the electrical distribution networks in such a way as to provide power to their customers at an agreed quality. At present, power quality requirements are based on national standards although a common European position is emerging described in British Standard (1995b). However, it should be noted that this document describes minimum standards of supply which a customer may expect and is not directly applicable to the connection of embedded generation.

The main parameters relevant to the connection of wind turbines are:

slow (or steady-state) voltage variations,

rapid voltage changes (leading to flicker),

waveform distortion (i.e., harmonics),

voltage unbalance (i.e., negative phase sequence voltage), and

transient voltage variations (i.e., dips and sags).

Therefore consideration of whether a wind generation scheme may be connected to a distribution circuit is based on its impact on other users of the network. Similar considerations apply to the connection of any load. The steady-state voltage variations are generally considered assuming conditions of minimum network load with maximum generation and maximum network load with minimum generation. These are rather onerous and conservative assumptions to ensure that steady-state voltage limits will never be violated. In some cases, local agreements have been reached between the network and generator operators so that some wind turbines are constrained off at times of low network load or abnormal network conditions. This then allows a larger wind farm to be connected than would otherwise be the case. Considerations of power quality (i.e., maximum instantaneous real and apparent power, reactive power, voltage flicker and harmonics) were considered recently by an IEC Working Party and a draft standard has now been issued (IEC, 2000b).

For wind farms of relatively large capacity connected to weak distribution networks the calculations required can be involved. Hence some countries have adopted an assessment approach to allowing connection based on the ratio of wind farm capacity (in MW) to symmetrical short-circuit level, without the wind farm connected (in MVA). This is sometimes known as the short-circuit ratio. Typical values chosen range from 2 percent–5 percent based on studies and on experience (Gardner, 1996). However, such simple rules can be too restrictive and lead to refusal of permission to connect or to excessive reinforcement of the distribution system. Table 10.5 gives data on two large wind farms in successful commercial operation in the UK with much higher ratios of wind farm capacity to short-circuit level. Both wind farms have operated successfully for some years but it should be noted that the number of turbines on each site is large and so the impact of any individual machine is small. Also, both sites are connected to 33 kV circuits from which no other customers are supplied directly.

Figure 10.11 shows the calculated variation of voltage with wind farm output at one site. The generator voltage of the turbines was 690 V and each turbine had a local 690/33 kV transformer. The 33 kV curve refers to the voltage at the point of connection of the wind farm to the public distribution network. With zero wind-farm output the system voltage is close to nominal as the distribution network is lightly loaded. As the wind farm output increases the voltage rises but, once rated power is reached, then the reactive power drawn by the generators increases rapidly and so the voltage drops. At 140 percent output power the load flow calculation fails to converge indicating that the network voltage is likely to collapse.

Table 10.5 Ratio of Site Capacity to Connection Short-circuit Level for Two Large Wind Farms

Site capacity (MW)	Number of turbines	Voltage of connection (kV)	Short circuit level of connection (MVA)	Ratio of site capacity to short-circuit level (%)
21.6	36	33	121	18
30.9	103	33	145	21

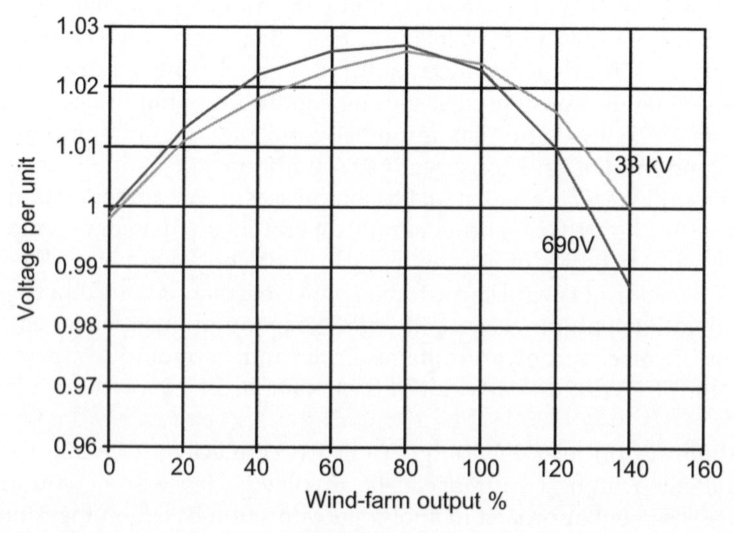

Figure 10.11 Variation of Wind-farm Voltage with Output Power (Generator Voltage 690 V, Connection Voltage, 33 kV)

Although such overloads are unlikely on wind farms with many turbines, individual machines can produce up to 200 percent of their rated output and so this phenomenon of voltage instability can be significant when considering the connection of large wind turbines to weak networks.

10.4.7 Power system studies

Power system studies are often required in order to assess the impact of embedded wind generators on the network and to ensure that the network conditions are such as to allow the wind generators to operate effectively. In the past simple manual calculations were used but sophisticated computer programs are now available at relatively low cost. However, considerable care is necessary when using these programs as a number of them were designed primarily to investigate systems with conventional generation using synchronous machines.

Load flow (or power flow) programs take as input the network topology and parameters of the lines, cables and transformers. The customers' load and generator outputs are then used to calculate the steady-state performance of the network in terms of voltages, real and reactive power flows and losses. Generally a balanced three-phase system is assumed. For wind-energy applications it is important that the load flow package includes a good model of the tap changers on the network transformers and an effective representation of induction generators. A simple load-flow gives a 'snap-shot' analysis of the network but some commercially available programs allow the use of daily load and generator profiles to repeat the calculations many times and so show the network performance over a period of time.

Fault calculators use the method of symmetrical components (Wagner and Evans, 1933) to calculate the effect of short-circuits on the power system. Both balanced and unbalanced faults may be investigated. Simple fault calculators tend to be of limited use when considering embedded wind generation as they often do not have good models of the induction generators or power electronic converters. Fault calculators give as output the fault flows and network voltages at particular times after the fault is applied. Some of the more sophisticated programs use analytical methods to estimate the decay of fault currents with time.

Transient stability programs allow investigation of how embedded wind generators will respond to disturbances on the network. A balanced network is assumed as the calculation is based on fundamental frequency phasors with an integration algorithm to calculate how the network conditions change over time. Traditionally, transient stability programs were used to investigate the angle stability of synchronous generators and so a good model of the wind turbines is required for useful results. In particular, the representation of induction machines in transient stability programs is often simplified and so this class of tools is not good for investigating voltage stability.

Electromagnetic programs are among the more sophisticated tools and unlike the previous programs do not assume perfect, fundamental frequency waveforms. Detailed representation of both the induction generators and power electronic converters is possible. These programs use detailed time-domain simulations which allow reconstruction of the distorted wave forms generated by power electronic converters as well as investigation of high-frequency transients such as those due to lightning. They may be used to investigate almost any condition on the power system but at the cost of considerable complexity. Electromagnetic simulations are not routinely used for wind farm design but only to investigate particular problems.

There are of course many more specialized calculation techniques and associated tools including probabilistic power flow, optimal power flow, harmonic load flow and various earthing/grounding codes. However, their use requires specialist knowledge and so they are not in common use for small wind-farm projects.

10.5 Power Quality

Power quality is the term used to describe how closely the electrical power delivered to customers corresponds to the appropriate standards and so operates their end-use equipment correctly (Dugan, McGranaghan and Beaty, 1996). Thus, it is essentially a customer-focused measure although greatly effected by the operation of the distribution and transmission network. There are a large number of ways in which the electrical supply (i.e., current, voltage or frequency) can deviate from the specified values. These range from transients and short-duration variations (e.g., voltage sags or swells) to long-term waveform distortions (e.g., harmonics or unbalance). Sustained complete interruptions of supply are generally considered an issue of network reliability rather than power quality. In the UK, complete interruptions lasting more than 1 min are recorded and used as one indicator of the *quality of supply*, as distinct from *power quality*, provided by the distribution utility. The growing importance of power quality is due to the increasing use of sensitive customers' load equipment including computer-based controllers and power electronic converters as well as the awareness of customers of the commercial consequences of equipment mal-operating due to disturbances originating on the power system.

The issues of power quality are of particular importance to wind turbines as individual units can be large, up to 2 MW, feeding into distribution circuits with a high source impedance and with customers connected in close proximity. For variable-speed designs, which use power electronic converters the issues of harmonic distortion of the network voltage must be carefully considered while the connection of fixed-speed turbines to the network needs to be managed if excessive voltage transients are to be avoided. During normal operation wind turbines produce a continuously variable output power. The power variations are mainly caused by the effects of turbulence, the wind shear, tower shadow, and the operation of the control systems. These effects lead to periodic power pulsations at the frequency with which the blades pass the tower (typically around 1 Hz for a large turbine) which are superimposed on the slower variations caused by changes in wind speed. There may also be higher-frequency power variations (at a few Hz) caused by dynamics of the turbine. Variable-speed operation of the rotor has the advantage that many of the faster power variations are not transmitted to the network but are smoothed by the flywheel action of the rotor. However, fixed-speed operation, using a low-slip induction generator, will lead to cyclic variations in output power and hence network voltage.

It is essential that wind turbines do not degrade the power quality of the distribution network as otherwise permission for their connection or continued operation will be refused by the distribution utility. The particular importance of the influence of wind turbines on power quality has been recognized in the creation of an international standard (IEC, 2000b). This standard (which is presently only in draft form) lists the following data as being relevant for characterizing the power quality of a wind turbine:

- maximum output power (10 min average, 60 s average and 200 ms average values),
- reactive power (10 min average) as a function of active power,
- flicker coefficient for continuous operation as a function of network source impedance phase angle and annual average wind speed,
- maximum number of wind turbine starts within 10 min and 120 min periods,
- flicker step factor and voltage step factor at start-up as a function of network source impedance phase angle,
- harmonic currents during continuous operation as 10 min averages for each harmonic up to the fiftieth.

It is general experience that the main power quality issue when connecting wind farms with a large number of generators (>10) is steady-state voltage rise while for individual large turbines on weak networks the limiting factor is often transient voltage changes. Power quality is an important consideration in the design and development of new large wind turbines and is one factor in the increasing popularity of variable-speed operation and 'active stall' control as both of these features reduce the transient variations in output power.

Figure 10.12 shows a representation of how power quality issues may be viewed with respect to wind generation. Figure 10.12(a) shows the various effects which may be considered to originate in the transmission and distribution networks and which can effect the voltage to which the wind turbines are connected. Voltage sags (i.e., a decrease to between 0.1 and 0.9 per unit voltage for a period of up to 1 min which is usually caused by faults on the transmission/distribution network) are of particular concern as they will cause fixed wind turbines to overspeed as the load on the generator is removed. This in turn can lead to a high demand for reactive power which further depresses the network voltage. The depressed voltage may also cause contactors to open and voltage-sensitive control circuits to operate. Voltage swells (an increase in voltage over 1.1 per unit) are less common and tend not to be a major problem for wind turbines. Ambient harmonic voltage distortion is increasing in many power systems, due to the power supplies of TVs and PCs,

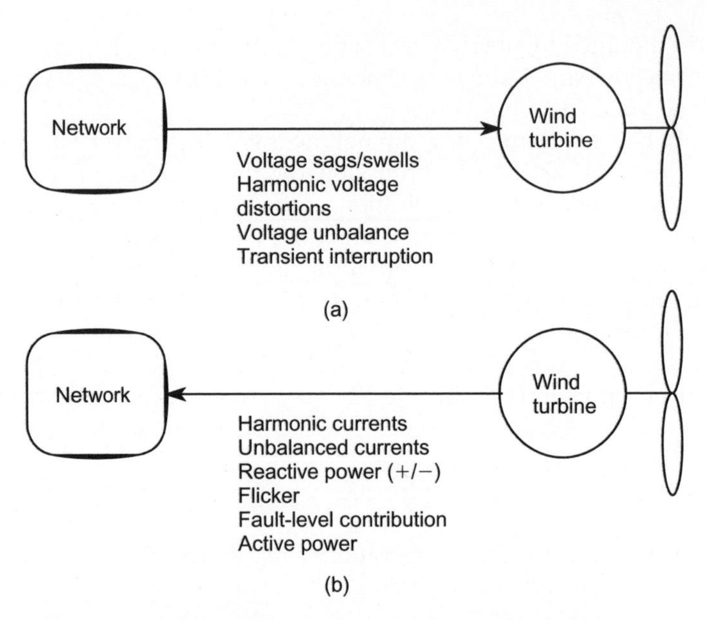

Figure 10.12 Origin of Power Quality Issues

and in the UK it is not uncommon to find levels at some times of the day in excess of that which is considered desirable by network planners (Electricity Association, 1976). Harmonic voltage distortions will lead to increased losses in the generators of wind turbines and may also disturb the control systems and harmonic current performance of power electronic converters. It is common practice to use power factor correction capacitors with induction generator (fixed-speed) wind turbines and these will, of course, have a low impedance to harmonic currents and the potential for harmonic resonances with the inductive reactance of other items of plant on the network. Transient interruptions (caused by the use of auto-reclosure on distribution networks) can potentially be very damaging due to the possibility of out-of-phase reclosure on to induction generators which are still fluxed, and hence developing a voltage. It is usual to apply fast acting loss-of-mains protection against this condition so that the turbine is isolated once the supply is interrupted.

Network voltage unbalance will also effect rotating induction generators by increasing losses and introducing torque ripple. Voltage unbalance can also cause power converters to inject unexpected harmonic currents back into the network unless their design has included consideration of an unbalanced supply. Figure 10.13 shows the negative phase sequence equivalent circuit of an induction machine (Wagner and Evans, 1933). It may be seen that during normal operation (i.e., when the slip (s) is close to zero) the effective rotor resistance to negative sequence (unbalanced) voltages is only $R_r/2$. Thus any small unbalance in the applied voltage will lead to high unbalanced currents. This has been found to be a particular problem for wind turbines connected to 11 kV distribution networks in the UK. It is common on rural 11 kV public supply networks to connect transformers and small spur lines between two phases only to supply farms and other small loads. These single-phase connections tend to unbalance the voltages (up to 2 percent voltage

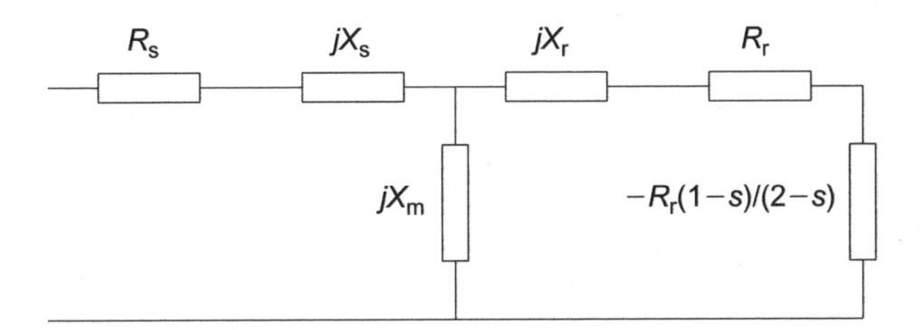

Figure 10.13 Negative Phase Sequence Equivalent Circuit of an Induction Machine (s = slip (negative for generator); R_s = stator resistance; jX_s = stator reactance; jX_m = magnetizing reactance; jX_i = rotor reactance (referred to the stator); R_r = rotor resistance (referred to the stator))

unbalance is not uncommon) and so generate high unbalanced currents (up to 15–20 percent) in wind turbines connected to the main 11 kV three-phase circuit. The wind-turbine control systems will detect excessive unbalanced current and cause a shut-down. Nuisance tripping often occurs in the night when heating loads in the farms are switched in automatically to take advantage of reduced price electrical energy tariffs.

Figure 10.12(b) indicates how an embedded wind generator might introduce disturbances into the distribution network and so cause a reduction in power quality. Thus, a variable-speed turbine using a power electronic converter may inject harmonic currents into the network. Similarly, unbalanced operation will lead to negative phase sequence currents being injected into the network which, in turn, will cause network voltage unbalance. As has been discussed earlier, variable-speed wind-turbine generators can either produce or absorb reactive power while exporting active power and, depending on the details of the network, load and generation, this may lead to undesirable steady-state voltage variations. Voltage flicker refers to the effect of dynamic changes in voltage caused by blade passing or other transient effects. Fixed-speed wind generators will increase the network fault level and so will effect power quality, often to improve it. There is considerable similarity between the power quality issues of wind turbines and large industrial loads and, in general, the same standards are applied to both.

There is extensive literature on power quality of wind turbines (Heier, 1998). Fiss, Weck and Weinel (1993) wrote an important early paper setting out the policy of the North German utility Schleswag at that time. Rather than carry out individual studies on each proposed installation Schleswag applied the following rules for connection based on short-circuit level.

For voltage change:

$$\sum S_{\mathrm{WKA}}\left(\frac{1}{S_{\mathrm{KE}}}-\frac{1}{S_{\mathrm{KSS}}}\right) \leqslant \frac{1}{33}$$

For voltage fluctuation

$$\sqrt{\sum\left(\frac{P_{\mathrm{WKA}}}{S_{\mathrm{KE}}}\right)^2} \leqslant \frac{1}{25}$$

For light flicker

$$\sqrt{\sum\left(\frac{P_{\mathrm{WKA}} P_{\mathrm{ST}}}{S_{\mathrm{KE}}}\right)^{2}} \leqslant \frac{1}{25}$$

where S_{WKA} is wind generator apparent power, P_{WKA} is wind generator real power, P_{ST} is short-term flicker severity, S_{KE} is short-circuit level at connection point, and S_{KSS} is short-circuit level at MV transformer station busbar.

The limit on the voltage change was based on the impedance of the circuit between the point of connection and the MV transformer busbar together with the apparent power of the wind-turbine generators. The voltage fluctuation limit was simply based on the root of the sum of the squares of the short-circuit ratio of the wind turbines while light flicker limit was the same calculation but with the value of P_{ST} measured for that type of wind turbine included. Simple rules such as these are now of historical interest only but they do serve to illustrate the dominant influences on aspects of power quality.

Davidson (1995, 1996) carried out a comprehensive, two-year, measurement campaign to investigate the effect of a wind farm in Wales on the power quality of the 33 kV network to which it was connected. The 7.2 MW wind farm of 24 × 300 kW fixed-speed induction generator turbines was connected to a weak 33 kV overhead network with a short-circuit level of 78 MVA. Each wind turbine was a WEG MS3-300 which are two-bladed, pitch-regulated machines with a teetering hub. A pole amplitude modulated (PAM) induction machine was used to give a rotor speed of 32 r.p.m. in low winds and 48 r.p.m. in high winds. An anti-parallel thyristor soft-start unit was used to connect each generator to the network and power factor correction capacitors were connected to each unit once the turbine started generating. Each wind turbine fed the wind-farm electrical system *via* a local 0.66/11 kV transformer. Power was collected by two underground cable feeders and passed to the 33 kV network *via* a conventional 11/33 kV distribution transformer with an on-load tap changer. The wind turbines were located along a ridge at an elevation of 400 m in complex upland terrain and so subject to turbulent winds. Measurements to assess power quality were taken at the connection to the distribution network and at two wind turbines. The results are interesting as they indicate a complex relationship of a general improvement in power quality due to the connection of the generators, and hence the increase in short-circuit level, and a slight increase in harmonic voltages caused by resonances of the generator windings with the power factor correction capacitors.

The operation of the wind farm raised the mean of the 33 kV voltage slightly but reduced its standard deviation. This was expected as, in this case, the product of injected active power and network resistance was approximately equal to the product of the reactive power absorbed and the inductive reactance of the 33 kV connection. The effect of the generators was to increase the short-circuit level and so reduce the variations in network voltage but with little effect on the steady-state value. The connection of increasing numbers of induction generators caused a dramatic reduction in negative phase sequence voltage from 1.5 percent with no generators connected to less than 0.4 percent with all generators operating. This was, of course, at the expense of significant negative phase sequence currents flowing in the generators and associated heating and losses, although in this wind farm nuisance tripping was not reported. The wind farm slightly reduced the voltage flicker measured at the point of connection using a flickermeter. This was a complex effect with the generators raising the short-circuit level but also introducing fluctuations in current. There was a slight increase in total harmonic voltage distortion with the wind farm in operation, mainly caused by a rise in the fifth and seventh harmonics when the low-speed, high-impedance winding of each generator was in service. This increase was probably associated with a parallel resonance of the high-impedance winding of the generators and the power factor correction capacitors and the high levels of ambient harmonics on the utility network. The most significant transient event was the energization of the entire wind farm including 24 turbine transformers, but without the wind turbines connected. This led to a transient 8 percent dip on the 33 kV voltage. Energization of the wind farm in this manner is rare and occurs only after disconnection by the grid interface protection.

Davidson identified three main sources of power fluctuation: wind turbulence, blade passing, and blade pitching. The major effects were found to be at wind speeds of 9–10 m/s, mainly due to wind turbulence, and above 20 m/s due to blade-pitching effects, although the pitch control on these turbines was not completely effective in high wind speeds. The highest wind speeds (>20 m/s) led to the greatest flicker. As might be expected siting turbines in turbulent locations significantly increased power fluctuations. Individual wind turbine starts/stops and speed changes made little impact on network voltage flicker.

These results are typical of the experience of connecting wind farms, with a large number of relatively small induction generators, on to rural distribution circuits. In a number of studies some aspects of power quality have been shown to be improved by the effective increase in short-circuit level. However, the effect of embedded generation plant on distribution network will vary according to circumstances and each project must be evaluated individually. In particular, the connection of large single generators, as opposed to a group of smaller generators, may be constrained by consideration of flicker.

Craig (1995) undertook a similar investigation on another WEG MS3-300 wind turbine connected to an 11 kV network in Northern Ireland over a 2 month period. A particular feature of this study was the very low network short-circuit level which varied between 8–15 MVA on the 11 kV point of connection depending on how the 11 kV utility circuits were arranged. The X/R ratio of the source impedance was approximately 0.5 illustrating the highly resistive nature of the long 11 kV overhead circuits. The site had a mean wind speed of 9 m/s and a turbulence intensity of 12.5 percent. Measurements were taken to investigate the voltage disturbances due to wind turbine starts (with and without an anti-parallel thyristor soft-start), steady-state voltage variations, dynamic voltage variations and flicker and also voltage unbalance.

The conclusions of the study were that, under certain circumstances (particularly high wind speed cut-out and connection without the soft-start) the dynamic voltage variations exceeded recommended limits. Steady-state voltage variations could be predicted reliably from load-flow calculations and were within limits. The wind turbine made some contribution to voltage flicker but this was masked during working hours by a local quarry which used stone-cutting machinery that created flicker well in excess of the recommended levels. The background voltage unbalance on the network exceeded 1.5 percent at times and this was reduced when the turbine was connected and drew negative phase sequence (unbalanced) current.

10.5.1 Voltage flicker

Voltage flicker describes dynamic variations in the network voltage which may be caused either by wind turbines or by varying loads (Bossanyi, Saad-Saoud and Jenkins, 1998). The origin of the term is the effect of the voltage fluctuations on the brightness of incandescent lights and the subsequent annoyance to customers (Mirra, 1988). Human sensitivity to variations of light intensity is frequency dependent and Figure 10.14 indicates the magnitude of sinusoidal voltage changes which laboratory tests have shown are likely to be perceptible to observers. It may be seen that the eye is most sensitive to voltage variations around 10 Hz. The various national and international standards for flicker on networks are based on curves of this type. The use of voltage flicker as an indicator of acceptable dynamic voltage variations on the network is rather unusual as the assessment is based on the experimentally measured effect of changes of incandescent lamps on the human eye and brain. It is, of course, possible to change incandescent bulbs to other types of lamp with a different response and the use of flicker as a measure of disturbance does not consider explicitly the effect of voltage variations on sensitive equipment, e.g., computers. However, the flicker standards are generally used to characterize transient voltage variations and are of considerable significance for embedded wind generation which: (1) often uses relatively large individual items of plant, (2) may

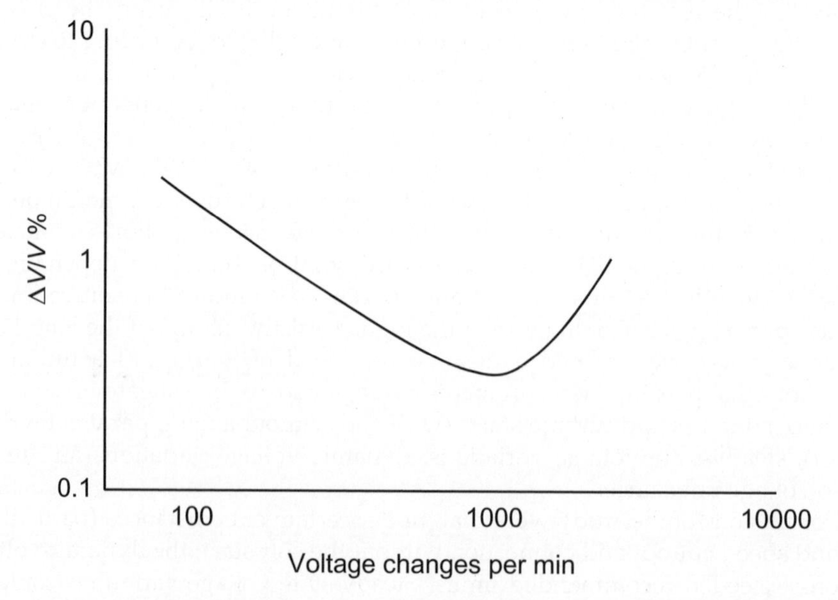

Figure 10.14 Influence of Frequency on the Perceptibility of Sinusoidal Voltage Changes

start and stop frequently, and (3) is subject to continuous variations in input power from a fluctuating energy source.

Flicker is usually evaluated over a 10 min period to give a 'short-term severity value' P_{st}. The P_{st} value is obtained from a 10 min time series of measured network voltage using an algorithm based on the nuisance perceived by the human eye in fluctuating light. This is illustrated in Figure 10.15 which indicates how flicker is measured. P_{st} is linear with respect to the magnitudes of the voltage change but, of course, includes the frequency dependency indicated in Figure 10.14. Twelve P_{st} values may then be combined using a root of the sum of the cubes calculation to give a 'long-term severity value' P_{lt} over a 2 h period (Electricity Association, 1989) although IEC 614000-21 makes no distinction between P_{st} and P_{lt} limits (IEC, 2000b).

If a number of wind turbines are subject to uncorrelated variations in torque then their power outputs and effect on network flicker will reduce as

$$\frac{\Delta P}{P} = \frac{1}{\sqrt{n}} \frac{\Delta p}{p}$$

where n is the number of generators, P and p are the rated power of the wind farm and wind turbine respectively and ΔP and Δp are the magnitude of their power fluctuation.

There is some evidence (Santjer and Gerdes, 1994) that on some sites wind turbines can fall into synchronized operation and in this case the voltage variations become cumulative and so increase the flicker in a linear manner. The cause of this synchronous operation is not completely clear but it is thought to be due to interactions on the electrical system caused by variations in network voltage. However, experience on most UK sites, where the winds are rather turbulent, has been that any synchronized operation lasts only a short while before being disrupted by wind-speed changes. This may not be the case offshore when the turbulence of the wind is likely to be lower.

A range of permissible limits for flicker on distribution networks is given in national and international standards. Engineering Recommendation P28 (Electricity Association, 1989) specifies an absolute maximum value of P_{st} on a network, from all sources, to be 1.0 with a 2 h P_{lt} value of 0.6. However, extreme caution is advised if these limits are approached as the risk of complaints increases at between the sixth and eighth power of the change in voltage magnitude once the limits are reached and the approximate assessment method proposed in the same document is based on P_{st} not exceeding 0.5. British Standard (1995), which specifically excludes embedded generation from its scope, is significantly less stringent specifying that over a 1 week period P_{lt} must be less than 1 for 95 percent of the time. Gardner (1996) describes P_{st} limits from a number of utilities in the range 0.25–0.5.

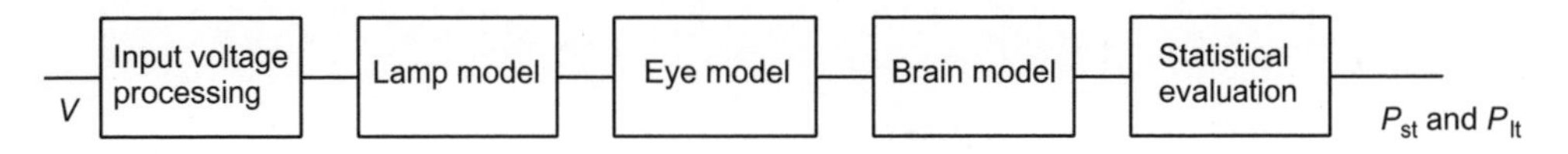

Figure 10.15 Principle of Flicker Measurement

10.5.2 Harmonics

Only variable-speed wind turbines inject significant harmonic currents into the network. Fixed-speed wind turbines, particularly those with power-factor correction capacitors, alter the harmonic impedance of the distribution network and, in some circumstance, create resonant circuits. This may be important if fixed- and variable-speed wind turbines are installed in the same wind farm. It is noted in IEC (2000b) that harmonic currents have been reported from a few installations of fixed-speed, induction-generator wind turbines but there is no known instance of customer annoyance or equipment damage due to harmonic currents from fixed-speed wind turbines.

Thyristor soft starts (Figure 10.16) are commonly used to connect the induction generators used on fixed-speed wind turbines. Their mode of operation is initially for the thyristors to be fired late in the voltage cycle and then the firing angle advanced (over several seconds) until the entire voltage wave is applied to the generator. Thus the network voltage is applied gradually to the generator and the current drawn by the generators controlled to reduce any voltage variations on the network. When operating, these devices produce varying harmonics as the firing angle of the thyristors is altered. Generally the soft-start units are only used for a few seconds during the connection of the induction generator and for this short period the effect of the harmonics is considered to be harmless and may be ignored (IEC, 2000b). If the anti-parallel thyristors are not by-passed, but left in service, then their harmonic currents need to be assessed. This continuous use of the soft-start unit has been proposed in order to reduce the applied voltage and hence the iron losses in the induction generator at times of low generator output. However, as well as the complication of trying to deal with harmonics which change with the thyristor firing angle there are also potential difficulties with the variation of the applied voltage to the generator altering the dynamic characteristics of the drive train.

Modern variable-speed wind turbines use network side voltage source converters, normally based on the Graetz bridge topology (Figure 10.17). These use insulated gate bipolar transistors (IGBTs) switching at several kHz to synthesize a sine wave and so eliminate the lower order (e.g., < 19th) harmonics. The switching techniques used can vary from rather simple carrier modulated techniques which compare a reference signal with a trigger signal to those based on space-vector

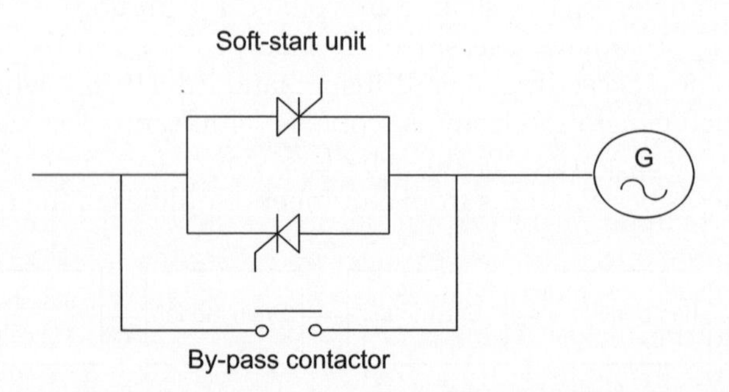

Figure 10.16 Soft-start Unit for an Induction Generator (One-phase only shown)

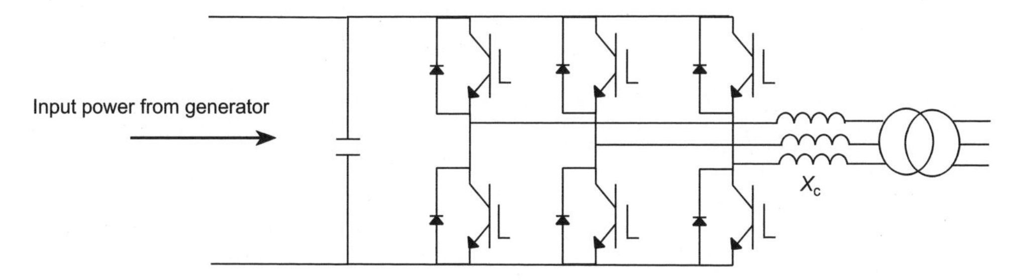

Figure 10.17 Six-pulse Two-level IGBT Voltage Source Converter (Graetz Bridge)

theory with some form of two-axis transformation (Jones and Smith, 1993). The rapid switching gives a large reduction in low-order harmonic currents compared to the classical line commutated converter but the output current will have significant energy at around the switching frequency of the devices (i.e., in the 2–6 kHz region). Although these high-frequency currents are relatively easy to filter they may influence the selection of the size of the coupling reactance X_c which can be used to form part of the filter to block harmonic currents around the switching frequency.

10.5.3. Measurement and assessment of power quality characteristics of grid connected wind turbines

Determination of the power quality of wind turbines and prediction of their performance in service is not straightforward and IEC 614200-21 (IEC, 2000b) has been written to provide guidance. There are a number of difficulties when assessing the power quality of wind turbines as their performance will depend on:

- the design of the entire wind turbine (including the aerodynamic rotor and control system),
- the conditions of the electrical network to which it is connected, and
- the wind conditions in which it operates.

For example, simple measurement of voltage variations at the terminals of a test turbine is not satisfactory as ambient levels of flicker in the electrical network and the X/R ratio of the source impedance at the test site will obviously have a great impact on the outcome. Hence, for evaluating flicker, a procedure is proposed where current measurements are made of the output of a test turbine and used to synthesize the voltage variations which would be caused on distribution networks with defined short-circuit levels and X/R ratios of their source impedance. This is illustrated in Figure 10.18 and is referred to in the Standard as simulation using a 'fictitious grid'. These voltage variation's are then passed through a flicker algorithm to calculate the flicker which the test turbine would cause on the defined networks. When the installation of the particular turbine is considered at a point on the real distribution network these test results are then scaled to reflect the actual

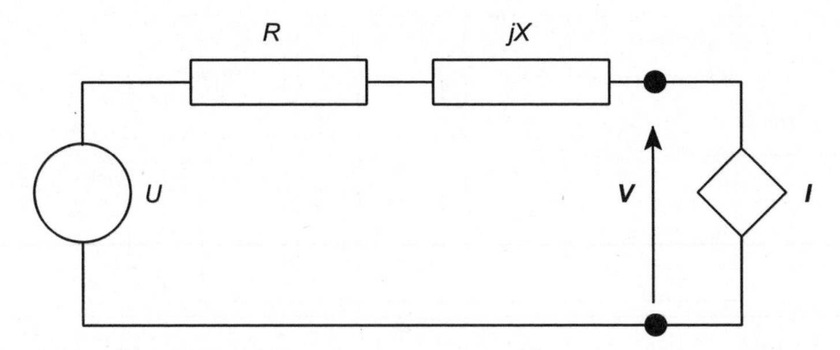

Figure 10.18 Use of 'Fictitious Grid' to Establish Voltage Variations for Various Potential Networks ($\boldsymbol{I}$ – measured current (complex quantity), R – resistance of fictitious grid, jX – inductive reactance of fictitious grid, U – ideal source voltage, $\boldsymbol{V}$ – synthesized voltage $|\boldsymbol{V}|$ passed through flicker algorithm)

short-circuit level and interpolated for the X/R ratio of the point of connection. A weighting factor, based on an assumed Raleigh distribution of wind speed, is also applied to provide flicker coefficients which may be used on sites with various average annual mean wind speeds.

The Standard also defines methods to evaluate the impact of wind turbine start-up at cut-in and rated wind speeds and during speed changing of two speed generators. Again currents are measured, combined with the 'fictitious grid' to provide a voltage time series and then passed through a flicker algorithm. For variable-speed wind turbines, harmonic currents are measured over 10 min observation periods.

10.6 Electrical Protection

All parts of a high-voltage power system are protected by relays which detect abnormal conditions and circuit breakers which then open to isolate the faulty circuits. Some lower-voltage circuits are protected by fuses, as these are cheaper but do not give the degree of control offered by relays and circuit breakers. However, fuses do have the advantage of operating very rapidly and so limiting the energy transferred into the fault. On a distribution network the protection system is designed primarily to detect excess currents caused by insulation failure in the circuits. Failure of insulation, e.g., breakdown of air or of solid insulation, leads to excess currents either between phases or between phases and ground. These high currents are only allowed to persist for up to about 1 s or so in order to limit the hazards which include: (1) risk to life caused by excess voltages as the high currents flow into the ground impedance, (2) risk to plant caused by the destructive heating and electromagnetic effects of the high currents, and (3) risk to the stability of the power system.

The electrical protection of wind turbines and wind farms follows the same general principles which are applied to any electrical plant (ALSTOM, 1987) but there are two significant differences. Because wind farms are frequently connected to the periphery of the power system it is common to find that the fault currents which will flow in the event of insulation failure are rather small. Although this is desirable from the point of view of reducing hazards, this lack of current can pose significant difficulties for the rapid and reliable detection of faults. In particular, some designs of high-voltage fuses rely on the energy within the arc for their correct operation. Hence they cannot be relied on to interrupt small fault currents when the arc energy is low. Secondly, fixed-speed wind turbines use induction machines and variable-speed wind turbines are interfaced to the network through voltage source converters. Neither induction generators or voltage source converters are a reliable source of fault current and so voltage or frequency-sensing relays are needed to detect abnormal conditions which are fed from wind turbine generators.

Figure 10.19 shows the current of an induction generator with a three-phase fault applied to its terminals. It may be seen that the fault current dies away rapidly as the stored magnetic energy in the electrical machine decays. There is no sustained fault current as an induction generator draws its magnetizing current from the network or local capacitors and this is not possible with the voltage collapsed by a three-phase fault at the terminals. Some asymmetrical faults (e.g., two-phase faults) can lead to sustained fault currents of two to three times that of full output (Jensen, 1990) but again these are not usually relied on to operate protective relays. However, it may be necessary to ensure that fault current supplied by wind turbines does not lead to mal-operation of relays on the distribution system by altering the flow of fault current in the network and, in effect, providing local support of the network voltage.

The protection of a wind farm has many similarities to the protection of a large industrial load equipped with machine drives which may back-feed into the network. This is a useful analogy, for the purposes of protection, as a wind farm can be thought of as a collection of large electrical machine drives with torque

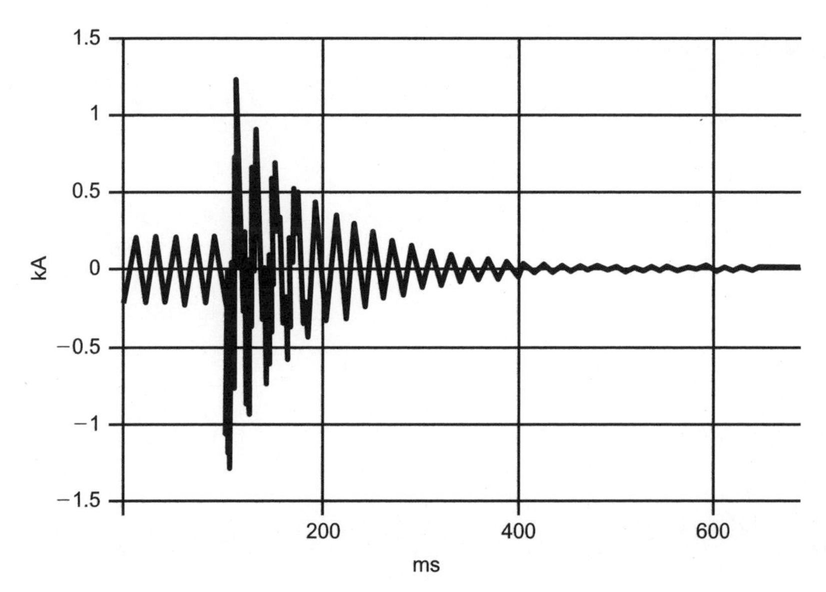

Figure 10.19 Fault Current of Induction Generator with a Three-phase Fault Applied to Terminals (Phase Shown with Minimum DC Offset)

applied to the machine shafts rather than taken to drive mechanical loads. As in the case of industrial machine drives, the terminal voltages and frequency of the electrical machines are determined by conditions on the network. The distribution network provides a reliable source of fault-current which can be used to detect insulation failure while there is also the possibility of a fault-current contribution from the rotating machines. In addition, on a wind farm there is also the hazard of prolonged generation with abnormal frequency or voltage when the wind farm is disconnected from the rest of the power system. This is known as *islanding* and is an important consideration for all embedded generation schemes.

10.6.1 Wind farm and generator protection

Figure 10.20 shows a typical protection arrangement for a wind farm of fixed-speed wind turbines with generator voltages of 690 V and with a collection circuit voltage of 11 kV. The 11 kV circuit is fed from a 33/11 kV Delta/Star wound transformer with the 11 kV neutral grounded either directly or through a resistor. The 11/0.69 kV transformers are also wound Delta/Star and so the 690 V neutral points of each circuit may be directly grounded. The neutral point of the generators is not connected to ground.

There are a number of zones of protection. At the base of the wind turbine tower a 690 V circuit breaker (usually a moulded case type as shown in Figure 10.1) will be fitted to protect the pendant cables and the generator. This is shown as Zone D.

Zone C in Figure 10.20 is the 690 V cables running from the turbine transformer to the tower-base cabinet. Fuses or another moulded case circuit breaker may be

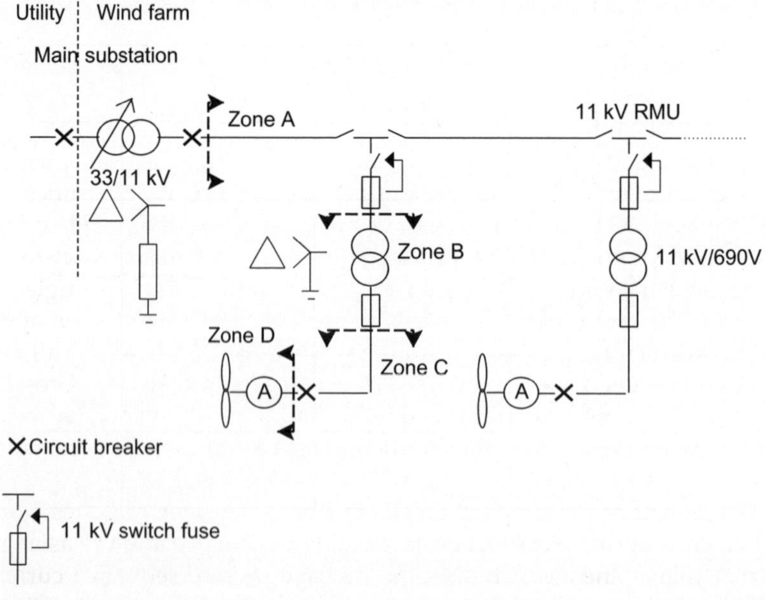

Figure 10.20 Protection of a Wind Farm with an 11 kV Connection Circuit (RMU – Ring Main Unit)

fitted to the 690 V side of the turbine transformer to provide protection of the cables and also a point of isolation so that all the electrical circuits of the turbine may be isolated without switching at 11 kV. In some designs of wind turbine, the main incoming busbar at the bottom of the electrical cabinet at the tower base consists of exposed conductor. The electrical protection of this area needs careful consideration as there is the possibility of dropping tools or other equipment on to these busbars. Zone B is the 11 kV/690 V transformer including the region around its 690 V terminals. This is a particularly difficult zone to protect as the 11 kV fuses must be robust enough to withstand the magnetizing inrush current of the transformer while sensitive enough to detect faults on the 690 V terminals when the fault current will be limited by the impedance of the transformer. This problem is common to all 11 kV/400 V transformers used on the public distribution network and a typical UK solution is to use an 11 kV combination fuse-disconnect (sometimes known as a switch-fuse). Faults on the 11 kV winding of the transformer will lead to high fault currents which are cleared by the 11 kV fuse. However, for faults on the low-voltage terminals, the fuse is unable to clear these low fault currents effectively and so, once the fuse operates, a striker pin operates a mechanism to open the disconnect switch to clear all the phases. A more expensive solution adopted on some industrial installations is to use so-called restricted earth fault protection to detect currents leaking to ground from the low-voltage winding and terminal area but this requires an 11 kV circuit breaker. There is obvious commercial pressure to reduce the costs of transformer protection on a wind farm as there is a transformer for each wind turbine. However, safety considerations require that all credible faults can be detected and cleared.

Zone A is the 11 kV cable circuit and this is protected in the conventional fashion by overcurrent and earth fault relays operating an 11 kV circuit breaker. The 33/11 kV transformer is protected in a similar manner to a transformer of the same rating used on the public distribution system.

As wind-turbine ratings increase and larger wind farms are constructed, 11 kV circuits cease to be cost-effective and so a number of large wind farms have been constructed with a collection voltage of 33 kV. Figure 10.21 shows a typical arrangement where a 33 kV wind farm is connected directly to a 33 kV public utility network. This arrangement poses a number of additional difficulties for the electrical protection. 33 kV switch fuses are not readily available and so it can be difficult to provide comprehensive protection of the 33 kV/690 V transformer for the full range of prospective fault currents. Effective protection for a single phase to ground fault on the low-voltage terminals is particularly difficult. One attempt to overcome these difficulties is to use a source protection relay to detect all faults on the 33 kV circuit, transformer and low-voltage terminals (Haslam, Crossley and Jenkins, 1999). Although a suitable technique has been shown to be technically successful this type of relay is not in production as the commercial demand for it is too low.

In the arrangement of Figure 10.20 the 33/11 kV transformer provides two useful features when considering electrical protection. Its impedance allows easier grading of overcurrent relays and, as it blocks the passage of zero-sequence current, fast acting earth fault relays can be applied to the wind-farm circuit. With direct connection to the utility circuit these desirable features are not available and, in

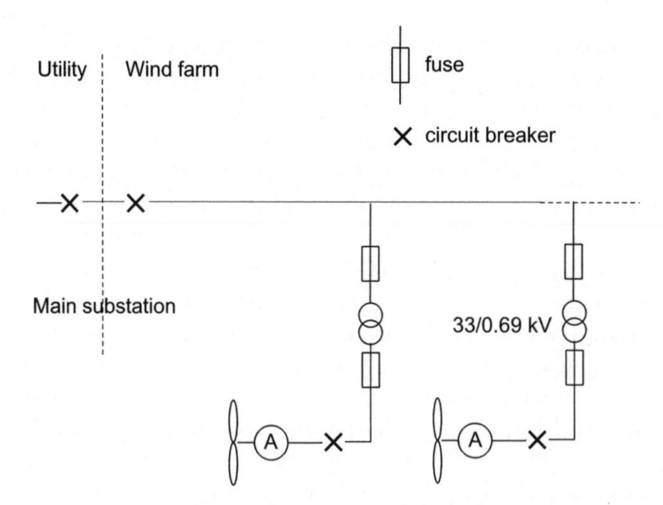

Figure 10.21 Protection of a Wind Farm with a 33 kV Connection Circuit

some countries, all embedded generation schemes are required to install an interface transformer even if its voltage ratio is 1:1. This is not an economic solution but does illustrate the difficulty of connecting a wind farm without a main transformer.

Electrical protection of wind farms follows conventional distribution engineering practice with the network as the main source of fault current. In the protection of a public supply network it is important only to isolate the faulty section or circuit and so maintain supply to as many customers as possible. This discrimination of the protection is less important in a wind farm as only some loss of generation will occur if correct discrimination is not achieved and so simpler and lower cost protection may be appropriate. The difficulty remains, however, of ensuring that effective protection is installed to detect all credible fault conditions even with limited prospective fault currents.

10.6.2 Islanding and self-excitation of induction generators

Fixed-speed wind turbines use induction generators to provide damping in the drive train and, as there is no direct access to the field of an induction generator, the magnetizing current drawn from the stator leads to a requirement for reactive power. In order to reduce the reactive power supplied from the network it is conventional to fit fixed-speed wind turbines with local power factor correction (PFC) capacitors. As long as the induction machine is connected to a distribution network its terminal voltage is fixed and so the PFC capacitors merely reduce the reactive power drawn from the network. However, once the induction generator is isolated from the network then there is the possibility of a resonant condition, known as self-excitation, leading to large over-voltages (Allan, 1959, Hindmarsh, 1970). When the generator is isolated from the network it will tend to accelerate as its load is removed. This increase in rotational speed, and consequently of frequency, adds to the possibility of self-excitation. There have been a number of cases reported of sections of distribution network, to which wind turbines are connected, becoming isolated and over-voltages resulting in damage to customers' equipment.

Figure 7.28 showed the conventional equivalent circuit of an induction machine with the PFC capacitors connected at the terminals (McPherson, 1981) but with no connection to the network. At normal running speed and frequency, the slip (s) is small and so the impedance of the rotor branch is high and, to a first approximation, can be considered an open circuit. As the stator impedance ($R_s + jX_s$) is much smaller than the magnetizing reactance jX_m then the equivalent circuit may be simplified to a simple parallel connection of the PFC capacitors and the magnetizing reactance. Thus the equivalent circuit of Figure 7.26 may be reduced to that of Figure 10.22. This is a conventional LC parallel circuit but in this case X_m is a non-linear function of voltage due to magnetic saturation. Energy is added to the circuit from the wind turbine. An indication of the likely voltage which will occur as the wind-turbine rotor accelerates can be gained by considering the intersection of the capacitive and inductive voltages as the current is identical in each element of this simplified model. This is shown in Figure 10.23. Point X indicates the self-excitation voltage at frequency f_1, say 50 Hz, while point Y illustrates the self-excitation voltage at an increased frequency, f_2, say 55 Hz. Such simple calculations are

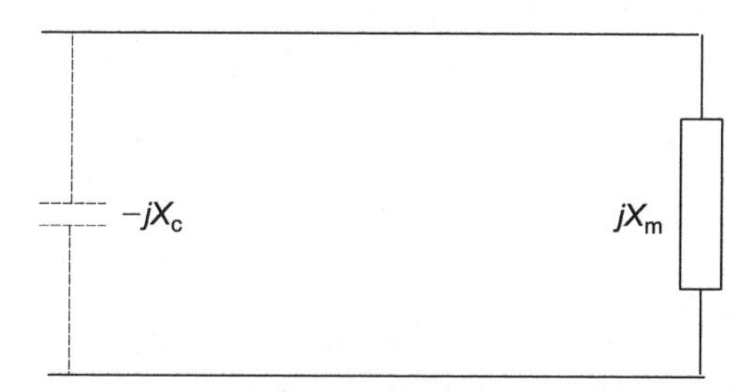

Figure 10.22 Reduced Equivalent Circuit to Illustrate Resonance Leading to Self-excitation

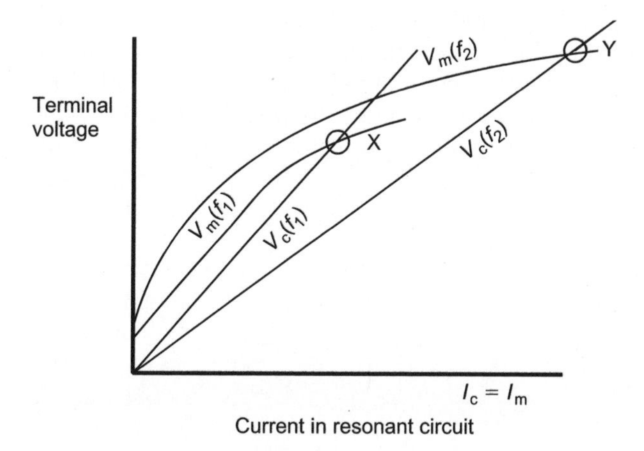

Figure 10.23 Illustration of Self-excitation at Two Frequencies f_1 and f_2 (I_c – Current in Capacitors, I_m – Current in Magnetizing Reactance of Induction Machine)

indicative only as they are based on a very simple representation of the induction generator and even then it is difficult to obtain reliable data for the saturation characteristic of the machine. However, as self-excitation is a most undesirable condition for grid-connected wind turbines, precise calculations of the voltage rise are seldom required.

Self-excitation can be avoided by limiting the PFC capacitance, including any capacitance of the distribution network, to a value which will not lead to the resonant condition at any credible frequency which will be experienced during overspeed. Alternatively, the potential for this condition must be recognized and fast-acting over-voltage and over-frequency protection arranged to stop the wind turbine if islanding occurs.

10.6.3 Interface protection

Section 10.6.1 considered the protection of the wind turbines from the effects of insulation failure and subsequent high fault currents, which were predominantly supplied by the distribution network. However, protection is also required to ensure that the wind farm does not feed into faults on the distribution network or attempt to supply an isolated section of network.

The problem is illustrated in Figure 10.24. For faults on the network, the difficulty is that wind turbines are not a reliable source of fault current and so circuit breaker B cannot be operated by over-current protection. Thus, for the fault shown, the current-operated protection on the network is used to open circuit breaker A. This then isolates the wind turbine which begins to speed up as the wind input remains but it is no longer possible to export power to the network. In fact, this acceleration begins as soon as the fault occurs, and before circuit breaker A trips, as the fault depresses the network voltage and so inhibits the export of power. Circuit breaker B is then opened by over-frequency relays which detect the increase in rotational speed of the wind turbine, and hence the increase in frequency of the output, or by under/over- voltage relays which detect that the connection to the network has been lost by monitoring the change in voltage. It is common for the over/under-frequency and -voltage relays to be time delayed by, for example 500 ms, to reduce spurious operation. Hence in high winds it is possible that the wind turbine will shut down on mechanical over-speed protection which is set to limit the speed of the aerodynamic rotor.

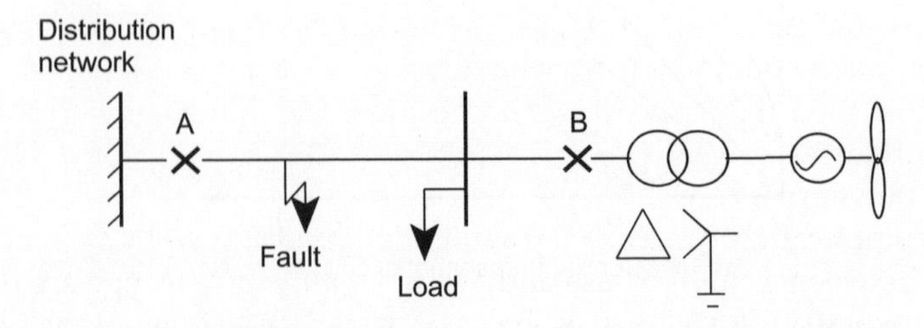

Figure 10.24 Protection of the Distribution Network from a Wind Turbine

This sequential tripping using voltage- and frequency-sensitive relays which operate once the generator has been isolated is not generally considered to be good practice in the protection of conventional power systems. However, with the use of induction generators or voltage source converters there is a little option and so the arrangement is generally accepted for wind farms.

The issue of *islanding* is the possibility that the output of the wind turbine (in terms of both real and reactive power) matches precisely the local load. In this case even though circuit breaker A opens there will be no change in the voltage or frequency of the wind turbine and so circuit breaker B will not operate. Most distribution utilities are extremely sensitive to possible islanded operation for a number of reasons:

- the possibility that customers may receive supply outside the required limits of frequency and voltage,
- the possibility that part of the network may be operated without adequate neutral earthing,
- the danger associated with out-of-phase reclosing,
- it is against the regulations governing operation of the distribution network, and
- the potential danger to staff operating the distribution network.

From the point of view of the wind farm operator the main danger is in out-of-phase reclosing. Many distribution circuits have automatic reclosing to allow transient faults, particularly on overhead lines, to be cleared without prolonged interruption to customers. Thus circuit breaker A may be arranged to reclose up to a few seconds after opening. If the circuit breaker B is still closed then very high currents and torques will occur as the network voltage is applied out-of-phase to the wind turbine. This is the main technical driver for fast islanding protection as, particularly with wind turbines, it is very unlikely that the required matching of generator output (both real and reactive power) to the load will be sustained for any length of time.

Considerable efforts have been made to devise robust protection systems to detect islanding. However, it may be seen to be very difficult as, by the principle of superposition, if there is no current flowing in circuit breaker A then it makes no difference to the network conditions at the wind turbine whether it is open or closed. The commonly used relays are the so-called rate of change of frequency (r.o.c.o.f.) devices or vector-shift relays which measure the jump in the voltage vector when islanding occurs. Both relays rely on some current flowing in circuit breaker A and can be adjusted to varying levels of sensitivity. However, if they are set too sensitive then they tend to be prone to spurious nuisance tripping due to external disturbances on the power system (e.g., the tripping of a large remote conventional power plant).

A further complication occurs if the fault shown in Figure 10.24 is not between phases but is between a single phase and ground. Faults of this type are particularly common on overhead lines. It can be seen that the turbine transformer (e.g., 33/0.69 kV) has a Delta winding on the high voltage side. This has no accessible neutral point and no connection to ground is made. Thus, there is no path in which the ground current may flow and so, in principle, the single-phase earth fault will remain indefinitely with no current flowing. In practice, some stray capacitive currents will flow but these will not be enough to operate conventional earth fault protection and will, in fact, lead to intermittent arcing. The conventional solution is to use a so-called neutral voltage displacement relay to detect that one phase of the circuit is connected to ground and so the neutral is displaced. The disadvantage of this scheme is its cost as a complex (five limb) voltage transformer is required on the high-voltage circuit. Although this can be accommodated for a large wind farm the expense is sometimes difficult to justify for individual wind turbines.

The requirements for interface protection vary widely between different countries (CIGRE, 1998, CIRED, 1999). Some countries favour the use of transfer tripping whereby when any upstream utility circuit breaker is opened this action is communicated to the wind farm circuit breaker which is then immediately opened. Although this provides a guarantee against islanded operation it can be expensive to implement as communication channels from a number of remote circuit breakers are required. It is interesting to note that in Holland, where almost all the distribution system is underground and so auto-reclose is not used to reclose circuits after overhead line transient faults, there is no requirement for loss-of-mains protection. In Denmark, positive phase sequence under-voltage relays are used extensively and appear to be effective in detecting islanded operation. No doubt over time practices will converge but, at present, very considerable national differences remain.

10.7 Economic Aspects of Embedded Wind Generation

Until 1990 most modern power systems were operated as vertically integrated monopolies with a single overall organizational structure for the generation, transmission, distribution and supply of electrical energy. Usually a national or regional government would effectively exercise control over the entire public electricity system. There were, of course, detailed differences in how the various parts of the power system were owned and administered but generally it was considered desirable to have a high degree of government control in order to provide electrical energy which was seen as of strategic national importance. Recently, however, there has been a move in almost all countries of the world to disaggregate or break up the power system into two major constituent parts: (1) the generation and supply of electrical energy (i.e., creation and sale of kWh), and (2) the transport facilities (i.e., the transmission and distribution networks) required to deliver the power to the customers. Generation and supply of electrical energy is seen as a competitive activity, in many ways similar to trading any other commodity, while the transport infrastructure is a natural monopoly and so requires regulation. The reasoning usually given for this change is that it allows competition to develop in at least parts of the electricity supply chain which should lead to lower costs and prices. Privatization of the electricity supply system also allows governments to relinquish their responsibility for providing capital funding and raises considerable revenue at the time of the sale.

In a deregulated power supply system it is important that all those involved in competitive activities (i.e., generators and suppliers of electrical energy) have equal access to the transmission and distribution monopoly networks in an open and non-discriminatory manner. This concept of 'open-access' is potentially very important for embedded wind generation as it guarantees that a wind farm can be connected to the network. A corollary of this free-market approach is that any change in the costs of the network caused by the wind farm need to be recognized and suitable payments made.

However, a major difficulty with open-access to the distribution network is the complexity necessary to ensure fairness and economic efficiency. There is no escape from this complexity as approximations and simplifications tend to introduce distortions which the participants, in what is essentially a market, are quick to exploit. Open access has been implemented effectively in transmission systems which although carrying very large amounts of power are, in fact, much simpler than distribution networks with many fewer busbars and circuits. In distribution networks the operators, who previously did not have to concern themselves with embedded wind generation at all, are now faced with both technical and commercial issues of considerable conceptual and practical difficulty. It is fair to say that appropriate commercial tariffs and arrangements to ensure open-access to the distribution network for embedded wind generation have yet to be put into place in most countries of the world.

10.7.1 Losses in distribution networks with embedded wind generation

Wind generation embedded in the distribution network will alter the flows of real and reactive power and so change the electrical losses within the network. This is illustrated in Figure 10.25. A load of real power, P, and reactive power, Q, is fed from a distribution feeder of resistance R. Now the electrical power losses in a circuit are by definition:

$$W = I^2 R$$

which is approximately equal to:

$$W = (P^2 + Q^2)R/V_0^2 \tag{10.3}$$

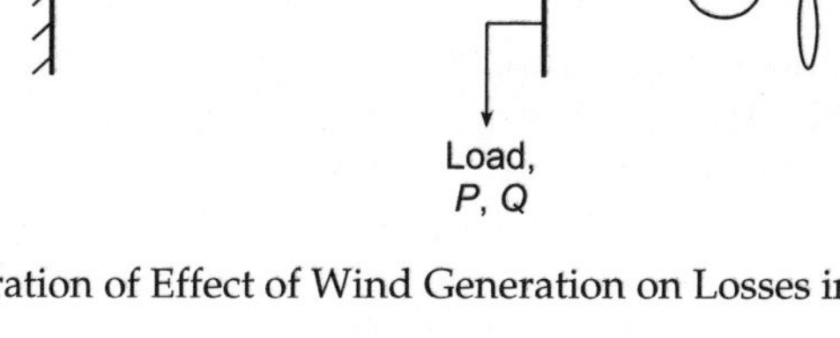

Circuit resistance, R

$p, -q$

Load, P, Q

Figure 10.25 Illustration of Effect of Wind Generation on Losses in a Distribution Network

and so the losses in the feeder may be estimated from Equation (10.3). If a wind turbine is then connected to the busbar and exports real power output p and imports reactive power q the losses will now be approximately:

$$W = [(P - p)^2 + (Q + q)^2]R/V_0^2 \tag{10.8}$$

Thus it may be seen that depending on the relative magnitudes of real and reactive load and the generation of the wind farm the network losses may either be reduced or increased. The flows in the feeder depend on the correlation of wind-farm output and load and so there may be times of the year when losses are increased and others when losses are reduced.

Because the electrical losses in the network are altered by the presence of the embedded wind farm it is necessary to calculate how the losses are effected so that, if they are reduced, the wind farm may be rewarded and if they are increased the wind farm must pay for them. One conceptually simple way of doing this is to calculate the network losses without generation and then with generation and so determine the effect of the wind farm. The calculation may be carried out for a range of typical network loadings and generator outputs. This is the so-called 'substitution method' which is used to calculate loss adjustment factors in the UK. Loss adjustment factors are used to gross up demand/generation to the point at which the distribution network is connected to the transmission system and so account for distribution losses. It has been shown recently by Mutale, Strbac and Jenkins (2000) that the substitution method produces inconsistent results which will lead to cross-subsidies between users of the network. Alternative rigorous techniques have been developed but these have not yet been adopted in practice. In fact, in many countries the impact of embedded wind farms on distribution network losses has been ignored as, whatever calculation technique is used, it is necessary to take account of both the location- and time-dependent nature of the load and generation. Typically loss adjustment factors on UK medium voltage circuits are in the range 5–10 percent and so they can be of significant commercial consequence. It should also be noted that flows, and hence losses, in the transmission system will also be modified by embedded wind generation.

10.7.2 Reactive power charges and voltage control

Equation (10.8) also shows that it is desirable to operate as near to unity power factor as possible (i.e., $q = 0$) if network losses are to be minimized. This preference to reduce reactive power flow is reflected in tariffs applied by many distribution companies to charge customers either for reactive energy (kVArh) or for the peak reactive power (kVAr) drawn during a period. Such tariffs are usefully applied to loads where both real and reactive power flows are in the same direction and so any reduction in reactive flows leads to smaller voltage variations on the network (voltage drop in the case of loads). However, they can have perverse consequences when applied to generation where it may be desirable for network voltage control to draw some reactive power and so limit voltage rise. In this case the generator wishes to draw reactive power to control the voltage rise caused by the real power export but will be charged for this reactive demand.

It is anticipated that, in time, reactive power/energy tariffs will cease to be applied to embedded wind generation and will be replaced by the concept of embedded generation taking part actively in the voltage control of distribution circuits perhaps through the commercial mechanism of an ancillary services market.

The design of distribution networks tends to be driven by voltage considerations rather than by thermal limits on plant. Therefore the connection of embedded wind farms is often limited by considerations of voltage variation. At present it is usual to evaluate these voltage limits by considering: (1) the maximum output of the wind farm in conjunction with the minimum network load, and (2) zero output of the wind farm with maximum network load. This rather simplistic approach can lead to a connection for a wind farm being rejected for conditions which only persist for a few hours per year. Thus, there is a move towards probabilistic assessment of voltage using so-called probabilistic load flows which allow calculation of the duration of the times in which voltage limits are violated. Probabilistic load flows may be based either on analytical calculation techniques or use some form of Monte Carlo simulation. The input is a probabilistic description of the network loads and generation and the output is a similar representation of network voltages and flows. Results of such studies may be used to predict the cost of lost revenue if generation is curtailed during periods of low network load or the increased charges for drawing higher levels of reactive power.

10.7.3 Connection charges 'deep' and 'shallow'

When a wind farm owner wishes to connect a project to the distribution network there are clearly costs associated with doing so and it is entirely reasonable that all the appropriate connection costs are borne by the wind farm. This is a similar situation to the connection of any large load. There are two main philosophies in charging for connection of either generation or load to a power system, i.e., 'Deep' or 'Shallow' charging.

In deep charging the embedded wind-farm project will pay for *all* costs incurred on the power system due to the wind farm. These costs will include the local circuit from the wind farm to the appropriate point of connection on the public distribution network but also for any uprating of plant (e.g., transformers or switchgear) at any point in the distribution network. Deep charging is conceptually simple and is presently applied on UK distribution networks. However, it has a number of significant problems. Perhaps the most important is that it operates on the basis of historical precedent, i.e., first-come, first-served.

As an example, consider that two 30 MW wind farms are to be connected to a public utility circuit with an available capacity of 50 MW. The first wind farm is to be constructed 1 year before the second. As there is adequate capacity, the first wind-farm developer will not pay for the capital assets of this existing circuit but will be allowed to connect. However, if a second wind farm, also of 30 MW, applies to connect to the same circuit then this second developer will incur all the expense of uprating the 50 MW circuit to at least 60 MW and any additional costs for increases in capacity elsewhere in the system. The first wind farm developer will continue to enjoy access to the network purely on the basis of his prior application to connect. The situation is complicated by the fact that distribution network capacity can only be added in discrete amounts and so the second developer may pay for excess capacity which is then available for any subsequent application, perhaps by the first developer expanding his wind farm!

A similar problem arises in the case of wind farms connected to parts of the network where the short-circuit levels are approaching the switchgear rated values. If a wind farm wishes to connect to part of the network and this results in the switchgear ratings being exceeded then the wind farm will be expected to pay for the uprating of the equipment even though most of the short-circuit capacity is being used by other, existing generators. Although a rather unusual problem for wind farms, which tend to be located in rural areas where short-circuit capacities are low, this issue of switchgear uprating causes major difficulties for other embedded generation (e.g., co-generation plant) which is located in towns and cities.

Clearly such arrangements are not equitable and *ad hoc* agreements are often entered into in order to resolve difficulties. These may include limiting charges to costs incurred one voltage level above that used for the connection or by claw back arrangements so that if a developer pays for excess capacity which is used subsequently by another network user than some repayment is made. However, the basic principles of deep charging lead automatically to such problems.

In contrast, 'shallow charging' involves the generator in paying directly only for that section of network necessary to connect the wind farm to the nearest point on the network. The extent of the network which must be paid for is generally defined by that which is used *solely* by the generator. Other costs incurred in reinforcing the distribution system are recovered by use-of-system charges. These charges are applied by the owner of the network on all users of the network and should reflect their use of the network assets. Shallow charging has the advantages that it recognizes that the monopoly network owner has a duty to manage the network assets for the benefits of the users and avoids the anomalies of deep charging. However, the difficulty with shallow charging is that as the users change, or the capital base of the network develops, then use-of-system charges may alter. In the UK shallow charging is applied on the transmission (400 kV and 275 kV) system and can result in a generator either paying charges, if they are increasing the overall requirement for transmission assets, or indeed, being rewarded if they are reducing the need for transmission plant. The resulting use-of-system charges are reviewed periodically. Embedded generators do not, at present, pay or receive distribution use-of-system charges in the UK.

10.7.4 Use-of-system charges

Use-of-system charges are applied to recover the cost of the capital assets required for the transport of electricity in either the distribution or transmission network. In general these are likely to be more significant than operating charges (i.e., losses or reactive power charges) as the costs of transport of electricity are dominated by the value of the assets required.

In the UK, distribution use-of-system charges for loads are generally calculated based on the long-run marginal cost of expanding, maintaining and operating the distribution system. The calculation is usually performed by considering the cost of adding 500 MW of load, evaluating the costs including those of providing system security and then allocating these costs to the various consumer types and voltage levels. The resultant tariff is generally in two parts: (1) a maximum demand element (£/kVA), and (2) a unit charge (£/kWh). The maximum demand charge is used to account for costs incurred at voltage levels close to customers as these are likely to be related to the peak demand of the customers. The unit charge is used for the costs of the higher-voltage distribution networks as it is considered that the individual maximum demand of a customer is less likely to occur at the same time as the system maximum.

Use-of-system charges are also applied to both loads and generators on the England and Wales transmission network. At present these charges are based on the peak power flows which occur at exit and entry points to the transmission network. Transmission use-of-system-charges are arranged so that in areas of the country where generators are remote from load centres, and so increase the requirement for transmission system assets, then the generator will pay charges to the transmission network operator while a generator will be rewarded for locating in an area of high load and so reducing the requirement for transmission system. Under present practice wind farms embedded in the distribution network can assist in reducing transmission use-of-system charges by reducing the peak power flows at the exit and entry points of the transmission network. However, the administrative arrangements for recovering this benefit are involved.

There is an argument that embedded wind farms should receive a benefit, in the form of negative use-of-system charge payment for the reduction in the distribution assets required due to the location of the wind generation. Clearly, if the distribution network was entirely radial then this would not be appropriate as intermittent wind generation would not reduce the need for any circuits or plant. However, as distribution networks are designed on the basis of system security and/or performance, rather than merely transport, it may be that in some cases embedded wind farms do actually reduce the need for distribution plant or improve the quality of supply. Such calculations can only realistically be carried out if the present deterministic security standards are replaced by probabilistic approaches which would quantify the contribution of intermittent wind farms to security and quality of supply. This is not the practice at present. It should also be noted that embedded wind generators may actually increase the requirement for distribution network assets in some cases and so would be required to pay use-of-system charges.

Robust techniques for the application of use-of-system charges for wind generators embedded on distribution networks have yet to be worked out and applied. However, these arrangements are required in order to provide to wind-farm developers clear pricing signals where, on the network, it is desirable to connect new projects. The costs of connection and the consequent charges for using the network are, of course, only one component to be considered when choosing a wind-farm site but equitable, transparent rules and genuine 'open-access' to the network are essential to encourage projects to be built.

10.7.5 Impact on the generation system

The connection of wind generation acts to displace conventional, often fossil fuelled, plant with the highly desirable consequence of reducing gaseous emissions. However, significant penetrations of wind energy will alter the operation of the conventional generators and their costs. Wind farms will respond to the wind resource and so the rest of the generation system needs to be able to accommodate the variations in their output. At present in the UK this is not considered to be a major issue as on a typical winter day conventional generation is scheduled to meet a load increase of up to 12 GW over the morning load pick-up period which lasts some 2 h. This need for flexible operation is much greater than any immediate requirement due to wind farms. Operational experience from Denmark (Falck Christensen *et al.*, 1997) is that the worst-case change in total wind power over 15 min is of the order of 10–15 percent of wind output at that time. This includes shut-downs of entire wind farms due to high winds but excludes the effect of system disturbances and faults. Translating this experience to the UK leads to a figure of some 10 GW of wind-farm capacity being required before the rate of change in output of wind farms exceeds that of the national load. This compares with a UK maximum demand of some 50 GW and minimum demand of approximately 20 GW.

In the 1980s (e.g., Grubb, 1988) there was considerable research carried out to investigate how very large wind farms would alter the costs of energy generated from conventional plant. This was based on an assessment of how wind generation would alter the optimal generating plant mix and its operation. At that time, the major utilities (e.g., the Central Electricity Generating Board in the UK) had well-defined procedures to determine the plant which needed to be constructed and how it should be operated. The research studies were undertaken by considering how wind generation would change the results of these assessments. For reasonable (up to 15–20 percent) penetration of wind power, the additional costs due to increased cycling of thermal plant or constraints on wind generation were likely to be small. However, the difficulty in predicting the output of wind farms was identified as a particular issue which would lead to higher reserve costs. This is also recognized in Denmark where considerable efforts have been made to provide effective prediction systems for the output of wind farms. Figure 10.26 (CIGRE, 1998) shows the expected development of wind farms and non-dispatched CHP in Denmark and it may be seen that these forms of generation will dominate the electricity supply system by 2015. However, it may be noted that Denmark is connected to other European countries by both AC and DC transmission links which will provide considerable frequency stability and the technical ability to provide generation reserve.

Figure 10.27 shows the measured capacity factor of a UK wind farm. The capacity factor is the energy delivered during a period of time expressed as a fraction of the energy which would have been supplied if the plant had operated at its rated

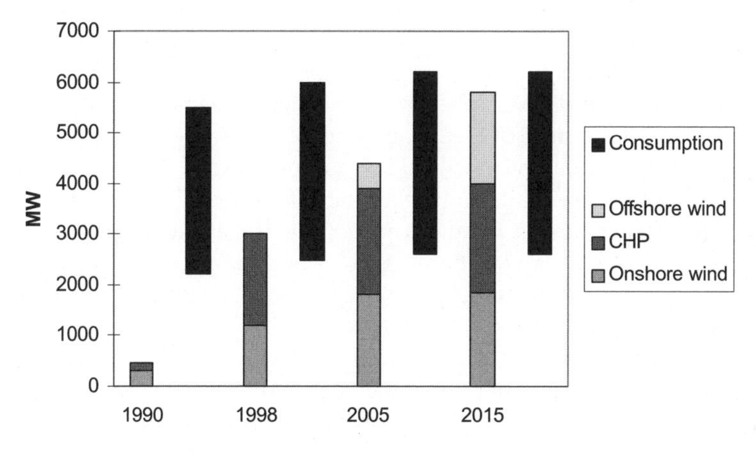

Figure 10.26 Anticipated Growth of Wind Generation in Denmark (after, CIGRE, 1998 and personal communication)

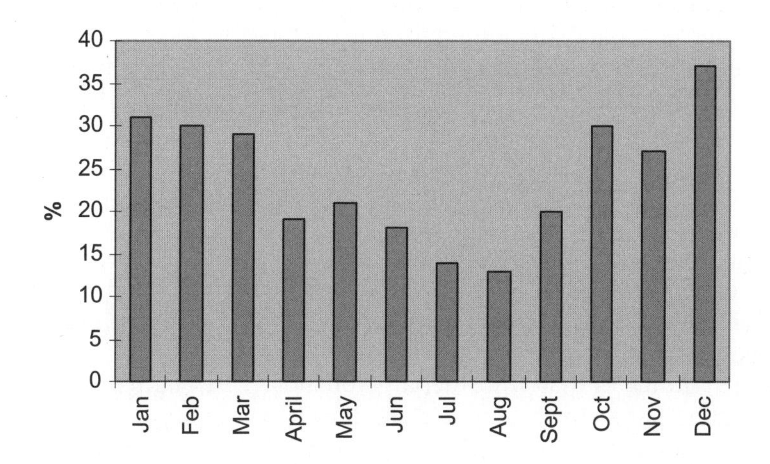

Figure 10.27 Monthly Variation in Capacity Factor of a UK Wind Farm

capacity. It may be seen that the wind-farm output is significantly greater during the winter months when, of course, the electrical load is highest. In the past, there has also been considerable study of the extent to which wind generation could be used to defer the construction of conventional generating plant. This question has been posed in terms of establishing the 'capacity credit' of the wind farm. Capacity credit is calculated by evaluating the capacity of conventional generation that need not be constructed while maintaining the same level of generation reliability (Milligan and Parsons, 1999). If these calculations are carried out carefully using conventional power system reliability techniques it is possible to establish long-term capacity credits as being approximately equal to the measured capacity factors of the wind farms providing overall wind energy penetration into the power system is low. However, it should be emphasized that these are statistically based calculations and at particular times of peak demand (e.g., on clear cold winter days in Northern Europe) there may be very little wind generation.

The studies of the 1980s have now been largely superseded by the change to a market structure for generation and supply of electrical energy and developments in the conventional plant mix. There is no longer any central control over the construction of generating plants in the UK and the energy trading system is based on bilateral agreements with only the final balancing mechanism being under the control of the network operator. These arrangements are such as to reward predictability and controllability in generating plants and, over time, a market price should become visible to indicate the extent of the additional costs of wind-energy generation.

Conventional generators, as well as being a source of electrical energy (kWh), are also used to provide a range of so-called ancillary services including: (1) system frequency control, (2) reserve capacity, and (3) black start capability. These services are essential for the stable operation of the power system and if they are not provided by conventional thermal, usually steam, plant then alternative sources need to be found. In a deregulated environment, these services are purchased by the network operator. For example the National Grid Company in England and Wales estimates that 100 MW of additional reserve for a year can be obtained for £10 M (House of Lords, 1999). Using this figure, and assuming that for large penetrations of wind energy into the UK system it will be necessary to provide 15 percent of the wind output as additional reserve, a cost of 0.17 p/kWh may immediately be calculated.

Synchronous generators also provide reactive power which is essential to support the voltage of a power system. It is difficult to transmit reactive power long distances in a power system and so local reactive sources are required to support the network voltage. Preliminary work from Denmark (Bruntt, Havsager and Knudsen, 1999) indicates that with the large offshore wind farms planned, which will use induction generators, there is a significant possibility of voltage collapse. This is now being addressed and possible remedial measures, including reactive compensators considered.

It is sometimes suggested that large penetrations of wind energy in a power system will lead to a requirement for energy storage. Although electro-chemical systems (flow batteries and fuel cells) are now becoming commercially available it will be very difficult for them to compete with conventional plant, particularly existing pumped storage, merely to provide generation ancillary services. It is likely the energy storage systems will have to use their distributed nature and location in the network to provide additional value, e.g., in terms of overcoming transmission constraints, as well as taking part in energy trading activities.

As wind power becomes a more significant component of the power system it will be necessary to integrate its operation more closely with that of the conventional plant. At present renewable generation is operated on a 'must run' basis with reactive power and voltage control being the responsibility of the network operator. This is entirely appropriate when wind energy is not significant compared with conventional generation but clearly alternative arrangements will be needed in the future.

References

Allan, C. L. C., (1959). 'Water turbine driven induction generators', *Proc. IEE, Paper No. 3140S.*

ALSTOM, (1987). 'Protective relays application guide'. ALSTOM Protection and Control, Stafford, UK.

ANSI/IEEE, (1986). 'IEEE Guide for safety in AC substation grounding'. *Standard 80.*

Bossanyi, E., Saad-Saoud, Z. and Jenkins, N., (1998). 'Prediction of flicker caused by wind turbines', *Wind Energy*, **1**, 1, 35–50.

British Standard, (1999). 'Code of practice for protection of structures against lightning'. *BS 6651.*

British Standard, (1995). 'Voltage characteristics of electricity supplied by public distribution systems'. *BS EN 50160.*

Bruntt, M., Havsager, J. and Knudsen, H. (1999). 'Incorporation of wind power in the East Danish power system'. *IEEE Power Tech '99.*

CIGRE, (1998). 'Impact of increasing contribution of dispersed generation on the power system', *Study Committee No. 37 Working Group 37-23 Final Report* (to be published).

CIRED, (1999). 'Preliminary report of Working Group WG04: Dispersed Generation'. *CIRED Conference.*

Copper Development Association, 'Earthing practice'. *CDA Publication No 119.* Copper Development Association, Potters Bar, UK.

Cotton, I. *et al.*, (2000). 'Lightning protection for wind turbines'. *International Conference on Lightning Protection*, pp 848–853.

Craig, L. M. and Jenkins, N., (1995). 'Performance of a wind turbine connected to a weak rural network', *Wind Engng*, **19**, 3, 135–145.

Davidson, M., (1995). 'Electrical monitoring of a wind farm in complex upland terrain'. *Proceedings of 17th British Wind Energy Association Conference.* Mechanical Engineering Publications, Bury St Edmunds, UK.

Davidson, M., (1996). 'Wind-farm power quality and network interaction'. *Proceedings of the 18th British Wind Energy Association Conference*, pp 227–231. Mechanical Engineering Publications, Bury St Edmunds, UK.

Dugan, R. C., McGranaghan, M. F. and Beaty, H. W., (1996). *Electrical power systems quality.* McGraw-Hill, New York, USA.

Electricity Association, (1976). 'Limits for harmonics in the UK electricity supply system', *Engineering Recommendation G5/3.*

Electricity Association, (1989). 'Planning limits for voltage fluctuations caused by industrial, commercial and domestic equipment in the UK'. *Engineering Recommendation P.28.*

Falck Christensen, J. *et al.*, (1997). 'Wind power in the Danish power system'. *CIGRE Symposium on Dispersed Generation.*

Fiss, H.-J., Weck, K. H. and Weinel, F., (1993). 'Connection of wind-power generating facilities to the power distribution system and potential effects on the system'. *Proceedings of the European Community Wind Energy Conference*, pp 747–750. H. S. Stephens and Associates, Bedford, UK.

Gardner, P., (1996). 'Experience of wind farm electrical issues in Europe and further afield'. *Proceedings of the 18th British Wind Energy Association Conference*, pp 59–64. Mechanical Engineering Publications, Bury St Edmunds, UK.

Grubb, M., (1988). 'The economic value of wind energy at high power system penetrations; an analysis of models, sensitivities and assumptions'. *Wind Engng*, **12**, 1, 1–26.

Haslam, S., Crossley, P. and Jenkins, N., (1999). 'Design and evaluation of wind-farm protection relay'. *IEE Proceedings Generation Transmission and Distribution*, **146**, 1, 37–44.

Heier, S., (1998). *Grid integration of wind energy conversion systems*. John Wiley and Son, Chichester, UK.

Hindmarsh, J., (1984). *Electrical machines and their application*. Butterworth Heinemann, UK.

House of Lords, (1999). 'Electricity from renewables'. *Select Committee on the European Communities. HL Paper 78.*

IEC, (1998). 'Short-circuit current calculations in three phase a.c. systems'. *IEC 909.*

IEC, (2000a). 'Lightning protection for wind turbines'. *Committee Draft, IEC 61400-24.*

IEC, (2000b). 'Measurements and assessment of power quality characteristics of grid connected wind turbines'. *Committee Draft, IEC 61400-21.*

IEEE, (1991). 'IEEE Recommended practice for the electrical design and operation of windfarm generating stations'. *IEEE Std. 1094*, (no longer current).

International Energy Agency, (1997). 'Recommended practices for wind-turbine testing and evaluation -9: Lightning protection for wind-turbine installations'.

Jenkins, N., (1995). 'Embedded generation – Part 1', *IEE Power Engineering Journal*, 145–150.

Jenkins, N., (1996). 'Embedded generation – Part 2', *IEE Power Engineering Journal*, pp 223–239.

Jenkins, N. and Vaudin, A., (1995). 'The electrical systems of five wind farms'. *Proceedings of I. Mech. E. Part A*, **209**, 195–202.

Jensen, K. K., (1990). 'Grid connection of wind turbines and wind farms'. *DEFU Report No. 77.*

Jones, R. and Smith, G. A., (1993). 'High-quality mains power from variable-speed wind turbines'. *IEE Conference on Renewable Energy – Clean Power 2001*, pp 202–206.

Lakervi, E. and Holmes, E. J., (1995). *Electricity distribution network design*, Second edition. Peter Peregrinus, Bristol, UK.

McPherson, G., (1990). *An introduction to electrical machines and transformers, Second Edition.* John Wiley & Sons, New York, USA.

Milligan, M. and Parsons, B., (1999). 'A comparison and case study of capacity credit algorithms for wind power plants'. *Wind Engng*, **23**, 3, 159–166.

Mirra, C., (1988). *Connection of fluctuating loads*. International Union of Electroheat, Paris, France.

Mutale, J., Strbac, G. and Jenkins, N., (2000). 'Allocation of losses in distribution networks with embedded generation', *IEE Proceedings Generation, Transmission and Distribution*, **147**, 1, 7–14.

Santjer, F. and Gerdes, G., (1994). 'Grid interference caused by grid connected wind energy converters'. *DEWI Magazine*, Number 5 (in German).

Tagg, G. F., (1964). *Earth resistances*. George Newnes Ltd, London, UK.

Thomson, M., (2000). 'Automatic voltage control relays and embedded generation'. *IEE Power Engineering Journal*, **14**, 2, 71–76.

Wagner, C. F. and Evans, R. D., (1993). *Symmetrical components*. McGraw-Hill, New York, USA.

Weedy, B. and Cory, B., (1998). *Electric Power Systems, fourth edition*, John Wiley and Sons, Chichester, UK.

Index